GLACIAL LANDSYSTEMS

Edited by

David J.A. Evans

Hodder Arnold

A MEMBER OF THE HODDER HEADLINE GROUP

First published in Great Britain in 2003 by
Arnold, a member of the Hodder Headline Group,
338 Euston Road, London NW1 3BH

www.hoddereducation.com

Distributed in the United States of America by
Oxford University Press Inc.,
198 Madison Avenue, New York, NY10016

British Library Cataloguing in Publication Data
A catalogue record for this book is available from the British Library

Library of Congress Cataloging-in-Publication Data
A catalog record for this book is available from the Library of Congress

ISBN–10: 0 340 80666 4 (pb)
ISBN–13: 978 0 340 80666 1

1 2 3 4 5 6 7 8 9 10

Production Controller: Anna Keene
Cover Design: Terry Griffiths
Cover photo credits: outlet glaciers on Ellesmere Island, arctic Canada and Oksfjordjokelen plateau icefield, artic Norway by D.J.A. Evans and passive continental margin glacial landsystem by T.O. Vorren.

Typeset in 10 on 12 pt Garamond by Phoenix Photosetting, Chatham, Kent
Printed and bound in Malta

CONTENTS

PREFACE

Since 1998 an informal group of graduate students and researchers have collected together twice a year somewhere in the British landscape, usually centred on a fine hostelry or even a fixed caravan site, to sample the landforms and sediments of glaciated basins. This informal gathering is the Glacial Landsystems Working Group. It reports to no parent body and it receives no formal funding but for me has served as an important testing ground for the landsystems concept and its application to glaciated terrain. The 'GLWG', as it has been endearingly termed, has provided a vehicle for field-based discussion and critique. The range of expertise on hand during any one meeting pools the knowledge of landform-sediment associations process-form relationships, modern analogues and theory. It rapidly became evident that the glacial community might benefit greatly from a suite of examples or templates of glacial landsystems that were informed by our ever-improving understanding of modern glacier systems.

When I took on the role of convenor for the Subglacial Processes Working Group of the INQUA Commission on Glaciation in 1999 it was one of my goals to compile a collection of glacial landsystems based upon modern research in both modern and ancient glacial settings. This book represents the fruits of my labours since then, and it is timely that it will be in production as we assemble for the INQUA Congress in Reno in 2003. The seventeen chapters of this book, written by specialists working in a wide range of glaciated environments, expand upon the glacial landsystems concept previously developed in textbooks by Nick Eyles in 1983 and by Doug Benn and myself in 1998. It is becoming increasingly evident that a better understanding of glacial landscapes stems from the integration of landform-sediment assemblages over large areas and that process-form models are best informed by modern process observations. The landsystems approach allows us to develop both of these research arenas and, with their integration, to contribute significant proxy data towards palaeoclimatic reconstruction. The chapters of this book offer a wide ranging, but by no means comprehensive, collection of glacial landsystems models. They are intended to act as catalysts for future research on glaciated basins, highlighting the value of holistic approaches to landform development.

In completing this book I have been aided by the considerable efforts of Yvonne Finlayson, Mike Shand and Les Hill, the technicians in Geography and Geomatics at the University of Glasgow, who I am sure from time to time wish that glacial geomorphology was not so diagram intensive! Liz Gooster at Arnold has been extraordinarily patient with our tardy completion rates and I'm sure will be glad to see this one on the bookshelves. Finally, Tessa, Tara and Lotte have given me the time and space to follow this project through when it was invading our home as well as my office—my mind will no longer be miles away when I'm reading those bedtime stories.

David J.A. Evans
Glasgow 2003

LIST OF CONTRIBUTORS

Colin K. Ballantyne, School of Geography and Geosciences, St Andrews University, St Andrews, Fife KY16 9ST, Scotland, United Kingdom.

Douglas I. Benn, School of Geography and Geosciences, St Andrews University, St Andrews, Fife KY16 9ST, Scotland, United Kingdom.

Vanessa Brazier, Earth Sciences Group, Scottish Natural Heritage, 2 Anderson Place, Edinburgh EH6 5NP, Scotland, United Kingdom.

Chris D. Clark, Department of Geography, University of Sheffield, Sheffield S10 2TN, United Kingdom.

Lee Clayton, Wisconsin Geological and Natural History Survey, 3817 Mineral Point Road, Madison, Wisconsin 53705, USA.

Patrick M. Colgan, Department of Geology, Northeastern University, 14 Holmes Hall, Boston, Massachusetts 02115, USA.

Paul M. Cutler, The National Academy of Sciences, 2101 Constitution Avenue NW (HA-372), Washington DC, 20418 USA.

Arthur S. Dyke, Terrain Sciences Division, Geological Survey of Canada, 601 Booth Street, Ottawa, Ontario K1A 0E8, Canada.

John England, Department of Earth and Atmospheric Sciences, University of Alberta, Edmonton, Alberta T6G 2H4, Canada.

David J.A. Evans, formerly at the Department of Geography and Topographic Science, University of Glasgow; new address, Department of Geography, University of Durham, South Road, Durham DH1 3LE, United Kingdom.

Sean J. Fitzsimons, Department of Geography, University of Otago, PO Box 56, Dunedin, New Zealand.

Neil F. Glasser, Centre for Glaciology, Institute of Geography and Earth Sciences, University of Wales, Aberystwyth SY23 3DB, United Kingdom.

Michael J. Hambrey, Centre for Glaciology, Institute of Geography and Earth Sciences, University of Wales, Aberystwyth SY23 3DB, United Kingdom.

Mark D. Johnson, formerly at the Department of Geology, Gustavus Adolphus College,

Minnesota; new address, Earth Sciences Centre, Goteburg University, Box 460, SE-405 30, Goteburg, Sweden.

Martin P. Kirkbride, Department of Geography, University of Dundee, Dundee DD1 4HN Scotland, United Kingdom.

David M. Mickelson, Department of Geology and Geophysics, University of Wisconsin, Madison, Wisconsin 53706-1692, USA.

Colm Ó Cofaigh, formerly at the Scott Polar Research Institute, University of Cambridge; new address, Department of Geography, University of Durham, South Road, Durham DH1 3LE, United Kingdom.

Lewis A. Owen, formerly at the Department of Earth Sciences, University of California; new address, Department of Geology, University of Cincinnati, PO Box 0013, Cincinnati, Ohio 45221-0013, USA.

Ross D. Powell, Department of Geology, Northern Illinois University, DeKalb, Illinois 60115, USA.

Brice R. Rea, formerly at the Department of Geology, University of Leicester; new address, Department of Geography, University of Aberdeen, Elphinstone Road, Aberdeen AB9 2UF, Scotland, United Kingdom.

Chris R. Stokes, Department of Geography, University of Reading, Reading RG6 6AB, United Kingdom.

James T. Teller, Department of Earth Sciences, University of Manitoba, Winnipeg, Manitoba R3T 2N2, Canada.

Frederik M. Van der Wateren, Landforms Research and Training, Amstel AB to 256, 1011PX Amsterdam, The Netherlands.

Tore O. Vorren, Institute of Biology and Geology, University of Tromso, N-9000, Tromso, Norway.

ACKNOWLEDGEMENTS

Copyright permissions have been included where pertinent throughout this book. In addition we would like to acknowledge the following for permission to use figures from previously published sources:

Blackwell Publishing for Figure 9 from Benn D.I. (1994) Fluted moraine formation and till genesis below a temperate glacier: slettmarkbreen, Jotunheimen, Norway, in *Sedimentology* 41, 279–292 and for Figure 12 from Kjaer & Kruger (2001) The final phase of dead ice moraine development: processes and sediment architecture, Kotlujokull, Iceland, in *Sedimentology* 48, 935–952;

Elsevier for Figures 85 and 150 from Brodzikowski & van Loon (1991) *Glacigenic Sediments* and for Figure 2 from Eyles et al. (1985) Models of glaciomarine sedimentation and their application to the interpretation of ancient glacial sequences, in *Palaeogeography, Palaeoclimatology, Palaeoecology* 51, 15–84;

The Canadian Journal of Earth Sciences for Figure 10 from Barnett et al. (1998) On the origin of the Oak Ridges Moraine, in *Canadian Journal of Earth Sciences* 35, 1152–1167 and Figure 10 from Sharpe & Cowan (1990) Moraine formation in northwestern Ontario: product of subglacial fluvial and glaciolacustrine sedimentation, in *Canadian Journal of Earth Sciences* 27, 1478–1486;

Geological Society of America for Figure 2 from Johnson & Hansel (1999) Wisconsin episode glacial landscape of central Illinois: a product of subglacial deformation processes?, in *Glacial Processes Past and Present* (eds. Mickelson & Attig), *GSA Special Paper* 337, 121–135 and Figure 5 from Clayton et al. (1999) Tunnel channels formed in Wisconsin during the last glaciation, in *Glacial Processes Past and Present* (eds. Mickelson & Attig), *GSA Special Paper* 337, 69–82;

The Geological Society for Figure 5 from Thomas (1984) Sedimentation of a sub-aqueous esker-delta at Strabathie, Aberdeenshire, in *Scottish Journal of Geology* 20, 9–20;

Scandinavian University Press for Figure 7 from Fyfe (1990) The effect of water depth on ice-proximal glaciolacustrine sedimentation: Salpausselka I, southern Finland, *Boreas* 19, 147–164 and Figure 11 from Matthews et al. (1995) Contemporary terminal moraine ridge formation at a temperate glacier: Styggedalsbreen, Jotunheimen, southern Norway, in *Boreas* 24, 129–139.

CHAPTER

INTRODUCTION TO GLACIAL LANDSYSTEMS

David J.A. Evans

1.1 LANDSYSTEMS

The landsystems concept was initially popularized in the reports of the Commonwealth Scientific and Industrial Research Organization for the mapping of large, sparsely populated areas in Australia in the 1940s (e.g. Christian and Stewart, 1952). Following the pioneering work of Bourne (1931), Unstead (1933) and Veatch (1933), these surveys were stimulated by the desire to evaluate the agricultural potential of large expanses of land. They sought to classify land based upon a landsystems approach, which recognized a landsystem as an area with common terrain attributes, different to those of adjacent areas. The areal coverage of a landsystem was therefore dictated by the size of the terrain attributes and could range from tens to hundreds of kms^2. A recurring pattern of topography, soils and vegetation was regarded as characterizing a landsystem. Theoretically, each landsystem should contain a predictable combination of surface features (landforms) and associated soils and vegetation types. During the preliminary stages of mapping, the topography and/or geomorphology is usually the most significant criterion employed in differentiating landsystems.

A landsystem is divided into smaller components called units (or facets) and elements (Lawrance, 1972). In early reports the land units were often depicted in three-dimensional sketches from which immediate impressions could be gained of relative relief and patterns and densities of land elements. Aerial photograph stereopairs were also used to exemplify typical units of the landsystem. The developmental history of landsystems mapping, as it pertains to terrain classification, is reviewed in more detail by Mabbutt (1968), Mitchell (1973), Ollier (1977), King (1987), and Cooke and Doornkamp (1990). Applications of the landsystems concept to assessments of glaciated terrain have been less of a quantitative mapping exercise for regional development purposes and more for landscape characterization, useful for reconstruction of palaeo-glaciation and predicting the presence of specific sediments. Moreover, because glaciated terrains are often blanketed with depositional features, glacial geomorphologists and geologists have emphasized the properties, structures and distributions of materials that lie beneath the surface in their landsystems models (Eyles, 1983a). This allows the landsystems approach to be employed as a holistic form of terrain evaluation, by not only linking the geomorphology and subsurface materials in a landscape, but genetically relating them through process-landform studies. Moreover, it provides a powerful tool for the reconstruction and interpretation of former glacial environments and ice dynamics.

1.2 GLACIAL LANDSYSTEMS

Mapping in glaciated terrains traditionally involved the grouping of landforms according to common origin and age. An excellent early example of this approach is that of Speight (1963) in the Lake Pukaki area of South Island, New Zealand. Suites of glacial features grouped as 'landform associations' essentially constituted inset sequences of glaciated valley landsystems (see below). The linking of process and form over a large area of diverse glacial landforms was pioneered by Clayton and Moran (1974) in their assessment of the spatial distribution of glacial features in North Dakota, USA (Fig. 1.1). They reconstructed changes in ice dynamics during a cycle of glacial advance and recession, emphasizing both spatial and temporal landform-sediment assemblages and recognizing that glacial features can be added to a landscape as a series of layers (stratigraphy). Important advances represented by this work included the focus on glacial stratigraphy as a prerequisite to studying glacial geomorphology, the commitment to assessing groups of landforms rather than individual forms in isolation, and the recognition of process continuums. This approach, therefore, acknowledged the complexity of glacial depositional systems and highlighted the superimposition of landform-sediment assemblages ('preadvance, subglacial, superglacial and postglacial elements') in spatially coherent 'suites' (Fig. 1.1).

Glacial landsystems were first compiled by Fookes *et al.* (1978) to provide engineers with process-form classifications of glacigenic landform-sediment assemblages. In order to simplify the predominantly complex sequences of landforms and sediments in glaciated basins, three

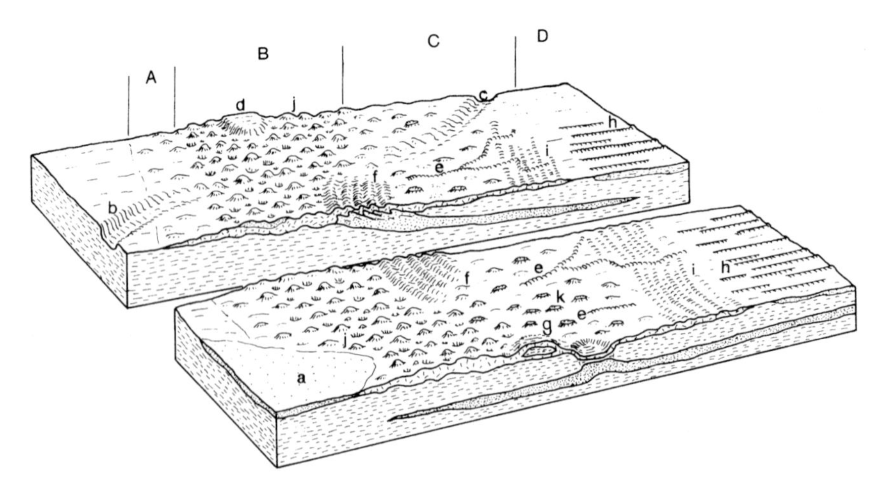

Figure 1.1 A process-form model for the glacial features of North Dakota, emphasizing the zonal nature of particular glacial landform assemblages. A = fringe suite, B = marginal suite, C = transitional suite, D = inner suite; a = fluvial plain, b = meltwater channel, c= partly buried meltwater channel, d = ice-walled lake plain, e = eskers, f = transverse compressional features, g = thrust mass, h = longitudinal features/flutings, i = recessional push moraines, j = hummocks, k = disintegration ridges. (After Clayton and Moran, 1974.)

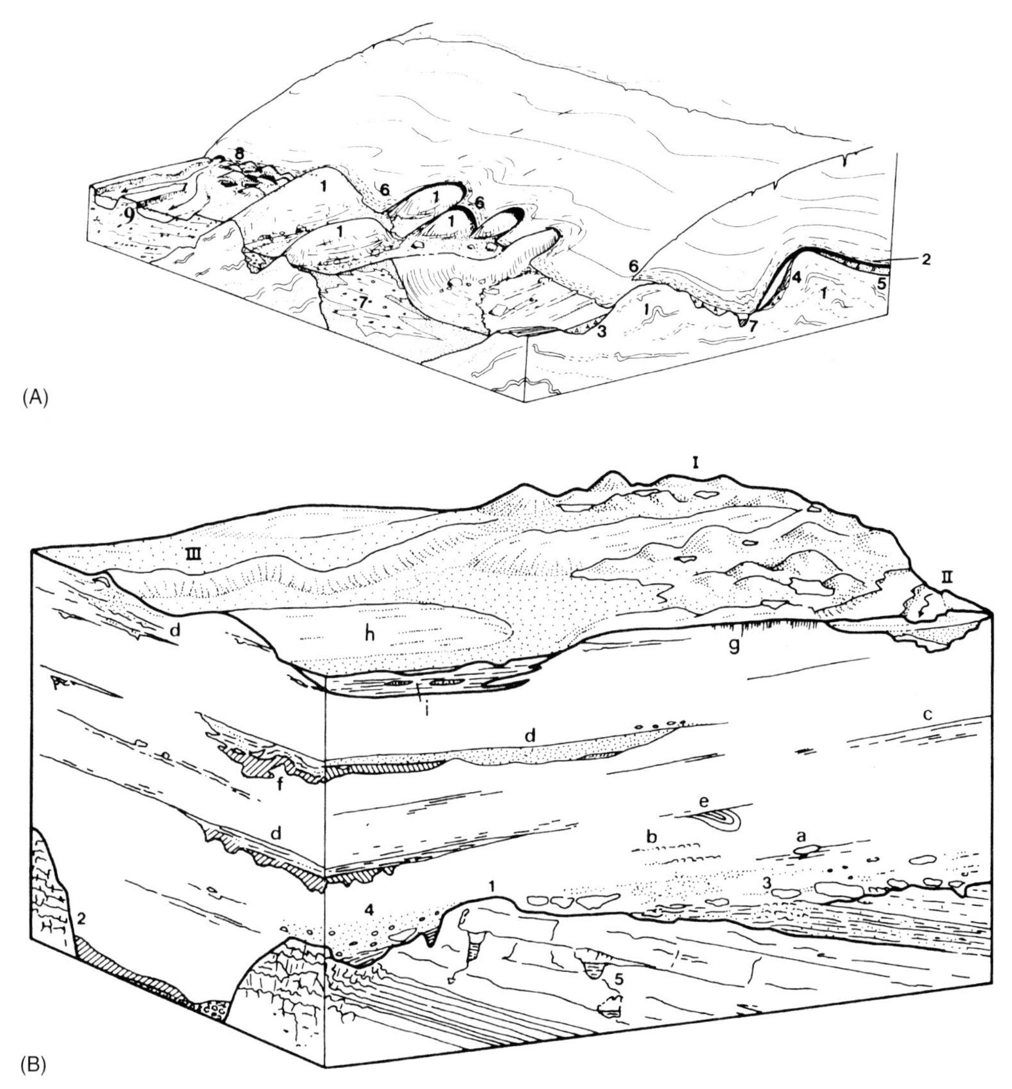

Figure 1.2 The subglacial landsystem. A) An area with hard substrate: 1 = rock drumlins, 2 = basal debris, 3 = subglacial lodgement/deformation till, 4 = lee-side cavity fill, 5 = basal till, 6 = supraglacial debris, 7 = esker, 8 = kettled outwash, 9 = proglacial sandur. B) An area of limestone substrate covered by stacked tills: I = hummocky kame and kettle topography, II = terraced outwash incised into subglacial sediments, III = esker, a = aligned faceted clasts, b = shear lamination, c = slickensided bedding planes, d = stratified subglacial cavity or canal fills, e = folded and sheared-off channel fill, f = diapiric till intrusion into cavity fill, g = vertical joints produced by pedogenesis, h = drumlin, i = interdrumlin depression filled with postglacial sediment, 1 = striated rock head, 2 = buried channel or valley, 3 = glacitectonized rock head producing rock rafts and boulder pavements in till, 4 = lowermost till with local lithologies, 5 = cold water karst. (From Eyles, 1983a).

landsystems were identified. The first was the 'till plain landsystem' which encompassed subglacially deposited tills and drumlinized surfaces. Second, the 'glaciated valley landsystem' referred to the materials produced by melting glacier ice (e.g. end, lateral and medial moraines) but excluded glacifluvial sediments. Finally, the 'fluvioglacial and ice-contact deposit landsystem' included all those landforms and sediments that are of ice-contact glacifluvial origin. These landsystems are differentiated not only by the landform-sediment assemblages that they contained but, more specifically and also on a practical level, by the engineering properties of their component materials. The Fookes *et al.* (1978) landform-sediment assemblages are, therefore, directly comparable with the 'elements' of Clayton and Moran (1974).

The component parts of landsystems, the elements and units, were placed in the context of glaciated basins by Fookes *et al.* (1978) and subsequently by Eyles (1983a). Elements are simply individual landforms that can be mapped at large scales. This includes features like drumlins, flutings, eskers, kame terraces and moraines. Such features can then be grouped as units, which constitute relatively homogeneous tracts of land that are distinct from surrounding surfaces. A typical land unit may comprise a drumlin or fluting field, an area of Rogen moraine, a suite of recessional push moraines or an outwash plain. The landsystem is a recurrent pattern of genetically linked land units. Eyles (1983a, b), Eyles and Menzies (1983) and Paul (1983) deviated from the initial landsystems classifications of Fookes *et al.* (1978) by incorporating all of the landform and sediment types associated with particular glaciation styles. For example, the subglacial landsystem included subglacial tills and streamlined landforms in association with glacifluvial features like eskers (Fig. 1.2). Similarly, the supraglacial landsystem contained stratified glacifluvial sediment accumulations associated with ice stagnation in addition to chaotic hummocky moraine (Fig. 1.3). The glaciated valley landsystem was re-packaged by Eyles (1983b) to include all of the landforms and sediments produced by the glaciation of mountain valleys and was not restricted to

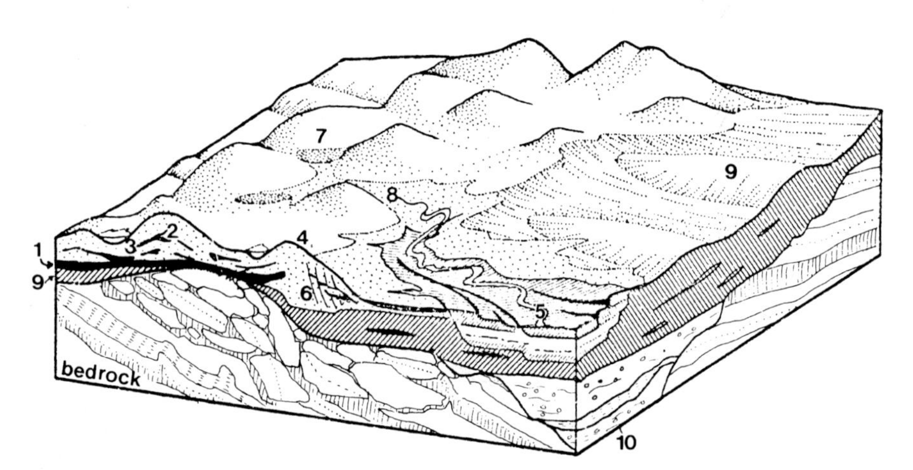

Figure 1.3 The supraglacial landsystem. 1 = melt-out till, 2 and 3 = flowed supraglacial diamictons, 4 = hummocky terrain composed of kames and chaotic hummocky moraine, 5 = terraced glacifluvial sediments, 6 = failure of strata by normal faulting due to the melting of adjacent ice cores, 7 = kettle holes and small ponds located between hummocks, 8 = outwash fan fed by melting ice cores in hummocky terrain, 9 = drumlinized subglacial till surface visible through a window in the supraglacial landsystem, 10 = buried valley or infilled tunnel valley. (After Eyles, 1983a).

supraglacial features (Fig. 1.4). Glaciated terrain was then mapped according to the dominant landsystems (Eyles and Dearman, 1981; Eyles *et al.*, 1983a), providing a predictive tool for rapid assessments of subsurface materials by those involved in resource and engineering project management.

Since the introduction of these primary glacial landsystems, intensive research around modern glacier margins has led to an expansion of the landsystem concept. Brodzikowski and van Loon (1987, 1991) maintained the separation of landform-sediment assemblages according to environment of deposition (Fig. 1.5), preferring not to emphasize the influence of glacier morphology and dynamics. However, glacial depositional environments are extremely complex, and variability in landform-sediment assemblages is dictated not only by the location of deposition but also by the 'style' of glaciation. Glaciation 'styles' are a function of climate, basement and surficial geology and topography, and consequently a wide range of glacial landsystems have been compiled for different ice masses and dynamics (Benn and Evans, 1998, chapter 12). For example, it has been recognized that the details of glaciated valley landsystems will reflect the relative relief and climatic regime of mountainous terrain in which they are

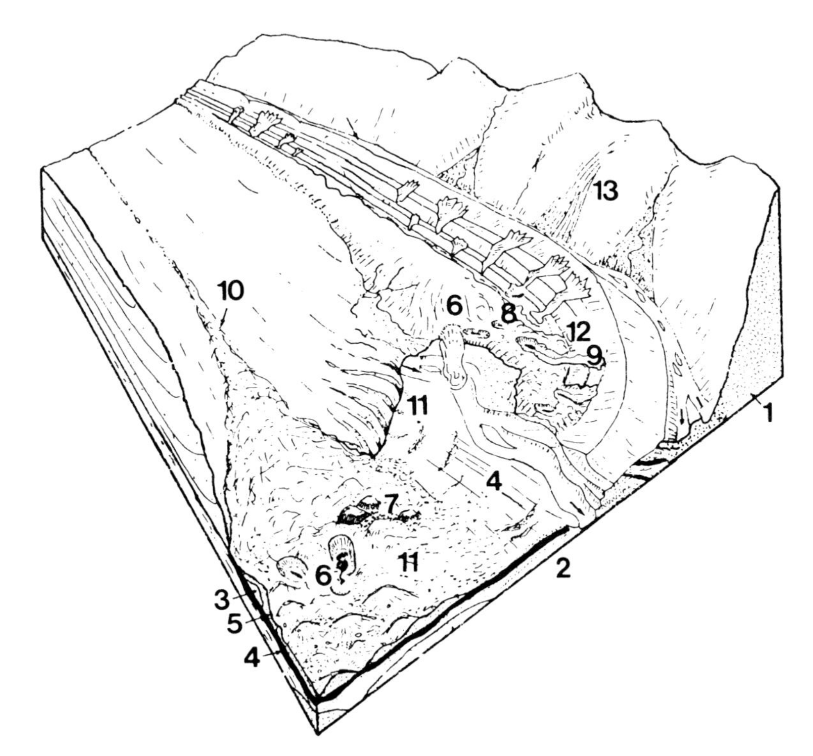

Figure 1.4 The glaciated valley landsystem. 1 = oversteepened bedrock slope, 2 = buried valley, 3 = melting ice core, 4 = subglacial till with fluted surface, 5 = hummocky supraglacial debris, 6 = supraglacial lateral moraine, 7 = large angular clasts derived from rockfall onto the glacier surface, 8 = interbedded glacifluvial sediments and flowed diamictons deposited in lateral positions, 9 = debris flows caused by melt-out of ice cores, 10 = medial moraine, 11 = minor ridges of supraglacially dumped debris, 12 = lateral moraine ridge, 13 = paraglacial fans. (From Eyles, 1983a).

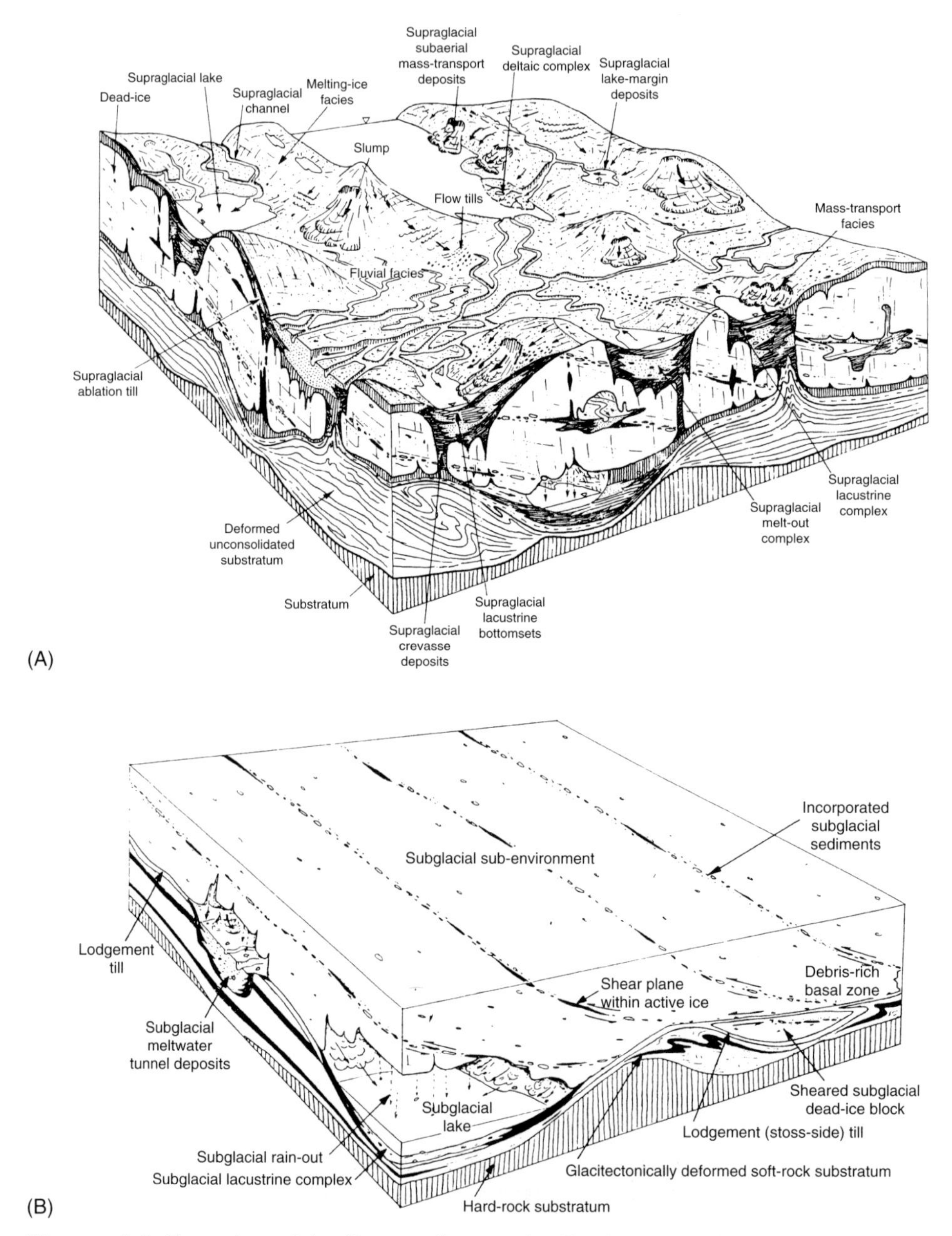

Figure 1.5 Examples of landform-sediment classification according to location. A) The supraglacial sub-environment. B) The subglacial sub-environment. (From Brodzikowski and van Loon, 1991).

located (e.g. Boulton and Eyles, 1979; Owen and Derbyshire, 1989; Evans, 1990a, b; Spedding and Evans, 2002). Additionally, some researchers have grouped the primary glacial landsystems in models that encompass the whole range of glacial processes active at particular glacier margins (e.g. Krüger, 1987, 1994a; Fig. 1.6). This is a particularly powerful approach where landscape evolution can be monitored (e.g. Kjær and Krüger, 2001; Fig. 1.7). Furthermore, it

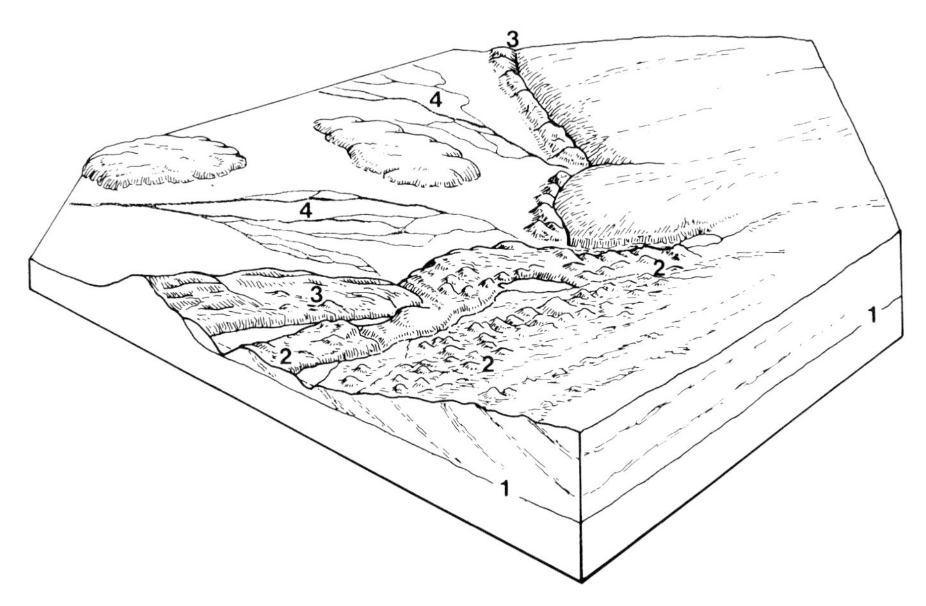

Figure 1.6 An example of applying primary glacial landsystems to an individual glacier margin. Lowland temperate glacier snout, Iceland. 1 = subglacial, 2 = supraglacial, 3 = terminoglacial, 4 = proglacial. (From Krüger, 1994a).

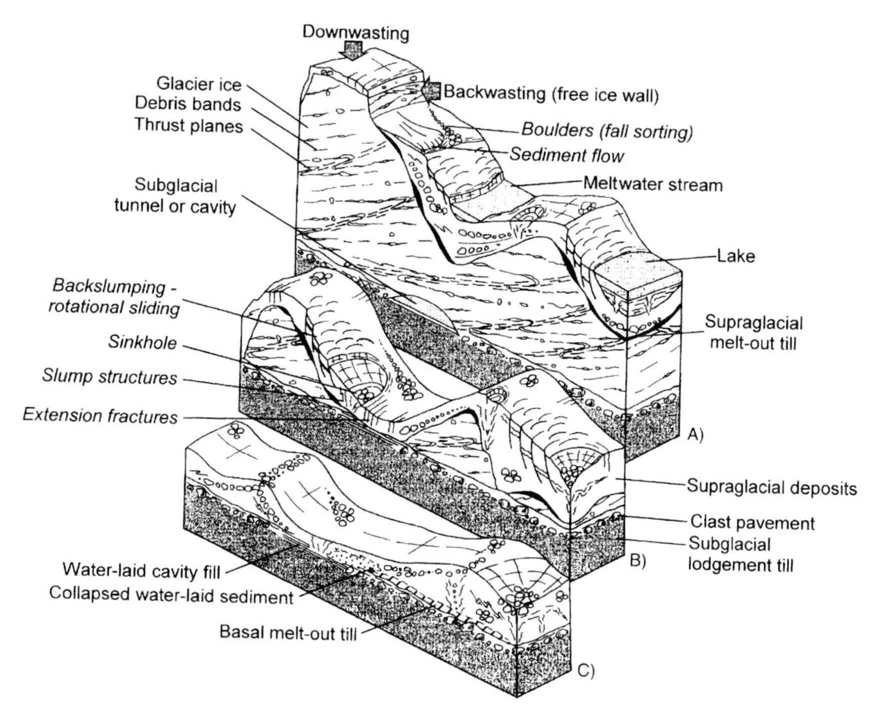

Figure 1.7 A model of moraine development due to the melt-out of debris-covered glacier ice. (From Kjær and Krüger, 2001).

has led to the identification of landform-sediment suites indicative of specific styles of glaciation (Benn and Evans, 1998) or the impacts of certain ice dynamics (e.g. surging glaciers, Evans and Rea, 1999; ice streams, Stokes and Clark, 1999). Once a landform-sediment suite pertaining to a single period of glacier occupancy or activity can be identified, it often becomes possible to differentiate overprinted signatures (e.g. Dyke and Morris, 1988; Clark, 1993; Krüger, 1994a; Evans *et al.*, 1999a; Fig. 1.8). Subaqueous depositional environments were recognized in the early landsystems models as extensions of the primary landsystems (e.g. Eyles and Menzies, 1983). Additionally models of the wide specturm of glacimarine processes and depositional features have appeared in reviews of offshore environments (e.g. Eyles *et al.*, 1985; Fig. 1.9) and have recently been given separate treatment in a landsystem context (e.g. Benn and Evans, 1998). This is in recognition of the genetic linkage between sediments from terrestrial glacier systems that terminate offshore and landforms in deep water settings (Powell, 1984; Brodzikowski and van Loon, 1987, 1991).

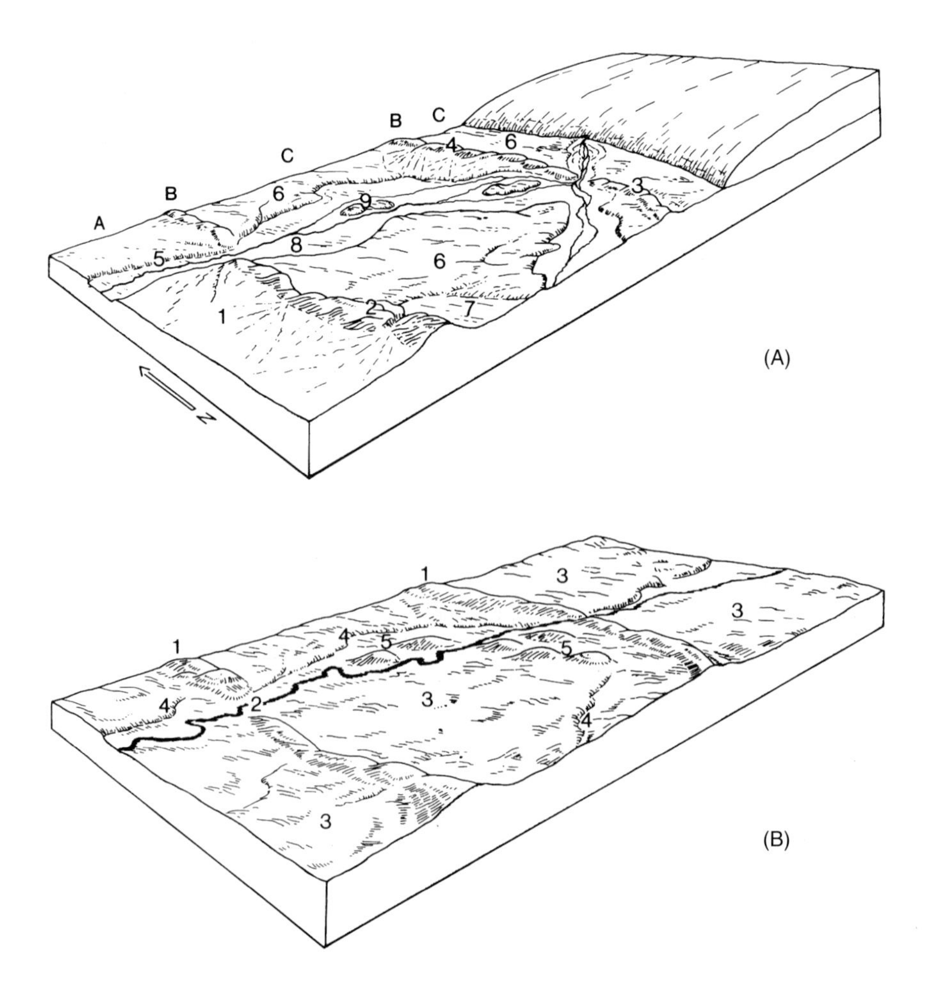

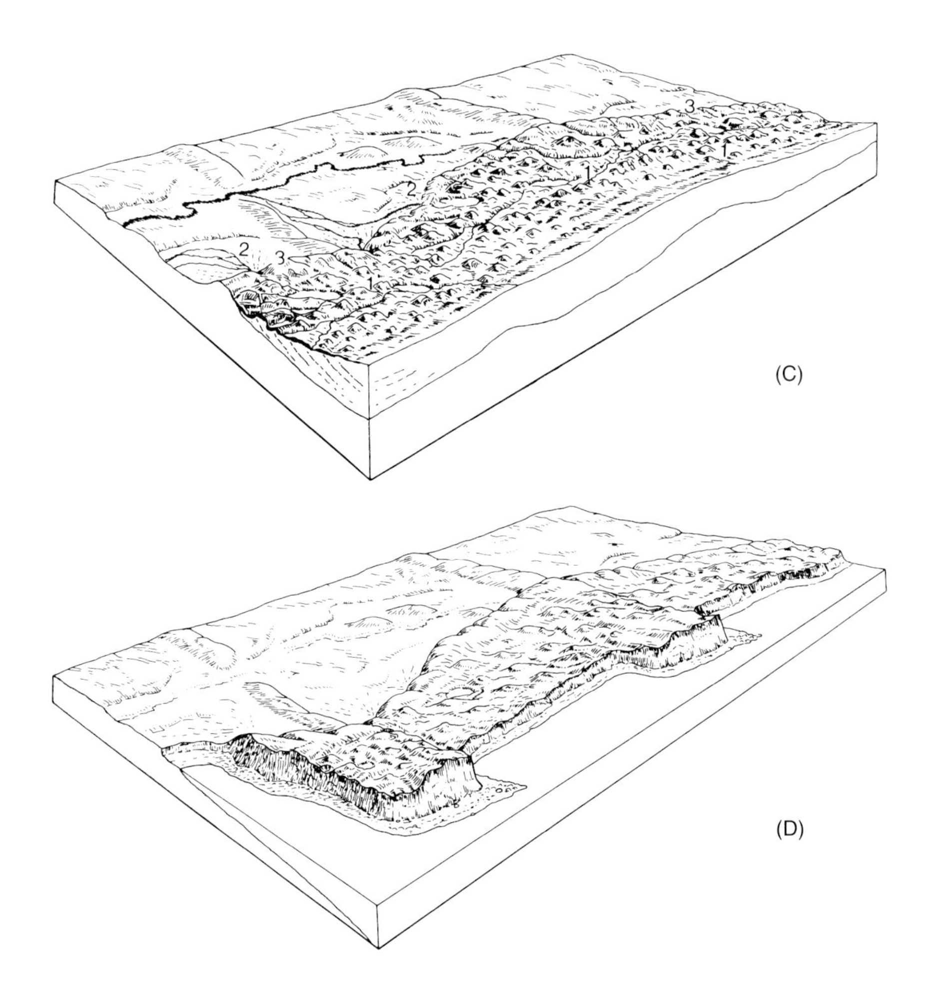

Figure 1.8 The overprinting of landform suites due to repeated glacier advances. A) Ice recession towards the east, leaving ice-contact fans (1) and an end moraine (2); a readvance then deposits a further moraine of stacked till slabs (3) and bulldozed outwash fan sediments (4); the foreland comprises A = proglacial landsystem, containing ice-contact fans and erosional terraces (1 and 5), B = terminoglacial landsystems, containing push moraines (2–4), C = subglacial landsystems, containing till surfaces and localized lake sediments (6 and 7) with areas of glacifluvially reworked material and isolated islands of subglacial till (8 and 9). B) The same terrain after a further glacier advance from the southeast. Because the glacier readvance only deposited a carapace of till over pre-existing landforms, those landforms still form the majority of the relief. Overridden push moraines and intervening meltwater corridors have been smoothed (1 and 2) and outwash plains and till surfaces have been flattened (3); some large glacifluvial terraces are still visible (4); some upstanding ridges have been smoothed into drumlins (5). C) A further, less-extensive glacier readvance from the south. Hummocky moraine develops as debris-covered ice stagnates (1) and ice-contact fans (2) are fed by supraglacial streams; an end-moraine ridge (3) marks the edge of the stagnating snout. D) The final deglaciated landscape after partial erosion by the sea or a proglacial lake. (From Krüger, 1994a).

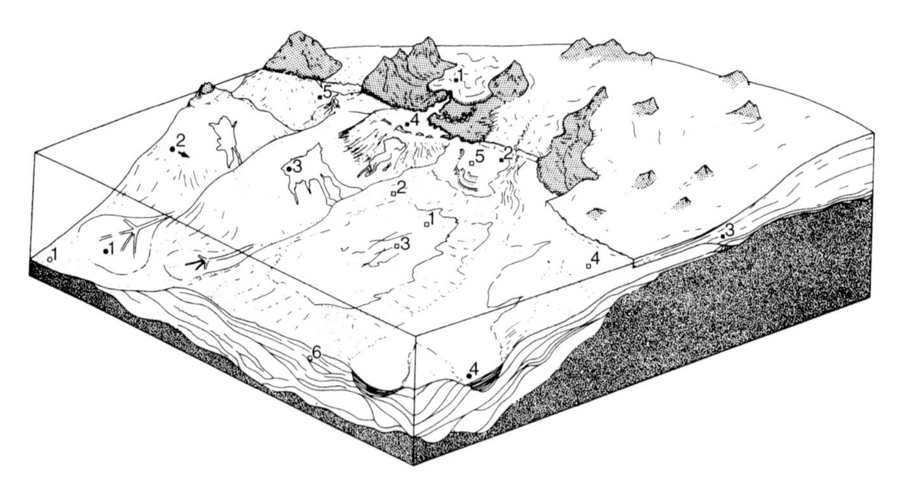

Figure 1.9 Model of glacimarine environments along a glaciated continental margin. Basin margin (■): 1 = grounded terrestrial ice margin, 2 = grounded marine ice margin and subaqueous fans, 3 = ice shelf grounded below sea level and subaqueous fan, 4 = glacier-fed marine delta, 5 = glaciated fjord. Shelf with active ice rafting (□): 1 = shallow bank carbonates, palimpsest sediments, 2 = deep bank mud and diamicton drapes, 3 = iceberg scour, 4 = sediment starved shelf, 5 = moraines left by receding ice lobe, 6 = stratiform sequences of diamicton, mud and channelized re-sedimented facies. Continental slope rise and canyons (•): 1 = glacially influenced submarine fan, 2 = upper slope contourites, 3 = downslope re-sedimentation from debris flows to turbidites, 4 = canyon fill and feeder to fans. Abyssal plain (○): 1 = pelagic sediments, ice-rafted debris, turbidites. (From Eyles *et al.*, 1985).

1.3 AIMS AND SCOPE OF THIS BOOK

The expanding research on glaciers and glaciation has led to the production of considerable volumes of data on glacial processes and their resultant forms and sediments. These data are presented in ever more complex formats for use in science, education and commercial enterprise (e.g. Barnett *et al.*, 1998; Fig. 1.10). This information can be parochial or reductionist in nature until it is applied through more holistic assessments of glacial environments and put to practical use, either in resource management/development or in regional through continental to global scale models of palaeo-climate reconstruction and future climate prediction. Due to the great spatial and temporal complexity of glacial processes and forms, glacial geomorphologists and geologists have found it appropriate to compile process-form models that relate to specific glaciation styles and dynamics. This book aims to provide a reasonably comprehensive and up-to-date selection of such models (landsystems) pertinent to both modern and ancient glacier systems and to a continuum of scales ranging from ice sheets to valley glaciers. These landsystems are useful for anyone embarking upon research or engineering surveys in glaciated basins and should be regarded as broad templates for interpreting glacigenic landform-sediment assemblages produced by different glaciation styles and ice dynamics in differing climatic, geologic and topographic settings.

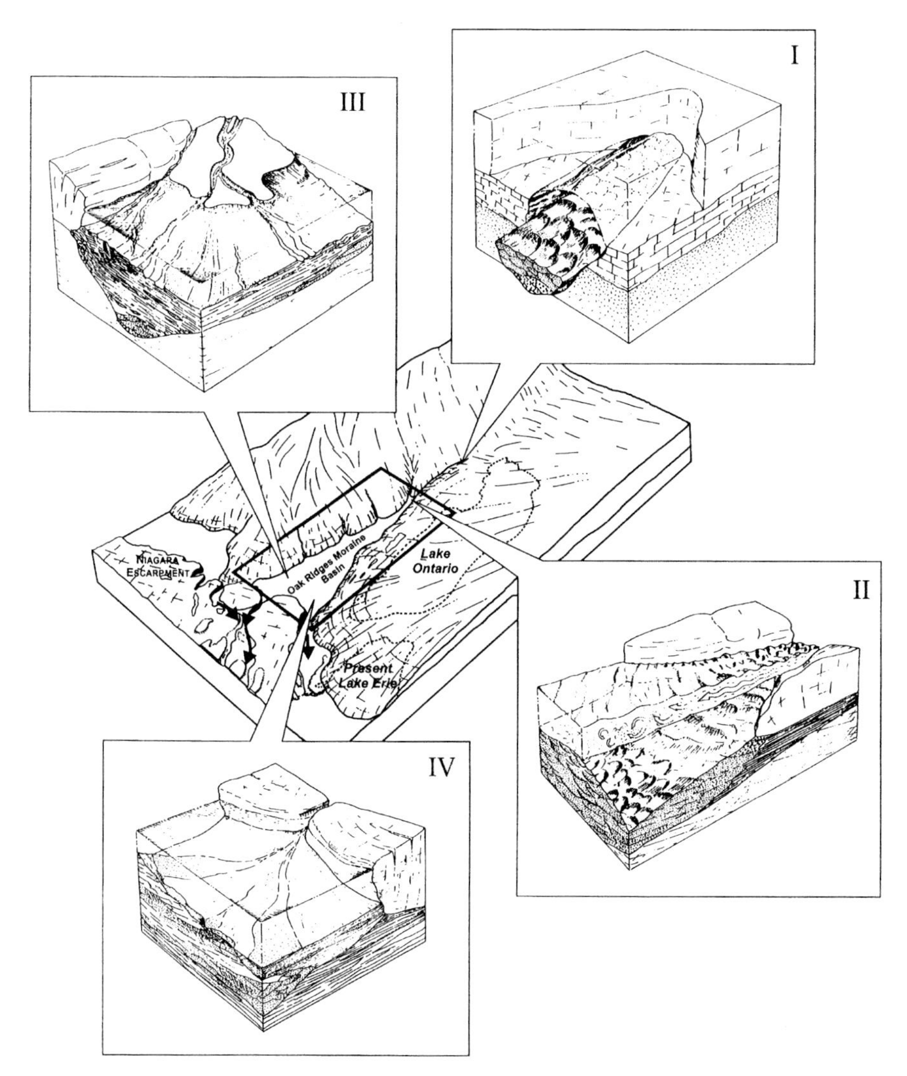

Figure 1.10 Reconstruction of the evolution of the Oak Ridges Moraine based upon extensive three-dimensional data. I = subglacial sedimentation, II = subaqueous fan sedimentation, III = fan to delta sedimentation, IV = ice-marginal sedimentation. (Drawings by J.R. Glew, from Barnett *et al.*, 1998).

CHAPTER

2

ICE-MARGINAL TERRESTRIAL LANDSYSTEMS: ACTIVE TEMPERATE GLACIER MARGINS

David J.A. Evans

2.1 INTRODUCTION

Temperate glacier margins are mainly wet-based for at least part of the year and are located in terrain that contains discontinuous or no permafrost. Such glaciers are considered as active when they are capable of forward momentum even during overall recession. This is manifest in the small winter readvances that characterize receding outlet glaciers in places like Iceland (e.g. Sharp, 1984; Boulton, 1986; Krüger, 1995). Cold winter conditions result in the penetration of a seasonal cold wave from the atmosphere through the thin ice. This produces a narrow marginal frozen zone thought to be significant in the production of some frontal moraines. In some settings the marginal frozen zone may persist for several years and is therefore technically discontinuous permafrost. The processes and major landform-sediment associations of active temperate glacier margins have been studied in great detail (e.g. Sharp, 1982, 1984; Harris and Bothamley, 1984; Krüger and Thomsen, 1984; Krüger, 1985, 1993, 1994a, 1997; Boulton, 1986; Boulton and Hindmarsh, 1987; Benn, 1995; Evans and Twigg, 2002), thereby informing models of landform production (e.g. Boulton and Eyles, 1979; Gustavson and Boothroyd, 1987; Krüger, 1987; Benn and Evans, 1998).

The impact of any glacier on landform development relies heavily upon its ability to transport debris. Although debris-rich basal ice sequences are typically thin or absent beneath temperate glaciers (Hubbard and Sharp, 1989), concentrations of debris are observed in the basal ice facies of glaciers whose subglacial meltwaters are influenced by supercooling in overdeepenings (e.g. Alley *et al.*, 1998, 1999; Lawson, *et al.*, 1998; Evenson *et al.*, 1999). Any debris-rich ice that does occur in a temperate glacier snout can be transported to englacial and supraglacial positions by compressive flow (Fig. 2.1). In relative terms, however, the volume of material transported through an active temperate lowland glacier by these mechanisms is small and so, consequently, supraglacial landforms and widespread ice stagnation topography are uncommon compared with temperate glaciers in high-relief settings. Exceptions to this rule occur in situations where unusually high concentrations of debris-rich ice are produced (e.g. through freezing of englacial drainage systems and/or supercooling, Spedding and Evans, 2002), or where complex marginal oscillations have led to proglacial thrusting and the re-incorporation

(A)

(B)

Figure 2.1 Debris entrained in active temperate glacier snouts. A) Debris-rich ice in the snout of Kvíárjökull. This originates either by freezing-on, by the process of supercooling over a subglacial overdeepening, or the freezing of sediment-charged englacial drainage networks (Spedding and Evans, 2002) and/or by freezing in fractures produced by pressurized meltwater discharges (e.g. Roberts *et al.*, 2000, 2001; Ensminger *et al.*, 2001); the sediments are then exposed on the glacier surface by compressive flow; (ice axe in foreground for scale). B) Typically sparse debris concentrations in the snout of Flaajökull. This view also shows the large amount of debris being transported as a subglacial deforming layer and used to construct push moraines in the foreground.

of stagnating ice in a fashion similar to the surging glacier landsystem (e.g. Kotlujökull, Iceland; Krüger, 1994a). Because later chapters develop the theme of temperate glacier lobes in mountain topography, this chapter concentrates on lowland outlet glaciers and uses mainly Icelandic case studies for illustration.

2.2 LANDFORM-SEDIMENT ASSEMBLAGES IN ACTIVE TEMPERATE GLACIER FORELANDS

Previous studies of landform-sediment associations on the recently deglaciated forelands of active temperate glaciers have highlighted three dominant depositional domains with characteristic landform-sediment assemblages. First, areas of extensive, low-amplitude marginal dump, push and squeeze moraines (Price, 1970; Krüger, 1987; Evans *et al.*, 1999a; Evans and Twigg, 2000, 2002) are derived largely from material on the glacier foreland and often record annual recession of active ice (Sharp, 1984; Boulton, 1986; Krüger, 1995). Short periods of readvance or stability can produce larger push moraines (Krüger, 1993). This can be a product of stacking of sub-marginally frozen sediment slabs (Humlum, 1985; Krüger, 1985, 1987, 1993, 1994a, 1996; Matthews, *et al.* 1995), or dump, squeeze and push mechanisms operating at the same location over several years. The incremental thickening of ice-marginal wedges of deformation till has also been proposed (e.g. Johnson and Hansel, 1999). Second, subglacial landform assemblages of flutings, drumlins and overridden push moraines dominate the areas between ice-marginal depo-centres (Krüger, 1987; Benn, 1995; Evans *et al.*, 1999a, b). These features have been linked to subglacial deforming layers, particularly in previous Icelandic research on active temperate glaciers (Boulton, 1987; Boulton and Hindmarsh, 1987; Benn, 1995; Hart, 1995; Benn and Evans, 1996). Their ubiquity in temperate glacier forelands attests to the sparsity of supraglacial sediments let down onto the substrate during deglaciation. Third, large areas of glacifluvial forms such as recessional ice-contact fans (Boulton, 1986; Price, 1969) and hochsandur fans (Krüger, 1997) cut through, or are directed by, moraine ridges. These outwash forms are often associated with simple and complex, anabranched eskers (Price, 1966, 1969) and small areas of pitted outwash (Price, 1969, 1971). Within the enclosed depressions on the foreland, proglacial lakes will expand and contract in response to the evolving drainage networks, acting as temporary storage for glacifluvial sediments. In addition to the three major depositional domains, anomalies are to be expected due to the signatures of erratic/non-cyclic or site-specific processes, and these are reported elsewhere. For example, jokulhlaup (Maizels, 1997; Russell and Knudsen, 1999; Fay, 2002) and surge (Evans and Twigg, 2002) features may be represented in the landform record of active temperate glaciers. The three major depositional domains are now reviewed in turn by concentrating on their diagnostic landform-sediment associations.

2.2.1 Marginal Morainic Domain

The most extensive and sharply defined moraines on active temperate glacier forelands are the small push moraines, often clearly linked to annual advances (Price, 1970; Boulton, 1986; Fig. 2.2). They comprise a wide range of sediments including:

- subglacially derived diamictons with large numbers of striated and faceted clasts
- glacifluvial sediments that have been reworked from pre-advance outwash
- glacitectonized slabs of laminated sands and muds originating in proglacial lakes.

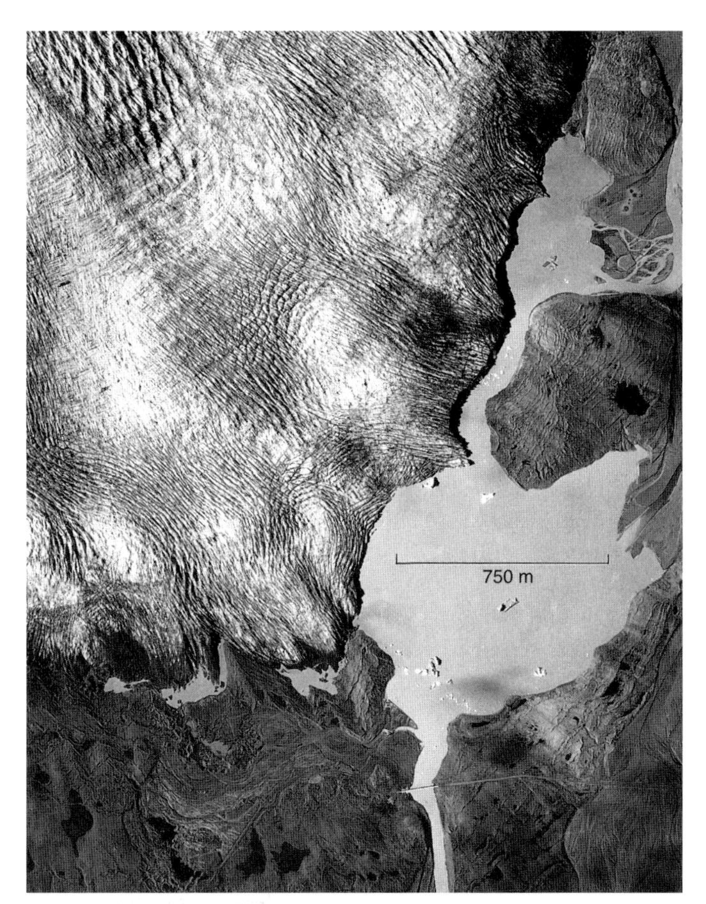

(A)

Figure 2.2 Push moraines at the margins of Icelandic active temperate glaciers. A) Aerial photograph of push moraines at the margin of Fjallsjökull, showing saw-tooth plan forms and partial overriding (Landmælingar Islands and University of Glasgow, 1965); (scale bar represents 750 m). B) Large composite push moraine at the margin of Flaajökull, constructed by the snout during a period of ice-marginal stability in the early 1990s.

(B)

Their crenulate, lobate or saw-tooth plan-forms mimic the indentations of the snout margin produced by the melting back of longitudinal crevasses (Price, 1970; Matthews *et al.*, 1979). At some locations push moraines may be coalescent or partially superimposed, indicating that one winter readvance was more extensive than that of the previous year. A stationary glacier snout may construct larger and more complex moraine ridges in this fashion (Krüger, 1985, 1993).

The local characteristics of push moraines indicate that most are produced either by ice-marginal squeezing of water-soaked subglacial sediment (e.g. Price, 1970) or pushing of proglacial materials (Howarth, 1968). The occurrence of flutings on the proximal slopes and crests of many push moraines (Fig. 2.3) prompted Evans and Twigg (2002) to propose that subglacial bedform production and push moraine formation are genetically linked on the Breiðamerkurjökull/Fjallsjökull foreland. Moreover, the fact that the flutings are arranged in

Figure 2.3 Part of an aerial photograph of the Breiðamerkurjökull foreland, Iceland (Landmælingar Islands and University of Glasgow, 1998), showing flutings arranged in strips between push moraines and terminating at the moraine crests. Note that some larger spindle drumlins are draped by push moraine (p^{i}) and terminate at the older push moraine (p^{ii}); (scale bar at bottom left represents 500 m).

strips separated by push moraines and aligned at right angles to the push moraine at their down-glacier end indicate that they clearly record individual flow events associated with the ice margin responsible for moraine deposition. Flutings are therefore the product of subglacial erosion and transport processes that advect sediment to the ice margin, where squeezing and pushing combine to create the moraine ridge (e.g. Sharp, 1984; Johnson and Hansel, 1999; Fig. 2.4). At locations where glacier snouts are thin and winters are relatively cold the construction of annual push moraines may involve the penetration of the winter cold wave down through ice marginal tills. For example, Krüger (1995) suggests that the annual moraines produced by the recession of northern Myrdalsjökull are initiated when the sub-marginal till is frozen on to the glacier snout and then pushed forward as a frozen slab in the winter (Fig. 2.4).

The role of a marginal zone of frozen sediment, possibly even discontinuous permafrost (see van der Meer *et al.*, 1999), is central to the explanation of larger moraines produced by stationary glacier snouts (Krüger, 1993, 1994a, 1996; Matthews *et al.*, 1995; Fig. 2.4). The model of Matthews *et al.* (1995) involves the winter freeze-on of supraglacial waterlain sediments and subglacial till that accumulate on and under the snout behind a large moraine. The moraine volume is increased each year as the glacier advances over its proximal slope and then melts out, draping the moraine surface with a layer of subglacial till and supraglacial outwash. Krüger (1993, 1994a, 1996) proposes a mechanism of incremental stacking of frozen subglacial till slabs. If the glacier snout reaches the same point each year then it will produce a large moraine composed of imbricate slabs of till interbedded with debris flow deposits.

Common features on the forelands of active temperate glaciers are the more subtle arcuate, low-amplitude ridges aligned parallel to recessional push moraines (Fig. 2.5). These ridges have clearly been overridden and moulded by the glacier as indicated by flutings that continue uninterrupted from their up-ice to down-ice slopes. They are also draped in some locations by recessional push moraines. Based upon these characteristics the arcuate ridges are interpreted as overridden push moraines, initially deposited during an earlier phase of glacier advance (Krüger, 1987; Evans *et al.*, 1999b; Evans and Twigg, 2000, 2002).

Due to the sparsity of supraglacial and englacial debris over large areas of temperate glacier snouts, hummocky moraine, defined as the product of melt-out of debris-mantled glaciers (Benn and Evans, 1998), is often sparsely developed over deglaciated forelands. Narrow, elongate zones of rubble or rubbly, low-amplitude hummocks can document the lowering of medial moraines onto the substrate (e.g. Dawson, 1979; Levson and Rutter, 1989a, b; Evans and Twigg, 2000, 2002; Fig. 2.6). However, such features are rarely reported, being difficult to identify among the stronger imprints of other landform-sediment accumulations. In addition, debris-charged glacier snouts, produced by marginal freeze-on and the development of debris-rich ice facies, may melt-out to form low-amplitude hummocky moraine (e.g. Kötlujökull, Krüger, 1994a; Kjær and Krüger, 2001; see Chapter 1).

2.2.2 Subglacial Domain (Tills and Associated Landforms)

The former beds of active temperate glaciers typically comprise areas of striated and polished bedrock with roches moutonnées, indicating that abrasion and quarrying is widespread at the ice/bedrock interface, covered or at least partially covered by subglacial sediments and landforms such as flutings and drumlins (Fig. 2.7). Subglacial experiments conducted beneath Breiðamerkurjökull demonstrated that the till was emplaced by deformation and that it

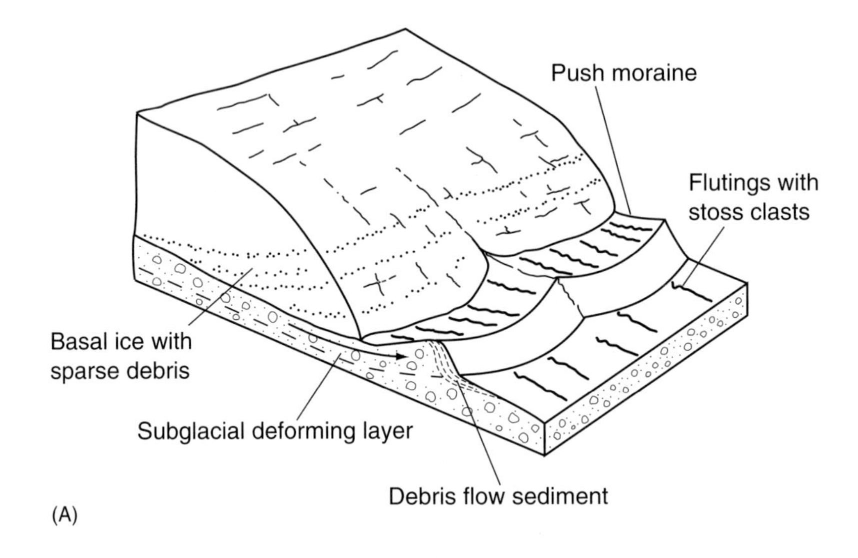
Push moraine
Flutings with stoss clasts
Basal ice with sparse debris
Subglacial deforming layer
Debris flow sediment
(A)

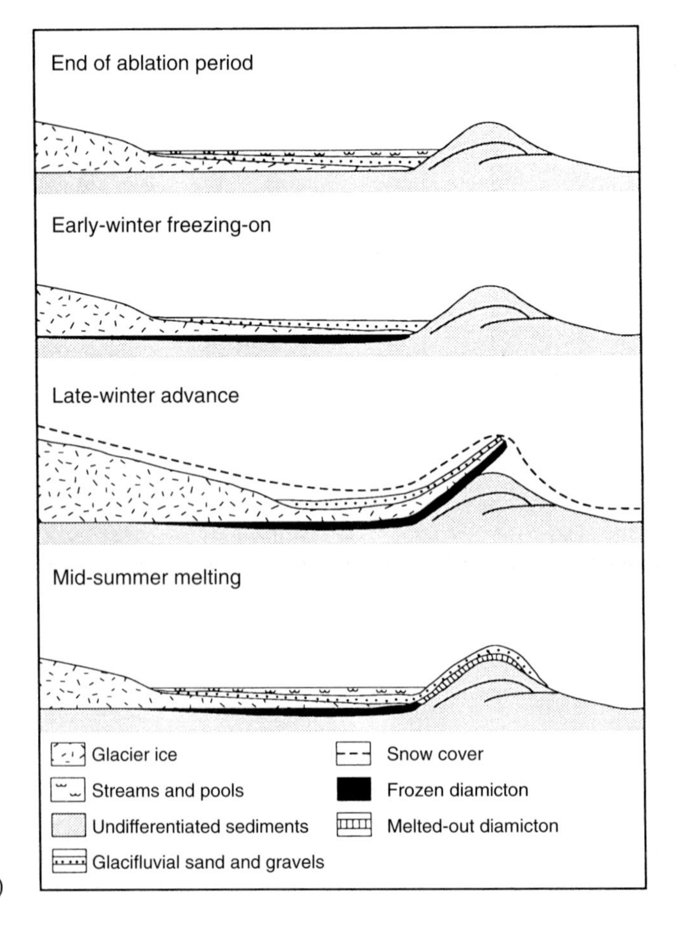
End of ablation period
Early-winter freezing-on
Late-winter advance
Mid-summer melting
Glacier ice
Streams and pools
Undifferentiated sediments
Glacifluvial sand and gravels
Snow cover
Frozen diamicton
Melted-out diamicton
(B)

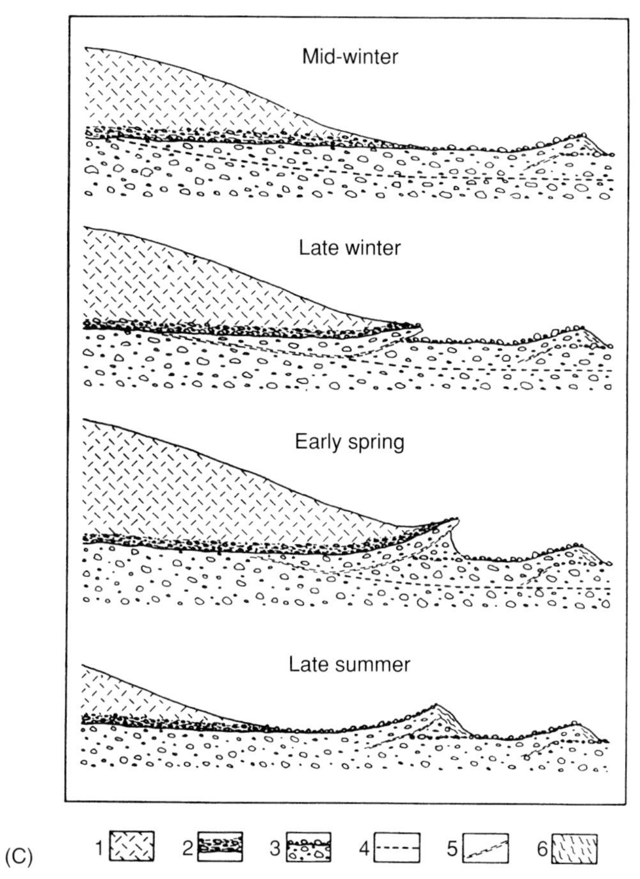

Figure 2.4 The main modes of push moraine formation. A) Subglacial deformation and ice-marginal squeezing. (After Price, 1970; Sharp, 1984; Johnson and Hansel, 1999). The subglacial deforming layer is extruded or squeezed out at the glacier margin as depicted by the arrow. B) 'Double-layer annual melt-out' model of Matthews *et al.* (1995). Although the example shows the production of a large moraine by a stationary glacier snout, individual recessional moraines could also be produced in this way. C) Pushing of a frozen till slab (From Krüger, 1993, 1994a, 1996): 1 = glacier ice, 2 = debris-rich glacier ice, 3 = clast paved lodgement till, 4 = 0 °C isotherm, 5 = thrust plane, 6 = mass movement deposits.

possessed a two-tiered structure (Boulton and Dent, 1974; Boulton, 1979; Boulton and Hindmarsh, 1987; Boulton and Dobbie, 1998; Fig. 2.8). This structure is thought to be the product of ductile flow of an upper dilatant layer (A-horizon) and brittle or brittle-ductile shearing of a lower stiff layer (B-horizon). The two-tiered structure has been observed in Icelandic tills by Dowdeswell and Sharp (1986) and Benn (1995). More recent process experiments conducted by Boulton *et al.* (2001a) suggest that tills may undergo a variety of responses to stress, these being driven by temporal variations in water pressure/effective pressure and concomitant vertical variations in the locus of plastic failure. The result would be cumulative, distributed net strain in response to localized failure events. They also suggest that the emplacement of tills by deformation is not necessarily an instantaneous process, but rather, cumulative whereby till is lost from the base of the deforming horizon. Specifically, the

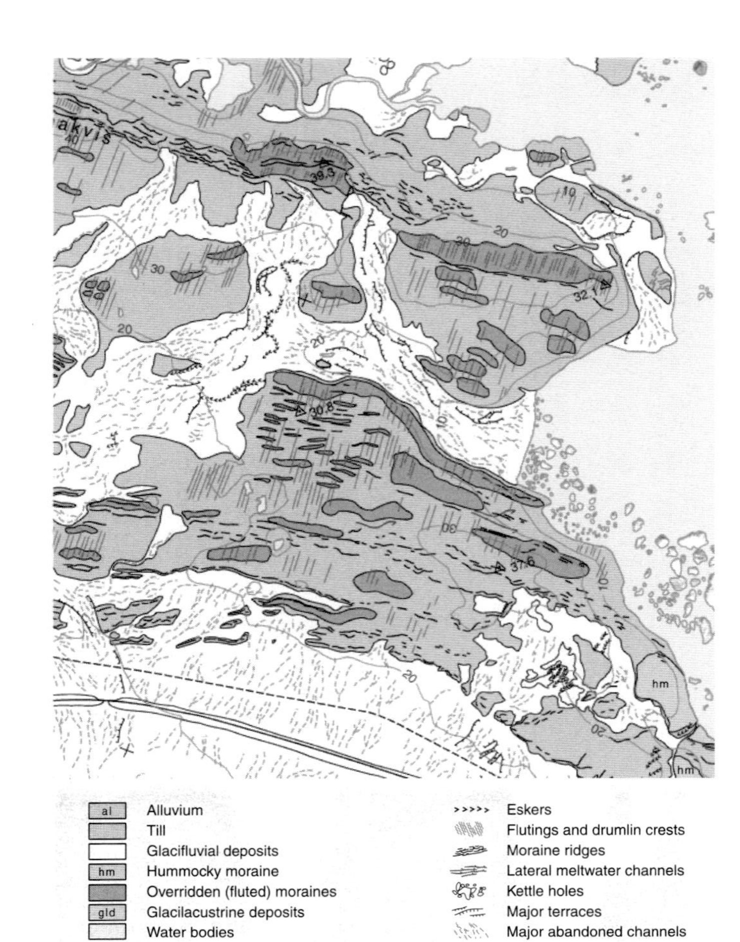

(A)

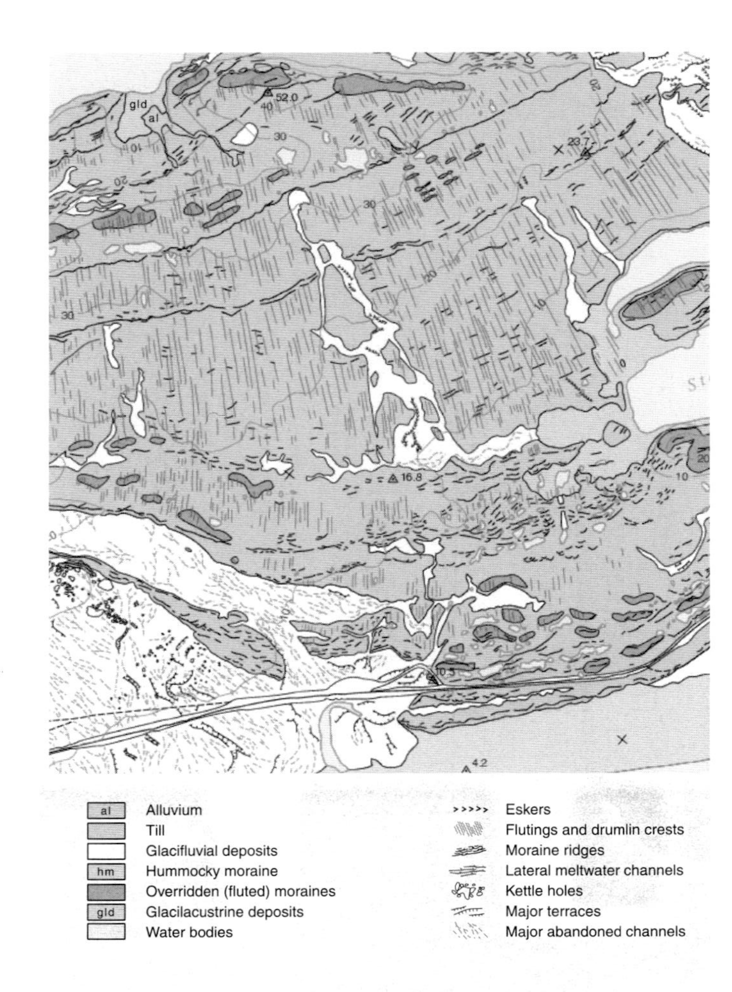

Figure 2.5 (Part 1)

(A)

Figure 2.5 (Part 2)

(B)

Figure 2.5 A) Portions of an aerial photograph (Landmælingar Islands and University of Glasgow, 1998) and surficial geology and geomorphology map of Breiðamerkurjökull, Iceland (from Evans and Twigg, 2000) showing overridden push moraines and their adornments of flutings and recessional push moraines. B) Portion of an aerial photograph (Landmælingar Islands, 1989) of the foreland of Skalafellsjökull (S) and Heinabergsjökull (H), Iceland, showing overridden push moraines (o) and superimposed recessional push moraines (p) and flutings.

cessation of deformation produces a single deformation till horizon but in most instances deformation tills probably accumulate by increments as material is lost from the base of the deforming horizon to produce 'tectonic/depositional slices'. Later-stage deformation features may also be superimposed on earlier structures, although this would be difficult to detect at higher strains. Changes in subglacial water pressure and associated stick-slip motions measured at the base of Breiðamerkurjökull are explained by Boulton *et al.* (2001a) as the result of a change from basal sliding to subglacial deformation (Fig. 2.9).

The numerous flutings that usually characterize the till surfaces of the forelands of active temperate glaciers (Figs. 2.3 and 2.10) have been explained as the products of till squeezing into cavities that develop on the down-glacier sides of lodged and striated boulders (Boulton, 1976;

Figure 2.6 Bouldery veneer draping the foreland of Breiðamerkurjökull, Iceland, and documenting the lowering of the medial moraine visible on the glacier snout in the distance.

Benn, 1994). Such boulders are commonly observed at the heads of flutings in temperate glacier forelands. Therefore, the occurrence of lodged clasts and their associated flutings, composed of deformed sediment and ice-flow parallel clast fabrics, represent landform-sediment evidence for subglacial deformation. However, recent proposals that fluted surfaces are the product of

Figure 2.7 The wind-deflated surface of a formerly thin till cover overlying striated bedrock. Southeast of þorisjökull, Iceland.

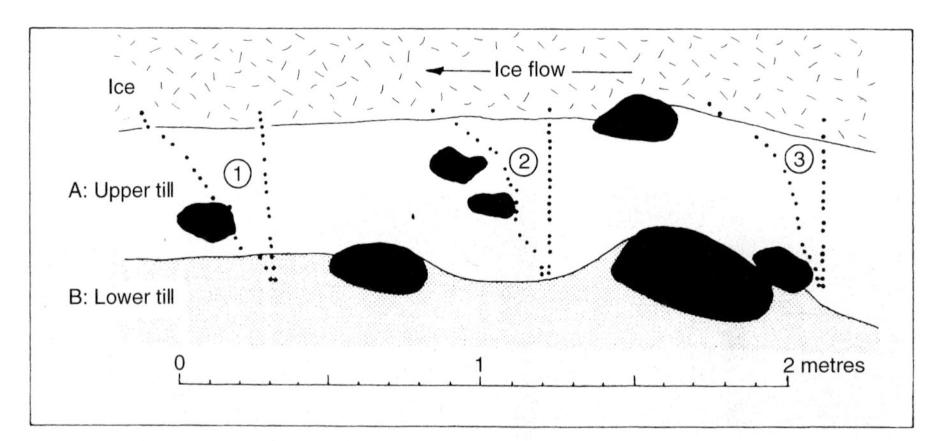

Figure 2.8 Subsole deformation and the occurrence of a two-tiered till at the base of Breiðamerkurjökull based upon the displacement of rods placed in the deforming layer. Although the convex upward displacement profile has been interpreted as the product of pervasive, viscoplastic deformation in the 'A' horizon it may also be reflective of distributed shear in a Coulomb plastic, as demonstrated by Iverson and Iverson (2001). (After Boulton and Hindmarsh, 1987).

ploughing of soft sediment beds by rough glacier soles may prove to be equally valid for the evolution of the subglacial bedforms of active temperate snouts (Tulaczyk *et al.*, 2001). This process can be equally as effective at advecting deformable sediment to the glacier margin as the deformation processes envisaged by Boulton and Hindmarsh (1987), Boulton and Dobbie (1998) and Boulton *et al.* (2001).

Exposures through the subglacial sediments of Icelandic glacier forelands typically reveal a sequence of glacitectonized pre-advance materials overlain by till, although localized outcrops

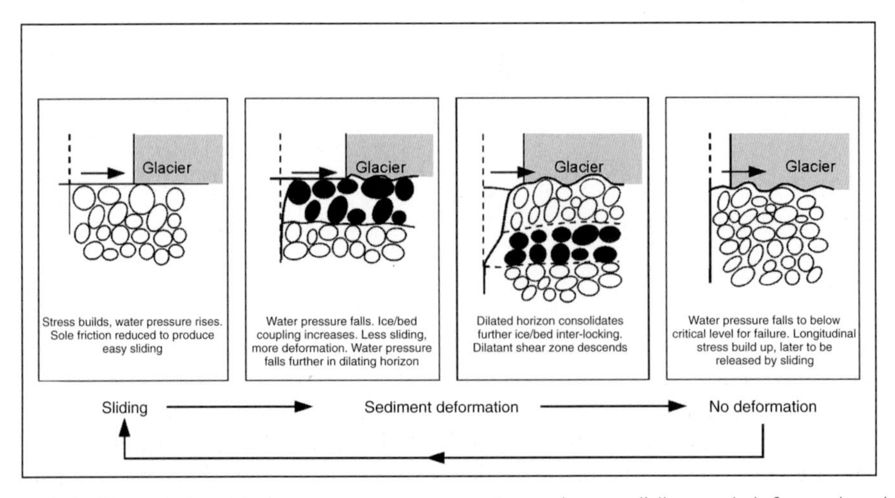

Figure 2.9 The relationship between water pressure cycles on sliding and deformation (stick-slip cycles) postulated by Boulton *et al.* (2001a) based upon data collected from Breiðamerkurjökull.

(A)

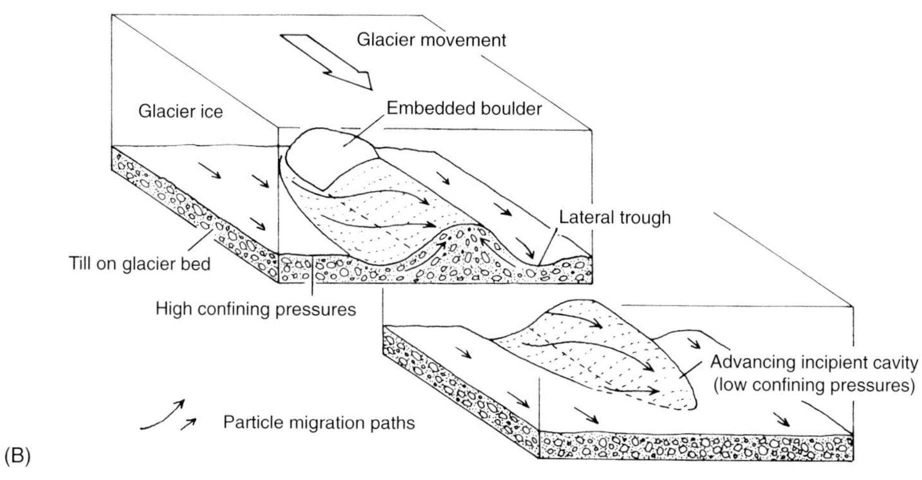

(B)

Figure 2.10 A) Lodged and striated boulder and lee-side fluting on the foreland of Skalafellsjökull, Iceland. B) Explanation of a fluting forming in the lee of a lodged boulder (from Benn, 1994).

near the outer limits of the Little Ice Age advance contain complex interbedded sequences of tills and stratified sediments. Often underlying the surface tills are stratified sediments deposited initially in ice-contact fans, braided streams and localized proglacial lakes (Howarth, 1968; Boulton, 1987; Benn and Evans, 1996; Evans, 2000; Evans and Twigg, 2002; Fig. 2.11). These sediments are truncated by the overlying tills and are characterized by internal glacitectonic disturbance in many locations. Evans and Twigg (2002) report a vertical continuum comprising a basal undisturbed zone of horizontally bedded sediments and peat beds overlain by a shear faulted and locally fluidized and hydrofractured zone in which original sedimentary structures are still discernible (non-penetrative glacitectonite of Benn and Evans, 1996). This is overlain by an intensely deformed zone or penetrative glacitectonite (Benn and Evans, 1996; Evans *et al.*, 1998). The whole sequence is capped by a till containing smeared inclusions of the stratified sediments and peat. Large numbers of striated and stoss-and-lee clasts in the tills on the foreland indicate that abrasion and lodgement is taking place in addition to widespread deformation. Additionally, till fabrics record clear clast alignments that are consistent with the former ice flow directions as recorded by local flutings. Therefore, the vertical sediment continuums probably record the local cannibalization of pre-advance sediments and the plastering of lodgement and deforming bed materials after their advection from up-glacier.

Conspicuous features in the subglacial tills of temperate glaciers are vertical dykes or injection structures composed of non-till sediment. On the foreland of Slettjökull, Iceland such features have been interpreted by van der Meer *et al.* (1999) as the products of subglacial meltwater discharges and related specifically to water escape. Previous research on clastic dykes suggests that they form by hydrofracturing due to the escape of pressurized groundwater (Mandl and Harkness, 1987; Boulton and Caban, 1995). The tensional cracks so formed are infilled by the sediment fluidized by the escaping water (Lowe, 1975; Nichols *et al.*, 1994; Rijsdijk *et al.*, 1999). Important in the interpretations of van der Meer *et al.* (1999) is the occurrence of an apparently perennially frozen outer glacier margin. This marginal belt of permafrost acts to force increased discharges of subglacial meltwater into sub-till outwash, thereby producing downglacier dipping water escape structures. Although the occurrence of permafrost is not

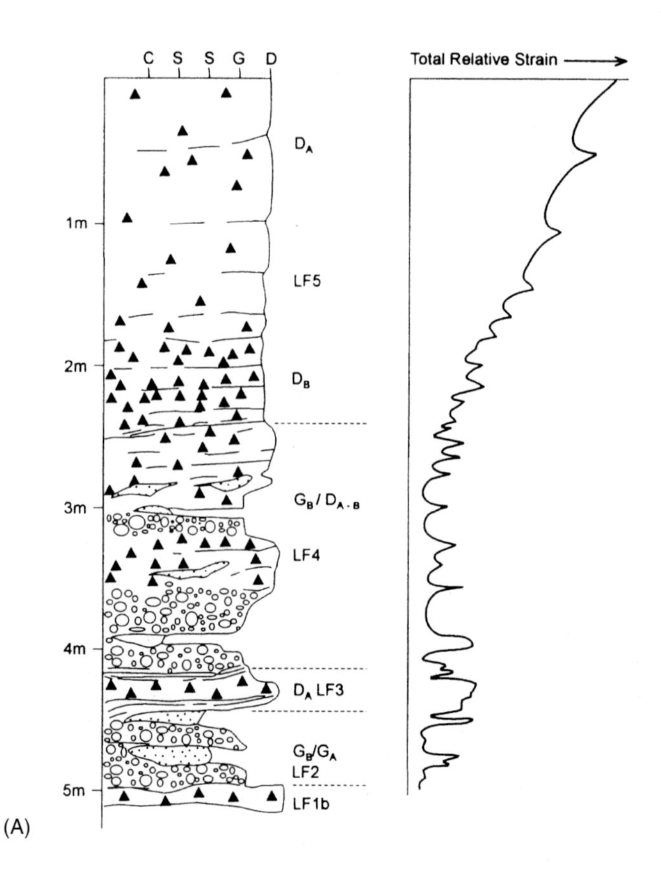

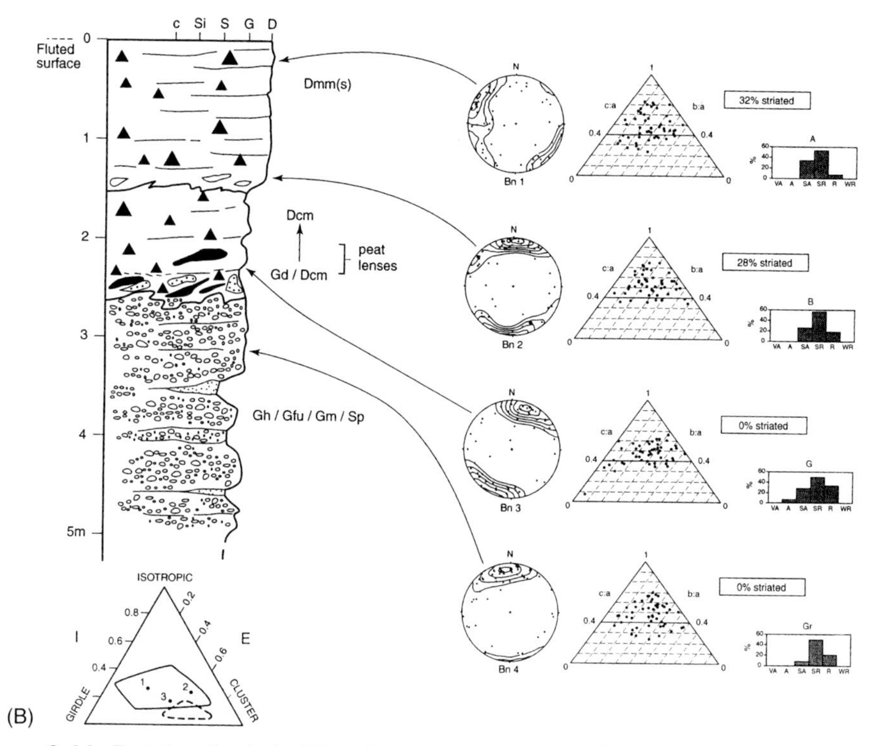

Figure 2.11 Details of subglacial sediment sequences at the margins of Icelandic glaciers showing typical vertical transitions from glacitectonized stratified sediments to overlying tills. A) Composite stratigraphic log from Skalafellsjökull showing various lithofacies and interpreted relative strain curve (from Evans, 2000a). B) Vertical profile log of section on the south shore of Breiðarlon on the Breiðamerkurjökull foreland showing a till with a possible A and B horizon superimposed on the B horizon of an eroded underlying till (from Evans and Twigg, 2002). Both tills have sheared rafts of underlying materials at their bases including peat from a nearby outcrop of *in situ* organics. The fabric shape triangle (Benn, 1994) includes envelopes for upper and lower tills (A and B horizons, respectively) based upon data collected by Dowdeswell and Sharp (1986) and Benn (1995). The cannibalization of underlying outwash by the deforming layer is apparent in the inheritance of clast shapes and lack of striae on clasts in the lower till.

necessary to the development of subglacial fracture fills, the possibility of a frozen bed at the glacier margin for significant periods of time clearly has implications for subglacial processes such as those envisaged by van der Meer *et al.* (1999) at Slettjökull. Just as supercooling in overdeepenings may produce abnormal quantities of debris-rich basal ice for an active temperate glacier, so short-lived ice-marginal permafrost may impart its own diagnostic structures on subglacial deformation tills beneath temperate ice. The exact significance of shallow permafrost at the snouts of temperate glaciers, whatever its age, is still not fully understood but its geological impact at the margins of Icelandic glaciers appears to be restricted to secondary subglacial till structures and the production of large push moraines by stacking of frozen till slabs (see above).

Generally, the characteristics of tills on temperate glacier forelands strongly suggest a subglacial deformation and lodgement origin. They comprise materials derived from pre-existing stratified sediments in addition to localized abrasion of rock surfaces and are generally thin (<2 m). Thicker sequences are constructed by the sequential plastering of several till layers onto stratified sediments and bedrock (e.g. tectonic/depositional slices of Boulton *et al.*, 2001). This is similar in depositional style to the rheologic superposition proposed by Hicock (1992), Hicock and Dreimanis (1992), and Hicock and Fuller (1995), and the till/stratified interbed successions of Eyles *et al.* (1982), Evans *et al.* (1995), and Benn and Evans (1996). Complex till sequences, often associated with larger push moraines, are constructed where glacier margins become stationary for substantial periods. It is evident that the construction or stacking of such complex till sequences at contemporary glacier snouts is a sub-marginal process and this knowledge may be applied to the interpretation of till architecture at regional scales (e.g. Boulton, 1996a, b; Johnson and Hansel, 1999). Although subglacial deformation and lodgement (and perhaps ploughing) are predominant till deposition processes in temperate glaciers, the delivery of subglacial material to the glacier margin and the construction of moraines involve the freeze-on of till slabs for at least part of the year. Where frozen sub-marginal conditions persist for long periods, possibly perennially, discontinuous permafrost must be considered in models of marginal till delivery and moraine construction even though lodgement and deformation processes still dominate the subglacial environment immediately up-ice.

2.2.3 Glacifluvial and Glacilacustrine Domain

The monitoring of glacifluvial landforms and proglacial lakes at temperate glacier margins provides valuable insights into process/form relationships, particularly during glacier recession (e.g. Welch, 1967; Howarth, 1968, 1971; Price, 1969, 1971, 1973, 1980, 1982; Price and Howarth, 1970; Churski, 1973; Gustavson, 1975; Smith and Ashley, 1985; Boulton, 1986; Evans and Twigg, 2000, 2002). Proglacial outwash streams drain either away from glacier margins to produce sandur fans or parallel to the margin due to topographic constraints to produce ice margin-parallel outwash tracts and kame terraces (Fig.s 2.12–2.15). Additionally, glacier recession uncovers large depressions in which proglacial lakes can evolve (e.g. Thorarinsson, 1939; Howarth and Price, 1969; Price and Howarth, 1970; Bjornsson, 1976; Price, 1982). The nature and size of depo-centres and sediment thicknesses in such lakes vary according to their longevity and ratio of glacier recession to sedimentation rate (Shaw, 1977c; Teller, Chapter 14).

Sandur fans are prograded from subglacial or englacial meltwater portals, often resulting in linkages between fan apices and eskers (Price, 1969; Fig. 2.13). The ice contact nature of sandur fans results in the development of pitted ice-proximal fan surfaces and steep ice-contact faces. During extended periods of glacier marginal stability, extensive and thick sandur fans may coalesce. An excellent example is Breiðamerkursandur located beyond the Little Ice Age maximum moraines of Breiðamerkurjökull where the routeways of proglacial streams have not been channelized by pre-existing topographic high points (Evans and Twigg, 2000, 2002). Such thick ramps of sediment are difficult for later proglacial meltwater and subglacial erosion to rework and hence proglacial drainage and then glacier flow is diverted around them. Stratigraphic evidence indicates that temperate glaciers often advance over considerable tracts of glacifluvial outwash without disrupting their general form. For example, Boulton (1986, 1987) and Evans and Twigg (2002) have reported that the forms of pre-advance sandur fans are still

(A)

(B)

Figure 2.12 Landforms produced in association with ice-margin parallel outwash tracts. A) High-elevation kame terraces and lower ice-margin parallel outwash tracts at the northern margin of Sandfellsjökull, Iceland. B) Push moraines composed of gravels derived from ice-contact outwash on the foreland of Heinabergsjökull, Iceland.

evident in the proglacial area of Breiðamerkurjökull. This indicates that they are relatively stable forms that become adourned with subglacial features (drumlins and flutings) during glacier overriding and push moraines during glacier recession but remain dominant as topographic features in the deglaciated terrain.

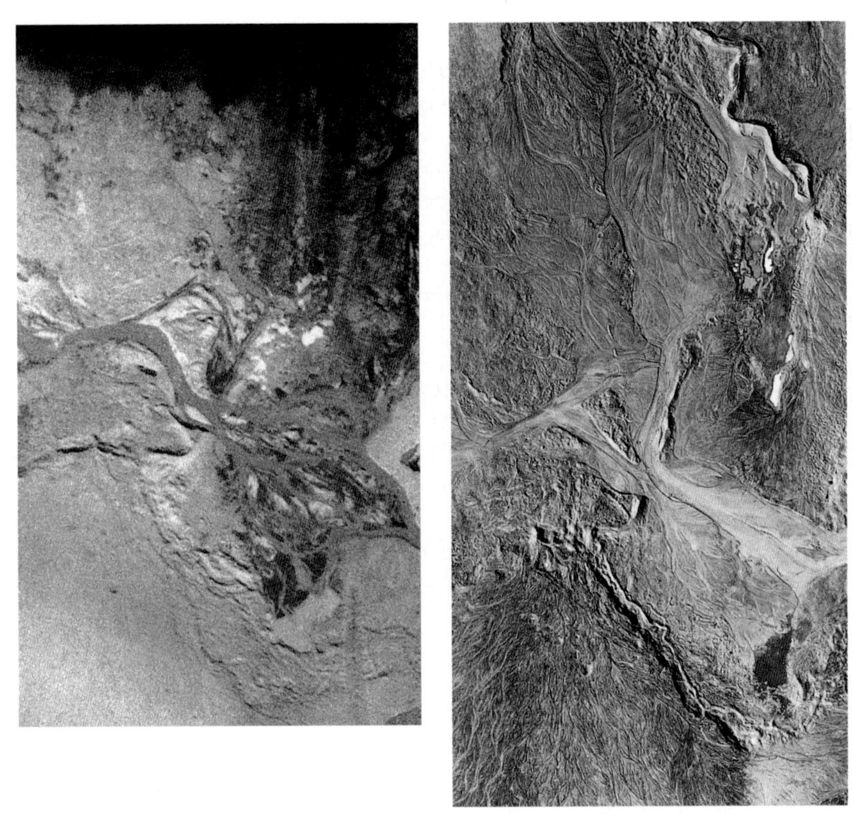

Figure 2.13 Portions of aerial photographs (Landmælingar Islands and University of Glasgow, 1945 (left), 1965 (right)) of the Breiðamerkurjökull foreland, Iceland, showing the development of eskers and pitted sandur fan during glacier recession. Ice marginal ponds are visible as dark patches to the top of the 1945 image. Note how the ice-contact face of the sandur fan and the esker which lies parallel to that face become clearer between 1945 and 1965 as buried glacier ice melts-out.

The progradation of outwash will not produce sandur fans in certain topographic settings. Rather the outwash will be deposited between the glacier margin and proglacial topographic high points to produce ice margin-parallel outwash tracts and kame terraces (Fig. 2.12a). Wherever localized glacier readvances impact upon such outwash tracts the resulting push moraine ridges will comprise glacifluvial sands and gravels (Fig. 2.12b).

Minor outwash fans composed of fine grained material ('hochsandur fans') are common at advancing or stationary temperate glacier margins where sufficient englacial and supraglacial debris is made available to meltwater streams (Gripp, 1975; Heim, 1983, 1992; Krüger, 1997). They are common at the margins of Icelandic glaciers where large quantities of englacial tephra bands melt-out and provide fine grained sediment for meltwater reworking in the ablation zone. Because hochsandur fans are produced during glacier advance they are unlikely to contain the pitted surfaces characteristic of outwash prograded over receding snouts.

Conversely, the predominantly flat surfaces of sandur fans and ice margin-parallel outwash tracts often contain numerous enclosed pits (Figs. 2.13 and 2.14) and are therefore termed pitted or kettled outwash (sandar). The nature of the pitting on the ice margin-parallel features is usually dominated by large individual kettle holes. Pitted sandur fans, on the other hand, are often characterized by numerous small kettle holes located at their apices (Fig. 2.13). This is a characteristic of jokulhlaup sandar reported from a number of Icelandic glacier snouts (e.g. Howarth, 1968; Churski, 1973; Galon, 1973; Klimek, 1973; Bodéré, 1977; Olszewski and Weckwerth, 1997; Fay, 2002). The development of kettle holes at the fan apex documents the melt-out of individual ice blocks originally deposited by flood waters on the fan (Maizels, 1977, 1992).

Observations on the evolution of the ice margin-parallel features at Breiðamerkurjökull by Welch (1967), Howarth (1968) and Price (1969, 1971, 1973) indicated that the kettle holes were opening up above extensive exposures of buried glacier ice (Fig. 2.14). Consequently the Breiðamerkurjökull ice-marginal features were explained as the product of melt-out of the shallow glacier margin previously buried by glacifluvial sediment. Melting of buried glacier ice in these settings explains the unusually large size of some individual kettle holes. This type of pitted outwash is not extensively developed in front of actively receding temperate glaciers due to the absence of large tracts of stagnating ice. It occurs at the lateral margins of temperate glaciers where streams are forced to flow over and through the ice, often due to topographic constraints.

More linear and complexly terraced sandar may evolve in the spillways that develop in association with proglacial lake growth. This is well illustrated by the forelands of

Figure 2.14 The pitted surface of an ice-margin parallel outwash tract at Breiðamerkurjökull, produced by the melt-out of extensive buried glacier ice. (Photograph by R.J. Price (1965).)

Breiðamerkurjökull and Fjallsjökull in Iceland. The recession of the two glaciers over the last 100 years has uncovered large depressions in which the lakes Fjallsárlón, Breiðárlón, Jökulsárlón and Stemmulón have evolved (Howarth and Price, 1969; Price and Howarth, 1970; Price, 1980, 1982; Evans and Twigg, 2000, 2002). Spillways from these lakes have routed substantial long-term discharges, which have constructed large, terraced outwash corridors. The most impressive example of a corridor lies between Breiðárlón and Fjallsárlón (Fig. 2.15), where the diversion of meltwater draining from west Breiðamerkurjökull after 1960 resulted in long-term progradation and incision of glacifluvial sediments around moraine and till-covered topographic high points (Evans and Twigg, 2000, 2002).

Ice-dammed and proglacial lakes are also common features around the receding margins of lowland temperate glaciers, particularly where the glacier has uncovered overdeepenings produced by long-term glacial erosion (Howarth and Price, 1969; Bjornsson, 1996; Bennett *et al.*, 2000c; Evans and Twigg, 2002). These lakes are significant sediment sinks on the forelands of temperate glaciers and give rise to the accumulation of thick sequences of glacilacustrine sediments and shorelines and deltas (Shaw, 1977c; Teller, Chapter 14). The surfaces of these glacilacustrine features can be heavily pitted or extensively deformed by ice melt-out in glacier-contact settings. Lake sediments also constitute ideal material for push and thrust moraine development (Howarth, 1968; Evans *et al.*, 1999b; Bennett *et al.*, 2000c) and the production of glacitectonite and deforming bed tills (Benn and Evans, 1996; Evans *et al.*, 1998, 1999c; Evans and Twigg, 2002).

The internal drainage networks of temperate glaciers are often documented in the landform record by eskers. The evolution of several esker systems at Breiðamerkurjökull, Iceland and Casement Glacier, Alaska has been documented by survey and mapping since the 1940s (Price, 1964, 1965, 1966; Petrie and Price, 1966; Welch, 1967; Howarth, 1968, 1971; Price, 1969, 1973, 1982; Evans and Twigg, 2002). The eskers are typically sharp-crested with steep sides and are composed predominantly of coarse gravels. They are arranged into complex forms comprising single ridges and multiple, anabranched sections often resembling fans. Aerial photographs taken since 1945 show that the largest of the eskers at Breiðamerkurjökull have developed at the same location as the medial moraines of the glacier, indicating that their location may be dictated by sediment availability.

The termination of some eskers at the apices of sandur fans clearly illustrates the importance of englacial and subglacial drainage systems to the construction of large areas of proglacial outwash. Based upon surveys of the eskers at Breiðamerkurjökull and Casement Glacier, Petrie and Price (1966, 1969), Price (1966), Welch (1967) and Howarth (1968, 1971) report considerable surface lowering of the original landforms. This indicates that large parts of the eskers have been deposited in englacial or supraglacial positions, an interpretation that is supported by observations of esker ice cores and eskers emanating from the wasting glacier surfaces. If drainage at Breiðamerkurjökull is subglacial at the snout (as suggested by Boulton *et al.*, 2001a) then the streams must deposit debris in englacial tunnels before they reach the bed at the outer glacier margin. This implies that eskers are deposited englacially in increments and in a zone just up-ice of the glacier margin. As the margin recedes the drainage migrates from englacial tunnels to subglacial pathways, leaving englacial tunnel fills to melt-out on the glacier surface. Aerial photographs record the evolution of large tracts of 'kame and kettle topography' (Welch, 1967; Howarth, 1968; Price, 1969; Evans and Twigg, 2000, 2002) associated with the Breiðamerkurjökull eskers (Fig. 2.16). Evans and Twigg (2002), incorporating the observations of Price (1969), demonstrate that eskers and a pitted outwash surface visible on 1945 aerial

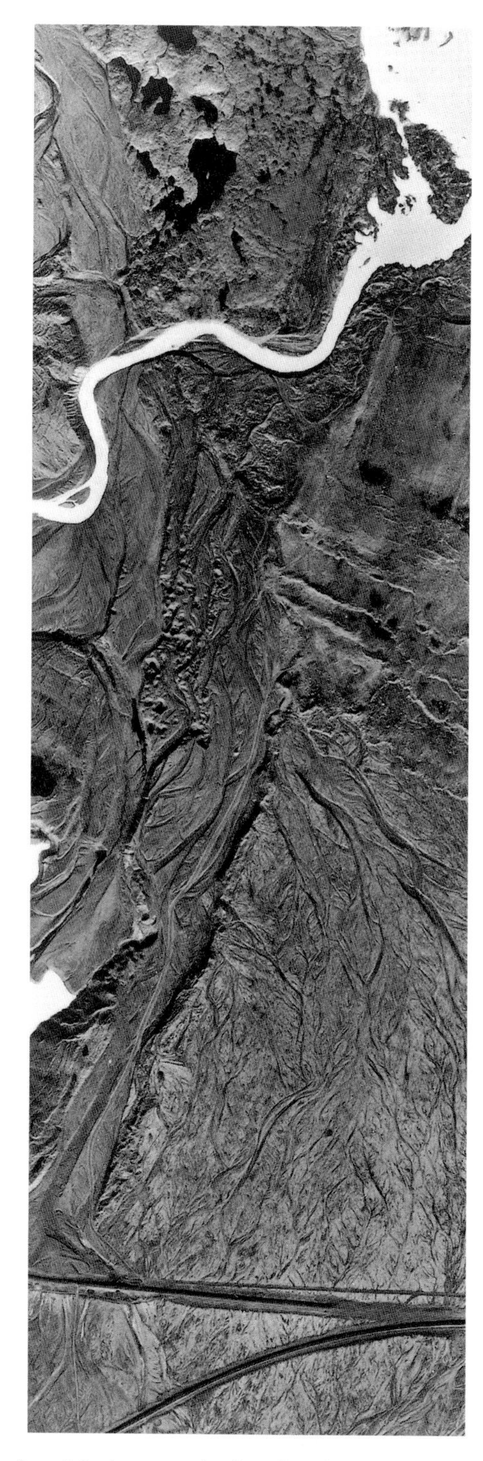

Figure 2.15 Portion of aerial photograph (Landmælingar Island and University of Glasgow, 1998) of the terraced outwash corridor located between the proglacial lakes Breiðárlón (top right) and Fjallsárlón (centre left) on the forelands of Breiðamerkurjökull and Fjallsjökull, Iceland.

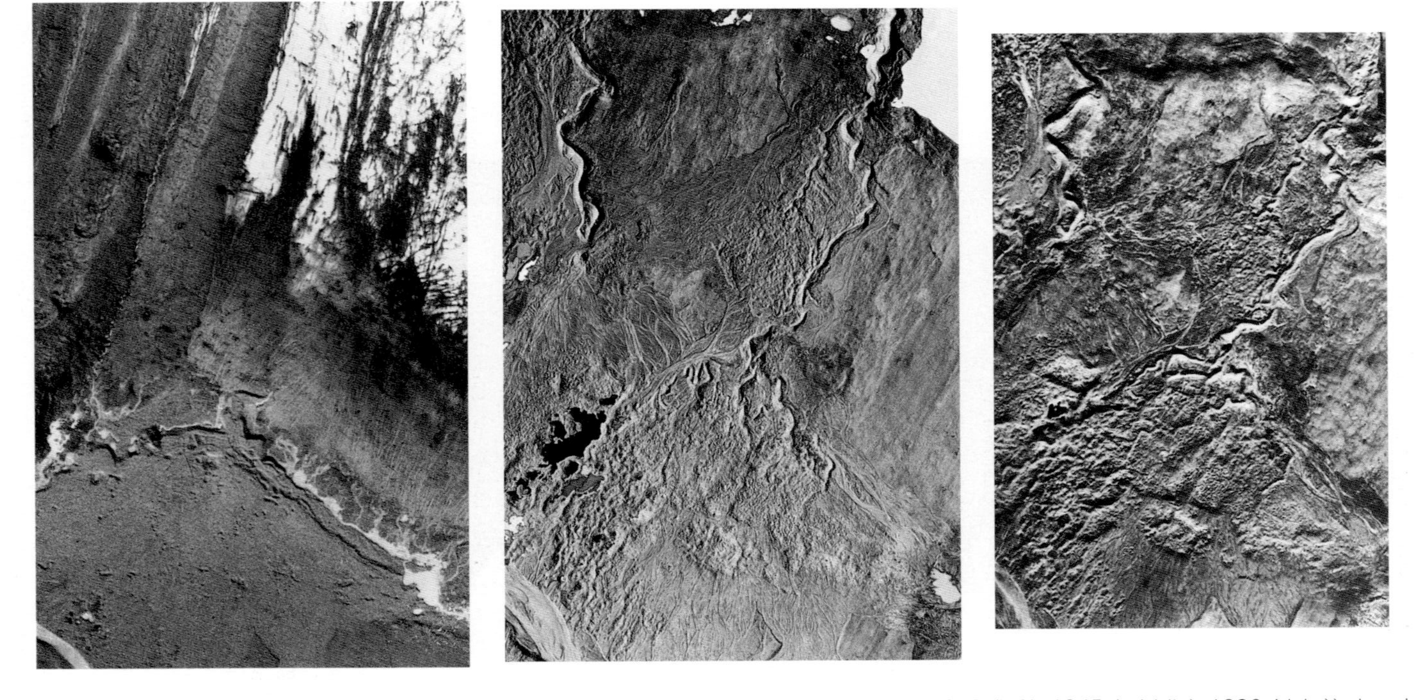

Figure 2.16 Portions of aerial photographs (Landmælingar Island and University of Glasgow, 1945 (left), 1965 (middle), 1998 (right)) showing the evolution of a fan-shaped arrangement of anabranched eskers from pitted outwash.

photographs evolve into eskers and kame and kettle topography by 1965, and then by 1998 into a complex anabranched system of esker ridges arranged in a fan shape. This indicates that a considerable expanse of glacier ice lay below a thin cover of glacifluvial outwash and that the larger volumes of esker sediment within the ice ensured their preservation after melt-out. These complex esker networks clearly develop in parts of the glacier snout that occupy topographic depressions, where a thin cover of glacifluvial sediment can accumulate, although landform evolution is difficult to reconcile with predominantly subglacial drainage. At the location of one of the Breiðamerkurjökull medial moraines in 1945, outwash was forming a fan that emanated from a prominent single esker. On 1965 aerial photographs it is clear that the fan apex was also the apex of a fan of eskers within the buried glacier ice. It appears that the outwash fan was thin and had developed where the englacial stream responsible for esker deposition became supraglacial rather than subglacial.

Although their models are based upon examples at the Malaspina Glacier with its marginal stagnant ice, Gustavson and Boothroyd (1987) have provided explanations for the variety of glacifluvial and glacilacustrine depositional environments that exist at active temperate glacier margins. Fig. 2.17 depicts modified versions of their original diagrammatic explanations. In each example the source of debris for the meltwater streams is assumed to be the glacier bed but, based upon the Breiðamerkurjökull esker and medial moraine associations, this need not be the case at every location. Fig. 2.17a depicts a sandur fan emanating from a fountain fed by subglacial meltwater. Such fountains are common at temperate glacier margins. A sandur fan fed by subglacial meltwater draining from beneath the glacier is depicted in Fig. 2.17b. Note that englacial tunnels will have the potential to rework debris from medial moraines. Fig. 2.17c depicts a subglacial tunnel connected to an ice-contact proglacial lake. Debris-charged meltwater will deposit a subaqueous fan by the progradation of density underflow deposits. This may be complicated where meltwater issues from an englacial tunnel (Fig. 2.17d) and deposits a subaqueous fan over glacier ice. In some circumstances englacial meltwater may exit the snout and travel a short distance over the glacier surface to feed an ice-contact delta (Fig. 2.17e). Ice-cored eskers that terminate at sandur fan apices demonstrate that englacial meltwater must exit the glacier and drain supraglacially even in the absence of ice-dammed lakes. Moreover, this englacial meltwater need not penetrate to the glacier bed (Price, 1969). In examples 2.17b to 2.17e, note that englacial sediment-filled tunnels may become supraglacial during glacier recession, thereby explaining ice-cored eskers and outwash. Additional complexity arises in the landform record where meltwater drainage pathways change location. This is well illustrated at Breiðamerkurjökull where a single large esker ridge terminates at a sandur fan apex but a smaller esker runs along the base of the ice-contact slope, at right angles to the larger esker (Boulton, 1986; Evans and Twigg, 2002). The larger esker clearly originally fed sediment to the fan surface but could no longer do so once the glacier margin had ceased to press against the fan's ice-contact slope. Although meltwater continued to be delivered to this part of the glacier snout it began to flow sub-marginally along the base of the ice-contact slope of the fan, thereby producing a smaller esker running parallel to the glacier margin.

2.3 ANCIENT RECORDS OF ACTIVE TEMPERATE GLACIER MARGINS

As active temperate glaciers provide us with a clear impression of their former passage, their impacts on ancient glaciated terrains should be identifiable. One excellent example in western

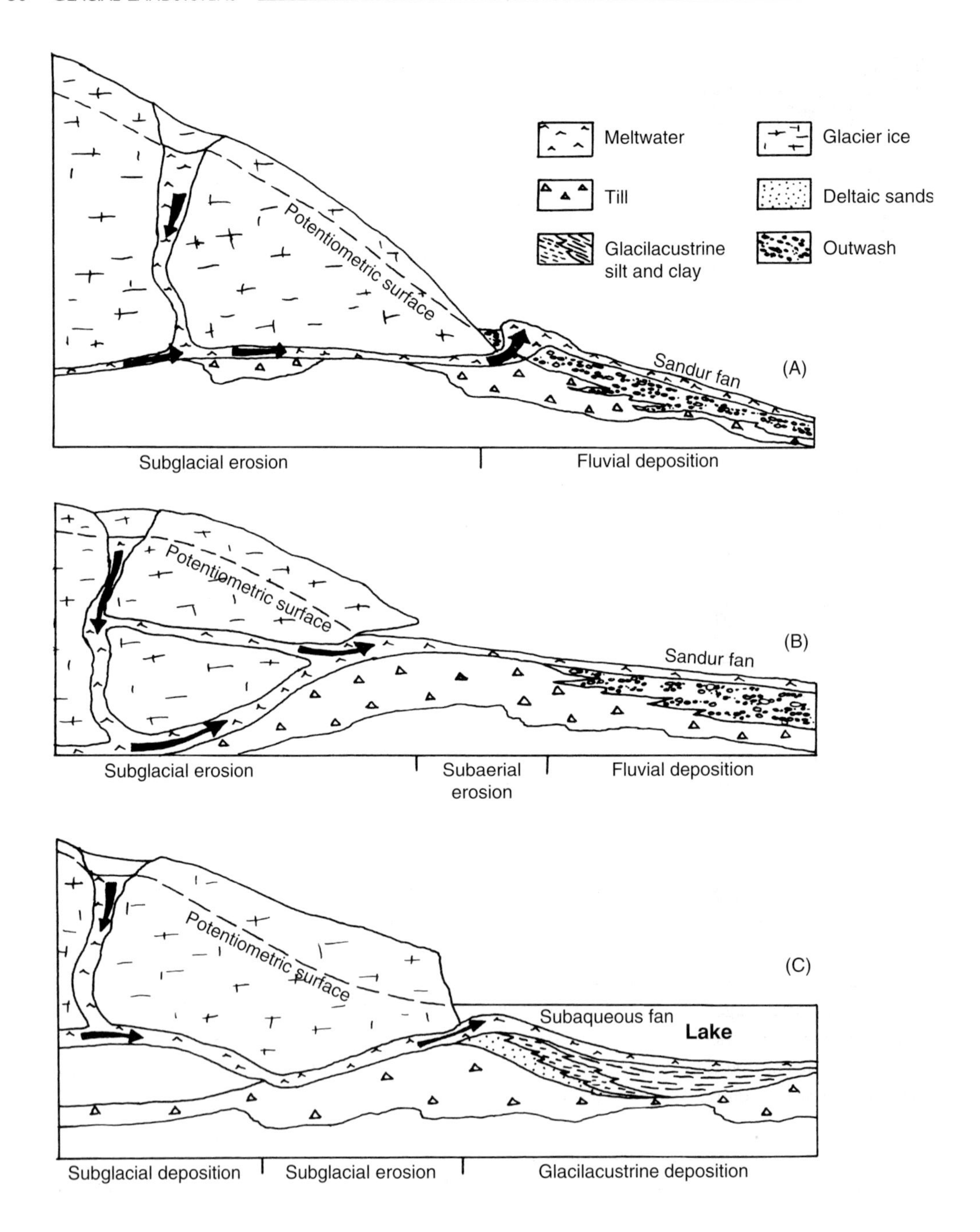

Canada has been identified by Evans *et al.* (1999a). In an extensive area deglaciated by the westernmost ice lobe of the southwest Laurentide Ice Sheet, between Calgary and Granum, Alberta, numerous closely spaced, low-amplitude moraine ridges drape fluted terrain and demarcate active ice-marginal recession over a distance of 90 km (Fig. 2.18a, b). Similar active recession appears to have characterized the margin of the adjacent 'central' lobe in southern Alberta. Here a 50 km wide belt of low-amplitude moraine ridges stretch largely unbroken over

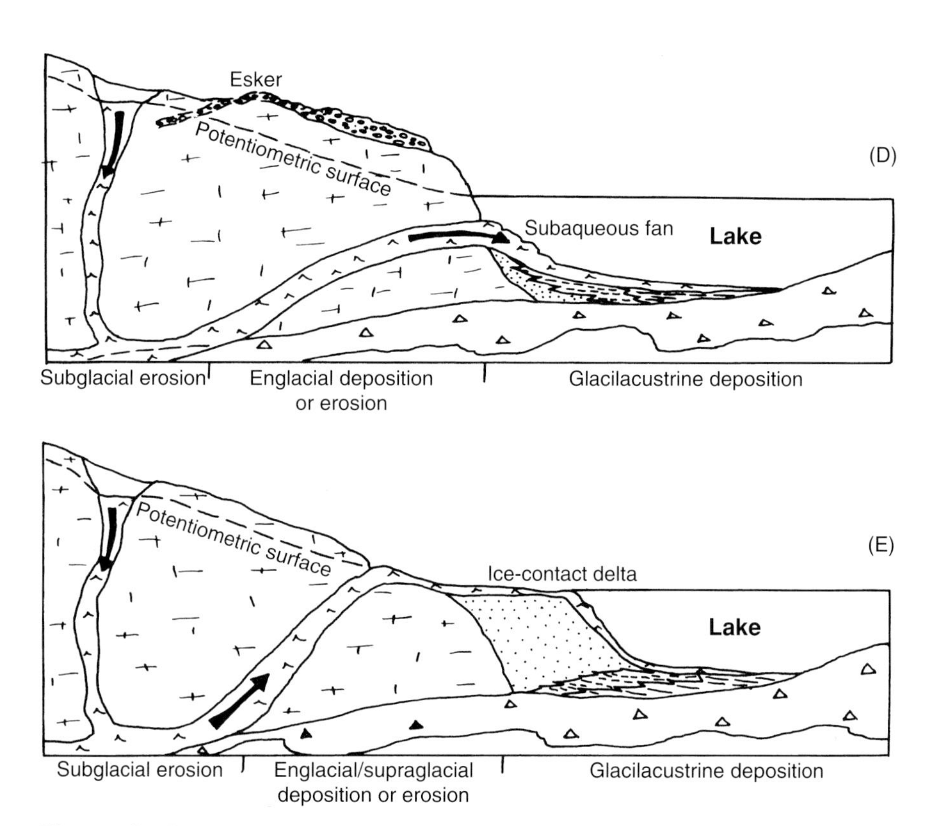

Figure 2.17 Models of glacifluvial and glacilacustrine deposition based upon the Malaspina Glacier, Alaska; (see text for explanation). (After Gustavson and Boothroyd, 1987).

130 km through the Milk River drainage basin and lie inside a broad arc of overridden thrust moraines (Fig. 2.18c). Similar recessional push moraines have been identified in Wisconsin, USA by Ham and Attig (2001), who make comparisons with Icelandic push moraines and, as a consequence, propose that the former Wisconsin Valley glacier lobe underwent active recession during the initial stages of deglaciation. Landform records of active temperate snouts exist in the lowlands bordering the mountain terrain of the Southern Alps in New Zealand. Here glacier ice has left a series of low-amplitude push moraines superimposed on a larger ridge of glacitectonized lake sediment that dams the southern shore of Lake Pukaki (Fig. 2.19).

2.4 CONCLUSION

Research has shown that the active temperate glacial landsystem is composed of three depositional domains (Fig. 2.20). First, areas of extensive, low-amplitude marginal dump, push and squeeze moraines are derived largely from material on the glacier foreland. These moraines often record annual recession of active ice but some may be superimposed during periods of ice margin stability. The continuation of flutings over the proximal slopes and crests of many push moraines indicates that subglacial bedform and push moraine production is genetically linked. A subglacial deformation/ploughing origin of flutings thereby implies that push moraines are

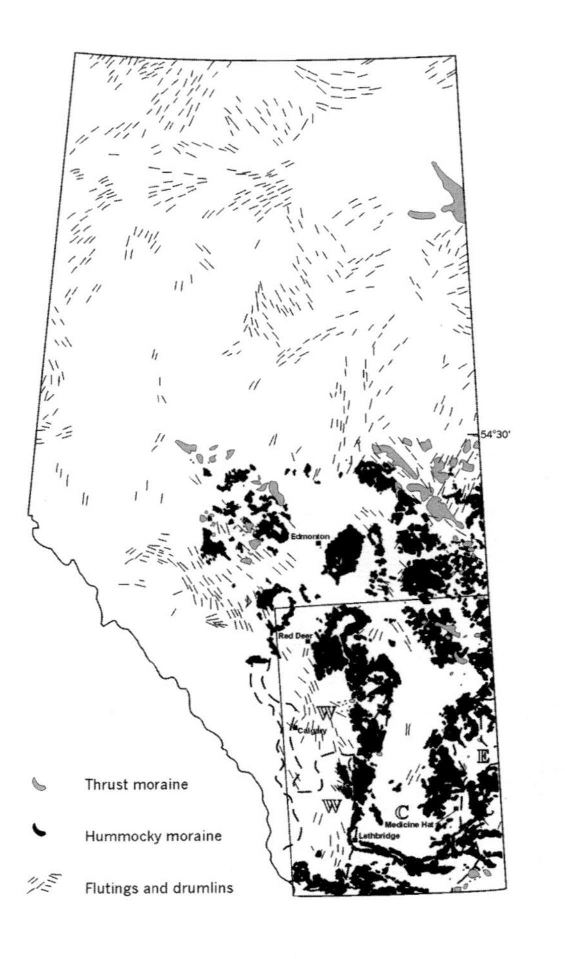
54°30'
Edmonton
Red Deer
Calgary
W
C
E
Medicine Hat
Lethbridge
Thrust moraine
Hummocky moraine
Flutings and drumlins

(A)

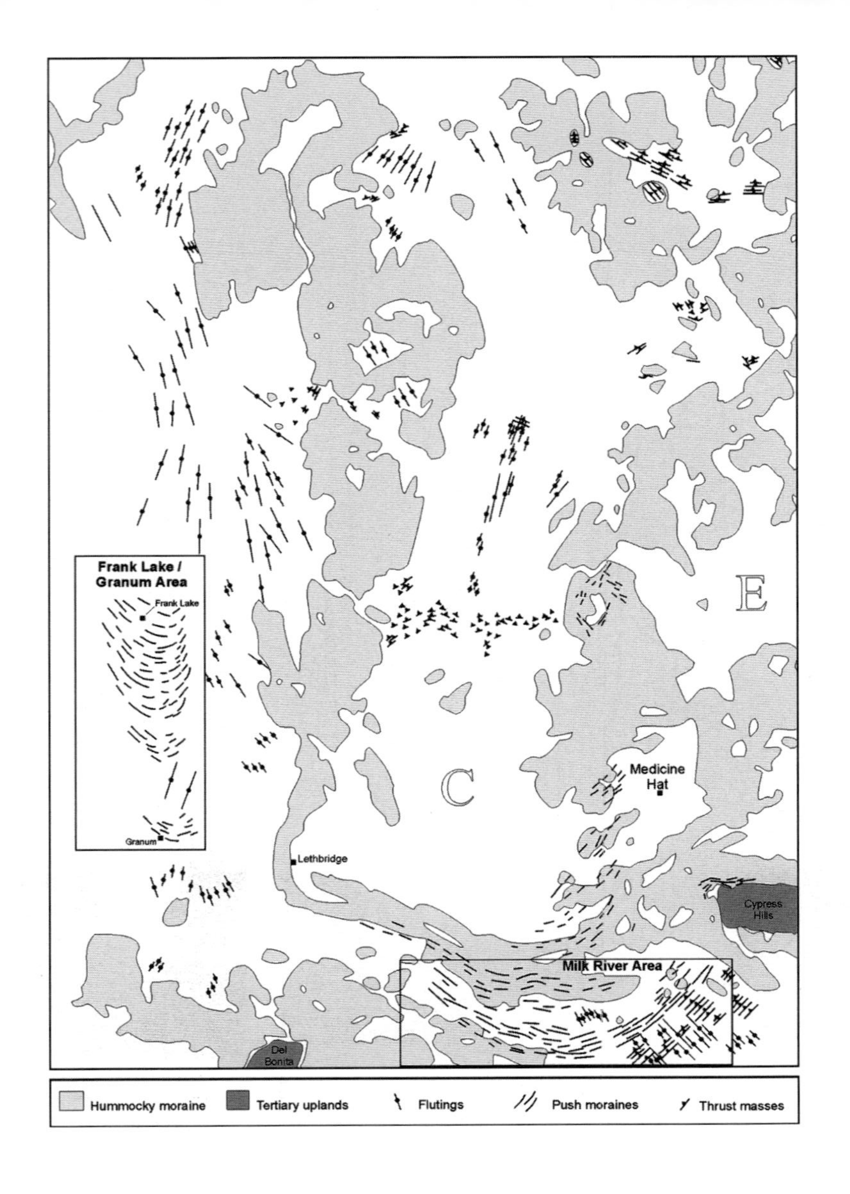
Frank Lake / Granum Area
Frank Lake
Granum
E
C
Medicine Hat
Lethbridge
Cypress Hills
Milk River Area
Del Bonita
Hummocky moraine
Tertiary uplands
Flutings
Push moraines
Thrust masses

(B)

(C)

Figure 2.18 Active temperate glacial landsystems of southern Alberta, Canada. A) Maps of southern Alberta showing major moraine systems and locations of three ice sheet lobes. (After Evans *et al.*, 1999a). W = west, C = central, E = east lobes; the Frank Lake/Granum and Milk River areas are boxed. B) Part of an aerial photograph mosaic of the Frank Lake area of southern Alberta, showing prominent recessional moraine ridges typical of the southern margins of the west lobe. Eskers and flutings are also well developed in the area (e.g. Evans *et al.* 1999a). C) Part of an aerial photograph of the Milk River drainage basin near Pendant d'Oreole, Alberta, Canada, showing numerous extensive and closely spaced recessional push moraines. In the same area more widely spaced push moraines are associated with flutings and drape overridden push moraines.

the product of the advection of subglacially deforming sediment to the ice margin where squeezing and pushing then combine to create the moraine ridge. Larger push moraines can be constructed by stationary glacier margins. A variety of processes may be involved in the production of larger moraines including the stacking of frozen sediment slabs, the prolonged

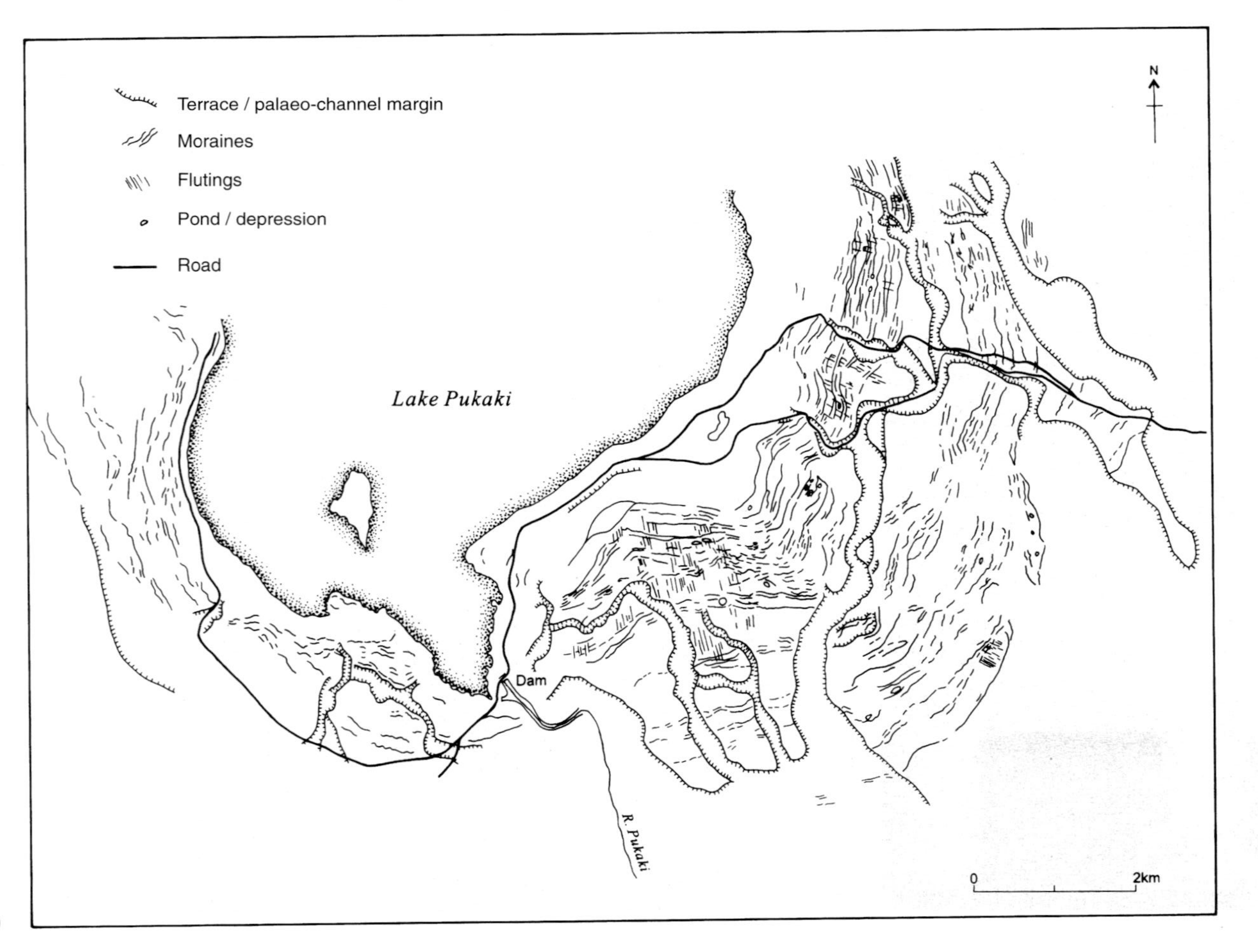
Terrace / palaeo-channel margin
Moraines
Flutings
Pond / depression
Road
N
Lake Pukaki
Dam
R. Pukaki
0
2km

(A)

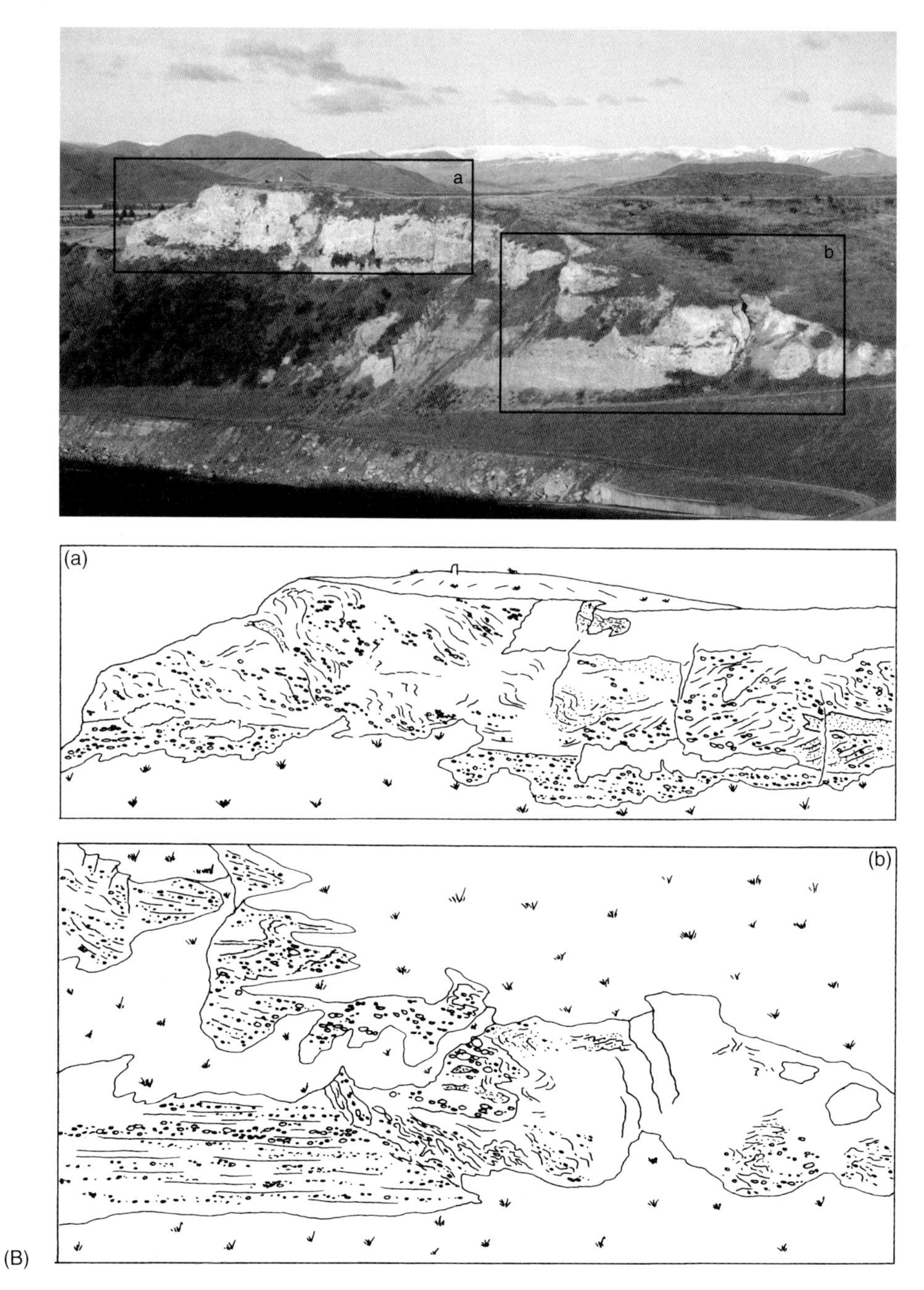

Figure 2.19 A) Map of recessional push moraines, many with saw-tooth plan forms, on the south shore of Lake Pukaki, New Zealand. Small areas of flutings are visible between moraine ridges, and channelled sandur are partially routed between moraines. B) Photograph and detailed section sketches (located by boxes on the photograph) of the glacitectonized fluvial and lacustrine sediments and overlying subglacially deformed materials (glacitectonite) that constitute the main ridge over which the recessional moraines at Lake Pukaki have been draped. Note the undeformed core of coarse gravels at the base of the section.

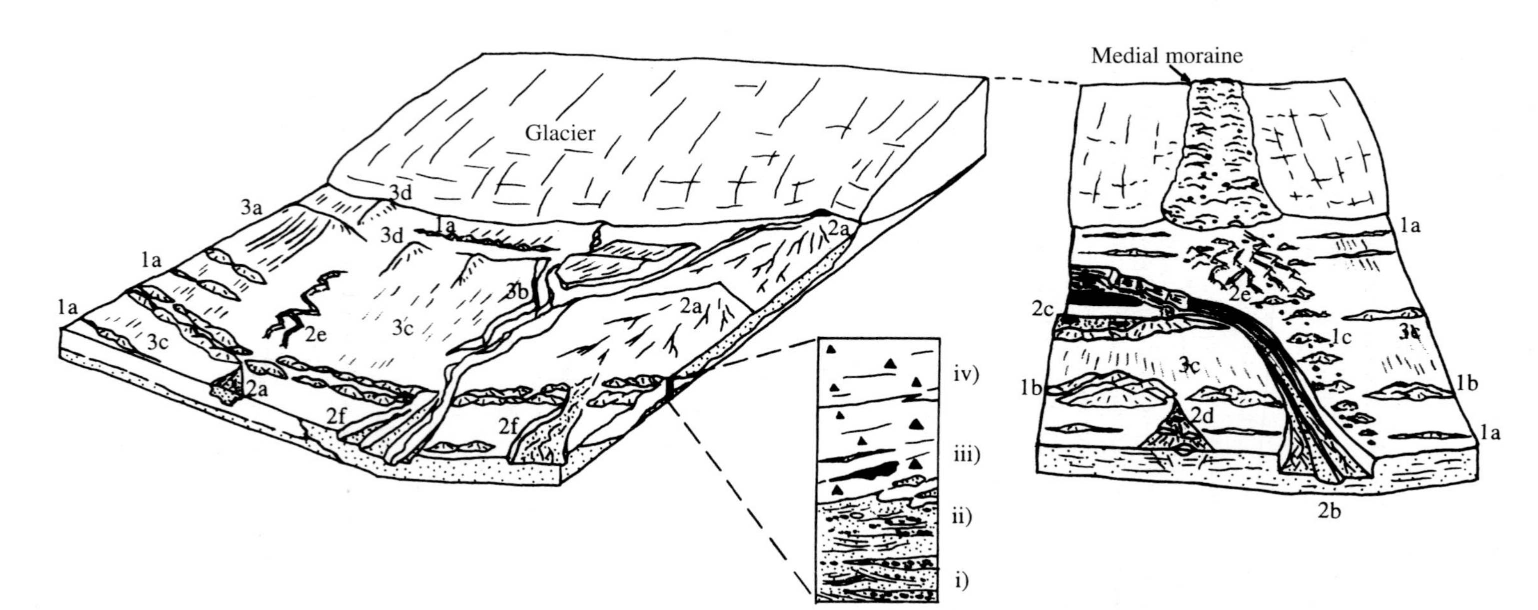

Figure 2.20 The active temperate glacial landsystem. Landforms are numbered according to their domain: 1 = morainic domain, 2 = glacifluvial domain, 3 = subglacial domain, 1a = small, often annual push moraines, 1b = superimposed push moraines, 1c = hummocky moraine, 2a = ice-contact sandur fans, 2b = spillway-fed sandur fan, 2c = ice margin-parallel outwash tract/kame terrace, 2d = pitted sandur, 2e = eskers, 2f = entrenched ice-contact outwash fans, 3a = overridden (fluted) push moraines, 3b = overridden, pre-advance ice-contact outwash fan, 3c = flutes, 3d = drumlins. The idealized stratigraphic section log shows a typical depositional sequence recording glacier advance over glacifluvial sediments, comprising: I = undeformed outwash, II = glacitectonized outwash/glacitectonite, III = massive, sheared till with basal inclusions of pre-advance peat and glacifluvial sediment, IV = massive sheared till with basal erosional contact. (After Kruger 1994a; Evans and Twigg, 2002).

impact of dump, squeeze and push mechanisms at the same location, or the incremental thickening of an ice-marginal wedge of deformation till. Wide and arcuate, low-amplitude ridges that are draped by flutings and recessional push moraines are interpreted as glacially streamlined push moraines overridden by glaciers during major readvances. The general lack of supraglacial sediment in active temperate glaciers generally precludes the widespread development of chaotic hummocky moraine, although low-amplitude, bouldery hummocks are produced by the melt-out of medial moraines and by the melting of debris-charged glacier snouts in settings where marginal freeze-on produces debris-rich ice facies.

Second, subglacial landform assemblages of flutings, drumlins and overridden push moraines dominate the land surfaces between ice-marginal depo-centres because the veil of supraglacial sediments is generally very thin and discontinuous. Although tills are thin over topographic high points in areas of hard bedrock, the stratigraphy of subglacial materials often displays a vertical continuum comprising glacitectonized stratified sediments of outwash and/or glacilacustrine origin capped by lodgement/deformation till containing rafts of underlying material and the products of localized abrasion. Complex till sequences thicker than 2 m are constructed by the sequential plastering of several till layers similar to rheologic superposition (Hicock, 1992; Hicock and Dreimanis, 1992; Hicock and Fuller, 1995) and the till/stratified interbed successions of Eyles *et al.* (1982), Evans *et al.* (1995), and Benn and Evans (1996). The ubiquitous flutings of the forelands are traditionally explained as the products of till squeezing into cavities on the down-glacier sides of lodged boulders (Boulton, 1976; Benn, 1994). Larger drumlins have been explained by Boulton (1987) as the streamlined remnants of coarse-grained sandur fans based upon the surface forms and internal stratigraphy of the Breiðamerkurjökull foreland.

Third, glacifluvial landforms are often extensive and include sandur fans (both ice-contact and spillway fed), ice margin-parallel outwash tracts and kame terraces, topographically channelized sandar, pitted sandar (ice-marginal and jokulhlaup types), and eskers of single and more complex anabranched forms. Although hummocky terrain located at receding glacier margins is often referred to as kame and kettle topography it can evolve through time due to melt-out of underlying ice into complex networks of anabranched eskers. The lowering of esker surfaces through time clearly demonstrates that they largely originated englacially or supraglacially.

The clear landform-sediment signatures of active temperate glacier recession have been recognized in some ancient glaciated terrains and can potentially provide glacial and Quaternary researchers with invaluable information about glacier dynamics and their linkages to climate change. In ancient glaciated basins where the landsystems approach has enabled us to recognize the imprints of active temperate glaciation, we are dealing with a landform-sediment record that possesses a clear regional palaeo-climatic signal rather than a glacial legacy that is dictated purely by localized physiographic controls and/or changes in the internal dynamics of the glacier system.

CHAPTER

3

ICE-MARGINAL TERRESTRIAL LANDSYSTEMS: SUB-POLAR GLACIER MARGINS OF THE CANADIAN AND GREENLAND HIGH ARCTIC

Colm Ó Cofaigh, David J.A. Evans and John England

3.1 INTRODUCTION

The Canadian high arctic comprises the Queen Elizabeth Islands, located north of Parry Channel (Fig. 3.1), a geologically and physiographically complex archipelago with widespread glaciers and ice caps above 1000 m above sea level (Miller *et al.*, 1975). The greatest relief occurs in the fretted mountains of northern and eastern Ellesmere Island, eastern Devon Island and central Axel Heiberg Island, where glaciers are at their most extensive (Koerner, 1977; Hodgson, 1989). This terrain grades into ridge and valley and dissected plateau landscapes on the remainder of Ellesmere, Axel Heiberg and Devon islands. Elsewhere, the smaller islands of the central and western parts of the archipelago are characterized by lowland and plain topography, and glacier ice cover is minimal, being restricted to small, scattered masses like the Meighen ice cap (Koerner, 1989).

The climate of the Queen Elizabeth Islands is dominated by anticyclonic air masses centred over the Arctic Ocean and cyclonic activity over Baffin Bay (Maxwell, 1980; Edlund and Alt, 1989). Mean annual temperature ranges from –16 °C to –19 °C, and the northern part of the archipelago is subject to extreme aridity (Bovis and Barry, 1974; Edlund and Alt, 1989). Permafrost is ubiquitous throughout the region and reaches depths of greater than 500 m, even at coastal sites (Taylor *et al.*, 1983). Although some ice caps are large (e.g. Agassiz Ice Cap and Prince of Wales Icefield; Fig. 3.1), the surface morphology of glaciers in the region is strongly controlled by the underlying bedrock topography and their margins are characterized by outlet valley/fjord glaciers or piedmont lobes. Moreover, due to their smaller accumulation zones, glacier velocities in the Queen Elizabeth Islands are relatively low compared with Greenland and Antarctic outlet glaciers (Iken, 1974; Koerner, 1989; Reeh, 1989). Beyond the terrestrial margins and fjord-outlet glaciers of the Greenland Ice Sheet, northern Greenland is characterized by small plateau ice fields and large areas of ice-free terrain (Dawes, 1987; Funder, 1989; Reeh, 1989; Weidick, 1995).

We now highlight the glaciological processes common to the sub-polar glacier margins of the Canadian and Greenland high arctic and link them to characteristic landform-sediment

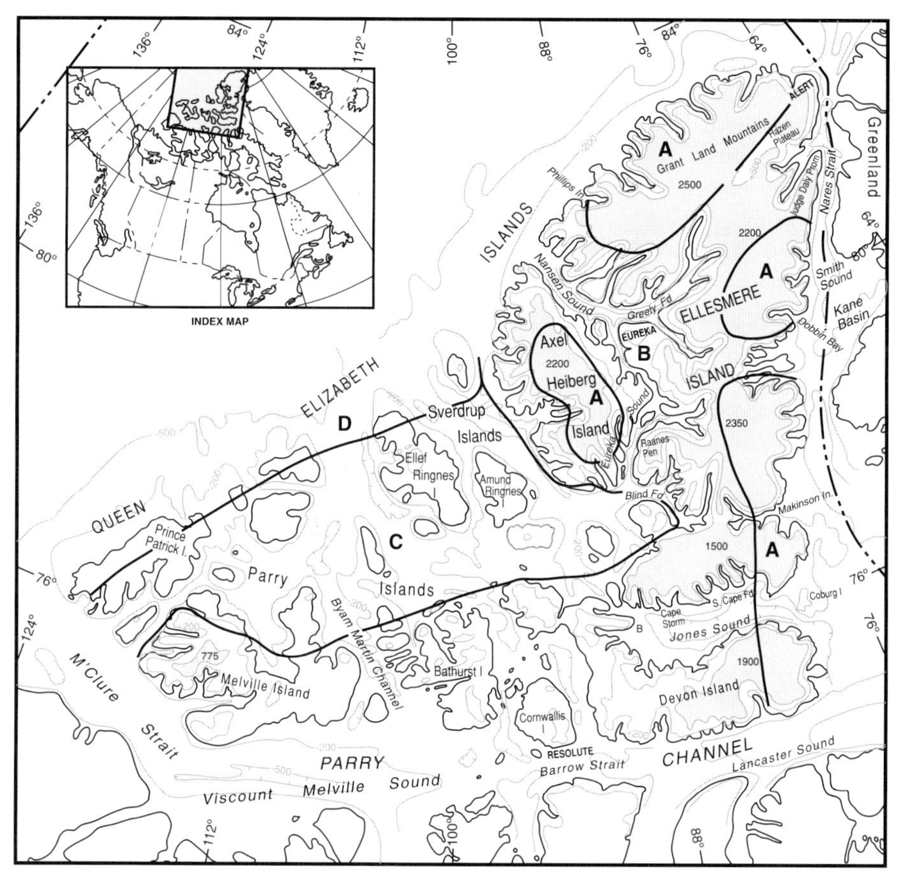

Figure 3.1 Location map of the Canadian and Greenland high arctic including physiographic zones from Hodgson (1989) and glacier ice cover. Physiographic zones are: A = fretted mountains, B = ridge and valley and dissected plateau terrain, C = lowland, D = plain.

assemblages (landsystems). Although emphasis is placed upon terrestrial margins, fjord depositional settings are also discussed.

3.2 SUB-POLAR GLACIERS IN THE CANADIAN AND GREENLAND HIGH ARCTIC

3.2.1 Glaciology

Due to the severity of the regional climate, much of the glacier ice in the Canadian and Greenland high arctic is below the pressure melting point, particularly where mean annual air temperatures and glacier thickness are low (Koerner, 1989). In the northern part of the region, this results in some glaciers terminating in the sea as ice shelves (Koenig *et al.*, 1952; Hattersley-Smith, 1957; Crary, 1958, 1960; Dunbar, 1978; Jeffries, 1987; Lemmen *et al.*, 1988; Higgins, 1989; Weidick, 1995). Because melting is largely restricted to the glacier surface and englacial drainage routes, and because glacier margins are predominantly cold-based, the glaciers of the region are classified as sub-polar (Paterson, 1994; Benn and Evans, 1998). However, many larger outlet glaciers reach pressure melting point over extensive areas of their beds due to high ablation rates at their snouts, the trapping of geothermal heat and the initiation of strain heating by glacier flow. Such glaciers are polythermal in nature (Blatter, 1987). For example, the vertical temperature profile through the snout of White Glacier, Axel Heiberg Island, indicates melting temperatures at the bed (Müller, 1976; Blatter, 1987; Blatter and Hutter, 1991). Additionally, studies on John Evans Glacier, eastern Ellesmere Island, report meltwater at the glacier bed (Skidmore and Sharp, 1999; cf. Iken, 1972, 1974). Such results demonstrate that even the snouts of some glaciers in the region are partly warm-based for short periods during the summer. However, with respect to debris entrainment and glacial geomorphology, ice thickness and activity dictate the larger-scale trends in thermal regime and landform development. Specifically, glaciers with relatively high-mass turnover (e.g. southeast Ellesmere Island bordering Baffin Bay) and/or strong converging flow will contain the most extensive zones of warm-based ice. The thinner, slower moving ice of sub-polar glacier margins will be cold-based. Evidence for cold-based glaciers within the Queen Elizabeth Islands is manifested by the recent retreat of glaciers on southeast Ellesmere Island where intact dead plant communities are being exposed (Falconer, 1966; Bergsma *et al.*, 1984).

Long-term observations on the margin of the Greenland Ice Sheet at Nunatarssuaq record the response of ice cliffs to mass balance changes (Fig. 3.2). The change in the snout profile from a steep cliff during periods of positive mass balance, to a low-angle, debris-covered ramp during periods of recession, has implications for both debris entrainment by apron overriding and for supraglacial debris release and the production of buried glacier ice (Goldthwait, 1960, 1961, 1971). Glacier recession and snout thinning exposes large areas of debris-rich folia parallel to the ice-margin, resulting in the formation of supraglacial controlled moraines (see below).

Surging glaciers have also been reported from the Canadian and Greenland high arctic (e.g. Hattersley-Smith, 1969a; Jeffries, 1984; Higgins and Weidick, 1988) and, based on our own observations, many glacier surfaces here contain looped and contorted moraines, which are usually interpreted to indicate surge activity. However, the geomorphic impact of such surging has never been investigated in the region. It is therefore possible that surge signatures are contained in the landform record of the region and this should be entertained in the explanations of process-form relationships outlined below.

3.2.2 Debris Entrainment Processes and Implications for Landform Development

Although fretted mountains are widespread in the eastern Queen Elizabeth Islands, supraglacial debris accumulations are sparse on most outlet glaciers. Where debris accumulates by rockfall below bedrock cliffs, this process may produce supraglacial lateral moraines, medial moraines and occasional ice-cored rubble cones (Evans, 1990a). Elsewhere, the aridity of the climate constrains periglacial slope processes and, therefore, the supply of extraglacial debris. Additionally, many ice fields have accumulated on plateaux and therefore their outlet glaciers cannot produce supraglacial debris (Rea *et al.*, 1998; Rea and Evans, Chapter 16). The most extensive supraglacial debris cover on sub-polar glaciers occurs at the snout where debris-rich basal ice is exposed by glacier marginal thinning. This debris cover plays a crucial role in the preservation of buried glacier ice because it often exceeds the active layer thickness. In the Canadian high arctic, buried glacier ice may remain

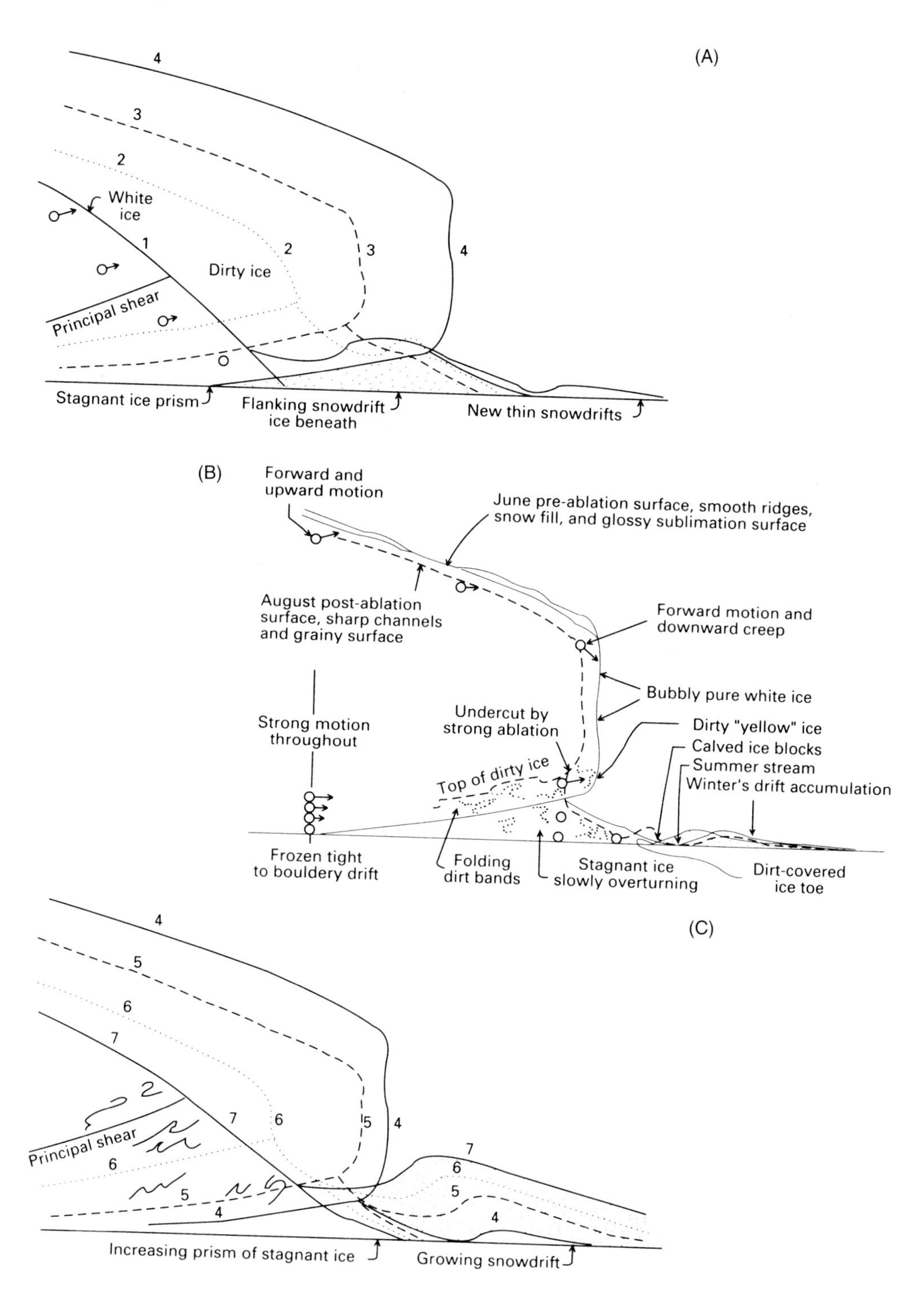

Figure 3.2 The evolution of the marginal profile of the Greenland Ice Sheet at Nunatarssuaq. A) Formation of a cliffed margin and overriding of apron. B) An ice cliff and associated processes during equilibrium phase. C) Thinning of an ice cliff during period of negative mass balance. (After Goldthwait 1971).

continuous with the glacier snout, forming a marginal supraglacial ramp up to the present ice margin. In addition, debris aprons may form where debris is exposed along shear planes and extrudes from the ice face. Such ramps are subject to fluvial incision and the development of glacier karst, resulting in the repositioning of supraglacial and englacial debris in the buried ice (Fig. 3.3). Glacier readvance involves re-incorporation of marginal ramps and their associated debris, contributing to the production of debris-rich, basal ice facies (Evans, 1989a).

The most prominent debris accumulations in sub-polar glacier snouts are those that comprise the basal ice facies. The thermal regime of a sub-polar glacier is critical to its ability to entrain and transport debris. Wet-based and sliding areas of the bed can erode and transport material towards the glacier margin in regelation ice, or perhaps in a deforming layer. Debris-rich basal ice is then produced in the snout due to net adfreezing, a process that is driven by the loss of heat through conduction at a rate that outstrips the provision of geothermal heat (Weertman, 1961; Hubbard and Sharp, 1989). Ice deceleration at the frozen margin of a sub-polar glacier induces compressive flow, which in turn thickens and elevates the debris-rich basal ice. Marginal debris, dry-calved ice blocks, buried glacier ice/marginal ramps and alluvium may also be entrained during glacier advance through a process known as apron entrainment (Goldthwait, 1960, 1961; Hooke, 1970, 1973a; Shaw, 1977a; Lorrain *et al.*, 1981; Evans, 1989a; Evans and England, 1992). Englacial folding and thrusting can thicken debris-rich basal ice (Hooke, 1973b; Hudleston, 1976; Hambrey and Müller, 1978). Furthermore, debris-rich basal ice can often be traced up-glacier for up to 0.5 km where it pinches out, suggesting that in some cases, the origin of this debris may be due to glacier readvance and overriding of pre-existing moraines during the Neoglacial or Little Ice Age. The resulting end products are the thick and complex debris-rich basal ice sequences that are observed at the margins of most sub-polar glaciers (Fig. 3.4). Temporal variability in basal thermal conditions is likely to be

Figure 3.3 A marginal supraglacial ramp with its extensive supraglacial debris cover, inner Dobbin Bay, eastern Ellesmere Island.

Figure 3.4 The cliff margin of a sub-polar glacier in Phillips Inlet, northwest Ellesmere Island, showing debris-rich basal ice facies and apron of dry calved ice blocks and debris.

significant, particularly where outlet glaciers thicken and occupy lowland areas where basal ice reaches pressure melting point and subglacial meltwater is evident (Skidmore and Sharp, 1999). Therefore, geomorphic signatures may reflect warm-based conditions in certain depositional settings. This is demonstrated by van Tatenhove and Huybrechts (1996) through their modelling of the west Greenland ice sheet margin through the Holocene.

3.3 GLACIAL GEOLOGY AND GEOMORPHOLOGY

3.3.1 Glacitectonic Landforms

Some of the most impressive landforms produced at the margins of sub-polar glaciers of the Canadian and Greenland high arctic are thrust-block moraines or composite ridges (Fig. 3.5) (Kalin, 1971; Evans, 1989b; Evans and England, 1991; Lemmen, *et al.* 1991; Lehmann, 1992). These landforms are constructed by proglacial glacitectonic disturbance of glacilacustrine, raised glacimarine or glacifluvial sediments on valley floors where the compressive stresses in the glacier snout are transmitted to unconsolidated sediments. Most thrust-block moraines of the region occur below local marine limit and in many cases record late Holocene readvances into recently emerged marine silts and/or recently deposited glacifluvial sediments (Blake, 1981; Evans and England, 1992). Because the moraines occur well below marine limit, the emergence and re-aggradation of permafrost likely date from the mid- to late Holocene. Occasionally, thrust-block moraines are constructed in glacilacustrine sediments that record former ice-marginal lakes.

Thrust-block moraines of the region are typically of the composite ridge form (Aber *et al.*, 1989), comprising relatively intact blocks of sediment displaced in en-echelon arcs by proglacial

Figure 3.5 Thrust-block moraine incised by proglacial meltwater channels, Axel Heiberg Island. The prominent arcuate moraine marks the terminus position contemporaneous with thrusting. Small lakes have become ponded on the glacier-proximal side of the thrust-block ridges.

thrusting. Thrust blocks are commonly tens of metres high (sometimes as great as 70 m), hundreds of metres wide, and en-echelon they may be hundreds of metres long. Bedding in the displaced blocks generally dips back towards the glacier snout, suggesting that they are imbricately stacked scales or deep-seated blocks partially rotated during thrusting. However, some moraines comprise blocks with bedding dipping away from the snout, indicating that the glacier was responsible for deep-seated wedging of the proglacial materials (Fig. 3.5) (Evans and England, 1991).

The role of permafrost in the thrusting process has long been debated (cf. Mathews and Mackay, 1960; Mackay and Mathews, 1964; Klassen, 1982). Furthermore, the coincidence of thrust-block moraines and former deep-water sediments or low-altitude outwash suggests that the sea level history of the region may also be significant in moraine construction. It has been shown that during the last glaciation of the Canadian high arctic, permafrost was removed by geothermal heat beneath warm-based glaciers in most valleys where the glaciers were undergoing extending flow (Dyke, 1993). During subsequent deglaciation, the sea transgressed these valleys to marine limit. Due to the high heat capacity of water bodies, permafrost development was retarded. Thus, permafrost re-aggradation was prevented until glacioisostatic emergence occurred and re-aggradation is progressively less from marine limit to modern sea level. Permafrost thickness ranges from hundreds of metres near marine limit to tens of metres at lower elevations. Therefore, decollement surfaces suitable for the mass displacement of large sediment bodies occur at the base of the aggrading permafrost. These surfaces would provide natural planes of weakness. It has also been suggested that thrust-block moraines are a major landform of the surging glacier landsystem (Evans and Rea 1999, Chapter 11).

Recent glacier advances in some valleys have resulted in the thrusting of former ice-contact deltas deposited at marine limit during Late Wisconsinan deglaciation. An excellent example of this process is the margin of Hook Glacier, Makinson Inlet, Ellesmere Island, which has proglacially thrust an existing delta and is presently incorporating the topset gravels by apron overriding (Fig. 3.6). The role of thrust blocks in providing sediments for later entrainment by the overriding glacier has also been stressed by Evans (1989a, b) and by Evans and England (1991), although most examples in the Canadian high arctic have been only partially overridden. This usually gives rise to a zone of controlled moraine ridges (see below) being superimposed on the inner blocks of thrust-block moraines during downwasting of the glacier snout (Fig. 3.7). Thrust-block moraines may also act to dam small lakes into which glacilacustrine sedimentation can take place (Fig. 3.5).

3.3.2 Glacial Debris-Release Processes and Moraine Deposition

Ice-marginal recession in the Canadian and Greenland high arctic appears to be characterized by thinning and the release of debris-rich basal ice, as suggested by Goldthwait (1960, 1961, 1971). In piedmont lobes and valley outlet glaciers this process results in the supraglacial melt-out of debris-rich folia until the snout is covered by a debris layer that exceeds the active layer thickness (≤0.5 m). The release of sediment from debris-rich basal ice of a receding and thinning glacier snout often produces transverse supraglacial debris concentrations referred to as controlled moraine (Benn and Evans, 1998). Debate continues as to the origin of the debris-rich folia, centring on the transfer of the material through the ice by shearing ('shear moraine'; Goldthwait, 1951; Bishop, 1957). The shearing mechanism was rejected by Weertman (1961) and Hooke (1968) prompting the names 'Thule-Baffin moraine' and 'ice-cored moraine' (Ostrem, 1959, 1963; Hooke, 1970).

The exposure of the complex debris-rich basal ice in many sub-polar glacier snouts, often with intense folds and thrusts, during snout downwasting leads to the construction of numerous transverse septa which ultimately control the pattern of differential ablation, meltwater flow and sediment reworking. The preservation potential of these moraines is low due to sediment redistribution during melt-out. At best the moraines may be represented in the landform record by discontinuous transverse ridges with intervening rubble hummocks. Even in situations where the buried glacier ice becomes part of the permafrost (see Dyke and Evans, Chapter 7), sediment reworking in the active layer will remove much of the inherited englacial structure. Hummocky till veneers interspersed with glacifluvial outwash tracts and occasional kames occur on valley floors where piedmont glaciers have receded onto surrounding uplands, leaving buried glacier snouts at lower elevations. Such buried snouts may be completely detached or remain connected to the outlet glacier via a debris-covered ramp.

At the margins of ice fields and upland outlet glaciers, particularly those associated with plateaux (Rea *et al.*, 1998; Rea and Evans, Chapter 16), debris turnover is low and moraines are rare. Recent glacier recession in the eastern Canadian arctic is documented by the occurrence of lines of boulders and associated rubble veneers that form conspicuous trimlines (Fig. 3.8). Such features attest to very low debris turnover in these glacial systems, as does the lack of thick unconsolidated sediments in the valleys in which the glaciers terminate (Evans 1990a). However, the major outlet glaciers that occupied the fjords and trunk valleys of the region during the last glaciation have deposited extensive lateral moraines (e.g. Lemmen *et al.*, 1991, 1994a; England *et al.*, 2000) and widespread till sheets (e.g. Bednarski, 1998). This is predominantly a function of the thermal

Figure 3.6 Main photograph: Holocene ice-contact delta being proglacially thrust by the advancing margin of Hook Glacier, Makinson Inlet, southwest Ellesmere Island. Inset top photograph: view across thrust blocks of delta showing contorted bedding of fine-grained bottomsets and gravelly foresets. Inset bottom photograph: detail of topset gravels in state of partial entrainment by apron overriding of the thrust-block moraine.

Figure 3.7 Controlled moraine ridges superimposed on the inner blocks of a thrust-block moraine at the margin of the Eugenie Glacier, Dobbin Bay, Ellesmere Island. This area was overlain by glacier ice containing discrete ice margin-parallel debris-rich folia on 1959 aerial photographs. The melt-out of the buried ice and retrogressive flow sliding at this site is gradually destroying the controlled moraine ridges.

characteristics of larger glaciers, the soles of which reached pressure melting point in most fjord/trunk valley systems.

3.3.4 Glacifluvial Processes and Forms

The most extensive evidence of glacier recession in the Canadian and Greenland high arctic are the numerous inset lateral meltwater channels cut along glacier margins. The nested patterns of such channels excavated in bedrock document the successive recessional positions of glacier snouts confined by topography (e.g. Hodgson, 1985; England, 1986, 1990; Lemmen, 1989; Evans, 1990a, b; Dyke, 1993; Bednarski, 1998; Smith, 1999; Ó Cofaigh *et al.*, 1999, 2000; England *et al.*, 2000) (Fig. 3.9a). Channel gradients are related to the gradient of the former ice margin. For example, low-gradient channels indicate similar low-gradient ice-surface profiles, probably related to rapid ice retreat and extensional flow, common in fjords, which would act to flatten the glacier profile (Fig. 3.9b). Steeper channels record a steeper glacier snout related to slower retreat, and are common inland of fjord heads where glacier margins become terrestrially based (Lemmen *et al.*, 1994a; Ó Cofaigh, 1998).

Although subglacial meltwater has been reported in the sub-polar glaciers of the region, eskers are rare. This probably reflects both the restricted nature of subglacial drainage and the sparsity of debris available for meltwater transport. Conical mounds of gravel in some valley bottoms are interpreted as kames. Because large volumes of meltwater are directed along the frozen margins of sub-polar glaciers, any sediment carried by such meltwater can be deposited as kame terraces allowing the reconstruction of former ice margins (e.g. Lemmen, 1989; Evans, 1990b; Smith, 1999). Where the retreating glaciers are in contact with the sea or lakes, meltwater draining through these channels routinely forms deltas (Ó Cofaigh, 1998; England *et al.*, 2000).

Figure 3.8 A trimline moraine comprising a line of boulders and a weakly developed moraine ridge, near Dobbin Bay, eastern Ellesmere Island. Inset photograph shows detail of the bouldery veneer that comprises the trimline moraine.

3.3.5 Rock Glacierization

Piedmont or tongue-shaped rock glaciers occurring at the base of cirque and valley glaciers, have not been reported from the Canadian and Greenland high arctic probably because of a lack of sufficient debris from surrounding slopes. Talus-foot or valley side rock glaciers are common, however, and have been subdivided by Evans (1993) into glacier ice-cored (glacial) and permafrost-related (periglacial) categories. The glacial ice-cored rock glaciers represent the former margins of outlet glaciers. Specifically, the lateral margins of outlet glaciers occupying major valleys are often characterized by supraglacial lateral moraines. These moraines later form discontinuous rock glaciers when the valley becomes deglaciated (England 1978). Paraglacial activity can also contribute to the production of rock glaciers in such settings where talus buries parts of ice margins during glacier downwasting (Fig. 3.10). Indeed, this relationship is so clear that extensive talus foot rock glaciers are often employed in the reconstruction of former glacier margins where they are thought to represent rock glacierized lateral moraines (England, 1978; Evans, 1990a, b, 1993).

3.3.6 Ice-Contact Glacimarine and Glacilacustrine Landforms

In the Canadian high arctic, research on modern marine-terminating glaciers and their sedimentary processes has been largely overlooked (Lemmen, 1990). Consequently, our understanding of glacimarine and glacilacustrine sedimentation from sub-polar glaciers in this region is largely based on investigations of emergent Holocene glacimarine, and to a lesser extent glacilacustrine, sediments (e.g. Bednarski, 1988; Evans, 1990a; Stewart, 1991; Ó Cofaigh, 1998; Ó Cofaigh *et al.*, 1999; Smith, 2000). It is likely that many modern outlet

(A)

(B)

Figure 3.9 Examples of lateral meltwater channels formed by meltwater erosion along the frozen lateral margins of valley/fjord glaciers. A) Nested lateral meltwater channels, Phillips Inlet, Ellesmere Island. B) Low-gradient lateral meltwater channels recording rapid Early Holocene retreat of fjord glacier with cold-based margins, Blind Fiord, Ellesmere Island. Note that postglacial streams have cut gorges oblique to the meltwater channels.

Figure 3.10 Glacier ice buried by talus in inner Dobbin Bay, eastern Ellesmere Island.

glaciers produce subglacial meltwater where they enter the sea given the evidence for subglacial meltwater in their terrestrial counterparts (Iken, 1972; Skidmore and Sharp, 1999). By contrast, much more is known about contemporary glacimarine sedimentation in the fjords of East and West Greenland (e.g. Dowdeswell *et al.*, 1994; Gilbert *et al.*, 1998; Ó Cofaigh *et al.*, 2001; Syvitski *et al.*, 2001). Studies of glacimarine sedimentation associated with fast-flowing outlet glaciers in East Greenland demonstrate that deposition by subglacial meltwater is significant in this environment (Ó Cofaigh *et al.*, 2001).

Where sub-polar glaciers terminate in marine or lacustrine environments as grounded or floating margins, englacial and subglacial meltwater emanating from these ice masses often constructs subaqueous depo-centres in the form of grounding line fans, ice-contact deltas and morainal banks. Spatially, the location of such ice-proximal glacimarine depo-centres exhibits a strong relationship to fjord bathymetry in that the most abundant sediment accumulations occur at topographic constrictions or areas of shallower water. Such areas acted as pinning points allowing retreating fjord glaciers to temporarily stabilize and deposit sediment. By contrast, between pinning points, glacimarine sediments are often sparse or absent, reflecting more rapid glacier retreat (e.g. England, 1987a; Lemmen *et al.*, 1994a; Ó Cofaigh, 1998).

Grounding-line fans form where sediment-laden meltwater enters deep water from englacial, or more typically, subglacial conduits. The ice-proximal location of these fans dictates that their lithofacies are texturally and sedimentologically heterogeneous. Meltwater deposits formed by the settling of suspended sediment from turbid overflow plumes are characteristic of grounding-line fans in the Canadian high arctic (Evans, 1990a; Stewart, 1991; Ó Cofaigh *et al.*, 1999). These sediments comprise various rhythmically interlaminated sand-mud (clay and silt) facies that have a drape-like geometry in section and contain variable amounts of ice-rafted debris (Fig. 3.11a). Mass-flow deposits are also a characteristic sedimentary component of grounding-line

Figure 3.11 Glacimarine sediments in Early Holocene grounding-line fans, Ellesmere Island. A) Horizontally laminated sand-silt couplets with occasional small dropstones, interpreted as suspension deposits from turbid overflow plumes with background iceberg-rafted debris. B) Channelized normally graded gravel with sharp erosional contacts, interpreted as the product of deposition from a high-concentration turbidity current, Ellesmere Island.

fan sequences in high-arctic fjords (Fig. 3.11b). Resedimentation is common due to high sedimentation rates, which result in oversteepening, failure and downslope transport. Mass-flow deposits range from channelized units of massive or variably graded, gravel and sand recording deposition from high-density turbidity currents, to finer-grained (mud and sand) laminated and stratified turbidites (Stewart, 1991; Gilbert *et al.*, 1998). Subglacial tills extruded at the ice margin or ice-rafted diamicts may also undergo resedimentation by cohesive debris flow to produce crudely stratified and massive matrix-supported diamict facies.

Subaqueous morainal banks are transverse, elongate landforms deposited along grounding lines during intervals of glacier terminus stability (Fig. 3.12a). They are commonly composed of coalescent grounding-line fans and generally range in height from about 5–30 m. Morainal bank size is controlled by the duration of grounding-line stability, sedimentation rate and availability of debris for entrainment. The elongate morphology of morainal banks reflects their origin by deposition from a series of point sources along the ice front, as well as by ice-marginal fluctuations that act to bulldoze and squeeze sediment along the grounding line. Depending on the availability of debris for entrainment and sedimentation rates, emergent morainal banks in the Canadian high arctic range from massive diamict veneers (<5 m thick) over striated bedrock to thicker accumulations of mud, sand and diamict. These include rhythmically laminated muds deposited by suspension sedimentation, and diamict facies ranging from massive tills and weakly-graded subaqueous debris flow deposits (Fig. 3.12b), to massive iceberg-rafted diamicts with *in situ* marine macrofauna (England, 1987b; Evans, 1990a; Stewart, 1991; Ó Cofaigh, 1998).

Grounding-line fans and morainal banks can aggrade to sea level and form marine limit deltas. These deltas commonly form isolated flat-topped hills in the landscape, often located on topographic highs, which acted as pinning points during glacier retreat. Distinguishing characteristics are steep ice-proximal slopes and pitted surfaces (kettle holes) due to the melt-out of buried ice (e.g. Evans, 1990b; Ó Cofaigh, 1998; England *et al.*, 2000; Fig. 3.13a). Where the ice margin has retreated above sea or lake level, it is commonly separated from the sea or lake by a braided outwash plain, which forms a delta where it enters standing water (Fig. 3.13b). Such glacier-fed deltas are common at many fjord heads throughout the Canadian and Greenland high arctic (e.g. Gilbert, 1990b; Gilbert *et al.*, 1993, 1998; England *et al.*, 2000).

In general, ice-contact and glacier-fed high-arctic deltas have a tripartite internal structure consisting of topsets, foresets and bottomsets (Fig. 3.13c). Delta topsets are essentially braided river deposits, and consist of sub-horizontally-bedded, massive gravel, with an a-axis transverse, b-axis imbricate, clast fabric. Foreset beds range from gravel to sand facies, depending on the depositional mechanism, sedimentation rate and proximity to the glacier. Commonly, however, mass-flow sediments dominate foreset beds (Fig. 3.13d). High sedimentation rates and mixing during downslope remobilization means that some of these units may be very poorly sorted. With increasing distance downslope, foresets become finer-grained and eventually grade into proximal bottomset beds of massive and graded sands and silts (Fig. 3.13e). Distal bottomset beds are composed of finer-grained silts and clays that in the marine environment often contain well-developed macrofaunal assemblages. Ice-proximal slopes are often characterized by a range of gravitational and soft-sediment deformation structures, slumps and normal faults, which are related to melt-out of buried ice and/or collapse associated with glacier retreat. In addition, the ice-proximal sides may be glacitectonized and contain subglacially-remoulded sediments in the form of glacitectonite and deformation till.

(A)

(B)

Figure 3.12 A) Arcuate morainal bank inset between fjord side and ice-moulded bedrock highs, Ellesmere Island. B) Stacked beds of massive, matrix-supported diamict with outsized clasts at bed tops, deposited by subaqueous cohesive debris flows.

Ice-dammed lakes are common in the region due to cold-based glacier margins (Fig. 3.14) (Maag, 1969, 1972), and raised shorelines often record their drainage (e.g. Hattersley-Smith, 1969b; Maag, 1969). Glacilacustrine sediments are also routinely documented from recently deglaciated terrain in the high arctic. However, studies on glacilacustrine landform development and sedimentation are rare (e.g. Smith, 2000). Sediment input to ice-dammed lakes is dictated by the debris-content of the glaciers in the catchment, and typical landforms produced in lakes with sufficient debris supply are ice-contact deltas, beaches and incised terraces, the latter recording the downcutting by meltwater streams through lake sediments during and after lake drainage. Because lake water tends to drain over cold-based glacier barriers, rather than through or under them, the most prominent landforms produced in association with ice-dammed lakes are erosional spillway channels (Maag, 1969, 1972; Smith, 1999).

(A)

(B)

(C)

(D)

(E)

Figure 3.13 A) Raised marine ice-contact delta, western Ellesmere Island. B) Braided outwash plain entering the sea, western Ellesmere Island. C) Detail of internal stratigraphy of ice-contact Gilbert-type raised marine gravel delta, western Ellesmere island. Note the horizontally bedded topset gravels which unconformably overlie planar crossbedded foresets. D) Delta foreset bed composed of normally-graded open-work gravels deposited by high-concentration turbidity current, Ellesmere Island. E) Delta bottomsets composed of normally-graded sands with silty mud cap, western Ellesmere Island.

As sub-polar glacier margins are prone to float upon contacting deep water they have the potential to deposit horizontal moraines or ice-shelf moraines at their margins (England *et al.*, 1978). However, ice-shelf moraines tend to be rare in the Canadian and Greenland high arctic and have (to date) only been documented from northeast and east Ellesmere Island (England, 1978, 1999; England *et al.*, 2000) and northwest Greenland (England, 1985).

3.4 OVERLAPS WITH OTHER GLACIGENIC LANDSYSTEMS

Important questions are how unique is the sub-polar glacier landsystem and to what extent can it be differentiated from other terrestrial and subaquatic glacigenic landsystems? This is important for understanding the nature of glaciers and ice sheets in the geological record, particularly in terms of their basal thermal regime and, hence, past climate. The sub-polar glacier landsystem contains elements that are common to several other landsystems discussed in this book. Thrust-block moraines have been described from modern surge-type, temperate glaciers in Iceland (Evans and Rea, Chapter 11). Temperate, marine-terminating tidewater glaciers in environments such as southeast Alaska, deposit subaquatic morainal banks, grounding-line fans and ice-contact deltas (Powell and Molnia, 1989; Powell, 1990, Chapter 13). Additionally, plateau ice fields in the

Figure 3.14 Ice-dammed lake, produced by the blocking of a valley by a plateau-outlet glacier (see also Chapter 16), Viking Ice Cap, northeastern Ellesmere Island. (Photograph T404L-56, Energy, Mines and Resources, Canada).

Canadian and Greenland high arctic are characterized by a suite of landforms and sediments that are diagnostic of the plateau ice field landsystem (see Rea and Evans, Chapter 16). The mountainous setting of much of the Canadian and Greenland high arctic, in which outlet glaciers extend towards sea level through valleys, means that the processes and landforms characteristic of the high-relief glaciated valley landsystem (see Benn *et al.*, Chapter 15) may also occur in the sub-polar glacier landsystem. These include rock glacierization, melt-out of supraglacial and englacial debris to produce buried glacier snouts (see Dyke and Evans, Chapter 7) and subaerial debris flow deposition.

As has been stressed for the other landsystems in this volume, it is apparent that no single criterion is diagnostic of the sub-polar glacier landsystem other than the exceptional detail provided by nested lateral meltwater channels. Rather it is important to consider the criteria collectively, particularly in investigations of the geological record. It is the combination of evidence for both a warm-based glacier bed with frozen glacier margins that is diagnostic of the sub-polar glacier regime. For example, the juxtaposition of thick accumulations of meltwater-derived glacimarine sediments related to the subglacial drainage system, with nested suites of lateral meltwater channels in the same fjord system, suggests the former presence of a sub-polar glacier margin rather than a temperate southeast Alaska-type regime. In the latter, the temperate glaciological regime would facilitate rapid penetration of supraglacial meltwater to the bed and preferential evacuation to the glacier terminus via the subglacial meltwater system.

3.5 CONCLUSION

Figure 3.15 presents a summary model of the sub-polar glacier landsystem in the Canadian and Greenland high arctic. Sub-polar glaciers of the Canadian and Greenland high arctic are characterized by a marginal zone of cold-based or frozen ice that passes up-glacier into a zone(s) of wet-based ice. This thermal regime means that sub-polar glaciers are characterized by both subglacial and supraglacial/lateral meltwater systems, and by significant compressive stresses where the zone of wet-based ice passes into the frozen bed of the terminus. These characteristics are reflected geomorphologically and sedimentologically in the sub-polar glacier landsystem. We emphasize that it is the presence of evidence for both wet- and frozen-based thermal conditions that are diagnostic of the sub-polar glacier landsystem, including the abundance of lateral meltwater channels that demarcate former glacier margins.

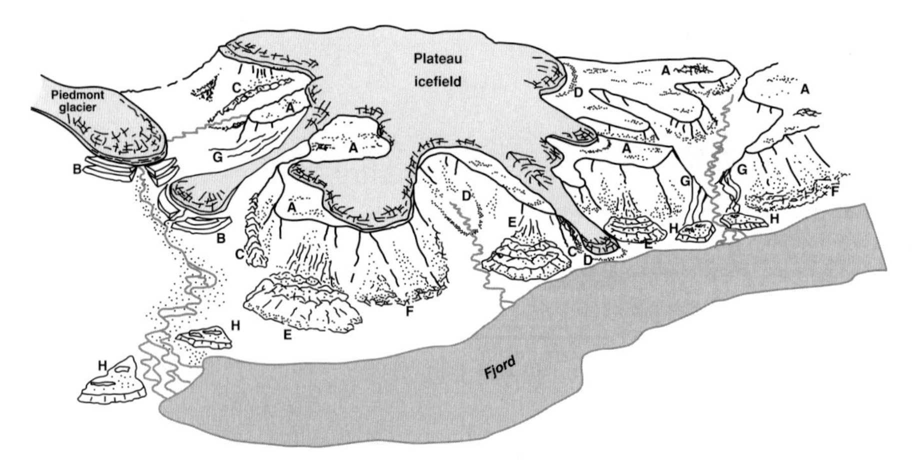

Figure 3.15 Landsystems model for sub-polar glaciers in the Canadian and Greenland high arctic, concentrating specifically on plateau icefields and piedmont lobes. A = blockfield/residuum, B = thrust-block moraine, C = ice-cored lateral moraine, D = trimline moraine, E = glacier-ice cored protalus rock glacier, F = periglacial protalus rock glacier, G = lateral meltwater channels, H = raised, former ice-contact deltas.

CHAPTER

4

ICE-MARGINAL TERRESTRIAL LANDSYSTEMS: SVALBARD POLYTHERMAL GLACIERS

Neil F. Glasser and Michael J. Hambrey

4.1 INTRODUCTION AND RATIONALE

The aim of this contribution is to describe the landform-sediment assemblages at the margins of polythermal glaciers in Svalbard and to present a landsystem model for terrestrially-based glaciers in this maritime high-arctic setting. The discussion is restricted principally to non-surge-type glaciers as the landform-sediment assemblages at surge-type glaciers are considered in a separate chapter (Evans and Rea, Chapter 11). The structural glaciological controls on landform development are described, together with the dominant landform types and their sedimentary facies.

The Svalbard archipelago (77°N to 80°N) lies at the northern extent of the Norwegian Current, a branch of the Gulf Stream, and enjoys a climate that is relatively mild and moist for its northern latitude. On the western coast the average annual temperature is –6 °C, the average temperature of the warmest month (July) is +5 °C, and in the coldest month (January) it is –15 °C. Although there are contrasts between the maritime west coast and the interior, precipitation in Svalbard at sea level varies typically from 400 to 600 mm annually. Orographic effects increase precipitation in the highland regions, but even on the glaciers snowfall of more than 2–4 m is rare (Hagen *et al.*, 1993). Ice-free land areas are underlain by permafrost to depths of between 100 and 460 m (Hjelle, 1993).

The archipelago is 60 per cent glacierized, the largest volumes of ice being accounted for by the highland ice fields and ice caps of northwestern, northeastern and southern Spitsbergen and Nordaustlandet (Fig. 4.1). These ice masses cover the highland areas, and their outlets are divided into individual glaciers by mountain ridges and nunataks. Many glaciers reach the sea, forming wide calving fronts. Smaller cirque glaciers are also common, particularly in the more alpine terrain of western Svalbard (Fig. 4.2). Many of the larger glaciers in Svalbard are polythermal, with extensive areas of temperate ice in the accumulation area, but with their termini frozen to the bed (Hagen and Saetrang, 1991; Hagen *et al.*, 1991; Ødegård *et al.*, 1992; Björnsson *et al.*, 1996). Most of the smaller cirque glaciers are probably cold-based throughout. The hydrology of polythermal glaciers is poorly understood compared with alpine glaciers (Bamber, 1989; Hagen

Figure 4.1 The Svalbard archipelago, showing location of glaciers mentioned in text.

and Saetrang, 1991; Hagen *et al.*, 1991; Vatne *et al.*, 1996), although recent advances have been made in this field (Hodgkins, 1997).

Until recently, the relationship between ice structure and debris distribution in polythermal glaciers was poorly understood in comparison with temperate glaciers (Weertman, 1961; Swinzow, 1962; Boulton, 1970, 1972b, 1978; Hooke, 1973a; Clapperton, 1975; Hambrey and

Figure 4.2 Norsk Polarinstitutt vertical aerial photograph S90-6526 of Brøggerhalvøya, on the southern side of Kongsfjorden in northwest Spitsbergen. Valley glaciers shown are (from left to right): Vestre Lovénbreen, Midtre Lovénbreen, Austre Lovénbreen, Pedersenbreen and Botnfjellbreen. These glaciers are typical of many Svalbard valley glaciers, with multiple accumulation basins feeding a single glacier tongue. The snout of Kongsvegen is just visible at the head of Kongsfjorden, in the lower right of the photograph. The large glacier at the bottom of the photograph is Uvêrsbreen.

Müller, 1978). Recent studies have clarified this relationship, confirming the importance of thrusting in elevating basal debris within polythermal glaciers (Hambrey and Huddart, 1995; Bennett *et al.*, 1996a and b; Hambrey *et al.*, 1996; Murray *et al.*, 1997). These studies have also highlighted the significance of folding of debris-rich stratification in re-organizing both supraglacial, basal and glacifluvial debris (Hambrey and Dowdeswell, 1997; Glasser *et al.*, 1998a, 1999; Hambrey *et al.*, 1999; Glasser and Hambrey, 2001a and b).

Estimates of the percentage of surge-type glaciers in Svalbard range from 13 per cent (Jiskoot *et al.*, 1998, 2000), to 35 per cent (Hamilton and Dowdeswell, 1996), and even as high as 90 per cent (Lefauconnier and Hagen, 1991). These glaciers are prone to dramatic increases in velocity and rapid frontal advances, followed by periods of quiescence during which velocities are generally low. Surge-type glaciers in Svalbard typically have relatively long quiescent phases between surge events (Dowdeswell *et al.*, 1991). Surges have been documented at numerous Svalbard glaciers including Usherbreen (Hagen, 1987, 1988), Bakaninbreen (Murray *et al.*, 1997), the Kongsvegen/Kronebreen tidewater complex (Melvold and Hagen, 1998; Bennett *et al.*, 1999), Seftstrømbreen (Boulton *et al.*, 1996), Holmströmbreen (Boulton *et al.*, 1999), Bråsvellbreen (Solheim and Pfirman, 1985) and Fridtjovbreen (Glasser *et al.*, 1998b; see Fig. 4.1 for locations of glaciers).

Most glaciers in Svalbard are currently receding from their Neoglacial maxima, achieved circa 1890–1900 (Fig. 4.3). Mass-balance measurements have been made at two Svalbard glaciers, Austre Brøggerbreen and Midtre Lovénbreen, since the 1960s (Hagen and Liestøl, 1990; Lefauconnier *et al.*, 1999). Statistical analysis of these records and of associated climatic data suggests that the net mass balance of these glaciers has been negative in the majority of years since 1900 (Lefauconnier and Hagen, 1990). Consequently, most glaciers terminating on land in Svalbard have receded 1–2 km since that time (Hagen *et al.*, 1993). Volume losses since 1900 have been substantial, possibly as much as 33 per cent, based on former ice-marginal positions and trim lines. Historical and photographic records show that, at the Neoglacial maximum, many Svalbard glaciers had near-vertical fronts with thick debris layers (Liestøl, 1988). This trend of overall recession means that many Svalbard glaciers have extensive zones of exposed sediments and landforms between their Neoglacial maxima and current snouts (Fig. 4.4).

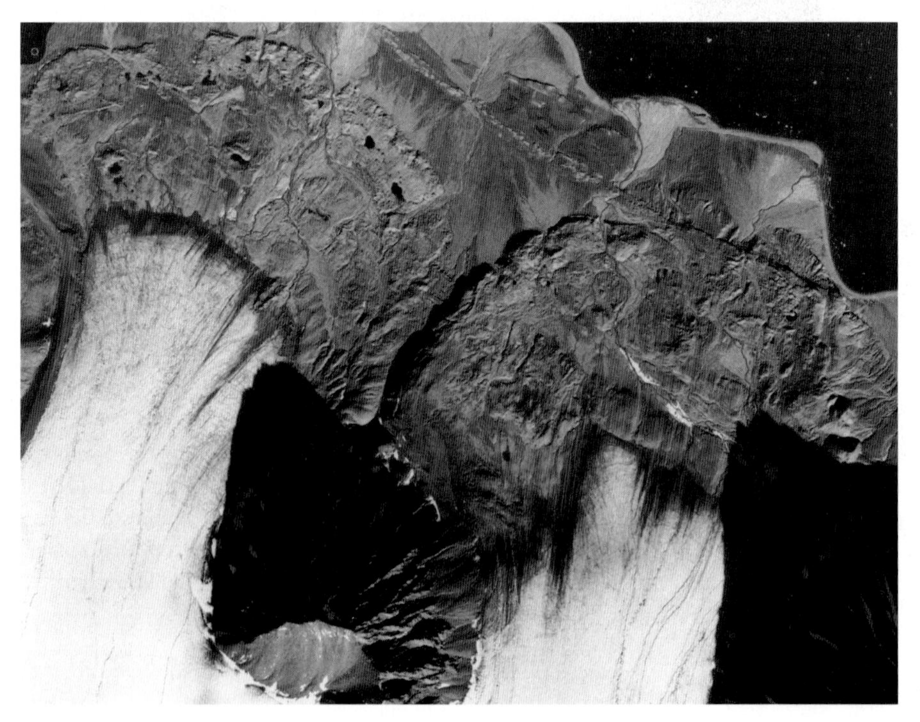

Figure 4.3 Norsk Polarinstitutt vertical aerial photograph S90-5788 of Midtre Lovénbreen (left) and Austre Lovénbreen (right) showing the recent (post *c.* 1890) recession of these two valley glaciers. Both have a prominent outer moraine ridge, within which are moraine-mound complexes, glacifluvial facies and linear debris stripes composed of supraglacial debris.

Figure 4.4 Part of Norsk Polarinstitutt vertical aerial photograph S95-1087 of Austre Brøggerbreen. Although heavily modified by glacifluvial activity and partly flooded by proglacial lakes, linear debris stripes are still a strong component of the landform-sediment assemblage at Austre Brøggerbreen.

4.2 STRUCTURAL GLACIOLOGICAL CONTROLS ON DEBRIS ENTRAINMENT AND TRANSPORT

An understanding of structural glaciology and its role in controlling debris entrainment and transfer is essential to the interpretation of the landsystem developed at Svalbard glacier margins (Hambrey and Lawson, 2000). Debris entrainment in Svalbard glaciers takes place by a variety of mechanisms that can be linked to the evolution of ice structures (Hambrey *et al.*, 1999). The most significant structures include stratification, folded stratification, foliation, the deformed basal ice zone, thrusting and thrust-related recumbent folding. These modes of debris entrainment are outlined below.

4.2.1 Incorporation of Rockfall Material

Primary stratification is inherited from snow accumulation and superimposed ice in the accumulation area, and is sometimes supplemented by rockfall material from the glacier headwalls (Hambrey *et al.*, 1999). This material is typically angular, and becomes incorporated englacially as a result of burial and folding of the stratified sequence. Converging flow, where multiple accumulation basins supply a narrow tongue, promotes folding (Fig. 4.5a). Component flow units may be reduced in width by more than 50 per cent. Fold styles typically range from open 'similar' to less common chevron and isoclinal types, each commonly associated with an axial-planar foliation. Fold axes tend to be parallel to flow and plunge gently up-glacier. As debris-laden folded stratification intersects the glacier surface in the ablation area, 'medial moraines' emerge from a single point-source or multiple point-sources, producing down-glacier widening spreads of surface debris (Fig. 4.5b, c). As in temperate glaciers (e.g. Meier and Post, 1969; Lawson, 1996), this relatively simple structure can be complicated by surges, giving rise to 'looped' moraine structures in which the debris layers originate from the stratification (Hambrey and Dowdeswell, 1997).

4.2.2 Entrainment of Debris at the Bed

Svalbard glaciers carry a large basal debris load, commonly reaching a thickness of several metres, with debris concentrations in some zones approaching 100 per cent. Although poorly understood at present, the principal processes of entrainment are considered to be regelation, water flow through the crystal vein system, bulk freezing-on, folding and shearing (Knight, 1997). Where observed in sections, layers of ice facies are commonly subjected to repeated shearing and isoclinal folding. The typical coarse-clear ice crystal character of basal ice, combined with varying proportions of debris allow it to be distinguished from coarse-bubbly ice derived from snowfall, when glacitectonically transferred to higher level positions within the glacier. Research into the composition and glacitectonic transport of the basal debris layers of Svalbard glaciers is an ongoing area of research.

(A)

(B)

(C)

(D)

(E)

Figure 4.5 Structural glaciology and debris transport in Svalbard valley glaciers. A) Oblique aerial photograph of a typical Svalbard cirque glacier with a single glacier tongue fed by multiple accumulation basins. Note the prominent debris stripes on the glacier surface and their continuation onto the forefield. B) Angular debris emerging from stratification on the surface of Sagabreen. The debris thickens down-glacier to form supraglacial debris stripes. C) Supraglacial debris stripes and their continuation onto the forefield at Midtre Lovénbreen. Individual stripes tend to consist of a single lithology, which in some cases can be traced to source areas in the accumulation basins. D) Debris-poor up-glacier dipping structures interpreted as thrusts in a lateral ice cliff on Kongsvegen. Note how the features rise asymptotically from the bed. The leftmost feature is 'blind', and does not reach the glacier surface. E) Debris-rich structure

(F)

interpreted as a thrust on the surface of Midtre Lovénbreen. The feature contains well-sorted sand and gravel, retaining original depositional characteristics indicative of a subglacial derivation. Glacier flow is from right to left and the plane of the thrust can be seen immediately above the ice axe. F) Small debris-rich structure interpreted as a thrust on the surface of Kongsvegen. The feature is composed of basal material, with high proportions of subrounded and striated clasts. Glacier flow is from left to right.

4.2.3 Association of Debris with Longitudinal Foliation

Longitudinal foliation is a structure common to all glaciers, being the product of tight folding, or simple or pure shear. In glaciers dominated by converging flow, longitudinal foliation pervades the width of a glacier (Hambrey and Müller, 1978). Svalbard glaciers commonly have pervasive near-vertical foliation throughout their widths, although this foliation varies in intensity, being strongest at the zones of confluence of adjacent flow units and at the margins. Debris is associated with foliation in two ways.

- Moderately well-sorted angular debris of supraglacial origin, which has been folded so tightly that the original stratification has been transposed into foliation (Fig. 4.6). Alternatively, the debris is dispersed around the hinge of a more open fold, which is intersected by an axial planar foliation (Hambrey *et al.*, 1999).
- Debris showing basal characteristics (i.e. a wide variety of clast shapes, clast surface features such as striations and facets on appropriate fine-grained lithologies, and poorly sorted texture), disseminated through, or layered within, coarse-clear ice. This debris is isoclinally folded or sheared within foliation, particularly in ice-marginal areas and occasionally within the lower reaches of medial moraines. The folding mechanism that allows basal debris to reach the surface is unclear, but it is possible that in the thicker parts of the glacier a deformable bed of debris is folded within the overlying ice in the zone of converging flow (Hambrey *et al.*, 1999). Again, the folding is in axial-planar relationship with the foliation.

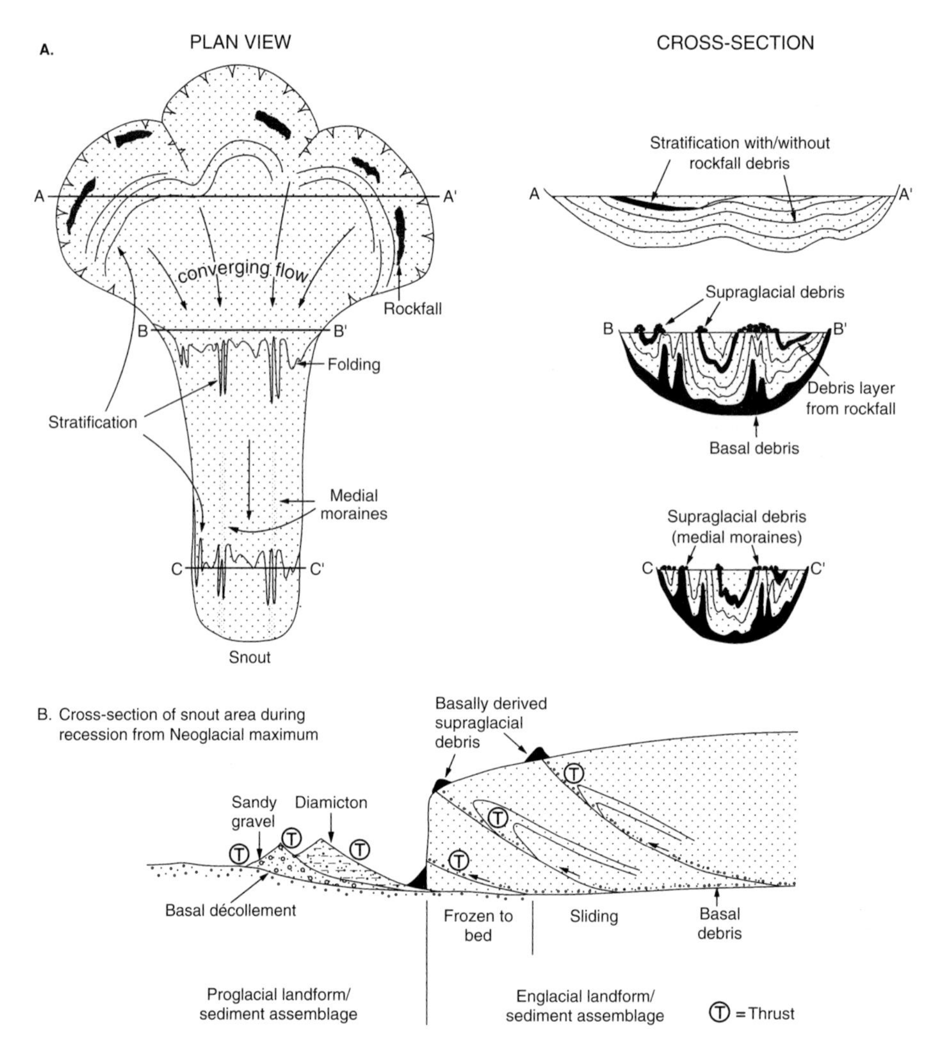

Figure 4.6 Schematic diagram illustrating the origin of structures in a typical polythermal Svalbard valley glacier.

4.2.4 Debris Entrainment by Thrusting

Thrusting is the most controversial mechanism of debris entrainment. Swinzow (1962) suggested that this was a valid mechanism for entrainment in the margin of the West Greenland ice sheet, but an alternative view is that incorporation of debris is a passive process as ice flowlines turn upwards towards the ice-frontal margin (Weertman, 1961; Hooke, 1973). Examination of structural relationships and mapping of numerous Svalbard glaciers suggests that thrusting is indeed a valid process for debris entrainment in polythermal glaciers (e.g. Hambrey *et al.*, 1996, 1999; Hambrey and Dowdeswell, 1997; Murray *et al.*, 1997; Glasser *et al.*, 1998a).

Thrusts in Svalbard glaciers tend to be new structures, unrelated to pre-existing structural inhomogeneities. They are particularly common on the glacier surface within several hundred metres of the snout (Fig. 4.5d, e, f,). Few thrusts are actively-forming today, except in glaciers that are actively surging (Glasser *et al.*, 1998b; Hambrey *et al.*, 1999). Thrusts incorporate whatever material lies on the bed or within the basal zone, including basal till, glacifluvial and glacimarine sediment. The sediment thickness associated with thrusting may reach several metres, and original sedimentary structures can be preserved. At the opposite end of the spectrum, thrusts may be marked simply by thin layers of debris-rich ice of basal origin or films of fine debris along a well-defined plane. Displacements along thrusts are variable, and these are often difficult to evaluate due to the lack of clear marker horizons in the glaciers. Where these marker horizons exist, measured displacements range from only a few centimetres to several metres. However, larger displacements of tens of metres are required to raise basal debris to the ice surface.

Geometrically, thrusts show an asymptotic relationship with the bed and emerge at the glacier surface at angles ranging from 15° to 70°. Most layers associated with prominent debris-ridges dip at ~30° or less. Thrusts tend to be arcuate in plan, mirroring the geometry of the ice margin. In places, many thrusts can be seen to intersect at low angles. Thrusts may be laterally continuous for several tens of metres, although significant quantities of debris usually only occur on a fraction of this length. Intersecting thrusts tend to promote accumulations of debris on the glacier surface (Bennett *et al.*, 1999). Debris-bearing thrusts are particularly well-developed in polythermal glaciers, where the thrusting process is facilitated by the transition from sliding bed conditions to frozen bed conditions at the margin and snout (Clarke and Blake, 1991; Hambrey *et al.*, 1999). At such locations, high water pressures can facilitate the upward displacement process of englacial and subglacial debris (Glasser *et al.*, 1999). Thrusting is not confined to polythermal glaciers, although large-scale debris-incorporation in temperate glaciers generally requires particular topographical conditions, such as ice-flow against a reverse slope (Glasser and Hambrey, 2002).

4.3 GEOMORPHOLOGY AND SEDIMENTOLOGY OF RECEDING SVALBARD GLACIERS

4.3.1 Moraine Complexes

Modern glaciers in Svalbard are invariably associated with large end-moraine complexes formed during the Neoglacial maximum in the late nineteenth or early twentieth century (Fig. 4.7). Some glaciers have receded 1–2 km from these complexes, leaving smaller moraine systems in the intervening zone. Some authors have referred to these moraine complexes as 'push moraines' (e.g. Croot, 1988a; Hagen, 1987, 1988; Van der Wateren, 1995; Boulton *et al.*, 1996, 1999) although not all Svalbard moraine complexes are produced by ice-push and in many cases thrusting of sediment and ice is the primary mechanism (Bennett, 2001). The descriptive term 'moraine-mound complex' should therefore be used to describe the large moraine systems in front of receding Svalbard glaciers (Hambrey and Huddart, 1995; Huddart and Hambrey, 1996; Hambrey *et al.*, 1997; Bennett *et al.*, 1999). The genetic term 'thrust-moraine complex' should only be used where this mechanism has been clearly demonstrated.

4.3.2 Moraine Complexes Produced by Thrusting

4.3.2.1 Morphology

Svalbard moraine-mound complexes commonly comprise arcuate belts of aligned hummocks or mounds comprising a wide variety of morphological types. They include linear ridges some 100 m long, short-crested ridges several metres long, and near-conical mounds reaching elevations of several metres. Irrespective of size, before degradation, they show ice-proximal rectilinear or curvilinear slopes with consistent angles of around 30°, and irregular distal slopes that are commonly steeper, formed by the stacking of different sedimentary facies (Fig. 4.7a). Stacking indicates thrusting in proglacial, ice-marginal and englacial positions (Hambrey *et al.*, 1997). Ice cores may be present, leading to the degradation of many mounds by mass-movement processes (Fig. 4.7b). Thrusting and stacking of thick sediment wedges promotes the survival of the initial moraine morphology, particularly if the sediments are not prone to subsequent reworking by glacigenic sediment flows (e.g. free draining gravels). However, large-scale degradation is likely if

(A)

(B)

(C)

Figure 4.7 Landform-sediment assemblages at Svalbard glaciers. A) Imbricate stacking of moraine mounds in front of Kronebreen (figure for scale). Thrust planes between the individual mounds are shown up by shadows between the mounds. Former ice flow from left to right. B) Moraine-mound complex ('hummocky moraine') in front of Kronebreen. Here the moraine-mounds are more degraded. Note the reworking of moraines by glacifluvial processes in the foreground of the photograph. C) Group of small flutes at the margin of Midtre Lovénbreen. Former glacier flow is from bottom left to top right of the photograph. Note moraine-mound complex beyond the flutes.

substantial quantities of ice are buried during formation, and if the sediments are composed of clay-rich diamicton that is prone to flowage when wet.

4.3.2.2 Sedimentary Facies

The sedimentary facies in moraine-mound complexes are as varied as the material over which the glacier flows (Fig. 4.8). Most of the terrestrial glaciers flow over ground occupied by extensive sheets of glacifluvial sediment and diamicton of basal glacial origin. These facies dominate the moraine assemblages at Uvêrsbreen (Hambrey and Huddart, 1995), Pedersenbreen (Bennett *et al.*, 1996a) and Midtre Lovénbreen (Table 2.1; Glasser and Hambrey, 2001a). Tidewater glaciers, such as Comfortlessbreen (Huddart and Hambrey 1996) and Kongsvegen/Kronebreen (Bennett *et al.*, 1999) tend to rework glacimarine deposits, ranging from ice-contact facies (diamictons, coarse gravels), through ice-proximal laminites (cyclopsams and cyclopels) to distal muds with dropstones up to boulder-size. Mixing with terrestrial sediments occurs where tidewater glaciers advance from the sea onto adjacent land. The wide variety of facies, however, is often organized systematically. A single ridge is commonly composed of one facies, but may be stacked as an inclined slab on another ridge of a different facies.

Preservation of sedimentary structures, for example cross-bedding and grading in sands and gravels or laminations with dropstones in fine-grained distal sediments, allows determination of the

(A)

(B)

(C)

(D)

(E)

(F)

(G)

Figure 4.8 Sedimentary facies produced by Svalbard valley glaciers. A) Clast-rich diamicton on the forefield of Kronebreen, interpreted as a basal till. B) Diamicton with anastomosing planar fabric interpreted as a shear fabric in front of Austre Lovénbreen. C) Muddy cobble/boulder gravel in front of Midtre Lovénbreen. This lithofacies is interpreted as an ice-marginal facies, created by the reworking of glacifluvial deposits and subsequent mixing with basal glacial material. D) Reworking of sediments on the forefield of Austre Lovénbreen by proglacial streams. E) Glacifluvial facies in moraine mounds at Finsterwalderbreen. F) Glacimarine sand-mud laminites in moraine mound, Comfortlessbreen. G) Glacimarine shelly mud in moraine mound, Comfortlessbreen.

original sedimentary facies. Intact and broken shells are a feature of reworked glacimarine sediment. Diamicton, on the other hand, being more susceptible to ductile deformation rarely preserves its original fabric. Typical basal till fabrics, with strong preferred alignment of clasts, are often (though not always) destroyed during thrusting.

4.3.2.3 Internal Structure

The internal structure of individual mounds has been documented in a number of cases, indicating evidence of deformation (Hambrey and Huddart, 1995; Bennett *et al.*, 1999). In mounds containing mud and sand, typically structures are low-angle thrust-faults, various normal faults and recumbent folds with sheared-off lower limbs. Such a combination of structures develops under strong longitudinal compression. Sometimes, however, the moraines show an earlier phase of longitudinal extension, such as boudinage. In mounds dominated by reworked sand and gravel, the degree of internal deformation is limited. Sedimentary bedding may be rotated, but otherwise survives intact within thrust slices, as in the gravels of Uvêrsbreen. Alternatively, bedding may be strongly modified, as in the fine-grained glacimarine sediments of Kronebreen and Comfortlessbreen.

Lithofacies (and interpretation)	Abundance	% clasts	Matrix (%)			Sorting coefficient	Sorting category
			Sand	Silt	Clay		
Clast-poor intermediate diamicton (basal glacial)	*	5	48	33	19	4.05	Extremely poorly sorted
Clast-rich sandy diamicton (basal glacial)	**	30	75–90	8–20	2–8	1.5–2.87	Very poorly sorted
Clast-rich intermediate diamicton (basal glacial)	***	25–35	41–83	5–43	11–20	2.91–4.27	Very poorly sorted
Clast-rich muddy diamicton (basal glacial)	*	30	15	78	7	1.22	Poorly sorted
Sandy gravel (glacifluvial)	**	10–40	92–98	1–6	1–2	0.87–1.61	Poorly sorted
Gravel (type 1) (fluvial)	**	80–95	75–90	6–19	2–8	1.78–2.96	Very poorly sorted
Gravel (type 2) (subglacial fluvial)	*	100	–	–	–	–	Well sorted
Gravel with sand (glacifluvial)	*	70–90	90–98	1–7	1–3	0.73–2.82	Moderately sorted to poorly sorted
Sand and mud (lacustrine)	**	0–5	76–99	1–19	1–5	0.88–1.12	Moderately sorted to poorly sorted

Table 4.1 Summary of lithofacies identified on the forefield of Midtre Lovénbreen. Key to abundance: *** = dominant, ** = prevalent, * = rare.

4.3.3 Moraine Complexes Resulting from Deformation of Permafrost

Drawing on data from two Svalbard valley glaciers, Usherbreen and Erikbreen, Etzelmüller *et al.* (1996) have suggested that the deformation of permafrost is important in the formation of ice-cored moraines and push moraines. Usherbreen is a surge-type glacier (Hagen, 1987), while Erikbreen is not. Stresses beneath the advancing glaciers are transmitted to the proglacial sediments and can be sufficient to cause proglacial deformation of the permafrost layer. Folding, thrust-faulting and overriding of proglacial sediments are possible under these conditions (Fig. 4.9). The nature of the deformation is controlled by the mechanical properties of the sediment, which is influenced by the water content and thermal condition (frozen/unfrozen). Typical landform/sediment associations are:

- an outermost push-moraine system
- an arc of ice-cored moraines, and
- an innermost area dominated by glacigenic sediment flows.

The deformed material can consist of a wide variety of sedimentary facies, including glacifluvial, glacilacustrine, glacimarine deposits and subglacial deposits.

Elsewhere in Svalbard, notably at Uvêrsbreen (Hambrey and Huddart, 1995) and Comfortlessbreen (Huddart and Hambrey, 1996) the outer parts of the moraine complexes represent deformation beyond the ice margin as stress was transmitted into frozen glacifluvial sediments in front of the glacier during the Neoglacial maximum. In such circumstances a basal décollement surface must have propagated at depth to form the outermost thrust.

4.3.4 Other Constructional Landforms

At least four other types of constructional landforms exist in the proglacial areas of Svalbard glaciers: linear debris stripes, foliation-parallel ridges, geometrical ridge networks and streamlined ridges/flutes.

Linear debris stripes are produced by the folding of supraglacially-derived debris layers in the ice, as described above. These are derived from folded stratified layers and emerge at the glacier surface as medial moraines as a result of ablation near the snout (Hambrey *et al.*, 1999). Debris is released from the ice as regular stripes of angular debris extending for considerable distances across the proglacial area (see Figs. 4.3 and 4.4). Commonly, linear debris stripes drape moraine-mound complexes. Individual debris stripes can often be traced to their source areas in the headwall of a glacier, where they are fed by rockfall material. These debris stripes are recognizable by their angular, unilithological nature and lack of fine matrix. Debris stripes survive as prominent features on the forefield following deposition because the large blocks and lack of associated fine sediment fails to support extensive vegetation.

Foliation-parallel ridges are ridges of basally-derived debris (Bennett *et al.*, 1996b; Glasser *et al.*, 1998a). They are particularly well-developed at surge-type Kongsvegen, where significant quantities of basal material are observed on the glacier surface parallel to longitudinal foliation. Other degraded examples occur at Vestre Lovénbreen, a non-surge-type glacier. Although the ridges are typically 1–2 m wide and up to 1.5 m high, the source debris layers in the ice below are rarely greater than 0.1 m wide. The dispersion of material associated with these features is a result of the melting of their ice core. Some ridges are composed of a clast-rich sandy diamicton,

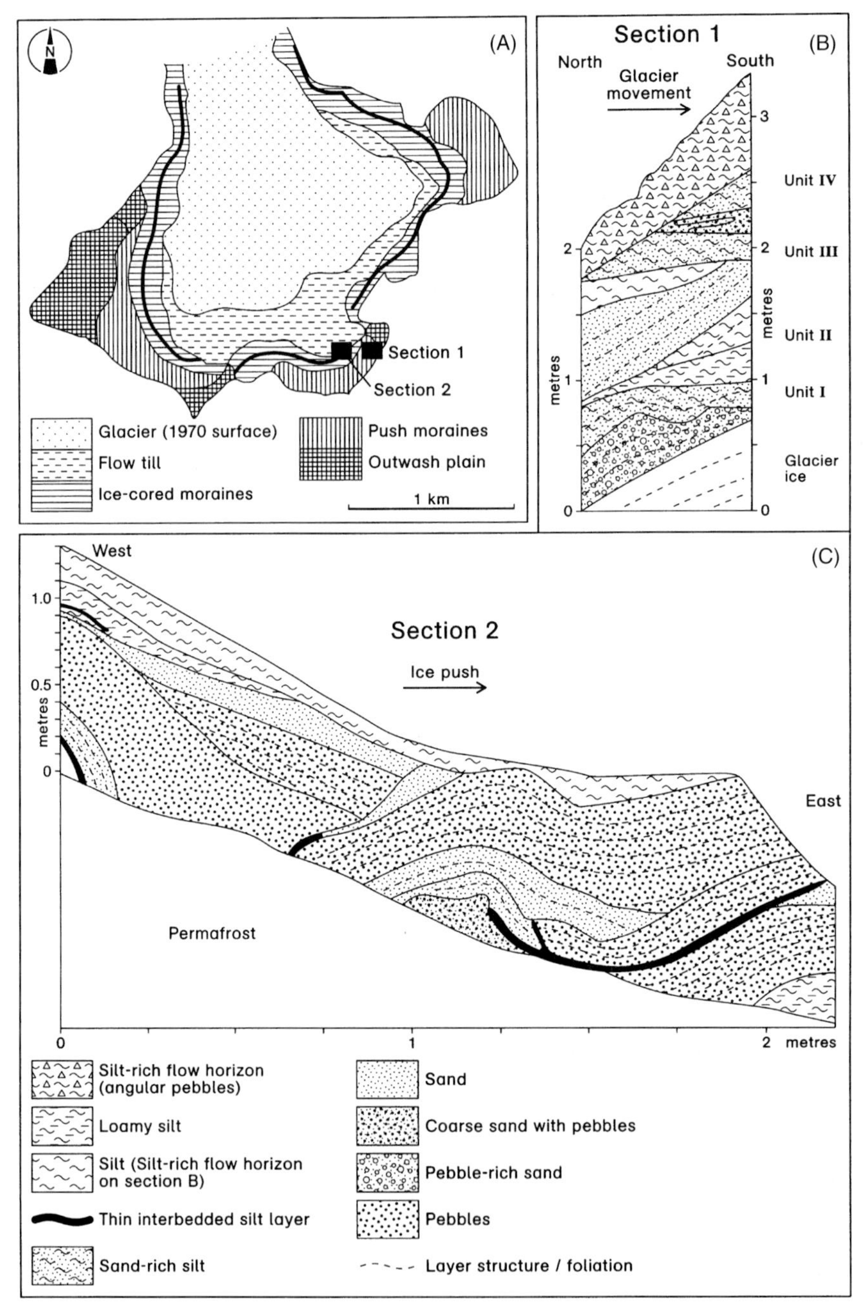

Figure 4.9 Structures and sedimentary facies associated with the deformation of permafrost at Erikbreen: A) The snout and proglacial area of the glacier. Solid rectangles indicate the locations of Sections 1 and 2. B) Section 1, showing an example of an ice-cored moraine. Unit I = foliated stratified sediments of glacifluvial origin, Unit II = stratified sands of glacifluvial origin, Unit III = horizontally stratified sandy silts, Unit IV = glacigenic sediment flow. C) Section 2, showing folded and thrust glacifluvial and glacimarine sediments in a push moraine. (Modified from Etzelmüller *et al.* (1996).)

characterized by subangular and subrounded clasts, which are occasionally striated. Lithologically, these foliation-parallel ridges are highly variable. The ridges can often be traced onto the glacier forefield as low (<1 m high) ridges. The foliation-parallel ridges are important as the incorporation of basal debris along longitudinal foliation is not a universally acknowledged process, and similar ridges elsewhere may have been mistaken for flutes. The mechanism invoked to explain this process is one where lateral compression of ice leads to the development of a transposition foliation parallel to flow, combined with the incorporation of basal debris-rich ice or soft basal sediment in the fold complex (Glasser *et al.*, 1998a; Hambrey *et al.*, 1999; Fig. 4.6). The base of the deforming layer represents a décollement surface that may represent the contact with bedrock. Incorporation of debris must take place where the ice is wet-based. The incorporation of debris by this process clearly precedes most thrusting, as foliation-parallel ridges are truncated by thrust moraines in the proglacial area. Their preservation potential appears to be low, because of destruction or burial by mass-movement and fluvial processes.

Geometrical ridge networks are created when both longitudinal and transverse debris accumulations melt out of the glacier and become superimposed. This landform-sediment assemblage has been described in the proglacial area of Kongsvegen as a result of the 1948 surge (Bennett *et al.*, 1996b). Here, small (4–8 m high) thrust ridges intersect low (<1 m) debris ridges or foliation-parallel ridges to form a complex of cross-cutting ridges on the glacier forefield. Although so far only observed at this glacier, there is no reason to suppose that they should not also occur at non-surge-type glacier margins. However, the preservation potential of these networks is probably low, as they are continually modified by slope processes and glacigenic sediment flows.

Flow-parallel ice structures can also incorporate large quantities of glacifluvial sediment, helping to redistribute sediment within the glacier. On Marthabreen, for example, two types of longitudinal debris-rich structures have been described on the glacier surface: 'longitudinal sediment structures' and 'longitudinal ridge accumulations' (Glasser *et al.*, 1999). Longitudinal sediment structures are ridges of sand and gravel, commonly 1–6 m long and 0.5 to 1.0 m high. They vary in width from 0.05 to 0.15 m and are always sub-parallel to the foliation. Their sediment fill consists of fine sand and granule gravel, and they are interpreted as former englacial channels formed parallel to longitudinal foliation (Glasser *et al.*, 1999). Longitudinal ridge accumulations are larger ridges that attain dimensions of between 20 and 30 m in length and 1–3 m in height. They have crests of *in situ* sand and gravel, while the ridge flanks contain slumped debris. These ice-cored sediment ridges frequently occur downstream of sediment structures or other debris pinnacles and have been interpreted as the product of sediment reworking by englacial or supraglacial streams flowing sub-parallel to the longitudinal foliation (Glasser *et al.*, 1999). The preservation potential of both types of structurally-controlled supraglacial and englacial fluvial deposits is probably low and their landform manifestation has yet to be observed in the proglacial areas of Svalbard glaciers.

Streamlined ridges have been described from the forefield of both Austre and Midtre Lovénbreen (Glasser and Hambrey, 2001a). They are between 25 and 50 m wide, up to 200 m in length, and reach 7 m in height. The ridges are elongated in the direction of glacier flow and, at the glacier margin, the ridges emerge from beneath the receding glacier. They generally comprise, from bottom to top: muddy-sandy gravel, diamicton, variable gravel and angular gravel. Flutes and fluted surfaces are also developed on the flat areas of the forefields at Midtre and Austre Lovénbreen (see Fig. 4.7c). Close to the ice margins, the ridges form low (<0.5 m high) and elongated (>10 m) fluted ridges composed of diamicton. The majority of these flutes commence in the lee of boulders, indicative of a subglacial origin. Further from the glacier margins, the flutes degrade rapidly and lose much of their surface relief. These small and relatively fragile landforms probably have a low preservation potential in the landform record.

4.4 A LANDSYSTEM MODEL FOR SVALBARD GLACIERS

From observations of modern Svalbard glaciers and the published literature we suggest the following landsystem model (Fig. 4.10). A typical receding Svalbard glacier has three zones within its forefield:

1. an outer moraine ridge
2. a moraine-mound complex, often draped by supraglacial debris stripes
3. an inner zone comprising various quantities of foliation-parallel ridges, supraglacial debris stripes, geometrical ridge networks, streamlined ridges/flutes and minor moraine mounds.

Outer moraine ridges are arcuate ridges rising steeply from the surrounding topography to heights of up to 15–20 m. They are commonly ice-cored, and the degree of degradation depends greatly on their constituent sediments. Some may be the product of permafrost deformation, while others represent former englacial, or proglacial thrusts. Some glaciers (e.g. Midtre Lovénbreen, Kongsvegen) are also flanked by large ice-cored lateral moraines (see Fig. 4.3).

Moraine-mound complexes commonly comprise arcuate belts of aligned hummocks or mounds of a wide variety of morphological types. They include linear ridges up to 100 m long, short-crested ridges several metres long, and near-conical mounds; all reaching elevations of several metres. Irrespective of size, they show ice-proximal rectilinear or curvilinear slopes with consistent angles of around 30°, irregular distal slopes that are commonly steeper, and are formed of stacked units of a variety of sedimentary facies. The rectilinear slopes and stacking indicate thrusting in proglacial, ice-marginal and englacial positions. There is a continuum of forms of moraine complex, based on displacement characteristics of the glacier bed or the décollement surface. Where friction at the décollement surface is high, then thrust-dominated moraine complexes form (e.g. Comfortlessbreen, Uvêrsbreen, Kronebreen). In permafrost areas, proglacial deformation may also occur, particularly during rapid advances into seasonally unfrozen sediments (Erikbreen, Usherbreen). Where friction at the glacier bed is very low, as when ice moves over saturated muddy sediments on the sea floor, the deformation in the sediment is represented by polyphase folding alone (Sefstrømbreen). The application of these principles to Pleistocene moraine complexes will be of value in assessing the nature of terrain over which the glacier moved.

The inner zone is located between the moraine-mound complex and the modern glacier snout. Foliation-parallel ridges, supraglacially-derived stripes of debris, geometrical ridge networks and streamlined ridges/flutes are formed in variable quantities in this zone. Sedimentary facies are predominantly diamicton (previously deposited as basal till), preserved in the geometrical ridge networks and streamlined ridges and flutes, commonly undergoing reworking by proglacial streams.

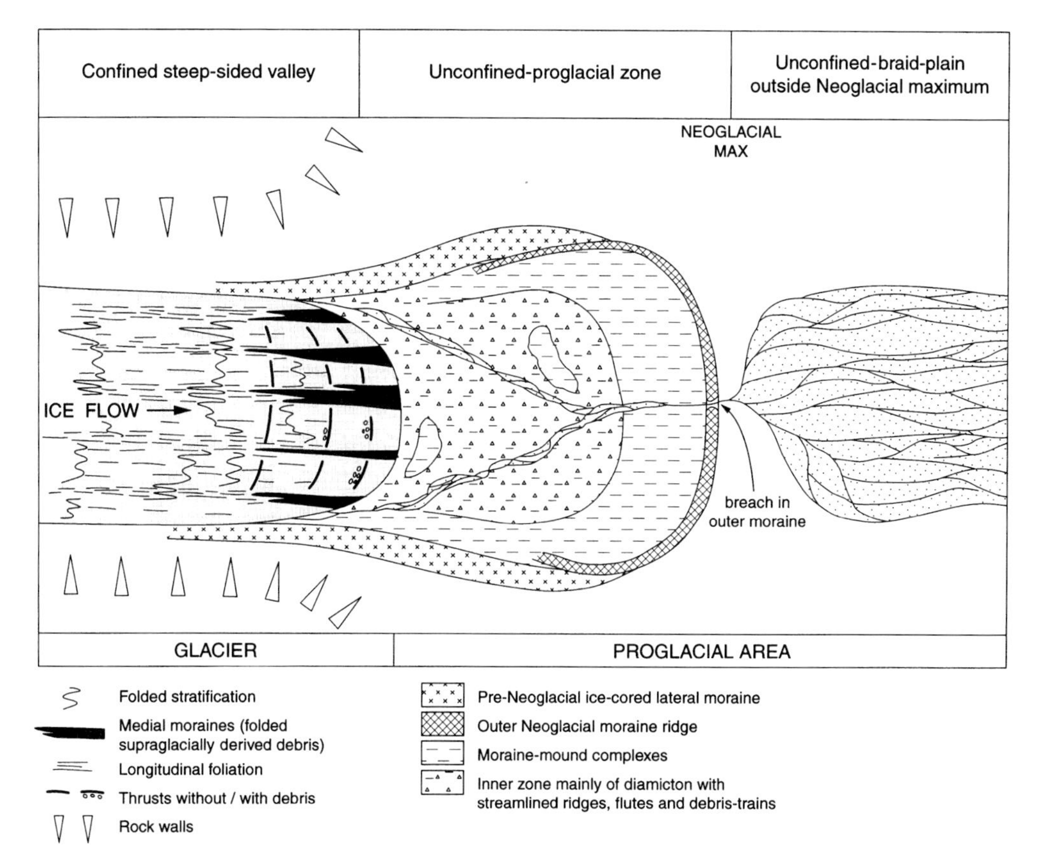

Figure 4.10 A landsystem model for a typical terrestrial Svalbard polythermal glacier.

4.5 APPLICATION OF THE MODEL TO PLEISTOCENE MORAINE COMPLEXES

Application of the Svalbard polythermal glacier landsystem model to the interpretation of Pleistocene landform-sediment assemblages is still in its infancy. In Great Britain, the model has been applied only to Younger Dryas moraine complexes in Cwm Idwal, North Wales (Hambrey *et al.*, 1997; Graham and Midgley, 2000), and in the Valley of a Hundred Hills (Coire a' Cheud-chnoic), in the Northwest Highlands of Scotland (Bennett *et al.*, 1998, but see also Wilson and Evans, 2000). These new comparisons stress the similarities between moraine-mound complexes formed by Svalbard glaciers and British 'hummocky moraine' in terms of moraine-mound morphology, sedimentology and facies variability. Understanding of the genesis of similar landforms in formerly glaciated areas of Scandinavia and North America may also benefit from these comparisons.

4.6 CONCLUSIONS

The landform-sediment assemblages associated with polythermal glaciers in Svalbard differ from those in temperate and cold-glacier systems. In particular, the deformation of glacier ice during debris entrainment and transport is a primary control on the nature of the Svalbard valley glacier landsystem. Debris entrainment and transport at Svalbard glaciers takes place by:

1. incorporation of angular rockfall material within the stratified sequence of snow/firn/superimposed ice. This debris takes an englacial path, becomes folded in zones of converging flow with axes parallel to ice flow, and is usually associated with the formation of an axial planar foliation. Near the glacier snout, the debris emerges at the surface along the hinges and limbs of the folds, producing medial moraines, which on deposition give rise to flow-parallel linear debris stripes.
2. entrainment of debris at the bed to form the basal ice layer, involving primarily regelation, bulk freezing-on, folding and shearing. These processes can result in a debris layer several metres thick within the glacier and a sheet of basal till on deposition.
3. incorporation of debris of basal character within longitudinal foliation. The foliation is the product of shearing or very tight folding within the ice. The resulting landforms, foliation-parallel ridges, have a low preservation potential.
4. thrusting, in which basal glacial sediments (including regelation ice) and subglacial sediments are uplifted into an englacial position, sometimes emerging at the ice surface. Thrusting is a dynamic process, and in polythermal glaciers may be linked mainly to the transition from sliding to frozen bed conditions. Thrusting may also be promoted by compressive flow against a reverse bedrock slope.
5. subglacial upright folding with transverse axes and faulting of sediment as a result of a glacier overriding a deformable bed.
6. reworking of thrust- or fold-derived glacifluvial material to produce longitudinal debris ridges within and on the ice. The resulting landforms are, however, ephemeral.

The most widespread deposit on the forefields of receding Svalbard polythermal valley glaciers is diamicton, which would typically be interpreted as a lodgement or melt-out till. Sediment release is controlled by the thermal regime of the glacier, the distribution of debris in the basal ice layer, and the nature of the bed beneath the glacier (deformable or non-deformable). Once released, glacifluvial processes rework much of this debris, although to a lesser extent than at temperate glaciers. The principal landforms in the Svalbard glacier landsystem resulting from ice-deformational processes are moraine-mound complexes (also referred to as thrust or push moraines where their genesis is known). The sediments in these complexes are highly variable and include diamicton of subglacial derivation, sand and gravel of glacifluvial origin, and mud with scattered gravel clasts and laminites of marine origin. Application of the Svalbard glacier landsystem to formerly glaciated areas is still at an early stage.

CHAPTER

5

ICE-MARGINAL TERRESTRIAL LANDSYSTEMS: POLAR CONTINENTAL GLACIER MARGINS

Sean J. Fitzsimons

5.1 INTRODUCTION

Research on landform and sediment assemblages formed by glaciers is dominated by studies of temperate glaciers in which sedimentary products reflect the influence of basal sliding, subglacial sediment deformation and subglacial hydrological systems (e.g. Boulton, 1972a and b). In contrast there have been relatively few studies of landform and sediment assemblages of polar and polythermal glaciers (e.g. Fitzsimons, 1997a; Ó Cofaigh *et al.*, 1999). The objective of this chapter is to synthesize recent investigations of polar continental glacier margins and move toward a depositional model for ice-marginal environments that links our understanding of glaciology and geomorphology. This review is based on field observations in East Antarctica (Vestfold Hills, Bunger Hills, Larseman Hills and Windmill Islands) and in south Victoria Land (McMurdo dry valleys and Ross Island). This chapter begins with a definition and review of the physical conditions that control depositional processes at polar glacier margins. The review is followed by a summary of the morphology and structure of modern polar glacier margins and associated landforms and sediments in low- and high-relief environments. The chapter concludes with a synthesis of ideas that form the basis of a depositional model for polar continental glacier margins.

5.2 POLAR ICE-MARGINAL ENVIRONMENTS

Polar continental glaciers constitute the bulk of ice on earth (Table 5.1). Polar glaciers constitute over 95 per cent of the glacier-covered area and over 97 per cent of ice volume (excluding the Ross and Ronne-Filchner ice shelves). Although polar continental glaciers dominate the earth's glacial systems, sediment-landform associations produced by these glaciers are the least well known and understood. The primary reason for this is that there is little land area beyond current ice margins and the glaciers are largely inaccessible. Consequently there is little terrestrial evidence of the growth and decay of the glaciers.

Our knowledge of the geomorphology and sedimentology of large polar continental ice masses is mainly derived from studies of small land areas that fringe Antarctica and Greenland. These areas

	Area (km^2)	%	Volume (km^3)	%
Ice caps, ice fields, valley glaciers	680,000	4.24	180,000	0.55
Greenland	1,784,694	11.06	2,620,000	7.96
Antarctica				
East Antarctic Ice Sheet	10,153,170	63.27	26,039,200	79.11
West Antarctic Ice Sheet	1,918,170	11.96	3,262,000	9.90
Antarctic Peninsula	446,690	2.79	227,100	0.69
Ross Ice Shelf	536,070	3.34	229,600	0.70
Ronne-Filchner Ice Shelf	532,200	3.33	351,900	1.07
Total	16,051,094	100	32,909,800	100

Table 5.1 Estimated areas and volumes of glaciers (From Williams and Ferrigno, 1993)

provide limited access to glacier margins and to landscapes that have experienced the advance and retreat of glaciers. These areas, often called oases, are cold deserts characterized by low mean annual temperatures (–10 to –20 °C), light precipitation the vast bulk of which falls as snow, and strong winds (typically mean monthly wind speeds of 2–9 $m.s^{-1}$).

The main controls on the nature and location of glacial deposition are glacier mass balance, thermal regime, bed configuration, the properties of the material being deposited and the climate near the ice margin (Andrews, 1975; Lawson, 1979). Studies of glacial deposits forming at the margin of glaciers have stressed the role of thermal regime in determining the processes involved in their deposition (Boulton, 1972b, 1975). Three different thermal boundary conditions have been recognized in glaciers on the basis of englacial temperature gradients (Weertman, 1961):

1. a temperature gradient that is sufficient to conduct all heat from the glacier bed, in which case there is no melting and the ice remains frozen to the bed
2. a temperature gradient that is just sufficient to conduct heat from the bed, in which case there is an approximate balance between melting and freezing
3. a temperature gradient that is insufficient to conduct heat from the bed, in which case there is melting and sliding.

These boundary conditions define two types of ice that are often called 'temperate' and 'cold' ice and recognizes the possibility that the state of the ice may change in space and time. When applied to whole glaciers the scheme has yielded a three-part classification of thermal or glaci-dynamic basal regimes that can be identified at modern glacier margins: 'temperate' glaciers, 'subpolar' or polythermal glaciers and 'polar' glaciers. The geographic terminology is regrettable because the distribution of glacier types is not simply determined by latitude. Consequently the terms 'wet-based', 'polythermal' and 'dry-based' or 'cold-based' are preferred and are used in this chapter. Most polar continental ice masses are of the polythermal type: where the ice is thin, such as the ice margins, they are dry-based and where the ice is thick or flowing rapidly the base of the ice is at pressure melting point and therefore wet-based. Thin glaciers in particularly cold environments may be entirely dry-based.

It has been argued that each thermal regime can be associated with diagnostic landform and sediment assemblages (Boulton, 1972b; Boulton and Paul, 1976; Eyles *et al.*, 1983b). Eyles *et al.* recognized a polar arid sediment assemblage based primarily on the work of Shaw (1977a, b) who examined depositional processes in the McMurdo dry valleys. The title of this sediment assemblage 'polar arid' encapsulates the problem of a thermal regime-based classification because this sediment assemblage is differentiated by the climate of the terminus area rather than basal thermal regime. Glacier thermal regime exerts a fundamental control on glacier behaviour by determining ice motion and erosion processes. Thermal regime is determined by the englacial temperature gradient, which is influenced by climate and the generation of heat close to the glacier bed. The indirect role of climate in controlling thermal regime contrasts strongly with the direct influence of climate on depositional processes at glacier margins. The use of both thermal regime and climate to distinguish sediment raises several interesting questions including:

- Can the roles of thermal regime and climate in glacier sedimentation at the terminus area be differentiated?
- If the roles of thermal regime and climate can be differentiated, which is the higher-order control in polar continental environments?
- Is glacier thermal regime a satisfactory basis for defining landform and sediment assemblages?

5.3 ICE MARGINS IN LOW-RELIEF LANDSCAPES

5.3.1 Glacier Margins

In East Antarctica the majority of the ice margin terminates in the sea. Relatively small parts of the ice margin terminate on land in small coastal oases. The largest oases in East Antarctica are the Vestfold Hills and Bunger Hills (Fig. 5.1). Recent investigations of the Quaternary history of these areas has suggested that the ice margin during the last glacial maximum was thinner and less extensive than previously thought (Colhoun *et al.*, 1992) and that deglaciation was almost complete by 10,000 years BP (Fitzsimons and Domack, 1993). These conclusions are clearly controversial as they contradict data from the Ross Embayment (Denton *et al.*, 1989) and marine seismic and core data in East Antarctica (Domack *et al.*, 1991). As the mode and pattern of ice advance and retreat have implications for the interpretation of palaeo-climate and ice dynamics, it is vital to have appropriate depositional models for landforms and sediments.

In Vestfold Hills the edge of the continental ice sheet runs from north to south, and the southern limit of the ice-free area is formed by the Sørsdal Glacier, which is the major outlet glacier of the area and forms a small ice shelf. The hills consist of a complex low-relief topography composed of valleys at and below sea level and ridges up to 158 m in altitude. Glacial sediments and landforms are absent from most of the ice-free area and are concentrated close to the glacier margin.

The mean annual temperature of the Vestfold Hills is –10.2°C (Schwerdtfeger, 1970) which is, on average, warmer than Antarctic stations of similar latitude (Burton and Campbell, 1980). Although no precipitation data are available, snowfall is light (probably <250 mm per year) and rainfall is very rare. Melting of snow and ice is restricted to the short summer (December to February). There is a strong diurnal component to the melt activity, which usually ceases between

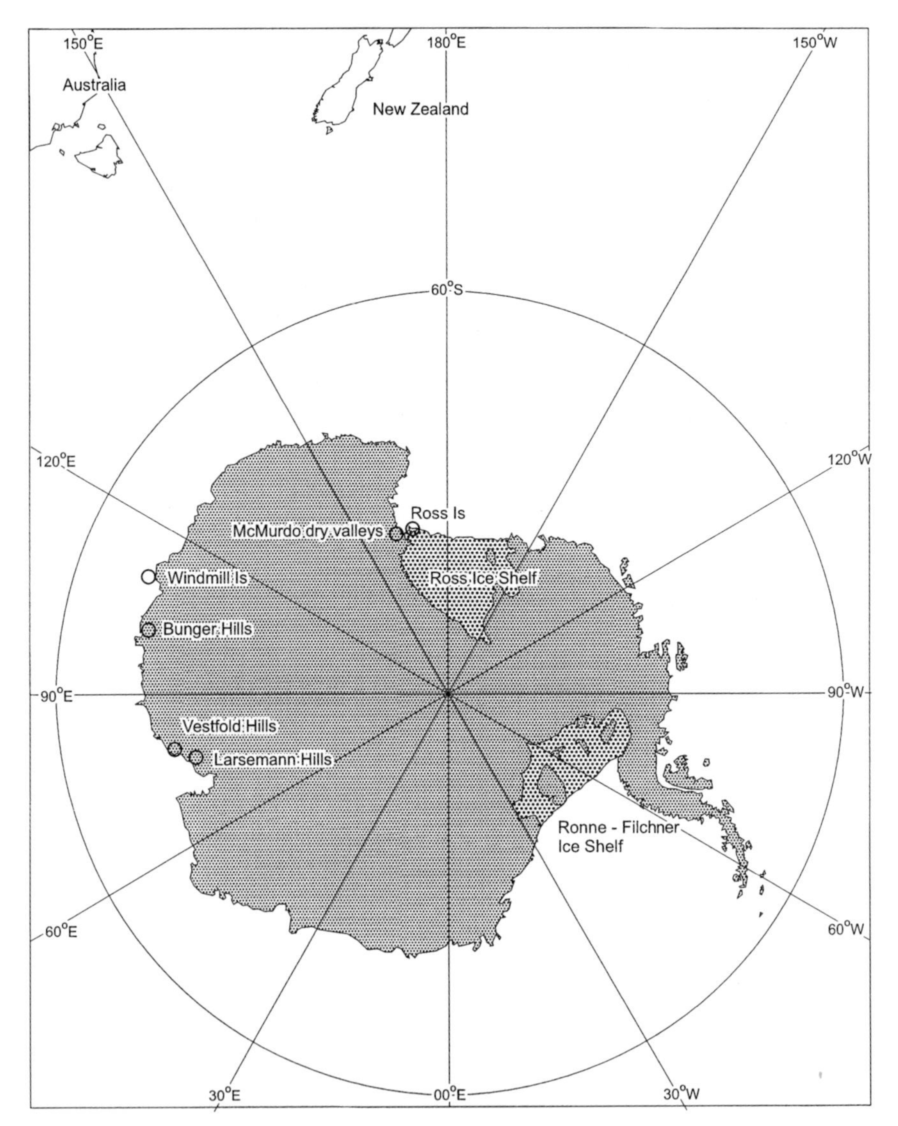

Figure 5.1 Map showing the location of the Vestfold, Larsemann and Bunger hills and the McMurdo dry valleys.

9 pm and 10 am when air temperatures are below or close to 0 °C and the sun has a low angle of incidence.

In many East Antarctic coastal oases the ice-sheet margin has a complex form, and distinct features such as an ice cliff are difficult to recognize. Consequently it is necessary to define some terms used in this chapter: 'ice edge' is used to describe the terminus of a glacier where it is sharp and easily recognizable (Fig. 5.2a) and 'ice margin' is used to describe an ice-terminus that it is not clearly recognizable (Fig. 5.2b). Within an ice margin an apparent ice edge is often recognizable as an ice cliff beyond which an area of ice-cored moraine occurs (Fig. 5.2b). The width of the ice-cored moraine or debris-covered glacier is highly variable and can range from tens of metres to several kilometres. The term outer ice edge is used to define the actual glacier terminus where ice movement ceases (Fig. 5.2b).

The three pre-eminent characteristics of the ice margin at Vestfold Hills are its variable shape, the presence of a large sinuous ice-cored moraine (Figs. 5.2 and 5.3), and the abundance of large snow drifts (Fig. 5.3). The ice-sheet margin has a convex form and descends rapidly from 300 m above sea level within 2 km of the margin to approximately 100 m above sea level at the margin. Where the ice flows into the sea, the ice margin forms 20–40 m high cliffs. On land the margin is considerably more complex, often with multiple cliffs and snow drifts (Fig. 5.4a).

The sinuous ice-cored moraine that dominates the ice margin at Vestfold Hills is a broad, discontinuous ridge of coarse debris, 100–300 m wide and about 20 km long that occurs inside the ice margin (Fig. 5.3). The debris is, on average, less than 0.5 m thick but accumulations up to 1.5 m thick occur on the sharp-crested ridges. The sinuous inner moraine contrasts with other moraines that occur in front of ice cliffs which have sharp-crested ridges. Moraine ridges beyond the ice margin are much higher (up to 20 m) and much shorter than the inner moraine ridges (less than 1 km long). Most are ice-cored and unstable, as indicated by the occurrence of numerous sediment flows, slumps and other mass movements (Fitzsimons, 1990).

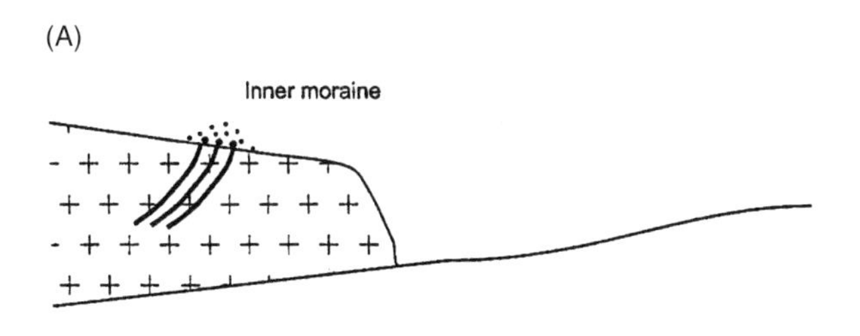

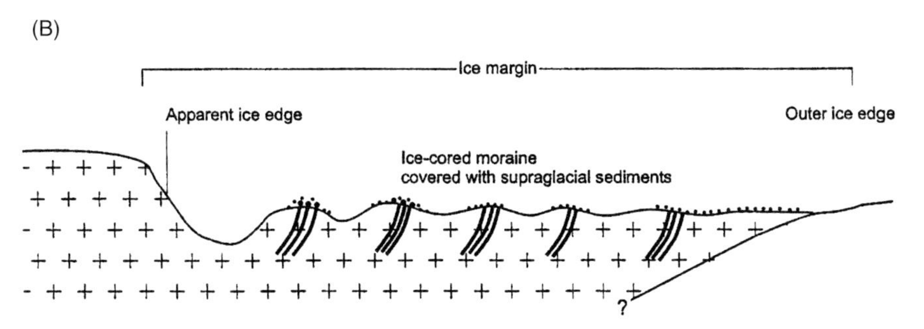

Figure 5.2 Ice-margin nomenclature. A) Simple ice margin with an ice cliff and inner moraine formed by basal ice cropping out on the glacier surface. B) Wide ice margin with an apparent ice margin separated from an outer ice edge by numerous outcrops of basal debris.

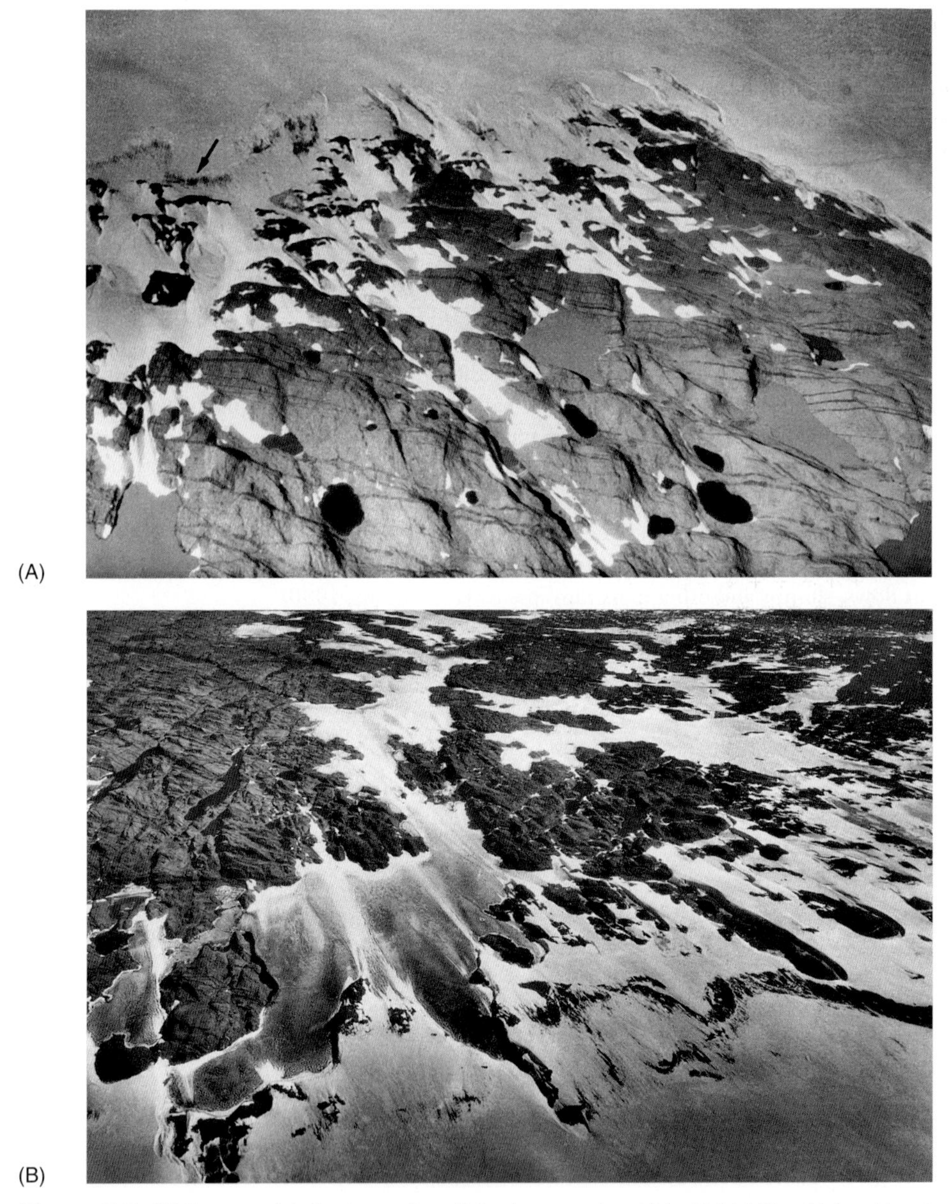

Figure 5.3 Oblique aerial photographs of the ice margin at Vestfold Hills. A) Looking toward the ice margin with Sørsdal Glacier at right. Note the large sinuous inner moraine (arrowed) and snow drift partially concealing the ice margin. The light, turbid lakes are connected to the proglacial drainage system and the dark ones are not. The ice margin is about 10 km long. B) Ice margin looking toward the coast. Note the deep snow drifts downstream of the inner moraine and partly frozen lakes and fiords. The numerous dolerite dykes belie the lack of unconsolidated sediment over the landscape.

In other parts of the Vestfold Hills the ice margin is buried by snow drifts and forms in a low-angle ramp. Where the margin is not buried, cliffs up to 30 m high occur near the ice-cored moraines and at the heads of fjords (Fig. 5.4a). In these steeper sections the debris is concentrated below the ice cliffs (Figs. 5.4a and b) forming narrow, sharp-crested ridges up to 10 m high.

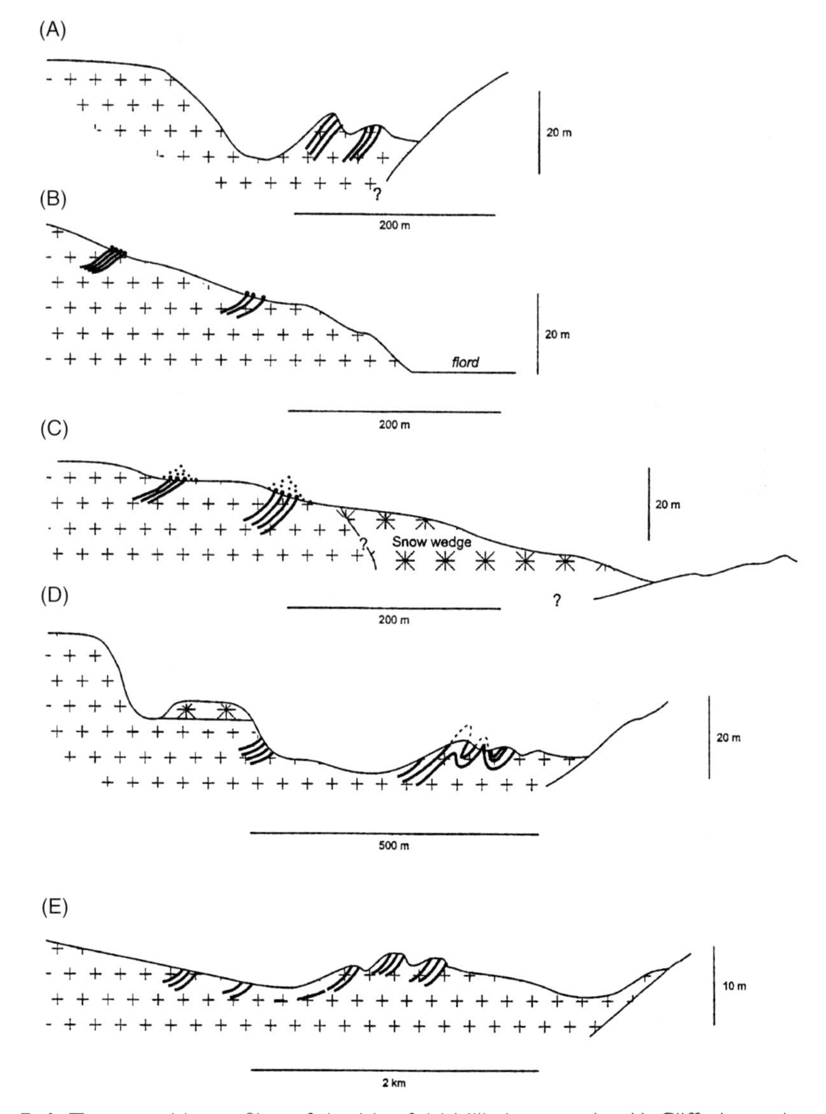

Figure 5.4 Topographic profiles of the Vestfold Hills ice margin. A) Cliffed margin with inner moraines beyond the apparent ice edge. B) Ramp margin with two sets of inner moraines. C) Ramp margin with a large snow wedge and two inner moraines. D) Multiple ice cliffs and snow wedge remnants with a folded inner moraine beyond the apparent ice edge. E) Ramp margin with numerous outcrops of basal debris.

In the southeastern corner of the hills, where the Sørsdal Glacier forms a distinct outlet glacier, the ice margin has a convex profile and has multiple ice cliffs (Fig. 5.4d). The slightly deformed, basal debris zone is unconformably overlain by clean white ice. This unconformity appears to record a former ablation surface that has been buried by ice which accumulated *in situ*. A zone of ice-cored moraine with numerous sharp-crested ridges parallel to the ice edge occurs beyond the main ice cliffs.

The structure of the ice margin at the Vestfold Hills is revealed by exposures of the basal debris zone in ice cliffs and gullies that cross the ice margin. Deformation structures range from relatively undeformed debris bands to intense deformation characterized by recumbent folds and shear structures (see Fig. 4 in Fitzsimons, 1990). Deformation structures in the basal zone of the ice cap can be divided into large-scale features, which involve the entire basal debris zone, and small-scale features which occur within the basal zone. The most prominent large-scale deformation structure is the upwarping of the basal debris zone to crop out on the surface of the glacier and form a large, sinuous ice-cored moraine (Figs 5.3 and 5.4). Exposures of basal debris reveal structures that vary from slightly deformed stratified ice (Fig. 5.5a) to complex multi-phase folding and shearing (Fig. 5.5b). A section through an ice-cored moraine in the southeastern corner of the hills shows that moraine ridges can form along the axes of a series of large recumbent folds that have amplitudes over 15 m (see Fig. 4c in Fitzsimons, 1990).

Measurements of debris concentrations in the basal zone ice are consistently below 10 per cent by volume. Debris concentration in individual bands is highly variable with most of the debris concentrated close to the bed. Unusually high concentrations occur in rare debris lenses of sorted fluvial sediments that have been entrained by the glacier. Most of the debris consists of silt and sand-sized particles with larger clasts either dispersed or occurring in small lenses. Gravel clasts are dominantly subrounded, rarely angular.

5.4 ICE-CONTACT LANDFORMS AND SEDIMENTS

Three types of well-preserved ridges can be recognized at the ice edge in East Antarctic oases: inner moraines, ice-contact fans/screes and thrust-block moraines.

5.4.1 Inner Moraines

Inner moraines form at the margin of the ice sheet where basal debris crops out and accumulates on the ice surface. Beyond the present ice margins older inner moraines often form prominent end moraines. Where an ice core remains in these moraines exposures reveal large recumbent folds with an amplitude of up to 6 m and numerous smaller sheared folds providing evidence of intense compressive deformation within the basal debris zone close to the ice margin (Fitzsimons, 1990, 1997a).

Exposures of the inner moraines reveal massive, matrix-supported diamictons with rare layers of poorly sorted, sandy gravel. Pebble-fabric strengths of the diamictons measured from clast a-axes range from 0.51 to 0.81 and tend to be weaker closer to the surface of the ridges (Fig. 5.6). Directions of maximum clustering are perpendicular to the trends of the ridges and in a few cases oblique to the trends of the ridges (Fig. 5.6).

(A)

(B)

Figure 5.5 Basal ice at the edge of the ice sheet in Vestfold Hills. A) Slightly deformed stratified basal ice resting on gneiss and unconformably overlain by drift snow. B) Highly deformed basal ice showing a series of tight sheared folds.

The diamictons are accumulations of basal debris that have cropped out on the surface of the ice sheet and subsequently been remobilized by sediment flows. Remobilization has resulted in relatively poorly defined directions of maximum clustering, and slight textural variation is probably related to sorting of sediments in less viscous flows. Stronger pebble fabrics below 1 m depth in the excavations can be interpreted as melt-out till in which the fabric of the basal debris zone has been preserved. The formation of melt-out tills, and the preservation of basal debris fabrics that record ice-flow direction are more likely where the sediment cover exceeds 0.5 m, after which melting slows and the debris is less likely to become saturated and flow.

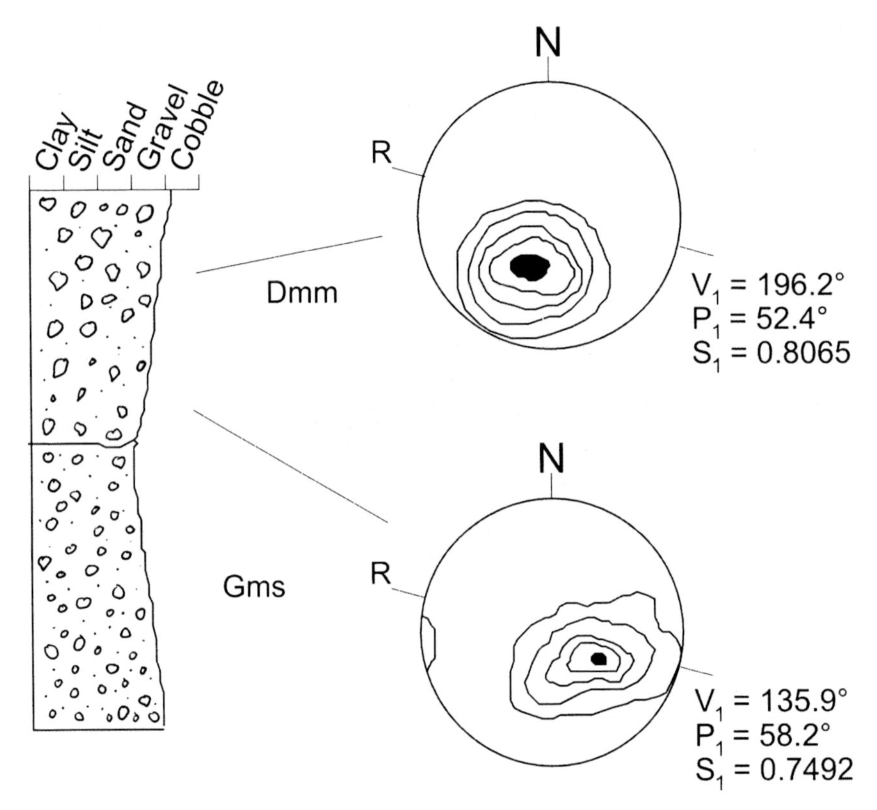

Figure 5.6 Sedimentary logs of sediments from the crests of inner moraines. The contour interval of the Schmidt nets is two standard deviations. V_1 and P_1 and give the azimuth and plunge of the principal eigenvector, S_1 gives the strength of clustering about the principal eigenvector and R shows the trend of the moraine ridge.

5.4.2 Ice-Contact Fans and Screes

Ice-contact fans and screes form sharp-crested cuspate ridge segments up to 20 m high and 500 m long. They form at ice cliffs where melting and sublimation of basal debris results in the fall and/or flow of debris at the foot of the cliff (Fig. 5.7a). Most of these ridges have asymmetrical profiles (Fig. 5.7a) characterized by proximal slopes between 25 and 15° and distal slopes between 15 and 25° (Fitzsimons, 1997b).

Sediments exposed at the crests of ice-contact fans and screes show a range of sedimentary facies, including massive and stratified gravels, horizontally laminated and cross-bedded sands, bouldery gravels with lenses of fine-grained sediment, massive matrix-supported diamictons, stratified diamictons and muds (Figs 5.7b and 5.8). The sediments range from moderately sorted to very poorly sorted, but on average are moderately sorted. Particles up to 0.8 m in diameter are common and occur in a chaotic mixture of diamicton, gravel and well-sorted and stratified sand. Most exposures show that the sediments are well stratified with dips down the distal slope of the moraines at angles of between 5 and 20°. The pebble fabric of diamictons and massive gravels are transverse or oblique to the trend of the ridges (Fig. 5.8) and the clustering about the mean axis ranges from moderate to strong (S_1 0.54–0.86).

(A)

(B)

Figure 5.7 A) An ice-contact scree forming at the ice margin (left) and two ice-cored ice-contact screes adjacent to the ice margin. Note the supraglacial stream emerging from the contact between basal ice and drift snow. B) Poorly sorted gravel overlain by laminated sand and gravel, and a clast supported diamict exposed in the crest of the ice-contact scree.

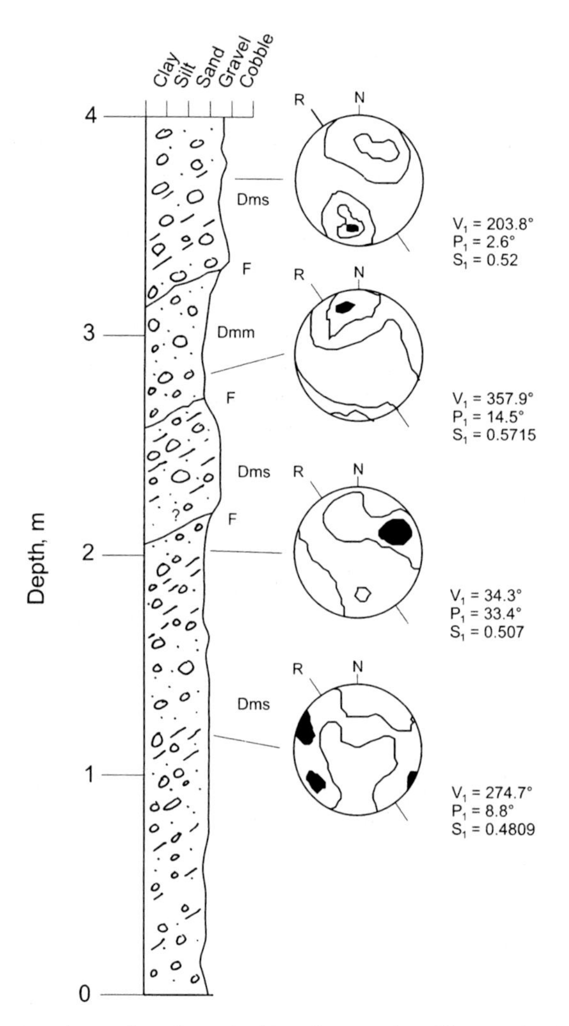

Figure 5.8 Sedimentary logs of sediments from the crests of ice-contact screes. The contour interval of the Schmidt nets is two standard deviations. V_1 and P_1 give the azimuth and plunge of the principal eigenvector, S_1 gives the strength of clustering about the principal eigenvector and R shows the trend of the moraine ridge.

The association of diamicton, gravel, sand and the bouldery facies suggests that both alluvial and colluvial processes are important during the formation of the ridges (Fig. 5.7b). The chaotic bouldery lithofacies is interpreted as the product of simultaneous accumulation of alluvial and mass-movement deposits (i.e. large particles fall or roll into alluvial deposits and sediment flows).

5.4.3 Thrust-Block Moraines

Thrust-block moraines form along the lateral margins of outlet glaciers, where ice flows across marine inlets or lakes. The ridges are up to 20 m high with proximal slopes of around 30° and distal slopes of around 25°. As the ice core melts, large tension cracks develop along the ridge crests.

Sediments in thrust-block moraines (Fig. 5.9a) consist of stratified diamictons (Fig. 5.9b), massive diamictons and rare layers of horizontally laminated sands (Fig. 5.9). Many exposures display low-angle thrust faults and sheared zones that consistently dip in an up-glacier direction at angles of between 10 and 25°. The pebble fabric of the diamictons can be divided into a group characterized by weak fabrics associated with stratified diamictons (S1 0.45–0.57) and a group of stronger fabrics adjacent to low-angle faults (S1 0.67–0.85). Massive diamictons frequently contain abundant shell fragments and stratified diamictons occasionally contain beds of shells, some in growth position (Fitzsimons, 1997b).

The distinctive fabric, lamination, and preserved marine shells, sometimes in growth position, suggests the diamictons are glacimarine sediments. Pebble fabrics of attenuated diamictons (faulted and sheared) have similar strengths to deformed lodgement tills described by Dowdeswell and Sharp (1986). The increased fabric strength is interpreted as a consequence of attenuation by shearing either as the blocks were detached or deposited. Preservation of beds of shells and laminations within the diamictons suggests at least some of the sediment may have been frozen during entrainment and transportation and/or that the strain was relatively low.

Low-angle faults, together with slickensides and attenuated diamicts adjacent to the faults, show that the glacimarine sediment has been entrained as a series of blocks with an average thickness of about 0.5 m (Fig. 5.10). The moraines have accumulated as successively older glacimarine sediments were eroded from the floor of the fjord and then deposited on the distal shore.

5.5 ICE MARGINS IN HIGH-RELIEF AREAS

5.5.1 Glacier Margins

The high-relief area that is described here is the McMurdo dry valleys, which are often called the McMurdo oasis. Glaciers of the McMurdo dry valleys can be divided into four groups: outlet glaciers, ice shelves, piedmont glaciers and alpine glaciers. Ice from the East Antarctic ice sheet flows through the Transantarctic Mountains to form outlet glaciers, such as Ferrar and Mackay glaciers, which reach the coast and form small floating ice tongues. Other outlet glaciers, such as Taylor Glacier, terminate on land. However, it could be argued that Taylor Glacier is not strictly an outlet glacier of the East Antarctic ice sheet because it flows from a local ice dome (Taylor Dome). North of the margin of the Ross Ice Shelf, ice streams that flow through the Transantarctic Mountains form outlet glaciers that feed small ice shelves. The largest ice shelf in the area is the Ross Ice Shelf, which is fed primarily by ice streams from the West Antarctic Ice Sheet. Although the Ross Ice Shelf does not directly impinge on the dry valleys today, during the Late Pleistocene the ice shelf grounded and flowed up the valleys. Consequently the ice shelf had a profound impact on the geomorphology of several valleys in the McMurdo oasis. In coastal areas of the McMurdo oasis, the slightly higher precipitation results in broad piedmont glaciers at the seaward margins of the Victoria and Wright valleys. Between the coastal piedmont glaciers and the inland glaciers, small alpine glaciers form a remarkable landscape in which bare rocky slopes contrast strongly with glacier ice. Most of these glaciers are no more than 15 km long.

Although ice margins in the McMurdo dry valleys range from gently sloping ice ramps to steep ice margins the most common and distinctive form is a 15–20 m high ice cliff. These

(A)

(B)

Figure 5.9 A) A series of thrust-block moraines on an island adjacent to the margin of an outlet glacier. B) Stratified glacimarine sediments exposed in the crest of a thrust-block moraine.

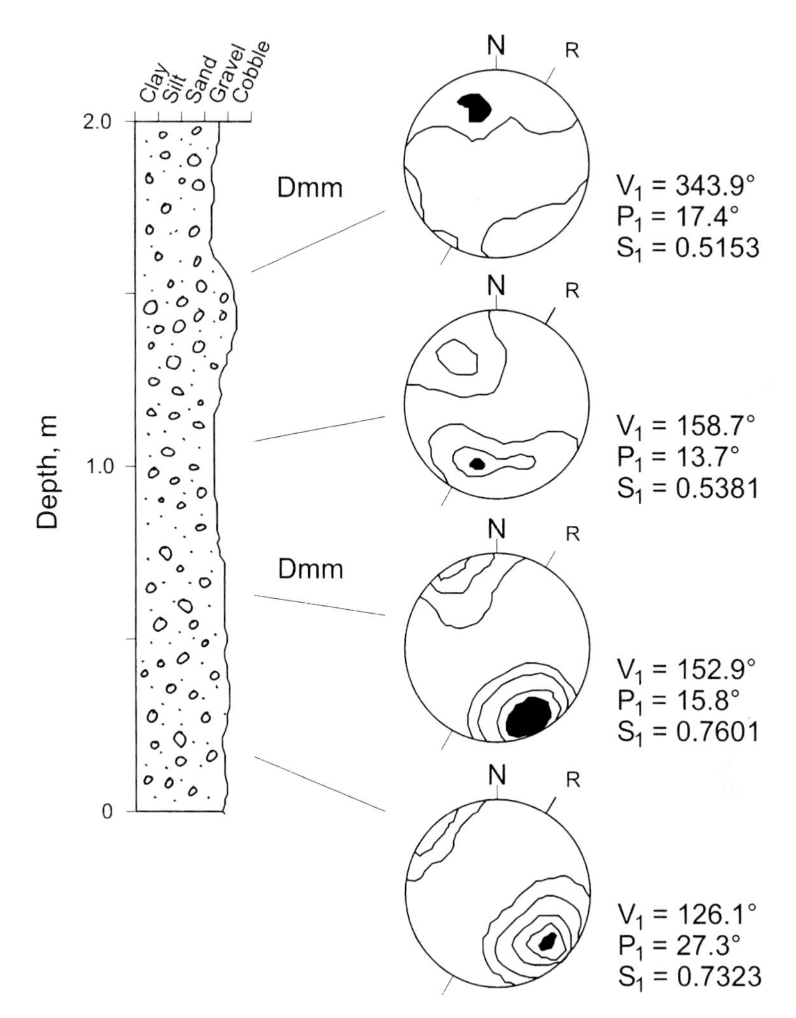

Figure 5.10 Sedimentary logs of sediments from the crests of thrust-block moraines. The contour interval of the Schmidt nets is two standard deviations. V_1 and P_1 give the azimuth and plunge of the principal eigenvector, S_1 gives the strength of clustering about the principal eigenvector and R shows the trend of the moraine ridge.

distinctive cliffs have been attributed to changes in the rheological properties of ice at around 20 m thickness (Chinn, 1991) and to a strong reduction in ablation from the foot of the cliffs to the glacier surface (Fountain *et al.*, 1998). Supraglacial debris is absent from most of the glaciers and the only visible debris is often restricted to small outcrops at the foot of ice cliffs where the basal zone of glaciers is exposed (Fig. 5.11). Although the ice margins do not have the thick snow accumulations that characterize the low relief landscapes described above, they are often characterized by an accumulation of ice at the foot of the cliffs produced by episodic calving (Fig. 5.11).

All glacier margins in the McMurdo dry valleys are dry-based with basal temperatures between –16 and –18 °C, which is very similar to the mean annual temperature (–19.8 °C at Vanda

(A)

(B)

Figure 5.11 Glacier margins in the McMurdo dry valleys. A) Clean white ice and marginal cliff characteristic of glaciers in the dry valleys. Note the ice apron produced by episodic calving (Hart Glacier, Wright Valley). B) Stratified basal ice at the margin of Suess Glacier.

in the Wright Valley). However, in the case of outlet glaciers such as Taylor Glacier, the ice is at pressure melting point within a few kilometres of the margin where the glacier is thicker (Robinson, 1984). The velocities of the glaciers are generally low. In the case of fully dry-based glaciers surface velocities are less than 1 $m.a^{-1}$ and 3 m above the bed velocities of around 250 $mm.a^{-1}$ have been measured (Fitzsimons *et al.*, 1999). The outlet glaciers move considerably faster.

Exposures of the basal zone of the glaciers show that debris concentrations are generally low and highly variable. In the case of Suess Glacier in the middle part of the Taylor Valley, debris concentrations range from less than 0.1 per cent to more than 70 per cent by volume with an average of less than 5 per cent. Debris concentrations in Taylor Glacier, which is at pressure melting point upstream of the terminus (Robinson, 1984), are considerably greater.

Ice-marginal landforms are absent or very small at most ice margins suggesting that the cold-based glaciers are not particularly effective agents of erosion. However, a few glaciers exhibit well-developed end moraines and have been the subjects of recent investigations. These landforms are described below.

5.5.2 Ice-contact Landforms and Sediments

Several types of moraines are recognized at the margins of glaciers in the dry valleys although there is considerable uncertainty about their origin. Given this uncertainty, for the purposes of this chapter they are divided into constructional and structural features.

5.5.2.1 Constructional Landforms

Constructional moraines are formed at ice margins where the glacier has been sufficiently stable to concentrate debris, usually at the foot of an ice cliff, by ablation of the basal debris zone. Chinn (1991) has argued that the outcrops of basal debris at the foot of ice cliffs are the equivalent of inner moraines that are commonly seen at dome-shaped polar ice-margins.

In the McMurdo dry valleys, constructional moraines, often covered with an ice and debris apron, occur at the margins of numerous glaciers. These features are formed by debris released from the basal zone together with sparse supraglacial debris. Shaw (1977b) has argued that an advancing glacier may override the ice and debris aprons thereby incorporating debris into the basal zone of the glacier. This 'apron entrainment' mechanism is similar to processes described in sub-polar glaciers in the Canadian arctic (Evans, 1989a).

5.5.2.2 Structural Landforms

Small structural moraines occur at the margin of several glaciers in the McMurdo dry valleys. These features appear to consist of either sediment blocks eroded from the base of the glacier and/or marginal sediments that have been deformed *in situ* (Fig. 5.12). Fitzsimons (1996a) described moraines that formed at the margins of Suess Glacier as thrust-block moraines. This paper posed the hypothesis that the moraines were produced by accretion of ice and debris as the cold-based glacier margin advanced into a proglacial lake (see Fig. 4 in Fitzsimons, 1996a). Subsequent investigations have demonstrated that the formation of moraines in this location is more complex than initially thought. Isotopic analysis of the basal ice exposed at the foot of the ice cliff has shown convincingly that some basal ice has formed as water froze onto the base and/or margin of the glacier (Lorrain *et al.*, 1999). However, excavation of a tunnel in the right side of the

(A)

(B)

Figure 5.12 Structural landforms at the margins of glaciers in the dry valleys. A) Moraines at the left margin of Suess Glacier form multiple ice-cored ridges up to 10 m high. Note the textural contrast between the small moraines in the foreground which are the product of deformation of proglacial fluvial sediments and the coarser, larger moraines produced by subglacial erosion. B) Ice-cored moraine at the margin of Wright Lower Glacier. Note the sedimentary stratification in the moraine. At least part of the moraine has been formed by deformation of the adjacent delta. The ice cliff is 18 m high.

glacier 100 m upstream of the moraine has demonstrated that the sediment blocks that feed the moraine have been entrained at least 100 m upstream of the glacier terminus (Fitzsimons *et al.*, 1999).

The new evidence shows that there is indeed support for the hypothesis that at least some of these features are thrust-block moraines formed as ice and debris was accreted and thrust at the glacier margin. However, the absence of the isotopic signature of ice accretion upstream of the glacier terminus suggests that a large proportion of the debris in the moraines has been entrained subglacially. The mechanism for erosion and detachment of the blocks of sediment are currently unknown. The evidence for subglacial entrainment casts doubt on whether the moraines are thrust-block moraines (*sensu* Kalin, 1971), which are features that are formed in a proglacial position as the foreland of a glacier is deformed.

5.5.2.3 Glacifluvial and Glacilacustrine Landforms

Outwash surfaces are absent from the proglacial areas of most glaciers in the McMurdo dry valleys. Their absence is a consequence of the low production of meltwater, low ephemeral discharges in streams, and low debris concentrations in and on the glaciers. The largest stream in the dry valleys and Antarctica is the Onyx River which flows from the Wright Lower Glacier and Lake Brownworth into Lake Vanda. Lake Vanda like many other lakes in the dry valleys has no outlet to the sea and water losses occur through sublimation and evaporation. Most lakes have a 4–6 m-thick perennial ice cover although some are frozen to their beds. The majority of these lakes receive little sediment from the glaciers again because of the low production of meltwater, low ephemeral discharges in streams, and low debris concentrations. Even the lakes in contact with glacier margins (Figs 5.13 and 5.14) are not strongly influenced by the presence of the glaciers. Divers and remotely operated vehicles operating beneath the ice cover of the lakes have revealed clear water and a lake bed covered with algae up to the contact with the glacier cliff.

5.5.2.4 Late Pleistocene Landforms and Sediments

During the Late Pleistocene the configuration of the glaciers in the McMurdo oasis was substantially different from the glacial systems described above. The alpine and piedmont glaciers are thought to have receded because their precipitation source was greatly diminished by the presence of a much larger Ross Ice Shelf. In the Taylor Valley, lacustrine strandlines and lacustrine deltas provide evidence of a large glacial lake that occupied much of the valley (Fig. 5.13). Glacial Lake Washburn is thought to have formed as the Ross Ice Shelf thickened and advanced into the valley. The grounded ice shelf deposited the younger Ross Sea Drift (Stuiver *et al.*, 1981), which extends westward into the valley as far as Canada Glacier (Fig. 5.14). The drift consists of numerous eskers 1–5 m high and up to 2 km long, and numerous small moraines which drape the eskers in a washboard-like structure (Fig. 5.14). The eskers and moraines are overlain by marine sediments and lacustrine deltas.

After withdrawal of the Ross Ice Shelf and the draining of Glacial Lake Washburn, alpine glaciers such as Canada Glacier advanced and reached their maximum positions on top of and abutting the Ross Sea Drift (Fig. 5.14). The relationship between the alpine glaciers and the Ross Sea Drift suggests the alpine and piedmont glaciers fluctuate out of phase with the grounded ice in McMurdo Sound.

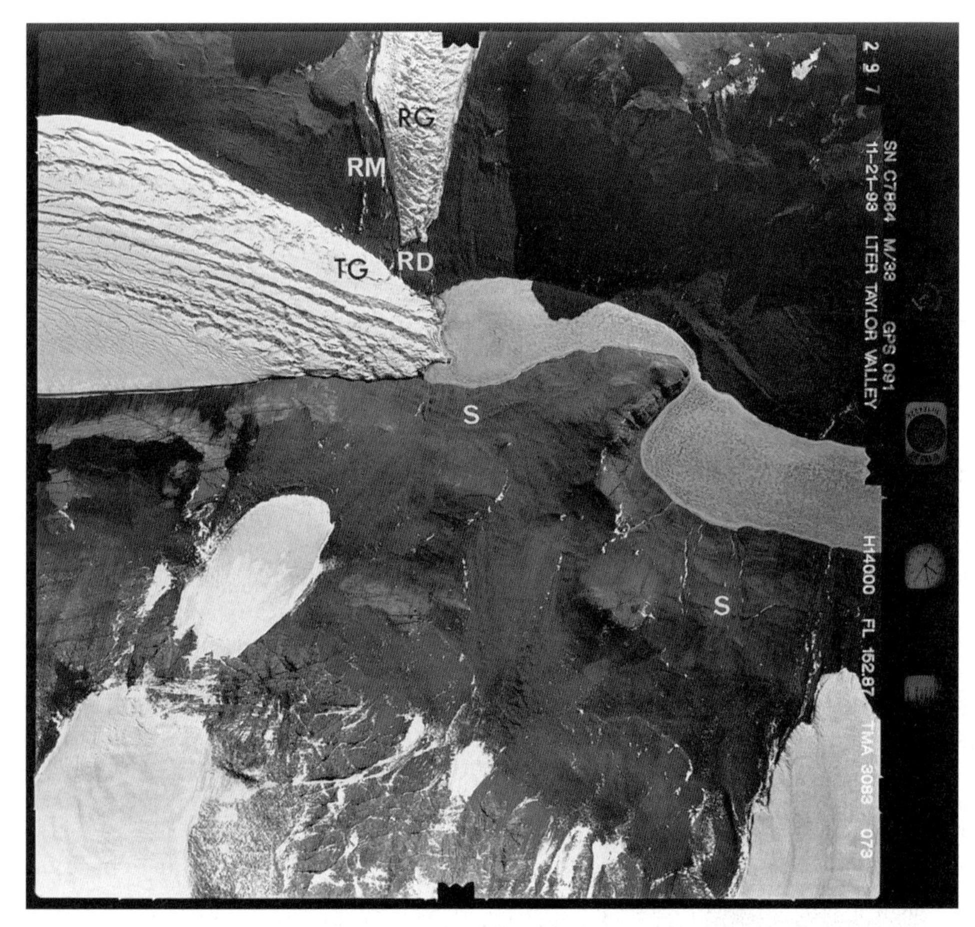

Figure 5.13 Aerial photograph of Taylor Glacier (TG), an outlet adjacent to the perennially frozen Lake Bonny. The adjacent Rhone Glacier (RG) is a small alpine glacier fringed by an older latero-terminal moraine (RM). Numerous strandlines (S) from Glacial Lake Washburn are evident on both sides of the valley. Several perched deltas (RD) deposited by streams that flowed from Rhone Glacier are evident.

5.6 TOWARDS A DEPOSITIONAL MODEL

Our knowledge of polar continental landform and sediment assemblages is incomplete. Consequently a comprehensive depositional model cannot yet be assembled. However, several elements of a model can be identified.

1. Relatively low volumes of sediment are produced by polar glaciers. Consequently landform and sediment assemblages have modest volumes and the preservation potential of ice-contact landforms is low.
2. A variety of constructional moraines form at stable ice–margins. The most common constructional landforms are ice-contact fans and screes. These form adjacent to steep ice margins and inner moraines (*sensu* Hooke, 1973a), which form where basal debris crops out on the glacier surface.

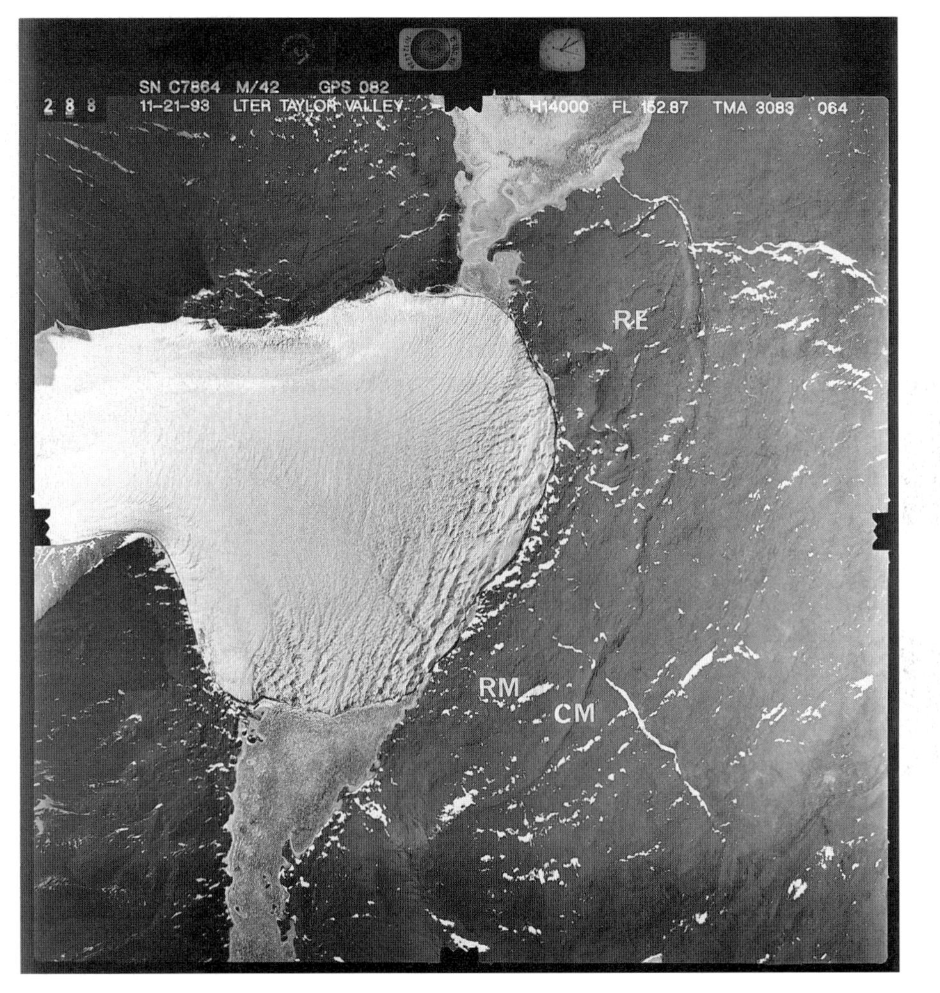

Figure 5.14 Aerial photograph of Canada Glacier showing a prominent end moraine loop (CM) and small moraines (RM) and eskers (RE) of the Ross Sea Drift deposited as the Ross Ice Shelf grounded and advanced into the Taylor Valley.

3. Thrust-block moraines and push moraines form where cold ice contacts saturated unfrozen sediment.
4. In some circumstances dry-based glaciers appear to be capable of bed deformation and production of structural landforms
5. Glacifluvial landforms are generally poorly developed elements of the depositional landscape.

Three high-order controls on the nature of polar landform and sediment assemblages are glacier thermal regime, climate of the terminus area and the topography of the landscape. Comparison of landform and sediment assemblages in polar maritime environments, such as Vestfold Hills, with polar continental environments, such as the McMurdo dry valleys, suggests that the availability of meltwater is the primary control on depositional processes in ice-marginal landscapes. If the summer is sufficiently warm and/or long enough for moderate quantities of meltwater production, glacial deposits are strongly influenced by remobilization after release from the ice. Given the critical role of meltwater, the wide climatic variability within polar environments, and the realization that the elements of the model summarized above are not inherently different from many other glacial environments, it seems that subdivision of landform and sediment assemblages based on glacier thermal regime is unsatisfactory.

This review points to a striking gap in our knowledge of polar landform and sediment assemblages. The gap in our knowledge concerns subglacial processes and resultant landform-sediment assemblages. Very little is known about polar subglacial landform and sediment assemblages because subglacial landscape elements such as streamlined forms or eskers are not preserved in the land areas that fringe polar ice masses. The main exception is the special case of Taylor Valley where subglacial landforms and sediments associated with an expanded and grounded Ross Ice Shelf are preserved. It appears that the absence of subglacial landform and sediment assemblages could be due to two factors. First, it is likely that subglacial landscapes are eliminated or modified because of the destructive thermal transition from warm-based to cold-based marginal areas and, second, it is possible that we do not yet recognize the sedimentary imprint of subglacial processes, particularly those associated with cold-based ice, which could be quite subtle. The main prospects for improving our understanding of polar subglacial landscapes are direct observations and measurements of basal processes under thick ice where the bed is at pressure melting point (e.g. Engelhardt and Kamb, 1998) and where glaciers are thin and dry-based (e.g. Fitzsimons *et al.*, 1999).

Acknowledgements

This work was supported by the Australian Antarctic Science Advisory Committee and the Marsden Fund (New Zealand). Logistical support was provided by the Australian Antarctic Division and Antarctica New Zealand. I thank Damian Gore, Massimo Gasparon, Roland Payne, Marcus Vandergoes, Regi Lorrain, Sarah Mager and Paul Sirota for assistance in the field, Sarah Mager and Dr C. Ó Cofaigh for critical comments on the text, and Bill Mooney for drawing the diagrams.

CHAPTER

ICE-MARGINAL TERRESTRIAL LANDSYSTEMS: SOUTHERN LAURENTIDE ICE SHEET MARGIN

Patrick M. Colgan, David M. Mickelson and Paul M. Cutler

6.1 INTRODUCTION

During the last glaciation, the Laurentide Ice Sheet (LIS) created the spectacular glacial landscape of northern North America. This landscape preserves a detailed record of the former ice-sheet size, subglacial bed conditions and ice-sheet behaviour. This chapter discusses the distribution of glacial landsystems that were created along the southern margin of the LIS in the northern USA during the late Wisconsin Glaciation (Fig. 6.1). As we show, landforms and sediments along the southern margin of the LIS are not randomly distributed but are arranged in patterns, which suggest that climate, topography, bed conditions and the resulting ice-sheet dynamics combined to yield distinct landsystems.

A difficulty in interpreting glacial features is the time-transgressive nature of the landscape. Ice-sheet conditions evolved and we must distinguish between landforms developed during cold conditions of ice advance and stability during the Last Glacial Maximum (LGM), and those that formed during deglaciation, when water was abundant and the ice margin was more dynamic. We define a landsystem as 'a genetically related set of landforms and sediments within a distinct region'. Our definition of landsystem contains numerous depositional environments (such as subglacial and ice marginal) that are created by a single ice lobe during a restricted time interval (such as a glacial phase). As genetic processes change with time and environments migrate laterally, one landsystem is superimposed on another. Therefore, to speak of a particular area as containing one specific landsystem is misleading as numerous depositional environments, during multiple phases, produce a suite of landforms and sediments. Commonly, an area is dominated by one landsystem, but includes older (palimpsest) or younger features (superimposed). Our experience and our landsystem maps reflect this, in that we find that two or three landsystems are commonly superimposed even in a small area such as is covered by a 7.5 × 7.5 minute USGS quadrangle. Nevertheless, the purpose of thematic mapping is to generalize and show patterns within the complexity of nature.

We seek to understand the southern LIS by:

1. mapping the distribution of landforms and sediments and classifying them (as landsystems)

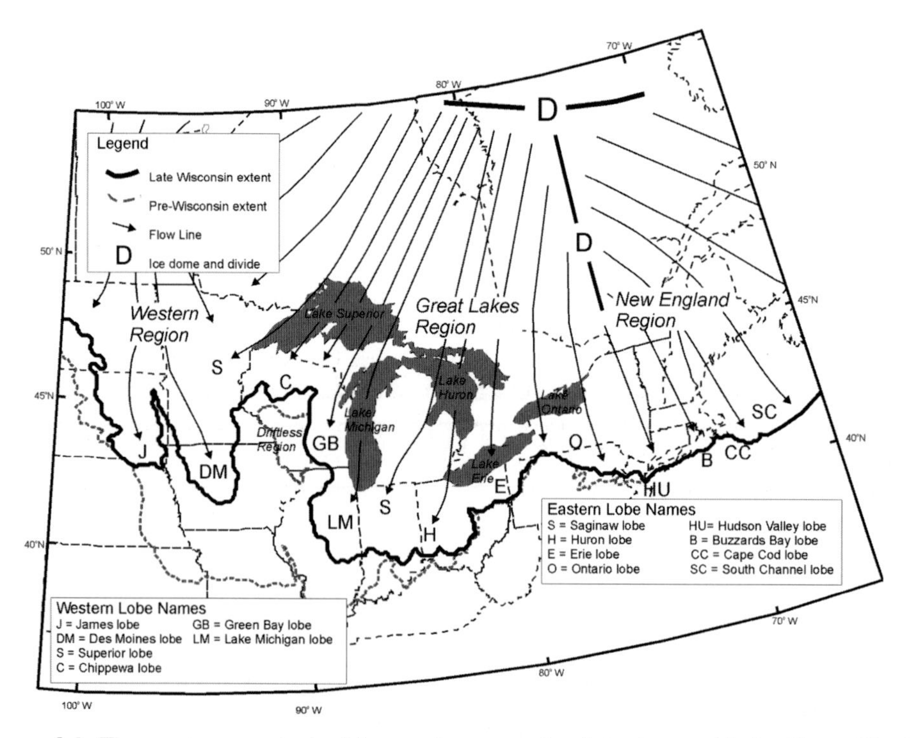

Figure 6.1 The maximum extent of the southern margin of the Laurentide Ice Sheet. The Last Glacial Maximum extent was probably never a continuous ice margin because each lobe reached its maximum position at slightly different times and began retreating at different times. For example the Des Moines lobe and other lobes in the western region reached their maximum extent sometime after 14,000 ^{14}C years BP, whereas most of the lobes in the Great Lakes region reached their maximum positions earlier at about 21,000 ^{14}C years BP. Lobes in New England may have reached their maximum even earlier by about 23,000 ^{14}C years BP. Ice flow directions and ice divides are after Dyke and Prest, 1987; Veillette *et al.*, 1999.

2. interpreting conditions of landform-sediment genesis, and
3. deducing characteristics of the ice sheet from those features.

We restrict most of our discussion to landforms created during the maximum extent of the LIS. We also discuss how in many areas, landforms and sediments created during deglaciation changed significantly from those created during advance and while ice was at its maximum position. This indicates that glacier-bed conditions also changed significantly between the LGM and deglaciation.

6.2 Physical Setting and Timing of Glaciation

Ice advanced into the northern USA after 26,000 ^{14}C years BP, yet the LGM extent was reached at different times in different places (Mickelson *et al.*, 1983). Lobes in the Great Lakes and New England regions reached their maximum before 21,000 ^{14}C years BP. Rapid decay of the ice sheet began after 14,500 ^{14}C years BP. Although lobes to the west of the Great Lakes also advanced before 21,000 ^{14}C years BP, they reached their maximum extent at about 14,000 ^{14}C years BP, out of phase with the rest of the ice-sheet margin (Hallberg and Kemmis, 1986). Readvance of lobes, some perhaps as surges, are recorded all along the southern LIS margin at 13,000 ^{14}C years BP, 11,800 ^{14}C years BP and 9,800 ^{14}C years BP. After about 9,800 ^{14}C years BP ice retreated out of the northern USA (Mickelson *et al.*, 1983).

The southern margin of the LIS stretched from Montana to Maine (Fig. 6.1). In the west the margin was highly lobate as it enlarged the bedrock-controlled lowlands of preglacial river valleys (Fig. 6.2). The Des Moines and James lobes advanced to 42°N into central Iowa and eastern South Dakota, respectively, after being split by the Prairie Couteau as ice moved south through the Red River lowland. The Superior and Chippewa lobes advanced southwestward out of the Superior lowland. The Green Bay lobe, Wisconsin Valley and Langlade sublobes were fed by ice advancing over the eastern end of the Superior lowland. The Lake Michigan lobe flowed down the axis of Lake Michigan, and into southern Illinois, to nearly 38°N. The Huron and Saginaw lobes advanced out of the Lake Huron lowland into Michigan and Indiana. In Indiana and

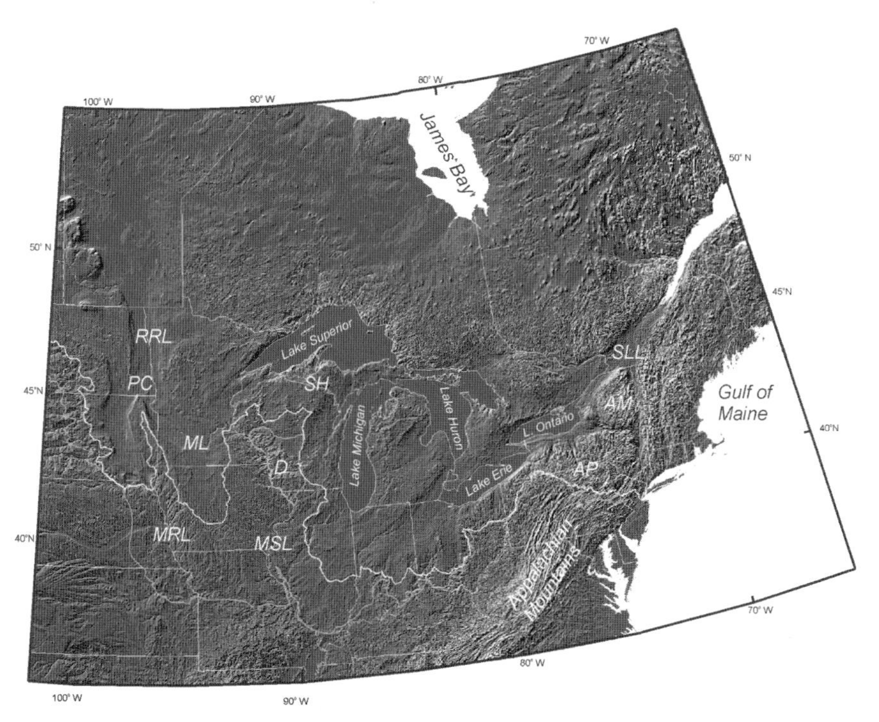

Figure 6.2 Shaded relief image showing the first-order topography of the northern USA. The major physiographic features of the study area discussed in the text are labelled. RRL = Red River Lowland, PC = Prairie Couteau, ML = Minnesota Lowland, MRL = Missouri River Lowland, D = Driftless Region, MSL = Mississippi River Lowland, SH = Superior Highlands, AP = Appalachian Plateau, AM = Adirondack Mountains, SLL = Saint Lawrence Lowland. The white line indicates the maximum extent of the Late Wisconsin Laurentide Ice Sheet. The grey line shows the maximum extent of pre-Wisconsin advances. (Data are from US Geological Survey ETOPO5 database.)

Ohio, sublobes of the Lake Huron and the Erie lobes advanced to nearly 39°N. The ice margin in Pennsylvania and New York was less lobate and formed a major re-entrant as it encountered the Appalachian Plateau and the Ridge and Valley Province. Small tongues of ice did project down narrow river valleys, but these were smaller than the lowland lobes to the west. Farther to the east, the ice-sheet margin remained much less lobate with the exception of the Hudson River and the Cape Cod lobes. Ice in eastern New England advanced well into the Gulf of Maine and onto the exposed continental shelf.

6.3 METHODS

We compile information about the distribution of glacial landforms and sediments in northern USA from North Dakota to Maine. Our compilation aims to minimize interpretations of genesis in the data collection because we do not fully understand the details of the genesis or conditions of formation of all landforms.

Unfortunately, past mapping has been non-standardized in different areas and the comparison of features from published sources is difficult. Ridges called end moraines in one area are not called end moraines in another (e.g. central Iowa). Similarly, different criteria are used to recognize streamlined features. This makes it difficult to compare areas using only existing glacial maps. Because of this, we examined the raw data sources (topographic maps, etc.) from which these maps have been created. We have used published maps and reports for information about sediment types. These are supplemented locally with high-resolution digital elevation models. Thus, we use a combination of published reports and maps and interpretation from topographic maps, digital elevation models and aerial photographs to add to the compilation.

6.4 MAPPING AND CLASSIFICATION OF THE DATA

Our database consists of sedimentologic and geomorphic information. Data are compiled in a pseudo-raster format with each cell dimension consisting of 7.5 minutes of latitude by 7.5 minutes of longitude (Fig. 6.3). The study area is divided into 18 zones (each 4 by 6 degrees). Each zone is represented with a grid of 1536 cells (Fig. 6.3). Some zones contain less than this number of cells because they border Canada or the ocean. In each zone, cells are represented by vector polygons, and each is linked to an attribute table, which lists the features found in that cell. For example if drumlins are present in a given $7.5' \times 7.5'$ area the attribute table notes this with the integer '1' in that attribute column for that cell. If drumlins are not located in this area then a '0' is entered. In other attribute columns the type of sediment that lies at the centre of that cell is recorded by an integer value. By this method the presence or absence of glacial features can be recorded, or the type of sediment, sediment thickness or other attribute of the area can be entered into the database.

To create our current maps we used nine attributes for which we have complete coverage (Fig. 6.4). These nine attributes are used as input to create the landsystem maps. From these nine attributes we defined seven different landsystems. Figure 6.5 illustrates how the nine attributes were used to create the seven landsystems shown in Figure 6.6. Some landforms are considered more important than others in this classification scheme. This is the case for features that we

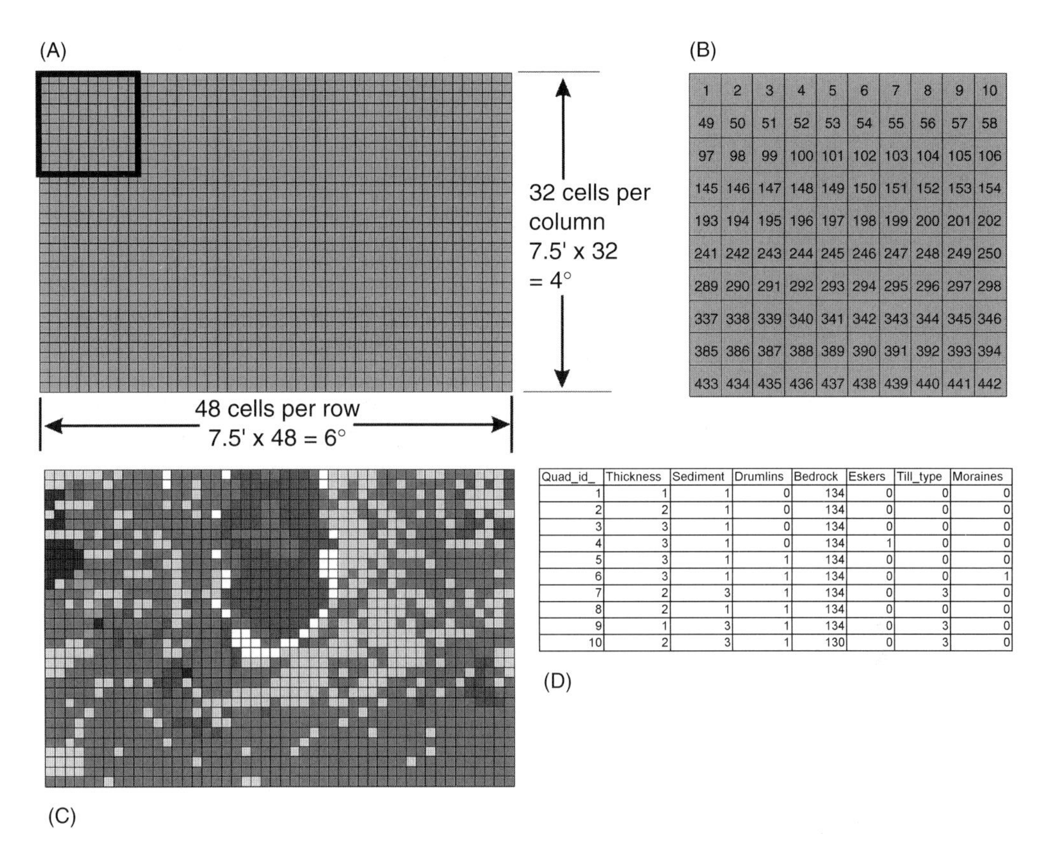

Quad_id_	Thickness	Sediment	Drumlins	Bedrock	Eskers	Till_type	Moraines
1	1	1	0	134	0	0	0
2	2	1	0	134	0	0	0
3	3	1	0	134	0	0	0
4	3	1	0	134	1	0	0
5	3	1	1	134	0	0	0
6	3	1	1	134	0	0	1
7	2	3	1	134	0	3	0
8	2	1	1	134	0	0	0
9	1	3	1	134	0	3	0
10	2	3	1	130	0	3	0

Figure 6.3 The structure of our geologic database. A) Grid of 1536 cells. We used part of 18 of these grids to cover the study area. B) A close-up of the numbering of the cells that are highlighted in black in A. Each cell is numbered and linked to a database table shown in D. C) An example of a map showing the distribution of sediment types in each cell. D) An attribute table showing the feature attributes for each cell. Though we only show seven attributes here, the database contains more than two dozen attributes for each cell area.

believe are the defining characteristic of a landsystem such as drumlins, till plains with thick glacial sediment, and low-relief hummocky plains. For example, any cell in which drumlins occur is classified as landsystem B (see Fig. 6.6a). Many cells contain landforms and sediments of more than one landsystem because landsystems created during deglaciation were superimposed upon LGM landsystems or because the size of each cell encompasses two or more distinct areas. Nevertheless, our maps show the regional-scale distribution of the dominant landsystems.

6.5 LANDSYSTEMS OF THE SOUTHERN LAURENTIDE ICE SHEET

Our maps show seven different landsystems (Fig. 6.6). We restrict our discussion here to the four landsystems that we believe are linked to bed conditions and glacier dynamics during the LGM:

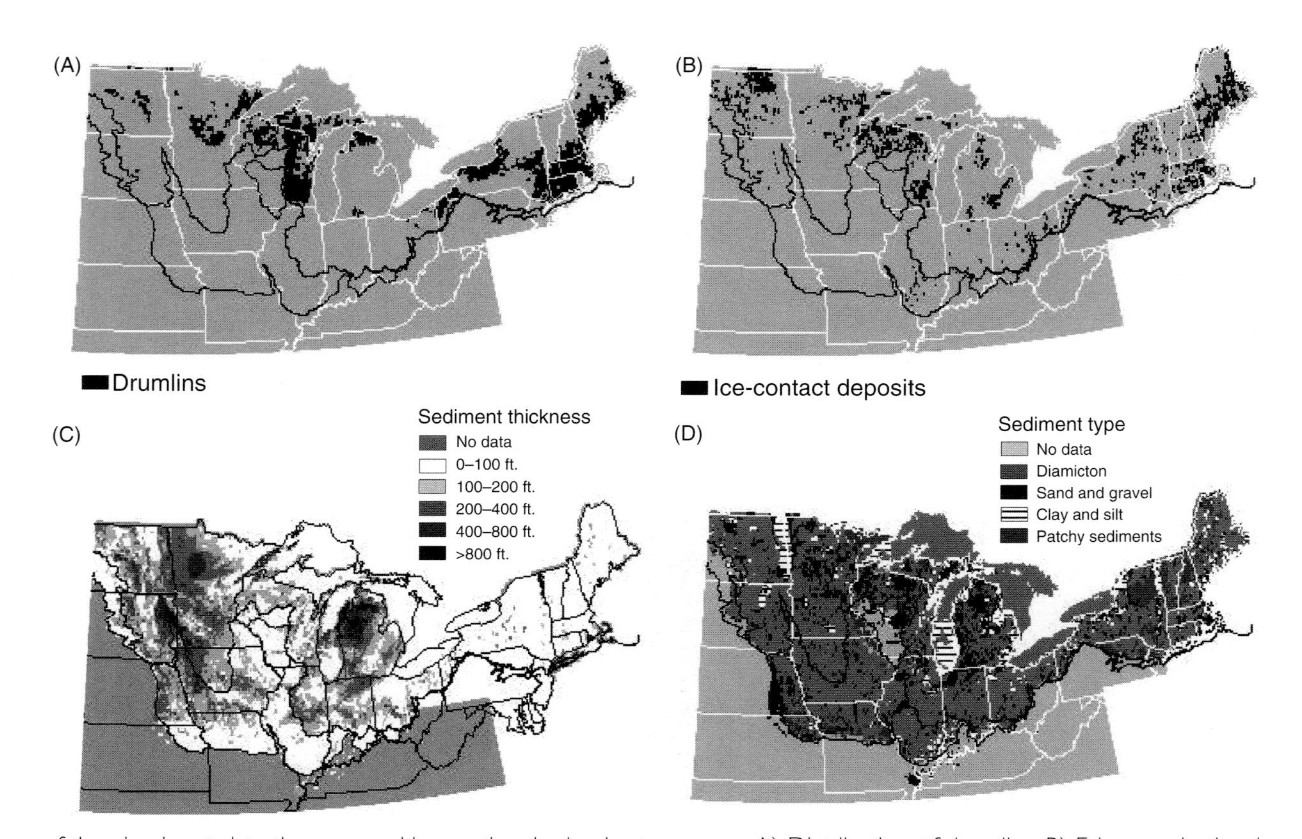

Figure 6.4 Four of the nine input data themes used in creating the landsystem maps. A) Distribution of drumlins. B) Eskers and other ice-contact deposits. C) Glacial sediment thickness from Soller and Packard (1998). D) Surficial sediment cover, also from Soller and Packard (1998).

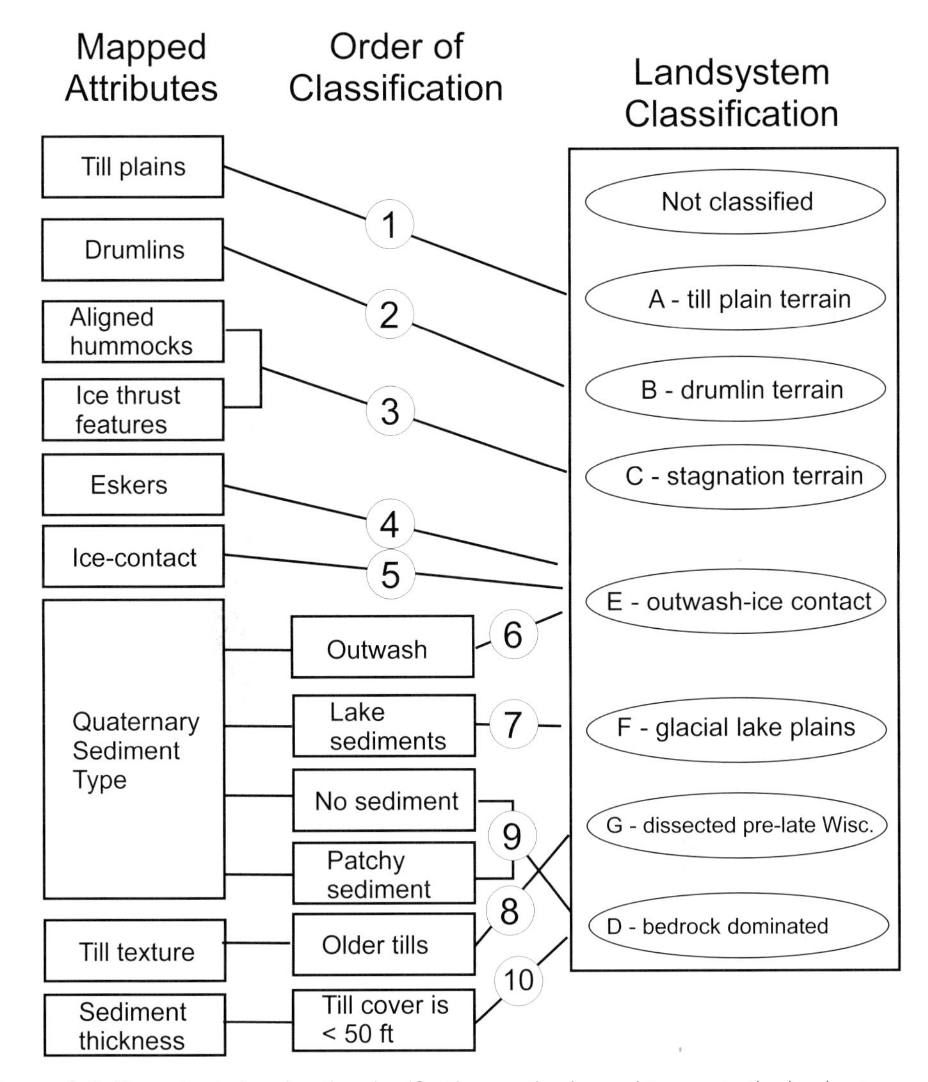

Figure 6.5 Flow-chart showing the classification methods used to create the landsystem maps (see Fig. 6.6). Input layers on the left were classified into the seven landsystems shown on the right. The order and type of classification are shown in the middle. The numbers show the order in which the classification was carried out. Classification was done by using a series of structured-query-language commands on the database tables.

1. low-relief till-plains (landsystem A)
2. drumlins and high-relief moraines (landsystem B)
3. low-relief hummocks and ice-thrust terrains (landsystem C)
4. bedrock-dominated glacial landscapes (landsystem D).

Glacifluvial (landsystem E) and glacilacustrine features (landsystem F) dominate two additional landsystems. Landsystems D and G were primarily created before the LGM, although there has clearly been continuing evolution of the landscape up to the present.

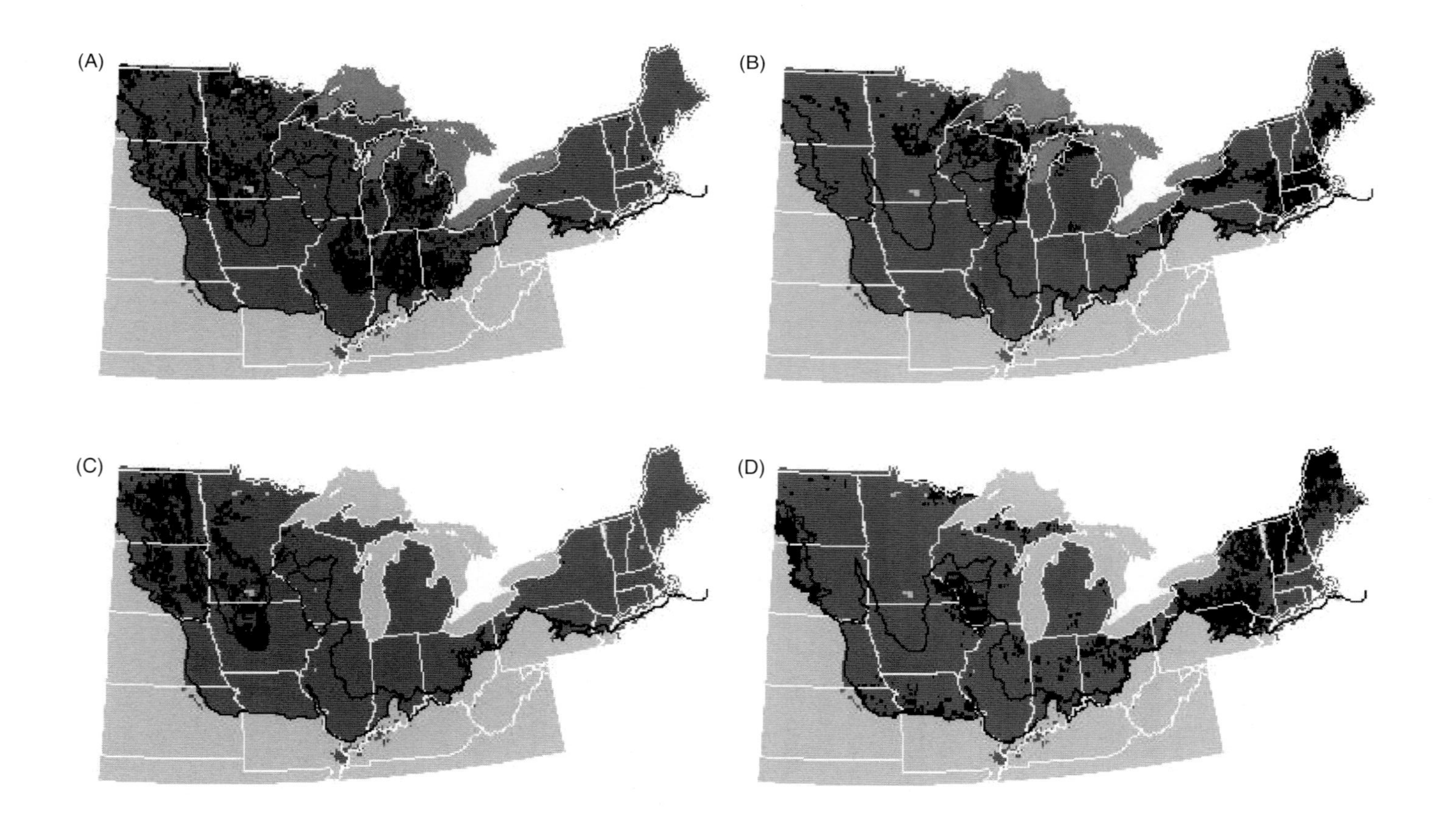
(A)
(B)
(C)
(D)

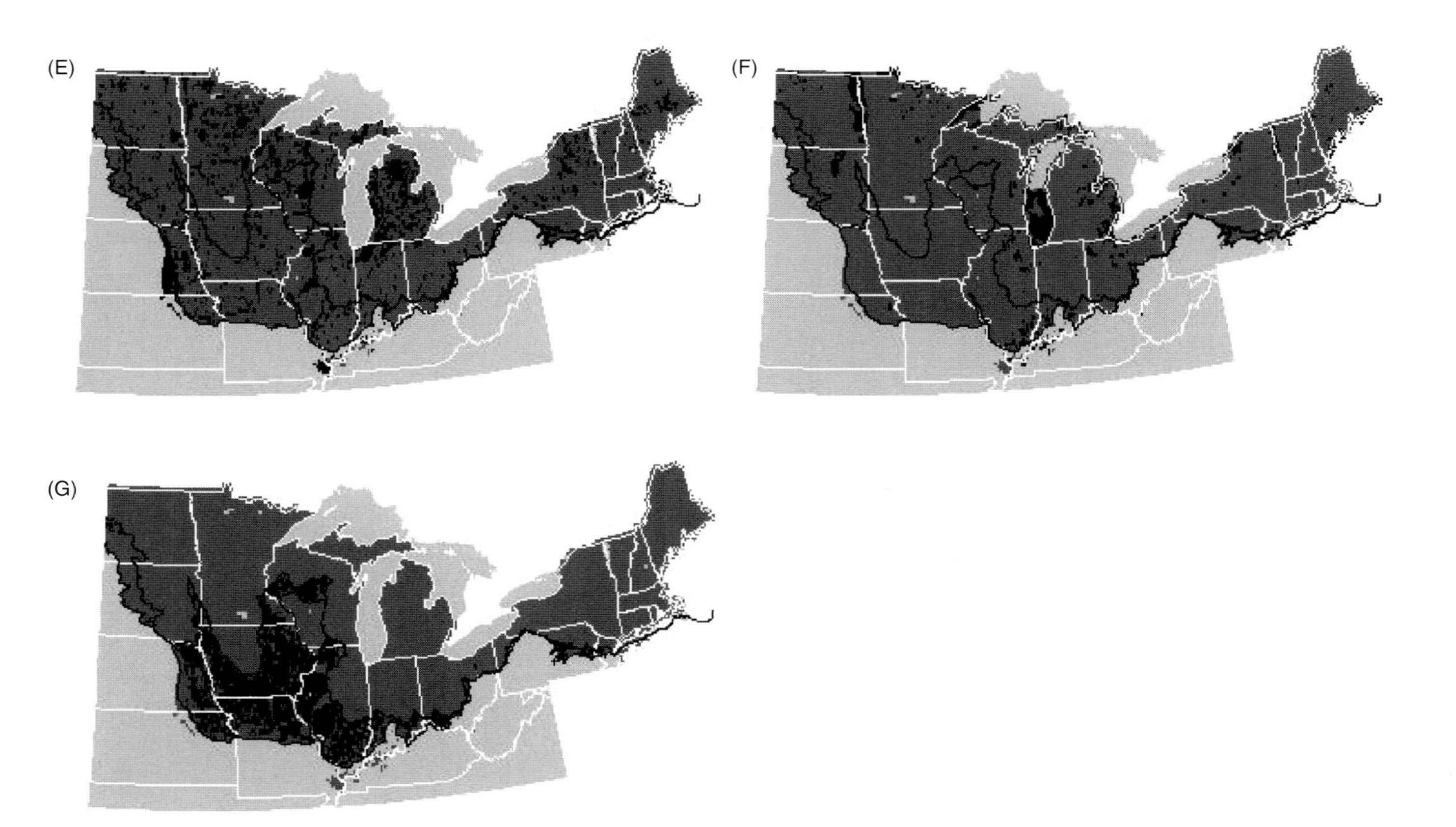

Figure 6.6 Landsystem maps of the area covered by the southern margin of the Laurentide Ice Sheet. Each cell (7.5' × 7.5') corresponds to a classification set shown in Fig. 6.5. Dark line shows the extent of Late Wisconsin glacial maximum. Light grey line shows the maximum extent of Pleistocene glaciation.

6.5.1 Landsystem A – Low-Relief Till Plains and Low-Relief End Moraines

Landsystem A occurs extensively in Ohio, Indiana, Illinois, Michigan, Minnesota, Iowa, North and South Dakota. It formed when ice was at or near its maximum southern extent and during retreat (Fig. 6.6A). We divide landsystem A into two zones: ice-marginal and subglacial (Figs 6.7 and 6.8). We interpret that landsystem A primarily reflects subglacial sediment transport and subsequent deposition as basal till.

6.5.1.1 Ice-Marginal Zone

This zone is composed of a ramp of diamicton inclined up-glacier. Steeply sloping fans composed of outwash and debris flow sediment occur along the outer edge of this zone. Proglacial outwash deposits merge into valley train deposits. Along the outermost margin of the ramp, older glacial and non-glacial deposits are deeply dissected by stream erosion. Fine-grained sorted sediment is present in major stream valleys.

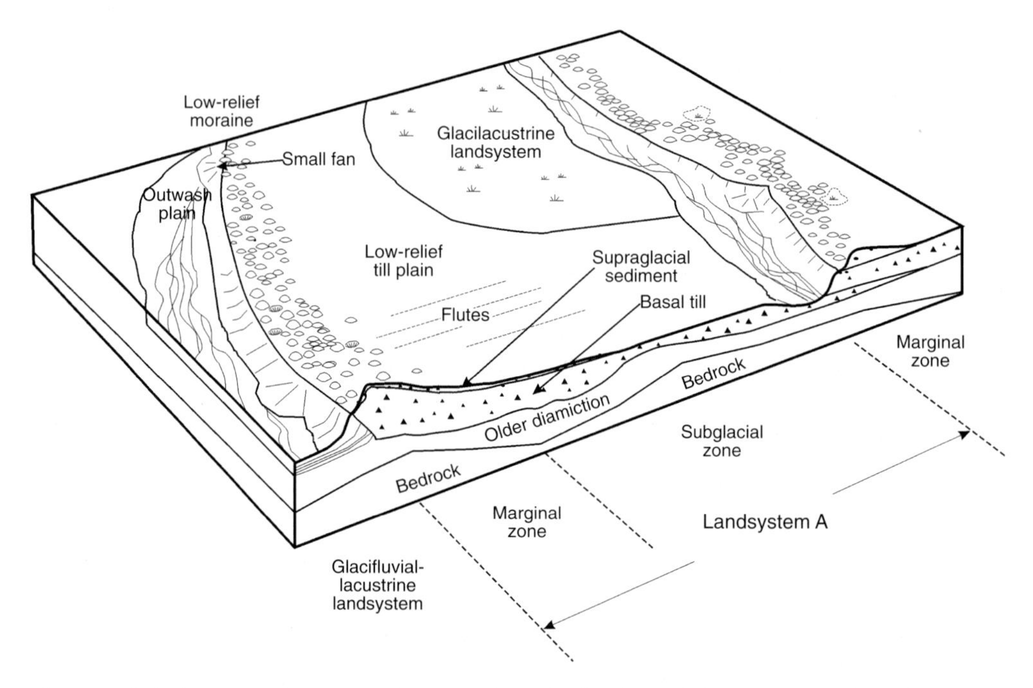

Figure 6.7 Schematic diagram showing the distribution of landforms and sediments in landsystem A. The marginal zone consists of a broad, low-relief end moraine composed primarily of basal till over older glacial drift with a cover of supraglacial sediment. The subglacial zone consists of till plain also composed of basal till with few features except for rare flutes. In the subglacial zone supraglacial sediment may be completely absent. Glacifluvial (landsystem E) and glacilacustrine (landsystem F) landforms and sediments fronted this landsystem in the proglacial area or were superimposed on this landsystem during deglaciation. Bedrock in this area only rarely crops out, and glacial deposits are generally greater than 15 m thick. During retreat, ice-marginal zones of landsystem A were superimposed on older landsystems as is shown on the right of the diagram. In the southern Great Lakes region in Illinois, Indiana, and Ohio each readvance deposited a wedge-shaped sheet of basal till on top of older sequences during retreat.

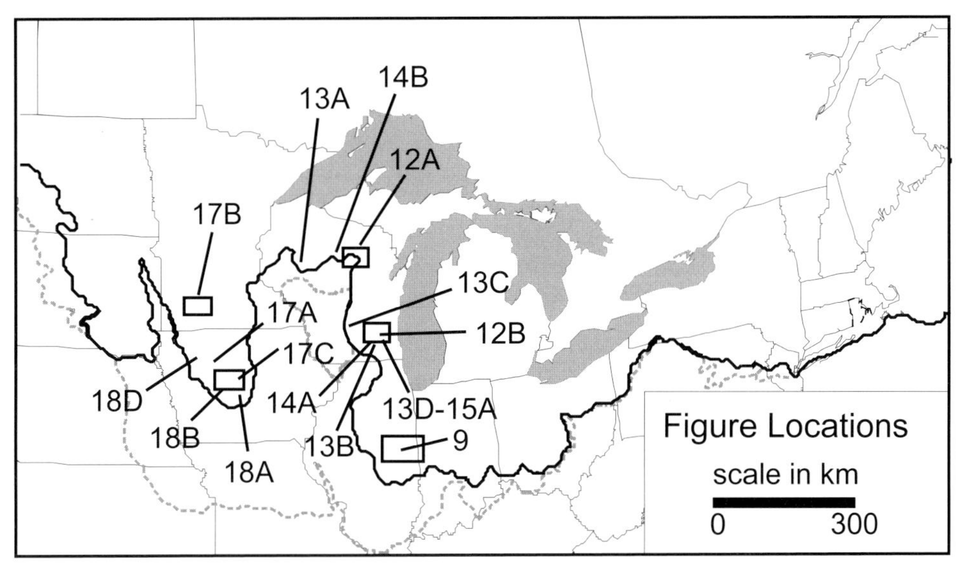

Figure 6.8 The location of aerial photographs and images cited in the text.

End moraines in landsystem A are 2–20 km wide, with asymmetric cross sections (Figs 6.9 and 6.10). The distal sides of the moraines have 1–2 per cent slopes, whereas the proximal slopes are less than 1 per cent (Hansel and Johnson, 1999). The moraine surface is smooth, except for postglacial gullies cut into surfaces (Fig. 6.9). Low-relief hummocks and closed depressions are rare and the internal moraine relief is generally less than 3 m. Moraines are mostly composed of homogeneous diamicton interpreted to be basal till capped with a thin layer (<2 m) of heterogeneous diamicton that may be supraglacial sediment. Although some of the material in the moraines is local, it has been suggested that most of the sediment originated a considerable distance up-ice and was transported into the area with little modification except clast abrasion (Hansel and Johnson, 1999). Many of the moraines have far-travelled erratic boulders concentrated at or near the surface.

6.5.1.2 Subglacial Zone

Behind the end moraine lies a flat or gently undulating till plain (Figs 6.7 and 6.9). The plain has few distinct landforms and generally a local relief less than 3 m. Rarely, a few low-relief flutes are present (Hansel and Johnson, 1999). Much of the diamicton is basal till, although locally there is a thin veneer of supraglacial sediment and a loess cap.

Stacked sequences of late Wisconsin diamicton and sorted sediments make up both moraines and till plains (Johnson and Hansel, 1990; Hansel and Johnson, 1996, 1999). Many of the moraines are composite forms and were modified during more than one advance or period of stability during the late Wisconsin. Palaeosols have been preserved between diamicton sheets. Diamicton units are wedge shaped, with their thickest end in the moraine zone and thinning to a wedge under the till plain. Sub-diamicton contacts are generally clear and show little thrusting or folding-in of older sediment. In some places, sub-bed material has been incorporated into the overlying diamicton. At a few sites, clast concentrations and clast pavements have been reported, but these appear to be neither continuous nor widespread.

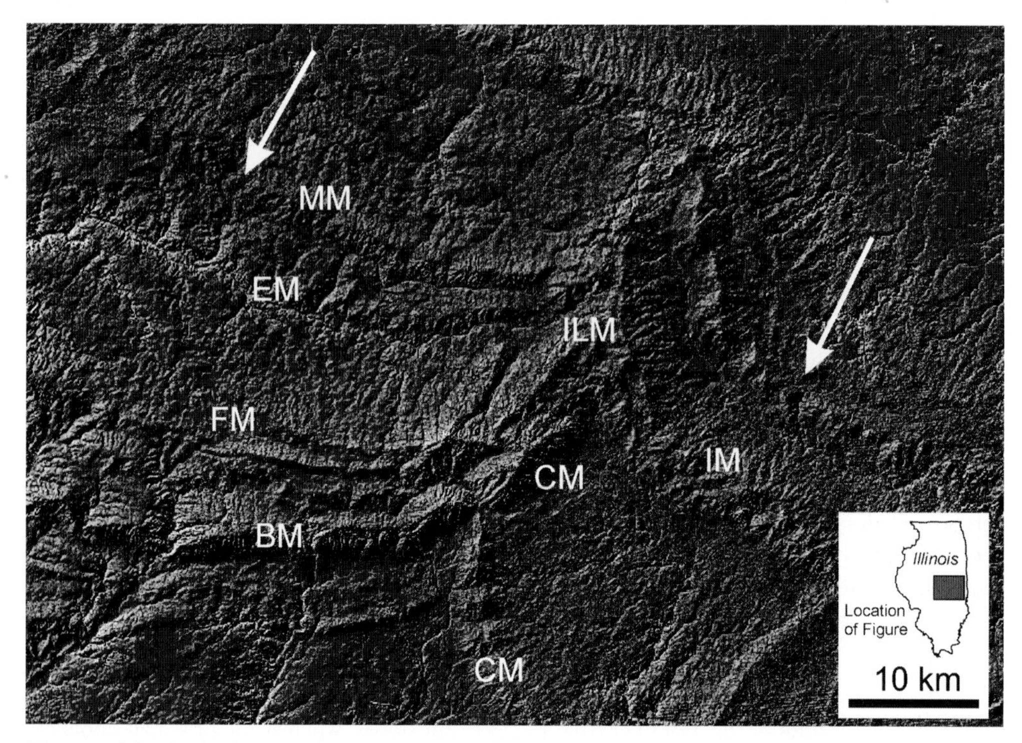

Figure 6.9 A shaded-relief image derived from a 30 m US Geological Survey DEM of the southern part of Lake Michigan lobe area. Moraine names are from Hansel and Johnson (1996). MM = Minonk moraine, EM = El Paso moraine, FM= Fletchers moraine, BM = Bloomington moraine, CM = Champaign moraine, IM = Illiana moraine, ILM = area of interlobate moraine. Each end moraine has a steep distal margin and a ramp-like proximal margin. Note that much of the surface roughness is the result of postglacial fluvial erosion as gullies and streams. Both the marginal and subglacial zone have very little original roughness or internal relief.

6.5.2 Interpretation of Landsystem A

The characteristics of landsystem A, with its till plains and broad low-relief moraines, suggest the following:

1. subglacial sediment transport as englacial and subglacial deformation till
2. little accumulation of debris on the ice surface and deposition as supraglacial sediment
3. ice motion dominated by some combination of sliding, ploughing and subglacial deformation
4. active ice with progressive retreat of the terminus.

There appears to have been very little, and perhaps no subglacial erosion of pre-existing sediments in the marginal areas. Widespread paleosols, present on older units, indicate limited erosion. Certainly, abrasion was taking place, and striated surfaces, rare clast-pavements and striated boulders are present throughout the area. The wedge-shaped till sheets suggest erosion by later advances but probably only at their up-ice ends.

Unlike landsystem B, drumlins are absent. With the exception of a few low-relief flutes, there appear to be no streamlined features. Tunnel channels, a common feature in landsystems B and C, are also absent. Likewise, ice-thrust masses and composite ridges, a common feature of landsystem C, and to a lesser extent landsystem B, are also absent. Eskers are nearly absent. The distribution of outwash along the former ice margins suggests that water was delivered to the ice margin in small subglacial or englacial tunnels as opposed to being concentrated in large discharge events or esker tunnels as in landsystems B and C.

Permafrost may have been present in Illinois, Indiana and Ohio (Johnson, 1990), but it was probably thinner than that in areas farther north. Wood preserved in tills in Illinois, Indiana and Ohio suggests that the Lake Michigan, Saginaw, Huron and Erie lobes advanced into areas with trees so it is likely that permafrost was discontinuous (Mickelson *et al.*, 1983; Ekberg *et al.*, 1993; Hansel and Johnson, 1999).

We support the conclusions of others that till in this landsystem was transported and deposited at the base of the ice, perhaps as a deforming bed (Boulton and Jones, 1979; Beget, 1986; Alley, 1991a; Clark, 1992; Hansel and Johnson, 1999). Ice motion was probably accommodated by a combination of sliding, ploughing and subglacial deformation. What is not clear is whether transport prior to deposition was as a subglacial deforming bed of dimensions metres-thick, or as melt-out of debris-rich ice in slabs detached near the terminus.

In summary, landsystem A with its till plains and broad low-relief moraines suggests:

1. predominantly subglacial sediment transport
2. little accumulation of debris on the ice surface
3. ice motion dominated by some combination of sliding, ploughing and subglacial deformation
4. initial advance over an unfrozen subglacial bed followed by progressively retreating active ice during deglaciation.

6.5.3 Landsystem B – Drumlins and High-Relief Hummocky End Moraines

Landsystem B is located in North Dakota, Minnesota, Wisconsin, northern Michigan, Pennsylvania, New York, New England and as a small window in southern Michigan (Fig. 6.6A). It is dominated by drumlins, sandy diamicton and high-relief, hummocky end moraines. We interpret that this landsystem reflects subglacial erosion and supraglacial deposition near an active ice margin.

6.5.3.1 Ice-Marginal Zone

This zone is a landscape that has always been mapped as end moraine (Figs 6.11 and 6.12). The zone is 2–20 km wide and consists of moderate- to high-relief hummocks in end moraines. The end moraine stands 5–40 m above the surrounding landscape (Fig. 6.10B). Distal and proximal slopes of the end moraine exceed 2 per cent. The internal relief of hummocks is much higher than in the smooth moraines of landsystem A. In most places the moraines are narrower (<2 km), but are about the same height as moraines in landsystem A. Within wider hummocky end moraines are large flat plateaux of sorted sediments and glacilacustrine sediment that have been interpreted as ice-walled lake plains (Johnson *et al.*, 1995; Ham and Attig, 1996). Ice-walled lake plains are found in Minnesota, North Dakota, Wisconsin, Michigan and in southern New England (as the 'high kames' of Stone and Peper, 1982).

Moraine composition is different from that in landsystem A. There is a much larger component of glacifluvial sand and gravel and much thicker accumulations of diamicton interpreted to be supraglacial sediment. Layers of sand and gravel, supraglacial sediment and basal till are stacked-up in this zone (Lundqvist *et al.*, 1993; Johnson *et al.*, 1995). Some of the gravel in the moraine was deposited in subglacial tunnels close to the ice margin (Lundqvist *et al.*, 1993). Erratic boulders are scattered on the moraine and were carried long distances from the Canadian Shield. In moraine exposures, beds of gravel and till are steeply dipping in the up-ice direction, indicating ice push and localized ice-thrusting (Oldale and O'Hara, 1984; Mooers, 1990).

Unlike landsystem A, the distribution of proglacial outwash outside the margin is concentrated in a few places. Large volumes of outwash were deposited as alluvial fans from the mouths of tunnel channels. Tunnel channels intersect the outermost moraine about every 5–10 km in Wisconsin and Minnesota. Most of the sediment in the outwash plains in front of the ice margin was delivered by these tunnel channels (Wright, 1973; Attig *et al.*, 1989; Mooers, 1989a; Patterson, 1994; Clayton *et al.*, 1999; Cutler *et al.*, 2001; Cutler *et al.*, 2002). Tunnel channels are about 0.5–1 km across where they intersect the moraine. Sometimes a steep-sided channel is present within the moraine, but often it is filled with hummocky sand and gravel, which collapsed as ice melted. Mapping of the channels in the Green Bay lobe indicates that water was driven up as much as 80 m in elevation (Fig. 6.10B)

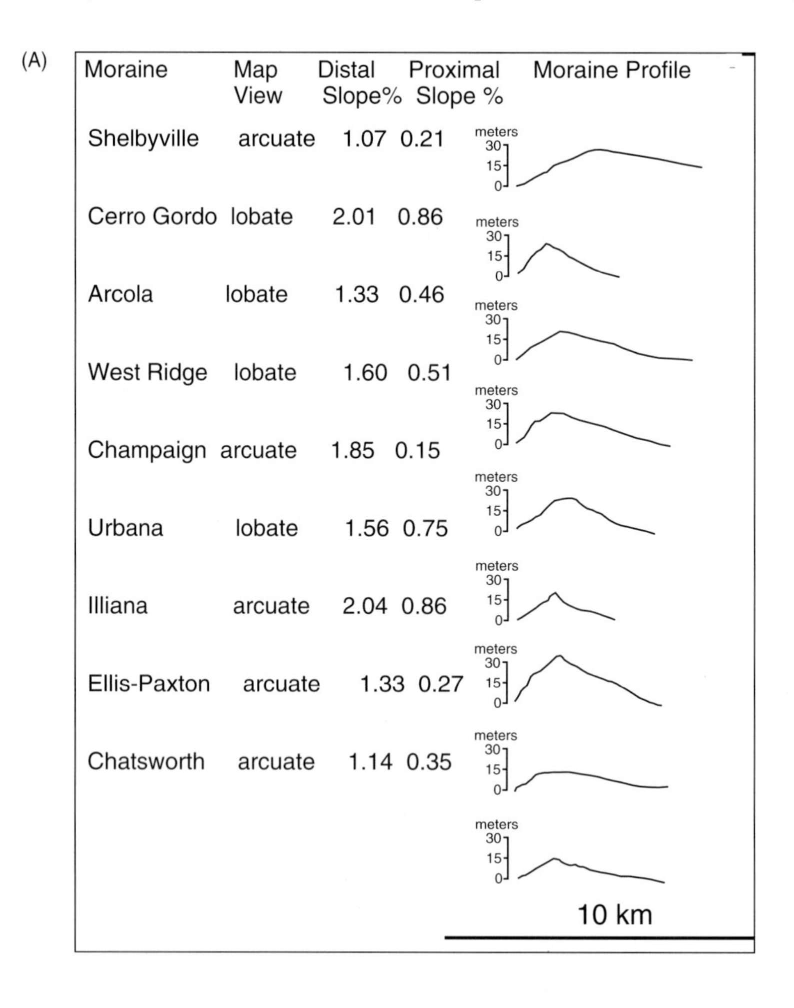

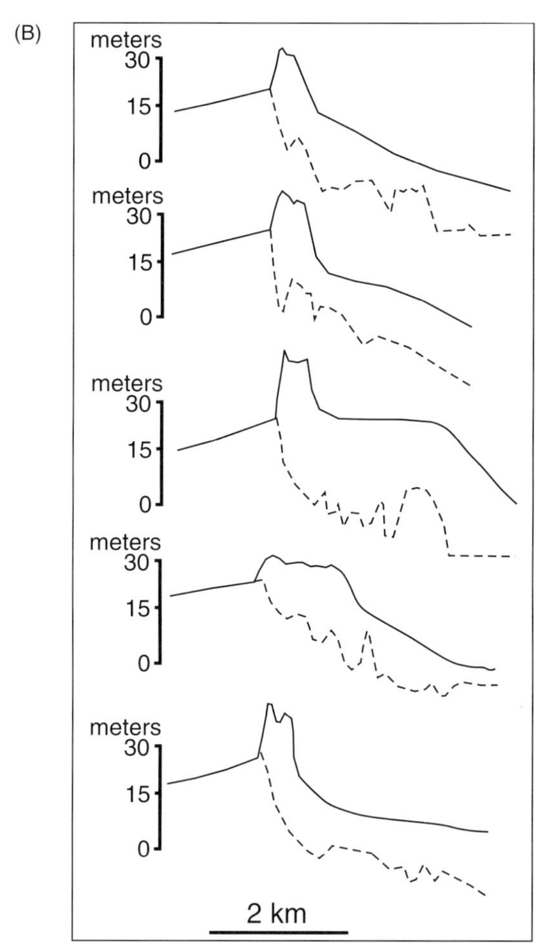

Figure 6.10 A) Distal and proximal slopes of Lake Michigan lobe moraines. (From Hansel and Johnson, 1999). B) Topographic profiles through end moraines of the Green Bay lobe showing moraine surface profiles in grey and tunnel channel bottom with the lower solid line. The dotted lines to the left of each profile show the slope of tunnel channel outwash fans. (From Clayton *et al.*, 1999). Note that the horizontal and vertical scales are different for each figure.

as it approached the ice margin (Clayton *et al.*, 1999). Some sediment within the fans is extremely coarse. Intermediate axes of boulders of up to 2 m have been described in the proximal part of the fans, grading into fine sand in the distal part of the fan (Cutler *et al.*, 2002).

A considerable volume of sand and gravel, and supraglacial sediment is present in interlobate areas and in the moraines themselves, indicating that there was more sediment on the ice surface than in landsystem A. During the retreat phase there was substantial sediment transport by supraglacial and englacial streams on and within the ice.

6.5.3.2 Subglacial Zone

The landforms in the subglacial zone are strikingly different from those in landsystems A and C. In landsystem B most of the surface has been streamlined by ice flow (Figs 6.12 and 6.13).

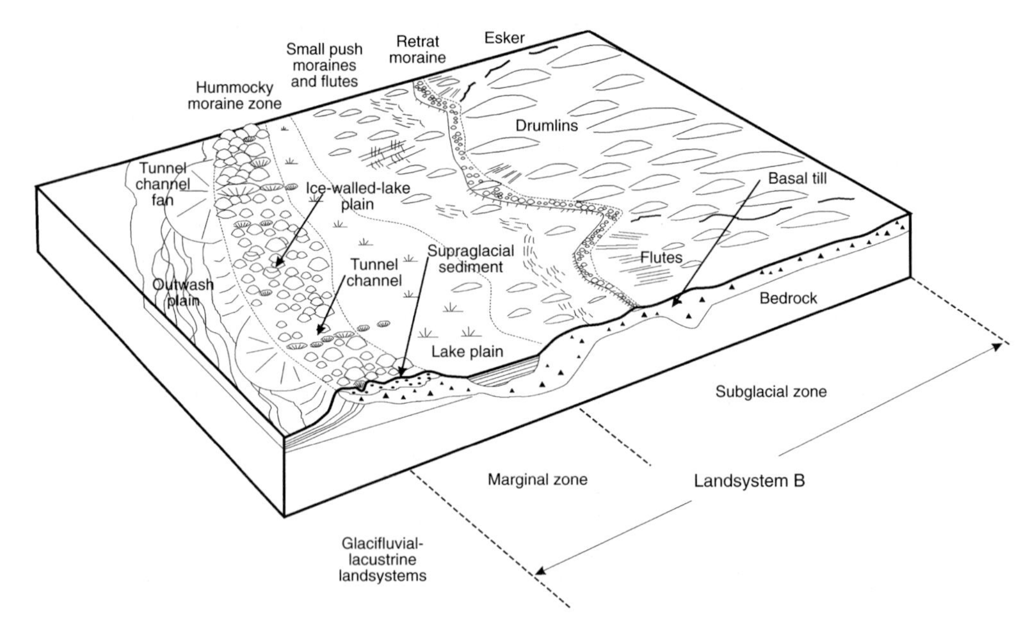

Figure 6.11 A schematic diagram showing the landforms and sediments in landsystem B. The marginal zone is dominated by high-relief hummocky topography. This zone contains a thick cover of supraglacial sediment, over basal till and sand and gravel. The subglacial zone is dominated by drumlins and flutes. This area contains less supraglacial sediment as a very thin cover sometimes existing as small moraines that were deposited during retreat. Eskers and other ice-contact material were also deposited during retreat. Bedrock in this area is usually covered with glacial sediments, but in many areas the glacial sediment cover is very thin (<5 m).

Drumlins and megaflutes up to 50 m high and as long as 30 km occur throughout this zone (Colgan and Mickelson, 1997). Thousands of drumlins occur in eastern Minnesota, in Wisconsin, Michigan, New York and in southern New England.

The composition of most of the drumlins is not known. In Wisconsin, New York and Massachusetts, some drumlins are composed of stratified sand and gravel (Upham, 1894; Alden, 1905; Fairchild, 1907; Whittecar and Mickelson, 1977, 1979; Stanford and Mickelson, 1985). Other drumlins are composed of diamicton, some have bedrock cores and some contain lake-sediment (Alden, 1905; Colgan and Mickelson, 1997). In Wisconsin and southern New England at least, the drumlins appear to be composed to a great extent of material that pre-dated the drumlin-forming phase (Whittecar and Mickelson, 1977, 1979; Stanford and Mickelson, 1985; Newman and Mickelson, 1994). Thus, drumlins appear to be partly erosional features, although certainly part of their height and length is attributable to growth beneath the ice from sediment moving toward the axis of the drumlin and deposition of diamicton and down-ice accretion of eroded material (Boulton, 1987). It has been argued (Mooers, 1989b) that drumlins formed at different times during retreat, but the largest fields formed when ice was near its outermost position. Smaller drumlin fields were then superimposed on larger forms as ice retreated and briefly stabilized at other positions (Colgan and Mickelson, 1997).

Throughout landsystem B, diamicton is fairly thin. In many places the diamicton of the last glacial maximum lies directly on bedrock. Basal till in the subglacial zone contains a high

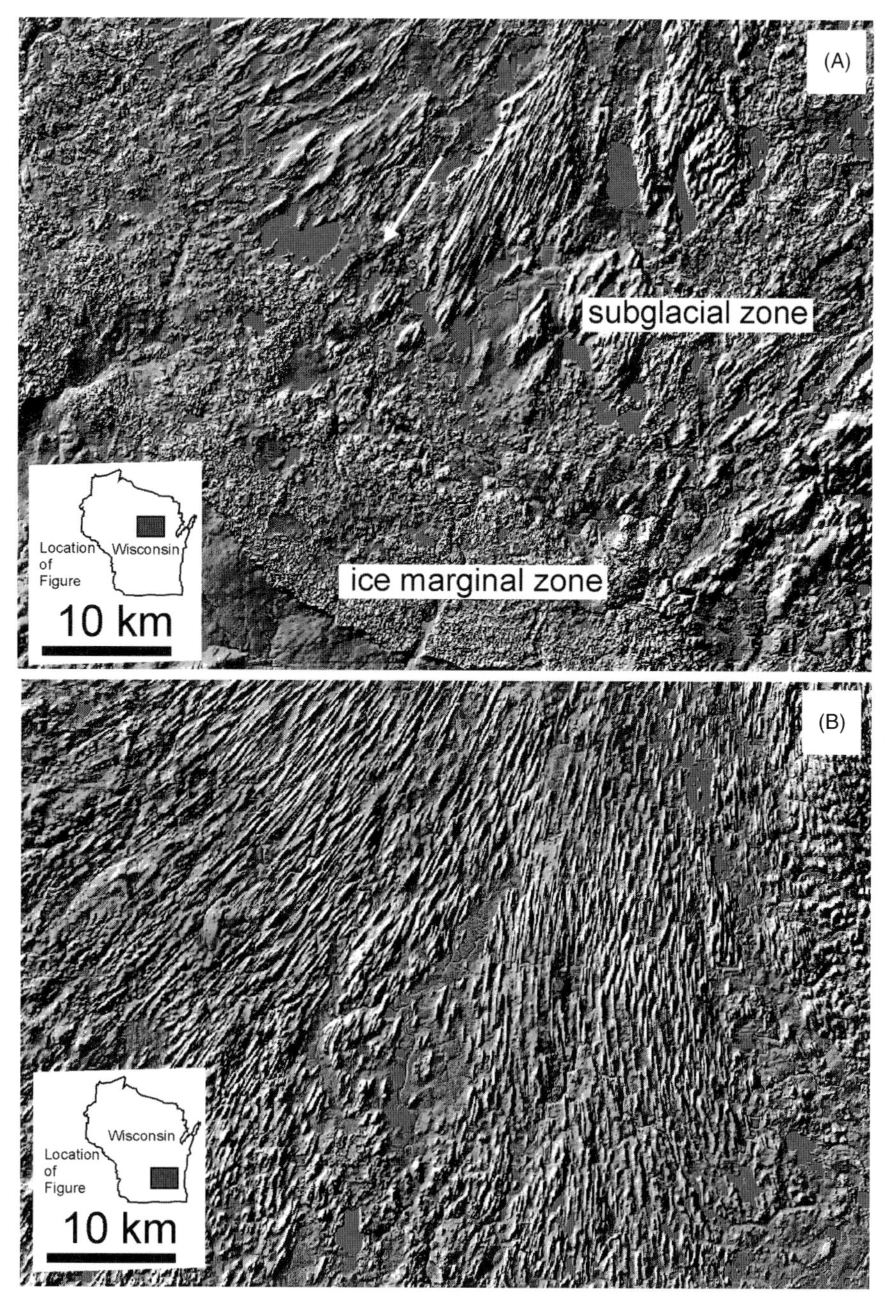

Figure 6.12 A) A shaded-relief image showing the topography of landsystem B in northern Wisconsin. This landsystem was created by the Langlade lobe from about 21,000–16,000 ^{14}C years BP. Note the distinct contact between proglacial glacifluvial deposits and hummocky end moraine. B) A shaded relief image showing the subglacial zone of landsystem B in southeastern Wisconsin created by the Green Bay lobe between 21,000 and 16,000 ^{14}C years BP.

Figure 6.13 Oblique aerial photos of landforms in landsystem B. A) High-relief hummocky topography and ice-walled lake plains deposited by the Chippewa lobe in northern Wisconsin. B) Green Bay lobe drumlins of the Madison drumlin field. This photo is of the same region shown in the DEM of Fig. 6.12B. C) Hummocky topography of the Johnstown moraine deposited by the Green Bay lobe. D) A single drumlin of the Madison drumlin field. (Courtesy of Donna Stetz).

percentage of local material. Far-travelled material is more common in supraglacial sediment in the marginal zone. This is different from the far-travelled nature of till in landsystem A and indicates much more local erosion. Striated rock surfaces are present even in the marginal areas, but striated boulder pavements are rare. Many bedrock exposures show indications of plucking and removal of large clasts as well as abrasion.

There are other major differences between landsystems B and A. The low-relief till plain of landsystem A is absent. Additionally, numerous small moraines are commonly superimposed on the drumlinized surface (Fig. 6.14). We believe that these are similar in origin to annual moraines formed in Iceland (Boulton, 1986; Krüger, 1994a), but different from the aligned hummocks found in landsystem C (see Colgan, 1996; Clayton and Attig, 1997; Ham and Attig, 2001).

6.5.4 Interpretation of Landsystem B

Most of the moraines in landsystem B were developed by the stacking of basal till slabs on top of layers of sand and gravel and supraglacial sediment (Lundqvist *et al.*, 1993). For some time

Figure 6.14 A) Aerial photograph showing small push moraines deposited by the Green Bay lobe during its retreat (photograph from USDA, WU-3P-26,1955, Dane County, Wisconsin). B) Aerial photograph showing small moraines deposited by the Wisconsin Valley lobe in northern Wisconsin (Ham and Attig, 2001).

when ice was at its maximum, extensive outwash was being generated along the ice margin as well as from tunnel channels. We interpret the tunnel channels as indicators of large discharges of subglacial water, perhaps stored as an extensive subglacial layer and then released during discrete events (Cutler *et al.*, 2002).

Wood is not present in the tills of landsystem B. We believe the absence of wood is an indication that continuous permafrost and tundra conditions existed when ice advanced to its maximum position in the area of landsystem B (Fig. 6.15). Both outside and inside the outermost margin of the ice in landsystem B there are indications of former permafrost. Patterned ground can be observed on aerial photographs (Fig. 6.15), particularly on outwash surfaces created during the last glacial maximum (Black, 1976b; Péwé, 1983; Ham, 1994; Johnson *et al.*, 1995; Colgan, 1996; Clayton and Attig, 1997; Clayton *et al.*, 2001).

The abundance of locally derived basal till suggests that material was eroded at the base of the ice and there was relatively little upward movement of basal sediment, except for stacking of debris-rich slabs near the terminus. Incorporation of local material and the prevalence of streamlined forms suggest that erosion was the dominant process occurring behind the terminus. This zone of erosion was time-transgressive and changed in size during advance and retreat because of the interaction of the thermal regime of the lobe and the underlying subglacial permafrost (Cutler *et al.*, 2000).

Another indication of the prevalence of subglacial erosion is the rarity of paleosols. There are a number of buried soils and organic deposits in Michigan associated with landsystem B, but most of these are in buried valleys where they have been protected by a thick cover of glacifluvial or glacilacustrine sediment.

In summary, the characteristics of landsystem B, with its drumlins and high-relief moraines, suggest:

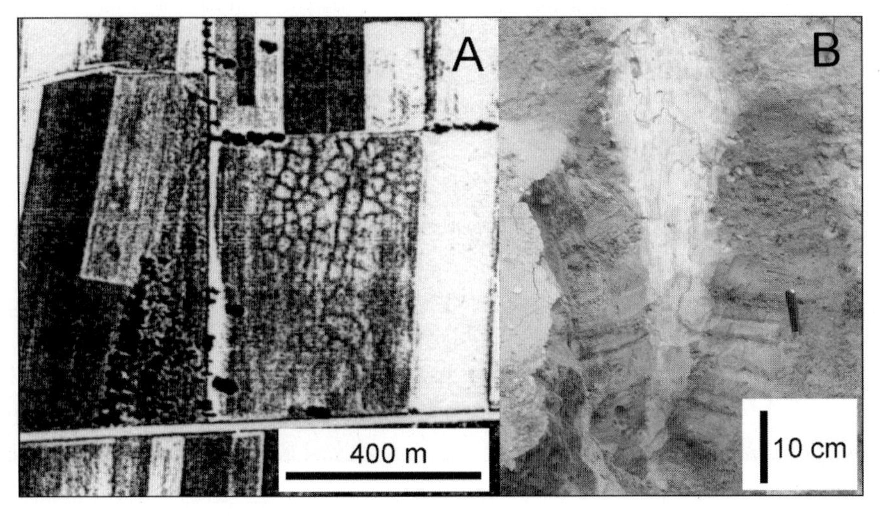

Figure 6.15 A) Aerial photograph showing relict ice-wedge polygons developed in outwash (USDA AX-5R-188, 1956, Dodge County, Wisconsin). B) Photograph of ice-wedge casts located in southern Wisconsin.

1. subglacial sediment transport and extensive subglacial erosion and deformation in the drumlin zone
2. extensive accumulation of debris on the ice surface in a narrow marginal zone 2–20 km wide
3. ice motion dominated by some combination of sliding and subglacial deformation
4. initial advance over a frozen subglacial bed followed by progressively retreating ice during deglaciation.

6.5.5 Landsystem C – Low-Relief Aligned Hummocks and Ice-Thrust Masses

Landsystem C occurs in North and South Dakota, western Minnesota, Iowa and in a small part of eastern Ohio (Fig. 6.6A). We interpret that this landsystem primarily reflects ice-lobe surges and widespread ice stagnation over large areas (Clayton *et al.*, 1985). While it is possible that glacier surges have occurred in areas where landsystems A and B are dominant we do not see evidence for large-scale ice stagnation in zones greater than 20 km wide as we do in landsystem C.

6.5.5.1 Marginal Zone

The marginal zone consists of a steep ramp of till and in places outwash, leading to the highest distal part of a hummocky end moraine (Figs 6.16 and 6.17). The hummocky end moraine

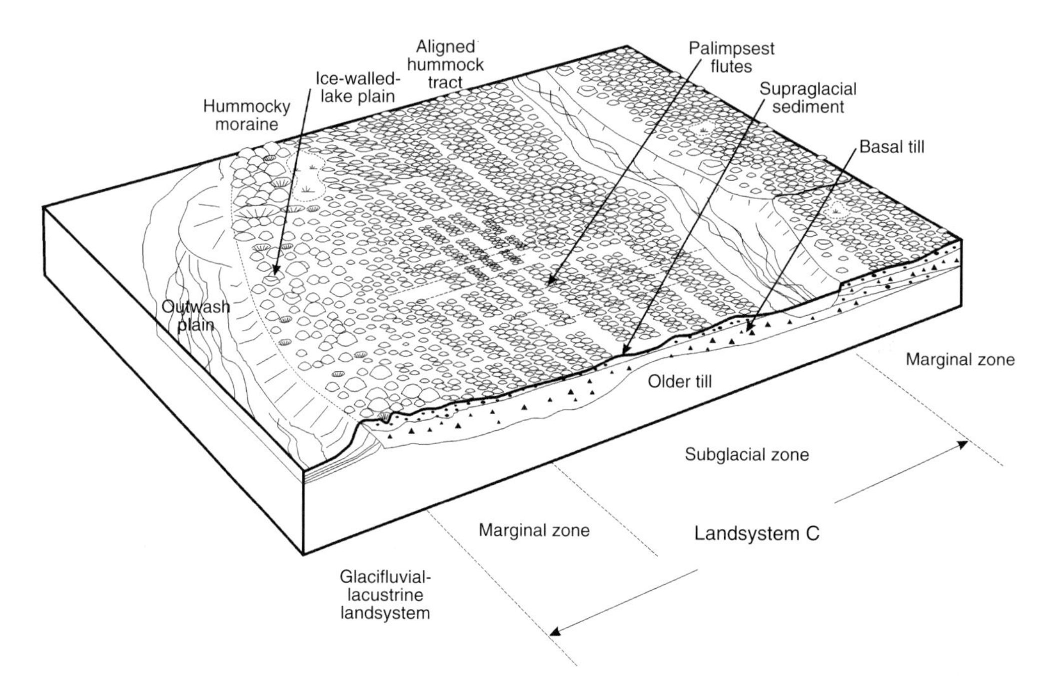

Figure 6.16 Schematic diagram showing landforms and sediments in landsystem C. End moraines are rare and consist of zones of thick supraglacial sediments (>3 m) over basal till. Hummocks are not well aligned in the end-moraine zone or may be superimposed on a broad end moraine. Aligned hummock tracts cover nearly the entire subglacial zone. In a few places, aligned hummocks may be seen to drape over flutes. In the subglacial zone, supraglacial sediments are thinner (<3 m). Older glacial deposits of loess and older tills underlie most of the area and can be quite thick (generally >30 m).

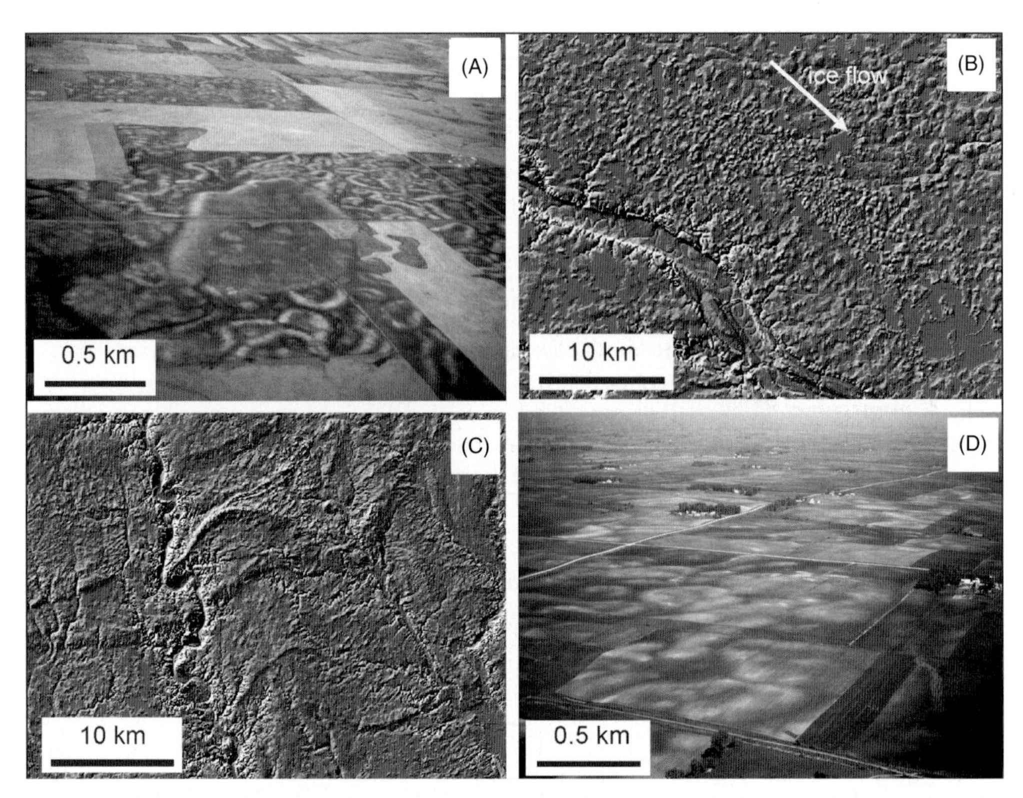

Figure 6.17 A) Oblique aerial photograph of low-relief hummocks in Iowa. (Courtesy of Tim Kemmis). B) DEM of area in southern Minnesota where aligned hummock tracks are superimposed on till plain of basal till. The hummocks are composed of thin supraglacial sediment and basal till. C) A DEM showing rare end-moraine ridges covered by aligned hummock tracts in central Iowa. The curved moraine ridges have been mapped as the Altamont moraine in Iowa. Aerial photographs show that this entire area is covered with aligned hummock tracts. D) Aerial photograph of aligned hummock tracts in central Iowa deposited by the Des Moines lobe. (Courtesy of Tim Kemmis).

slopes gently downward on the proximal side. In some areas hummocky end moraines are not present near the terminus and the subglacial zone of aligned hummocks begins at the terminus. Low-relief hummocky topography creates closed depressions that are generally less than 3 m deep (Fig. 6.18). End moraines are generally 2–20 km wide but are less continuous than in landsystem A and B. Another landform found in the marginal zone of landsystem C in North Dakota is the ice-thrust mass (Moran *et al.*, 1980; Bluemle and Clayton, 1984; Clayton *et al.*, 1985; Aber *et al.*, 1989). This mass appears as a hill of sediment or bedrock up to 1 km across and a few metres to a few tens of metres high. Generally, the thrust mass is paired with a depression of similar size up-glacier. These masses have moved down-ice with very little internal deformation except faulting and drag folding along faults at the base. These 'hill-hole pairs' generally formed within 1 or 2 km of the terminus. Far more common are larger thrust masses of Cenozoic and Cretaceous shale and sandstone that have been moved up to several km in the down-ice direction at the bed of the glacier (Moran *et al.*, 1980).

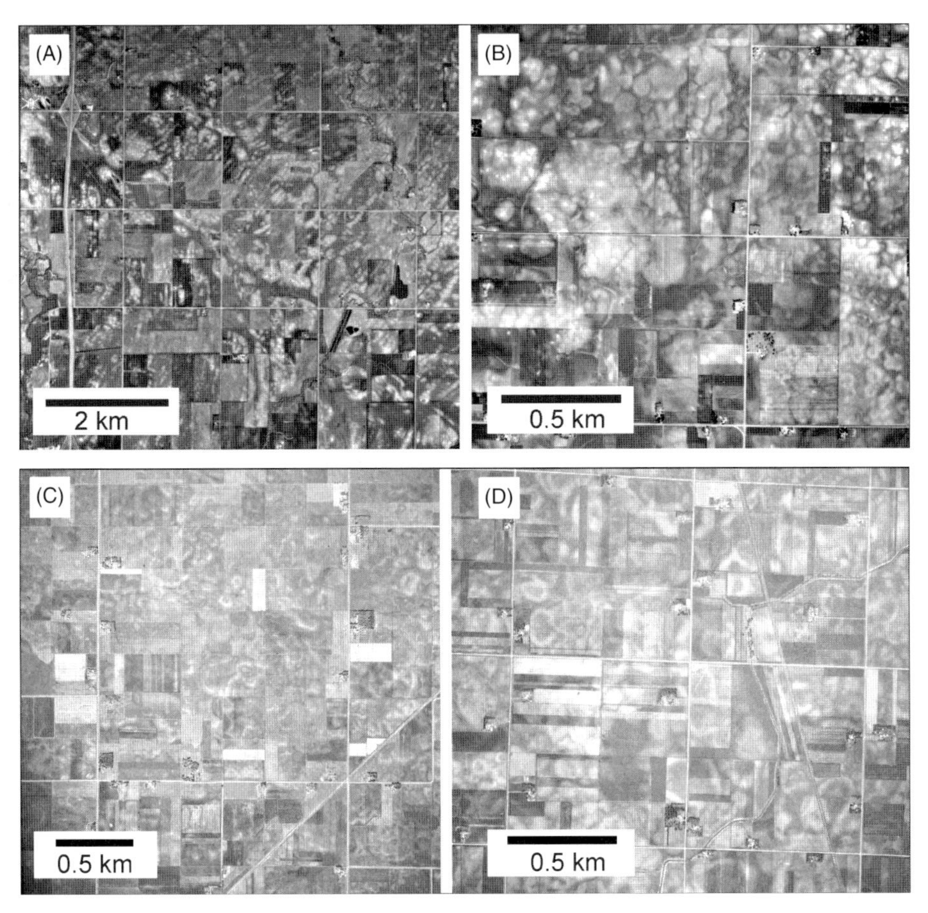

Figure 6.18 A) Aerial photograph of aligned hummock tracts in Story County, Iowa (NAPP photograph 2122-121, 1990). B) Circular aligned hummocks grading into disintegration rings in central Iowa (USDA BZI-2JJ-78, 1968). C) Disintegration rings in central Iowa (USDA photograph BZR-2V-28, 1958). D) Disintegration rings in central Iowa (USDA photo BZF-1JJ-243, 1968). These photographs show that aligned hummocks are transitional features to disintegration rings. Both features are a product of ice stagnation and deposition of supraglacial and englacial debris.

6.5.5.2 Subglacial Zone

A major difference between landsystems C and A occurs behind the moraine. Here instead of flat till plain, the surface in most places has small (1–3 m high) ridges that have been called 'washboard moraines' or 'minor moraines' (Gwynne, 1942, 1951; Elson, 1957; Foster and Palmquist, 1969; Kemmis *et al.*, 1981; Stewart *et al.*, 1988). Colgan (1996) called these features aligned hummocks because they consist of individual hummocks about 10–30 m in diameter, aligned in rows perpendicular to flow (Figs 6.17 and 6.18). We believe that they are not end moraines and did not form at the terminus, but instead formed in a zone of stagnant ice within 40 km of the terminus (Colgan, 1996). These features are transitional to hummocks superimposed on the end moraine of the marginal zone. In many places where no end moraine exists, these features extend to the former terminus of the lobe.

The subglacial zone is up to 40 km wide and consists mostly of aligned hummocks oriented perpendicular to the ice margin (Fig. 6.18). They grade into larger hummocks in the marginal zone. Spacing is about 200 m between groups of aligned hummocks. Detailed mapping of these features in central Iowa indicates that areas of aligned hummocks are grouped into coherent tracts up to 10–40 km behind former ice-margin positions (Colgan, 1996). In Iowa there are at least seven such ice-margin positions and associated aligned hummock tracts (Colgan, 1996). These features merge with disintegration doughnuts and other features that clearly suggest widespread ice-stagnation (Parizek, 1969; Clayton and Moran, 1974). Their close association with circular doughnuts and the transitions commonly seen between aligned hummocks and doughnuts suggest that they are genetically related as ice-stagnation features showing various degrees of structural control by stagnant ice features such as crevasses and ice foliation (Fig. 6.18). In at least a few places in Iowa and in North Dakota aligned hummocks are superimposed on megaflutes assigned by us to landsystem B (Bluemle *et al.*, 1993). Basal till is up to 10 m thick in the marginal zone (Kemmis *et al.*, 1981), and Clark (1991) showed that striated clast pavements are also common in this area at the base of till units.

Megaflutes underlie aligned hummocks in North Dakota and in parts of northern Minnesota. In North Dakota streamlined landforms are long and narrow, commonly about 0.1 km wide and 1–5 km long. Some are up to 27 km long (Bluemle *et al.*, 1993). Megaflutes appear to be erosional or ice-moulded features. They contain till, sand and gravel, lake clay, shale and other sediments that were at the bed at the time of the last glacial advance. Bluemle *et al.* (1993) report evidence of squeezing and flow of material into the megaflute form. This is a good example of how one landsystem can be superimposed over another. In this case, landsystem C has been superimposed over landsystem B.

6.5.6 Interpretation of Landsystem C

Several researchers have examined the origin of aligned hummocks in Iowa. Stewart and others (1988) called them 'corrugated moraines' and suggested they formed by thrusting of basal debris. Kemmis (1992) referred to the depressions between hummocks as 'linked depressions' and concluded that doughnut topography and associated aligned hummocks are mostly englacial sediment that accumulated in crevasses or other openings in stagnant ice. Other observations (Stewart *et al.*, 1988; Patterson, 1997b) suggest that aligned hummocks are subglacial in origin with only a drape of supraglacial sediment over them. Similar to this, Colgan (1996) hypothesized that they formed when basal debris-rich ice was extended and then compressed longitudinally within 40 km of the terminus during a surge event. After this, dead ice melted in this broad zone of compressed ice, aligned hummocks formed where concentrated basal debris-rich layers melted out and were then partially reworked as supraglacial sediment into patterns reflecting ice fabric in the zone of compression. In some areas farther up-ice where longitudinal extension of ice occurred some of the aligned hummocks may have directly formed as basal crevasse fills (Colgan, 1996; Patterson, 1997b). The discussion of the origin of these features has been somewhat confused in the past, probably because some of these features formed near the terminus (the aligned hummock tracks) related to compression of stagnant ice, while other features formed far up-ice as basal crevasse fills. The basal crevasse fills described by Patterson (1997b) in southwestern Minnesota are much more ridge-like and linear than are the aligned hummocks in central Iowa (Kemmis *et al.*, 1981; Kemmis, 1992; Colgan, 1996).

It has been suggested that megaflutes, ice-thrust masses and aligned hummocks are genetically related (Clayton and Moran, 1974; Moran *et al.*, 1980; Mickelson *et al.*, 1983; Bluemle and

Clayton, 1984; Evans *et al.*, 1999b). This could be the case, as these extremely long, streamlined features are very different from the drumlins found in landsystem B. It is clear that the megaflutes are palimpsest features and that aligned hummocks, crevasse fills and eskers are superimposed upon them, and this is what would be expected if supraglacial sediment was draping basal till. Megaflutes may have formed during ice streaming to the ice maximum or during surge events that produced ice-thrust masses, and then longitudinally deformed basal debris formed the overlying aligned hummocks as this material melted out. Presumably where supraglacial sediment was thick enough, megaflutes would be completely buried. In both Iowa and North Dakota, streamlined landforms can be seen to underlie a thinner layer of supraglacial sediment in a few places.

Ice-thrust features in North Dakota overlie sand and gravel aquifers in buried valleys. It has been suggested (Moran *et al.*, 1980; Bluemle and Clayton, 1984) that ice-thrust masses occur because of steep groundwater gradients in the terminal zone caused by the blocking of subglacial drainage by subglacial permafrost at the ice margin. They may also be associated with surges and share a common genesis with similar features in Iceland (Evans *et al.*, 1999b).

Although basal till is common it appears that more supraglacial sediment is present in landsystem C than in either landsystem A or B (Clayton *et al.*, 1985). The aligned hummocks and basal crevasse fills are composed primarily of basal till in their lower parts with varying amounts of supraglacial sediment deposited on them during wastage of stagnant ice (Kemmis *et al.*, 1981; Kemmis, 1992; Colgan, 1996).

We believe that the superposition of landsystem C over megaflutes in this zone suggests that glacier surges or fast flow dominated post-LGM readvances in this part of the ice-sheet margin (Clayton *et al.*, 1985). The timing of the maximum ice extent in this region (at 14,000 ^{14}C years BP) was out of phase with the rest of the southern LIS margin. This suggests that these lobes were advancing when more basal water was available during warmer post-LGM climatic conditions (Kemmis *et al.*, 1981; Clayton *et al.*, 1985). Megaflutes and boulder pavements could have formed during rapid basal sliding during the surge event or as fast steady flow fed by an ice stream up-ice (Patterson, 1998). Ice and basal debris were longitudinally compressed near the terminus perhaps during the propagation of a surge wave through the glacier. Ice thrusting of proglacial and marginal material and compression of basal debris-rich ice resulted in the formation of ice thrust masses, aligned hummocks and widespread ice disintegration features (Colgan, 1996).

In summary, the characteristics of landsystem C, with its aligned hummocks and broad low-relief moraines suggest:

1. predominantly englacial and subglacial transport and ice stagnation over large areas
2. extensive accumulation of debris on the ice surface in a broad zone 10–40 km wide, and
3. ice motion dominated by unsteady surge-type or fast ice flow followed by stagnation of ice over a broad area behind the terminus.

6.5.7 Landsystem D – Bedrock-Dominated Landscapes

This landsystem is common in New England, New York, Pennsylvania and New Jersey. It consists of high-relief bedrock terrain draped by till, with glacifluvial and glacilacustrine sequences filling most valleys. We make the interpretation that most of this region reflects glacial erosion of a pre-existing high-relief bedrock terrain during advance to the last glacial maximum.

Ice-contact and glacifluvial (landsystem E), and glacilacustrine/glacimarine deposition (landsystem F) dominated as a progressively retreating terminus developed local ice-stagnation zones near the margin. After about 13,000 ^{14}C years BP, ice probably become regionally stagnant in northern New Hampshire and Maine as it became isolated from the main body of the dwindling LIS. Features such as cols, ice-marginal channels, eskers and other ice-contact deposits suggest that as the ice wasted, high peaks were exposed as nunataks followed by isolation of ice in major valleys (Goldthwait and Mickelson, 1982). Detailed mapping of striations also suggests that ice flow reversed direction in northern Maine as ice was drawn down in the St Lawrence lowland (Lowell *et al.*, 1986).

Besides the well-formed and mapped end-moraine systems in southern New England that we have classified as landsystem B or E (Oldale and O'Hara, 1984), large end moraines are relatively rare in New England (Koteff and Pessl, 1981). Very small moraines in Connecticut, Maine and Massachusetts are present, but these features are most similar to the small push moraines present in the Great Lakes region (Stone and Peper, 1982; Goldsmith, 1987). Small recessional moraines in Maine occur in association with glacimarine deposits and may be similar in origin to DeGeer moraines (Kaplan, 1999). Most of this area where these moraines occur has been classified as landsystem B or F.

The dominance of glacifluvial deposition in bedrock valleys during deglaciation has been described in detail by Koteff and Pessl (1981), Mulholland (1982), and Stone and Peper (1982). Outwash sequences called 'morphosequences' (Koteff and Pessl, 1981) were created by a progressively retreating ice margin that stagnated near the terminus in zones 2–5 km wide (Mulholland, 1982). Each morphosequence includes glacifluvial and glacilacustrine deposits that are graded to an ice margin position and a local base level established briefly during a progressive retreat. A zone of stagnant ice several km wide formed as ice retreated over high-relief bedrock surfaces. In this zone ice-contact sand and gravel were deposited. Temporary lakes also formed in this bedrock-controlled landscape, and many of the valley sequences include deltaic and varved sediment (Ashley, 1975). Bedrock ridges oriented parallel to the ice margin were particularly effective at detaching zones of stagnant ice from active ice as the margin retreated northward over the terrain.

6.6 DISCUSSION

We hypothesize that the types of landforms and sediments within a particular region are the result of how the ice sheet responded to the fundamental influences of climate, bed geology and topography. These influences were filtered in a non-linear manner by the ice sheet. The combination of these factors produced different dynamic behaviours of the ice-sheet margin and different depositional environments in various regions of northern USA. Below we discuss the influence of each factor on glacier dynamics and the resulting landsystems in each region.

6.6.1 The Role of Climate in the Genesis of Landform-Sediment Landsystems

Climate is a first-order control on glacier behaviour. Besides determining mass balance and the extent of ice, climate also influences the temperature of ice as well as its bed near the margin. It also influences the amount and nature of water movement. During the LGM the climate along the southern LIS margin varied spatially (Kutzbach, 1987). In the western and northern Great

Lakes regions, and perhaps in northern New England, permafrost formed before, during, and even after the LGM (Péwé, 1983). It is likely that the ice-sheet margin advanced over this permafrost many times. It may have taken several thousand years for this overridden permafrost to melt, depending on the initial thickness of the permafrost and the thickness and temperature of the overlying ice (Cutler *et al.*, 2000). In the southern Great Lakes region of Illinois, Indiana and Ohio, permafrost was probably discontinuous (Johnson, 1990). Abundant wood in tills in this region (Ekberg *et al.*, 1993; Hansel and Johnson, 1999) suggests that ice advanced into a warmer region than that in Wisconsin, Michigan and Minnesota, where till contains no wood and sub-till organic material indicates tundra conditions. In New York and New England ice may have advanced over permafrost as relict ice-wedge casts and ice-wedge polygons are reported there also (Péwé, 1983). In southern New England the proximity to the Atlantic Ocean may have warmed this part of the margin compared with areas to the north or west (Gustavson and Boothroyd, 1987).

In Wisconsin and Minnesota this permafrost lasted until about 13,000 ^{14}C years BP in the south, disappearing northward about 3,000 years later (Clayton *et al.*, 2001). Not only was permafrost overridden, it must have reformed at times on the deglaciated surface, between 17,000 and 13,000 ^{14}C years BP. Discontinuous permafrost may have also been present in some areas of landsystem A (Johnson, 1990), but we believe this permafrost was not sufficiently extensive to have had the same impact as it did in the area of landsystem B. Throughout most of landsystem A ice advanced into spruce forest.

The marked difference between landsystem B and landsystems A and C is the presence of drumlins, tunnel channels and high-relief hummocky end moraines. We believe that these features are consistent with an interpretation that ice advanced onto permafrost and eroded pre-existing landforms and sediments. We suggest that when ice was at its maximum position, a zone of subglacial permafrost up to 100 km wide was present. Behind this zone, a zone of partially frozen bed may have been present back to the up-ice edge of the drumlin zone. Subglacial water was produced at the bed in this zone and probably accumulated behind the wedge of frozen ground near the terminus because insufficient water was able to drain through the groundwater drainage system. The trapped water was eventually released through tunnel channels that intersected the end moraine (Attig *et al.*, 1989; Clayton *et al.*, 1999; Cutler *et al.*, 2002). We attribute the mobility and deformation of sediments in the drumlin zone in part to the likely high pore pressure and the high water content of the subglacial sediments upstream from the frozen margin.

In the southern Great Lakes region, where landsystem A is dominant, we suggest that the glacier bed was thawed and that basal motion, through some combination of glacier sliding, ploughing and bed deformation, was the primary result of ice motion. We suggest that little erosion was accomplished in this region. Mickelson and others (1983) report radiocarbon dates on wood between 23,000 and 14,000 ^{14}C years BP in Illinois and Indiana. Much of the wood is spruce and this suggests that ice advanced into a spruce forest or spruce parkland throughout the late Wisconsin, with the exception of the period between 17,500 and 16,000 ^{14}C years BP, when there are no wood dates from Illinois. Permafrost features are not as abundant as they are to the north, although Johnson (1990) reports ice-wedge casts and other permafrost features on late Wisconsin deposits that may have formed during this interval after ice advance to the maximum position. We conclude that the absence of drumlins in this area is because of the prevailing thawed-bed conditions and not because of a difference in topography or till texture. Drumlins are present on Illinoian surfaces just outside the late Wisconsin margin in Illinois in till of similar texture to that of late Wisconsin till plains (Lineback *et al.*, 1983), indicating that texture was not a significant influence on the presence or absence of drumlins.

6.6.2 The Role of Bed Geology and Topography in the Genesis of Landsystems

In addition to climatic controls, the behaviour of the ice sheet was influenced by bed geology and topography. The distribution of bedrock lithologies was important to ice dynamics and hence landsystem distribution for several reasons. First, differences in the lithology, and therefore rate of bedrock erosion, have influenced the regional topography (Fig. 6.19) and hence regional patterns of ice flow. The lobate nature of the southern LIS was determined by bedrock topography, with lobes channelled down each of the basins in the Great Lakes region. For example, these and other major basins must have been present prior to the last glaciation, and were deepened by an unknown amount during the late Wisconsin. The style of ice-marginal behaviour differed between areas where the ice was advancing up a regional slope and those where it advanced down a regional slope. In addition, large lake basins significantly altered the mass balance and ice-surface profiles (Cutler *et al.*, 2001). In the high-relief areas of central and northern New England, the style of deglaciation and resulting deposits were greatly controlled by topography (Koteff and Pessl, 1981; Mulholland, 1982; Stone and Peper, 1982; Goldthwait and Mickelson, 1982).

Second, the rate of erosion of bed materials also influenced the amount of debris being carried by the ice, and topography determined if that material could be transported away from the terminus. Ice in northern Wisconsin and in southern New England retreated progressively up a surface that was gently sloping away from the ice margin. This produced detachment of stagnant ice from active ice in zones up to 20 km wide (Koteff and Pessl, 1981; Mulholland, 1982; Stone and Peper, 1982; Gustavson and Boothroyd, 1987; Ham, 1994; Johnson *et al.*, 1995). As these stagnant-ice zones melted, they produced large hummocky moraines with ice-walled lake plains (called 'high kames' in New England). In the Dakotas, Minnesota, Wisconsin, New York and northern New England, ice retreated into lake basins and lowlands. This produced ideal conditions for the formation of proglacial lakes that trapped fine-grained sediments. Subsequent readvances after the LGM incorporated large amounts of these glacilacustrine silts and clays (Acomb *et al.*, 1982; Clark, 1994b) and deposited these as fine-grained tills (Fig. 6.19).

Lithology also influenced the nature of subglacial fluvial deposits. For example, the texture of glacier deposits controls the present appearance of glacial landforms (Fig. 6.19). High relief and steep slopes tend to occur where till is sandy. Silt- and clay-rich tills tend to have low-relief landforms because of postglacial weathering and erosion (Clayton and Moran, 1974). Finally, the nature and abundance of glacifluvial deposits must in part be a function of the availability of coarse material. Eskers and outwash fans cannot occur unless there is sufficient coarse sediment to be deposited.

Finally, the hydraulic conductivity of the glacier bed controls subglacial drainage and therefore water pressure and basal motion (Arnold and Sharp, 1992; Clark and Walder, 1994). The effective stress at the bed is in part controlled by the ability of water to drain towards the ice margin. If flow is impeded, effective pressure declines and basal motion may be enhanced. Consequently, lithologically dependent hydraulic properties of bedrock and overlying sediments

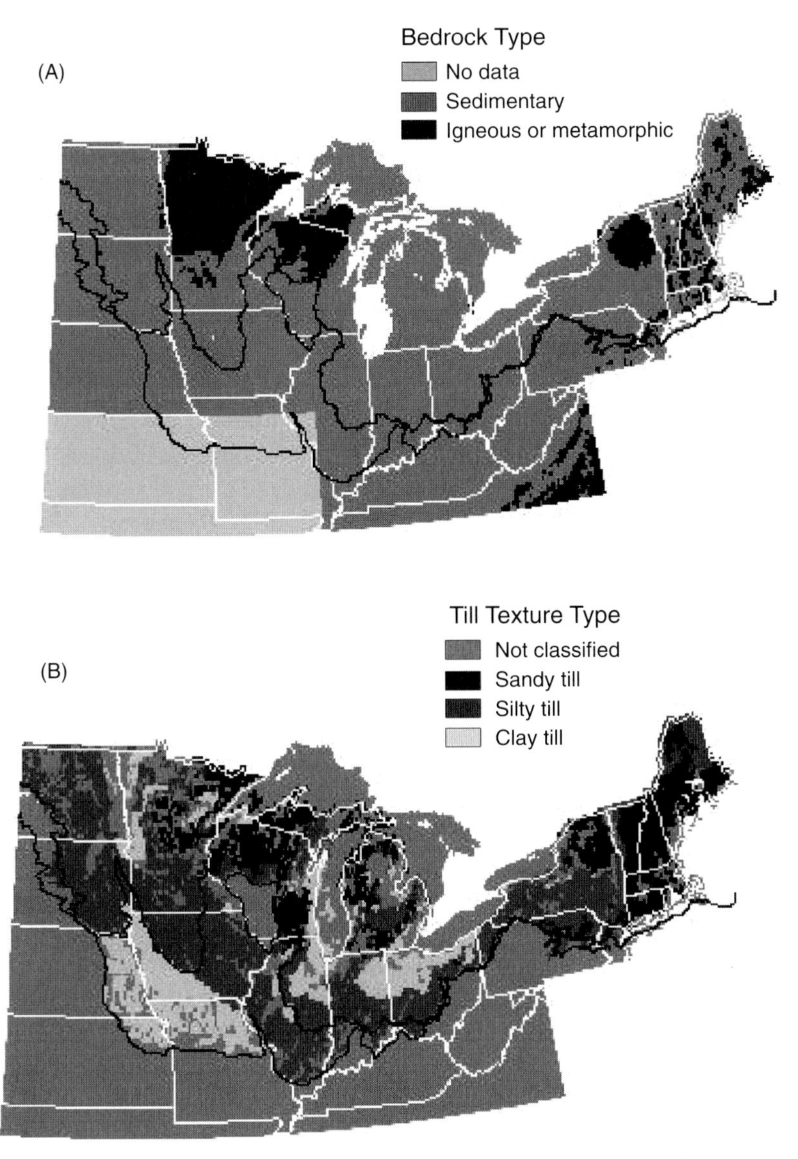

Figure 6.19 A) Map showing bedrock types in the northern USA (map derived from the digital version of the geologic map of the US, Schruben *et al.*, 1999). B) Map showing the dominant texture of till matrix in the northern USA (map derived from data of Soller and Packard, 1998).

exert a direct influence on spatial variations in ice dynamics. The distribution of landsystems found around the southern LIS reflects this influence. In the western region and the southern Great Lakes region where landsystems A and C dominate, the bed is composed of thick silty or clayey tills derived primarily from Palaeozoic sedimentary rocks. This bed would have been soft, smooth and easily deformed, with low hydraulic conductivity. Low basal shear stress would result in low-profile, unstable glacier lobes. Enhanced subglacial sliding, bed deformation or surging behaviour would be associated with this bed type. In the northern Great Lakes and New England where landsystem B dominates, the bed is composed of igneous or metamorphic rocks or sandstone. This bed would have been hard, rough, thinly covered by sediment, and had a higher hydraulic conductivity compared with soft-bedded areas. This may have caused ice lobes to be thicker and more stable than those advancing over a soft bed.

6.6.3 The Role of Glacier Dynamics in Influencing the Distribution of Landsystems

A final factor in producing various landsystems is the dynamic behaviour of the ice sheet during advance to its maximum extent, and during deglaciation (Fig. 6.20). Dynamic behaviour directly results from the climate and bed geology discussed above. The ice margin appears to have retreated progressively at rates of between 50 and 500 m/year along much of the southern margin of the LIS in the Great Lakes and New England regions (Andrews, 1973; Mulholland, 1982; Colgan, 1996; Ham and Attig, 2001). Evidence for this includes radiocarbon chronologies, small push moraines, numerous well-defined retreat moraines, and extensive and well-dated varve deposits in proglacial lakes (Ashley, 1975; Johnson *et al.*, 1999). In the western prairie region it appears that ice surged into lowlands several times and then experienced regional stagnation (Clayton *et al.*, 1985). Retreat rates in this region were higher than 700 m/year and could have been as high as 2000 m/year (Fig. 6.20). As these rates are much higher than advance-retreat rates of normal glaciers it has been suggested that they reflect regional stagnation following each surge of a lobe in the areas west of the Great Lakes (Clayton *et al.*, 1985).

Lobes in the western prairie region also advanced to their maximum advance position out of phase with the rest of the ice sheet margin (Fig. 6.20) (Clayton and Moran, 1982; Mickelson *et al.*, 1983; Clayton *et al.*, 1985). The Des Moines lobe reached its maximum at about 14,000 ^{14}C years BP when the rest of the margin was rapidly wasting (Kemmis *et al.*, 1981; Hallberg and Kemmis, 1986). We believe that this is an indication that these lobes surged. The resulting landforms and sediments (landsystem C), consisting of widespread ice-thrust and ice-stagnation features, are compatible with surging as a major process in this region after the LGM.

Reconstructed ice-surface profiles also provide information about ice-lobe behaviour that ties in with our interpretations of landform-sediment landsystems. Most of the major lobes have been reconstructed from ice-marginal features and moraine elevations for their LGM positions and for some deglaciation phases (Mathews, 1974; Beget, 1986; Ridky and Bindschadler, 1990; Clark, 1992; Colgan, 1996; Colgan and Mickelson, 1997; Socha *et al.*, 1999). These profiles show that glacier lobes that created landsystem B in the northern Great Lakes and New England regions had comparatively steep ice-surface profiles. Lobes that created landsystem A in the southern Great Lakes region had lower ice-surface profiles. Finally, the lowest ice-surface profiles have been reconstructed for lobes in the western region (landsystem C), and for lobes that readvanced out of lake basins during deglaciation.

These ice-surface profiles reflect a combination of lithologic and climatic influences, as well as, in the case of the lower-profile lobes, internal ice-dynamic behaviour related to surges. In contrast with steep ice-surface profiles that suggest high basal shear stress and progressively (steadily) advancing and retreating lobes, the extremely low profiles of the western region suggest that ice was very thin and stagnant following surges (Clayton *et al.*, 1985).

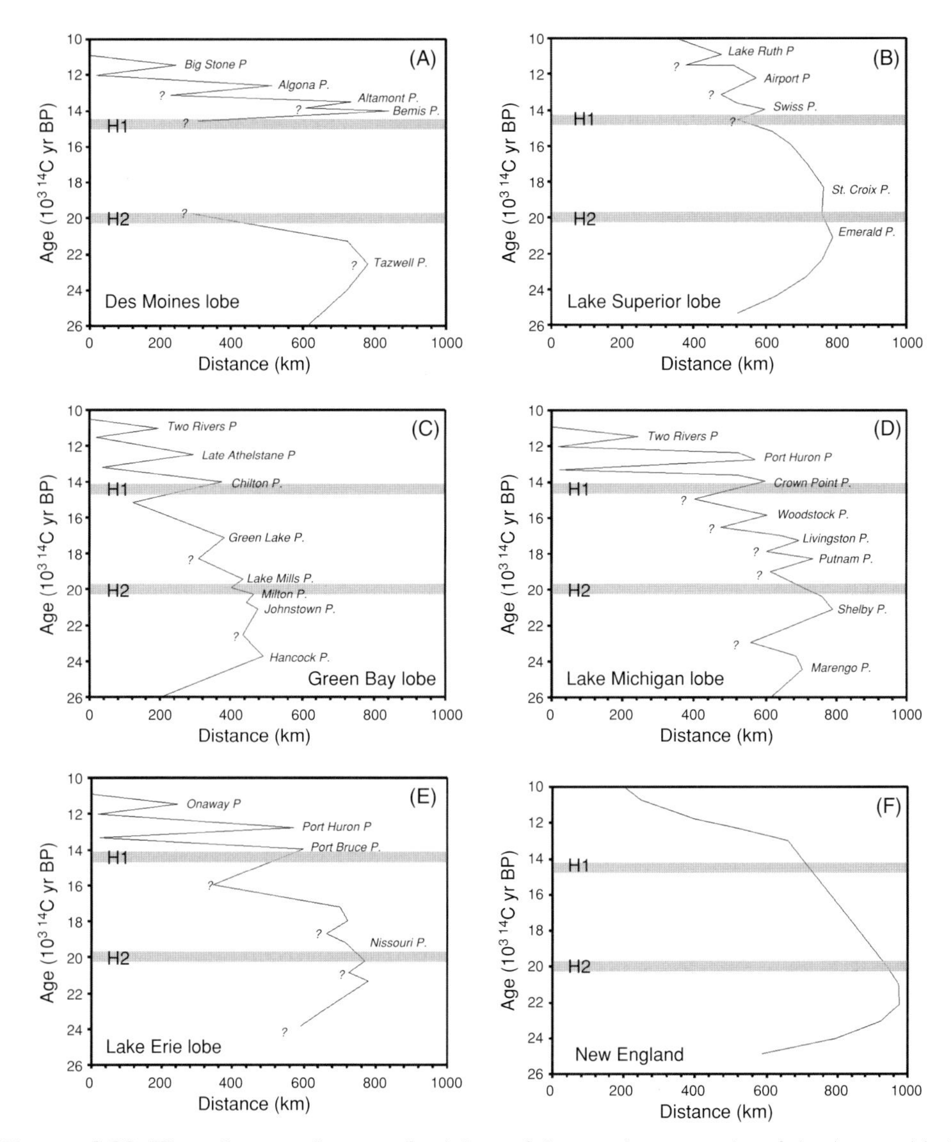

Figure 6.20 Time-distance diagrams for lobes of the southern margin of the Laurentide Ice Sheet. A) Des Moines lobe in the western region. A major advance occurred in phase with the rest of the margin at about 21,000 ^{14}C years BP, but this was not the maximum advance. The maximum advance occurred at about 14,000 ^{14}C years BP when the rest of the margin had retreated 100–300 km behind the Last Glacial Maximum margin (Kemmis *et al.*, 1981; Hallberg and Kemmis, 1986). B) Lake Superior lobe (Attig *et al.*, 1985). C) Green Bay lobe (Colgan, 1996). D) Lake Michigan lobe (Hansel and Johnson, 1999). E) Lake Erie lobe (Clark, 1994). F) New England (Clark, 1994). Grey shading shows Heinrich events at about 14,500 (H1) and 20,500 (H2) ^{14}C years BP.

6.7 CONCLUSIONS

The thrust of this paper is to point out that there are striking differences in glacial landforms and sediments along the southern margin of the LIS. We believe that these differences are significant from a palaeo-glaciological point of view, and we encourage further research on the genesis of glacial landsystems and their significance to reconstructions of former ice sheets and ice lobes.

It is clear that climate and bed geology were important controls on the landforms that developed. Another important control was temperature at the base of the ice and in particular whether the bed was frozen or unfrozen. We postulate that there was an area (landsystem A) along the southernmost margin of the ice sheet where wet-based ice advanced into a spruce forest, and that permafrost was probably discontinuous and formed later than the maximum advance and then quickly disappeared. This area was dominated by basal melting, and broad end moraines were developed almost entirely of basal till. Behind this, subglacial sliding, ploughing and deformation produced a flat till plain with only small-scale flutes on the surface. Subglacial sediment was brought to the ice surface in interlobate areas and a few locations where bed slope was upward toward the ice margin, particularly along deep basin margins. This was particularly true during the short-lived readvances out of Great Lakes basins when large volumes of lake sediment were moved short distances into end moraines.

Further north in landsystem B, we believe ice advanced over permafrost that was tens if not hundreds of metres thick and extended laterally tens of kilometres under the advancing ice margin. The thickness of the permafrost probably increased to the north. This permafrost wedge must have had a major influence on subglacial drainage, on subglacial pore pressures, and on subglacial processes in general. In the marginal zone in the southern part of landsystem B, where the subglacial permafrost zone was fairly narrow (tens of kilometres), significant upward thrusting and stacking of material took place. Further north, where the frozen zone was wider, an extensive area of high-relief hummocky topography now marks the location of that frozen bed. Tunnel channels, which probably formed by drainage of subglacial water dammed behind this frozen wedge, are common in landsystem B and northward, but are absent in landsystem A. Likewise thrust masses and drumlins also appear to require the presence of a permafrost zone near the ice margin for their formation. There are thousands of drumlins in landsystem B, but drumlins are absent in landsystem A and commonly palimpsest in landsystem C. Most eskers in the northern landsystems appear to be later features, formed when climate warmed and water was present at the bed in marginal areas.

To the west, where regional stagnation of lobe margins followed surges, vast areas of ice-stagnation topography were produced (landsystem C). Low topographic relief, the presence of large shallow lake basins, and the fine-grained nature of tills in the area may have predisposed these lobes to unstable dynamic behaviour, particularly after the LGM when the climate warmed and subglacial water was available. Similar features and behaviour seem also to have occurred around all the Great Lakes basins during deglaciation.

CHAPTER

ICE-MARGINAL TERRESTRIAL LANDSYSTEMS: NORTHERN LAURENTIDE AND INNUITIAN ICE SHEET MARGINS

Arthur S. Dyke and David J.A. Evans

7.1 INTRODUCTION

The 2000 km stretch of glaciated terrain between the north coast of Ellesmere Island and the centre of ice recession in Keewatin can be divided into several large concentric zones, each with characteristic ice-marginal and subglacial landform assemblages determined by basal ice thermal conditions. The entire region lies well north of the southern limit of continuous permafrost, which is several hundreds to more than a thousand metres thick and represents the normal interglacial ground temperature for this region. Permafrost either persisted through the last glaciation under continuously cold-based ice patches or, except where submerged, it reformed upon deglaciation, sometimes early in the deglacial cycle under thinning marginal ice. In this chapter, we describe the glacial landsystems that demarcate the former margins of the Laurentide and Innuitian ice sheets in the area located between northern Baffin Island and the Yukon coastal plain, including the Mackenzie Delta region, and between the islands of the Canadian Arctic Archipelago and central Keewatin (Fig. 7.1). We also discuss vying interpretive models of landform genesis.

7.2 THE REGION AND ITS GENERAL GLACIAL LANDSCAPES

The region is physiographically varied, ranging from the upland and lowland tundra on the Barren Grounds of the Arctic mainland west of Hudson Bay to the channels and islands of the Canadian Arctic Archipelago. Most of the region comprises undulatory plateaux and intervening wide valleys or lowlands, but extensively glaciated fretted mountains extend along the eastern margin of the region from Axel Heiberg and Ellesmere islands south to the Torngat Mountains of Labrador (Bird, 1967; Bostock, 1970; Dyke and Dredge, 1989; Hodgson, 1989; Fig. 1.7).

The larger of the alpine glacier complexes in these mountains persisted throughout the Holocene from Pleistocene precursors, whereas the smaller systems reformed during the Neoglacial. High parts of the plateau adjacent to the mountain rim also support ice caps, the largest of these, the

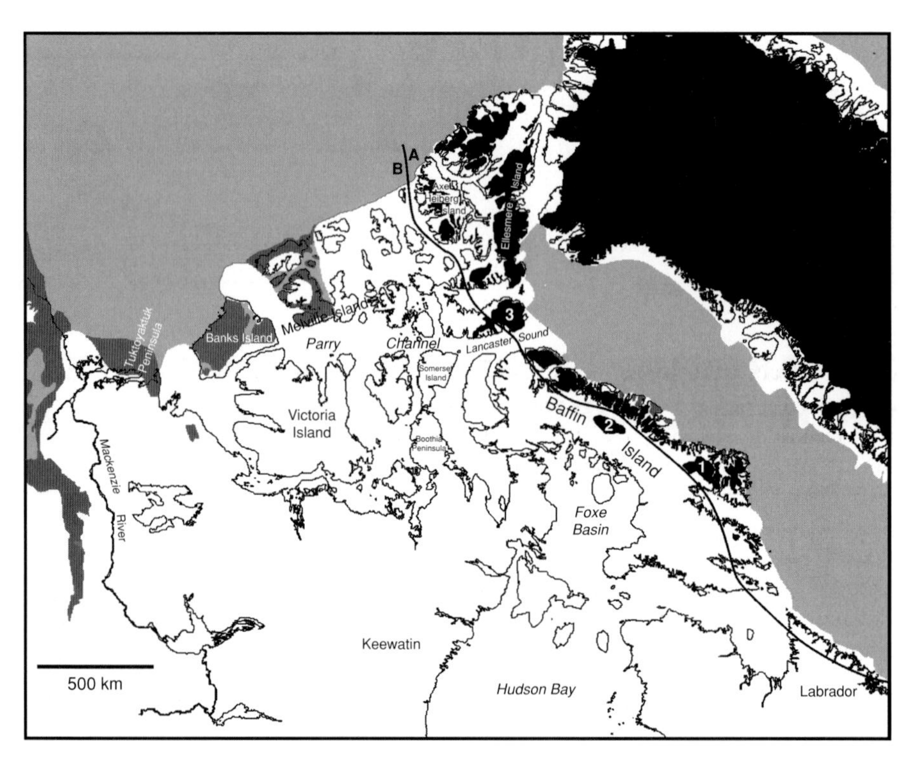

Figure 7.1 Map of northern Canada showing physiography, ice sheet limits and major place names. Region A is the mountainous eastern rim of Arctic Canada, heavily glaciated today (glaciers in black). Region B consists of medium-to-low elevation plateaux and lowlands. 1 = Penny Ice Cap, 2 = Barnes Ice Cap, 3 = Devon Ice Cap. (After Dyke *et al.*, 2002).

Barnes, Penny and Devon ice caps, also having persisted throughout the Holocene. The Barnes Ice Cap (*c.* 6000 km^2 and 600 m thick) on Baffin Island is a remnant of the Laurentide Ice Sheet, consisting in part of Pleistocene ice, and has fluctuated in size during the Holocene (Ives and Andrews, 1963; Dyke, 1974; Dyke and Hooper, 2001). The similarly sized Penny Ice Cap is also a remnant of the Laurentide Ice Sheet with Pleistocene ice at its base (Fisher *et al.*, 1998) but it probably functioned under its own ice divide even at the last glacial maximum (Dyke, 1979; Dyke *et al.*, 1982). The plateau-based Devon Ice Cap and the large alpine ice complexes on Ellesmere and Axel Heiberg islands are remnants of the Innuitian Ice Sheet and have re-expanded substantially in the late Holocene (Blake, 1981, 1989; Koerner, 1989). Thus ice-marginal features in this region range from those currently forming, through a multitude of Neoglacial forms, to those that formed close to the last glacial maximum.

The islands of the Canadian Arctic Archipelago are separated by 50–100 km wide channels and dissected by marine re-entrants ranging from large fjords to drowned lowlands. The striking continuity of some pre-Quaternary erosion surfaces from island to island has prompted suggestions that they developed on a once contiguous landscape that was fractured by regional faulting (Bird, 1967; Kerr, 1980). The extent of glacial dissection of the archipelago remains an open question (Dyke *et al.*, 1992), but it is inappropriate at present to apply a dominantly tectonic, fluvial, or glacial interpretation to physiographic evolution of the region.

During the Late Wisconsinan, the northern margin of the Laurentide Ice Sheet coalesced with local ice caps on northern Baffin Island and Somerset Island (Dyke, 1993; Dyke and Hooper, 2001) and with the Innuitian Ice Sheet in Parry Channel (Blake, 1970; Dyke, 1999). The northwest Laurentide ice margin abutted Banks and Melville islands where it was partly afloat as extensive areas of ice shelf (Vincent, 1982; Hodgson and Vincent, 1984; Hodgson *et al.*, 1984; Dyke, 1987; Dyke and Prest, 1987; Hodgson, 1994). In the far west the ice sheet terminated on the lowlands adjacent to the Mackenzie River, its margins occupying the Tuktoyaktuk and Yukon coastal plains, Mackenzie Bay, and the then dry Beaufort continental shelf (Dyke and Prest, 1987; Dyke *et al.*, 2002; Rampton, 1988; Vincent, 1989; Fig. 7.1).

The impact of Laurentide glaciation is manifest by up to 100 m thick drift and landforms such as flutings, drumlins, eskers and a variety of transverse ridges and belts of hummocky to ridged moraine, recording a strong radial flow of ice from former ice sheet dispersal centres in Keewatin (Sharpe, 1988; Dyke and Dredge, 1989) and Foxe Basin (Blackadar, 1958; Ives and Andrews, 1963, 1989; Andrews and Sim, 1964; Dredge 1995; 2000). Vigorous ice-stream flow is recorded in sets of subglacial streamlined landforms, which in places crosscut each other, and thereby document highly mobile dispersal centres and ice divides throughout the last glacial cycle (Dyke and Morris, 1988; Dyke and Dredge, 1989; Dyke *et al.*, 1992; Hodgson, 1994; Clark and Stokes, 2001). The cross-cutting flow sets indicate that ice streams continued to actively channel glacier ice to the ice sheet margin during its recession, thereby providing considerable volumes of sediment for deposition in glacimarine basins and for moraine formation.

In contrast to the Laurentide Ice Sheet, the flow imprint of the Innuitian Ice Sheet and of plateau ice caps is much less distinct, a fact that encouraged the protracted debate about the very existence of this ice sheet (Dyke, 1999). Innuitian ice streams and associated debris dispersal trains are now being recognized as having flowed along the marine channels (Blake, 1992; Bednarski, 1998; Dyke, 1999; England, 1999; Ó Cofaigh *et al.*, 2000). The large fjord systems (e.g. Nansen Sound, Eureka Sound, Greely Fjord, Baumann Fjord) extending back to the dispersal centres of the Innuitian Ice Sheet were probably the chief conduits for ice evacuation from its eastern, alpine sector. Early bathymetric work identified a large, hummocky drift belt on the polar continental margin that may represent the Late Wisconsinan Innuitian terminal moraine (Pelletier, 1966). In contrast, many upland plateaux appear to have escaped glacial scouring altogether, attesting to the persistence of cold-based ice beneath local ice divides (Dyke, 1993, 1999).

7.3 GLACIAL LANDSCAPE ZONES

Glaciated landscapes are systematically ordered in a north to south progression, from simple to complex in terms of landscape alteration. The 'complex' southern landscape is the one that will be most familiar to geologists working south of the permafrost limit. In the far north, lateral meltwater channels cut into terrain that either lacks other forms of glacial modification or is weakly glacially scoured and bears a thin till cover with sparse, faint flutings. Moraines are rare but include small end moraines (e.g. Craig Lake moraine; Smith, 1999), minor lateral moraines along fjords, and glacially tectonized bedrock masses (Dyke *et al.*, 1992; Dyke, 1999; 2000; Fig. 7.2). Ice-contact deltas are limited mainly to fjord heads. Along with prodelta muds and more complex accumulations of glacimarine sediment (morainal banks; Evans 1990a, b; Ó Cofaigh *et al.*, 1999),

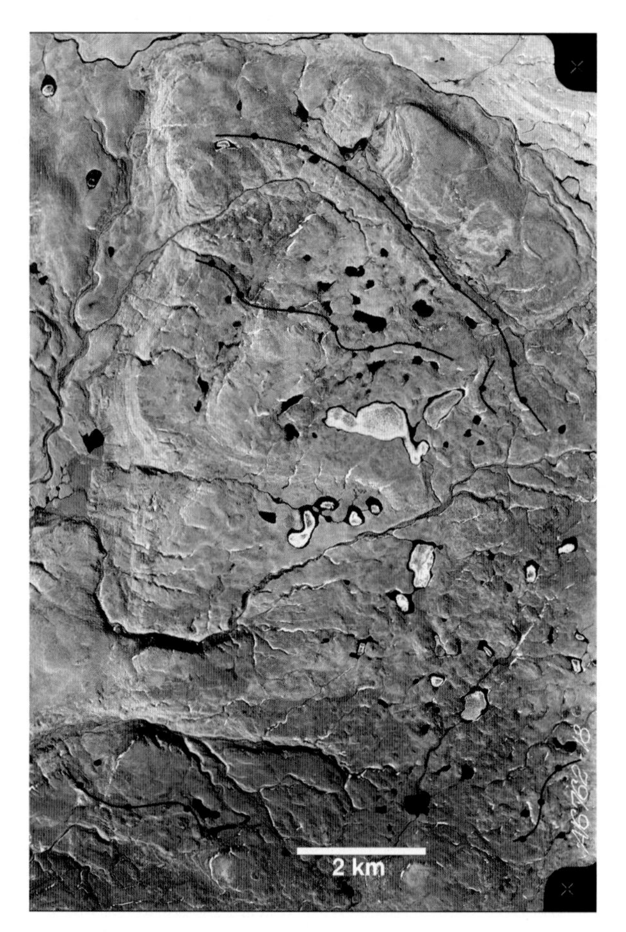

Figure 7.2 Glacially tectonized bedrock (ice-thrust moraine) on Devon Island. Morainal ridges are indicated, but the ice-thrust material extends throughout the rough terrain studded with small lakes. This material is exceedingly coarse, blocks being typically metres across. (NAPL A16762-18.)

they constitute the only thick glacigenic sediments in this vast region. This terrain was covered during the last glaciation by a coalescent complex of alpine glaciers and plateau ice caps, collectively referred to as the Innuitian Ice Sheet (Blake, 1970). South of Parry Channel, alpine and plateau glaciers were coalescent with the northeast margin of the Laurentide Ice Sheet but remained dynamically distinct from it (Fig. 7.1).

Within the limits of the Late Wisconsinan Laurentide Ice Sheet, there are terrains of limited extent that are similar to those described above (e.g. south-central Melville Peninsula; Sim, 1960; Dredge, 2000, 2002), north-central Boothia Peninsula (Dyke, 1984), northeast Prince of Wales Island (Dyke *et al.*, 1992) and parts of northwest Victoria Island (D.A. Hodgson, pers. comm.). Elsewhere, northern Laurentide terrain exhibits a further three-part, north–south (or outer–inner) zonation, which may be briefly described as follows (Fig. 7.3):

1. The outer Laurentide zone contains belts of ridged and hummocky moraine (Fig. 7.4) fronting well-fluted or drumlinized till. The streamlined till bordering the morainal belts is

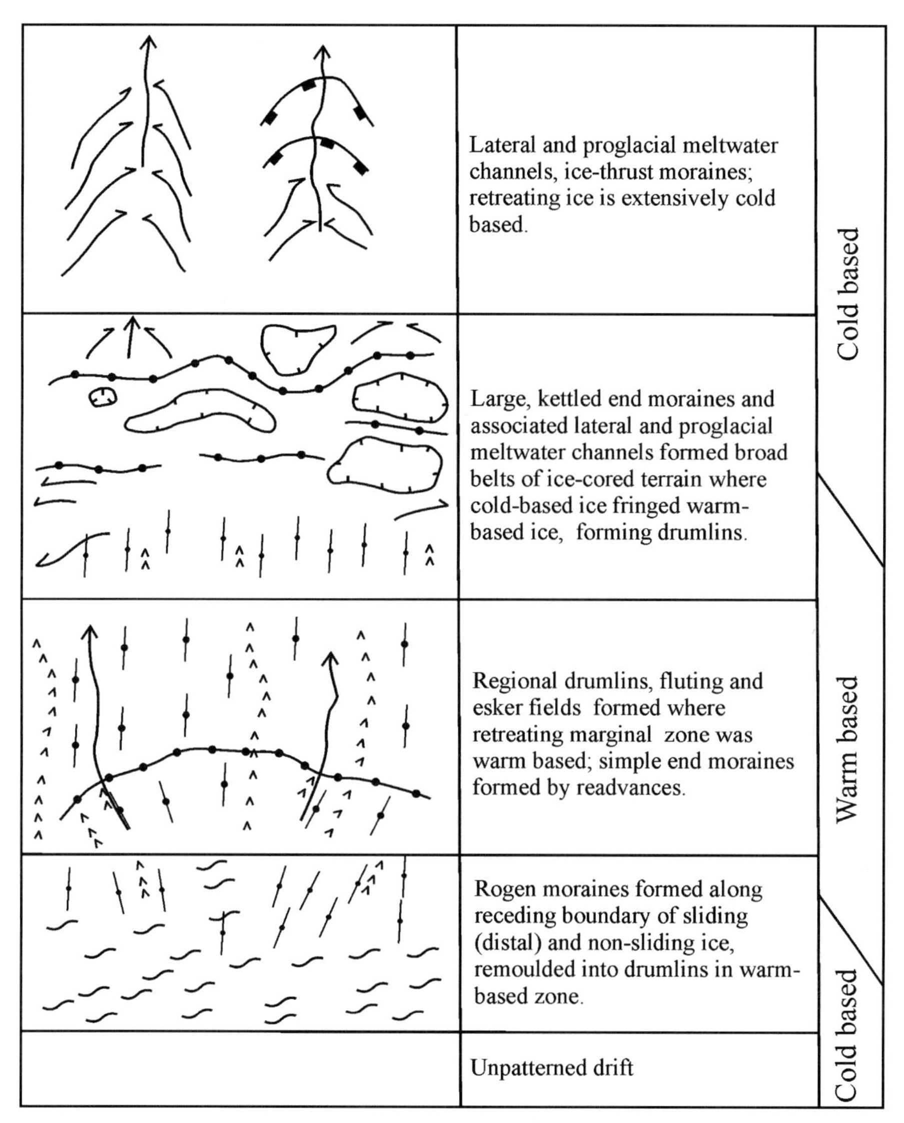

Figure 7.3 Schematic sketch of major glacial landscape zones described in the text. Sinuous arrows are proglacial meltwater channels; arrows with single barbs are lateral meltwater channels; lines with rectangular ornaments are ice-thrust moraines; lines with multiple dots are end moraines; closed depressions represent kettles; lines of 'Vs' are eskers; lines with single dots are drumlins and flutings; and attenuated 'Ss' are Rogen moraines.

inset by sparse, minor eskers or subglacial meltwater channels. Lateral meltwater channels tend to dominate patches of terrain where moraines have not formed, but also occur within the moraine belts, where they are cut into moraine ridges, till surfaces or bedrock. The largest Laurentide lateral meltwater channels occur within this zone (e.g. wrapping around the northern and eastern flanks of the Melville Hills; Klassen, 1971; Dyke *et al.*, in press; Fig. 7.5). Inwash kame deltas formed in and filled small ice-marginal lakes (Fig. 7.5B). Proglacial

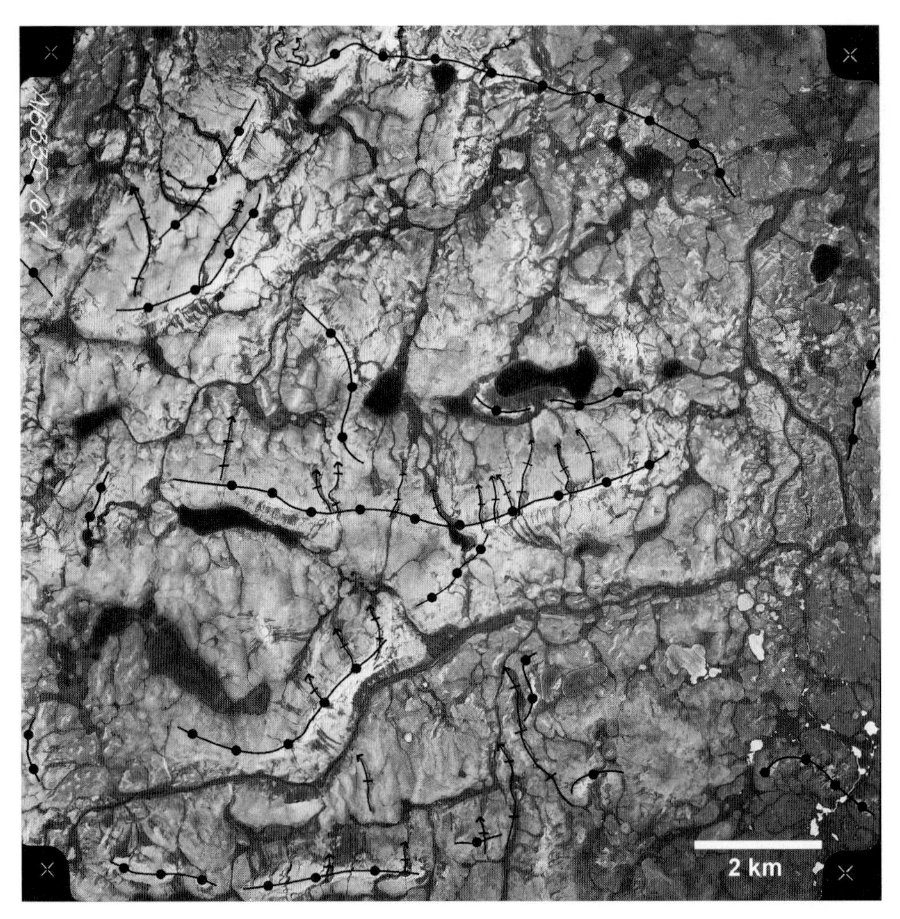

Figure 7.4 Part of ridged and hummocky end-moraine belt on Wollaston Peninsula, Victoria Island. The light-toned areas along the moraine crests (lines with dots) are sand and gravel kames. Note the proglacial meltwater channels (arrows with cross ticks) trending northward from each moraine. (NAPL A16335-167.)

meltwater channels and associated outwash deposits are fairly common. Where the ice front contacted the sea or a glacial lake, moraines typically consist of coarse, crudely bedded glacimarine or glacilacustrine sediment. Ice-contact deltas are locally significant as are DeGeer moraines. A significant ice-marginal assemblage, the Winter Harbour moraine (till) and associated marginal drainage features, formed at the grounded distal edge of a large ice shelf during a readvance (Hodgson and Vincent, 1984).

2. In the middle Laurentide zone, there are some regionally significant end moraines of various compositions. Prominent among these are the Melville moraine (Sim, 1960; Dredge, 1990; Dyke and Prest, 1987), the Chantrey moraine (Dyke, 1984; Dyke and Prest, 1987), and the MacAlpine moraine (Blake, 1963; Falconer *et al.*, 1965; Aylsworth and Shilts, 1989; Dyke and Prest, 1987). They are tens to hundreds of kilometres long with minor gaps and tend to form single ridges or tightly spaced parallel ridge sets (Fig. 7.6). These moraines front strongly developed fluting and drumlin fields that are inset, particularly in Keewatin, by impressively developed esker systems and associated subglacial channels and meltwater scour zones.

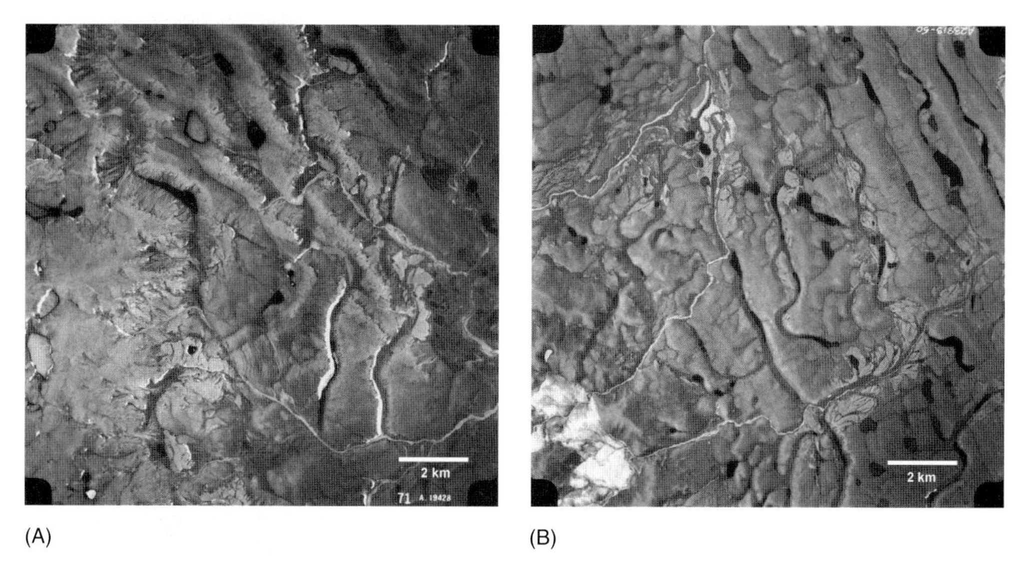

(A) (B)

Figure 7.5 A) Large lateral channels cut into the thick drift of the Bluenose Lake end-moraine complex. The deepest channels are about 100 m deep (northwest corner). Note 'inwash' kame deltas built at the mouths of three channels where small ice-marginal lakes were filled with glacifluvial materials. (NAPL A19428-71). B) Large lateral channels cut into the lower flank of the Bluenose Lake end-moraine complex and adjacent drumlin field. The moraine is to the west, overlain in places by wind-eroded (white) glacial lake silts. Drumlins trend SSE–NNW in the northeast part of the photograph. The lateral channels bend abruptly in places into sub-marginal chutes. Kame deltas were deposited at these bends in small ice-marginal lakes occupying thermally eroded notches at these locations. Descending ice margins and lake levels are well marked by the succession of delta pads along a chute in the southeast quadrant. (NAPL A23813-50).

Collectively these are the giant radiating systems illustrated on the Glacial Map of Canada (Prest *et al.*, 1968) and in later larger scale maps (Aylsworth and Shilts, 1989). Proglacial outwash and associated deltas are important here and DeGeer moraines are more common in formerly submerged areas than in the distal zone. Esker beads (or nodes, ranging up to esker deltas) mark debouchments of channelized subglacial meltwater into standing water. Regionally thick glacimarine deposits attest to abundant sediment delivery to the ice front. Lateral meltwater channels are not known in this zone.

3. The innermost Laurentide landscape is devoid of end moraines and is dominated instead by fields of Rogen moraine, commonly superimposed by flutings and eskers (Lee, 1959; Aylsworth and Shilts, 1989). However, subglacial meltwater features are far less prominent here than in the neighbouring middle zone.

7.4 GENERAL INTERPRETATION

The sequence outlined above reveals the following general trends:

1. a north-to-south decrease in the importance of lateral meltwater erosion

Figure 7.6 Low oblique aerial view of the Melville Moraine. The moraine has a gentle proximal and steep distal flank. It curves to the right into a valley where it has attached to it a gravel delta terrace marking relative sea level at time of formation, visible in the lower right corner with three snow banks on the delta riser. From the delta, the moraine continues left along the bottom edge of the photograph. (GSC photo library 205141F by L.A. Dredge, with permission).

2. a north-to-south increase in the importance of subglacial meltwater action
3. a north-to-south increase in the development of streamlined subglacial bedforms, with drumlin fields first appearing in the zone of minor subglacial meltwater action and significant lateral meltwater action
4. a north-to-south increase in the amount of debris in the ice sheets (ignoring recognizable influences of bedrock resistance)
5. a maximal concentration of large ridged and hummocky moraines in the zone of well-developed streamlined bedforms, minor subglacial meltwater action, and significant lateral meltwater action.

We interpret these trends in terms of the basal-ice thermal conditions in the marginal and near-marginal zones. For the moment, we attempt to explain only the generalized pattern.

1. In regions dominated by lateral and proglacial meltwater channels, where ice-marginal accumulations are either absent or consist of ice-thrust material, the ice-marginal zone – and perhaps most of the retreating glacier – was cold-based. Warm-based ice was limited mainly to deep, fjordic valleys, the only sites of significant ice-marginal deposition (see Dyke, 1993, for detailed arguments on the thermal evolution of ice caps growing on permafrost). Although these ice caps were capable of scouring in their outer zones while at maximal extent, they became cold-based during the early stages of deglaciation, probably as a response to ice thinning, reduced strain heating, and concomitant geothermal heat

dissipation. This favoured the development of lateral meltwater channels in such large numbers that they record recession at a very fine temporal resolution possibly of (annual?) melt-event scale (Fig. 7.7; see also Bednarski, 1998, 2002; Ó Cofaigh, 1998; England *et al.*, 2000).

2. The bulky Laurentide ridged and hummocky moraine belts and associated lateral and proglacial channel systems also formed in cold-based marginal zones above contemporaneous sea level. These cold-based zones were narrow, no more than 10–40 km wide, because the moraine belts are flanked up-ice by well-developed streamlined bedforms that splay toward the moraines, which we conventionally interpret as having formed under warm-based ice. The abundant debris delivered to the ice-marginal zone was generated in the warm-based zone up-ice. This fact strongly differentiates this landsystem from its more northerly neighbour, which is nearly devoid of ice-marginal debris. The trivial development of subglacial meltwater features directly inboard of the moraines indicates that only small amounts of subglacial water were available. Surface meltwater on the warm-based ice behind the cold-based marginal fringe probably was unable to reach the bed due to the low temperatures of the upper ice layers, cold upper ice being an inevitable consequence of the low mean annual air temperatures of the region. Furthermore, thick ice would be expected in the distal warm-based zone because of compressive flow against the cold-based fringe. There are few tunnel channels or eskers extending through these end moraine belts, from which we infer that there were few breaks in the marginal fringe of cold-based ice. However, on a broader scale, wide marine channels interrupt the moraine belts. It is improbable that the cold-based marginal fringe extended across these channels, except where fronted by an ice shelf, and hence probable that subglacial meltwater drained almost entirely along these corridors. The distribution of eskers and subglacial channels (e.g. Sharpe, 1992; St Onge and McMartin, 1995) is in accord with this view.
3. If there was a cold-based, marginal ice fringe in the next inner zone, dominated by

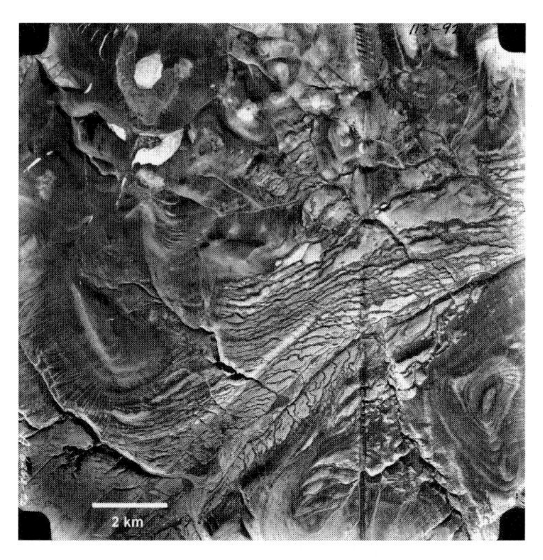

Figure 7.7 Lateral meltwater channels, northern Baffin Island. The channels formed along both sides of the valley that drains from southwest to northeast, but most profusely on the southeast-facing slope, where there are about 30 channels in the centre of the photograph. (NAPL A16263-92.)

drumlins and eskers, it was much narrower and perhaps discontinuous. Here there was abundant basal meltwater under at least the outer tens of kilometres of the ice sheet. With the exception of the end moraines noted above, which probably indicate readvances (Falconer *et al.*, 1965), only minor ice-marginal features occur in this zone. Continuous recession is thus inferred.

4. It has long been a mystery why Rogen moraine in North America is nearly limited to the final ice recession centres of Keewatin and Quebec-Labrador (e.g. Dyke and Prest, 1987; Aylsworth and Shilts, 1989). We suspect that the explanation of this distribution lies in changing basal thermal conditions similar to the proposition of Hattestrand (1997). Dyke *et al.* (1992) observed that Rogen moraines on Prince of Wales Island – one of the rare occurrences beyond the recession centres – are situated along the erosional contact around the head of a drumlin field. Up-flow from there an older drumlin field survived the younger flow phase under cold-based ice, unmodified by the younger flow. The implication of this relationship is that these Rogen moraines formed at the boundary between sliding and non-sliding (cold-based) ice. On a much larger scale, the Rogen moraine field of Quebec-Labrador is identically situated around the head of the Ungava Bay drumlin field (Prest *et al.*, 1968). The contact between the Rogen moraines (older) and the drumlins (younger) is an erosional unconformity commonly referred to as the Labrador Ice Divide (Veillette *et al.*, 1999). If the basal sliding/non-sliding boundary propagates up-ice during deglaciation, Rogen moraines will be preserved only near the final position of the boundary, which will normally be close to the final centres of ice recession. Those that formed earlier were probably reshaped into drumlins, a process that can be inferred from the common superposition of flutings and drumlins on Rogens and from the lateral transition from Rogens to drumlins where these features are arrayed in flow-aligned trains (Dyke and Dredge, 1989). Rogen moraines are bedforms that formed behind the margin. Some may nevertheless be difficult to distinguish from small recessional moraines. A useful distinguishing characteristic is that Rogens are apparently never kettled, whereas end moraines commonly are.

In summary, the salient and broadest characteristics of glacial landscape zonation, including the distribution and types of ice marginal landforms, can be understood in terms of the changing configuration of basal-ice thermal conditions between the last glacial maximum and deglaciation. There appear to have been:

a. extensive areas of receding cold-based ice wherein little debris had accumulated and where lateral meltwater channels formed profusely
b. long, cold-based margins, along which broad morainal belts formed, backed by warmed-based ice, and
c. broad, warm-based marginal zones flanking inboard cold-based central ice zones, along the contact of which Rogen moraines formed.

We interpret (a) above traditionally as a cold-based glacier landsystem. Similar interpretations employing similar principles are applied in Scandinavia (e.g. Kleman *et al.*, 1999). We consider in greater detail the more complex ice-marginal landform assemblages. Assemblages that were necessarily formed in a permafrost environment may provide useful analogues for glacial landforms produced in deglacial permafrost environments further south. However, there is much divergence of interpretation of moraine-like features within the permafrost zone. Therefore, we first outline some general restrictions that must on principle apply to interpreting deglacial features formed in permafrost areas and then discuss the three major conceptual models that are currently applied to interpretations of moraine-like belts. These are:

- pseudo-moraines consisting of thermokarsted regional ground ice sheets formed by ice segregation
- hummocky moraine formed by regional stagnation of a broad ice-marginal zone, and
- moraines with cores of glacier ice formed along active ice margins.

The term 'segregation' here refers to the formation of massive ground ice bodies by the process of soil water migrating to, and freezing along, a freezing front.

7.5 CONSTRAINTS ON INTERPRETATION OF GLACIAL LANDFORMS AND DEPOSITS IN THE CONTINUOUS PERMAFROST ZONE

The thick, continuous permafrost of northern Canada either survived under cold-based Wisconsinan ice, formed under the marginal fringe of the receding ice sheet, formed immediately after release from a warm-based ice margin, or formed after emergence from the sea or after drainage of glacial lakes. In all these cases, because the basal ice and water temperatures were necessarily higher than current mean annual air temperatures, the permafrost would have thickened postglacially towards equilibrium with air temperature. Permafrost aggradation was most rapid at the start and slowed exponentially thereafter (see below). In other words, it is safe to assume that permafrost has never disappeared during postglacial time within the zone of continuous permafrost, except in taliks due to local water bodies.

If we accept these conclusions, certain interpretations commonly applied to landforms and deposits south of the permafrost zone, and sometimes within the permafrost zone, are not tenable within the zone of continuous permafrost. For example, kettles in permafrost cannot result from the melting out of deeply buried ice blocks. In the continuous permafrost zone, a kettle can only form upon the melting of exposed ice that is clean or nearly clean. Because the equilibrium base of permafrost is hundreds of metres below the kettle and because permafrost aggraded rapidly in early postglacial time, the ice block could only have melted from the surface downward, and melting was arrested as soon as a surface debris cover equal to the active layer thickness had accumulated. In the continuous permafrost zone, kettle-like basins are prominent features of ridged and hummocky moraine belts (e.g. Fyles, 1963; Dyke and Savelle, 2000; Dyke and Hooper, 2001), typically occupying 10–30% of the surface area, in places up to 50%. Indeed, they are essentially restricted to such settings. Exceptionally, these basins are floored by bedrock, the kettle penetrating entirely through the moraine. Many others penetrate much of the apparent moraine thickness, and others are shallow. Kettles also form basins at the heads of outwash and in ice-contact deltas, but rarely in distal outwash. They have a wide variety of shapes and sizes, with elongations imparted only by bounding or adjacent ice-contact features. Depth seems roughly proportional to area. Similar basins in similar geological contexts south of permafrost are also interpreted as kettles.

These observations demonstrate that kettle-like basins are characteristic of ice-marginal accumulations in the zone of deep, continuous permafrost. The common interpretation of similar kettled terrain south of permafrost is that the kettling and associated processes occurred well after, often millennia after, deglaciation with the slow removal of ground ice. Some of these southern kettles probably did form in this way, but we suggest that many, particularly the large and deep ones, are essentially contemporaneous with deglaciation. Kettled terrain in both the south and the north is also typically interpreted as 'dead ice' or ice stagnation terrain. However, in the continuous permafrost zone, kettle-like basins are essentially restricted to the prominent morainal belts. If kettles signify stagnation, an inference with which we disagree, then most of the major end moraines of the northern Laurentide Ice Sheet were formed by stagnation. Below we develop the alternative case that these moraine belts were formed by active ice margins.

Permafrost imposes the further restriction that basal meltout till, characterized by the preservation of delicate englacial debris-rich ice structures due to slow basal melting (Shaw, 1982), cannot form. Such till can only form in the continuous permafrost zone if basal melting of permafrost progressed to the surface, passing up through the entire thickness of the putative buried or stagnant ice. There is no evidence that such an event ever occurred and every reason to believe that instead permafrost thickened rapidly in early postglacial time. Therefore, interpretations of glacial landform genesis within the zone of continuous permafrost that start with the inference that the features are comprised wholly or partly of basal meltout till (e.g. Sharpe and Nixon, 1989) are untenable.

7.6 PROCESS-FORM RELATIONSHIPS: TOWARDS A LANDSYSTEMS MODEL FOR THE PERMAFROST ZONE

Regional mapping of Quaternary geology by the Geological Survey of Canada has provided a comprehensive overview of the major landform-sediment assemblages associated with the last glacial cycle in northern Canada. This has enabled attempts to reconstruct ice sheet behaviour based upon landform zonation and process-form relationships, although interpretations of certain landform elements and their palaeo-glaciological implications may differ (cf. Hodgson and Vincent, 1984; Dyke, 1987; Sharpe, 1988; Rampton, 1988; Dyke and Savelle, 2000). For example, landform and surface material mapping generally has a high degree of reproducibility but the processes responsible for the landforms that are mapped can be the subject of dispute (e.g. Fyles, 1963; Sharpe, 1988, 1992; Dyke and Savelle, 2000; Rampton, 2001).

7.6.1 Glacially Deformed Permafrost

Glacially thrust and deformed beds of permafrost outcrop in numerous coastal sections along the western Canadian Arctic between the Tuktoyaktuk region and the glacial limit. All are evidently overlain by till and have been described mainly by J.R. Mackay (see summary and references in Mackay, 1971). Internal structures (folds and thrust faults) everywhere accord with glacier flow directions as inferred from up-ice flutes and end-moraine configurations. The largest known transported mass of pre-Wisconsinan permafrost comprises Herschel Island (170 m high and 110 km^2), part of the Laurentide terminal moraine just off the Yukon Coast. The island is directly down-ice from Herschel Basin, which is similar to the island in size and is therefore the likely source of the detached permafrost mass. Where large upstanding, deformed permafrost masses like Herschel Island and Nicholson Island form distinct ridges (Rampton 1988), they constitute thrust moraines. Other probable thrust moraines, which are known to contain deformed permafrost, are the so-called fingers of the Eskimo Lakes (see Fig. 62 in Mackay, 1963). The deformed permafrost, comprising a variety of fossiliferous terrestrial and marine sediments and massive ice bodies, is in effect deformation till or glacitectonite (Elson, 1981; Benn and Evans, 1996, 1998).

The widespread occurrences of deformed and displaced permafrost illustrate two important points. First, some of the relict frozen cores of moraines pre-date the last glaciation and are of nonglacial origin except in their deformation and displacement. To date, however, such occurrences are known only from the vicinity of the glacial limit. They have probably not survived elsewhere and could not have survived under warm-based ice. Hypothetically, similar features could have been generated by a readvance across permafrost during deglaciation, provided sufficient time was available for proglacial permafrost aggradation. However, no such occurrences are known from the permafrost zone of Canada. Second, the glacier ice that entrained and (or) deformed the permafrost bodies must have been cold-based from time of entrainment through to deglaciation. The extensive distribution of deformed permafrost in the vicinity of the glacial limit indicates that much (or all) of the marginal fringe was cold based.

7.6.2 The Major Belts of Ice-Cored Terrain: End Moraines Versus Thermokarst Terrain

As northern Canada lies within a zone of continuous permafrost, large bodies of glacier ice are likely to be preserved in the deglaciated landscape wherever melting snouts generated a surface debris cover thicker than the active layer (~1 m thick). The common occurrence of modern debris-covered glacier snouts, along with the fact that thousands of glaciers in this region and in the adjacent Cordillera formed ice-cored moraines during the Neoglacial period, makes it exceedingly unlikely that similar features did not form during Late Wisconsinan deglaciation.

Considerable exposures of ground ice have been observed, encountered in boreholes, or inferred from geophysical and geomorphic data in the lowlands of the western Canadian Arctic. Some ice exposures occur within features that have been mapped as end or hummocky moraine and have been interpreted as the remnants of glacier ice now entombed within the permafrost (e.g. Lorraine and Demeur, 1985; French and Harry, 1988, 1990; Dyke *et al.*, 1992; Sharpe, 1992; St Onge and McMartin, 1995, 1999). This interpretation implies that many morainic landforms are still supraglacial in character and that deglaciation continues albeit episodically today (Worsley, 1999; Dyke and Savelle, 2000). Other exposures of massive ground ice occur in a similar till-mantled terrain with a ridged and hummocky character in the Tuktoyaktuk region east of the Mackenzie Delta. However, these ice bodies have been interpreted as segregation ice (e.g. Mackay, 1971; Rampton, 1974, 1988). A serious interpretive difficulty thus arises in distinguishing kettled end moraines from extensive segregation ice-cored terrain that has later been subjected to thermokarst and is essentially identical morphologically. This is the case where a landform that looks like a moraine apparently is not. If correctly interpreted, these thermokarst terrains might be termed pseudo-moraines. Note that we use 'moraine' here in a morphological sense, rather than as a synonym for till. Rampton's use was the latter or both.

Mackay (1971) and Rampton and Walcott (1974) among others have convincingly shown that massive ice and ice-debris mixtures underlie much of the moraine-like terrain of the western Arctic mainland. Rampton (1974, 1988) hypothesized that most of the ice is not buried glacier ice, as might be expected in true end moraines, but is of segregation origin. Nevertheless, the isotopic composition of the ice (Mackay, 1983) indicates a Wisconsinan glacier ice origin. Rampton inferred, therefore, that segregation occurred by freezing of subglacial meltwater as it encountered the aggrading base of permafrost, located in his model at the ice margin and in the immediate glacier forefield during deglaciation. Thus the glacial isotopic signature was preserved in nonglacial ice (Fig. 7.8). In this model, the subglacial meltwater flowed beneath an existing till layer and generated a 20–30 m thick layer of segregation ice. This uplifted the till layer at or just behind the receding ice margin and thereby separated the till from underlying sediment. It is not evident why the beginning of the permafrost wedge is placed at the ice margin rather than behind it, nor therefore is it evident what might have prevented the subglacial meltwater from simply exiting at or escaping to the surface at the ice front. Similarly, it is not evident why the considerable pore water pressures, sufficient to cause tens of metres of ground uplift, did not rupture the warm, incipient, ice-marginal permafrost and allow water escape. Nevertheless, and taking all elements of the model at face-value, the rate of ice segregation kept pace with the rate of ice-marginal recession, and thus generated a regional sheet of massive ground ice. The massive ice was later pocked by thermokarst, which left a terrain resembling, and perhaps indistinguishable from, ice-cored hummocky and ridged moraine. Rampton mapped this terrain as 'rolling moraine'. His map shows that there is considerable internal ridging within this terrain and that it is bounded by prominent ice-front positions and ice-marginal deposits. The putative thermokarst lakes in the moraine-like region are identical in form to features interpreted as kettles elsewhere (Fig. 7.9).

In effect, Rampton invokes the formation of features that resemble morainal belts as proglacially formed ground-ice injection structures. In a broad sense, ice-cored terrains formed in this way might still be considered to be end moraines in that they formed precisely along former ice margins. It is important to bear in mind, however, that the hummocky and ridged moraine-like features in this model developed long after deglaciation and that no morainal relief will survive removal of the segregation ice core. Only an extensive thermokarst lake plain will remain, and hence the model has no potential application to moraines south of permafrost. Rampton's model has no basis in observations at modern ice margins. For example, regional ground ice sheets are not known to be forming and uplifting ground in front of modern glaciers and the model has not been applied to similar terrain elsewhere in arctic Canada. However, Rampton (2001) has recently argued its explicit application to the morainal belt on Wollaston Peninsula of Victoria Island (see below) in response to the contention by Dyke and Savelle (2000) that these moraines are cored extensively by glacier ice. Its applicability is, therefore, considered below. But first we need to consider the proposed chronology of development of moraine-like topography in the Tuktoyaktuk type area.

Rampton's model for the ridged and hummocky moraine of the Tuktoyaktuk region supports and may be essential to his interpretation of regional glacial history (Fig. 7.8). In that interpretation, the massive, segregated, ground ice sheet was formed during Early Wisconsinan deglaciation (*c.* 115–64 ka BP). However, the lake basins, and hence the moraine-like topography, developed in Late Wisconsinan time (mainly 13–10 ka BP), as shown by numerous radiocarbon dates on basal lake sediments. Before that time, the massive ice was apparently left undisturbed under the cold climate of intervening Late and Middle Wisconsinan time (64–13 ka BP). However, if the lake basins in the hummocky moraine in the Tuktoyaktuk region are instead kettles, formed by melting of glacier ice as outlined above, the radiocarbon dates on these lake basins indicate that the moraine belt is also of Late Wisconsinan age. Support for Rampton's model in its type area would therefore lie in a demonstration of an Early Wisconsinan age for the till overlying the massive ice. The lack of any convincing evidence that the till is Early Wisconsinan and the fact that correlative or older deposits overlie beds of Middle Wisconsinan age (e.g. Hughes *et al.*, 1981; Morlan *et al.*, 1990; Hill *et al.*, 1985), have allowed most reviewers of regional glacial history to prefer a Late Wisconsinan age for the till overlying the massive ice in the Tuktoyaktuk region (e.g. Hughes *et al.*, 1981; Denton and Hughes, 1981; Dyke and Prest, 1987; Dyke *et al.*, 2002). Nevertheless, demonstration of a Late Wisconsinan age for the till at or close to the glacial limit, as is the case at Tuktoyaktuk, would not invalidate the general process model advanced by Rampton; it could still be claimed that the ground ice is of proglacial segregation origin and that the lake basins and moraine-like topography formed well after deglaciation, early in Late Wisconsinan time.

However, any application of Rampton's model as an alternative interpretation of younger Late Wisconsinan moraine-like belts (e.g. Rampton, 2001) should take into consideration the finer chronological implications of the model. One should demonstrate, for example, that sufficient time was available to exhaust the latent heat released by freezing meltwater at the freezing front upward through the overlying permafrost during the postulated interval of ice segregation. Rampton did not develop or apply his model quantitatively and hence did not evaluate its chronological implications. Nevertheless, the accumulation of segregation ice as proposed in his model is a problem analogous to that of pingo growth, both processes involving addition of massive ice at the base of aggrading permafrost. If relief of moraine-like areas is entirely due to cores of massive segregation ice and if these areas become flat upon ice removal, as Rampton (1988, 2001) suggests, then ice under the moraine-like ridges (ignoring for the moment the inter-ridge depressions, which presumably originated by thermokarst) must have formed in lock-step with ice recession across the moraine-like belt. This stepwise formation of segregation ice is necessary because the plunging permafrost base extending down from the retreating glacier front (Fig. 7.8) would have prevented groundwater from reaching far into or below the glacier forefield.

The moraine-like belt on Wollaston Peninsula is well behind the Late Wisconsinan glacial limit and hence is of deglacial age. It is as much as 40 km wide (Sharpe, 1992b, Map 1650A) and a typical transect would cross 20 or so moraine-like ridges or very large hummocks of 20–100 m relief (Fig. 7.5). The simplest calculation of the rate of segregated ice formation by basal accretion is provided by Stefan's solution:

$$z = bt^{1/2}$$

where: z is depth of the freezing front
t is time
b is a constant (ice is 1.4 – incorporating latent heat of fusion, thermal conductivity, and temperature appropriate to the region; Mackay 1971, 1979).

The time taken to form an ice layer by basal accretion thus increases as the square of ice thickness. A 10 m thick layer would form in about 50 years (20 m in ~200 years; 30 m in ~460 years; 50 m in ~1,275 years; 75 m in ~2,870 years; 100 m in ~5,100 years). These are minima because the solution ignores geothermal heat flux, the lower conductivity of the capping till layer, and the insulating effect of any surface water or snow bodies, and because an unlimited supply of basal meltwater is assumed at all times. If moraine-like relief is due totally to excess segregated ice, the largest ridges would each have required thousands of years to form and the moraine-like belt probably would have required more than 10,000 years, bearing in mind that the basins in the topography were produced by removal of a volume of segregated ice by thermokarst at least equal to that remaining under the ridges. However, the entire moraine belt on Wollaston Peninsula formed in an interval of about 1,000 years, as shown by radiocarbon dating of marine deposits on distal and proximal sides of the belt (Dyke and Savelle, 2000; Dyke *et al.*, in press). This interval allows for an average of only 50

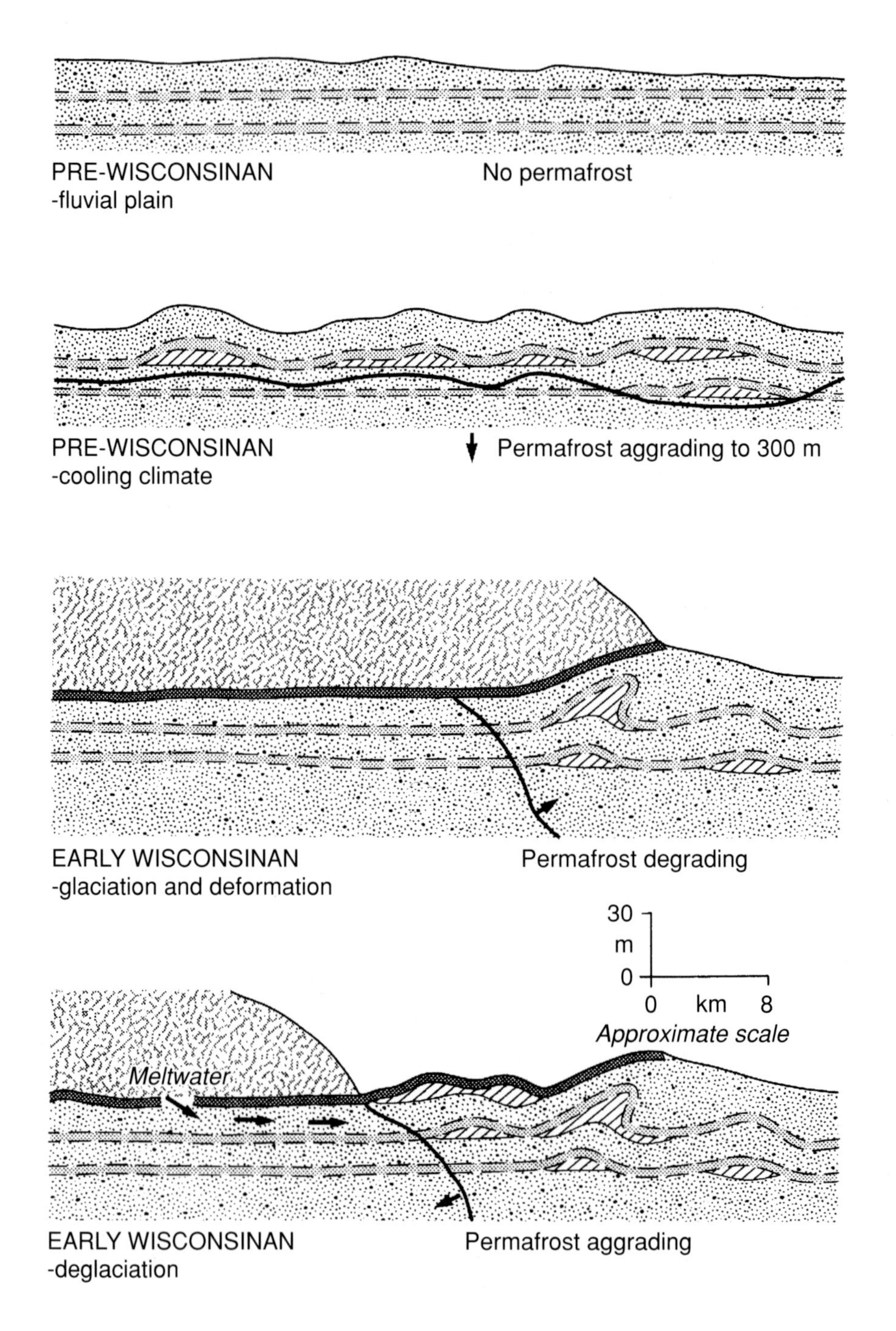

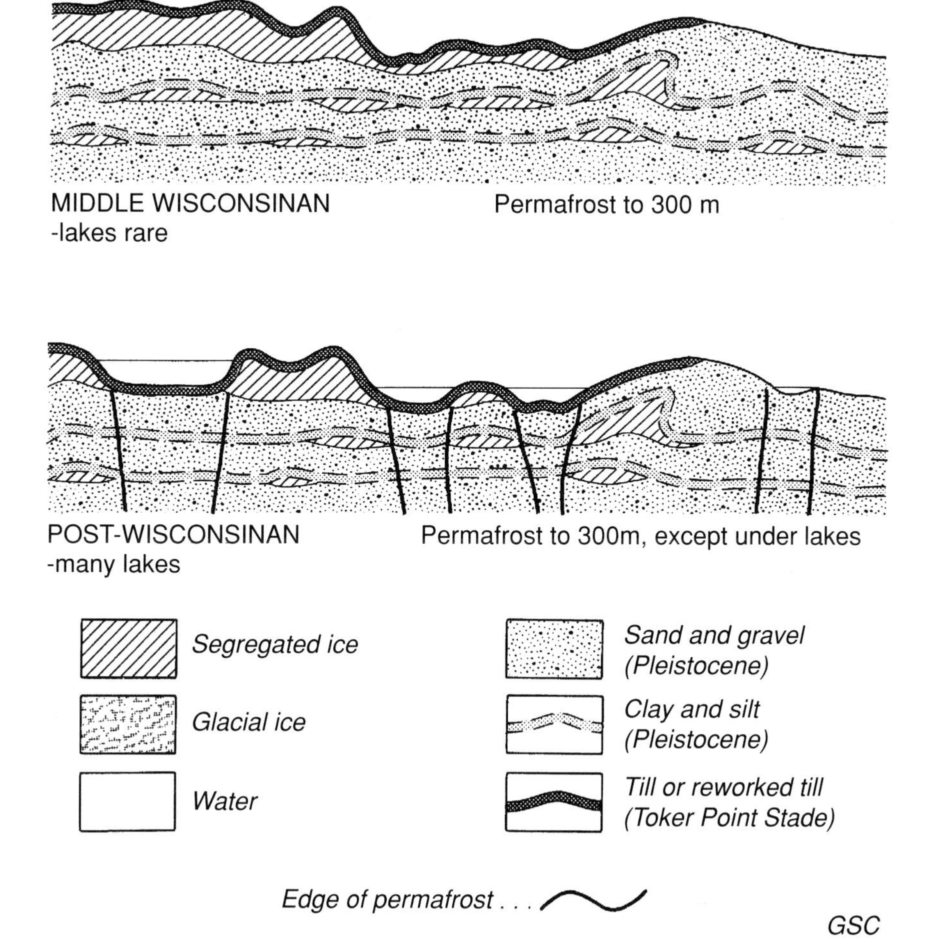

Figure 7.8 Rampton's model of thermokarst terrain development in the western Canadian Arctic (From Rampton, 1988).

years to form each of the estimated 20 ridges. Therefore, we suggest that Rampton's model for the origin of moraine-like belts is not applicable to Wollaston Peninsula, nor to any large moraine-like accumulation formed during deglaciation. Realistically it is not applicable to any individual moraine-like ridge more than about 10 m high. Moreover, the model embodies the irony that the incorporated deglacial process operates so slowly that end moraines would almost inevitably have formed by normal processes of sediment delivery to the ice margin in the long time intervals required for thick ice segregation.

We now consider whether there might not be simple geomorphic criteria to distinguish kettle lakes formed within the permafrost zone from thermokarst lakes. The distinction has significance beyond proper identification of moraine belts. For example, kettles that formed in permafrost terrain have no particular palaeo-climatic significance, having formed chiefly in response to the same warming that caused deglaciation. On the other hand, thermokarst lakes that formed long after deglaciation are taken as evidence of a warming event. The best-known

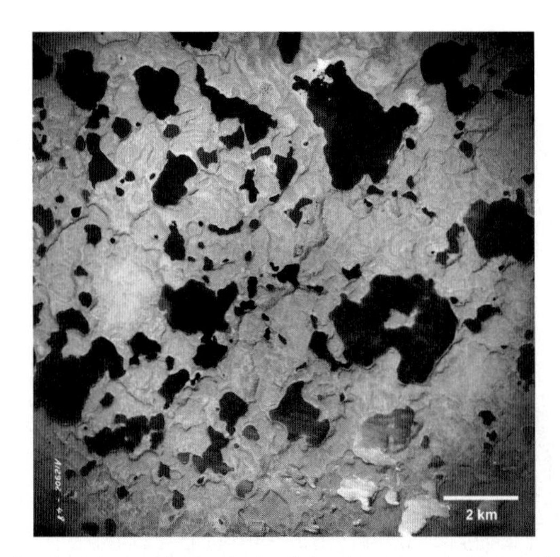

Figure 7.9 Extensively pitted terrain typical of the ice-cored areas of Tuktoyaktuk Peninsula, illustrated in part as figure 3 in Rampton (1988). The basins, interpreted herein as kettles in an ice-cored end-moraine belt, are interpreted as Late Wisconsinan thermokarst features that formed in Early Wisconsinan drift by Rampton. (NAPL A12902-48.)

inference of climatically induced thermokarst in North America is the formation of the numerous lake basins in the hummocky moraine-like belt near Tuktoyaktuk where this event is thought to reflect warmer conditions of the last Milankovitch insolation maximum (e.g. Rampton, 1988; Burn, 1997). If our interpretation is correct, this inference is incorrect because the lakes are kettles. Classic thermokarst lakes, on the other hand, are fairly shallow and are commonly wind oriented (e.g. Mackay, 1963). Lakes of this kind occur in the Tuktoyaktuk region (see Fig. 4 or 21 in Rampton, 1988) adjacent to, but evidently not in, the belt of moraine-like topography, where lakes are kettle-like. No explanation has been offered as to why thermokarst lakes in terrain that is non-morainal should have become wind-aligned whereas those in moraine-like terrain are not. We suggest that kettle-like lakes in areas of hummocky and ridged moraine-like topography are best interpreted as kettles unless a stadial-scale age difference between ground ice formation and lake basin development can be demonstrated.

7.6.3 Hummocky and Ridged Moraine Belts: Active Versus Stagnant Ice

In the largest morainal belts, the most common topography is hummocky or chaotically moundy in character, particularly where the surface material is ice-contact stratified drift and where the debris accumulated in the deep-cut recesses between major ice lobes. However, sub-parallel linear ridges impart a broad organization to the belts, strongly suggestive of multiple ice fronts or of large englacial structures or supraglacial debris-covered ridges parallel to a margin. The latter are commonly arranged in nested suites lying transverse to former glacier flow. The longest continuous moraine ridge on Wollaston Peninsula extends for 100 km with no break wider than a meltwater channel. Cross-cutting relationships of both ice-marginal and ice-flow features indicate that the more prominent ridges, both here and in the Bluenose Lake region on the adjacent mainland, represent culminations of readvances, several of which are recognized (Dyke *et al.*, in press).

(A)

(B)

Figure 7.10 A ground ice slump exposing massive ground ice on upper slope of an ice-cored moraine on western Victoria Island. The slump was triggered by heavy rain the day before the photograph was taken, and the ice face was inaccessible because material in the floor of the slump remained liquid. The capping colluviated till approximates the thickness of the maximum active layer development (~1 m).

Extensive areas of this moraine type are clearly underlain by buried ice that is undergoing a very slow melting from the top down. The dominant ablation-triggering process is the development of flow-slides (ground ice slumps) and active layer detachments that expose the ice cores (Fig. 7.10). Fresh slumps are rare compared with the numerous old slump scars. In all probability, slumping and ice-core degradation was at a maximum during earliest postglacial time, when moraine slopes were steepest and climate was generally warmer than present. However, even today an unusually warm summer or unusually heavy summer rain will trigger new slumps or reactivate older slumps. Exposures of the buried ice in places display considerable debris concentrations including large, commonly striated boulders, debris bands and folded folia (French and Harry, 1990). A considerable amount of ice, presumably of glacial origin, has also been detected beneath morainic topography by gravity profiling (e.g. Kotler *et al.*, 1998), and surface kettles and large ice wedge polygons on moraine surfaces have been used to infer ice cores (Dyke and Savelle, 2000). Additionally, Dyke and Savelle (2000; also Dyke *et al.*, 1992) point out that moraine volume is considerably reduced where a moraine extends below the marine limit or into a glacial lake basin. This reduction is explained by the fact that the water either prevented formation of an ice core or quickly destroyed one that did form. Debris-rich bands within glacier ice, regardless of origin (e.g. regelation, apron overriding, thrusting; see Ó Cofaigh *et al.*, Chapter 3), control the distribution of sediment melting out on the glacier surface and therefore often produce linear ridges lying transverse to glacier flow. The resulting moraine is therefore strictly defined as hummocky moraine and controlled hummocky moraine, being moraine 'deposited during the melt-out of debris-mantled glaciers' (Benn and Evans, 1998), though we stress that within the permafrost zone, melting occurs only from the top down. However, by most common definitions a moraine is a landform produced after complete removal of glacier ice and therefore most of the expansive areas of hummocky glacial terrain in northern Canada are in fact supraglacial accumulations of debris or debris-mantled, relict glacier snouts. Nonetheless, the buried glacier ice is now part of the permafrost of the region, and if the ice is now in equilibrium with the environmental conditions, that is as far as deglaciation and deposition progress.

The extensively kettled, ice-cored, ridged and hummocky moraine of northern Canada has, not unlike in many other regions, been interpreted as the product of regional glacier ice stagnation (e.g. Sharpe, 1988, 1992). However, several characteristics question this mode of formation. The moraine possesses numerous linear ridges arranged in wide belts, the belts often bordered or separated by large 'end moraines'. Dyke and Savelle (2000) consider the individual linear ridges within the belts of hummocky moraine to be a record of ice marginal, supraglacial debris accumulation, or ice-cored end-moraine formation, and therefore indicative of repetitive moraine building and readvances during the recession of active glaciers. Proglacial meltwater channels and outwash trains and fans emanate from numerous individual ice-marginal positions, which could only have formed in sequence as opposed to being formed randomly or contemporaneously by regionally stagnant ice. These characteristics of ice-cored moraine belts are important, because were the ice cores to be removed, these belts would more closely resemble vast hummocky terrains typically interpreted as products of regional ice stagnation, and successive ice-marginal positions would be more difficult to discern.

7.6.4 Ice-Shelf Landforms

Ice-shelf landforms warrant special mention because they are better documented from this region than anywhere else in glaciated North America. This limited distribution accords with the evident constraint that ice shelves can be sustained only where the feeding ice is cold-based, warm-based ice having too little tensile strength to prevent calving (Benn and Evans, 1998).

The simplest evidence of large former ice shelves comes from low-to-negligible moraine gradients over significant distances. Bearing in mind that low gradients are at least in part due to differential glaci-isostatic rebound, and hence could have been zero at time of formation, large ice shelves are postulated on this basis to have fringed the Late Wisconsinan Laurentide ice limit in Amundsen Gulf and M'Clure Strait (Vincent, 1982; Dyke and Prest, 1987). Similarly, a horizontal moraine on the north coast of Bylot Island (Klassen, 1993) may have been formed by an ice shelf extending to the mouth of Lancaster Sound. The latter was proposed as the last glacial maximum position by Dyke and Prest (1987) but is now thought to be of early deglacial age (Dyke *et al.*, 2002). Small horizontal moraines of few kilometres length in valleys on Ellesmere Island have similarly been attributed to ice shelves (England et al., 1978), and many other small horizontal moraine segments in the region may have a similar origin.

The Viscount Melville Sound Ice Shelf formed during a readvance of the northwest Laurentide margin in the late Pleistocene (Hodgson and Vincent, 1984), probably during Younger Dryas time. It was approximately 60,000 km^2 in extent and is recorded geologically by the distinctive Winter Harbour Till, which was deposited where the shelf edge grounded in water shallower than the freeboard of the ice shelf. The till is thin and nearly featureless, its distinctive characteristics being a near-horizontal elevational limit along hundreds of kilometres of coastline and its relationship to both pre- and post-Winter Harbour marine limits. Weakly inscribed striae below the till consistently trend normal to shoreline regardless of shoreline orientation. Far-travelled erratics were commonly deposited during the event, indicating that the ice shelf was laden with debris from its previous grounded phase or that debris was efficiently transferred through it. Associated morainal forms are rare and evidently of ice-push origin. Lateral drainage channels are also rare and possibly associated with lakes that were impounded along the shelf edge.

Viscount Melville Sound is a nearly ideal location for the formation of an ice shelf. Its broad eastern end was the access route for inflowing ice from the main body of the ice sheet, and it is otherwise nearly enclosed by land. Thus the ice shelf was of the 'confined' type, with grounded distal margins (Benn and Evans, 1998). The lack of similar evidence of deglacial ice shelves in other large marine basins in the region suggests that conditions suitable for ice shelf formation were rarely met at that time. Theoretically, conditions should have been more suitable for ice shelf formation during ice advance phases, and indeed this process may have been essential for regional ice sheet establishment in the Canadian Arctic Archipelago.

7.7 Discussion and Conclusion

We have reviewed the broader aspects of ice-marginal landform distribution in the continuous permafrost zone of northern Canada. The environment places constraints on possible basal ice thermal conditions in glacier marginal and sub-marginal zones throughout glacial cycles. Ice caps necessarily nucleated and expanded on pre-existing permafrost. The permafrost then either survived throughout the glacial cycle or reformed upon or prior to deglaciation. Ice-marginal landforms and bedforms in this region are zonally distributed along the ice sheet radius in a manner that can be interpreted in terms of the relative prevalence of cold-based and warm-based ice. Where the ice sheet was extensively cold-based, ice-marginal landforms are nearly limited to lateral and proglacial meltwater channels and ice-thrust moraines. The latter are most common and most voluminous where composed of disturbed unconsolidated sediment. Broad constructional morainal belts are extensively cored with glacier ice and backed by streamlined terrain. These belts formed along a cold-based marginal fringe backed by warm-based, debris-rich ice. Characterized by large kettles, these morainal belts can be misinterpreted as thermokarst terrains. They would lose much of their topographic organization and misleadingly resemble dead-ice terrain were they to lose their ice cores. Short of serious global warming, loss of ice cores is improbable in the continuous permafrost zone. Such ice-cored terrain probably survived through previous interglacials and likely provided large, regional debris sources to advancing cold-based ice during phases of glacial build-up. Where the ice sheet was extensively warm-based, few regionally significant end moraines formed except during readvances. Rogen moraines probably formed from previously deposited drift along the boundary between distal, sliding and proximal, non-sliding (cold-based) ice and hence are well preserved only near final ice recession centres.

We know of no extensive terrain south of permafrost in North America that resembles the northernmost terrain outlined above with its extensive marginal meltwater channels. Less-extensive occurrences have been described from the Cordillera (Tipper, 1971; Dyke, 1990) and from central Quebec-Labrador (Ives, 1960). However these are areas of discontinuous permafrost today, and it is therefore probable that the marginal zone was cold-based during deglaciation.

Considerations of former permafrost conditions during deglaciation have already been thoughtfully incorporated into conceptual models of landform genesis along the southern margin of the Laurentide Ice Sheet starting perhaps with the seminal model of Clayton and Moran (1974), which incorporated a cold-based marginal fringe (see Chapter 1). This model was applied to interpreting the landscape zonation of the James and Des Moines lobes of the Laurentide Ice Sheet. The broad morainal belts there can thus be seen as former ice-cored moraines, and the still-ice-cored moraines of similar size in the Canadian Arctic can serve as analogues of the evolutionary stage of the southern moraines after deglaciation but before loss of ice cores. More recently, Attig *et al.* (1989) and Clayton *et al.* (2001) have attributed the formation of hummocky end moraine in Wisconsin to compressive flow across a frozen toe zone creating broad belts of ice-cored moraine. The inferred ice-walled lakes within the ice-cored terrain (e.g. their Fig. 1.1) are equivalent to the extant kettles in the moraines of Arctic Canada. Ham and Attig (1996) present a similar model for end moraines further west in Wisconsin. These models are entirely compatible with our observations from the present permafrost zone. However, we would prefer to refer to the forming or detached zone of buried ice, a transient permafrost feature in southern areas but a permanent one in the north, as ice-cored moraine rather than stagnant glacier ice. This is because they formed along active, commonly readvancing, ice margins rather than by regional stagnation.

In the area described by Attig *et al.* (1989), broad end-moraine belts ceased to form once regional permafrost disappeared and drumlin formation extended to the ice margin. This is a reasonable inference because it is based in part on the distribution of palaeo-permafrost features (Clayton *et al.*, 2001). However, landscape zonation within the permafrost zone described above also shows that retreating ice sheets can be warm-based right out to the margin where permafrost is necessarily forming upon deglaciation.

These examples and more direct evidence demonstrate that permafrost was widespread along the southern ice margins at the last glacial maximum and early in deglaciation. Clayton and co-workers have incorporated permafrost into glacial process-form models, with the cold-based marginal fringe being the key element. We suspect, however, that this model is still under-applied in interpreting landscape zonation in presently non-permafrost regions.

Acknowledgements

Discussions with Larry Dyke, Geological Survey of Canada, about permafrost evolution through glacial cycles is particularly appreciated. The manuscript was reviewed by Greg Brookes, Geological Survey of Canada, who clarified several aspects.

CHAPTER

ICE-MARGINAL TERRESTRIAL LANDSYSTEMS: SOUTHERN SCANDINAVIAN ICE SHEET MARGIN

Frederik M. Van der Wateren

8.1 INTRODUCTION

During the Middle and Late Pleistocene, the Scandinavian Ice Sheets repeatedly covered the northern half of The Netherlands and Germany, most of Poland, Estonia, Latvia, Lithuania and Belarus, a small part in the north of the Czech Republic and the northern parts of Ukraine and Russia (Fig. 8.1). This chapter focuses on the southern part of the area glaciated by the Scandinavian Ice Sheets, between 49°N and 60°N, 3°E and 30°E.

The landscapes left behind by the Southern Scandinavian Ice Sheets show the typical characteristics of glaciated sedimentary basins, including interaction of the ice sheet with weakly lithified and unlithified sediments. Expanding from their core region, the Scandinavian shield – including Norway, Sweden and Finland – the ice sheets invaded the northern and northeastern European plains, underlain by major sedimentary basins filled with sequences of Mesozoic and Cenozoic sediments of up to 8 km in thickness (Ziegler, 1990). The distribution, morphology, structural geology and sedimentology of glacigenic deposits and landforms are strongly influenced by the geological structure of these basins. Existing crustal scale structures, such as the central European Variscan and Alpine orogens, rifts, major faults and salt structures, affected the geometry of aquifers over which the ice sheets advanced and the distribution of sediments of varying mechanical properties. This, in turn, controlled the hydrology, and thus the dynamics of the ice sheets, determining the location of ice sheet limits, ice streams, tills, thrust moraines and other moraines.

There has been considerable debate about the physical properties of the geological substrate over which the Fennoscandian Ice Sheets advanced and their consequences for ice sheet volumes, advance and retreat rates, sea level change and atmospheric circulation (e.g. Boulton, 1996b and references therein). Arguments for widespread deformable bed conditions beneath temperate ice sheets include features that are regarded as typical of subglacial deformation of saturated sediments. These range in scale from megaflutes and drumlins to smaller scale boudins, augen, 'pods', and tectonic laminations developed within tills (Boulton, 1987; Hart, 1994; Hart and Roberts, 1994). Arguments against a deformable bed model originate from the interpretation of laminated diamicts as glacimarine deposits and subaquatic, melt-out or lodgement tills (Eyles and McCabe, 1991; Piotrowski and Kraus, 1997; Piotrowski *et al.*, 2001; Piotrowski and Tulaczyk, 1999).

Boulton (1996b) argued that it would be highly unlikely for a temperate ice sheet not to deform the underlying sediments. Large volumes of meltwater are produced at the base of temperate ice sheets, which are drained through the bed. Over large areas subglacial drainage is likely to produce high pore water pressures in the subglacial sediments, reducing the effective stresses to levels low enough for these sediments to deform, or even fluidize. Identification of features resulting from subglacial deformation may therefore help to reconstruct the dynamics of former ice sheets.

A recent debate on the distribution and relative importance of subglacial deformation (Hart *et al.*, 1996, 1997; Piotrowski *et al.*, 1997, 2001; Piotrowski and Kraus, 1997; Piotrowski and Tulaczyk, 1999) made it clear that widely accepted criteria to distinguish sediments that have been deformed by subglacial simple shear from sediments, that are undeformed – or deformed by other mechanisms – are largely lacking. Part of the disagreement stems from the lack of understanding of the various deformation histories and the resulting till fabrics. It is also the consequence of not consistently applying a structural approach to the analysis of subglacial tills, as is clear from the statement by Piotrowski *et al.* (2001): "It is easier to show that some tills were not pervasively deformed, than to demonstrate that others were". Most likely the reverse is true.

In a recent paper, Van der Wateren *et al.* (2000) use a structural approach to analyse unlithified to weakly lithified sediments, which have been subglacially deformed. The primary objective is to establish a set of criteria to distinguish undeformed sediments from those that have been subglacially sheared or otherwise deformed, and produce reliable tools for the reconstruction of past ice movement directions. The former is of critical importance for studies of piston or drill cores where the structural and sedimentological context is much less clear than it is in a good outcrop. Structural analysis is not a stand-alone tool; a multi-scale approach including sedimentological, morphological and geophysical data is strongly preferred (Kluiving *et al.*, 1999).

This chapter reviews the glacial landsystems of the Southern Scandinavian Ice Sheets. The following section presents a brief history of Middle and Late Pleistocene glaciations in Northern Europe. Glacial landsystem analysis seeks to reconstruct the dynamics of past ice sheets. As the Fennoscandian Ice Sheets expanded across regions of greatly varying structure and lithology it is important to consider first how glacial landsystems are affected by subsurface geological conditions. The regional geology of the Northern European Plains is then introduced with a discussion of the distribution of glacial landforms and glacitectonic styles produced subglacially as well as at the ice margin, and how these styles can be used to reconstruct glacial landsystems. This is followed by a focus on one particular glacitectonic style, structures developed in response to subglacial deformation, as these have been a continuous source of misunderstanding. Aspects of glacifluvial outwash, ice-marginal valleys and lakes in relation to end moraines are then discussed and the chapter concludes with examples of glacial landsystems of the Northern European Plains with a focus on Central and Northern Germany, and The Netherlands.

(A)

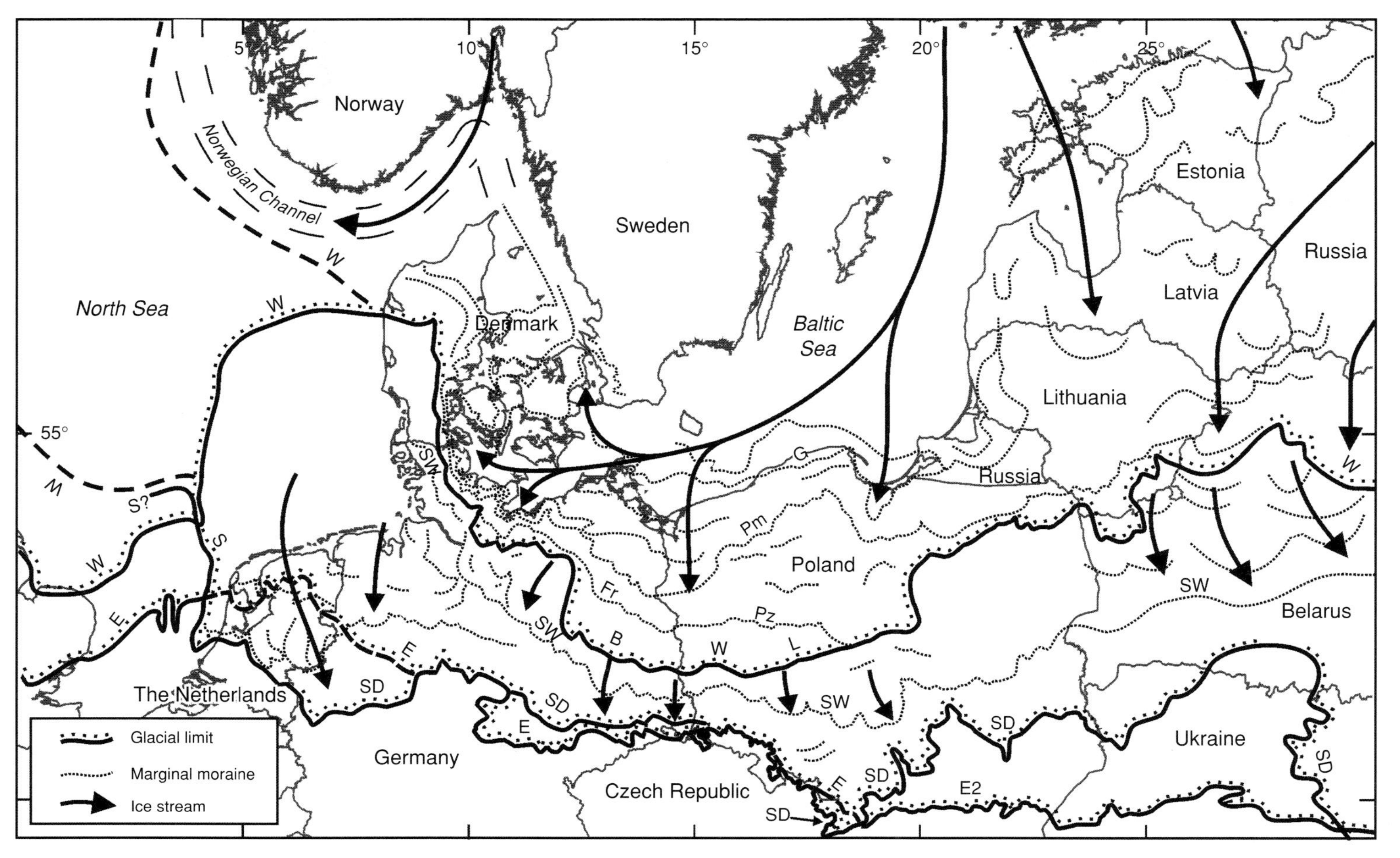
Norway
Norwegian Channel
Sweden
North Sea
Denmark
Baltic Sea
Estonia
Latvia
Russia
Lithuania
Poland
Belarus
The Netherlands
Germany
Czech Republic
Ukraine
5°
10°
15°
20°
25°
55°
Glacial limit
Marginal moraine
Ice stream

(B)

8.2 GLACIAL HISTORY

Northwest Europe contains evidence of continental glaciation from the latest three cold stages, the Elsterian (Marine Oxygen Isotope Stages 14 and 12), Saalian (Stages 8 and 6) and Weichselian (Stages 4 and 2) (Bowen *et al.*, 1986; Sibrava, 1986; Ehlers, 1996). The landscapes produced during these three cold stages differ markedly:

1. The Elsterian glaciation is characterized by extensive systems of very deep (up to 400 m) tunnel valleys. They form an anastomosing pattern of generally overdeepened and closed elongate depressions (Huuse and Lykke-Andersen, 2000). In The Netherlands and West Germany thrust moraines are absent. Elsterian thrust moraines are only known from the North Sea area (Bammens, 1986; Cameron *et al.*, 1989; Laban, 1995), Central Germany (Eissmann, 1995) and Poland (Brodzikowski, 1995). Elsterian tills are found throughout the area and are interrupted only where they have been removed by post-Elsterian erosion and reworked into younger sediments. They contain commonly large proportions of locally derived Tertiary material. The rarity of thrust moraines in Northwest Europe may be due partly to removal by the more extensive Saalian glaciation. Yet, it is at least striking that Elsterian push structures have been rarely described as part of younger structures. In the area east of Leipzig and north of Dresden in central Germany a series of probably Elsterian age thrust moraines occur, which have been overridden by the Drenthe and Warthe advances of the Saalian glaciation (Eissmann, 1994).
2. The Saalian glaciation is characterized by large thrust moraines, particularly those belonging to the oldest and southernmost Rehburg line. Several thrust moraine lines are arranged along

Figure 8.1 Middle and Late Pleistocene margins of the Southern Scandinavian Ice Sheet in the Northern European Plains. A) Shaded relief map showing Elsterian (E), Saalian (S) and Weichselian (W) glacial limits. In central Europe, Elsterian and Saalian ice sheets were bounded by Variscan mountain chains. The Saalian margin in The Netherlands coincides with the northern fringe of NW–SE trending Central High (see also Fig. 8.19). Morphology of the Northern European Plains is strongly dominated by chains of end moraines, thrust moraines and till plateaux. Many of the present day major river valleys within this area follow ice-marginal drainage systems. Numbers indicate areas shown in detailed maps: 1 = central Germany (Fig. 8.16), 2 = Mecklenburg (Fig. 8.18), 3 = The Netherlands and western Germany (Fig. 8.19). B) Glacial limits and major end moraines: E = Elsterian, SD = Saalian, Drenthe glaciation, SW = Saalian, Warthe glaciation. Weichselian end moraines: L, Br = Leszno/Brandenburg, Pz, Fr = Poznán/Frankfurt, Pm = Pomeranian, G = Gardno. Correlation of end-moraine lines between western and eastern Europe is often uncertain because of different stratigraphical and dating methods. The map compiles data from Houmark-Nielsen (1983, 1987, 1989), Sjørring (1983), Ter Wee (1983), Van der Wateren (1985, 1995), De Gans *et al.* (1987, 2000), Stephan (1987, 1995), Van den Berg and Beets (1987), Long *et al.* (1988), Brodzikowski (1995), Caspers *et al.* 1995), Eissmann (1995), Gozhik (1995), Khristophorova (1995), Knoth (1995), Macoun and Králik (1995), Matveyev (1995), Mojski (1995), Müller *et al.* (1995), Raukas (1995), Rühberg *et al.* (1995), Skupin *et al.* (1993), Sejrup *et al.* (1998), Huuse and Lykke-Andersen (2000), Marks (2002). Location of major ice streams within the Weichselian glaciated area is based upon Boulton *et al.* (2001b), and location of those within the Saalian glaciated area is based upon the assumption that lobate moraines are produced by fast-flowing ice streams.

series of marginal positions. Some of these lie along deep glacial basins, up to 150 m below mean sea level, for example the Gelderse Vallei and the Ijssel Valley in The Netherlands (De Gans *et al.*, 1987; Van der Wateren, 1981, 1985), the Nordhorn basin (Richter *et al.*, 1951) and the Quakenbrück basin (Meyer, 1987; Van der Wateren, 1987, 1995) in Germany. Except for the North Sea area and the Hunze valley in the northern Netherlands (Van den Berg and Beets, 1987) extensive systems of deep tunnel valleys are not known from the Saalian, although it may be argued that the overdeepened glacial basins on the upstream side of thrust moraines are similar structures (Boulton and Hindmarsh, 1987). Shallow dry valleys running parallel to the assumed ice flow direction and relatively few drumlins are known to occur in the Saalian till covered areas (Schröder, 1978; Ehlers and Stephan, 1983; Rappol, 1984).

3. The Weichselian glaciation is characterized by generally smaller thrust moraines than those of the Saalian. Tunnel valleys are not as numerous and on the average about half as deep as the Elsterian tunnel valleys. Drumlins have been reported from Denmark, Schleswig-Holstein (Stephan, 1987) and Poland (Karczewski, 1987).

In Germany subdivision of the glaciations had traditionally been based on morphostratigraphy, distinguishing various ice-marginal lines, which were assigned to 'stages'. Each of these, particularly within the Weichselian glaciated area, was subdivided into end moraines. According to Ehlers *et al.* (1995) and Ehlers (1996) the Brandenburg/Leszno, Frankfurt/Poznán moraines probably belong to the same ice (re)advance as they are represented by one single till unit. The end-moraine lines define different stages in the retreat after 20 ka BP. In other cases such end moraines could be shown to have formed during an ice sheet advance and not necessarily marked a major stagnation of the advance, for example the Older Saalian Rehburg line of thrust moraines. It must be borne in mind that some of these moraines may be diachronous, which makes correlation with similar end-moraine lines elsewhere problematic (Van der Wateren, 1995).

Figure 8.1a is a shaded relief map of Northwest, Central and part of Eastern Europe showing the Middle and Late Pleistocene margins of the Southern Scandinavian Ice Sheets in relation to topography. The high topography in the south comprises Variscan mountain chains – including the Teutoburger Wald, Harz Mountains, Thüringerwald, Erzgebirge and Sudetes – and the Alpine chain of the Carpathian Mountains. The Elsterian and Saalian ice sheets terminated against these natural barriers. The plains north of these mountains, underlain by Cenozoic sediments, are largely shaped by successive overriding ice sheets. Major end-moraine lines with extensive plateau-like till plains on their proximal side, stand out in the map. Large valleys have been carved by ice-marginal drainage, which originated from south-to-north running rivers that were redirected parallel to the ice margins. Examples of these are the Vistula River valley in Poland (~53°N, 15°–20°E), the Elbe River (~54°N, 10°E) and Weser River (~53°N, 10°E) in Germany and the Rhine and Meuse Rivers in The Netherlands (~52°N, 5°–6°E). The smaller Weichselian ice sheet terminated near the northern margin of the sedimentary basins, in the plains previously overridden by the Elsterian and Saalian glaciations. Except in Eastern Europe it was less confined by topographic barriers.

Figure 8.1b shows the margins of the ice sheets, as well as major end moraine lines within these margins (marking advances, or readvances during retreat), based upon a compilation of data from various sources. This map of the southern Scandinavian glaciated area must be regarded as tentative. The limits of the Elsterian, Saalian and Weichselian ice sheets are generally well established. However, detailed correlation of individual end-moraine lines between different countries – particularly between those of Eastern and Western Europe – remains problematic in places. This chapter therefore focuses on the western and central parts of the area glaciated by the Southern Scandinavian Ice Sheets.

The Elsterian glaciation completely changed the regional drainage of the Northern European Plains. Many rivers that previously drained northward into the Baltic basin were redirected along the ice-sheet margin into westward directions. Large ice-marginal lakes developed next to the Variscan highlands, for example the ice-dammed lake occupying the Elbe River valley south of Dresden in central Germany (Fig. 8.1b). This pattern repeated itself during subsequent glaciations, for example the dammed Weser River system in Lower Saxony alongside the Drenthe ice margin in western Germany (Van der Wateren, 1994b, 1995), and extensive tracts of ice-marginal valleys and lakes succeeding the last glacial maximum in Poland (Marks, 2002). Ice-marginal lake deposits are major components of the lithostratigraphy of many thrust moraines (Ruegg, 1981; Van der Wateren, 1995).

Boulton *et al.* (2001b), using a combination of satellite image analysis, existing geological data and numerical ice-sheet modelling, recognized patterns of flow-parallel lineations and ice-marginal features. These allowed them to make the following palaeo-glaciological inferences.

1. Shape and dynamic behaviour of the ice sheet show a strong spatial asymmetry.
2. Ice-sheet margins are commonly diachronous. Ice margins continued to advance in some areas while in others retreat set in.
3. As has been suggested previously (Ehlers, 1990), this asymmetry stems from a clockwise rotation of the main ice divide. According to Boulton *et al.* (2001b) this is the result of changing accumulation patterns across the ice sheet with relatively warm and moist conditions prevailing on the western flank, while cold and dry conditions dominate the areas to the east and on the leeside of the divide. Climatic warming leads to decay of the western part of the ice sheet while the eastern part may continue to grow.

8.3 REGIONAL GEOLOGY AND GLACIAL LANDSYSTEMS

The Scandinavian shield (Fig. 8.2), the core region of the Fennoscandian Ice Sheets, consists of Precambrian and Early Palaeozoic metamorphic and intrusive rocks (the same rocks that are found as erratics in all Pleistocene tills). These are covered at most with a sparse cover of Quaternary sediments. The Southern Scandinavian Ice Sheets covered areas almost completely underlain by large sedimentary basins. These basins separate the Scandinavian craton from the Variscan and Alpine orogens in the south. They are filled with up to 14 km of Palaeozoic to Cenozoic sediments (Ziegler, 1990). The upper few hundred metres consist of unlithified and weakly lithified sediments, which may deform when subjected to glacial stresses. Under the right conditions deformation by ice-marginal stress fields may produce thrust moraines. Subglacial shearing may produce tills, which are a mixture of local sediments and far-travelled Scandinavian material.

The margins of the Elsterian and Saalian ice sheets were located in the northern foothills of the Variscan uplands. The shaded topography map (Fig. 8.1a) clearly shows that major end moraines occur in narrow belts parallel to the uplands, for example the moraines within the Drenthe

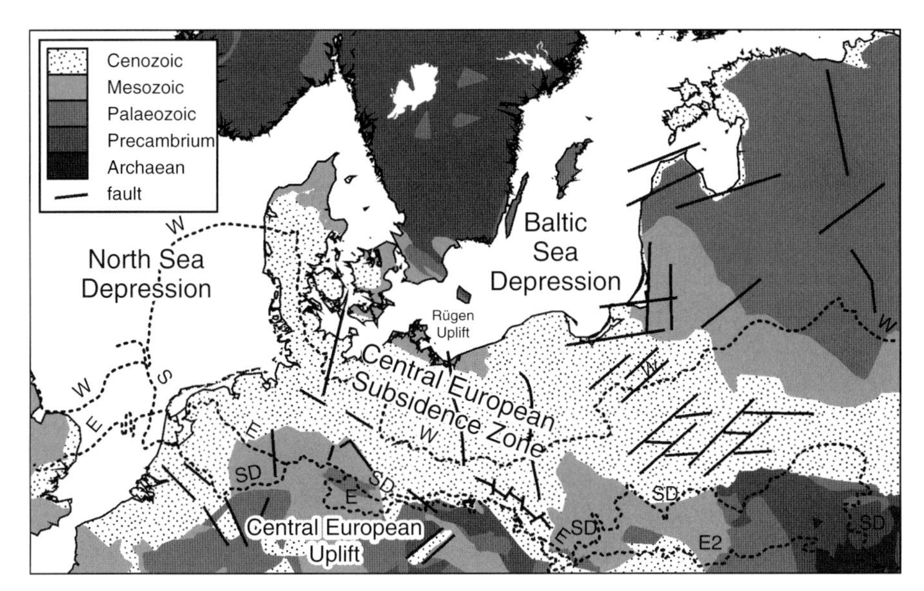

Figure 8.2 Geological sketch map of northwest Europe (data courtesy of Cornell University Interactive Mapping Tool; Kirkham, 1995) including Elsterian, Saalian and Weichselian ice sheet margins. Cenozoic sedimentary basins separate the Scandinavian Precambrian craton in the north from the Variscan and Alpine orogens in the south. They form an area of extended and thinned lithosphere and are filled with up to 14 km of Palaeozoic to Cenozoic sediments. Major neotectonically reactivated basement faults (sketched) and neotectonic structural subdivision are based upon Garetsky *et al.* (2001).

glaciated area (SD in Fig. 8.1b), or the prominent belt of Warthe (SW) end moraines, which runs from northern Germany across Poland into Belarus. These end moraines are mainly thrust moraines built of unlithified and weakly lithified preglacial sediments, deposited in Mesozoic/Cenozoic sedimentary basins. The coincidence of moraine lines with the shallow margins of these basins draws attention to a possible relationship between the formation and distribution of glacial landforms and geological structure. This correlation of substratum geology and glacial landforms has been noted by many workers in Europe as well as North America (see the Introduction in Van der Wateren, 1995).

In The Netherlands the distribution of Saalian thrust moraines and till sheets is evidently controlled by Cenozoic crustal scale structures in the substratum (De Gans *et al.*, 1987). The Saalian limit, which is marked by large thrust moraines, exactly follows the northern margin of major structural highs. Other thrust moraine belts north of the Saalian limit in The Netherlands can likewise be shown to correlate with subsurface structures. Most of the larger faults have been active throughout the Neogene up to the present. The large Saalian thrust moraines of the Rehburg line in Lower Saxony, west Germany, and those in central Germany occur in areas near the margin of sedimentary basins where Mesozoic and Tertiary clay and silt beds shallow and crop out. Within the Weichselian glaciated area, substratum control is less obvious, except for the Pomeranian end moraines, which tend to follow the –50 and –100 m contours of the Quaternary base in Germany as well as in Poland (Stackebrandt *et al.*, 2001).

In the area occupied by the Southern Scandinavian Ice Sheets, including the sedimentary basins of the North German Lowlands and adjacent Polish lowland, the Quaternary base has a relief of more than 1100 m, ranging from more than 600 m below sea level in the central parts of the basins to more than 500 m in the highlands south of them (Stackebrandt *et al.*, 2001). The deepest depressions are produced by Elsterian subglacial erosion and constitute tunnel valleys reaching up to 500 m in depth. Many of these, particularly those in the Central European Subsidence Zone (Fig. 8.2), follow NNE–SSW trending basement faults, which have been neotectonically active. Neotectonics in this area appears to have played a major role in channeling Elsterian ice stream flow (Stackebrandt *et al.*, 2001). In the transition zone between the Central European Subsidence Zone and the Central European Uplift (including the Variscan highlands) the Quaternary base contours are affected by another set of NW–SE trending neotectonically reactivated faults (Garetsky *et al.*, 2001; Stackebrandt *et al.*, 2001).

In Poland, Brodzikowski (1995) demonstrated that glacitectonic styles, intensity and dimensions of the glacitectonic deformation of the pre-Vistulian (pre-Weichselian) glaciated area in Poland is strongly related to the geological structure of the substratum. In Belarus, Matveyev and Nechiporenko (1995) showed that Pleistocene sediment thickness and the distribution of glacigenic landforms are strongly correlated with the structure of the pre-Quaternary basement as well as with neotectonic structures.

In all these cases the depth of fine-grained sediments (e.g. Cretaceous and Tertiary marine clays and silts, Tertiary lignites, glacial and interglacial lacustrine clays and silts), appears to be the main controlling factor for the formation of thrust moraines. These layers formed a low-friction décollement on which the pushed sediment sheets moved. In areas where they were lacking, or out of reach of the ice-marginal stress field, thrust moraines are absent or very small. A discussion of the conditions and mechanics of thrust moraine formation can be found by Van der Wateren (1994a, 1995, 2002a). Deformation of sediments in the glacier bed rather than formation of thrust moraines can be expected where the surface before the arrival of the Pleistocene ice sheets consisted of clay and silt. Here the décollement was very shallow, within the top layer of the bed immediately beneath the glacier sole. The tectonic style is dominated by a shear zone fabric of extremely high shear strain, as described below. The properties of glacial sediments and landforms are to a large degree controlled by the structure of the substratum. In the plains underlain by Mesozoic/Cenozoic sedimentary basins, thrust moraines form predominantly where a suitable décollement – in fine-grained clastic and/or lignite layers – is sufficiently shallow. The schematic profile across the Northern European Plains (Fig. 8.3) illustrates the association of subsurface geology and thrust moraine structure/lithostratigraphy.

Local geological conditions determine whether a décollement is provided by either relatively young, or old lithostratigraphic units. In the centre of a sedimentary basin, salt diapirs may bring up layers, which in other parts of the basin are out of reach of the glacial stress field. If these contain fine-grained sediments these may act as a décollement for a thrust moraine ('a' in Fig. 8.3). Shallowing of the basin towards its margin brings up a potential décollement in stratigraphically higher units (e.g. upper Neogene marine clay and silt and lower Pleistocene fluvial and lacustrine clays; 'b' in Fig. 8.3). Near the margin of the basin again stratigraphically deeper units may be incorporated in thrust moraines (e.g. Mesozoic clays and Tertiary lignites; 'c' in Fig. 8.3). Thrust moraines formed close to the highlands tend to be relatively small, because sedimentary units thin out towards the basin margin. The highlands are fringed with zones which are generally free of

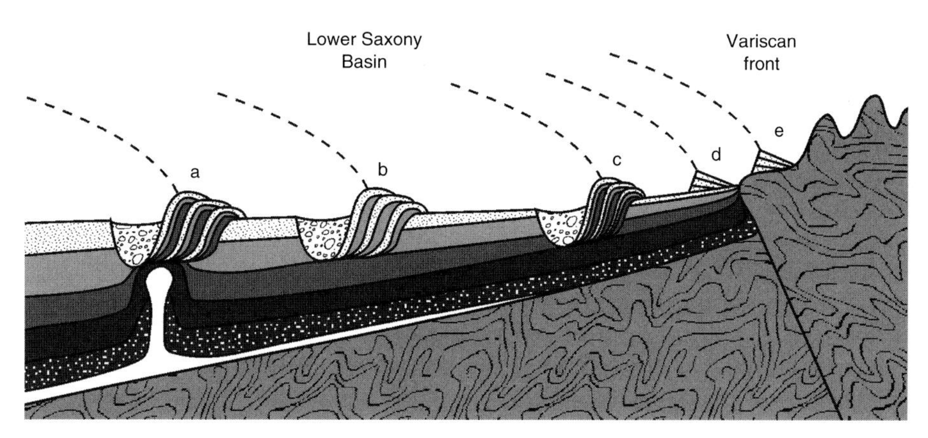

Figure 8.3 Schematic profile of the northern European plains north of the Variscan highlands showing the association of substrate geological structure and thrust-moraine structure/lithostratigraphy. In the plains underlain by Mesozoic/Cenozoic sedimentary basins, thrust moraines form predominantly where a suitable décollement in fine-grained clastic and/or lignite layers is sufficiently shallow. Depending on local geological conditions a décollement is provided by stratigraphic units of varying age. a = salt diapir brings up layers, which are too deep elsewhere, to act as a décollement for the thrust moraine, b = shallowing of the basin towards its margin provides a décollement in stratigraphically higher units (e.g. upper Neogene or lower Pleistocene clays), c = near the margin of the basin stratigraphically deeper units are incorporated in the thrust moraine (e.g. Mesozoic clays and Tertiary lignites). Thrust moraines formed close to the highlands (shallowest part of the basin) tend to be smaller than those further away from it. Dump end moraines (till and outwash) form at the ice margin where a suitable décollement is lacking ((d) and (e)). Dashed lines indicate successive ice sheet profiles. Vertical scale exaggerated.

thrust moraines. Here, basin sediments consist mainly of coarse-grained erosion products from the Variscan and Alpine ranges (e.g. Neogene and Early Pleistocene river and alluvial fan sediments), while fine-grained units that could produce a décollement are lacking. No thrust moraines are found in the highlands. In both areas, former ice margins are marked by accumulations of till and glacifluvial outwash (dump end moraines; 'd' and 'e' in Fig. 8.3).

The Dammer Berge, one of the largest Saalian thrust moraines (Fig. 8.4), illustrates how basin structure and (neo)tectonic activity combined to create the conditions necessary for its formation (Van der Wateren, 1995). Up to 50 m thick nappes and thrust sheets in the western half of this horseshoe-shaped thrust moraine are built largely of Miocene and Pliocene units, while Pleistocene units are relatively thin or even lacking. Tertiary clays invariably form the intensely deformed basal units on which the nappes moved. These sediments originated from the glacial basin where an originally Variscan anticline just north of this part of the thrust moraine (Fig. 8.4a) has been reactivated during the Alpine orogeny (Ziegler, 1990). This has lifted the Tertiary units to shallower levels (Fig. 8.4b). The lithostratigraphy in the eastern half of the Dammer Berge is dominated by Early and Middle Pleistocene fluvial deposits with a basal shear zone usually consisting of a Pleistocene, sometimes Tertiary, clay layer. Glacial overdeepening of the glacial basin from which the nappe units had been removed enabled the ice to cut into a Campanian-

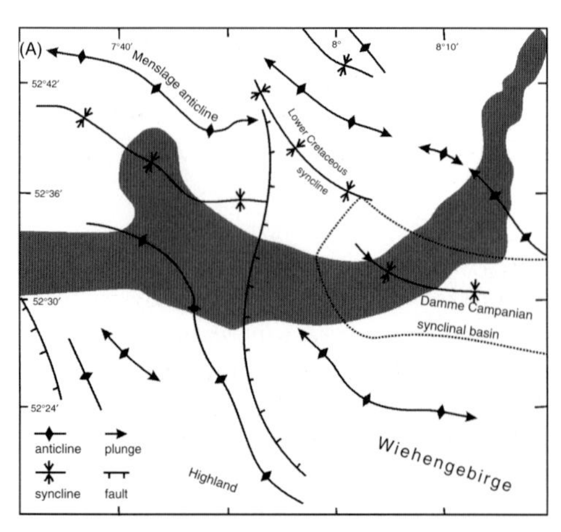

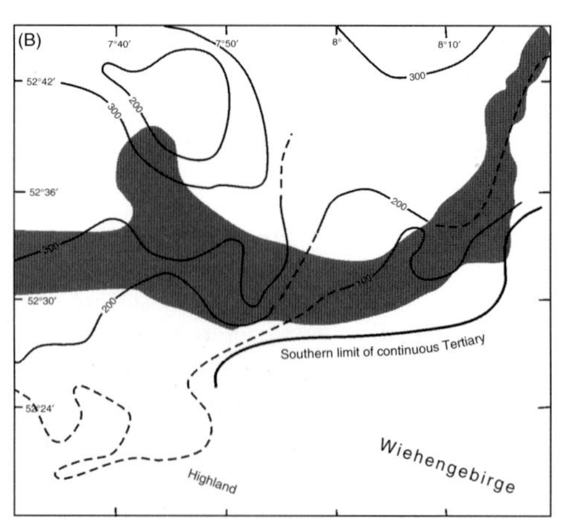

Figure 8.4 The Dammer Berge thrust moraine (Van der Wateren, 1995; grey shading, see Fig. 8.19a for location) north of the Variscan highlands in Germany. Neotectonic movements strongly affected the underlying Variscan folds and high-angle faults (A), which is reflected by the contours of the base of the Tertiary (B). The combination of shallowing of the Lower Saxony Basin towards the margin with the effect of (neo)tectonic movements carried upper Cretaceous and middle Tertiary clays close enough to the surface for them to be incorporated in the thrust moraine. The Campanian synclinal basin underlying the eastern part of the Dammer Berge is the source of the clay which has been reworked in some of the nappe shear zones and a clay-rich till (Van der Wateren, 1995). (Data courtesy of the Archive, Niedersächsisches Landesamt für Bodenforschung, Hannover.)

Maastrichtian clay. These sediments were moved to the highest flanks of the thrust moraine to produce a till, which is made up entirely of this clay with a few scattered Scandinavian erratic pebbles (Van der Wateren, 1987).

8.4 DISTRIBUTION OF GLACIAL LANDFORMS AND GLACITECTONIC STYLES

The discussion by Hart *et al.* (1996, 1997) and Piotrowski *et al.* (1997) is a clear illustration of how different interpretations of sedimentary structures, and whether or not they have been significantly deformed by subglacial shearing, may lead to widely disparate interpretations of glacial landsystems. The reconstruction of the dynamics of former ice sheets requires that it can be determined whether sediments have either been sheared by overriding ice, or deposited by lodgement or subglacial melting. Consequently, it is of critical importance to establish an unambiguous set of criteria that can be used to identify subglacial deformation.

Slater (1926), Gry (1942) and Banham (1975, 1977) were among the first to recognize the similarities between the deformation structures present in tills and those in regionally deformed metamorphic rocks (e.g. slates, schists) and shear zones (e.g. mylonites). Shear tests with clay by Maltman (1987) produced structures very similar to cleavages in low-grade metamorphic rocks (e.g. slates), while later studies of actively deforming accretionary wedges (Byrne, 1994; Maltman, 1994) demonstrated that these structures indeed form in unlithified sediments under natural conditions. Alley (1991a), Boulton (1996a, b), Boulton and Hindmarsh (1987) and Boulton and Jones (1979) showed that the dynamics of mid-latitude ice sheets are, to a high degree, controlled by deformation of the bed. As we have seen, outside their core regions the Southern Scandinavian Ice Sheets advanced over a bed of unlithified and weakly lithified sediments.

As a consequence of ice flow, sediment is continuously being transported from the internal parts of an ice sheet towards its margin. Under the right conditions basal shear stresses may cause the sediments in the bed to deform by simple shearing, which moves them towards the margin. Slowing down of the sediment flow in the marginal zone leads to horizontal compression and deposition. This pattern of deformation of the bed is similar to that in the overlying ice, where extending flow gives way to compressive flow near the margin. In sediments within a glaciated region we can therefore distinguish two glacitectonic regimes, the subglacial shear zone and the marginal compressive belt (Van der Wateren, 1995).

Two types of thrust moraine serve as models for the interpretation of structures forming at the ice margin. The first (Fig. 8.5) is based on the Holmströmbreen thrust moraine in Spitsbergen (Van der Wateren, 1995; Boulton *et al.*, 1999). The structure of this thrust moraine is dominated by folds and relatively thick fold nappes (aspect ratios of 5:1 to 8:1). This style is typical of thrust moraines where the bulk rheology is controlled by relatively ductile fine-grained sediments. A unit of coarse-grained sediments overlies a unit of roughly equal thickness of fine-

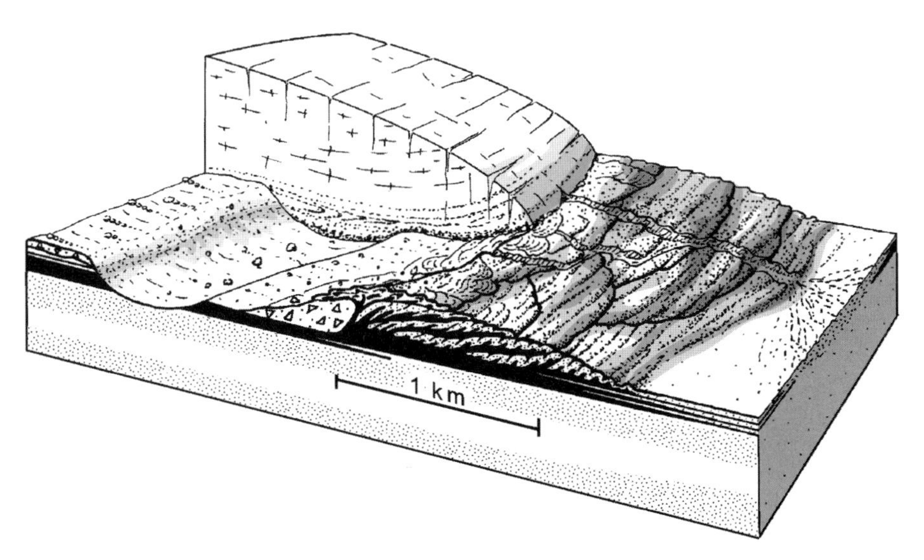

Figure 8.5 Model of a thrust moraine in which the glacitectonic style is controlled by relatively ductile fine-grained sediments. Nappes have low aspect ratios of 5:1 to 8:1 (up to 20 m in thickness and 100 m long in cross section). The model is based upon the thrust moraine which formed during a late 19th century surge of the Holmströmbreen glacier in Spitsbergen (Van der Wateren, 1995; Boulton *et al.*, 1999). Vertical scale is exaggerated. Volume of the glacial basin, from which the sediments have been removed, does not match the volume of the thrust moraine.

grained sediments – in this case, sand and gravel of several glacifluvial outwash fans, and silt and clay of a proglacial delta, respectively. This type of thrust moraine is characteristic of areas in central Europe where Tertiary lignites crop out. In the Holmströmbreen thrust moraine the entire structural succession is represented in a section from the undeformed foreland to the glacier margin (Van der Wateren, 1995). The central European thrust moraines are mainly built of one unit of folds and thrust sheets above a detachment surface, while piles of nappes are rare.

The second type (Fig. 8.6) is based on the Dammer Berge thrust moraine, Germany, Drenthe Stage, Saalian Glaciation (Van der Wateren, 1995). It comprises a stack of up to six subhorizontal nappes measuring up to 12 km^2. Their aspect ratios range from 20:1 to 50:1. They are composed of relatively thick coarse-grained sediments (typically 20–50 m, 150 m maximum) overlying a thin layer of silt and clay (0.5–3.0 m). This is characteristic of areas where glacifluvial outwash fans and fluvial sands and gravels overlie Pleistocene fluvial clays or Tertiary marine clays. Such sedimentary sequences are relatively stiff and favour transmission of stress over great distances. This leads to a wide spacing of folds and thrusts and the formation of relatively thin nappes, which each may contain several high-angle thrusts and folds.

Terrains within the marginal zone experience a different style of deformation resulting from horizontal shear stresses applied by the overlying ice (Fig. 8.7):

A. Undeformed foreland. Dump end moraines and outwash fans, consisting mainly of proglacial outwash and supraglacial flow tills, directly related to the latest pushing event, overlying older sediments.

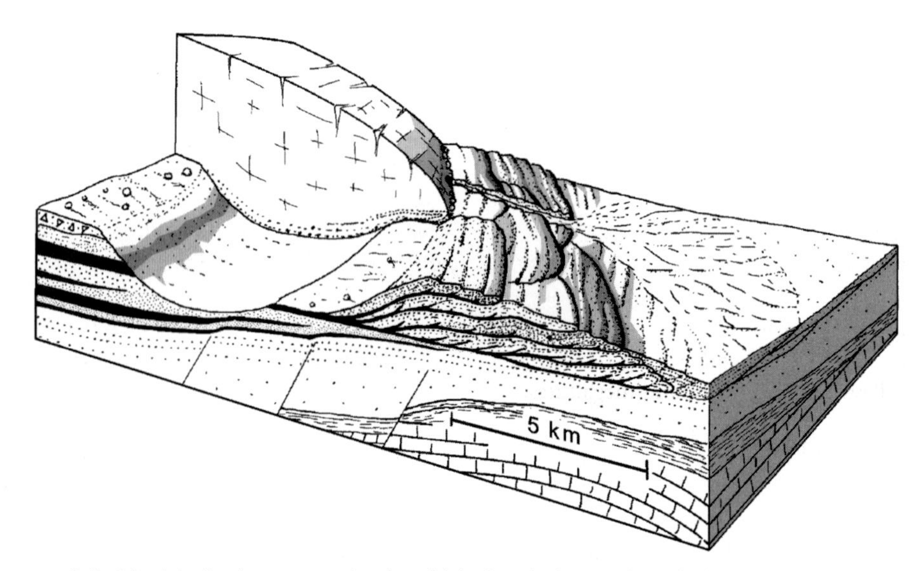

Figure: 8.6 Model of a thrust moraine in which the glacitectonic style is controlled by relatively stiff and brittle coarse-grained sediments. Based upon the Dammer Berge thrust moraine, Germany, Drenthe Stage, Saalian Glaciation (Van der Wateren, 1995). This type of thrust moraine is built of nappes and thrust sheets of high aspect ratios – between 20:1 and 50:1. Nappes are typically between 20 and 50 m in thickness and may measure up to 12 km^2. Vertical scale exaggerated and glacial basin not to scale.

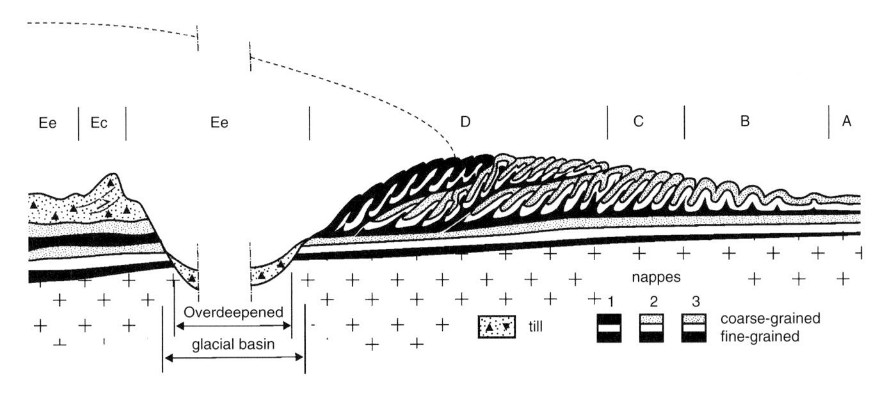

Figure 8.7 Synthesis of glacitectonic styles in relation with the ice margin (dashed line). A = undeformed foreland: dump end moraine, outwash fan, B = steeply inclined structures: concentric style folds, reverse faults, C = overturned and recumbent structures: thrusts, folds, D = nappes, E = extensional structures: deformation tills and other subglacial tills, Ec = compressive structures in E style terrain: stacked till sheets, Ee = strong extension in E style terrain: erosional features, tunnel valleys, overdeepened glacial basins. E-type structures are particularly widespread in the extensive till plateaux well inside the glacial margins. Till on top of the thrust moraine is not shown. It may consist of strongly deformed sediments from the thrust moraine or deformed sediments from stratigraphically lower levels subcropping in the overdeepened glacial basin (e.g. Dammer Berge, Van der Wateren, 1995).

B. High-angle structures. Jura-style concentric and box folds with vertical and steeply dipping axial surfaces. Thrusts are rare and steeply dipping, with small offsets. Minimal horizontal tectonic shortening.

C. Low-angle structures. Strongly asymmetric, overturned and recumbent folds. Low-angle thrusts. Medium shortening.

D. Nappes. Extensive horizontal and relatively thin thrust sheets. Maximum shortening. Nappes may contain a number of high-angle folds and reverse faults.

E. (Deformation) tills. Boudinage, boudinaged folds and folded boudins, subhorizontal shear planes, transposed foliation. Extremely high shear strain and horizontal extension. Subdivided into:
Ec (compression), comprising stacked till sheets and other compressive structures and
Ee (extension), with strong erosion of the substratum, comprising overdeepened (tunnel) valleys, drumlin fields and megaflutes.

Styles B, C and D represent terrains of horizontal compression, with increasing longitudinal strain from B to D, whereas E is characterized by strong horizontal extension as a result of progressive simple shear. As the flow rate in a subglacial deforming bed decreases towards the margin, compressive structures may be expected to occur in tills, which are deposited close to the margin. Hence the division of E-style terrains in a purely extensional zone and a compressive zone, Ee and Ec, respectively.

Terrains may be expected to show different overprinting relationships if they are produced by either an advance, or a readvance during a general glacial retreat. Two sequences may be distinguished (Fig. 8.8):

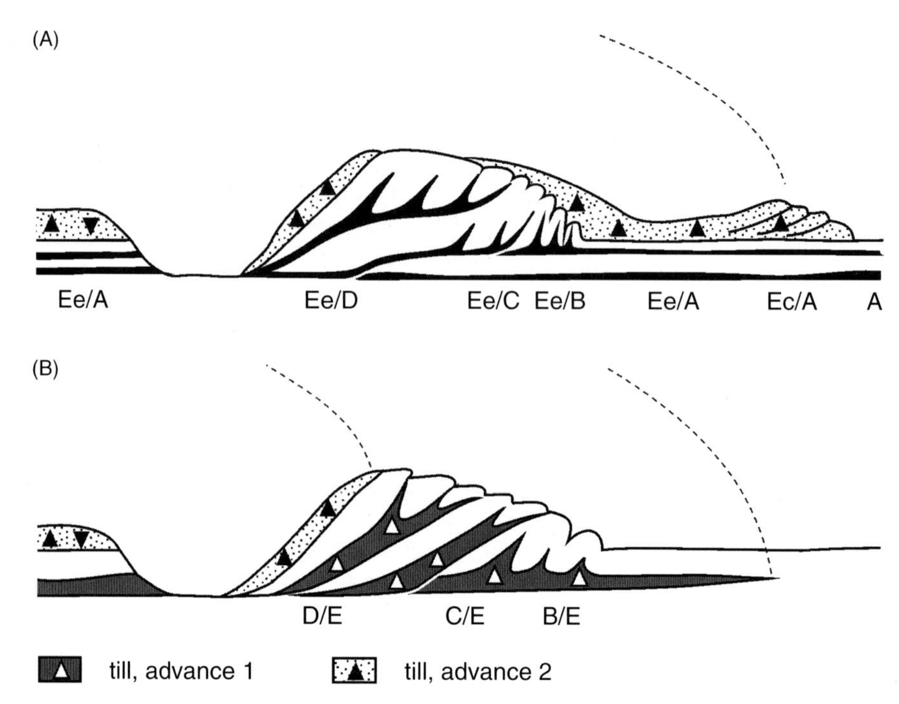

Figure 8.8 Overprinting of glacitectonic styles in A) advance sequence, and B) readvance sequence during retreat. Open triangles are older till, black triangles are younger till associated with the formation of the thrust moraine. Dashed lines indicate ice sheet limits. Using simple overprinting relationships makes it possible to distinguish moraine lines produced during a glacial advance from those produced during retreat.

1. Advance sequence (Fig. 8.8a), in which style E overprints compressive structures B, C, D and undeformed foreland, A. Boulton (1987) argued that drumlins cored by coarse-grained sediments may be the result of overriding and streamlining of glacial outwash (i.e. tectonic style E overprinting A (E/A)). Many drumlins in northwestern Germany originated as thrust moraines, which have subsequently been overridden (Stephan, 1987). The Rehburg thrust moraines in western Germany are overlain by deformation till and have been streamlined by the overriding ice sheet. Those in the northern part of The Netherlands have been pushed and subsequently overridden after a temporary stagnation of the glacial advance (Rappol *et al.*, 1989; Van den Berg and Beets, 1987). These are examples of style E/(B,C,D).
2. Readvance sequence (Fig. 8.8b). When a readvance occurs during a general retreat, tills and outwash dating from the last glacial advance may be incorporated in thrust moraines. They may be quite rare, as they are susceptible to erosion by meltwater and overriding ice. Thus B, C and D overprint E style structures. The Lamstedt thrust moraine in western Germany is an example of (C,D)/E.

Overprinting of glacitectonic styles can be observed on all scales, from the map scale to outcrop and microscopic scale, and may be a useful tool to unravel advance and retreat sequences. Overprinting relations provide a means to relatively date successive deformation episodes. This has proved to be successful in the Lamstedt and Wilsum thrust moraines, west Germany, and the adjoining area in The Netherlands (van Gijssel, 1987; Kluiving, 1994; Kluiving *et al.*, 1991).

Figure 8.9 shows the different map patterns ensuing from overprinting of different glacitectonic styles. These allow the identification of stable ice margins, advance and retreat sequences. A stable ice-sheet margin, or an episode of long-term stagnation during either an advance or retreat (Fig. 8.9a), is associated with dump end moraines (style A, usually large outwash fans) downstream of the overridden thrust moraine (E/D, E/C). A stable ice-sheet margin is also indicated by a line of B-, C- and D-style thrust moraines, which have not been overridden (Fig. 8.9b). An advance sequence (Fig. 8.9c) shows overridden thrust moraines (E/D, E/C) upstream from the advance limit, which is indicated by patterns such as in Fig. 8.9a and 8.9b. A readvance sequence (Fig. 8.9d) is indicated by stacked sheets of till (Ec) deposited during the previous advance or, alternatively, by a thrust moraine containing till from the previous advance.

8.5 STRUCTURES DEVELOPED IN RESPONSE TO SUBGLACIAL DEFORMATION

The developing discussion within the field of glacial research regarding deforming beds and their possible widespread occurrence (e.g. Boulton, 1987, 1996a; Boulton and Hindmarsh, 1987; Hart and Boulton, 1991; Hart and Roberts, 1994; Benn and Evans, 1996; Boulton *et al.*, 1996; Hart *et al.*, 1996, 1997; Piotrowski *et al.*, 1997, 2001; Piotrowski and Tulaczyk, 1999; Van der Wateren, 1999, 2002b; Van der Wateren *et al.*, 2000; Ó Cofaigh and Evans, 2001a, b; Stephan, 2002;) calls for reliable criteria to distinguish subglacially deformed from undeformed sediments.

Where meltwater drains subglacially, deformation of the bed is the most likely mechanism to produce tills in areas such as the Northern European Plains, which are underlain by unlithified and weakly lithified sediments (Boulton, 1996a). Identifying the products of subglacial deformation thus remains vitally important in the reconstruction of glacial landsystems. The simplest model to describe subglacial deformation is progressive simple shear (Van der Wateren *et al.*, 2000). This describes the deformation history of most subglacial tills with sufficient accuracy, while the geometry of its fabric is relatively easy to understand in terms of the model. In reality, both simple and pure shear (flattening and horizontal compression) usually act together. Compaction by dewatering and ice loading are examples of shear plane normal compression that result in volume loss. Flow within the deforming bed is generally not uniform. Large boulders or lenses of stratified sediment that move relatively slowly tend to have sectors of compressed sediment on their upstream side and extended sectors on their downstream side. Shear plane parallel/subhorizontal compression occurs at the ice-sheet margin, leading to stacking of till sheets. Although far from uncommon, these departures from the ideal simple shear model are minor compared with the vast amounts of shear strain in subglacial shear zones. Usually they do not present problems for the interpretation of shear zone structures. For a comprehensive treatment of the kinematics of subglacial shear zones and their structural characteristics, see Van der Wateren *et al.* (2000).

The genesis of laminated diamicts is the most controversial issue in the deforming bed debate. Layers, lenses and boudins of sorted sediments interbedded with diamict are common features of

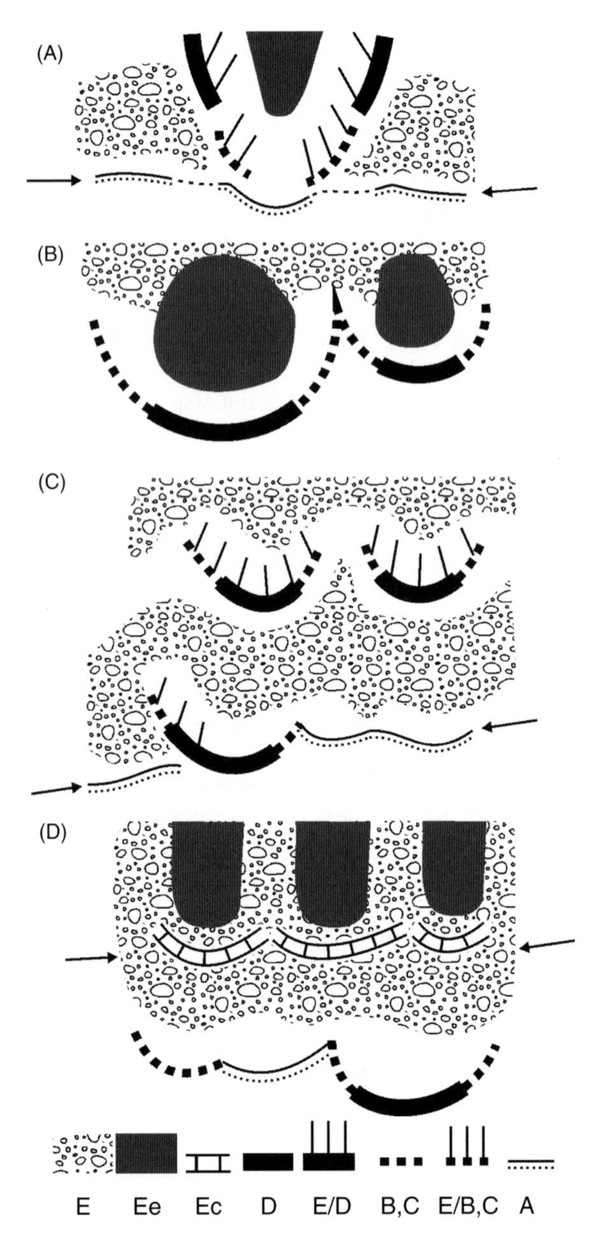

Figure 8.9 Examples of map patterns of glacitectonic styles indicating stable ice sheet margins, advance and retreat sequences. A) Stable ice sheet margin (arrows) located along the dump end moraines (A) downstream of the overridden thrust moraine (E/D, E/C). B) Stable ice sheet margin was located along the (B)-, (C)- and (D)-style thrust moraines which had not been overridden. C) Advance sequence. Overridden thrust moraines ((E/D), (E/C)) upstream from the advance limit (arrows). D) Readvance sequence (arrows: limit of readvance). Stacked sheets of till (Ec) deposited during the previous advance, which has its limit further downstream. See Fig. 8.19 for an application of this method of mapping and identifying different glacial margins.

subglacial tills. They produce a lamination, which may sometimes be confused with sedimentary layering. In this context, intraclasts of stratified and sorted sediment are commonly seen as evidence of subglacial meltwater discharge (Menzies and Shilts, 2002). Where these can be correlated with footwall sediments that are older than the diamict, subglacial melt-out and/or water flow cannot account for the laminations. In these, indeed very common, cases subglacial deformation of the sediments together with the diamict is the most obvious explanation (Van der Wateren, 1995, 1999, 2002a; Van der Wateren *et al.*, 2000).

8.6 STRUCTURAL STYLE OF SUBGLACIAL SHEAR ZONES

Structures in the deforming bed are produced by progressive simple shear, disregarding relatively minor deviations from this ideal deformation model. This deformation history leads to a monoclinic symmetry in three dimensions (Passchier and Trouw, 1996). This shows in two-dimensional view (in the plane of shearing) as a typically asymmetric structural geometry, which is the most reliable indicator of tectonic movement and therefore ice movement direction. Structural styles of brittle and ductile shear zones are summarized below. They are treated in greater detail in Van der Wateren *et al.* (2000).

Brittle- and ductile-style structures usually coexist in shear zones in unlithified sediments. Brittle-style structures usually form in the more competent coarse-grained sediments, while clays show more ductile behaviour. Generally this is a matter of scale, as macroscopically, ductile deformation may be the cumulative effect of numerous microscale discrete displacements. Figure 8.10 explains the ductile and brittle shear zone structures, which are common in tills on all scales varying from microscale to outcrop scale. The fabric is similar to that of many mylonites and cataclastic shear zones and a matching terminology to describe these features (Passchier and Trouw, 1996) is adopted here.

A ductile shear zone contains shear band surfaces with a characteristic asymmetry in the plane of shearing (Fig. 8.10a). Parallel-oriented platy and prismatic mineral grains define a penetrative cleavage of S surfaces, which is oriented roughly parallel to the direction of maximum finite extension. In thin section and in crossed polarized light, the S cleavage shows a dominant birefringence direction (illumination) of clay minerals. Rotation of skeletal grains (sand-to-pebble size) leads to preferred orientations toward the shear direction. The common method of measuring clast orientations to determine the till fabric is based on this principle. C and C′ planes are more widely spaced shear bands also known as extensional crenulation cleavages (ECC), with the C surfaces oriented parallel to the boundaries of the shear zone. As in mylonites (e.g. Passchier and Trouw, 1996) the two most common combinations of cleavages are termed S-C fabric (Fig. 8.10b) and S-C′ fabric (Fig. 8.10c).

Figure 8.10d explains the geometry of Riedel shear surfaces (Y, P, R and R′) in a brittle shear zone. Tension veins (T) are oriented perpendicular to the direction of maximum extension. They are filled with either fluidized sorted sediment or till and are equivalent with injection veins, till wedges and clastic dikes described in numerous till sections (e.g. Åmark, 1986; Dreimanis, 1992; Larsen and Mangerud, 1992; Dreimanis and Rappol, 1997; Rijsdijk *et al.*, 1999; Van der Wateren, 1999). The characteristic asymmetry in the plane of shearing makes it relatively easy to determine the shear direction – in this case dextral (clockwise) shearing. The fabric has a markedly higher degree of symmetry in sections perpendicular to the direction of shearing. Figure 8.11

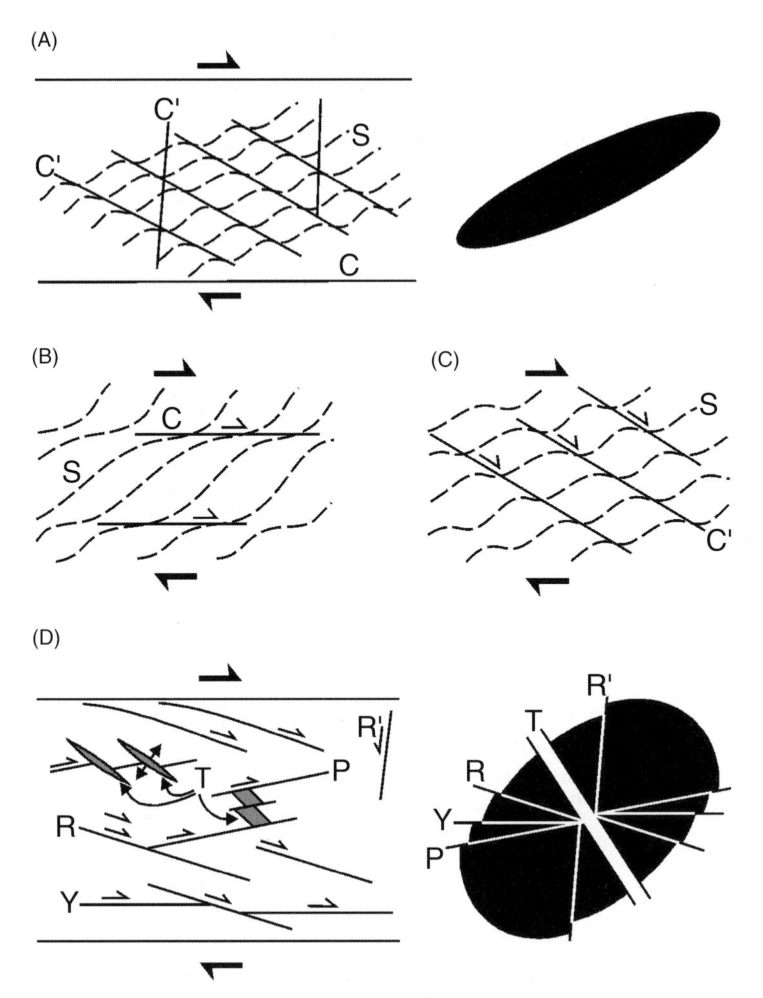

Figure 8.10 Ductile (A–C) and brittle (D) shear zone structures, which are common in tills on a microscale as well as on outcrop scale. Simple shear produces structures with monoclinic symmetry (asymmetric in the plane of shearing – see strain ellipse), which is an important indicator of the direction of shearing and, therefore, in tills of the ice flow direction. The fabric is similar to that of many mylonites and cataclastic shear zones: A) Geometry of shear band surfaces and symmetry of a dextral ductile shear zone. C and C′ planes also known as extensional crenulation cleavages (ECC). B) S–C fabric. C) S–C′ fabric. D) Geometry and shear direction of Riedel shears (Y, P, R and R′) in a dextral brittle shear zone. T indicates tension veins oriented perpendicular to the direction of maximum extension, which are filled with fluidized sediment. The diagram also shows a strain ellipse produced by brittle fracture.

summarizes structures produced by progressive simple shear, indicating how they can be used as kinematic indicators. For ease of comparison, all structures indicate dextral simple shearing. The cross section of sheath folds – strongly attenuated tube- or sock-shaped structures parallel to the shear direction – are in a plane perpendicular to the shear direction.

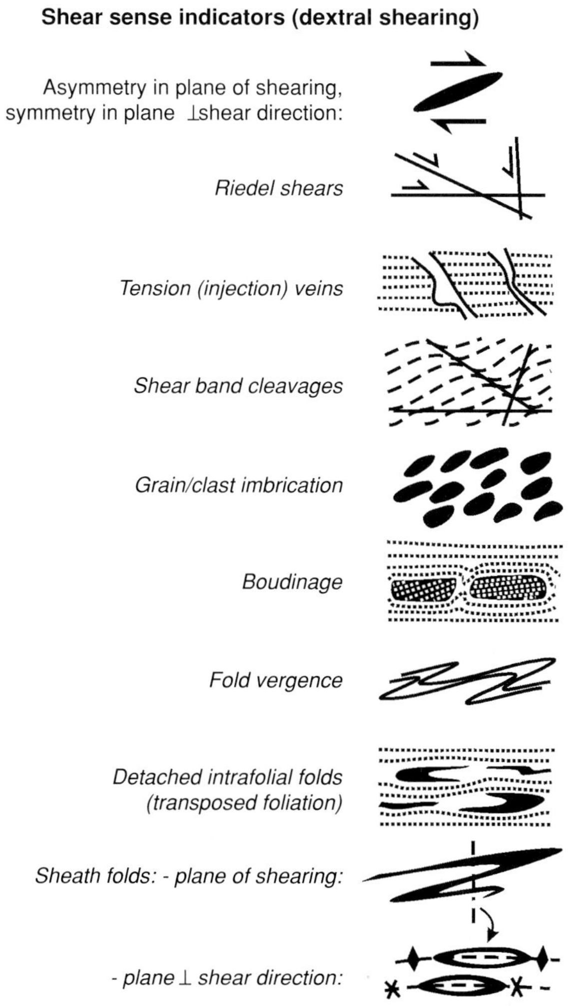

Shear sense indicators (dextral shearing)

Asymmetry in plane of shearing,
symmetry in plane ⊥shear direction:

Riedel shears

Tension (injection) veins

Shear band cleavages

Grain/clast imbrication

Boudinage

Fold vergence

Detached intrafolial folds
(transposed foliation)

Sheath folds: - plane of shearing:

- plane ⊥ shear direction:

Figure 8.11 Summary of structures produced by progressive simple shear. The marked asymmetry of these structures in sections parallel to the plane of shearing makes them reliable kinematic indicators. All structures indicate dextral simple shear. The cross sections of sheath folds at the bottom are perpendicular to the shear direction, illustrating striking difference in symmetry between the two sections.

8.7 Ductile Shear Zone Structures

A deformation till at Gliedenberg, Dammer Berge, Germany (Fig. 8.12) is a clear illustration of how subglacial shearing of sediments may produce a laminated till and ultimately a massive diamicton within just 100 m. This process is similar to the one described by Hart and Boulton (1991). The Dammer Berge (Van der Wateren, 1987, 1995) are a thrust moraine formed during the Drenthe advance of the Saalian glaciation. They were overridden when the ice expanded to the Saalian limit, 100 km further to the south. The till is a mixture derived from the four Tertiary and

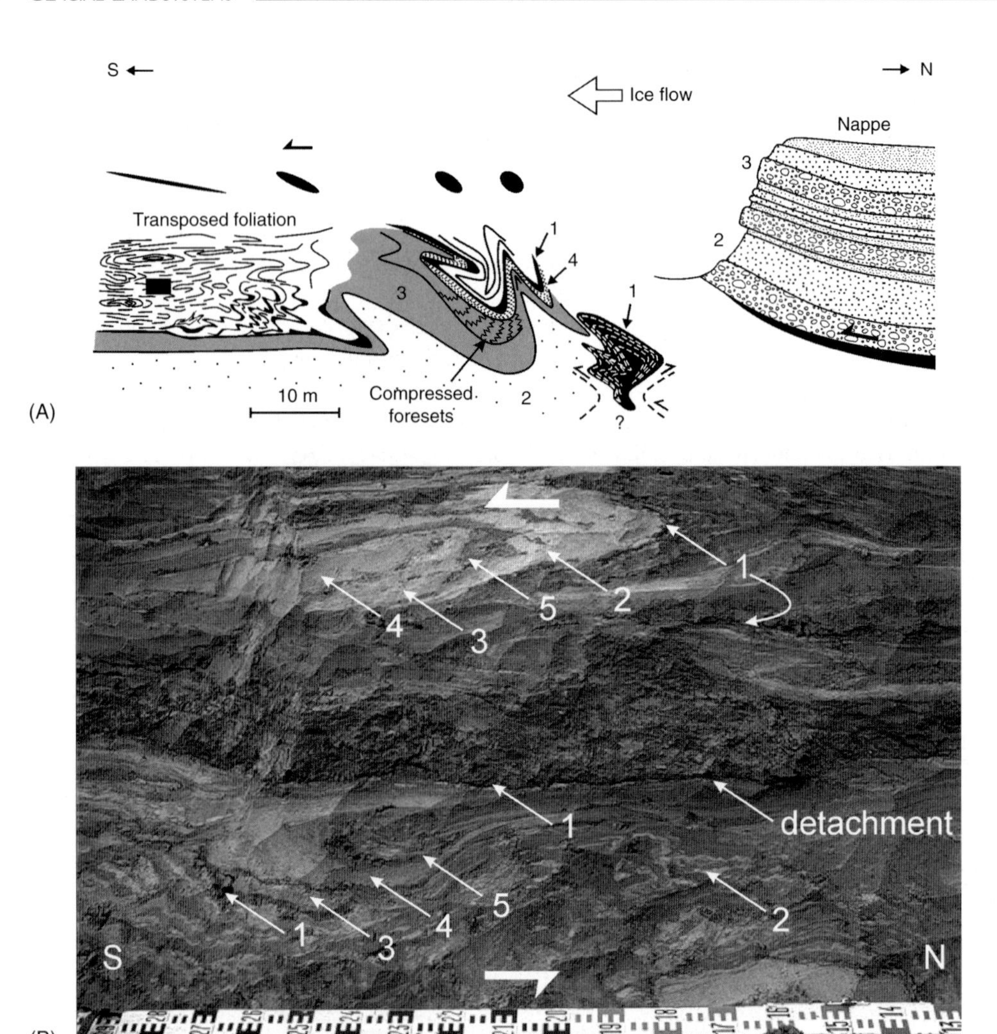

Figure 8.12 Deformation till at Gliedenberg, Dammer Berge, Germany (Van der Wateren, 1995). The Dammer Berge is a thrust moraine from the Saalian Drenthe advance and which had been overridden when the ice expanded 100 km further to the south. The till is a mixture derived from the four Tertiary and Pleistocene formations which comprise the thrust moraine. A nappe containing these sediments is exposed 50 m upstream (north) from the till outcrop. They are folded and extruded in a diapiric structure, in front of the nappe, cored with Tertiary clay (black). A) Schematic diagram showing deformation till and its source sediments at Gliedenberg, Dammer Berge thrust moraine, Germany (Van der Wateren (1995). Subglacial shearing of the source sediments produced a laminated diamict with numerous boudins and detached intrafolial folds. The lamination therefore is not a sedimentary layering but a transposed foliation, the product of glacitectonic deformation. Numbers refer to caption of Fig. 8.12b. B) The deformation till is composed of: 1 = Tertiary glauconitic clay, 2 = white Lower Pleistocene fluvial sand, 3 = pink Middle Pleistocene fluvial sand, 4 = green glauconitic glacifluvial sand (mixture of Pleistocene and Tertiary sediments), 5 = brown diamict.

Pleistocene formations, which build the thrust moraine. A thrust sheet containing these sediments is exposed 50 m upstream (north) from the till outcrop. As a result of loading by this structure, the sediments are folded and extruded in front of it in a diapiric structure, cored with Tertiary clay (Fig. 8.12a).

The structures downstream of this are the product of overprinting of the upright and steeply dipping folds and diapirs by intense horizontal simple shear. Shearing of the source sediments produced a laminated diamict with lenses, discontinuous layers, boudins and detached intrafolial folds in which the original lithology can still be distinguished (Fig. 8.12b). The lamination clearly is not a sedimentary layering but a transposed foliation, the product of glacitectonic deformation. The individual components of the deformation till are Tertiary glauconitic clay, Lower Pleistocene white quartz-rich fluvial sand, Middle Pleistocene pink Buntsandstein-rich fluvial sand, green glauconite-rich glacifluvial sand (mixture of Pleistocene and Tertiary sediments) and a brown diamict.

Only a few tens of metres downstream the sediment changes aspect from a laminated diamict to a macroscopically massive till in a manner described by Boulton (1996a) as 'dissolving' of the sorted sediment bodies into the unsorted matrix. On the microscale, the matrix appears less homogeneous, as it consists of clay-rich and sandy laminae. The structural symmetry of the S, C and C′ cleavages indicates sinistral shearing, consistent with the macrostructural geometry. Small boudins of muddy diamict and strongly oriented clay rims around rotated sand grains indicate very high finite shear strain, the sediment lenses and boudins forming 'islands of low strain' (Passchier and Trouw, 1996) in a high-strain matrix. At high finite strains (related to the transport distance within the till) they are only visible at high magnifications in thin sections (Van der Wateren, 1995, 2002a; Van der Wateren *et al.*, 2000). The example of the Gliedenberg till demonstrates that homogenization may occur within a distance of the order of 100 m.

8.8 Brittle Shear Zone Structures

Shear zones in unlithified sediments rarely contain either only brittle- or only ductile-style deformations, usually the two are associated. Clay, which typically deforms in a ductile way, is sometimes cut by brittle structures due to strain hardening associated with loss of pore water. This association of brittle structures and ductile structures is illustrated by a shear zone beneath one of the nappes in the Dammer Berge thrust moraine (style D; Van der Wateren, 1995). The sands and gravels of the footwall are cut by sets of Riedel shear planes, while the silts and clays at the base of the overriding nappe contain a transposed foliation of boudins, sand lenses and intrafolial folds (Fig. 8.13). This is a mixture of footwall and hangingwall materials.

This sub-nappe shear zone shows characteristic upward increasing shear strain, which is also common in many tills. At the interface between the glacier sole and the bed, strain rates are at maximum (Boulton, 1987; Boulton and Hindmarsh, 1987). The intensity of subglacial deformation therefore increases from the *in situ* footwall sediments to the top of the deforming layer, which carries the most allochthonous elements. Following Van der Wateren (1987), and similar to Hart and Boulton (1991), the shear zone can be subdivided in units Sr, Sb and Sh, grading from relatively intact to completely homogenized sediment (Fig. 8.13). Homogenized unit Sh at the base of the overriding nappe contains boudins of a Cretaceous clay. They contain a

Figure 8.13 Association of brittle structures (Riedel shears) and ductile structures (boudins, transposed foliation, intrafolial folds) in a shear zone beneath the uppermost nappe in the Dammer Berge thrust moraine (Van der Wateren, 1995). The shear strain increases upward, which is also representative of many tills. Unit S_r contains rooted structures (top of lower nappe). S_b in the lower half of the shear zone consists of a transposed foliation of boudins and detached intrafolial folds of footwall material in a more allochthonous matrix. S_h, the unit with the highest finite strain, is a completely homogenized mixture of local (footwall, nappe III) and far-travelled material (hangingwall, nappe IV). (After Van der Wateren, 1987). This unit contains boudins of a Cretaceous clay which has been transported from a level ~150 m below the outcrop.

Campanian-Maastrichtian fauna and originate from Cretaceous beds subcropping in the glacial basin north of the thrust moraine, ~150 m below the level of the outcrop.

Observations of till sections in Germany and The Netherlands indicate that the intensity of deformation quite commonly increases upwards from undeformed sediments to a strongly sheared and homogenized diamict at the top (Van der Wateren, 1987, 1995; Kluiving *et al.*, 1991). This produces a petrographic layering where higher levels contain clasts of distant provenance, whereas those from lower levels have a more local origin (Rappol and Stoltenberg, 1985; Rappol *et al.*, 1989; Boulton, 1996a). Subglacially deforming sediments are incorporated into the shear zone from below, and consequently particles higher up in the subglacial shear zone have travelled greater distances than have those in the bottom layers.

In a cliff section on the Baltic coast near Heiligenhafen, northern Germany, three Pleistocene tills can be distinguished on the basis of structural and petrographic characteristics. The Lower and Middle Tills had previously been ascribed to the Saalian, and the Upper Till to the Late Weichselian (Stephan *et al.*, 1983). In a more recent evaluation Stephan (2002) recognized three stratigraphically different tills below the Weichselian Upper Till.

An alternative explanation holds that the Lower and Middle Tills constitute one single subglacial shear zone with relatively autochthonous sediments in the bottom part and more exotic far-travelled material at the top (Van der Wateren, 1999, 2002b). This is corroborated by a parallel trend towards increasing deformation (finite strain) from the bedrock to the top of the section. The most obvious aspect of this latter trend is the massive appearance of the Middle Till, which can be interpreted as the result of homogenization by repeated folding and attenuation of sediment lenses that have been incorporated into the till.

8.9 GLACIFLUVIAL OUTWASH, ICE-MARGINAL VALLEYS AND LAKES

Meltwater plays an important role in ice-marginal landsystems. Where a glacial margin stagnates, sediment accumulates in the form of outwash fans (sandur) and glacifluvial deltas. A new advance may incorporate these sediment bodies into thrust moraines. Lakes form where glacial advances block ice-marginal valleys. These lakes may cause permafrost to decay over large areas, thus generating an important morphodynamic control on glacial advances. The area covered by the Rehburg advance during the Saalian may serve as a model to illustrate the morphodynamic relationships at an ice-sheet margin.

Rivers like the Rhine and Meuse in The Netherlands, Weser and Ems in western Germany, the Elbe in central Germany, and the Odra and Vistula in Poland, used to drain into the North Sea and Baltic Sea basins. Successive Pleistocene glaciations redirected these rivers to a westward course parallel to the ice-sheet margins. True ice-marginal valleys (in German Urstromtal, plural Urstromtäler) formed only during the Weichselian glaciation. From south to north (old to young) the major Weichselian ice-marginal valley systems are the Baruth-Głogów, Warsaw-Berlin, Warsaw-Toruń-Eberswalde and Kashubian-Pomeranian Urstromtäler (Ehlers, 1996; Marks, 2002). For the Elsterian and Saalian glaciations ice-marginal drainage can only be reconstructed by sediment-petrographical and provenance studies (Meyer, 1983; Zandstra, 1983). Valley incision into bedrock was largely inhibited because the ice margins spent relatively short intervals of time next to and in the highlands (Liedtke, 1981).

When an ice sheet advances across an Urstromtal, ice-marginal river and lake sediments may be incorporated into thrust moraines (Van der Wateren, 1994b, 1995; Fig. 8.14). The advancing ice lobe and emerging thrust moraines block river channels, producing ice-marginal lakes. Subaquatic fans build at the mouths of subglacial meltwater tunnels (Fig. 8.14b). Ice-marginal deltas and subaerial alluvial fans form on top of and adjacent to subaquatic fans, silting up the lake (Fig. 8.14c). The advancing ice lobe compresses these sediments into thrust moraines and the subaerial drainage forms a rectangular pattern between the outcropping nappe and fold structures. Subaquatic mass flows move down the partly submerged thrust moraine slopes into the lake.

Meltwater streams interact with the emergent thrust moraine forming syn- and post-tectonic terraces and channel fills (Boulton, 1986; Van der Wateren, 1987). In cross section these show as stacked sediment bodies of contrasting finite strain (Fig. 8.15). The terms pre-, syn- and post-tectonic refer to stages in the formation of a thrust moraine. The lowest of the stack, pre-tectonic folded/thrusted sediments – including glacifluvial outwash overlying preglacial sediments – have undergone the highest degree of deformation (finite strain). Syn-tectonic sediments fill the synclinal valleys on top of the thrust moraine while it was forming and therefore show varying amounts of finite strain: folding of the basal unconformities becomes progressively less tight in Fig. 8.15. They form terraces, glacifluvial deltas and channel deposits unconformably overlain by undeformed post-tectonic sediments (glacifluvial channel and alluvial fan deposits).

One of the most outstanding features of glacifluvial sediments in many thrust moraines in northwestern Europe is the large proportion of subaquatic sediments. Sedimentary sections

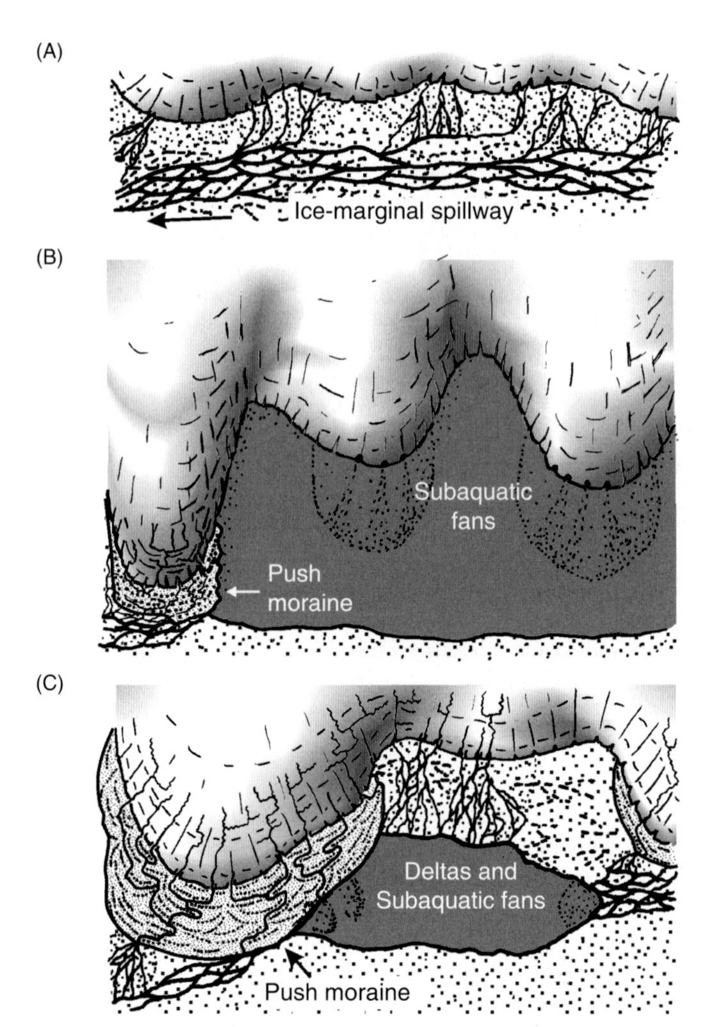

Figure 8.14 Model of pre-tectonic and syn-tectonic drainage of the area of the Rehburg thrust moraines based on sedimentological analyses of glacifluvial sediments in thrust moraines (Van der Wateren, 1994b, 1995). A) Pre-tectonic westward drainage of the ice-marginal braided system (Urstromtal). B) Formation of an ice-marginal lake by blocking of river channels. Subaquatic fans build at the mouths of meltwater tunnels. C) The lake fills up with subaerial alluvial fans, ice-marginal deltas, subaquatic fans and subaquatic mass flows moving down the thrust moraine slopes. These sediments make up a large proportion of thrust moraines. Note rectangular drainage pattern on the thrust moraine.

often contain structures that are typical of deltaic and lacustrine environments. Apparently, a large proportion of the glacifluvial sediments in thrust moraines has been deposited in lakes dammed off by ice or emerging thrust moraines. Glaciated lowlands form a suitable environment to produce lakes of a wide size range, silting, draining and falling dry in response to the continuously changing topography. As many of these lakes formed as a result of damming by advancing ice masses, it may be tentatively concluded that thrust moraines form preferentially along ice-marginal river systems.

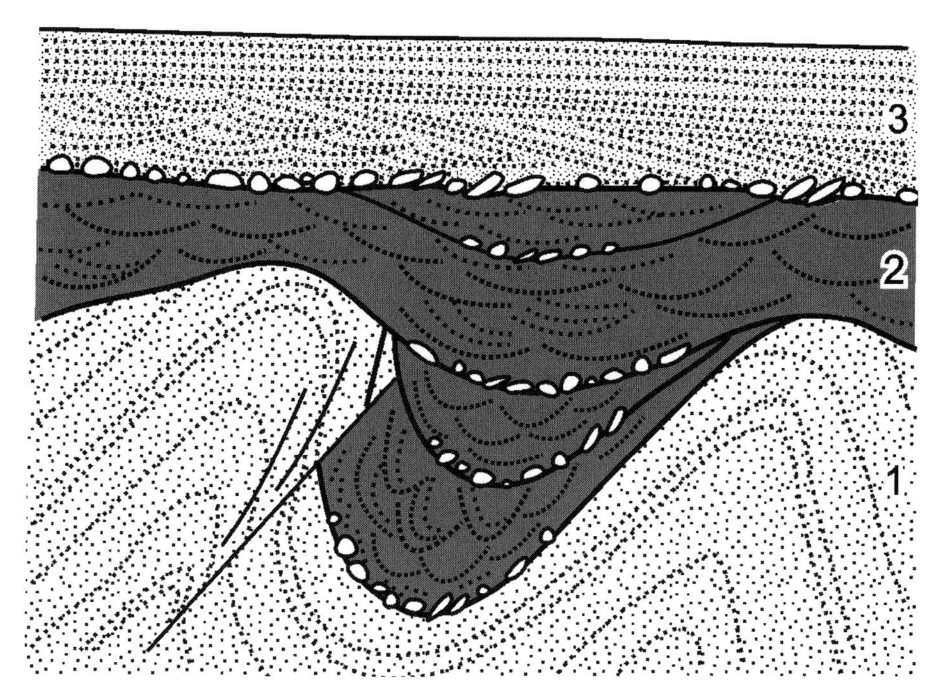

Figure 8.15 Pre-, syn- and post-tectonic sediments in a thrust moraine. (After Van der Wateren, 1987). 1 = pre-tectonic folded/thrusted sediments, 2 = syn-tectonic fill of synclinal valleys on top of the thrust moraine (glacifluvial delta and channel deposits, terraces), 3 = undeformed post-tectonic sediments (glacifluvial channel and alluvial fan deposits). Structures in unit 2 show decreasing finite strain in successively younger deposits.

The deltas usually contain onlapping lobes and bodies of unstructured fine loamy sands and sandy loams alternating with sets of climbing ripple and parallel-laminated fine-grained sands (Van der Wateren, 1994b). These fine-grained units alternate with trough-shaped sets of coarse-grained gravelly sands. A likely interpretation of these sequences is that they belong to fans and fan deltas at the mouths of subglacial and englacial tunnels draining into ice-marginal lakes. The alternation of coarse-grained trough-shaped units and fine-grained ripple laminated units reflects the strong variation in competence, which is typical of meltwater streams. In this way they are quite similar to eskers described by Banerjee and McDonald (1975), Rust and Romanelli (1975) and Saunderson (1975).

While large eskers are abundant in Scandinavia, only a few and usually small eskers have been described in the areas glaciated by the Southern Scandinavian Ice Sheets. The reason is, as explained by Clark and Walder (1994), that eskers, the sedimentary fill of englacial and subglacial meltwater tunnels (R-channels, Röthlisberger 1972), have a preference for a substratum of low permeability, such as bedrock. Where most of the area under consideration is underlain by sedimentary basins filled with permeable fluvial and glacifluvial sediments, esker formation is impeded. The few examples given on the regional maps in the next section are from areas where either bedrock or impermeable clay beds are shallow, thereby confirming Clark and Walder's (1994) ideas. In the area occupied by the Southern Scandinavian Ice Sheets, eskers have been described from the Hamburg area (Homci, 1974), the Münsterland area near the Dutch-German border (Van den Berg and Beets, 1987; Klostermann, 1995), the Barnim till plain northeast of Berlin (Chrobok and Nitz, 1995) and Mecklenburg (Schulz, 1963, 1970).

8.10 GLACIAL LANDSYSTEMS OF THE NORTHERN EUROPEAN PLAINS

The last section concluded that the palaeogeography and, more particularly, the distribution of ice-marginal drainage systems and ice-dammed lakes, is another strong controlling factor in the formation of thrust moraines. This section discusses three regions where these aspects combined in different ways with glacial dynamic influences. The first is an area in central Germany where the three Mid- and Late-Pleistocene glacial margins partly coincide in a small area on the north side of the Variscan highlands. The second is an area in northern Germany along the Baltic coast – the classical area of the Weichselian end moraines. The third is the area shaped by the Saalian glaciation in The Netherlands and adjacent western Germany.

8.10.1 Central Germany

In this central part of the marginal area of the Southern Scandinavian Ice Sheets (Fig. 8.16) the Elsterian, Saalian and Weichselian ice-sheet margins come together within less than 100 km of each other. The Elsterian was the most extensive glaciation, invading the highlands to elevations of up to 500 m above sea level (Eissmann *et al.*, 1995). The Saalian was almost as extensive as the Elsterian, their margins coinciding in the northern foothills of the Harz, but diverging in the Thüring Basin, and almost merging again near the German-Polish border (Fig. 8.16). The area comprises the Mesozoic/Cenozoic Altmark-Brandenburg Basin (Ziegler, 1990) north of the Harz, Thüringer Wald, Erzgebirge and Sudetes – the Variscan highlands alongside the German-Polish border. The distribution of end moraines, their size, glacitectonic style, lithostratigraphy and sedimentology are largely controlled by the geological structure of the area.

The maps in Fig. 8.16 (after Eissmann, 1995; Knoth, 1995) demonstrate that thrust moraines are limited to the central parts of the basins where Cenozoic sediments are sufficiently thick and suitable décollement layers (lignite, clay) are available. The glacitectonic deformations reached depths of up to 175 m below the surface (Eissmann *et al.*, 1995). Closer to and within the highlands, glacial margins are defined by dump end moraines consisting of bouldery and sandy till and glacifluvial outwash. In Poland there is a clear correlation of thrust moraine size and distance from the highlands, with the largest ones occurring in the central part of the basin (Brodzikowski, 1995). While the shape of the Elsterian and Saalian outer ice margins is

Figure 8.16 End moraines in central Germany. Within a relatively small area, just north of the central German Variscan highlands – Harz, Thüringer Wald, Erzgebirge – the Elsterian, Saalian and Weichselian ice sheet margins coincide and partly overlap. Thrust moraines are confined to the area underlain by unlithified Cenozoic sediments (compare with geological map, Fig. 8.2). Margins beyond this area are defined by dump end moraines consisting of bouldery and sandy till and glacifluvial outwash (sandur). Glacial limits: E1, E2 = Elsterian 1, 2, SD = Saalian, Drenthe Stage, SW = Saalian, Warthe Stage, WB = Weichselian, Brandenburg end moraine, WF = Weichselian, Frankfurt end moraine. (After Eissmann, 1995; Knoth, 1995).

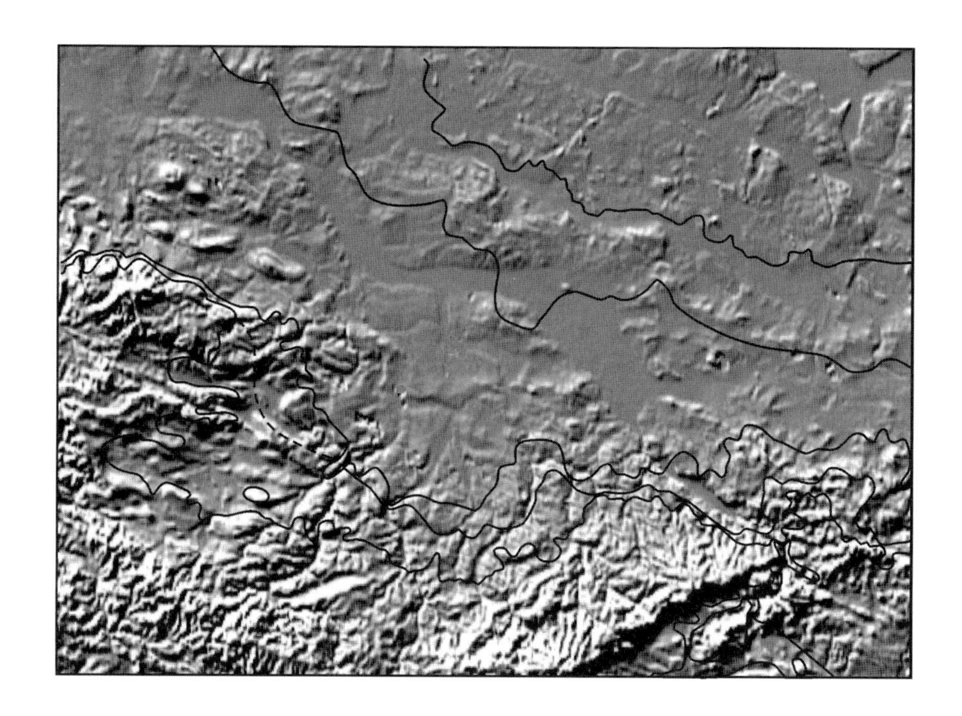

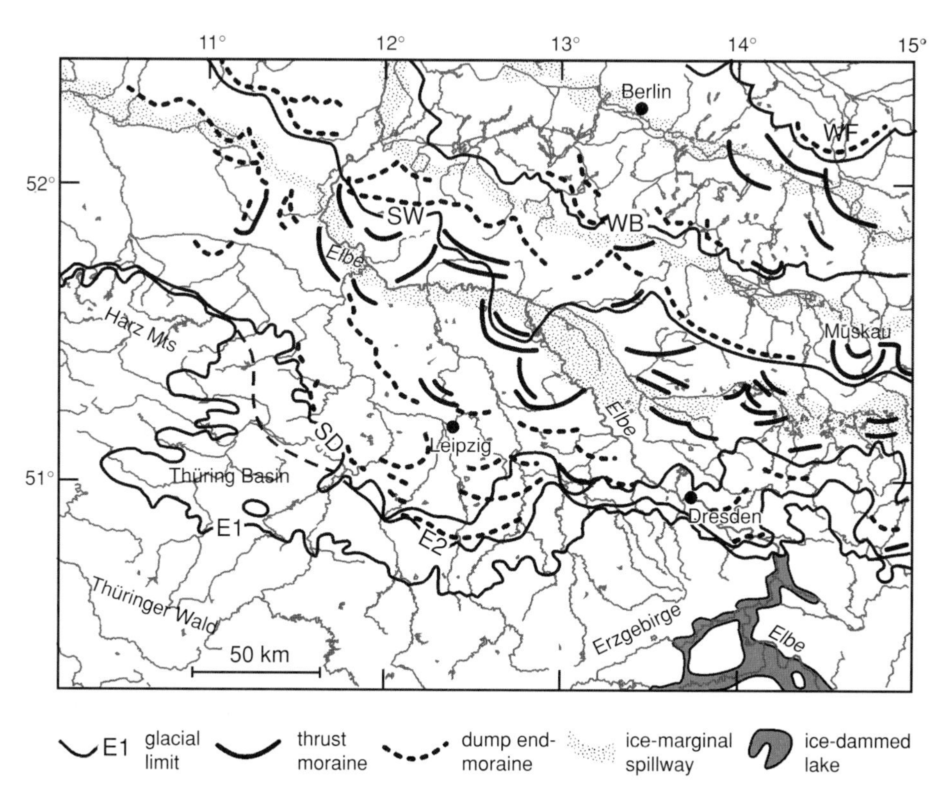
11°
12°
13°
14°
15°
52°
51°
Berlin
WF
SW
WB
Elbe
Harz Mts
Muskau
SD
Leipzig
Elbe
Thüring Basin
E1
E2
Dresden
Thüringer Wald
Erzgebirge
Elbe
50 km
E1 glacial limit
thrust moraine
dump end-moraine
ice-marginal spillway
ice-dammed lake

determined by the highland valleys and ridges, end-moraine lines in the sedimentary basins show remarkable lobate patterns. This is a positive sign of ice stream activity, which according to a theory developed by Boulton and Hindmarsh (1987) is promoted by a substratum of deformable sediments. The maps in Fig. 8.16 show a series of ice-marginal valleys, which are part of the Weichselian retreat sequence (Ehlers, 1996; Marks, 2002). The Elbe River, which also acted as a late Saalian ice-marginal system, was previously dammed by the Elsterian ice, producing a large lake south of Dresden (Fig. 8.16).

Most thrust moraines in the area to the east of Leipzig and north of Dresden originated during the Elsterian (Eissmann, 1995). They were subsequently covered and possibly modified to some extent by the Saalian Drenthe and Warthe glaciations. The Mużaków Hills/Muskauer Faltenbogen (Fig. 8.17) – along the Warthe end moraines – are the most prominent example of the inheritance of older landforms in younger landscapes.

Central Germany and bordering Poland are underlain by sedimentary basins of which the Upper Cenozoic sequence is dominated by Miocene and Pliocene lignites, and to a lesser extent clays, alternating with medium- and fine-grained shallow-marine and freshwater sands. As was explained above this kind of lithology promotes the formation of thrust moraines of the Holmströmbreen type (Fig. 8.5), characterized by concentric and inclined folds, occasional diapirs and folded nappes. Diapirs have formed as a result of two mechanisms (Eissmann *et al.*, 1995). In some areas loading by glacial ice of a saturated substratum produced diapir-like upright and steeply inclined folds. They sometimes developed from anticlines, or, alternatively, developed into antiforms. Many lignite diapirs occur in isolated positions and are interpreted as the result of glacial loading (Eissmann *et al.*, 1995). In areas outside of the influence of continental ice sheets diapirs have been ascribed to permafrost processes.

Straddling the Polish-German border, the Mużaków Hills (Brodzikowski, 1995), or Muskau thrust moraine (German: Muskauer Faltenbogen; Kupetz, 1997), is a representative example of the glacitectonic style in this region (Fig. 8.17). The structural style of this thrust moraine, one of the largest in Europe, is strongly controlled by the rheology of the sequence of evenly bedded Tertiary clays, sands and lignites prevailing in this region. The main detachment is approximately 150 m below the surface in Tertiary clays and lignites. Horizontal shortening is accomplished by steeply inclined concentric folds and reverse faults, which with increasing strain grade into overturned disharmonic similar style folds. Detachment surfaces in lignite beds at intermediate levels produce heterogeneous shortening, which increases from the lowest décollement to the highest level of the structure. In terms of glacitectonic style terminology (Fig. 8.7), this and similar thrust moraines from this region classify as style B and C. Horizontal shortening is considerably less than for style D thrust moraines, which are built of subhorizontal stacks of nappes.

8.10.2 Mecklenburg-Vorpommern

This type area of the classical Weichselian end-moraine landscapes (Woldstedt, 1925; Ehlers, 1996), lies along the southern Baltic coast, immediately north of the region discussed in the previous section (Fig. 8.18a, b, after Müller *et al.*, 1995; Rühberg *et al.*, 1995). All main late Weichselian end moraines from the European mainland are represented in this relatively small area. They mark readvances during the retreat after the last glacial maximum. Up to 100 m thick tills from the Saalian Drenthe and Warthe glaciations have been found in drillings underlying the Weichselian sediments (Müller *et al.*, 1995).

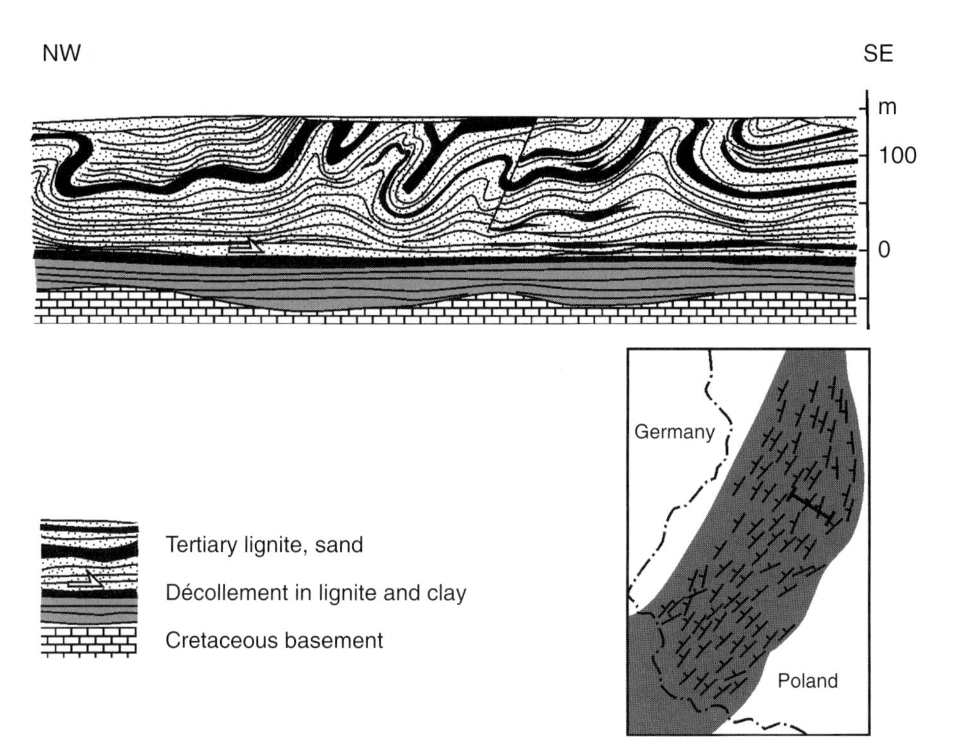

Figure 8.17 Profile of the Mużaków Hills on the Polish–German border. (After Brodzikowski, 1995). The German part of this thrust moraine, which dated from the Saalian Warthe stage, is called the Muskau thrust moraine (Kupetz, 1997). For location, see Fig. 8.16. The profile is based on drilling and mapping around brown coal mines and has an unknown, but less than 2-times, vertical exaggeration. The structural style of this thrust moraine is strongly controlled by the rheology of the sequence of Tertiary clays, sands and lignites prevailing in this region. In terms of glacitectonic style terminology (see Fig. 8.7) this and similar thrust moraines from this region classify as style B and C.

A landscape of hummocky moraines and glacifluvial outwash fans (sandur) occurs mainly between the outer margin (W1B, Brandenburg end moraine) and the Pomeranian end moraine (W2). Depressions in this landscape are occupied by numerous lakes and peat bogs. From the highest crests – the Pomeranian end moraine – the landscape slopes gently towards the Baltic coast. These proximal slopes are scattered with eskers, which run approximately parallel to the palaeo-ice flow direction. Thrust moraines occur mainly in the Pomeranian and Mecklenburg end moraines. They originated partly from Saalian advances (Rühberg *et al.*, 1995). The large thrust moraine on the Island of Rügen contains thrust sheets of Cretaceous chalk, similar to those on the Island of Møn, Denmark.

8.10.3 The Netherlands and West Germany

This area (Fig. 8.19a, b) has been shaped mainly by the Drenthe and Warthe stages of the Saalian Glaciation. Here, the various structural and palaeogeographic controls on the formation of Saalian glacial landsystems are well documented. The Elsterian margin runs ESE-WNW across the area, but none of its landforms survive at the surface. Drilling and mapping have

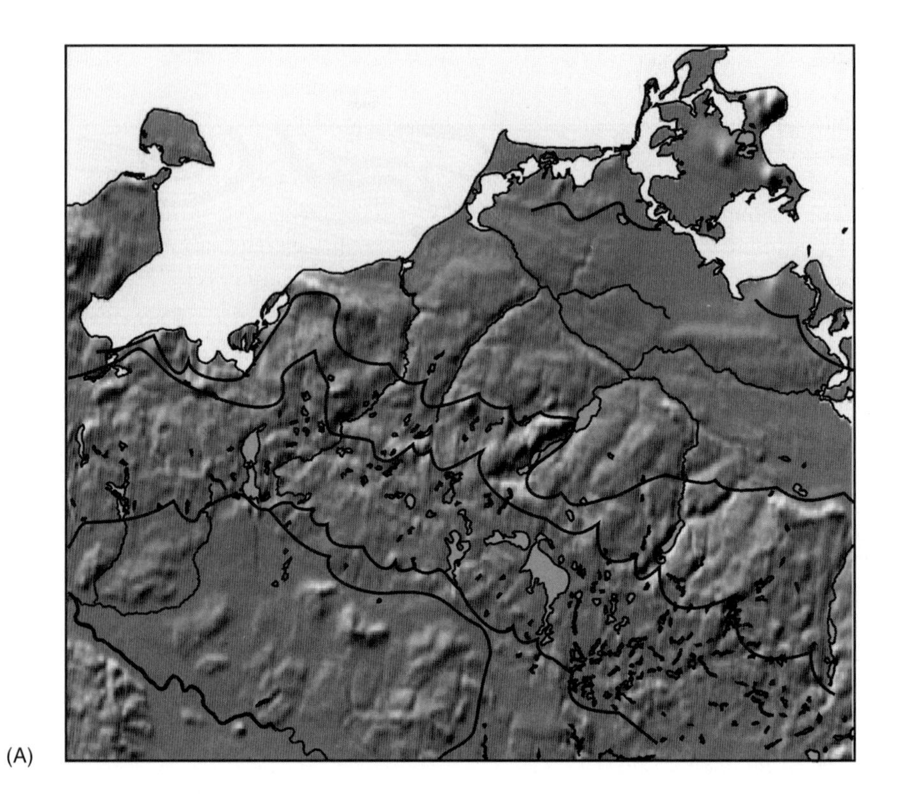

(A)

11°
12°
13°
50 km
Fehmarn
Baltic Sea
Rügen
push m.
esker
outwash
W3V
Rostock
54°
W3
W2
W2max
W1F
W3
W2
Elbe River
W1B
W1F
53°

(B)

demonstrated the existence of Elsterian deep tunnel valleys in the northern Netherlands and Germany (Kuster and Meyer, 1979; Ter Wee, 1983; Huuse and Lykke-Andersen, 2000). Unlike central Germany, this area does not contain evidence of inheritance of Elsterian landforms in Saalian glacial landsystems.

The landscape is dominated by vast till plateaux ('Geeste' in German) and thrust moraines. Few eskers are found in areas of shallow bedrock and some of the till plateaux are drumlinized (Fig. 8.19b). The largest thrust moraines occur in the E-W trending Rehburg line, Drenthe stage. The German Rehburg line probably correlates with a line of thrust moraines and stacked till sheets (style Ec) in The Netherlands at 52.5°N to 53°N (Van den Berg and Beets, 1987; Van der Wateren, 1995). The chain of large thrust moraines in the central Netherlands at the Saalian outer margin formed after overriding of the Rehburg line.

In Germany thrust moraines frame the Variscan highlands of the Teutoburgerwald and Wiehengebirge at the southern margin of the Lower Saxony Basin. In The Netherlands those at the Saalian margin follow a major NW-SE trending fault system delineating the uplifted flank of the Roer Vally Graben, the northwest branch of the Rhine Valley rift system (Fig. 8.19a). In The Netherlands the distribution of thrust moraines and till sheets is quite clearly controlled by Cenozoic structures in the substratum (De Gans *et al.*, 1987). To clarify this relationship Fig. 8.19a includes contours of the Quaternary base (after Zagwijn and Doppert, 1978) that are strongly influenced by Cenozoic faulting. Quite remarkably, the Cenozoic Zuiderzee Low is rimmed on three sides by thrust moraines, which is a clear indication of the influence of subsurface structural and hydrological conditions on glacitectonic processes.

Van den Berg and Beets (1987) demonstrated the close link of the various tectonic styles with zones in which strata of contrasting hydraulic conductivities are close to the surface. The Rehburg line in The Netherlands follows a zone where fine-grained Elsterian glacilacustrine deposits reach the surface. The detachment is in Holsteinian fluvial clays, while thrust moraines near the German border and further east regularly have their décollement in Tertiary and Cretaceous formations.

A reconstruction of the glacier bed just before the Drenthe ice sheet reached The Netherlands indicates it is likely that the Saalian ice sheet expanded in the northern and eastern parts of the country over a surface of fine-grained sediments. South of the line (53.0°N, 4.0°E – 52.7°N, 6.7°E), running eastward from the island of Texel, the surface consisted of coarse-grained

Figure 8.18 The Weichselian glaciated area of Mecklenburg. All main late Weichselian end moraines from the European mainland are represented in this relatively small area. A) Shaded relief map (GTOPO30). Numerous lakes occur on hummocky moraine and glacifluvial outwash landscape between successive Weichselian end moraines. From the crests of the end moraines the landscape slopes gently towards the Baltic coast. B) The various margins mark readvances during the retreat after the Last Glacial Maximum. They are associations of dump end moraines, thrust moraines, eskers and glacifluvial outwash fans. W1B = Brandenburg advance, W1F = Frankfurt readvance, $W2_{max}$ = limit of Pomeranian till, W2 = Pomeranian readvance, W3 = Mecklenburg readvance, W3V = Velgast readvance. (After Müller *et al.*, 1995; Rühberg *et al.*, 1995).

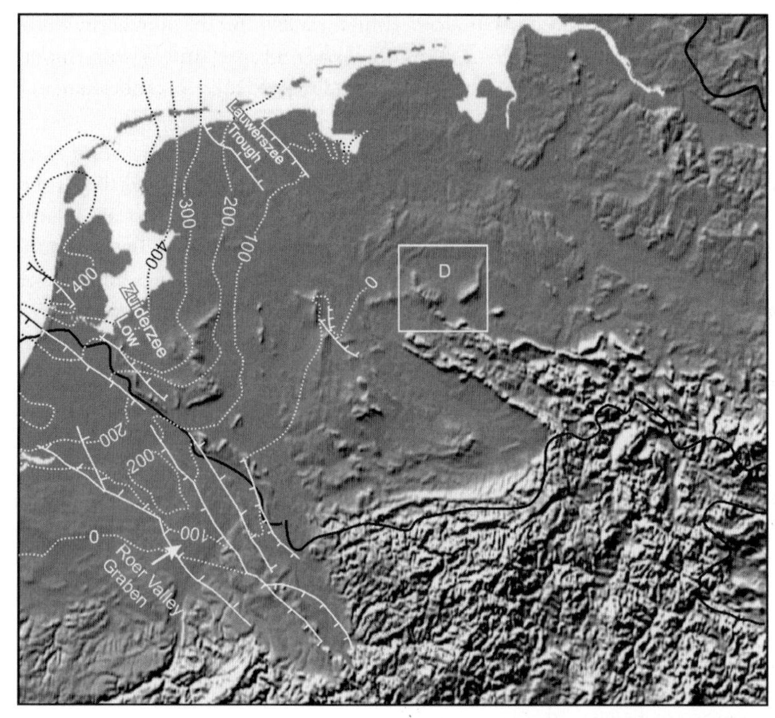

(A)

53°
52°
51°
5°
10°
W
H
L
SW
D
N
M
Wiehengebirge
Teutoburger Wald
SD
E

A; dump endmoraine
outwash fan
B, C
E/B,C
D
E/D
E; till plateaus
Ee, glacial basin
Ec
drumlin
esker
ice-marginal valley

(B)

sediments. A deforming bed of fine-grained deposits north of this line was reworked into thick accumulations of till (Rappol, 1984; Rappol and Stoltenberg, 1985). Thrust moraines formed once the ice reached areas covered with coarse-grained fluvial material and where a fine-grained layer in the lower substratum could act as décollement. These coarse-grained sediments filled ice-marginal valleys, once more stressing the impact of palaeogeographical conditions on glacial landsystem formation. Where a suitable décollement was lacking, stacked till sheets rather than thrust moraines formed along the stagnating ice margin (Fig. 8.19b).

A similar relation seems to exist between salt structures and pushed zones. Most of the domes in Zechstein salt originated as Hercynian anticlinal structures that intruded into the overlying strata (Jaritz, 1973). Their trends closely follow those of the Variscan and Alpine orogenies in the area. They remain usually well below the deepest level of glacial deformation and are only cut by the deepest Elsterian tunnel valleys (Kuster and Meyer, 1979). However, some have been mobile in the Tertiary and even up to the present time (Gripp, 1952; Jaritz, 1973; Picard, 1964), which is why in those areas fine-grained Mesozoic and Tertiary strata may be found on higher levels than elsewhere. This relationship is demonstrated by the Lamstedt thrust moraine (L in Fig. 8.19b) and the Altenwald thrust moraine to the northwest. These ridges run more or less parallel to a series of long N-S trending salt domes that crop out locally (Kuster and Meyer, 1979).

The conspicuous lobate shape of many of the Saalian thrust moraines is a clear indication of the activity of ice streams. These were particularly active following the advance beyond the Rehburg line to the Drenthe margin. An ice stream, which occupied the Nordhorn basin (N, Fig. 8.19b), advanced into the embayment south of the Teutoburgerwald (Skupin *et al.*, 1993). The orientation of drumlins and eskers in this area indicate a strongly diverging ice flow with even northeastward flow in the eastern part of the embayment (Fig. 8.19b).

Although thrust moraines are commonly considered to consist of steeply imbricated thrust sheets ('Schuppen'), in reality they are built of thin subhorizontal nappes (Van der Wateren, 1995). Figures 8.20 and 8.21 show two examples, the Blankenese thrust moraine west of Hamburg (after

---

Figure 8.19 Glacial landsystems in The Netherlands and western Germany – thrust moraines and till plateaux from the Saalian glaciation. The prominent moraines trending east–west across the middle of the map belong to the Rehburg line, Drenthe stage (SD). End moraines and till plateaux from the Saalian Warthe stage (SW) are in the northeastern part of this map. In Germany (eastern half of the map) thrust moraines rim the Variscan highlands, while in The Netherlands, these formed along a major NW-SE trending fault system. A) Shaded topography. Morphological features are strongly related to the structure of the substratum as shown by depth contours of the base of the Tertiary (stippled lines, from Zagwijn and Doppert, 1978) and major faults. Box outlines the area of the Dammer Berge (Fig. 8.4). B) Distribution of glacitectonic styles (see also Fig. 8.7). M = Münsterland esker ridge, N = Nordhorn glacial basin, D = Dammer Berge, H = Hamburg-Blankenese thrust moraine (Fig. 8.20), L = Lamstedter Berge thrust moraine (Fig. 8.21). The Lamstedter Berge (L) show a change in style from north to south from E/D to E/C to A/dump end moraine, implying that the strongest shortening occurred in the northern part of the thrust moraine. The Warthe advance overrode the thrust moraine, reaching its limit at approximately the same latitude as the southern tip of the thrust moraine.

Figure 8.20 Profile across the Blankenese thrust moraine west of Hamburg. The thrust moraine formed during the Warthe stage, Saalian glaciation. Décollement is in Upper Miocene micaceous clay (black) which is overlain by Pliocene, Elsterian and Holsteinian silts and sands (white, stippled lines indicate bedding). Till of the Drenthe stage (white triangles, grey shading) is incorporated in the nappe structures. The thrust moraine is covered on its proximal side by Warthe stage till (black triangles, stipple shading). Although this and other Saalian thrust moraines have been commonly cited to consist of steeply imbricated thrust sheets ('Schuppen'), the profile without vertical exaggeration shows they are really built of thin nappes extending more than 1 km (Van der Wateren, 1995). In terms of glacitectonic style terminology (Fig. 8.7) this and similar thrust moraines from this region classify as style D. Horizontal shortening is much greater than for style B and C thrust moraines. (After Wilke and Ehlers, 1983).

Wilke and Ehlers, 1983) and the Lamstedter Berge (after van Gijssel, 1987), respectively. Thrust moraines in this region are predominantly built of large nappe structures, glacitectonic style D (Fig. 8.19b). Investigations in the Dammer Berge (Van der Wateren, 1987, 1995) and the Itterbeck-Uelsen thrust moraine (Dutch-German border, (Kluiving, 1994) showed that these nappes may measure several km^2 up to 12 km^2. This style, contrasting with that of central European thrust moraines, is controlled by thick sequences of coarse-grained fluvial sediments derived from the ice-marginal valley systems across which the ice sheets advanced.

8.11 CONCLUSIONS

Mapping glacitectonic styles is a powerful way to reconstruct former ice sheets and to identify the glaciological conditions existing at the time, if the following reservations are made. Not all thrust moraines are equally useful for a reconstruction of ice-sheet margins (Van der Wateren, 1995). If the subsurface geology is the main factor controlling the position of a line of thrust moraines, individual ridges do not have to be synchronous and the use of this line for an ice sheet

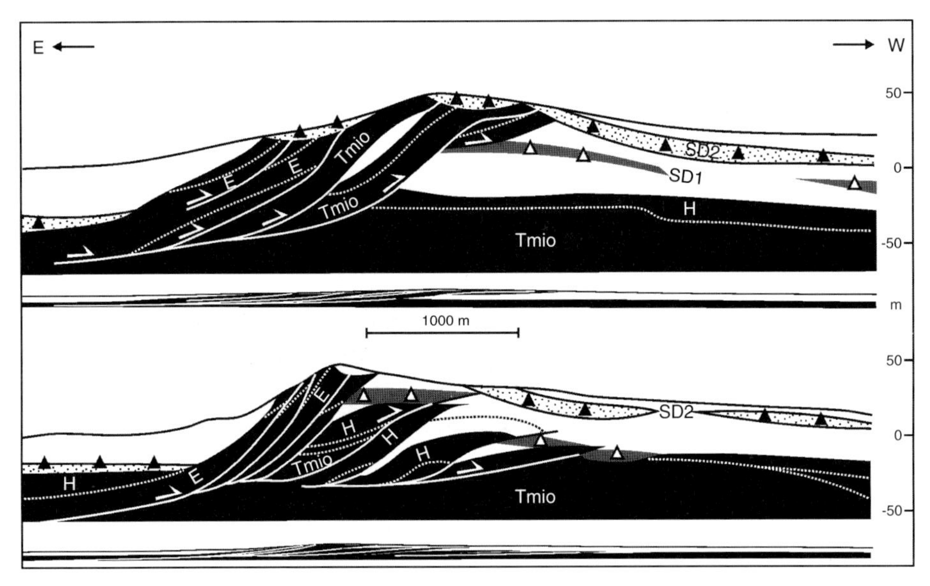

Figure 8.21 Profile across the Lamstedter Berge. Black: Clay and silt, Tmio = Miocene clay and silt, E = Elsterian (Lauenburg) clay, H = Holsteinian (interglacial) clay and silt; white: Saalian glacifluvial sand and gravel. Till of the Drenthe 1 stage (SD1, white triangles, grey shading) is incorporated in the nappe structures. The thrust moraine is covered by Drenthe 2 stage till (SD2, black triangles, stipple shading). Like the Hamburg-Blankenese thrust moraine the Lamstedter Berge are built of sub-horizontal nappes measuring several square-kilometres (style D). (After van Gijssel, 1987).

reconstruction is rather pointless. Stable margins are identified by the presence of stacked till sheets, outwash fans and dump end moraines, either undeformed or as main constituents of thrust moraines.

The combination of glacitectonic analysis with other methods, such as sedimentology, sediment petrology, fabric analysis and stratigraphy, offers great scope to improve regional reconstructions of ice movement. This method of kinetostratigraphy has been most successfully applied in Denmark (Berthelsen, 1978, 1979; Houmark-Nielsen and Berthelsen, 1981; Houmark-Nielsen, 1987, 1994).

Conditions promoting the formation of thrust moraines, as discussed in previous sections, are:

1. structure of the underlying Cenozoic sedimentary basins, particularly the depth of a detachment layer (clay, lignite)
2. palaeogeography, particularly the occurrence of ice-marginal valleys containing accumulations of coarse-grained sediments
3. glaciology – stagnation of the ice margin during (re)advance and an ice thickness sufficient to provide the potential energy to move thrust sheets (Van der Wateren, 1995).

Many of the largest European thrust moraines are horseshoe-shaped around lobate margins. Whether this is an indication of surging is debatable, but they clearly formed around the termini of ice streams. Map patterns of glacitectonic features suggest that most of these ice streams were rather long-lived and therefore not really surges.

Although permafrost has been commonly cited as a prerequisite for the formation of thrust moraines (e.g. Richter *et al.*, 1951; Boulton and Caban, 1995; Boulton *et al.*, 1999), this does not appear to be the case. Aber (1988) emphasized that thrust moraines and related structures may form under water as well as subaerially. Similar but larger structures, such as accretionary wedges, form in subduction zones below several thousands of metres of sea water (e.g. Byrne, 1994).

The observation that many thrust moraines contain subaquatic sediments from lakes in ice-dammed ice-marginal river systems is another argument against permafrost as a precondition for pushing. Although tempting, equalling the dimensions of thrust moraines with the thickness of the permafrost layer (Boulton *et al.*, 1999) may lead to erroneous palaeoclimatic conclusions.

Apparently massive as well as laminated tills in the area glaciated by the Southern Scandinavian Ice Sheets are mainly produced by deformation of the unlithified sediments that fill the sedimentary basins. This can be demonstrated most readily where the tills can be correlated with relatively *in situ* sediments cropping out nearby.

8.12 SUMMARY – LANDFORM-SEDIMENT ASSOCIATIONS OF THE SOUTHERN SCANDINAVIAN ICE SHEET

The morphology and distribution of landforms of the areas glaciated by the Scandinavian Ice Sheets are strongly controlled by the structure of the substratum. This effect is most prominent for the distribution of thrust moraines, which for their formation depend on the availability of a suitable décollement: usually Cretaceous, Tertiary or early Pleistocene clays or Tertiary lignites. These units of low shear strength, shallow near the margins of the sedimentary basins of the Northern European Plains and along salt diapirs or fault blocks. Coupling of the glacial stress field to the foreland is promoted by coarse-grained sediments in proglacial fans or in ice-marginal drainage systems, which subsequently make up the bulk of the thrust masses, sliding on a fine-grained décollement layer. This explains why the largest thrust moraines form chains parallel to the boundaries of the Variscan highlands and major neotectonic structures. In view of the large influence of structural and palaeogeographical conditions on their formation, we must be careful when using thrust moraines to reconstruct stable ice sheet margins.

The majority of the tills in the area outside the Scandinavian shield consists of reworked locally derived sediments with a relatively minor component of far-travelled erratic material. Subglacial tills are usually characterized by a fabric indicating high-strain simple shear, characterized by recumbent structures, which are strongly attenuated in the direction of shearing (= ice flow direction). In an ideal section, tills show an upward increase in finite shear strain and therefore of sediment transport distance. The lower portion of a till may be almost entirely made up of local sediment, grading from an undeformed footwall to a laminated till, which represents a glacitectonic lamination (transposed foliation) rather than sedimentary layering. The next higher unit, the more massive appearing part of the till, nearest to the former glacier sole, has undergone the highest amount of shear strain and contains the most far-travelled material. The resulting petrographic layering may resemble a stratigraphic sequence of different tills. Combining a careful structural analysis with petrographic and sedimentological observations helps to avoid this pitfall.

Thick till sequences forming plateaux inward of the major end moraines indicate temporary stagnation of the ice margin during an advance or a readvance during glacial retreat. Where thrust moraines are absent (lacking the necessary preconditions for their formation), till plateaux are rimmed by stacked till sheets produced by marginal compression. These may line up with dump end moraines and proglacial alluvial fans, thus reliably indicating a stable ice-sheet margin.

Landform-sediment associations, which can be used for regional mapping and palaeo-glaciological reconstructions, show a characteristic sequence of glacitectonic styles. From the foreland to the interior of an ice sheet these are:

A. undeformed foreland, including dump end moraines and outwash fans overlying older sediments
B. high-angle structures: Jura-style concentric folds, box folds and steeply dipping reverse faults; minimal horizontal glacitectonic shortening
C. low-angle structures; strongly asymmetric, overturned and recumbent folds; low-angle thrusts; medium shortening
D. nappes; extensive horizontal and relatively thin thrust sheets; maximum shortening
E. (deformation) tills; boudins, boudinaged folds and folded boudins, subhorizontal shear planes, transposed foliation; extremely high finite shear strain and horizontal extension; includes Ec (compression), comprising stacked till sheets and other compressive structures and Ee (extension), with strong erosion of the substratum, comprising overdeepened (tunnel) valleys, drumlin fields and megaflutes.

Thrust masses B, C and D are unconformably overlain by glacifluvial, glacideltaic and glacilacustrine sediments that show varying degrees of deformation, depending on when, during the evolution of the thrust moraine, they had been emplaced.

A sequence of glacitectonic styles that formed during a glacial advance shows overprinting of style E (tills) over styles B, C and D. In a readvance sequence styles B, C and D overprint style E in tills from the previous advance. These overprinting relationships thus offer a reliable method to relatively date successive advances and readvances. They can be identified on all scales, from the regional to outcrop and microscopic scale. Eskers are relatively rare in the area outside the Scandinavian shield. They are confined to areas where bedrock is shallow and where the glacial bed cannot accommodate large meltwater fluxes.

CHAPTER

PALAEO-ICE STREAM LANDSYSTEM

Chris D. Clark and Chris R. Stokes

9.1 Introduction and Rationale

Ice streams are spatially restricted regions in a grounded ice sheet, which flow much faster than the surrounding ice (Fig. 9.1). They are known to be one of the main regulators of ice sheets because of their ability to rapidly drain large volumes of ice. The processes that promote fast flow and restrict it to well-defined arteries are still poorly known. In recent times, the focus of research has been on investigating Antarctic ice streams to try to elucidate their processes of operation (e.g. Bell *et al.*, 1998). Meanwhile, researchers of Quaternary oceanography and palaeo-glaciology have discovered that ice streams were responsible for producing large-order ice sheet instabilities that were of great enough magnitude to force climate change on abrupt (millennial) timescales (Bond and Lotti, 1990). It follows that an understanding of ice stream operation is critical to both contemporary and palaeo-glaciology, and has implications for mechanisms of abrupt climate change.

The definition of ice streams as an artery of fast flow surrounded by slower flow, makes it relatively easy to identify contemporary examples from observed flow structures (e.g. Fig. 9.1) or velocity fields measured by satellite interferometry (Joughin *et al.*, 1999). For palaeo examples, it is much harder and we have to rely on evidence left behind that is indicative of the existence of fast and spatially restricted flow. This chapter tackles this issue and builds a landsystem model of what we expect the geological and geomorphological products of ice streaming to look like. This expectation is driven by the characteristics of contemporary ice streams. Reasons for wishing to find and investigate palaeo-ice streams are outlined below.

9.2 Significance of Palaeo-Ice Streams

9.2.1 Palaeo-Ice Streams and Ice-Sheet Reconstructions

The large flux within ice streams has a profound effect on ice sheet configurations, including drainage basin and ice divide locations, and local and regional ice sheet topography. This is demonstrated by the effect that ice streams have on the mass balance, flow configuration and ice

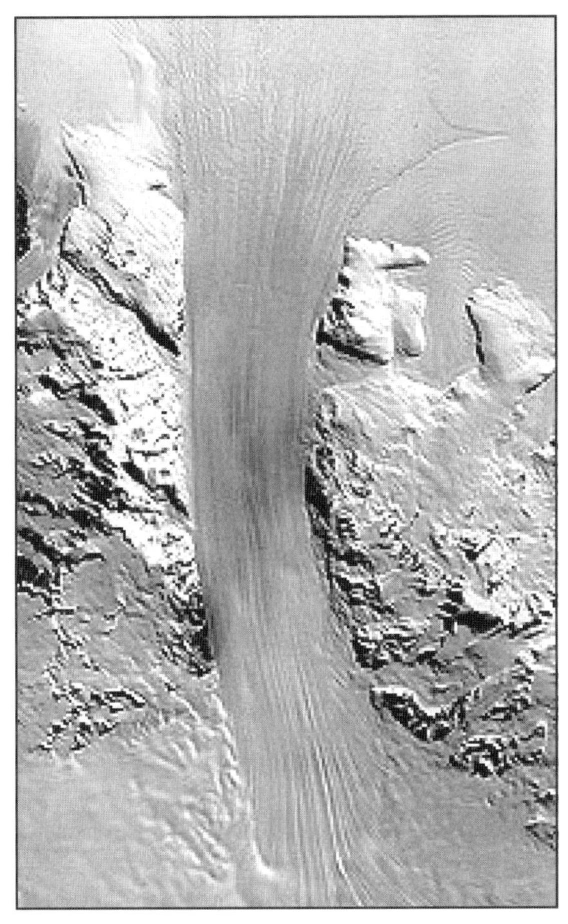

Figure 9.1 An Antarctic ice stream (Byrd Glacier) viewed from a satellite image. Flow is towards the top of the image. Note the upstream convergence of flow leading into the main trunk, and the sharp margins between fast- and slow-moving ice. At the top of the image, flow diverges as it spreads into the floating ice shelf. Image is about 100 km in width.

divide positions of the Greenland and Antarctic ice sheets. It is estimated, for example, that around 90 per cent of Antarctica's drainage is accounted for by ice streams (Morgan *et al.*, 1982) even though they occupy only around 13 per cent of the ice-sheet perimeter (Paterson, 1994). Clearly, the control that ice streams have is disproportionate to their size. Given that ice streams appear to be an intrinsic part of the ice-sheet system (i.e. will always tend to occur as 'release valves' within large ice sheets), then any reconstruction of former ice-sheet configuration and dynamics that neglects ice streams is likely to be seriously flawed. This applies both to reconstructions built by geomorphological inversion or by numerical modelling. For the former we need to be able to recognize former ice stream tracks, and for the latter we need numerically defined flow laws governing fast-flowing ice and must know the processes that initiate streaming flow and restrict its spatial extent. We regard that locating the positions of former ice streams is of paramount importance when reconstructing ice sheets, and their elucidation is likely to radically alter our views of ice-sheet geometries and their tempo and pattern of change.

9.2.2 Palaeo-Ice Streams and Climate Change

The marine geological record has provided important evidence of the role of ice streams in forcing 'abrupt' (decadal to millennial) climate changes. Episodes of ice streaming, particularly from the eastern margin of the former North American (Laurentide) Ice Sheet, were responsible for large iceberg discharge events into the North Atlantic between 60,000 and 10,000 years ago (Bond *et al.*, 1992). Evidence for these events comes from bands of ice-rafted debris found in ocean cores, known as Heinrich layers. Andrews and Tedesco (1992) have specifically linked the carbonate detritus associated with the two most recent Heinrich events to sedimentary rocks eroded by an ice stream in Hudson Strait. It has been postulated that the influx of freshwater resulting from massive iceberg discharge was sufficient to cause changes in sea surface temperature and salinity, which had a considerable impact on ocean circulation and northern hemispheric climate (Broecker and Hemming, 2001). Although the trigger for the ice streaming itself remains unclear, it is now recognized that ice streams are instrumental in driving abrupt changes in high latitude climate and oceanographic circulation (Bond and Lotti, 1990; Andrews, 1998).

In addition, climate modelling by Manabe and Broccoli (1984) has demonstrated the important coupling between ice-sheet elevation and atmospheric circulation, and Shin and Barron (1989) suggested that North Atlantic climate is highly sensitive to ice-sheet elevation. Because ice streams have the ability to rapidly drain large portions of ice sheets, they play a critical role in controlling overall ice-sheet thickness and therefore, may impact on atmospheric circulation.

9.2.3 Palaeo-Ice Stream Beds and Basal Processes

Current research in western Antarctica is striving to ascertain the basal characteristics and processes of the ice streams (e.g. Kamb, 2001). While providing invaluable insights into the nature of the ice stream bed, borehole investigations are limited by the scale of the investigation, and seismic studies suffer from a lack of detailed resolution. In an ideal world, we would like to view the whole ice stream bed at a variety of scales. Studying former ice stream tracks allows us to do just that, and if we can confidently find a former ice stream bed, we have an unprecedented opportunity to investigate its basal characteristics on a variety of scales, from large scale mapping of geomorphology to micromorphological till analysis.

Palaeo-ice stream beds can provide information with regard to the thermal conditions of the ice sheet, the basal topography, the bed roughness, the hard-bed-to-soft-bed fraction and the geology and lithology, all of which have been hypothesized to be critical with respect to ice-stream location and functioning. However, it should be noted that the bedform record of a palaeo-ice stream may only be related to the final stages of ice-stream operation (i.e. conditions immediately prior to shut down), and that this may have suffered from post-depositional modification, particularly during deglaciation.

9.2.4 Palaeo-Ice Streams and Sediment Transport

While some ice streams may only 'smear' or redistribute sediments, others are extremely powerful erosional agents. In particular, some marine-terminating ice streams may be the most important mechanism by which sediment is delivered to continental margins, and their fluxes have been compared with the efficacy of the largest fluvial systems (e.g. Mississippi delta), despite their far shorter duration (Elverhøi *et al.*, 1998). A consequence of this is that ice-stream location and vigour determines the distribution and volume of major accumulations of sediments, or fans. The location, deposition rate, sediment volume and facies architecture have strong practical implications for mineral exploration and geohazards because gravity-driven slumping may occur within these fans.

9.3 HYPOTHESES OF PALAEO-ICE STREAM LOCATIONS

As early as 1981, Denton and Hughes (1981) recognized the role ice streams likely played in the dynamics of former ice sheets and presented a hypothetical map of potential ice-stream locations (Fig. 9.2). These were predicted on the basis of obvious topographic troughs and on ice-sheet geometry. The map is highly speculative, as at the time there was little or no evidence for ice-stream existence, but it made the point that ice streams probably existed and significantly affected the topography, flow geometry and behaviour of the ice sheets. Since this early work, there has been a drive to find more substantial evidence for their existence and significant progress has been made. A wide range of reconstructions that argue for the existence of specific ice streams have been hypothesized, at various locations within the Cordilleran, Laurentide,

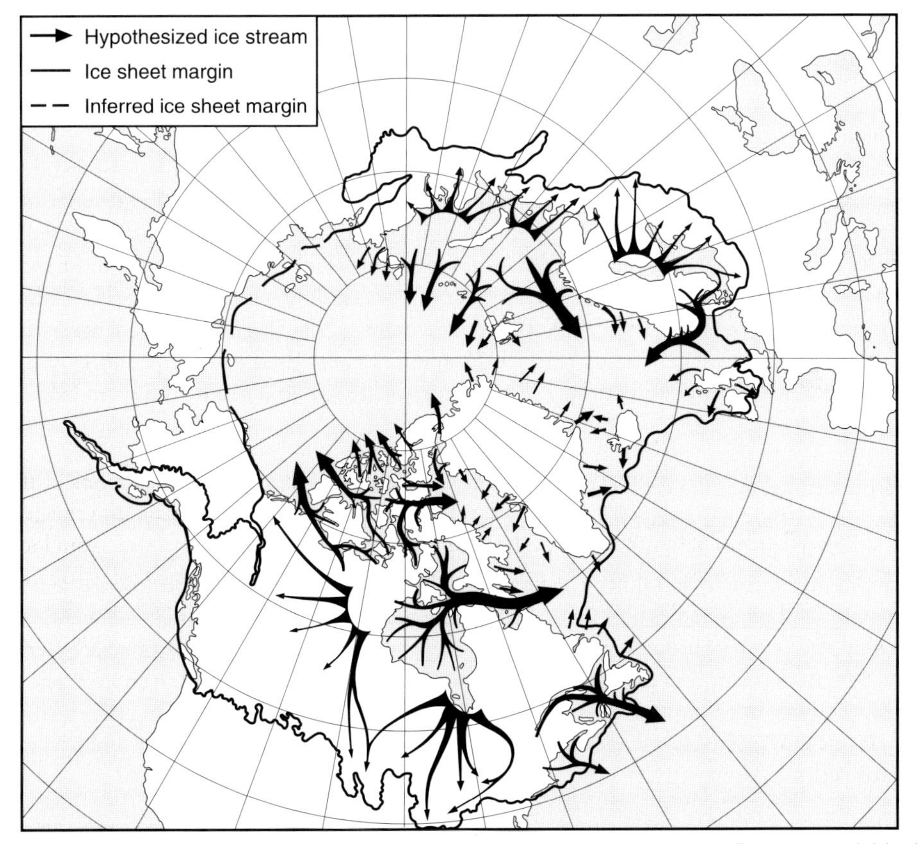

Figure 9.2 Given that contemporary ice sheets contain ice streams, Denton and Hughes (1981) hypothesized that ice streams were also present within former ice sheets. Their map, reproduced above, indicates the predicted locations.

British-Irish, Icelandic, Scandinavian, and Barents ice sheets. These have been reviewed by Stokes and Clark (2001), who comment that: 'unfortunately, hypothesised locations have tended to outweigh meaningful evidence because our understanding of ice stream geomorphology is limited' (Stokes and Clark, 2001, p. 1455). Evidence used to hypothesize specific ice streams ranges widely and includes: topographic troughs, intense glacial scour and streamlining of bedrock, specific erratic dispersal 'plumes', drumlin patterns indicative of fast flow or characteristic ice-stream flow patterns, mega-scale glacial lineations indicative of fast flow, sedimentary and tectonic evidence of pervasive deformation in glacial sediments, hummocky topography indicative of wastage following ice stream stagnation, large till deltas or trough mouth fans indicating spatially focused sediment delivery, and so on. Over fifty specific palaeo-ice streams have been hypothesized some of which have a strong basis in terms of evidence, but most of which are highly speculative.

To enable us to reliably identify a palaeo-ice stream there is a clear need for the various types of geomorphological and geological evidence to be synthesized into an idealized template of what they might leave behind. Rather than building such a landsystem model based on palaeo evidence, we firstly outline some of the problems in identifying palaeo-ice streams and turn to the characteristics of contemporary ice streams as a guide.

9.4 PROBLEMS IN IDENTIFYING PALAEO-ICE STREAMS

9.4.1 Terminology

Unfortunately, the term 'ice stream' is often used very loosely to describe a range of fast ice flow phenomena. What are the differences, for example, between an ice stream, a surge and a surging ice stream? Does a transient ice stream qualify as a surge? This has led to many ambiguities in the literature. Ice streams have been used to account for terrestrial lobes protruding from the ice sheet margin (e.g. Patterson, 1997a), and to describe zones of fast flow within a lobe protruding from the margin (e.g. Boyce and Eyles, 1991). The term 'ice stream' has also been interchanged with the term 'tidewater glacier' (e.g. Merritt *et al.*, 1995), and it has been used to describe short-lived surge behaviour in a marine environment (e.g. Kaufman *et al.*, 1993), and surging lobes on land (Eyles *et al.*, 1994).

It is recommended that the definition of an ice stream given by Paterson (1994) should be adhered to. That is: 'a region in a grounded ice sheet in which the ice flows much faster than in regions on either side'. This carries no implication as to whether the flow is transitory or continuous, or to the mechanism that promotes fast flow. The term 'surge' is usually taken to describe flow acceleration of a temporary (and non-steady state) nature, which in an ice sheet context is potentially confusing. It is probably best to restrict use of 'surging' to just glaciers, whereby a large part of the glacier experiences cyclical flow acceleration, and use the term 'ice stream' for zones of fast flow within an ice sheet. These ice streams may be in continuous operation or transitory. An 'ice-sheet surge' would thus be reserved for a dramatic flow acceleration and possible margin advance of the whole ice mass, and which happens at regular cycles.

9.4.2 Lack of Diagnostic Evidence

Little theoretical work has been carried out to predict the landforms and landform assemblages that we should expect an ice stream to produce. In essence, we have no clear criteria on which to base our assumptions. Mathews (1991) emphasized the need for diagnostic criteria when attempting to find palaeo-ice streams. There are two main problems:

1. a wide variety of evidence has been used to infer former ice streams (see above and Stokes and Clark, 2001), and
2. such evidence has rarely been scrutinized in detail, despite its often subjective nature.

Because ice streams are distinct features within an ice sheet, it seems logical to suggest that they will, in general, leave behind a distinct imprint of their activity, a 'geomorphological signature' or footprint. However, it seems that in some cases the identification of a distinct flow pattern (i.e. of drumlins) is regarded as ample justification for postulating the location of an ice stream. We would argue that the real challenge lies in demonstrating why the distinct flow pattern was produced by an ice stream.

9.4.3 Lack of Modern Analogues

All contemporary ice streams are marine-based, leaving us with the problem that we do not actually know what a terrestrial-terminating ice stream looks like. This poses a number of important questions when trying to reconstruct their configuration, not least of which is how does the ice stream evacuate ice so rapidly? Because contemporary ice streams feed ice shelves or terminate in open water, the removal of ice at the terminus is rapid and this maintains a high velocity. They can maintain a high velocity without advancing. In contrast, a terrestrial ice stream has a much less effective method for removing ice and presumably, therefore, has to advance producing a large splayed ice lobe at its terminus. This explains why many former ice streams have been postulated for the southern margin of the Laurentide Ice Sheet, the lobate configuration of which is thought to represent the distal portion of a number of palaeo-ice streams. However, is the fast flow restricted to a narrow zone along the central axis of the lobe, or does the divergence of flow result in a more uniform but slower regime of flow velocities? We presume it to be the latter case, with a lobe of slower moving ice (i.e. below typical ice stream velocities) protruding beyond the overall ice margin.

9.5 CHARACTERISTICS OF EXISTING ICE STREAMS

Retaining Paterson's (1994) broad definition of ice streams as spatially restricted fast flow surrounded by slower moving ice, then there seem to be two main breeds of ice stream. 'Topographic ice streams' are fixed in position, their location predetermined by bedrock troughs through which they flow. In comparison with others they have higher surface slopes and driving stresses (typically 50–200 kPa), higher basal shear stresses and attain higher velocities. A classic example is Jakobshavns Isbræ, in western Greenland, which extends for about 90 km upstream of the grounding line, with a steep surface gradient and high driving stresses (200–300 kPa) producing velocities of 4–7 km a^{-1} (Echelmeyer *et al.*, 1991). Pine Island and Thwaites glaciers are examples from Antarctica. In contrast, 'pure ice streams' are features with geometry and flow patterns that are not determined by bedrock topography. They are characterized by very low surface gradients and low driving stress with a concave-upwards surface profile (Bentley, 1987). Typically, velocities are of the order of 400 m a^{-1}. In order to flow so fast with such low driving stresses requires efficient lubrication at their base. Their flow geometry is idealized in Fig. 9.3. Ice streams A–E of the Siple Coast of West Antarctica are the most well known and widely studied of pure ice streams. Recent discoveries in the East Antarctic Ice Sheet suggest that there may be a

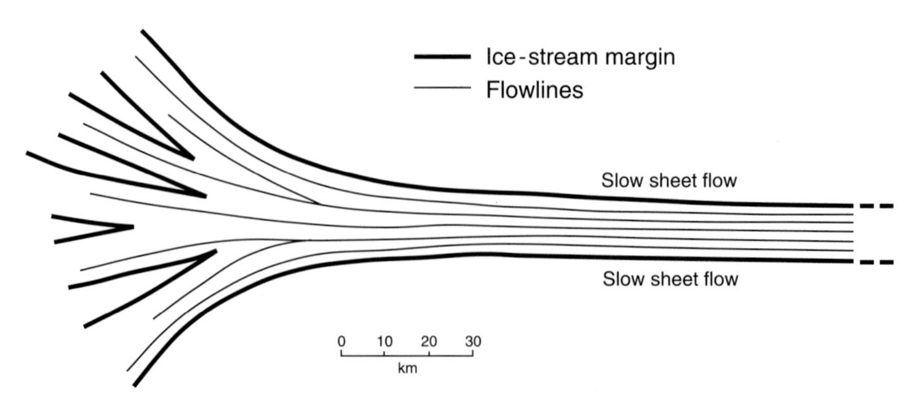

Figure 9.3 Idealized flow geometry of an ice stream. A highly convergent onset zone feeds the faster-moving ice stream trunk. The ice-stream margin is extremely abrupt. The onset zone may comprise a number of tributaries.

third breed of ice stream called 'fast flow features', which extend much greater distances into their parent ice sheets, and whose locations seem to be partly controlled by underlying topography (e.g. Bamber *et al.*, 2001).

Although the processes that permit these different categories to flow so fast are poorly known and may differ between different breeds, typical velocities are higher than can be achieved by internal deformation of the ice column alone. Table 9.1 summarizes some characteristics of the more well-known ice streams, in terms of their dimensions and flow velocities.

Given that ice often transports sediment englacially or subglacially and that higher ice fluxes are experienced within ice streams than adjacent to them, then it follows that we should expect higher levels of sediment transport along ice-stream corridors. This may lead to spatially focused sediment delivery at ice-stream termini. Large 'trough mouth fans' or 'till deltas' deposited on submarine continental slopes are examples of these (e.g. Alley *et al.*, 1989; Dowdeswell, *et al.*, 1996; Vorren and Laberg, 1997).

In summary, the basic characteristics of ice streams in the broadest sense are:

- large in dimension (>20 km wide, 150 km long)
- highly convergent onset zones feeding a main channel
- rapid ice velocities (>300 m a^{-1})
- abrupt lateral shear margins
- spatially focused sediment delivery.

9.6 GEOMORPHOLOGICAL CRITERIA OF ICE-STREAM ACTIVITY

Ice streams are discrete features within an ice sheet and so we may expect them to leave behind a unique suite of glacial landforms. By coupling the known characteristics of existing ice streams (outlined above) with traditional theories of glacial geomorphology, it is possible to predict several geomorphological products of palaeo-ice stream activity. While individual criteria are not

Ice Stream type	Name	Length (km)	Width (km)	Thickness (m)	Velocity (m.a^{-1})	Drainage basin area (km^2)	References
Pure Ice Streams	Ice Stream A	>200	~50	~1,000	217–254	No data	Rose (1979); Shabtaie *et al.* (1987); Echelmeyer *et al.* (1994)
	Ice Stream B	500 300	35 50 30–80	1,000–>2,000	450 >500 ~500	163,000	Rose (1979); Shabtaie and Bentley (1987); Alley *et al.* (1989); Scambos and Bindschadler (1993); Whillans and Van der Veen (1993); Engelhardt and Kamb (1998)
	Ice Stream C	<400	50–60	200–1,500	1–13 40 50 5	122,000	Rose (1979); Fastook (1987); Shabtaie and Bentley (1987); Whillans *et al.* (1987); Whillans and Van der Veen (1993)
	Ice Stream D	550	46–59 30–82	800 927–1,432	420–670 700 100–370	104,000	Rose (1979); Scambos and Bindschadler (1993); Scambos *et al.* (1994); Hodge and Doppelhammer (1996)
	Ice Stream E	~320	75–100	975–1,091	400–550 ~400	131,000	Rose (1979); Bentley (1987); Scambos and Bindschadler (1993)
Topographic Ice Streams	Pine Island Glacier	200 ~300	26 30	1,564 ~2,000	1300–2600	220,000	Bentley (1987); Lucchitta *et al.* (1995); Jenkins *et al.* (1997)
	Thwaites Glacier	c. 300	c. 83	~2,000	2,200–3,400	121,000	Bentley (1987); Rosanova *et al.* (1998); Ferrigno *et al.* (1998)
	Rutford Ice Stream	>150	18–26	1,640–2,250	360–400 302–377	~36,000	Doake *et al.* (1987); Frolich and Doake (1988)
	Jutulstraumen Ice Stream	c. 300	c. 40	2,000	443	124,000	Høydal (1996)
	Jakobshavns Isbræ	70–80 85–90	~6	2,500 1,900–2,600	800–7,000 8,360	10,000	Lingle *et al.* (1981); Echelmeyer and Harrison (1990); Echelmeyer *et al.* (1991); Fastook *et al.* (1995); Clarke and Echelmeyer (1996)

Table 9.1

necessarily exclusive to ice-stream activity, collectively they can be viewed as an idealized palaeo-ice stream imprint. The criteria are used to assemble a conceptual landsystem model of the bed of former ice streams, and it is envisaged that such models can provide an observational template upon which hypotheses of palaeo-ice streams can be better based.

9.6.1 Characteristic Shape and Dimensions

Palaeo-glaciological reconstructions rely heavily on the use of palaeo-flow indicators such as striae, roches moutonnées, flutes, drumlins, ribbed (Rogen) moraine, and mega-scale glacial lineations, to constrain ice-sheet flow geometry. If such indicators are found to record a pattern resembling the geometry and size of typical ice streams then this may be the most obvious clue as to the existence of an ice stream. Subglacial bedforms are the most widespread flow indicators and may cover around 70 per cent of the beds of former ice sheets. In contemporary ice streams, the onset of streaming flow (or onset zone) is characterized by a large convergence zone (Fig. 9.3), whereby slower moving ice is gradually incorporated into the ice-stream trunk (probably by tributaries). Thus, subglacial bedforms produced by an ice stream should exhibit a large degree of convergence in the onset zone, leading to a much narrower and well-defined trunk.

Any such pattern should be at a scale appropriate to ice streams, typically greater than 20 km wide and 150 km long. We acknowledge that contemporary ice streams may not be a representative sample of the population of ice streams that have ever existed, and so there is plenty of scope for an ice stream to be bigger or even smaller, particularly if the shape is consistent with ice-stream activity.

9.6.2 Bedform Signature of Fast Ice Flow

Mega-scale glacial lineations (MSGL) are elongate streamlined ridges of sediment (Fig. 9.4) produced subglacially and are similar to flutes and drumlins but much larger in all dimensions

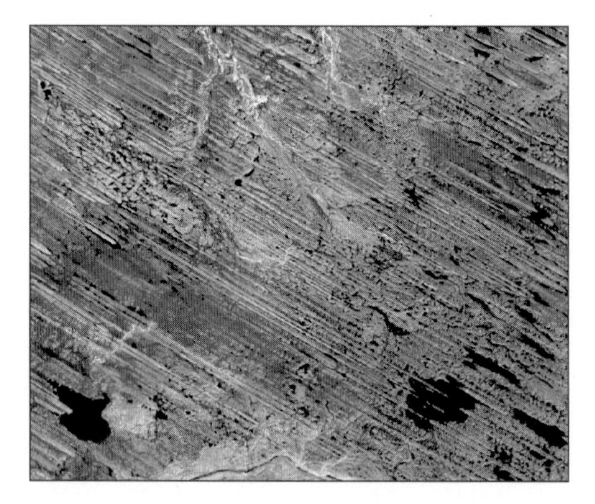

Figure 9.4 Mega-scale glacial lineations observed from a satellite image. Ice flow was towards the northwest. These ridge-groove systems, expressed in surficial sediment, extend for tens of kilometres, and are a record of fast ice flow. Note the highly parallel nature of landforms over great distances and with no discordances in flow pattern. Image is about 30 km across and is centred at 101° 51′ W, 64° 02′ N, in the Dubawnt Lake region of northern Canada.

(Clark, 1993). Typical lengths are 6–100 km, widths of 200–1,300 m and spacing of 200 m to 5 km. Given the association of MSGLs with inferred high strain rates and fast ice velocity it has been argued that they can be be used as an indicator of palaeo-ice stream location (Clark, 1993; 1999; Stokes and Clark, 1999). Until recently, these assumptions could not be tested because the foregrounds of contemporary ice streams are largely inaccessible. However, using swath bathymetry and high-resolution seismic investigations, MSGLs have, for the first time, been discovered on the Antarctic continental margin (Shipp *et al.*, 1999; Canals *et al.*, 2000). We take these remarkable finds as validation of the association between MSGLs and fast flow as they are found to lie distal to positions of existing or inferred ice streams. MSGL can thus be used as reliable indicators of palaeo-ice stream tracks.

9.6.3 Proxies for Ice-Stream Velocity Field

Ice-stream velocities have two main characteristics. First, the velocity of a marine-based ice stream steadily increases all the way to the grounding line. Second, ice-stream velocities remain high all the way across the ice stream until there is an abrupt decrease in the marginal areas. This unique characteristic is described as plug flow and is exhibited by all ice streams. Figure 9.5 illustrates this pattern of surface velocity across an ice stream, from the centre line to the slower-moving adjacent ice.

A manifestation of the rapid velocity of an ice stream may be highly attenuated streamlined bedforms. Elongation ratio (length divided by width) is a useful way of quantifying the degree of attenuation of subglacial bedforms (such as flutes, drumlins and MSGLs). Although there is no method for recovering former flow velocities from elongation ratios, there are many studies that report correlations between inferred fast ice flow and high elongation ratios (Boyce and Eyles, 1991; Clark, 1994a; Hart, 1999; Stokes and Clark, 2002b). Thus, swarms of highly attenuated drumlins and MSGLs could record the velocity field of an ice stream. We would expect to find (i) highest elongation ratios in the trunk rather than the convergent onset zone, and (ii) higher elongation ratios along the central axis of the trunk. For marine-terminating ice

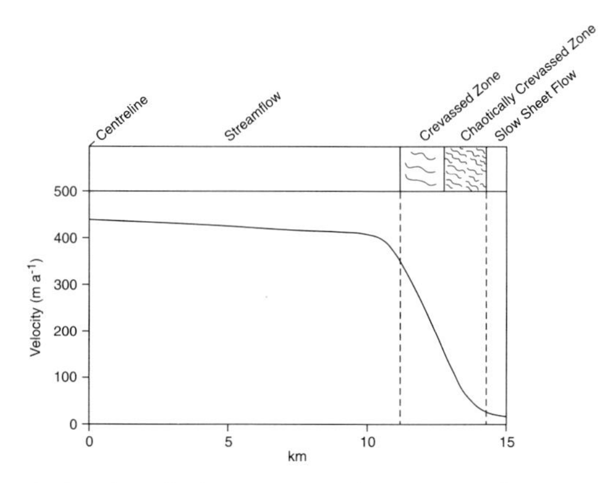

Figure 9.5 Typical variation in velocity across an ice stream, from the centre line to adjacent slower-moving 'sheet ice'. Note that the ice stream moves by plug flow, with slightly higher velocities in the centre, decreasing outwards and with a dramatic change across the shear margin. (Simplified from Echelmeyer *et al.*, 1994).

streams, elongation ratios should steadily increase downstream towards the grounding line. For terrestrially terminating ice streams elongation ratios should decrease towards the outer margins of the lobate terminus.

Further proxy evidence for velocity fields may be gleaned from specific erratic dispersal patterns. Dyke and Morris (1988) recognized two types of dispersal train, schematically represented in Fig. 9.6. The Boothia-type dispersal train forms when an abrupt lateral variation in ice velocity transports distinctive sediment from a large source area. Such a lateral variation in ice velocity is a unique characteristic of ice streams, and Boothia-type dispersal plumes may be a product of their activity. In contrast, the Dubawnt-type dispersal plume implies no lateral variation in velocity. Although it may appear similar to a Boothia-type dispersal plume, the source area of the sediment is the key control on the pattern and it can be formed by slow sheet-flow. Hence, it is important to identify the spatial extent of the source area from which the distinctive till is transported. There is not necessarily a blatant connection between ice streams and dispersal trains, but when found in conjunction with other criteria, it may be highly suggestive of ice-stream activity.

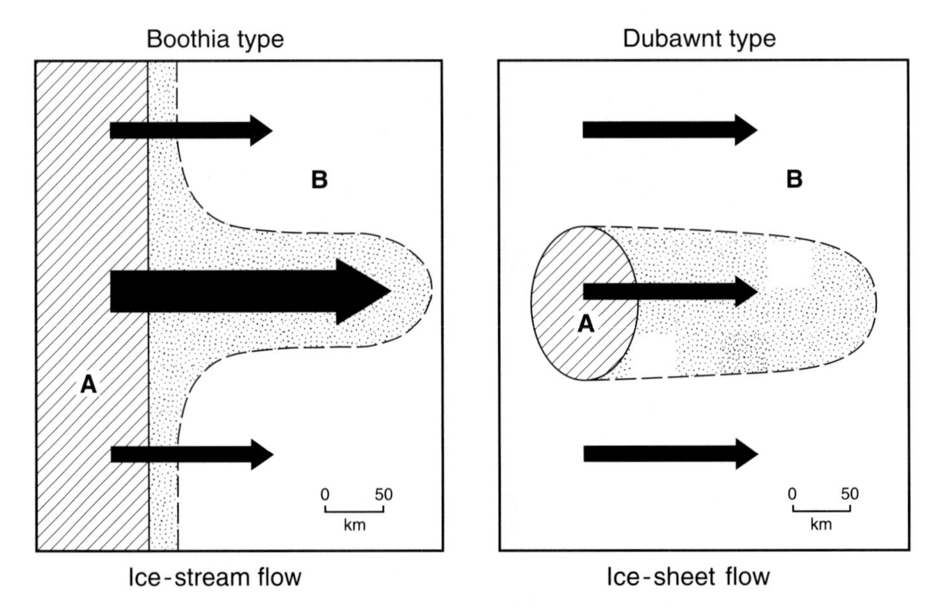

Figure 9.6 Specific erratic dispersal patterns may be used for a proxy of relative ice velocity in order to identify an ice stream. Two rock types (A and B) experiencing stream flow may give rise to a plume of erratic dispersal beneath the faster flowing stream ice, called 'Boothia-type' dispersal. This must not be confused with the more usual 'Dubawnt-type' dispersal that produces an apparent plume simply because of a restricted source area. (From Dyke and Morris, 1988).

9.6.4 Abrupt Lateral Margins

Ice streams are characterized by their abrupt lateral margins bordered by slower-moving ice (Figs. 9.1 and 9.5). The characteristic geomorphology inscribed by a former ice stream would be expected to exhibit an abrupt margin or an abrupt zonation of subglacial bedforms at the margin. We would expect a high density of bedforms within the ice stream to cease abruptly. This is in contrast to most drumlin patterns, which gradually diminish in drumlin density orthogonal to flow. For example, Hodgson (1994) noted a well-defined margin of a drumlin field when postulating the existence of an ice stream flowing northwards over the eastern portion of Victoria Island in the Canadian Arctic. Figure 9.7 illustrates the abrupt nature of this change, which can be taken to record the ice-stream margin. Dyke and Morris (1988) also noted an extremely abrupt margin to a bedform pattern formed by an inferred ice stream on eastern Prince of Wales Island. In addition, Kleman and Borgström (1994) and Kleman *et al.* (1999) suggested that an abrupt margin can be produced at the transition between cold- and warm-based ice. Given the fact that some ice-stream margins may well be characterized by such a transition, then abrupt marginal areas (<2 km) could be used to de-limit the width of a palaeo-ice stream (see Kleman *et al.*, 1999).

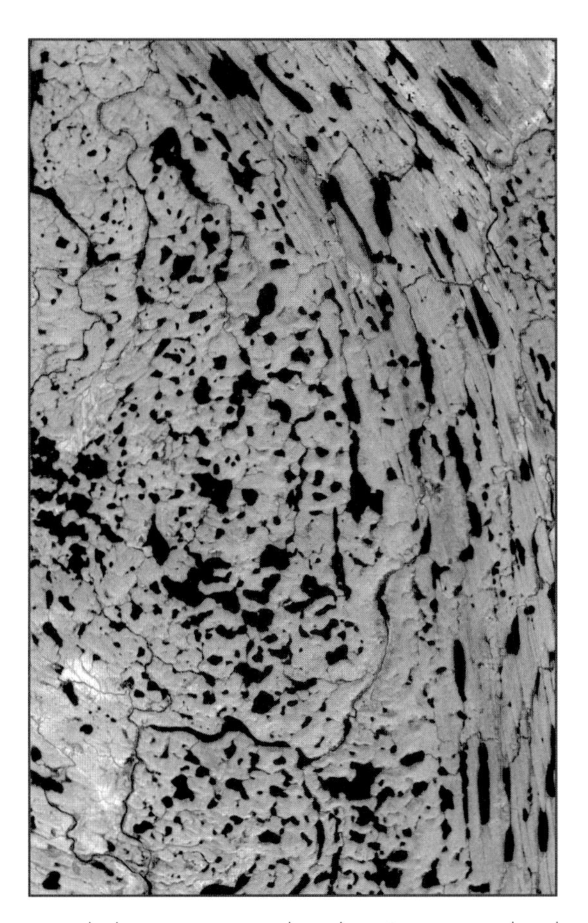

Figure 9.7 Bed geomorphology across a palaeo-ice stream margin, viewed from a satellite image. Note the contrast between hummocky and slightly drumlinized terrain in the west and the strong ice-stream signature (flow towards the north) in the east. The image is around 40 km across and the extreme abruptness of the margin (<2 km) is apparent. The margin forms the western limit of the M'Clintock Channel Ice Stream on Victoria Island, Arctic Canada. (From Clark and Stokes, 2001).

9.6.5 Subglacial Ice Stream Shear Margin Moraines

An abrupt lateral margin associated with a change in ice velocity (as described above) may also be conducive to the generation of characteristic landforms. We might expect a concentration of sediment accumulation at this boundary, producing an ice stream shear margin moraine. High melting rates should exist at ice-stream margins arising from strain-heating and the shear experienced between slow- and fast-moving ice. In addition, crevassed ice may also allow surface meltwater to penetrate to the ice-sheet bed. Sub- or englacial drainage of these waters may serve to transport and concentrate sediment in the marginal zone, and upon deglaciation, leave prominent moraines, composed primarily of glacifluvial material (e.g. Punkari, 1997). Alternatively, if sediment is eroded and entrained in the onset zones of ice streams, then downstream convergence of flow will concentrate this englacial debris into higher densities. In order to produce sediment accumulations at the margin, we hypothesize that elevated levels of strain-heating could produce sufficient melting for debris to be preferentially deposited here.

On southeastern Prince of Wales Island, Dyke and Morris (1988) identified a single, narrow ridge of till that delineated the western side of a drumlin field. This ridge can be traced for up to 68 km but is less than 1 km in width (Dyke *et al.*, 1992). In a discussion of its origin, the ridge was interpreted as being a 'lateral shear moraine', marking a shear zone at the side of an ice stream, separating fast-flowing ice from slower-flowing cold-based ice (Dyke and Morris, 1988).

Stokes and Clark (2002a) discuss ice stream shear margin moraines and report four examples associated with the former M'Clintock Channel Palaeo-Ice Stream (Hodgson, 1994; Clark and Stokes, 2001), Victoria Island, Arctic Canada. The moraines range in length from 11–22 km, maintain fairly constant widths of about 500 m and range in height from 10 to 50 m above the surrounding terrain. They are composed of carbonate drift of a similar composition to the ice-stream bedforms and have been laid down irrespective of local and regional topography. Two of the moraines display a lateral offset and are thought to reflect minor migrations of the ice-stream margin. Their mode of formation is explored and it is suggested that they occur when erosion at the ice-stream margin provides a surplus of sediment, which is 'smeared out' in the downstream direction.

Ice stream shear margin moraines are subglacial accumulations of sediment that form at the margins of active ice streams. They mark the shear zone between fast- and slow-moving ice, and may or may not coincide with an abrupt change in the basal thermal regime.

9.6.6 Spatially Focused Sediment Delivery

Although not necessarily indicative of ice-stream activity, focused accumulations of sediment on a continental shelf or slope may complement, and indeed strengthen terrestrial evidence for ice-stream flow. Vorren and Laberg (1997) have identified huge submarine till deltas, thought to have been produced by ice streams draining the northwestern part of the Fennoscandian Ice Sheet and the Barents Sea Ice Sheet.

Because a stable ice-sheet margin (i.e. without an ice stream) would not deliver such concentrated accumulations of sediment to the continental shelf, it is clear that offshore sediment accumulations can provide a valuable clue to the existence of marine-based ice streams. Clearly, if terrestrial evidence that suggests ice stream activity is available, the identification of offshore sediment accumulations serves to support such hypotheses. Furthermore, such fans, if well dated, can also be used as proxies for former ice velocities and discharges.

9.6.7 Summary

The above criteria are summarized in Table 9.2, which indicates how they are derived from the characteristics of existing ice streams. Flow indicators (e.g. drumlins and MSGLs) of a palaeo-ice stream need to have a distinctive well-defined pattern, or flow-set, that conforms to the characteristic shape and dimensions of ice streams, and display a large degree of convergence in the onset area. It is crucial that such bedforms are significantly different in morphology and elongation from surrounding flow indicators. Further verification may come from proxies for the velocity field such as the pattern of variation in bedform elongation ratios or by the distinctive 'Boothia-type' dispersal trains. For example, if we could find a pattern of distinct bedforms that conforms to the shape of an ice stream and if the bedform elongation ratios within it vary as would be expected both across and down-ice, then this is very strong evidence for a palaeo-ice stream. A manifestation of the sharply delineated margin found on ice streams may be an abrupt lateral margin to the bedform pattern and ice stream shear margin moraines. Evidence of a deforming till layer and sediment accumulation fans are not necessarily indicative of ice-stream flow but may provide substantial supporting evidence when found in conjunction with the other criteria.

The focus of this chapter has been on identifying 'pure' palaeo-ice streams from their bed imprint. Many topographic ice streams, discharging through large bedrock troughs, may be erosional in nature and leave an entirely different record. From erosional evidence (whalebacks, roches moutonées) it is probably much harder to demonstrate that these were definite ice-stream tracks, rather than simple topographic capture of flow leading to greater lengths of time of occupancy and slightly higher flow velocities. Evans (1996) argues that topographic ice streams are recognizable

Contemporary ice stream characteristic	**Proposed geomorphological signature**
A. Characteristic shape and dimensions	1. Characteristic shape and dimensions (>20 km wide and >150 km long) of distinct flow pattern
	2. Highly convergent flow patterns leading into a trunk
B. Rapid velocity	3. Bedform signature of fast flow; MSGLs and highly attenuated drumlins (length:width >10:1 100:1)
	4. Boothia-type erratic dispersal trains
C. Distinct velocity field (plug flow, downstream variation in velocity)	5. Expected spatial variation in MSGL and drumlin elongation ratios
	6. Boothia-type erratic dispersal trains
D. Sharply delineated shear margin	7. Abrupt lateral margins (<2 km)
	8. Ice stream shear margin moraines
E. Spatially focused sediment delivery	9. Submarine till delta or sediment fan

Table 9.2

from erosional evidence, but Mathews (1991) cautions against the simple equation that a glacially eroded trough equals an ice stream. For topographic ice streams the best lines of evidence are likely to be significant volumes of sediment deposited as trough mouth fans. If dating of these indicates high sediment efflux over short bursts of time, then the inference of an ice stream seems reasonable (i.e. Dowdeswell *et al.*, 1996; Vorren and Laberg, 1997).

9.7 A LANDSYSTEM MODEL FOR PALAEO-ICE STREAMS

It would be highly unlikely that all of the geomorphological criteria should be found in one location, produced by a single ice stream. This is because not all ice streams will leave a complete geomorphological signature and because preservation and modification often obscure the complete picture. However, the criteria outlined above can be thought of as comprising a characteristic 'landsystem' produced by a former ice stream. This illustrates the perfect, or unaltered, geomorphological signature of ice-stream activity. The more criteria we can find the more certain we can be of the existence of a palaeo-ice stream. We now bring together these criteria to assemble a series of 'fantasy' landsystem models.

9.7.1 Marine and Terrestrially Terminating Ice Streams

Ice streams can be broadly categorized as either terrestrial or marine-based, depending upon the environment in which they terminate. All contemporary ice streams are marine in nature but former ice sheets are likely to have been drained by terrestrial ice streams. In Fig. 9.8, terrestrial and marine-terminating ice streams are illustrated, and an important consideration is the means by which they can ablate large ice fluxes from their termini. For the marine ice streams, this is by iceberg calving directly into the ocean, or via floating ice shelves. Terrestrial examples may calve ice into large proglacial lakes, or if these are not present, we presume that the terminus must comprise a large splayed lobe extending beyond the adjacent ice-sheet margin. These lobes would present a large surface area below the equilibrium line altitude and thus facilitate efficient surface melting and mass loss. It is only by such a configuration that sufficiently high ablation losses could balance the high ice flux being delivered to the margin. Note that diverging flow within the lobe implies much lower ice velocities than in the trunk of the ice stream.

9.7.2 Bedform Signature

Given the distinct velocity field that ice streams possess, then we might expect to find an expression of this in the subglacial bedform signature. From the earlier discussion about the relationship that seems to exist between ice velocity (cumulative strain) and elongation ratio of

Figure 9.8 Simplified configurations of terrestrial and marine-based ice streams, and contemporary and palaeo examples of each. (a) For ice streams that terminate on land, there must be a method for rapidly removing ice mass. For this reason we presume that the margin must advance, producing a large splayed ice lobe which further lowers surface elevation below the equilibrium line, enhancing ablation losses. (b) Other terrestrial ice streams may drain into locally impounded glacial lakes and thus discharge their ice flux in the form of icebergs. All contemporary ice streams are marine-based and either drain directly into open water (c) or feed floating ice shelves (d).

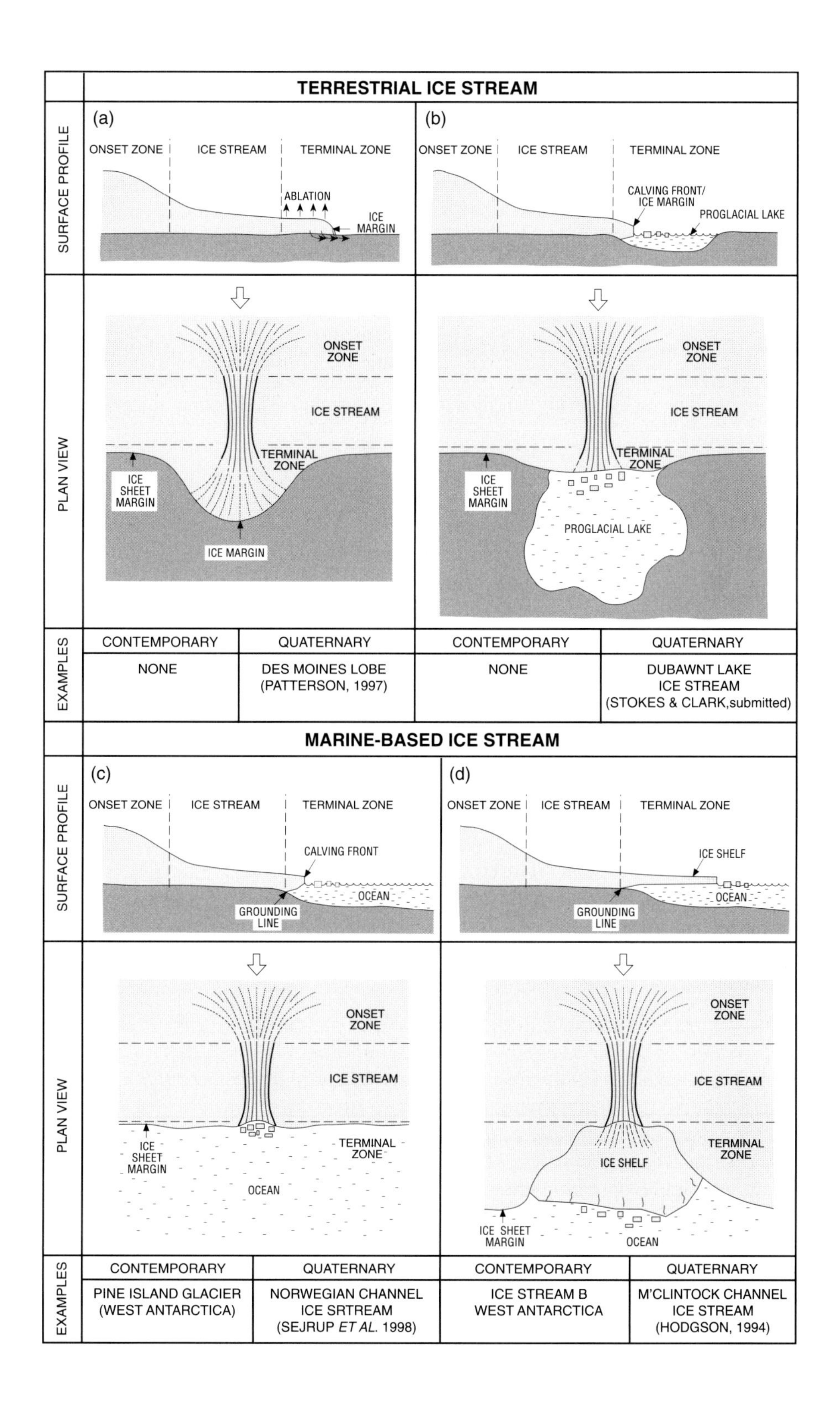
TERRESTRIAL ICE STREAM
SURFACE PROFILE
(a)
ONSET ZONE
ICE STREAM
TERMINAL ZONE
ABLATION
ICE MARGIN
(b)
ONSET ZONE
ICE STREAM
TERMINAL ZONE
CALVING FRONT/ ICE MARGIN
PROGLACIAL LAKE
PLAN VIEW
ONSET ZONE
ICE STREAM
TERMINAL ZONE
ICE SHEET MARGIN
ICE MARGIN
ONSET ZONE
ICE STREAM
TERMINAL ZONE
ICE SHEET MARGIN
PROGLACIAL LAKE
EXAMPLES
CONTEMPORARY
QUATERNARY
NONE
DES MOINES LOBE (PATTERSON, 1997)
CONTEMPORARY
QUATERNARY
NONE
DUBAWNT LAKE ICE STREAM (STOKES & CLARK,submitted)
MARINE-BASED ICE STREAM
SURFACE PROFILE
(c)
ONSET ZONE
ICE STREAM
TERMINAL ZONE
CALVING FRONT
OCEAN
GROUNDING LINE
(d)
ONSET ZONE
ICE STREAM
TERMINAL ZONE
ICE SHELF
OCEAN
GROUNDING LINE
PLAN VIEW
ONSET ZONE
ICE STREAM
ICE SHEET MARGIN
TERMINAL ZONE
OCEAN
ONSET ZONE
ICE STREAM
TERMINAL ZONE
ICE SHELF
ICE SHEET MARGIN
OCEAN
EXAMPLES
CONTEMPORARY
QUATERNARY
PINE ISLAND GLACIER (WEST ANTARCTICA)
NORWEGIAN CHANNEL ICE SRTREAM (SEJRUP ET AL. 1998)
CONTEMPORARY
QUATERNARY
ICE STREAM B WEST ANTARCTICA
M'CLINTOCK CHANNEL ICE STREAM (HODGSON, 1994)

subglacial bedforms, then we propose that, under ideal circumstances, bedform elongation ratios should vary in accordance with Figure 9.9.

9.7.3 'Rubber Stamped' Versus 'Smudged' Ice-Stream Imprints

Ice streams must initiate at some point in time, remain active and then cease to function. During all stages of their activity, they are likely to do geomorphological work, eroding, transporting and depositing sediment, and erasing and generating landforms. The imprint left may relate to a single phase or a complex combination of activity during the ice-stream life cycle. We hypothesize two end-member landsystems, shown in Figure 9.10.

The whole ice-stream imprint may have been produced isochronously leaving behind a snapshot of ice-stream activity at a point in time. We can think of this as a 'rubber-stamped' imprint. This may be produced by an ice stream that operates, then switches off, 'freezing' the geomorphology

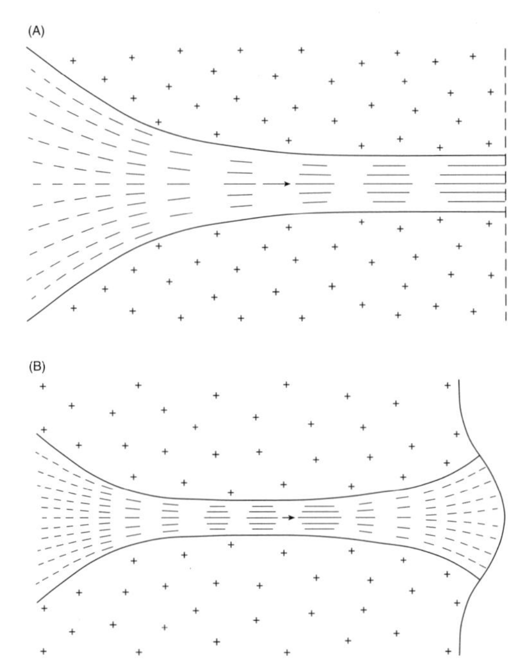

Figure 9.9 Given the distinctive velocity field that ice streams possess, and that elongation ratio of subglacial bedforms approximates (cumulative strain) velocity, then we propose that bedforms (drumlins and mega-scale glacial lineations) should, under ideal circumstances, conform to the above patterns. A) For a marine-terminating ice stream, velocity increases towards the grounding line and so it should be expected that lineations should progressively increase downstream. Short drumlins in the slower-moving convergence zone should grade into larger forms or mega-scale glacial lineations downstream. B) Terrestrially terminating ice streams will show a similar pattern, except that where ice diverges into a lobate terminus, the slower velocities should produce shorter lineations. For both cases, lateral variation in bedform elongation is likely to be slight, with longer forms along the central axis and shortening towards the shear margins, reflecting slower velocities here.

at that time, with the imprint remaining preserved during deglaciation. Preservation of landforms during deglaciation is not as problematic as once thought and is now known to be widespread (cf. Kleman and Borgström, 1994; Clark, 1999). On the other hand, an ice stream may operate throughout several cycles of advance and retreat, or be continuously operating during margin retreat, whereby the earlier imprints are modified and overprinted by the younger ice flow patterns. The other extreme therefore, is a 'time-transgressive' imprint that is continuously reorganized over time and may appear disjointed and complex (a 'smudged' imprint). Figure 9.10 illustrates the different patterns that we might expect. Clark (1999) specifically addresses how the geomorphological record may be used to distinguish between the two cases, and this is outlined in Figure 9.11.

9.8 APPLICATION OF LANDSYSTEM MODELS

Hypothesized palaeo-ice streams from the major ice sheets have been reviewed (Stokes and Clark, 2001), and elements of the criteria and landsystems models above are inherent to many of the arguments that authors have used for demonstrating the existence of ice streams. From the above consideration of marine and terrestrial ice streams, and those that leave a rubber-stamped or smudged imprint, we construct four landsystem models that we expect ice streams may leave behind as evidence (Fig. 9.12). It is hoped that these conceptual models will help in deciphering the bedform record of palaeo-ice streams. In the following, we apply our knowledge of the four main landsystems to actual examples of palaeo-ice streams.

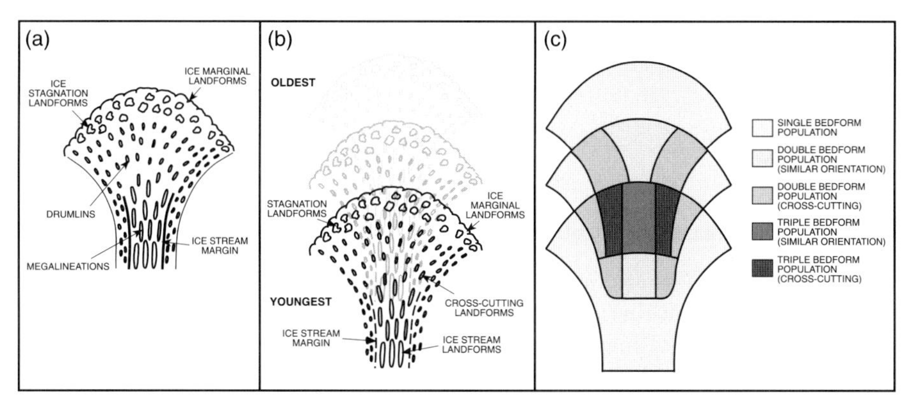

Figure 9.10 A terrestrial ice stream may leave behind a diagnostic landsystem of bedforms and ice-marginal landforms. If the ice stream switched off and retreated back without much remoulding of the landscape we would expect to find an imprint as in (a), the 'rubber-stamped' imprint. However, if the ice stream retreated, while continuously operating, then we should expect reorganization of the bed during retreat, producing a more 'smudged' or overprinted landsystem as in (b). This is schematized into fainter systems overprinted and modified by later margin positions and flow patterns, whereby some parts of the corridor exhibit singular flow directions (i.e. down the trunk) and other parts exhibit cross-cutting bedform populations, particularly on the lobe flanks (c).

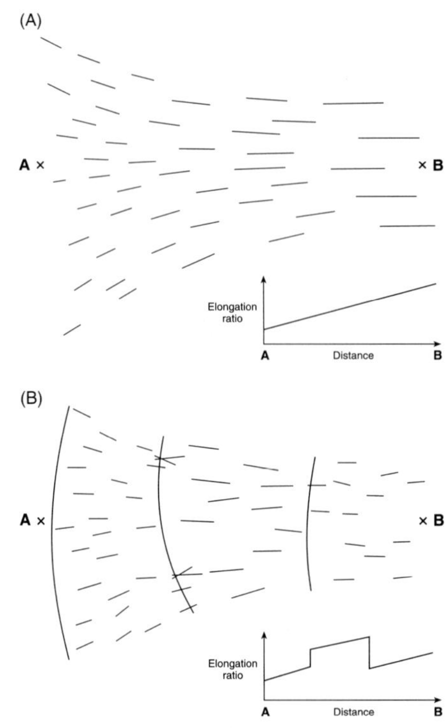

Figure 9.11 Distinguishing between rubber-stamped and smudged bedform imprints can be aided by careful observation of lineation morphometry, discordances in flow direction and cross-cutting relationships between bedforms. A) A rubber-stamped flow pattern should contain lineations of high parallel conformity, no cross-cuts and with gradual variations in morphometry. B) Conversely for the smudged or time-transgressive case, we expect low parallel conformity, possible occurrence of cross-cuts and abrupt discontinuities in lineation morphometry. Superimposed ice marginal landforms such as end moraines would provide further evidence.

9.8.1 Marine-Terminating, Rubber-Stamped, Palaeo-Ice Stream (Fig 9.12a)

The best example is the Western Bransfield Basin Palaeo-Ice Stream, discovered by Canals *et al.* (2000). It was found by marine geophysical investigation of the continental shelf just off the northern tip of the Antarctic Peninsula, and is about 100 km in length and up to 21 km wide. The imprint is remarkable for its clarity and completeness, and comprises:

- converging flow pattern
- appropriate dimensions (100 by 25 km)
  abrupt margins, and
- a bedform signature of MSGLs.

The well-organized pattern of MSGLs, without any sign of discordance or cross-cutting, is a strong indicator that this is a snapshot record of the glacier bed, which has survived deglaciation with little modification. It represents a classic example of the rubber-stamped imprint, and almost

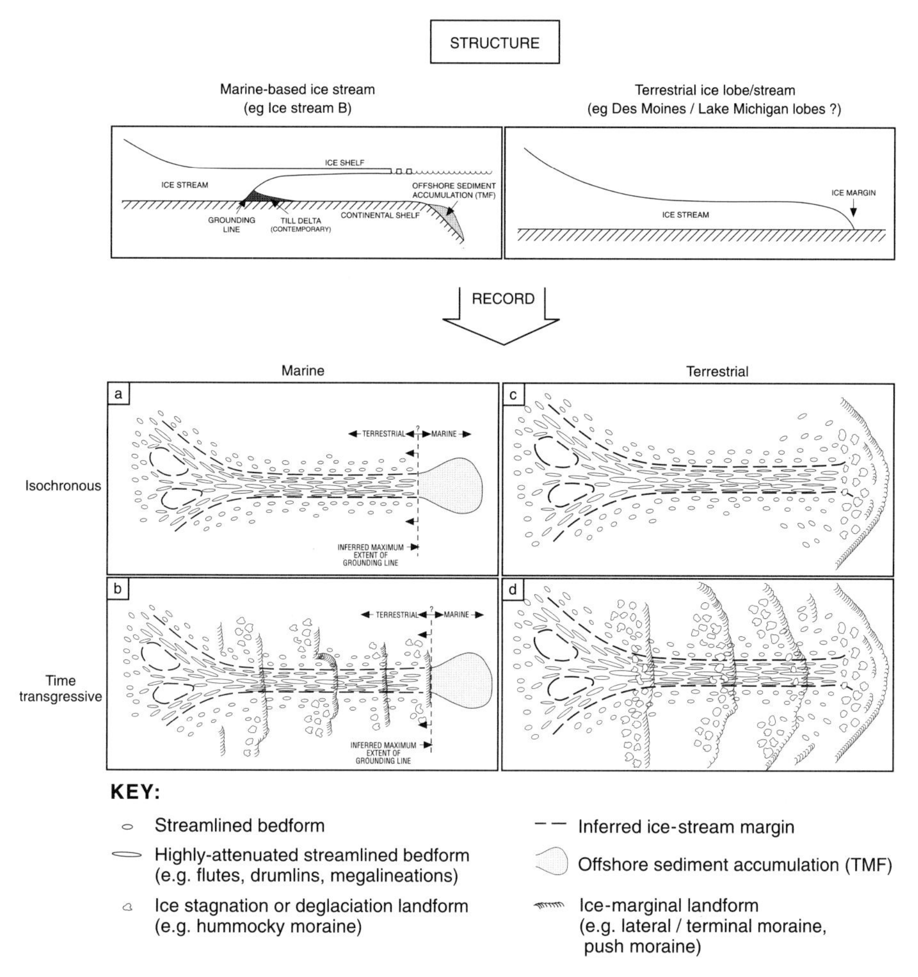

Figure 9.12 The four types of landsystem signature produced by palaeo-ice streams. Ice streams may be marine or terrestrially terminating (top diagrams), and may lay down their geomorphological imprint either isochronously (rubber-stamped) or time-trangressively (smudged). This yields four landsystem models (a, b, c and d).

looks as if the ice sheet has just been lifted off, preserving a perfect record. Indeed, on this latter point there is the possibility that it did in fact lift off as sea level rose.

9.8.2 Marine-Terminating, Smudged, Palaeo-Ice Stream (Fig. 9.12b)

Again from the seafloor surrounding Antarctica, a good example of this type is that reported in Shipp *et al.* (1999), where MSGLs are found that have been overprinted by moraines during ice-stream retreat. The M'Clintock Channel Palaeo-Ice Stream is a further example and is illustrated in Figures 9.13 and 9.14. This ice stream, reported in Clark and Stokes (2001), has left terrestrial evidence of its existence. It drained the northwestern portion of the Laurentide Ice Sheet during deglaciation (Fig. 9.13) and likely terminated in an ice shelf. Figure 9.14 illustrates some of the complexity of flow patterns recorded in subglacial bedforms, and shows a strong ice-stream

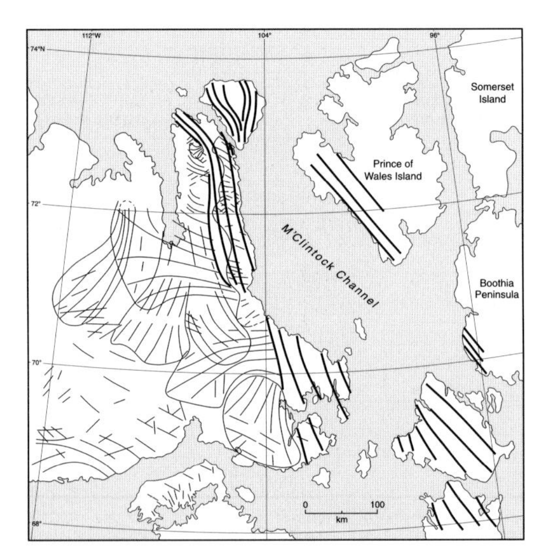

Figure 9.13 Extent of the M'Clintock Channel Ice Stream (marked in bold) in Arctic Canada and adjacent palaeo flow patterns. The ice stream was centred along the main trough and is reconstructed as 720 × 140 km in size with a cross-sectional area of 100 km^2, and is thought to have discharged into an ice shelf. (From Clark and Stokes, 2001).

signature that has in part been overprinted by landforms relating to a subsequent readvance and retreat. The ice-stream record comprises:

- appropriate shape and dimensions
- converging flow pattern
- abrupt lateral margins (see Fig. 9.7)
- shear margin moraines, and
- drumlins and MSGLs recording fast-flow velocities.

While the geomorphology has been somewhat smudged by later activity, the ice-stream signature remains well-enough preserved to see an isochronous record of the bed at the time of ice-stream shut down. Although the M'Clintock Channel Palaeo-Ice Stream is a (slightly) smudged example, it is likely that tracks exist elsewhere where the smudging from later flows is more severe and the main trunk may appear much more subtle.

9.8.3 Terrestrially Terminating, Rubber-Stamped, Palaeo-Ice Stream (Fig. 9.12c)

The Dubawnt Lake Palaeo-Ice Stream (Stokes and Clark, 2003) existed entirely terrestrially and drained the northwestern portion of the Laurentide Ice Sheet during late glacial times. Figure 9.4 displays part of the bed which comprises highly parallel MSGLs, and Figure 9.15 illustrates the extent of the ice stream. Criteria demonstrating that this flow imprint is an ice stream include:

- appropriate dimensions
- converging – trunk – diverging flow, as we would expect of a terrestrial ice stream
- abrupt lateral margins

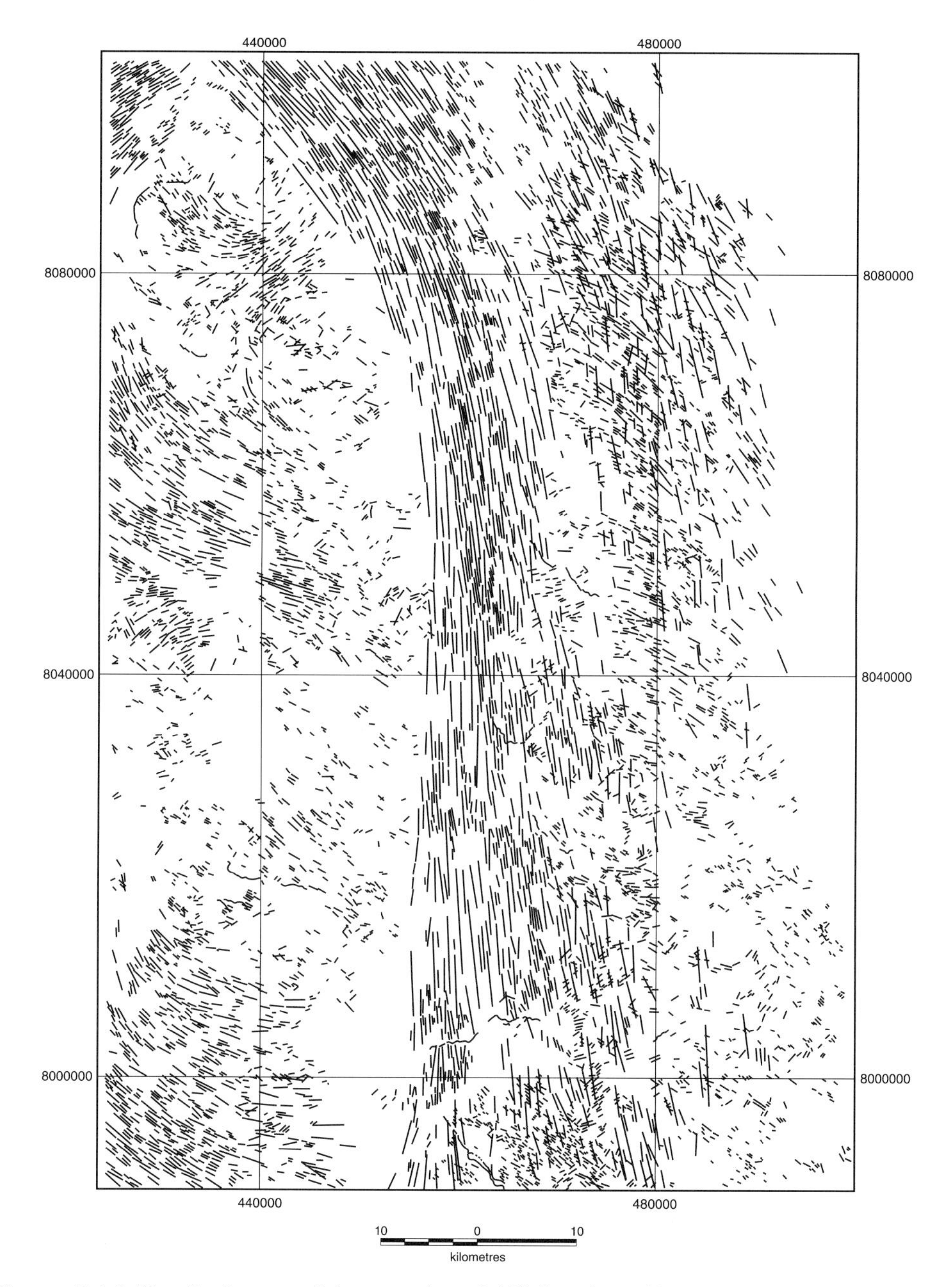

Figure 9.14 Detail of some of the mapping of drift lineations (drumlins and mega-scale glacial lineations), for part of the M'Clintock Channel Ice Stream on Victoria Island, Arctic Canada. In the west, drumlins record early flow patterns that were obliterated by the ice-stream signature (flowing northwards through centre of image), which have then been slightly remoulded, or 'smudged' by a drumlinization event subsequent to cessation of ice-stream operation. (From Clark and Stokes, 2001).

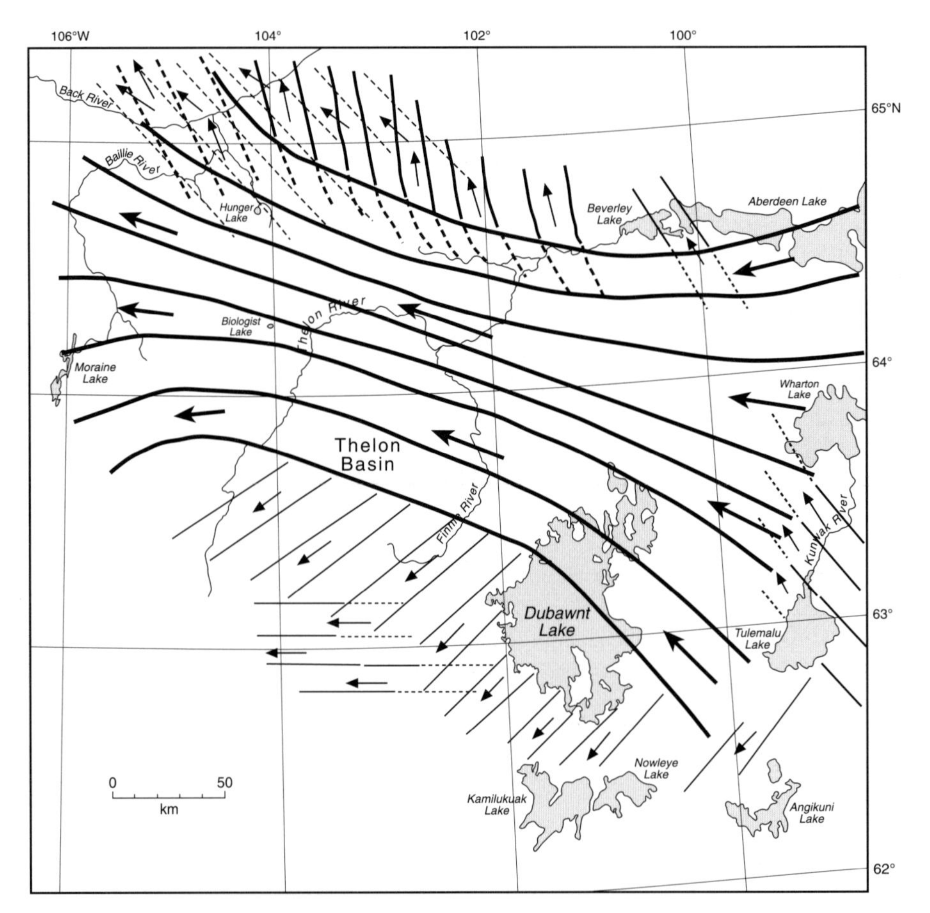

Figure 9.15 Extent of the Dubawnt Lake Palaeo-Ice Stream (marked in bold), Nunavut, Canada, as defined by the imprint of subglacial bedforms. This rubber-stamped bedform pattern overprints the adjacent and earlier flow patterns. (From Stokes and Clark, 2003).

- MSGLs, and
- importantly, an inferred velocity field (via elongation ratios) that exactly matches our expectation (i.e. Fig. 9.9, and Stokes and Clark, 2002b).

The ice-stream bedform signature is remarkably clear and consistent, with no discordances or cross-cuts, and is taken to be a rubber-stamped record of activity. The ice stream must have shut down, preserving the bed record, and from superimposed esker patterns at an oblique orientation to the ice stream axis, we infer that deglaciation 'unzipped' from a different direction. This example is the only known record of a rubber-stamped terrestrial ice stream; we presume them to be rare.

9.8.4 Terrestrially-Terminating, Smudged Palaeo-Ice Stream (Fig. 9.12d)

There are many examples of this type. If an ice stream operates while the ice-sheet margin is gradually retreating then we expect continuous reorganization of the bed as the ice stream changes position and its margin back-steps. This may give rise to a hint, or partly obscured record of the trunk but with numerous lobe-shaped imprints smudged on top. It has been argued that many of the lobe imprints of the southern margin of the Laurentide Ice Sheet (Boyce and Eyles, 1991; Paterson, 1997a; Evans *et al.*, 1999b) and the lobes of central Finland (Punkari, 1993, 1995, 1997; Dongelmans, 1996; Boulton *et al.*, 2001b), are examples of this type. These may provide good examples of the record that actively retreating ice streams should produce but some caution is required. This is because it is hard to distinguish between the record that a retreating ice stream should produce, and that of simple retreat of an ice lobe. We would never expect a retreating ice margin to be straight over large distances. Therefore, simply citing a lobate pattern as evidence for an ice stream is fundamentally flawed. The key, as ever, must be in demonstrating evidence for spatially restricted fast flow extending into the ice sheet, which can be adequately demonstrated by finding the trunk zone. If only smudged imprints of the marginal lobe pattern are recorded, then there is less certainty that this was produced by an ice stream.

9.9 Conclusion

Palaeo-ice streams have a profound impact on ice sheet mass balance, flow configurations and surface elevation. We need to know where and when they operated in order to reconstruct former ice sheets and understand how they interacted with climate and sea level. Hypotheses of palaeo-ice stream tracks are now abounding in the literature, but there has been some confusion as to what constitutes good evidence for their existence. In this chapter we have outlined reasons why it is imperative to discover and examine palaeo-ice stream tracks, and report a range of criteria that can be used to aid their identification. From these we have built landsystem models of the expected geomorphology that should arise from ice-stream activity, and report case studies that illustrate these.

The landsystem models presented in this chapter only include the geomorphological imprint and pay little regard to details of the sedimentology of deposits that should arise. A future challenge lies in taking a sequence-stratigraphic approach to extend these models to include styles of deposition and the nature of the stratigraphy that should be left behind. Once such models have been assembled it may be possible to gain extra diagnostics for identifying palaeo-ice streams.

CHAPTER

SUPRAGLACIAL LANDSYSTEMS IN LOWLAND TERRAIN

Mark D. Johnson and Lee Clayton

10.1 INTRODUCTION

Bands and tracts of hummocks commonly mark the former margins of the great ice sheets that covered Europe and North America during the Late Pleistocene. These landforms are best explained as forming by the collapse of abundant supraglacial debris during the melting of stagnant glacier ice, and hence are said to make up the supraglacial landsystem. It is also possible that subglacial squeezing of soft till contributes to landform development, and, in that case, this landsystem may be better described as the 'stagnant-ice landsystem.'

By 'lowland terrain' we refer to regions of moderately low relief at the margin of ice sheets. This refers primarily to the non-mountainous margins of the Laurentide and Scandinavian Ice Sheets, large areas of which are dominated by the supraglacial landsystem. The term also implies that the glaciers in these regions received little or no sediment directly from supraglacial sources: the vast bulk of sediment in these landforms was derived from subglacial sources and initially transported to the margin subglacially or englacially.

The most common landform in the supraglacial landsystem is the hummock, but this landsystem also includes ring forms, ice-walled-lake plains, dump moraines, outwash fans and disintegration ridges. In places, these landforms are superimposed on active-ice landforms. Regions of hummocky topography have been referred to as dead-ice moraine, hummocky moraine, stagnation moraine, ice-disintegration features and moraine-mound complexes, to name a few. They have also been referred to as end moraine where the hummocks occur in a distinct band, and ground moraine where hummocks are widespread and of low relief.

A supraglacial interpretation of this suite of landforms above was first developed in the 1940s and 1950s in Europe and North America (Milthers, 1948; Gravenor and Kupsch, 1959). Further studies in the 1960s in North America, Europe, and on modern glaciers (e.g. Clayton, 1964, 1967; Parizek, 1969; Boulton, 1967, 1968) strengthened a supraglacial interpretation. These studies were accompanied by investigations that pointed out the sedimentological complexity of the supraglacial environment (Hartshorn, 1958; Boulton, 1968, 1972a;

Marcussen, 1973; Lawson 1979; Eyles, 1979, 1983b and c), along with the corollary that several till layers of varying genesis could be deposited during one glacial cycle.

Early workers who supported a supraglacial explanation for these features also considered subglacial squeezing to be possible and important (Gravenor and Kupsch, 1959; Parizek, 1969). In recent years, the supraglacial theory has been challenged by proponents who think that subglacial pressing dominates the hummock-forming process or that hummocks are formed by subglacial meltwater erosion. These theories and others for hummock formation are listed below. Only one of the following theories (number 1) occurs in a supraglacial setting: 2 through 7 take place subglacially, 8 and 9 bear the stamp of englacial processes, and 10 and 11 occur proglacially. The theories include:

1. collapse of thick supraglacial debris lying on stagnant ice (Gravenor and Kupsch, 1959; Clayton, 1967; Boulton, 1967, 1972; Parizek, 1969; Clayton and Moran, 1974; Eyles, 1979, 1983b and c; Krüger, 1983; Paul, 1983; Sollid and Sørbel, 1988; Johnson *et al.*, 1995; Ham and Attig, 1996; Mollard, 2000)
2. subglacial pressing of stagnant ice blocks into a soft bed (Hoppe, 1952; Stalker, 1960; Aartolahti, 1974, 1975; Eyles *et al.*, 1999a; Boone and Eyles, 2001)
3. chaotic erosion by subglacial meltwater (Shaw, 1996; Munro and Shaw, 1997)
4. subglacial moulding by active ice, forming hummocks along with drumlins and Rogen moraine (Aario, 1977; Lundqvist, 1981)
5. subglacial glacitectonic thrusting forming 'cupola hills' (Aber *et al.*, 1989; Evans, 2000b)
6. subglacial accumulation on stoss side of patches of frozen bed (Kleman *et al.*, 1999)
7. patchy formation of ground ice beneath stagnant, cold-based glaciers (Aario, 1992)
8. melting of debris-rich stagnant ice containing extensive karst tunnels (Kemmis *et al.*, 1994)
9. deformation of supraglacial and englacial debris by rising diapirs of clean ice (Minell, 1979)
10. groundwater blowouts (Bluemle, 1993; Boulton and Caban, 1995)
11. partly or wholly by periglacial means (Bik, 1968; Mollard, 2000).

Of these, we consider 1 the theory that best explains the observed morphology and sedimentology of these features, although we consider 2 possible as well. Of the remaining, we regard 3 and 9 as being unlikely, for a variety of reasons. Theories 4 through 8 and 10 and 11 likely occur, but are of only local importance or they produce hummocks not associated with the suite of landforms described in this chapter.

10.2 LANDFORMS

10.2.1 Hummocks

A landform with the shape of a hummock can be formed in a variety of environments, some non-glacial. Even in glaciated landscapes other than the supraglacial landsystem, there are likely to be found examples of incompletely formed Rogen moraine or drumlins, the forms of which could best be described as hummocky. But geologists working in glaciated terrain have rarely used 'hummock' in a purely descriptive sense. Rather, the term has been reserved for the knobs, hillocks and mounds occurring in widespread tracts and interspersed with the other landforms mentioned below. These are the hummocks we describe here.

Individual hummocks are round to elongate, conical to flat topped, and interspersed with depressions, which are sometimes called dead-ice hollows, ice-block depressions, or kettles. Hummocks occur in groups of thousands of individuals and generally display no preferred orientation of slopes or elongate elements (Figs. 10.1, 10.2 and 10.3).

As reported in the literature, hummocks are 15–400 m in diameter and are spaced 150–500 m apart (Table 10.1). The relief of hummocks ranges from 2 to 70 m, but most are less than 20 m. Gravenor and Kupsch (1959) classified hummocks in classes of relief:

Figure 10.1 Aerial photographs of hummocks. These hummocks lack a preferred orientation of elements and are referred to as uncontrolled. A) A hummock tract 35 km north of Vilnius, Lithuania, from archives of the Geological Survey of Lithuania; ice flow direction was to the southeast. B) A stereopair showing high-relief hummocky topography with ice-walled lake plains (A) and some disintegration ridges (B) approximately 125 km southeast of Edmonton, Alberta, Canada (Photo number 160-5216: 1368-10 and -11, Location, T. 46, R. 12, W. 4th. Mer., from Technical Division, Department of Lands and Forests, Edmonton, Alberta). C) Hummocky topography from northwest Wisconsin (BRO-2AA-95, U.S. Department of Agriculture); ice-flow direction was approximately S 30 E, parallel to the elongate lakes.

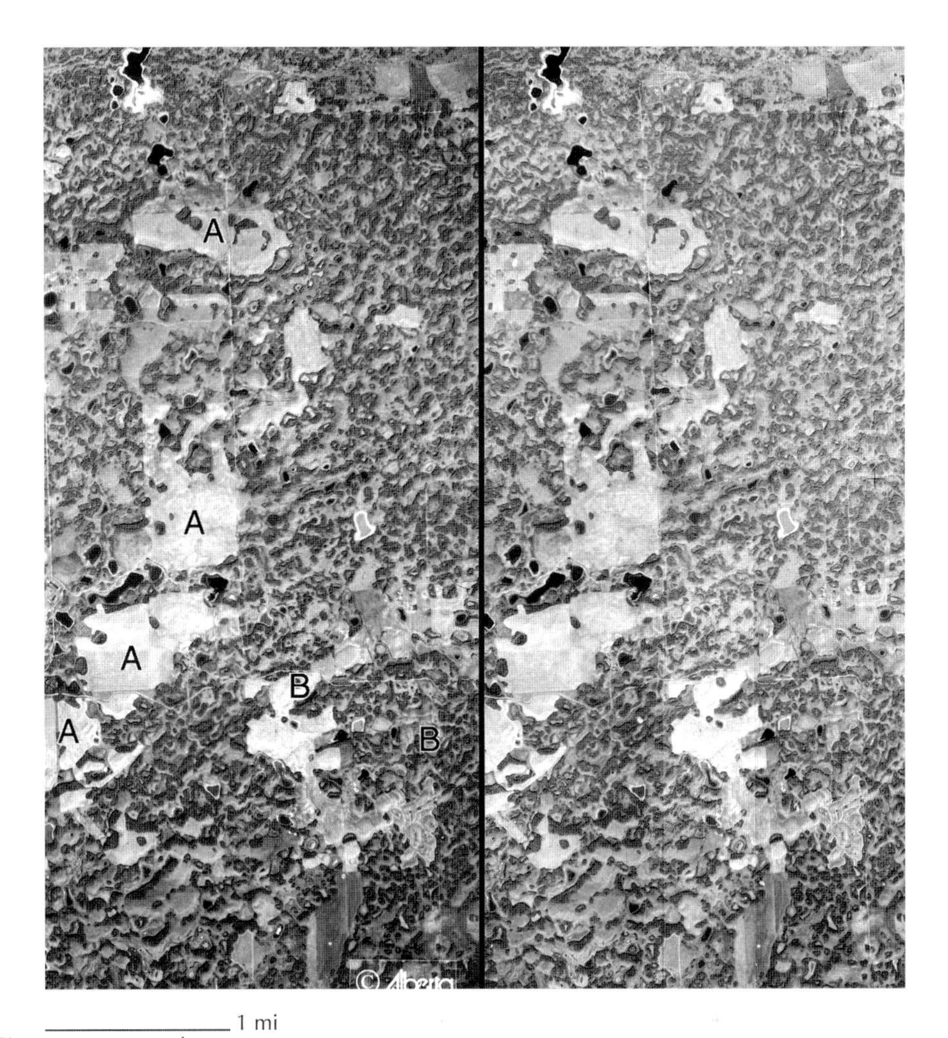

(B)

(C)

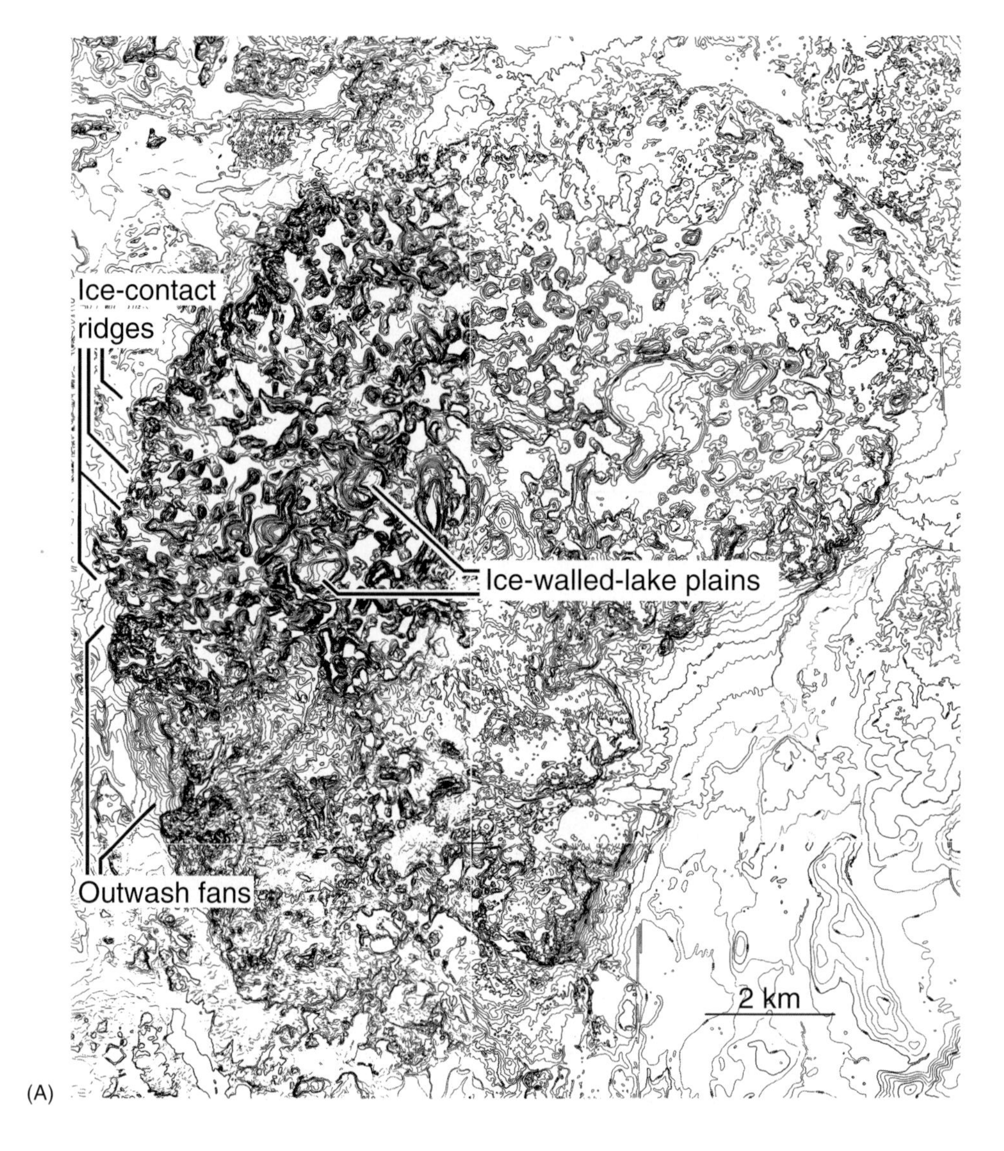

Figure 10.2 Topographic maps of hummocky landscapes. A) Mosaic of topographic maps showing the Harrison Hills of northern Wisconsin (Ham and Attig, 1997); ice-flow direction was approximately S 45 E. Contour interval is 10 ft except in the upper-right portion where it is 20 ft. The Harrison Hills are dominated by high-relief hummocks (up to 70 m) that were formed along the Late Wisconsin margin of the Wisconsin Valley Lobe. Numerous ice-walled-lake plains are present, two of which are identified. These ice-walled-lake plains are examples of the stable ice-walled-lake environment (see Fig. 10.12). Several ice-contact ridges and outwash fans are present, some of which are identified. B) Mosaic of topographic maps from western Wisconsin near the Late Wisconsin margin of the Superior lobe showing hummocks and three ice-walled-lake plains with rim ridges; contour interval 10 ft (Johnson, 2000). These ice-walled-lake plains are examples of the unstable ice-walled-lake environment (see Fig. 10.12); ice-flow direction was approximately S 50 E. C) Hummocks from southern part of the island of Sealand, Denmark, contour interval 2.5 m (Krüger, 1969).

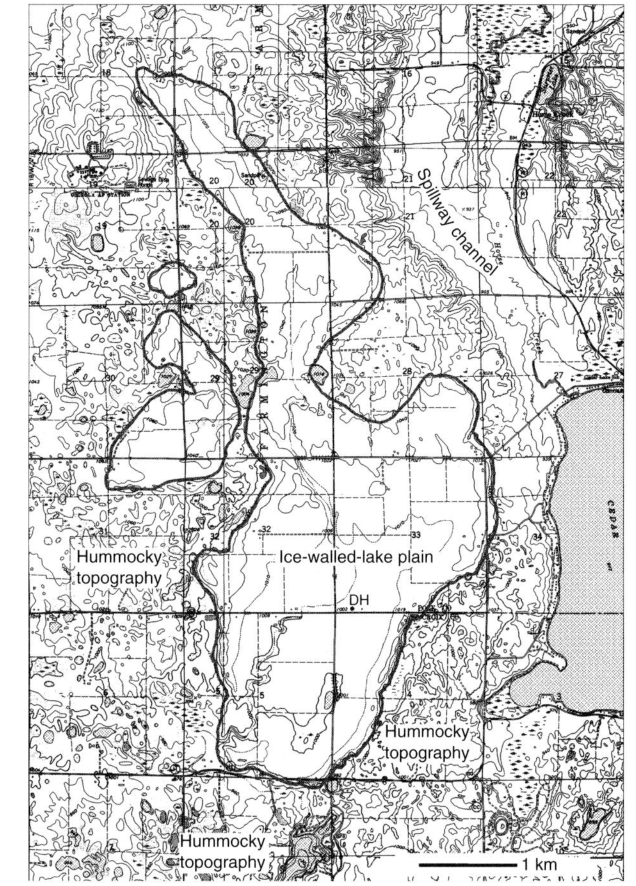
Spillway channel
Hummocky topography
Ice-walled-lake plain
DH
Hummocky topography
Hummocky topography
1 km

(B)

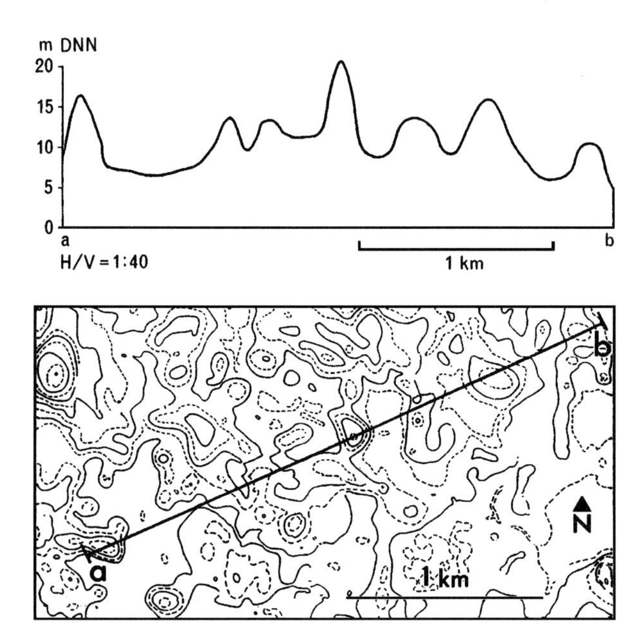
m DNN
20
15
10
5
0
a
b
H/V = 1:40
1 km
b
a
N
1 km

(C)

- low-relief hummocks – less than 10 ft (3 m)
- intermediate-relief hummocks – 10–25 ft (3–8 m), and
- high-relief hummocks – greater than 25 ft (8 m).

This distinction may be important in terms of genesis. For example, Ham and Attig (1996) describe high-relief hummocks in north-central Wisconsin, USA, that occur in distinct tracts amidst broad regions of lower-relief hummocks (Fig. 10.2A), implying that the two types of hummocks have had a different developmental history. Other than these rough measurements of hummock dimensions, we have not attempted a detailed analysis of the variation of hummock size, shape, slope and spacing. Considering the modern abilities of imaging technology this may be a rather easy and fruitful area of future geomorphic research.

Most authors emphasize the haphazard orientation of linear elements in hummocky terrains, which is the result of what Gravenor and Kupsch (1959) refer to as uncontrolled deposition (Fig. 10.1 and 10.2). However, in some regions, hummock tops are oval to elongate (they are

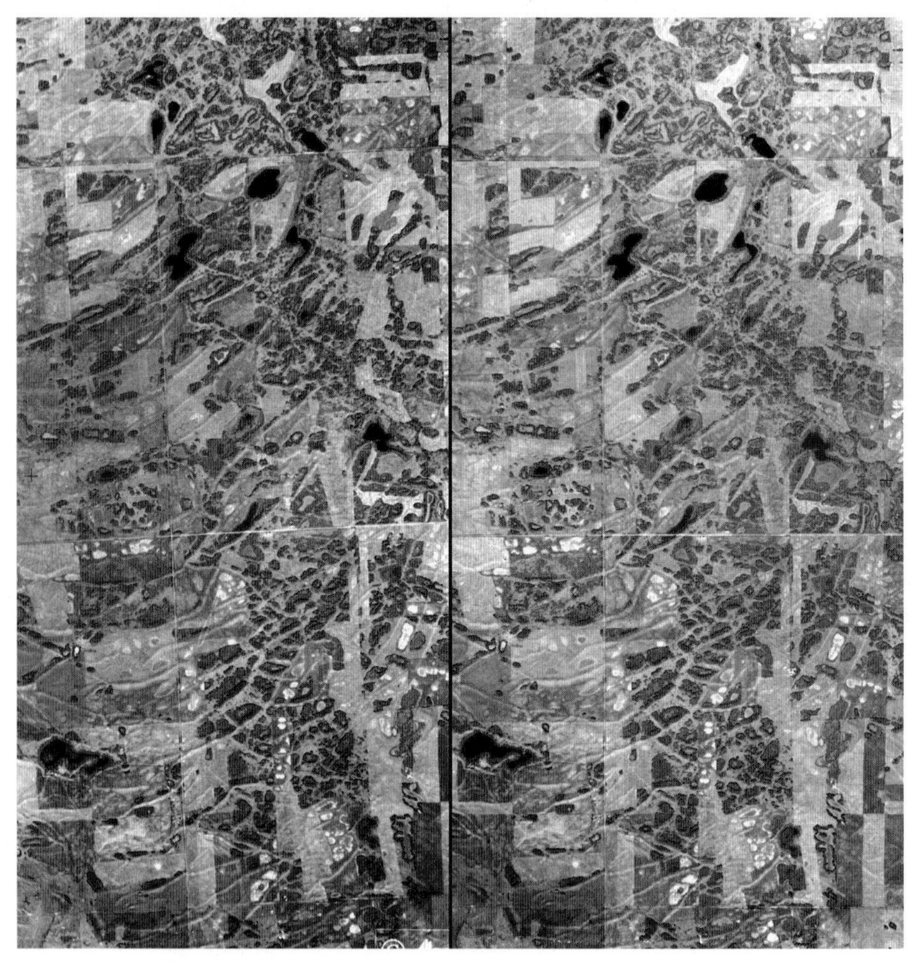

(A)

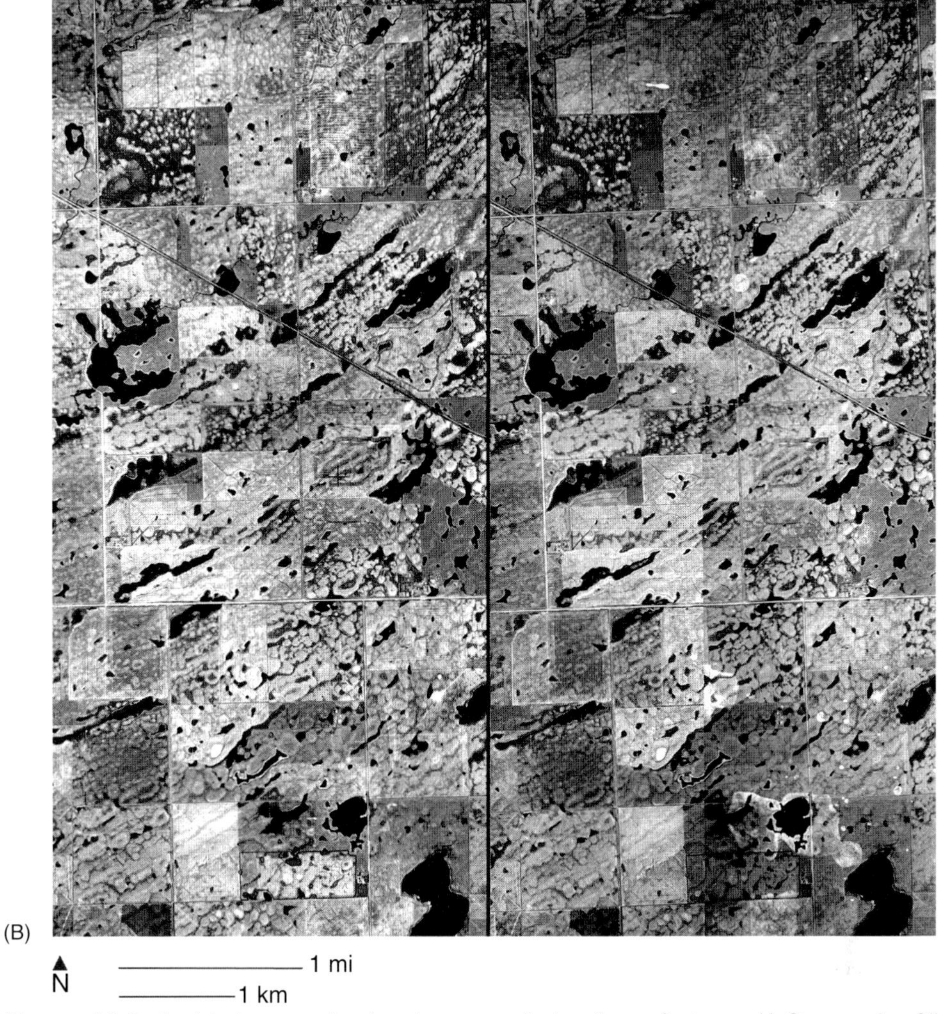

Figure 10.3 Aerial photographs showing controlled collapse features. A) Stereopair of linear disintegration ridges located about 200 km east of Edmonton, Alberta, Canada. Ridges are interpreted to represent the location of crevasses in the ice (ice flow slightly east of south) that were filled by supraglacial debris or squeezed subglacial till (Gravenor and Kupsch, 1959) (Photo number 160-5303:1333-36 and -37, Location: T. 48, R.1, W. 4th Mer., from Technical Division, Department of Lands and Forests, Edmonton, Alberta). B) Stereopair of aligned hummocks forming ridges about 65 km SE of Calgary, Alberta, Canada showing control by structures within the ice, presumably thrust planes near the ice margin (Gravenor and Kupsch, 1959); ice flow was to the southeast.

rarely exactly circular), or the hummocks are grouped in clear patterns. Such an inherited geometry is referred to as controlled deposition (Gravenor and Kupsch, 1959; Fig. 10.3). These hummocks are interpreted to have inherited their orientation from similarly oriented structures in the parent ice, mainly crevasses (Fig. 10.3A) and thrust planes (Fig. 10.3B). For example, controlled deposition has been explained as resulting from melting of debris-rich thrust zones to produce hummocks in Scotland (Hambrey *et al.*, 1997; Bennett *et al.*, 1998), the washboard

Reference	Location	Relief (m)	Diameter (m)	Spacing (m)	Slope angle
Gravenor, 1955	Canada (Alberta)	5	90		
Klassen, 1993	Canada (Saskatchewan)	5–20			
Schou, 1949	Denmark	5–25	50–200	150–300	
Aartolahti, 1975	Finland	15–25	200–300		
Okko and Perttunen, 1971	Finland	3–5	20–100		
Möller, 1987	Sweden	2–5	15–50		
Clayton, 1967	USA (North Dakota)	7–70			5–15°
Ham and Attig, 1996	USA (Wisconsin	5–70		300–500	
Johnson *et al.*, 1995	USA (Wisconsin)	5–20	25–400	250–500	2–14°
TOTAL RANGE		2–70	15–400	150–500	2–15°

Table 10.1 Dimensions of hummocks

moraines of the Canadian Prairies (Gravenor and Kupsch, 1959), and aligned hummock tracts of Iowa (Colgan, 1996). Additionally, direct deposition in crevasses produces disintegration ridges (see below), and this is also a type of controlled deposition. Controlled formation of hummocks may not be obvious with just a quick inspection of maps or aerial photographs, but may be revealed after measurement of linear elements in a hummock tract (Johnson *et al.*, 1995; Evans, 2000b).

Hummocks are composed entirely or partly of till, stream sediment and lake sediment. Hummocks composed of collapsed lake sediment that displays faults and folds are described in North Dakota (Clayton and Cherry, 1967) and Wisconsin (Attig, 1993). Collapsed stream sediment with hummocky topography is widespread in many glaciated regions and may even be the predominant feature in some areas. A collapse origin is shown by topography, sedimentology and normal faults that cut sedimentary structures.

Many hummocks contain a complex internal stratigraphy consisting of interbedded till and stratified sediment, often with normal faults, and these are well described from the margins of modern glaciers (e.g. Boulton, 1972a; Lawson, 1979; Krüger, 1994a) as well as in Pleistocene examples from northern Europe (Krüger, 1969; Stephan, 1980; Haldorson, 1982, Möller, 1987; Malmberg Persson, 1991) and North America (Sharpe, 1988; Attig and Clayton, 1993; Johnson *et al.*, 1995; Munro and Shaw, 1997) (Figs. 10.4 and 10.5). These sedimentological relationships are almost always interpreted as developed by the interbedding of supraglacial flow till or melt-out till with stream or lake sediment, which subsequently collapses from melting of underlying ice, forming faults and other disruptions of bedding.

However, the most common type of hummock in hummocky regions are those composed entirely or predominantly of till, and these have been described in Canada (Gravenor and Kupsch, 1959; Stalker, 1960) and the Upper Midwest of the USA (Clayton, 1967; Kemmis *et al.*, 1981; Mickelson, 1986; Hansel and Johnson, 1987; Johnson *et al.*, 1995; Ham and Attig, 1996). The origin of the till in these hummocks has been the subject of debate in the literature.

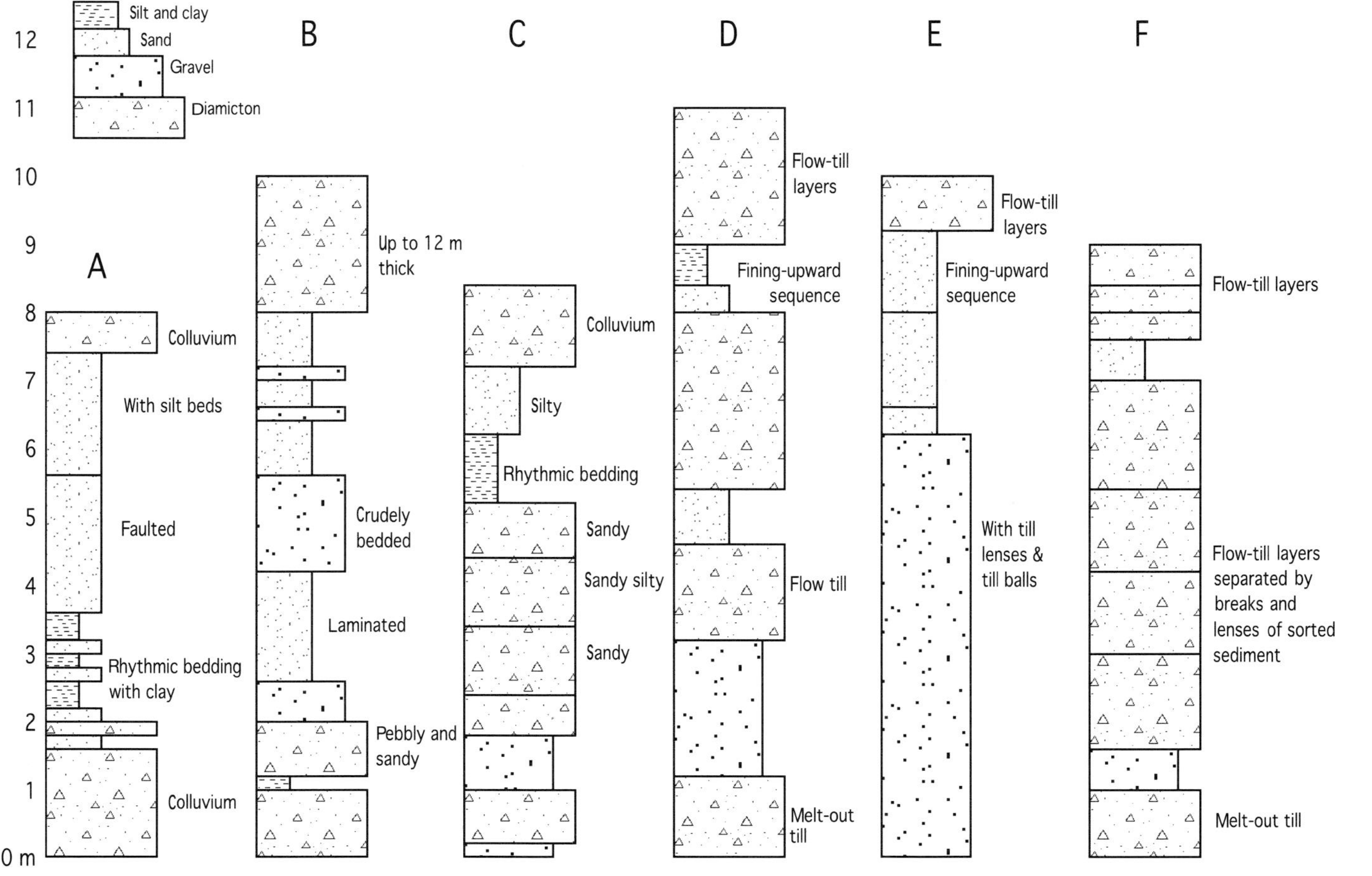

Figure 10.4 Vertical profiles of sediment described from Pleistocene hummocks on the Wollaston Peninsula, northern Canada, and from modern glaciers. These sequences show interbedded till and sorted sediment in various proportions. A and B = sequences from hummocks on the Wollaston Peninsula (Sharpe, 1988), C = Idealized section of hummock sediment from the Matanuska Glacier, Alaska (Lawson, 1981), D, E, and F = Idealized sections from several modern glaciers showing trough (depression) fillings on the glacier surface where troughs receive mixed flow till and sorted sediment (D), predominantly sorted sediment (E), or predominantly flow till (F). (Paul, 1983).

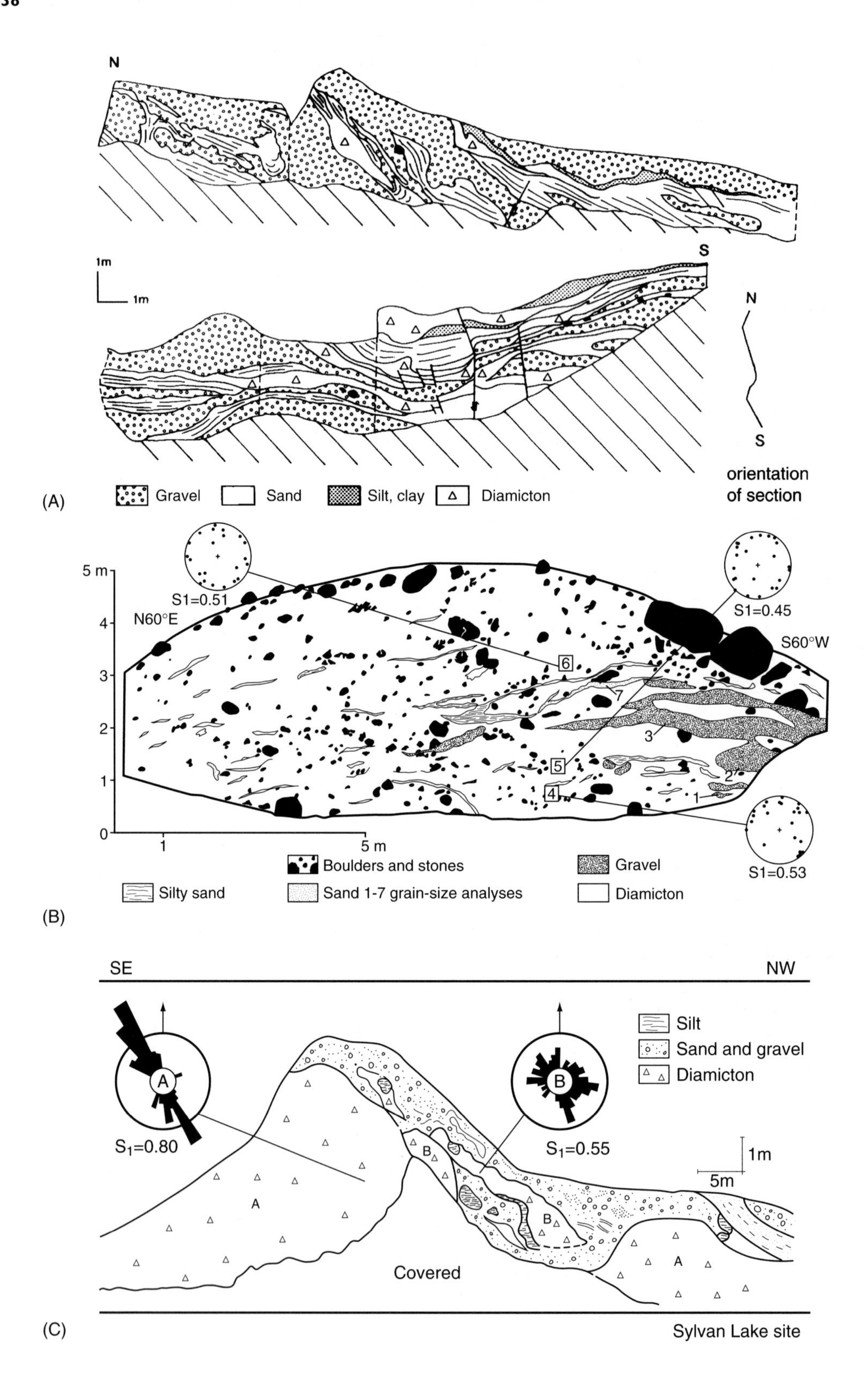

N
S
1m
1m
N
S
orientation of section
(A)
Gravel
Sand
Silt, clay
Diamicton
5 m
4
3
2
1
0
1
5 m
S1=0.51
N60°E
S1=0.45
S60°W
S1=0.53
6
5
4
7
3
2
1
Boulders and stones
Gravel
Silty sand
Sand 1-7 grain-size analyses
Diamicton
(B)
SE
NW
A
B
S1=0.80
S1=0.55
Silt
Sand and gravel
Diamicton
1m
5m
Covered
(C)
Sylvan Lake site

Because the topography of till hummocks is identical to hummocks composed of collapsed lake sediment and outwash, it is likely that most till hummocks formed in the same way, that is, by supraglacial collapse.

Another argument indicating that till hummocks are supraglacial collapse features is their close association with ice-walled-lake plains (described in more detail below). Fossils in ice-walled-lake plain sediment in North Dakota indicate that ice-walled lakes existed long after the period of initial ice stagnation into a time of warmer climate. Such a climate would cause rapid ablation unless the surrounding stagnant ice was covered with thick supraglacial debris (Tuthill, 1967; Clayton and Cherry, 1967). These ice-walled-lake plains are today surrounded by the hummocks produced when the insolating supraglacial debris was eventually let down by the melting of the underlying ice.

The few sedimentological studies that exist also indicate a supraglacial-collapse origin for till hummocks. For example, Ham and Attig (1996) described high-relief hummocks (Fig. 10.2A) with thick uniform till that they interpret to be flow till. They observed flow-till bedding dipping into the centres of hummocks, presumably from an adjacent supraglacial source, as well as pebble fabrics that are randomly oriented, suggesting flow till. Johnson *et al.* (1995) describe hummocks with thick uniform till that they interpret to represent supraglacial melt-out till. They cite till fabric measurements that parallel regional ice-flow direction (Fig. 10.5C), a relationship that would not be expected with a flow-till or a squeezed-till hypothesis. Supraglacial sandy melt-out till, like outwash, would be well drained and would collapse by faulting leaving the till between faults undisturbed. Recently, the till of hummocks on the Canadian Great Plains has been reinterpreted as subglacial squeezed till (Eyles *et al.*, 1999a; Boone and Eyles, 2001), but this interpretation is not based on sedimentological measurements.

The matrix grain-size composition of till has an effect on the geomorphic characteristic of hummocks. Boulton (1972a) noted that till with higher water content would flow more easily, producing hummocks with lower slopes. Thus, hummocks composed of poorly drained, clayey till would have lower slopes and generally lower relief than hummocks composed of well-drained, sandy till. The more poorly drained clayey sediment would tend to collapse and flow,

Figure 10.5 Sketches of exposures through hummocks containing mixed till and sorted sediment. A) Outcrop sketch from a hummock in southern Sweden showing interbedded till and sorted sediment. Weak fabrics are found in the till layers and interpreted to indicate flow till (Malmberg Persson, 1991) (reprinted with kind permission from the author and the Geological Society of Sweden). B) Outcrop sketch of till with some sorted sediment from a hummock in southern Sweden. Weak fabrics are interpreted to indicate flow till (Andersson, 1998) (S1 = primary eigenvalue) (reproduced from 'Genesis of hummocky moraine in the Bolmen area, southwestern Sweden,' by G. Andersson, from *Boreas*, www.tandf.no/boreas, 1998, volume 27, pages 55–67, by permission of Taylor and Francis AS). C) Outcrop sketch of a portion of a hummock in northwestern Wisconsin showing flow till (fabric B) and sorted sediment overlying thick, uniform till with strong, ice-flow-parallel fabric (fabric A) (S1 = primary eigenvalue) that is interpreted as melt-out till (reproduced from 'Composition and genesis of glacial hummocks, western Wisconsin,' by M.D. Johnson, D.M. Mickelson, L. Clayton, and J.W. Attig, from *Boreas*, www.tandf.no/boreas, 1995, volume 24, pages 97–116, by permission of Taylor and Francis AS).

where the better-drained, stiff sandy sediment would flow less often and fault, and produce a stable debris cover that would insulate the glacier, cause ice to melt slowly, and allow for the preservation of melt-out till. This difference in grain size explains why high-relief hummocks in Wisconsin are associated with sandy till (Johnson *et al.*, 1995; Ham and Attig, 1996), and

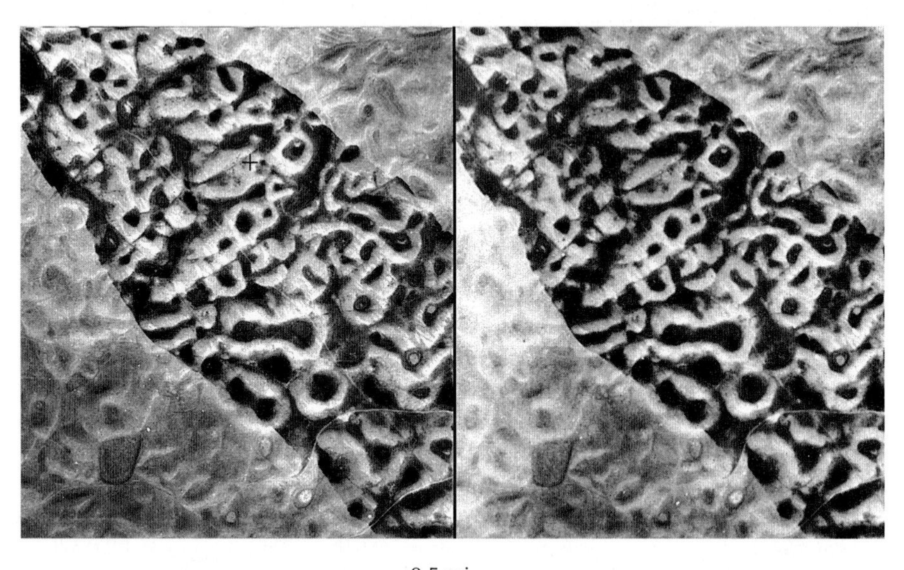

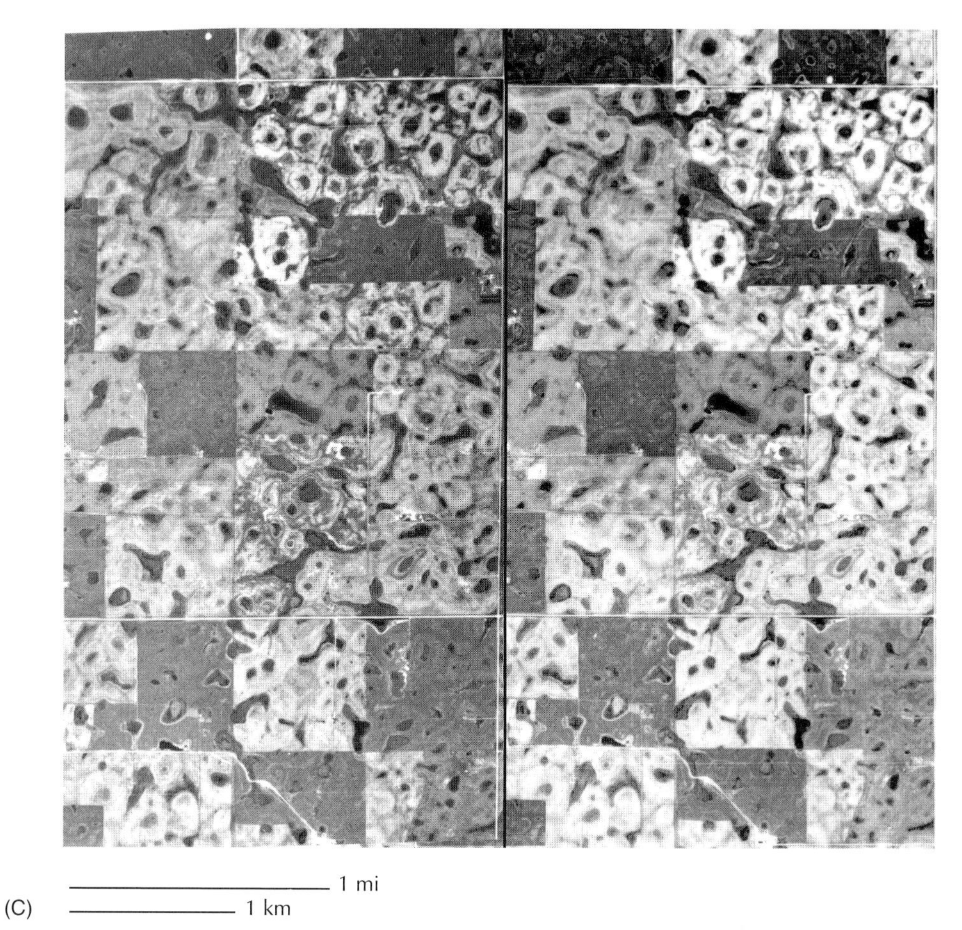

Figure 10.6 Photos showing ring forms. A) Stereopair of ring forms composed of thin lake sediment over clay till southwest of Long Valley, Saskatchewan, Canada, highlighted by a prairie-grass fire (A6729-12 and 13 (R41-12 and 13); Lat. 50° 42′, Long. 107° 03′). B) Stereopair of ring forms composed of clay till east of Kenaston, Saskatchewan, Canada (95675-08 68 and 69; Lat. 51° 33′, Long. 106° 08′). C) Stereopair of low-relief ring forms in clay till north of Steelman, Saskatchewan, Canada (A21749-14 and 15; Lat. 49° 22′, Long. 102° 37′).

low-relief hummocks of Alberta, Saskatchewan and North Dakota are associated with clayey till (Clayton, 1967).

10.2.2 Ring Forms

A type of landform common on the Great Plains of central North America are fields of low-relief rings; circular ridges with depressions in their centres (Fig. 10.6). These ring forms have also been referred to as doughnuts, circular disintegration ridges, closed ridges, rim ridges, rimmed kettles and humpies (Gravenor and Kupsch, 1959; Parizek, 1969; Mollard, 2000). They are essentially a type of hummock, but one with a central depression. Most of these ring forms are clearly glacial in origin, but some associated with lake sediment in cold regions likely have a periglacial origin.

Ring forms in North Dakota, Alberta and Saskatchewan (Fig. 10.6) are up to 200 m across, are generally of low relief, uniform size, and uniform spacing, and composed predominantly of clayey till. Nearly identical ring forms can form tracts consisting of hundreds or thousands of rings. Ring forms are much less common in sandy-till regions, though some composed of sandy till have been reported in northern Sweden (Melander, 1976). Ring forms can be formed in several ways (Fig. 10.7). According to several authors (Gravenor, 1955; Clayton, 1967; Mollard, 2000), these features form supraglacially by flowing of till into a glacier sinkhole (Fig. 10.7A, B). A subglacial-squeezing origin (Gravenor and Kupsch, 1959; Aartolahti, 1974; Eyles *et al.*, 1999a; Mollard, 2000) suggests that ring forms may be produced by till being squeezed into the sinkhole from below (Fig. 10.7C) or by squeezing up around individual, foundering ice blocks.

10.2.3 Ice-Walled-Lake Plains

Ice-walled-lake plains (Figs. 10.2, 10.8, 10.9, 10.10) form where sediment accumulates in broad, water-filled sinkholes in stagnant ice. After the ice completely melts, the sediment

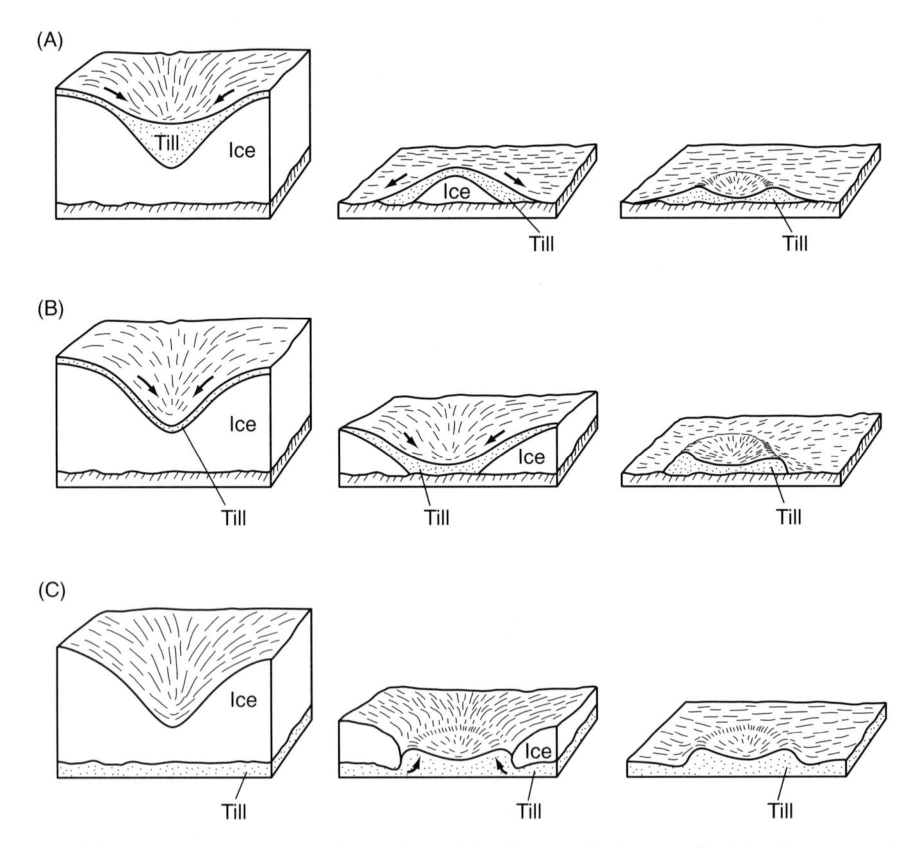

Figure 10.7 Mechanisms for the formation of ring forms. A) A supraglacial collapse mechanism where a block of ice is insulated by flow till in an ice depression (Clayton, 1967). B) A supraglacial collapse mechanism where flow till is deposited around the margins of sinkhole. C) A subglacial squeezing mechanism for the formation of hummocks and ring forms. See text for description of a fourth mechanism.

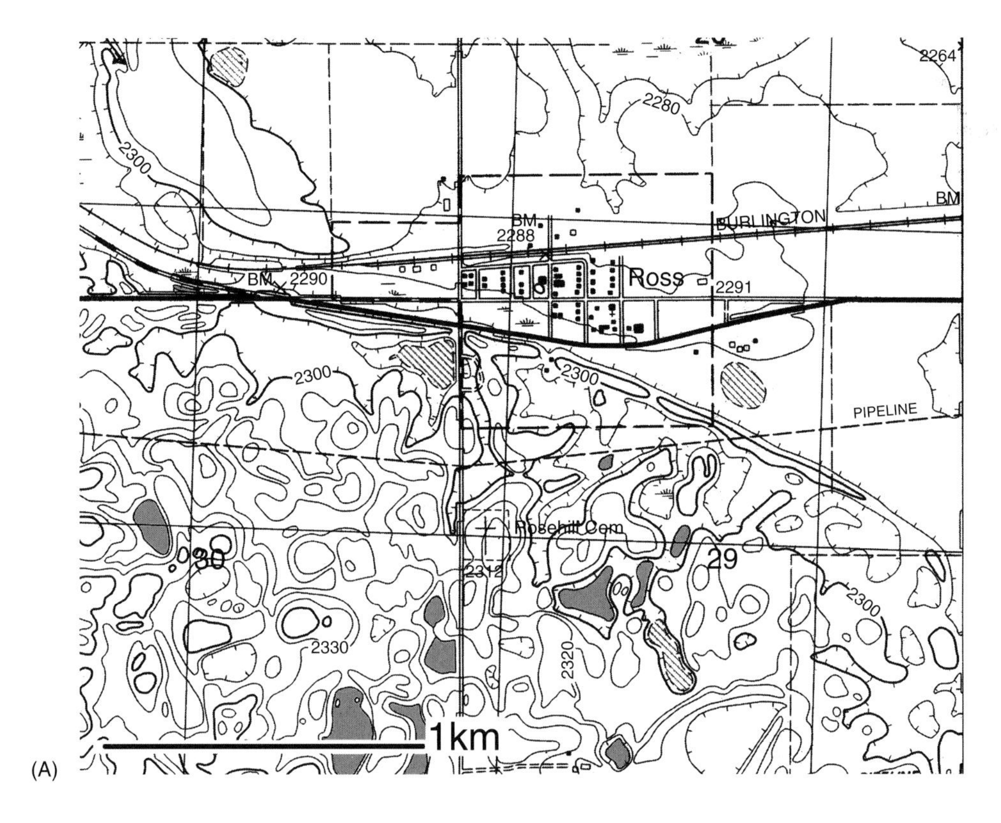

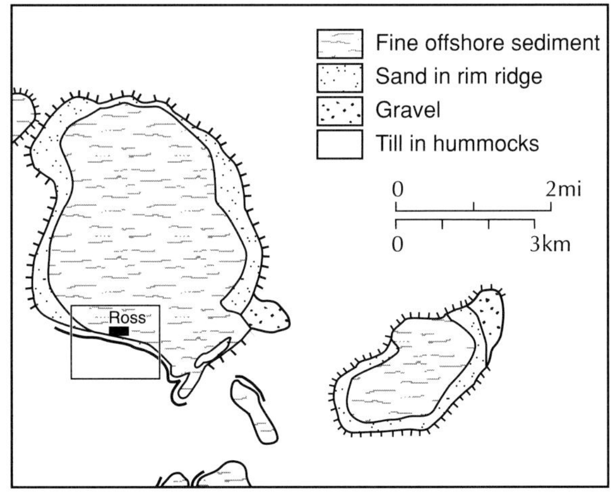

Figure 10.8 An ice-walled-lake plain in North Dakota, USA. A) Portion (shown in B) of the Ross 7.5′ series topographic map showing a rim ridge with ice-walled-lake plain to the north and hummocks to the south; contour interval 10 f. B) Geologic interpretation of the features in A.

remains as a roughly circular, flat-topped landform. Ice-walled-lake plains are rounded in map view, similar to the shape of lakes where headlands tend to be eroded and the lake shape smoothed out by erosion. They stand out amidst the surrounding hummocky terrain by being flat to dish-shaped, and they are often, but not always, marked on their edge by distinct rim

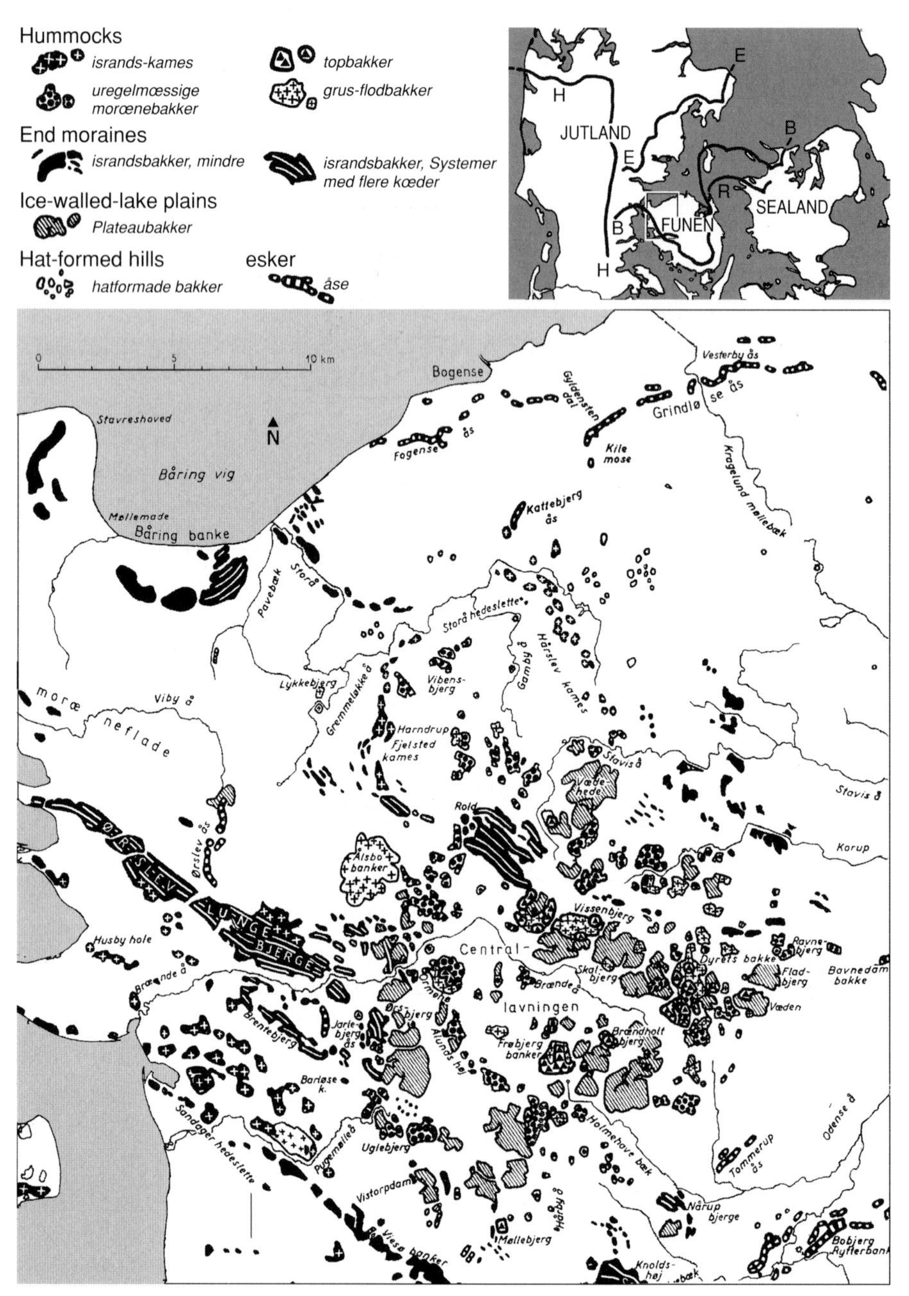

Figure 10.9 A portion of Smed's (1962) geomorphologic map of the island of Funen, Denmark. Only selected geomorphic features from Smed's map are shown, and we have identified these as hummocks, end moraines, ice-walled-lake plains, hat-formed hills, and eskers. Names in italics are Danish terms for these features. Among the features present in the area but not shown on the map include outwash plains, tunnel channels and drumlins, none of which are prominent in this area dominated by supraglacial landsystem landforms. These stagnant-ice features developed in ice left from the East Jylland advance. Prominent Danish ice-margins are shown in the inset: H = Main Stationary Line, E = East Jylland ice limit, B = Bælthav advance ice limit, R = Røsnes advance ice limit (from Houmark-Nielsen, 1983). End moraines shown on the map are associated with the Bælthav advance.

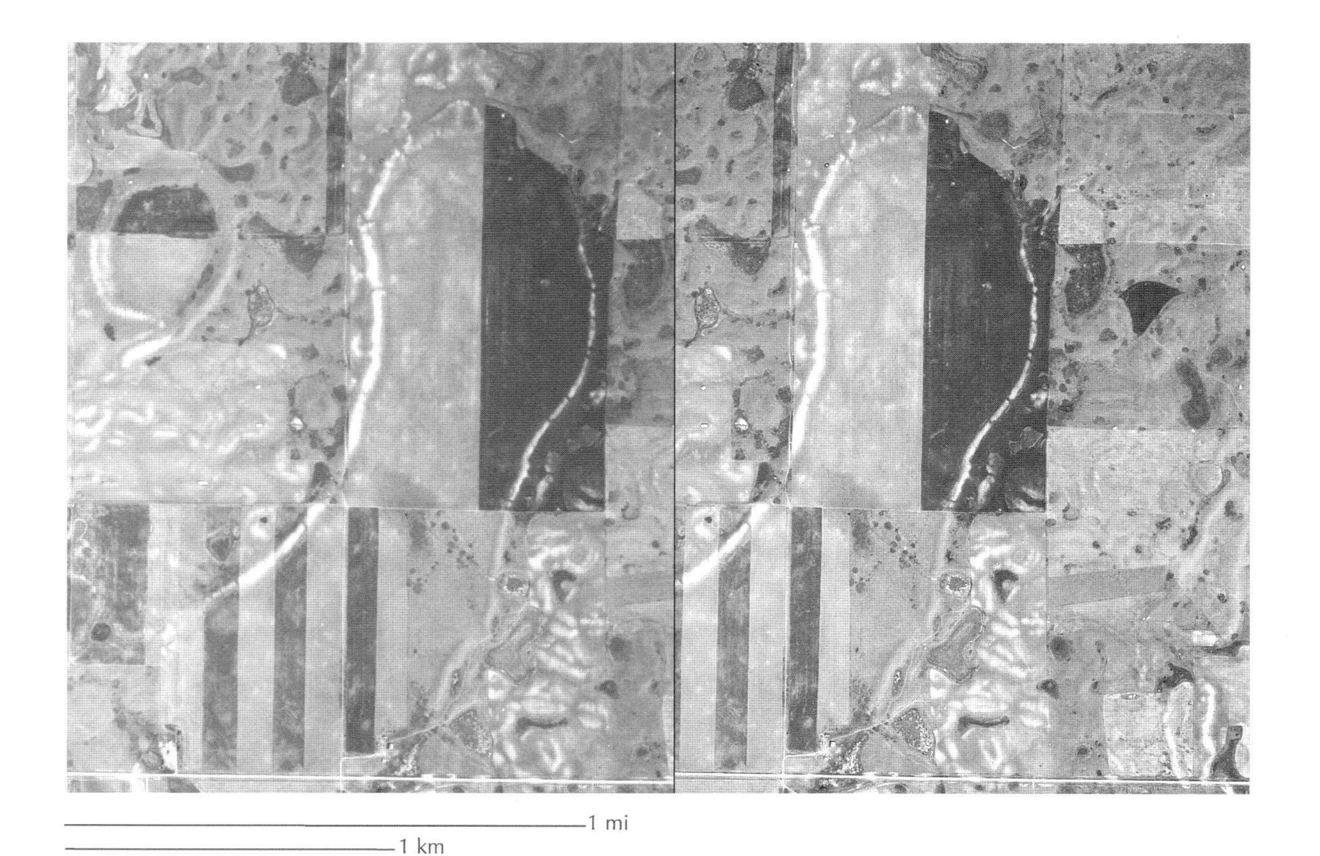

Figure 10.10 Stereopair of an ice-walled-lake plain in Mountrail County, North Dakota USA (BAL-4V-9 and 10, USDA). The ice-walled-lake plain is surrounded by hummocks and ring-forms.

Reference	Location	Relief (m)	Diameter (m)	Area (km^2)	Thickness of lake sediment (m)
Stalker, 1960	Canada (Alberta)	2–5	6–180 (90)	Up to 30	'Thin'
Parizek, 1969	Canada (Saskatchewan)	7–50		Up to 13	1–10
Klassen, 1993	Canada (Saskatchewan)	5–25	200–800		
Brehmer, 1990	Denmark	10	200–750		
Hansen, 1940	Denmark	10–15	150–4,000		
Schou, 1949	Denmark	8–15	300–700		
Smed, 1962	Denmark	25–35	500–2,000		2–5
Strehl, 1998	Germany	9–13	300–1,200	0.1–1.0	9–18
Bitinas, 1992	Lithuania	10–50		1–30	
Westergård, 1906	Sweden	15–30	500–4,500		1–3
Clayton and Cherry, 1967	USA (North Dakota)		400–2,500	0.5–13	
Ham and Attig, 1996	USA (Wisconsin)		100–1,500		15
Johnson *et al.*, 1995	USA (Wisconsin)	5–10	500–4,000	1–13	20
Syverson, 2000	USA (Wisconsin)	40–60	500–1,000		50 (stable environment)
Syverson, 2000	USA (Wisconsin)	10–35	1,000–1,500		23 (unstable environment)
TOTAL RANGE		2–60	100–4,500	0.1–30	

Table 10.2 Dimensions of ice-walled-lake plains

ridges that rise up to 10 m above the central parts of the ice-walled-lake plains. The rim ridges slope gently towards the plain centres, but have slopes close to angle of repose on their outer slopes. Ice-walled-lake plains are generally 1–15 km^2 in area but can be up to 30 km^2 (Table 10.2). They may occur as isolated forms but generally appear in clusters amidst their hummocky surroundings (Figs. 10.2B, 10.9).

Ice-walled-lake plains were referred to as plateau clay-hills by Milthers (1948) and Schou (1949) in Denmark, and as moraine plateaux by Stalker (1960) in Canada. The high kames of southern New England are likely ice-walled-lake plains (Stone and Peper, 1982) as are rimmed Veiki moraine plateaux of northern Sweden (Hoppe, 1952; Lagerbäck, 1988). Ice-walled-lake plains have also been described in Poland (Niewiarowski, 1963), Lithuania (Bitinas, 1992), Germany (Strehl, 1998), and Russia (Ekman *et al.*, 1981).

Ice-block depressions may occur in ice-walled-lake plains, and there are tracts of hummocky topography underlain entirely by collapsed lake sediment (Figs. 10.11A and B). However, as noted above, the majority of ice-walled-lake plains show little sign of collapse in their centres, and this is due to the ability of lakes, once formed on the ice surface, to melt completely through the stagnant ice due to the high heat content of water. Multiple rims positioned concentrically within an ice-walled-lake plain indicate progressive widening of the ice-walled lake, with each rim forming from ice-contact sedimentation (Fig. 10.11D).

Clayton and Cherry (1967) described two end-member types of ice-walled-lake plains. The unstable environment type (Figs. 10.2B, 10.12) is formed in ice with thin surface debris, resulting in ice-walled-lake plains of lower relief but having well-defined rim ridges. Stable environment settings occur where debris cover is thick, and the ice-walled-lake plains that result have higher relief, lack rim ridges, have thick till and are often among the highest points in the landscape (Figs. 10.2A, 10.12). Because of the ability of thicker till to insulate ice, stable environment lakes last a longer period of time and have the potential of preserving long sediment records.

Rim ridges on ice-walled-lake plains are composed of sand and gravel interbedded with layers of flow till (Fig. 10.11C). Deltaic sediment underlies rim ridges (Stalker, 1960; Johnson, 1986, 2000; Syverson, 2000), and indicates stream flow on the adjacent ice surface. The sorted sand and gravel in the ridges is interbedded with beds of flow till dipping towards the centres of the plains, indicating flowage from the adjacent ice surface. In a few places, till makes up the entire rim ridge (Parizek, 1969), some of which may be subglacial-squeeze till (Stalker, 1960).

The centres of ice-walled-lake plains contain finer sediment, which is clayey or silty, well bedded and even varved in places (Hansen, 1940; Smed, 1962; Stalker, 1960; Brehmer, 1990). In Denmark, an asymmetrical distribution of grain sizes in ice-walled-lake plain sediments (e.g. sandier on the southern half, finer on the northern half) has been interpreted to represent, for this example, dominantly northward flow of sediment-laden streams on the stagnant-ice surface (Smed, 1962; Brehmer, 1990).

There is a range of ice-walled-lake plains composed entirely of lake sediment to ones composed entirely of till, even within a given geographic region (Stalker, 1960). Most ice-walled-lake plains seem to contain sorted sediment that accounts for the total relief of the plain, and lake sediment can be 20–50 m thick (Fig. 10.13B) (Stalker, 1960; Clayton and Cherry, 1967;

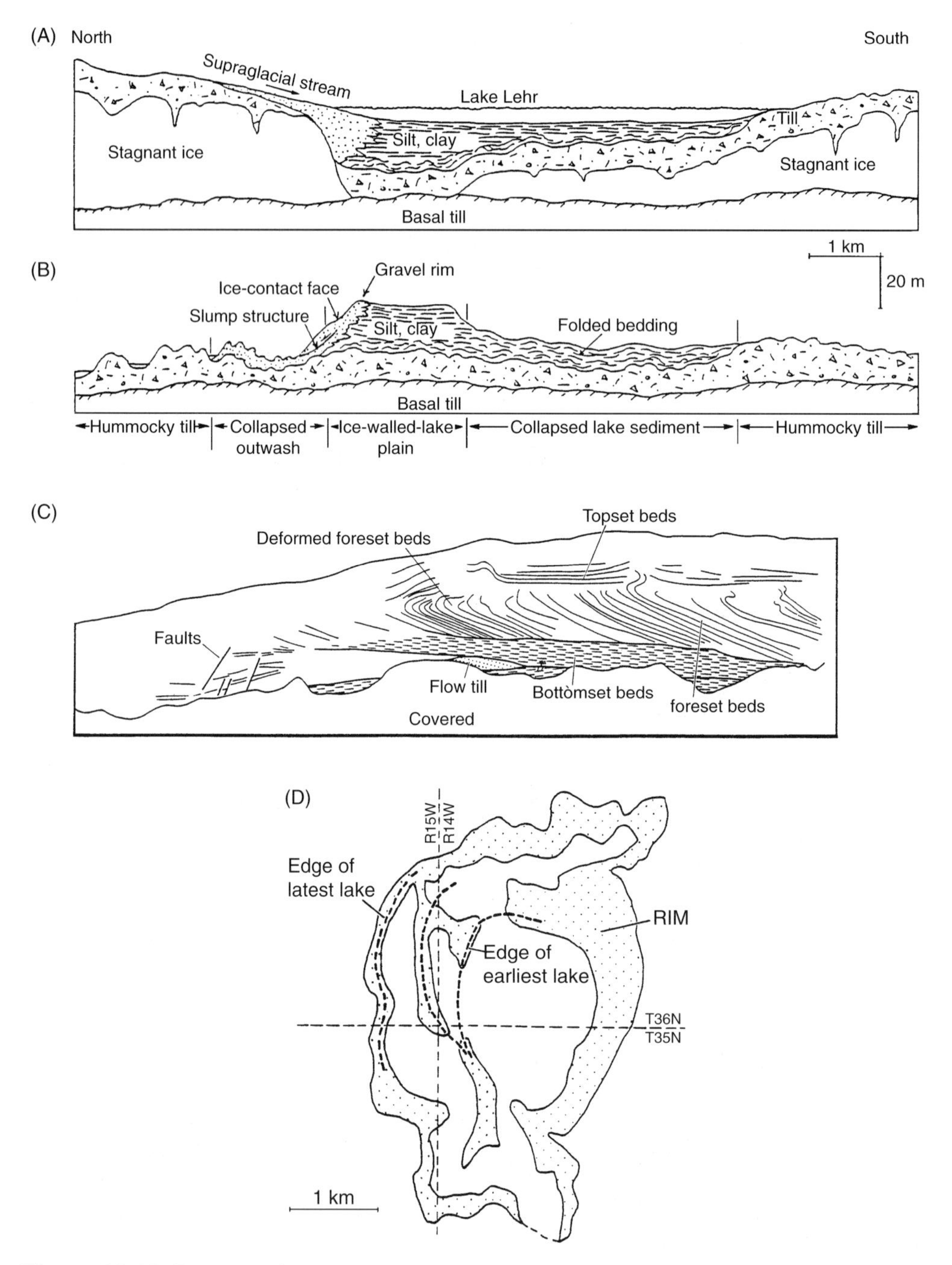

Figure 10.11 Features of ice-walled-lake plains. A) and B) Sketches showing the development of an ice-walled-lake plain near Lehr, North Dakota in which a portion of the ice-walled-lake plain collapses. C) Outcrop sketch from the rim of an ice-walled-lake plain in western Wisconsin showing deltaic sediment (Johnson, 1986). D) Map of ice-walled-lake plain in western Wisconsin showing multiple rims of different age (Johnson, 1986). The dashed lines are local township boundaries.

Unstable ice-walled-lake plain development

(A)

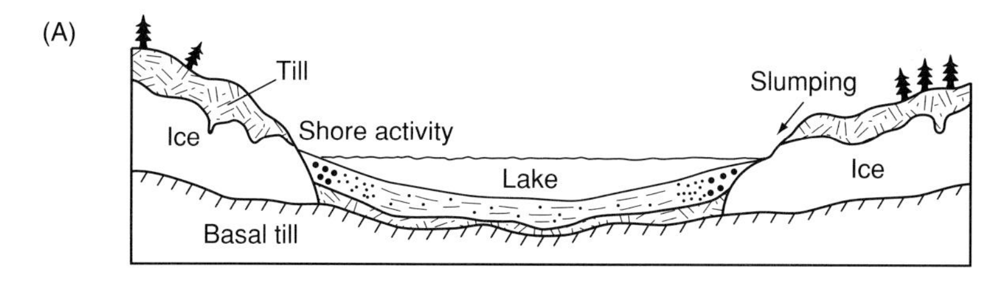

(B)

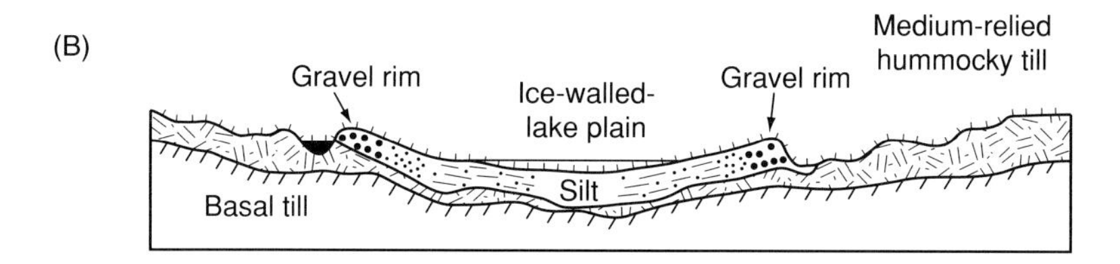

Stable ice-walled-lake plain development

(C)

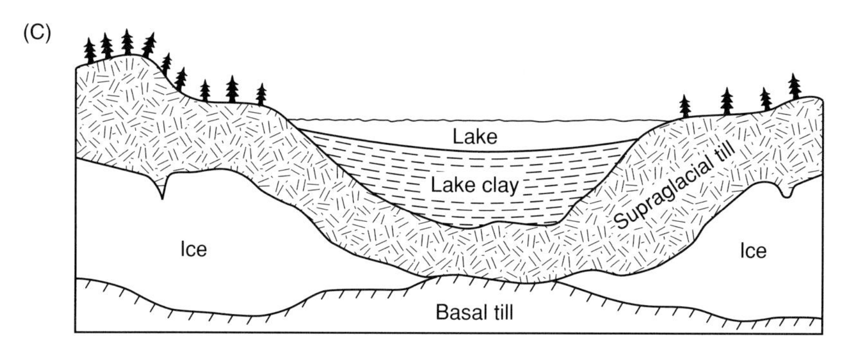

(D)

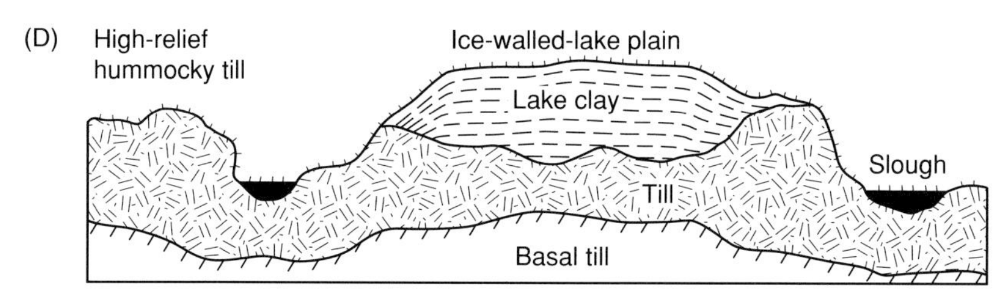

Figure 10.12 Figure showing the differences between unstable and stable ice-walled-lake plains. A) and B) illustrate the development of the unstable ice-walled-lake plain, where supraglacial debris is relatively thin and the surrounding thin ice melts fairly rapidly. The ice-walled-lake plain features well-developed rims and is surrounded by low-to-medium relief hummocks. C) and D) illustrate the stable ice-walled-lake plain, where supraglacial debris is thick, and the ice-walled lake exists for a comparatively longer period of time. The ice-walled-lake plain lacks well-developed rims and is surrounded by high-relief hummocks. (From Clayton and Cherry, 1967).

Johnson *et al.*, 1995; Ham and Attig, 1996; Syverson, 2000). However, in a few places, the thickness of lake sediment is greater than the relief of the ice-walled-lake plain (Strehl, 1998; Johnson *et al.*, 1995; Fig. 10.13C). This is significant, because such a relationship would not be expected if subglacial pressing is called on to explain the formation of surrounding hummocks.

Some ice-walled-lake plains are composed nearly entirely of till, with only a thin surficial cover of lake sediment (Fig. 10.13A; Table 10.2), and the term 'moraine plateau' has been applied by some to this type of ice-walled-lake plain. Ice-walled-lake plains with thin lake sediment have been described in Canada, the USA, Lithuania, Sweden and Denmark (Westergård, 1906; Hansen, 1940; Lagerbäck, 1988; Brehmer, 1990; Bitinas, 1992; Klassen, 1993). The till in these forms is likely flow till derived from the surrounding ice surface, although some melt-out till and squeezed till may occur towards the base of the form. In southern Minnesota, the flat-topped circular hills of Patterson (1997b) are made entirely of bedded till, which she interprets as indicating supraglacial flow till. It is unlikely that squeezing can account for all the till in these forms. Stalker's (1960) squeezing hypothesis was applied only to till in the rim ridges and not across the centre of the ice-walled-lake plains.

10.2.4 Disintegration Ridges

Crevasses open to the surface can localize deposition of supraglacial materials. As stagnant ice continues to disintegrate, crevasses extending through the ice to the bed may allow soft

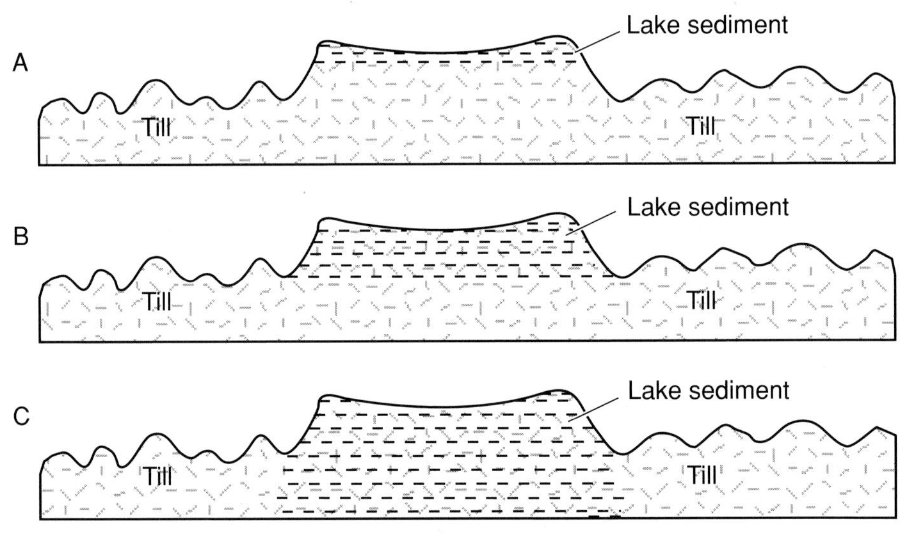

Figure 10.13 Sketches of ice-walled-lake plains containing a range of thickness of lake sediment from A) thin to B) as thick as the relief of the plain to C) a thickness greater than local relief. Examples of A) are described by Hansen (1940), Westergård (1906), Bitinas (1992), Stalker (1960), and Klassen (1993); examples of B) are described by Clayton and Cherry (1967), Johnson (1986), and Syverson (2000) and are thought to be the most common type; examples of C) are described by Johnson *et al.* (1995) and Strehl (1998). A), B), and C) can all form supraglacially, and the till in the lake plain would be flow till derived from the ice surface. Though the type in A) could be formed with subglacial pressing, subglacial pressing would be difficult to form type B) and impossible to form type C).

subglacial till to squeeze into crevasses from below. These processes may result in linear to hummocky landforms that have a complex internal structure. Gravenor and Kupsch (1959) call these features linear disintegration ridges, a type of controlled deposition (Fig. 10.1B, 10.3A). The features they describe are 1–10 m high, 8–100 m wide, and extending from a few metres to up to 10 km in length. The ridges may form two sets in a boxwork pattern resembling ice-crevasse patterns (Fig. 10.3B). They are made mostly of till, although crevasse sediments can contain stratified material as well. Gravenor and Kupsch prefer 'disintegration ridge' to 'crevasse fill' because they recognize that sediment may get into the ridge by subglacial squeezing. Subglacial squeezing into crevasses has been called on to explain radial till ridges in Finland (Aartolahti, 1995), minor moraines in Minnesota (Patterson, 1997b), and disintegration ridges in Spitzbergen (Boulton *et al.*, 1996). In addition to ridges associated with crevasses, the term 'disintegration ridge' has also been applied to sinuous, curved and irregular ridges of unknown genesis that are associated with other landforms of the supraglacial landsystem.

10.2.5 Ice-Contact Dump Ridges

Supraglacial debris transported off the terminal margin of an ice sheet as flow till or outwash may form a dump ridge. This ridge may form at the margin of an active glacier, in which case it is an end moraine, or at the peripheral margin of a stagnant ice mass. Melting of ice may collapse parts of the ridge producing hummocks. Dump ridges have been observed to form this way along modern glaciers (Boulton, 1968; Eyles, 1979, 1983b; Krüger, 1994a). Sediment in dump ridges flows from supraglacial positions close to the ice margin where englacial material had been released from the ice by ablation. Figure 10.14 shows a topographic map of such a dump ridge in central Minnesota, USA, together with an interpretation of its origin. Similar ridges have been described in Poland (Kasparek and Kozarski, 1989; Kozarski, 1981). Ham and Attig (1996) show examples of ice-contact ridges formed concentrically around a former stagnant ice mass (Fig. 10.2A).

10.2.6 Outwash Fans

Though the majority of glacial outwash deposits are derived from subglacial streams emerging at the ice margin, small outwash fans can develop derived solely from supraglacial material. These have been well described by Krüger (1997) who refers to these fans as hochsander fans, a term first used by Gripp (1975). Hochsander fans are finer grained than other outwash fans, being primarily composed of horizontally stratified sand with some gravel lenses: coarser material is often preferentially left on the glacier surface. Vertical variations in the sedimentology of the fan deposit reflect the character of surface processes and variations in rainfall and surface drainage. Similar fans of Pleistocene age occur in Denmark, Germany, and Poland (Krüger, 1997), as well as Wisconsin, USA, (Ham and Attig, 1996) where the source of sediment for the fans was a slowly melting debris-covered, stagnant-ice mass (Fig. 10.2A). Like the dump ridges described above, outwash fans need not form at an active ice margin.

10.3 THE SUPRAGLACIAL LANDSYSTEM MODEL

The landforms described above result as stagnant ice, covered with supraglacial debris, slowly melts. In the lowland-terrain setting, supraglacial debris is derived initially from the base of the ice: subglacial or basal-ice debris is moved upwards to englacial positions and, eventually,

(A)

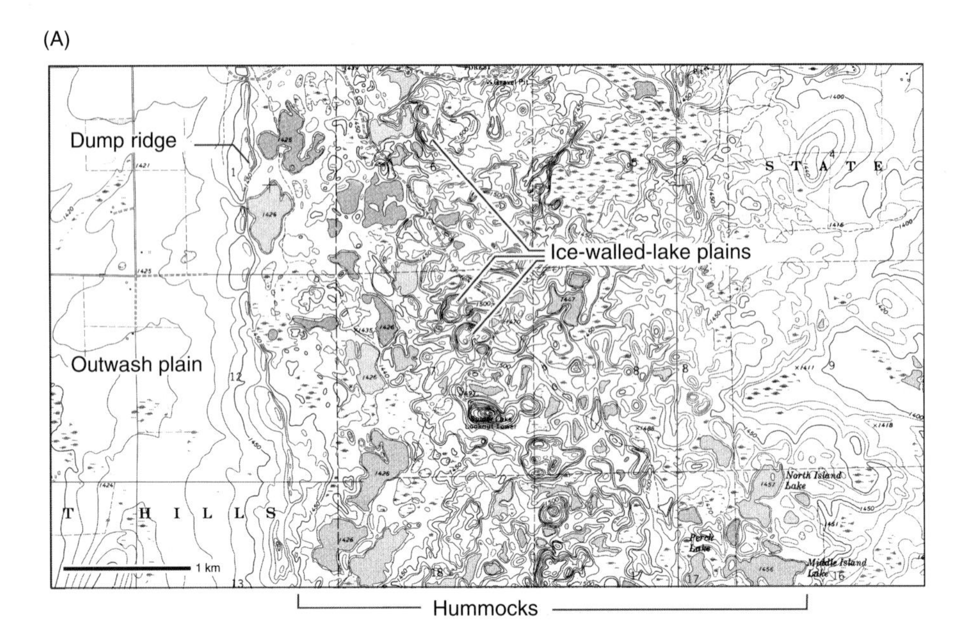

(B)

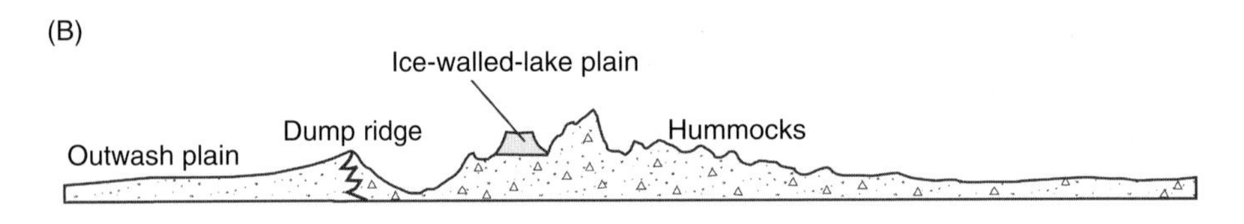

(C)

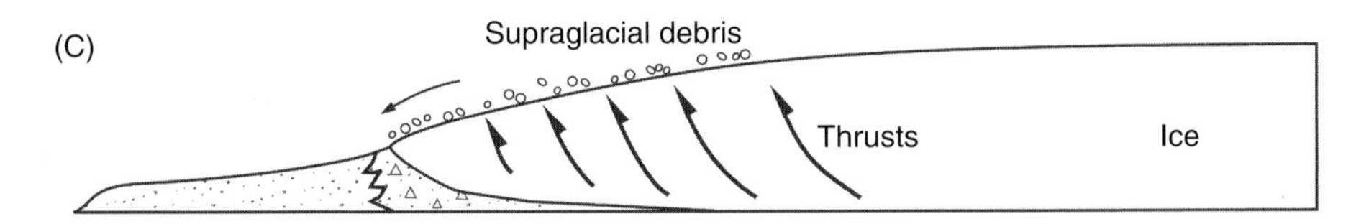

Figure 10.14 A dump ridge and an associated ice-marginal band of hummocky topography in central Minnesota. A) Figure of a portion of the Spider Lake and Bungo Creek, Minnesota 7.5′ topographic quadrangles, (USGS), contour interval 10 ft. The outwash plain at the left edge of the map is composed of outwash derived from an esker tube to the north of this map. B) and C) Hypothetical cross sections showing the ice of the Superior Lobe with thrust planes rich in debris melting out on the ice surface, and eventually forming the hummocks, dump ridge and ice-walled-lake plains.

supraglacial positions as the ice melts. Critique of the supraglacial model (Gravenor and Kupsch, 1959; Munro and Shaw, 1997; Eyles *et al.*, 1999b; Boone and Eyles, 2001) is founded on the supposition that it is difficult to transport enough debris to the ice surface from a subglacial position. Furthermore, these authors point out that supraglacial debris on modern glaciers has quite different characteristics from basal till, whereas much of the till in the landforms described in this chapter is essentially identical to the local basal till. Therefore, our model of the supraglacial landsystem must explain how sediment can get to the ice surface in the lowland setting and how it retains its basal characteristics.

10.3.1 Generating Supraglacial Debris

Observations from the margins of modern glaciers show that there are predominantly three ways that abundant basal debris can be moved to englacial and supraglacial positions:

1. by thrusting subglacial and basal debris upward along shear planes within active ice near its margin
2. by stacking of debris-rich ice, and
3. by the subglacial incorporation of debris by freezing-on followed by thrusting or release by top-down melting.

We consider these mechanisms sufficient to provide the supraglacial sediment necessary for the landforms we describe.

Thrusting along shear planes at the ice margin can carry debris-rich basal ice and unfrozen subglacial debris to englacial positions, and this has been noted on modern glaciers in Greenland, Iceland and Svalbard (Swinzow, 1962; Boulton, 1967, 1970; Hambrey *et al.*, 1999; Sletten *et al.*, 2001; Lyså and Lønne, 2001). This thrusting mechanism has been used by several authors to explain abundant supraglacial debris on Pleistocene glaciers (Gravenor and Kupsch, 1959; Clayton, 1967; Moran *et al.*, 1980; Ham and Attig, 1996; Bennett *et al.*, 1998; Fig. 10.3B). A striking example of this process is provided by St Onge and McMartin (1999) who describe buried extant Late-Wisconsin glacier ice underlying abundant supraglacial till in the Northwest Territories, Canada. The remnant glacial ice is debris rich, which they interpret to have been generated by marginal thrusting and stacking. They point out that, with a future climate change, the melting of this ice and supraglacial debris would produce a hummocky topography. Sharp (1985) and Hambrey *et al.* (1996) noted that this thrusting process is accentuated with surging glaciers and can cause debris to be lifted to positions as high as 200 m above the ice bed. Furthermore, there are many examples of glacitectonically deformed bedrock where glaciers have actively moved subglacial material up into the ice (Aber *et al.*, 1989; Moran *et al.*, 1980). This material can be moved to englacial positions, be deformed by glacier flow and eventually melt-out on the ice surface.

Stacking of debris-rich ice may also produce thick sequences of stagnant ice that later melt and generate abundant supraglacial debris. This may occur by the thrusting mechanism mentioned above, but also by glacier readvances or slight fluctuations of the ice margin in which active ice rides up over older, stagnant ice (Fig. 10.15C). Older ice may be reactivated (Eyles, 1983c; Sharp, 1985b) and incorporated into the active ice, bringing debris-rich ice to higher englacial positions. Because stagnant ice can remain for thousands of years following a glacier advance (Florin and Wright, 1969; Böse, 1995), it is not difficult for stagnant ice to become further buried by subsequent ice advances.

The thrusting and stacking mechanisms described above are accentuated where glacier margins have strong compressive flow. There are three situations in which compressive flow is accentuated at the ice margin and where thrusting can move large amounts of subglacial debris into englacial and supraglacial positions:

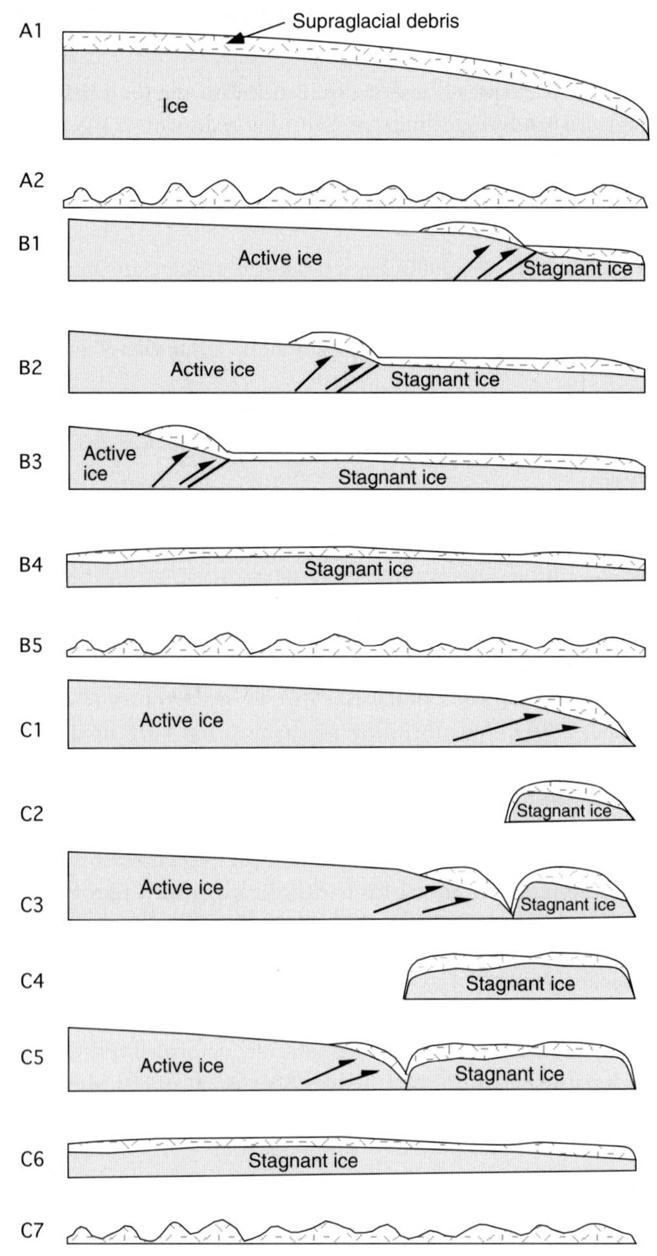

Figure 10.15 Large tracts of hummocky topography do not necessarily mean widespread ice stagnation. Here are three models for how active and stagnant ice can produce a large hummock tract. A) A large portion of the glacier is debris-covered and later stagnates in place to produce hummocky topography. This is a common interpretation for regions characterized by the supraglacial landsystem. However, we consider such a scenario unlikely. A similar suite of landforms can be produced by the mechanisms shown in B) and C). B) A marginal zone of ice is covered with debris brought from the glacier bed by thrusts, inhibiting melting and causing ice to stagnate and separate from the active ice. This marginal band of stagnant ice expands as the active-ice margin retreats. C) The debris-covered marginal zone is isolated as ice retreats (C1 and C2). Ice readvances (C3) only to melt back again, leaving a debris-covered marginal zone. This process can be repeated several times.

1. sub-polar glaciers, where warm ice underneath the thicker, central parts of a glacier flows into the marginal region where the glacier is frozen to the bed
2. glaciers with subglacial slopes near the ice margin that slope steeply up-glacier (Clayton, 1967; Hansel and Johnson, 1987), and
3. surging glaciers (Clapperton, 1975; Sharp, 1985b; Clayton *et al.*, 1985; Hambrey *et al.*, 1996).

Two or even all three of these mechanisms can act jointly. Additionally, surging ice lobes moving as plug flows can form levee-like accumulations of stagnant, debris-rich ice along their margins, which then melt out to form hummocks, as along the edges of the Des Moines Lobe in Minnesota and Iowa, USA (Kemmis *et al.*, 1994).

In addition to thrusting and stacking, the amount of englacial debris can increase by the freezing-on of subglacial material. This occurs on sub-polar glaciers where debris is frozen-on at the boundary between warm and cold ice near the ice margin (Boulton, 1972b). Sediment entrainment also occurs when supercooled subglacial water freezes-on near the ice margin (Evenson *et al.*, 1999). Supercooled freezing-on occurs where there is a steep subglacial gradient near the ice margin, a condition that would have been present near the margins of several parts of the Pleistocene ice sheets. Additionally, surging glaciers may cause an increase in freezing-on of subglacial sediment, as suggested by Clapperton (1975) who noted that frozen-on basal debris was more common on glaciers in Iceland and Svalbard that had surged than those that had not.

10.3.2 Characteristics of Glaciers With Abundant Supraglacial Debris

These observations indicate that sub-polar glaciers that surge or that flow against regional slopes are likely to have greater amounts of basal-debris-rich ice and englacial debris, which in turn can be readily carried to higher positions in the ice and eventually be released supraglacially. Pleistocene glaciers in lowland settings in North America and Europe matched these characteristics: they were sub-polar, based on climatic evidence (Black, 1976a and b; Attig *et al.*, 1989; Boulton *et al.*, 1995; Cutler *et al.*, 2000), many of them surged, as shown by radiocarbon datings (Clayton *et al.*, 1985), and many portions of the ice sheets flowed out of basins near the ice margin (e.g. the Winnipeg Lowland and Great Lakes region of North America and the Baltic basin of northern Europe). These observations support our contention that it is reasonable to explain these landforms in terms of abundant supraglacial debris.

The supraglacial landscapes we have studied in Wisconsin and North Dakota formed during a time when permafrost was present around the margins of the existing glaciers, and we find hummocky terrains to be best developed at the outer margins of the Late Wisconsin ice advances, at a time when permafrost was present. During ice recession, permafrost would not have had time to develop as deeply as during ice advance, and this explains in part why there is less hummocky terrain associated with younger ice margins. However, hummocks present at younger margins indicate that factors other than permafrost were important in supraglacial debris genesis (Attig and Clayton, 1993). Though the presence of permafrost creates conditions that favour the incorporation of englacial debris and its transference to higher positions within the ice, it is not a requirement for the supraglacial landsystem.

These mechanisms indicate why supraglacial till often looks like the regional basal till, because it is the regional basal till. That is, it is till derived from the base of the glacier and moved or frozen into englacial positions by the processes we outline above. The uppermost surface of the supraglacial debris will experience the sorting and winnowing commonly seen on modern glaciers, but these processes will not affect the entire thickness of supraglacial till thereby preserving the sedimentary characteristics produced englacially or at the glacier bed.

10.3.3 Supraglacial Processes as Stagnant Ice Melts

Englacial debris is exposed on the surface due to ablation. Ablation causes downwasting and backwasting of the debris-rich or debris-covered ice (Kjær and Krüger, 2001, Lyså and Lønne, 2001), the rate of which is controlled by climate, geothermal heat flow, meltwater drainage, and debris content in and on the ice. Several processes may occur on the debris-covered ice surface including flowage or sliding of debris, mechanical mixing of till with other supraglacial lake or stream sediment, frost action, winnowing of fines, concentration of clasts, and soil formation (Eyles, 1983c; Kjær and Krüger, 2001). Supraglacial sediment may be resedimented several times during ablation (Boulton, 1968, 1972a; Clayton and Moran, 1974; Lawson, 1979; Sletten *et al.*, 2001). As ice becomes more deeply buried beneath supraglacial debris, downwasting rates decrease and the remaining ice melts out slowly (Krüger and Kjær, 2000, Sletten *et al.*, 2001), often so slowly that the surface sediment becomes thick enough for vegetation to become established. As the supraglacial debris cover becomes thicker, the deeply buried portions of the supraglacial debris may escape extensive reworking and may produce melt-out till (Johnson *et al.*, 1995) or experience only a few resedimentation episodes.

Supraglacial lakes form on the surface and expand by melting downward and laterally with some marginal calving (Benn *et al.*, 2000). Expanding lakes may coalesce with others. Drainage of meltwater will take advantage of weaknesses and crevasses in the stagnant ice and form sinkholes. The water level in these sinkholes will be controlled by the groundwater system in the glacier. As sinkholes grow, sedimentation by till flowage and lake processes will initially be rapid.

The presence of permafrost will significantly slow the melting of buried ice (Driscoll, 1980), and may even affect geomorphic development of hummock tracts. For example, high-relief hummocks in northern Wisconsin (Fig. 10.2A) likely formed in two phases, with stable environment ice-walled-lake plains (Fig. 10.12) forming first when extensive buried ice was maintained by regional permafrost (Fig. 10.16). With climate warming and melting of permafrost, the buried ice melted and the surrounding hummocks formed by supraglacial collapse (Ham and Attig, 1996). A rare modern analogue for this model exists today in Siberia where ice-walled lakes occur in extant Late Weichselian glacial ice, maintained by permafrost (Alexandersson *et al.*, 2002; see also Chapter 7). With a climate warming, the melting of ice would produce ice-walled-lake plains surrounded by hummocks. Additionally, if permafrost slows the melt rate of stagnant, debris-covered ice, more melt-out till may be produced because slow drainage and slow ice melt would lessen the number of flow-till events (Eyles, 1983c).

10.3.4 Marginal or Widespread Stagnation?

The development of the supraglacial landsystem may take place in a zone of stagnation at the margin of an otherwise active ice lobe, or it may indicate widespread stagnation of a much larger portion of the ice body. Figure 10.14 shows a band of hummocky topography clearly associated with an individual ice margin. However, areas characterized by the supraglacial landsystem often cover hundreds of square kilometres, and it is not clear that they are associated

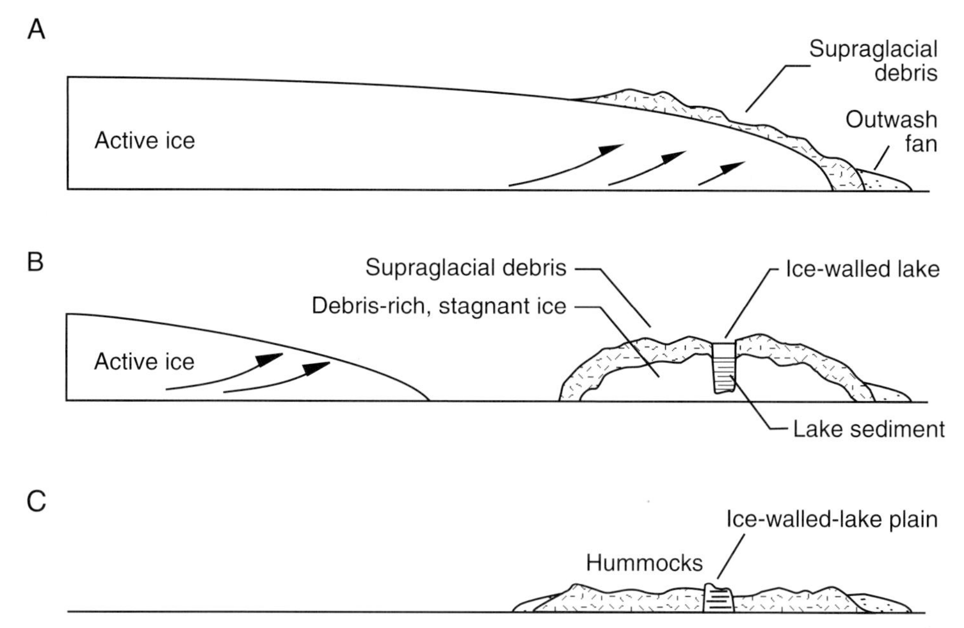

Figure 10.16 Two-step formation of hummocky topography in northern Wisconsin controlled by permafrost. A) Supraglacial debris accumulates in the marginal zone of a glacier. B) Ice retreats, but supraglacial debris and permafrost insulate and preserve much of the stagnant ice. Ice-walled lakes form, perhaps melting to the bottom of the stagnant ice. C) With the abatement of permafrost, the remaining stagnant ice collapses producing hummocks. Note outwash fans on both the proximal and distal sides of the hummock tract. (From Ham and Attig, 1997).

with specific ice-margin positions (e.g. Gravenor and Kupsch, 1959; Evans, 2000b). Such large regions have been traditionally interpreted to result from widespread, regional, en-masse stagnation (Parizek, 1969; Flint, 1971; Fig. 10.15A). However, progressive development of stagnation zones during overall retreat may produce a broad area of stagnant-ice deposits that may be misinterpreted as widespread, regional stagnation, when it actually represents a style of active-ice retreat or the stagnant ice from several advances (Figs. 10.15B and C). Incremental, marginal stagnation is described by Dyke and Savelle (2000) for a group of presently ice-cored Younger Dryas moraines on Victoria Island, northern Canada. They point out that if permafrost were to leave this area, the ice cores would melt, and the now well-defined ridges would be transformed into a chaotic hummocky landscape, leaving the landscape to appear no different from landscapes interpreted to have formed by widespread regional stagnation (see also Chapter 7). Additionally, Hambrey *et al.* (1997) and Bennett *et al.* (1998) describe hummocky landscapes produced from actively receding ice during which supraglacial debris (brought to englacial and supraglacial positions through thrusting) is slowly let down. Considering that marginal processes (thrusting, stacking, and freezing-on) are the processes we believe to generate much supraglacial debris, we regard widespread regional stagnation (Fig. 10.15A) as unlikely and suggest that most large hummock tracts result from the scenarios shown in Fig. 10.15B and C.

10.3.5 The Lack of Suitable Modern Analogues

A key dimension of glacial research in the past 30 years has been the attention paid to modern glaciers as providing modern analogues for Pleistocene sediment and landforms (Boulton, 1972a). As essential as this research is, we point out that there are still many Pleistocene features that lack modern analogues, and that landforms of the supraglacial landsystem are among them. Though the formation of hummocky topography has been noted on modern glaciers (Boulton, 1967, 1972a; Eyles, 1979, 1983c; Wright, 1980; Sharp, 1985b; Krüger, 1994a; Krüger and Kjær, 2000; Sletten *et al.*, 2001), none of these field studies note hummocks with the relief or internal structure of many Pleistocene hummocks. Descriptions of modern glacial-sinkhole deposits and the behaviour of modern supraglacial lakes (Rubulis, 1983; Benn *et al.*, 2000) bear little resemblance to the ice-walled-lake plains we describe. Additionally, the thickness of supraglacial debris described on modern glaciers (Boulton, 1968, 1972a; Lawson, 1979; Kjær and Krüger, 2001) is insignificant compared with the likely few tens of metres for the till hummocks of North America (Clayton, 1967). Though any model needs to be supported by careful fieldwork with description of geomorphology and sedimentology, it is clear that glacial geologists should not be restricted to models developed solely on modern glaciers.

CHAPTER

SURGING GLACIER LANDSYSTEM

David J.A. Evans and Brice R. Rea

11.1 INTRODUCTION AND RATIONALE

Glacier surging represents a cyclic flow instability that is triggered from within the glacier system rather than by external climate forcing. The active phase of a surge involves the transfer of ice from a reservoir area to the snout of a glacier and can produce ice flow velocities up to one thousand times the flow rate of intervening non-surge phases (Clarke *et al.*, 1984; Raymond, 1987). This may result in the rapid advance of the glacier front and a concomitant thinning of the reservoir area. Between surges, periods of slow flow, or quiescent phases, are characterized by snout stagnation and ice build up in the reservoir area. Although individual surging glaciers display uniform return periods there are large variations between glaciers and regions (Post, 1969; Clarke *et al.*, 1986; Dowdeswell *et al.*, 1991; Hamilton and Dowdeswell, 1996). A climatic linkage to surging was modelled by Budd (1975), who suggested that a continuously fast-flowing glacier is capable of discharging its annual mass balance, whereas a surging glacier has a total mass throughput that is too small to sustain fast flow but too large to be discharged by slow flow alone, thereby initiating a regular surging cycle. As the reservoir zone builds up, a thermal boundary may exist at the down-glacier end, and water storage increases at the bed. Thus, it should be noted that, while there is a climatic linkage, ultimately surging is the result of oscillations in the internal dynamics of the glacier. Specifically, the large changes in glacier velocity are driven by reorganizations in the subglacial drainage system (Clarke *et al.*, 1984; Kamb *et al.*, 1985; Clarke, 1987; Fowler, 1987; Kamb, 1987). In Iceland, surging is likely associated with geothermal activity as this can allow cyclical build up and release of large subglacial meltwater reservoirs (Björnsson, 1975, 1992). However, many of the surging glaciers may be predisposed to this type of behaviour. For example jökulhlaups drain regularly from Grimsvötn lake out beneath Skiedararjökull (intervals of 1–10 years) but these do not induce the glacier to surge, an event which occurs independently (Björnsson, 1998). Regardless of the trigger mechanism, the landform-sediment assemblages produced by surging glaciers appear to be consistent and predictable. Significantly, moraines deposited by surging glacier margins cannot be modelled as the products of steady-state glaciers in equilibrium with climate. Moreover, the identification of former surging within Pleistocene ice sheets remains a problem area that glacial geologists continue to tackle through the identification of diagnostic palaeo-surge signatures in the geomorphological, sedimentological and stratigraphical record.

Contemporary surging glaciers are instantly recognizable by their surface forms. For example, prior to the surge there may be surface bulging associated with the filling of the reservoir area which may be coincident with a thermal boundary (e.g. Trapridge Glacier). Associated with the passage of the surge front is extensive crevassing, thrusting and folding of the glacier surface, and the formation of looped medial moraines (Meier and Post, 1969; Clarke *et al.*, 1984; Raymond *et al.*, 1987; Clarke and Blake, 1991). Once such features have melted out on the glacier surface and become more subdued, or when the entire glacier has gone, the identification of palaeo-surges becomes problematic. Addressing this problem requires systematic investigations of the marginal areas of known surging glaciers and the integration of the landforms, sediments and stratigraphy into a diagnostic landsystems model. Previous attempts to identify the landform-sediment assemblages of Pleistocene ice sheet surging have been piecemeal, employing various combinations of diagnostic geomorphological and sedimentological criteria. For example, Clayton *et al.* (1985) identified extensive tracts of hummocky moraine as characteristic features of surging margins in the southwestern part of the Laurentide Ice Sheet. Dredge and Cowan (1989b) proposed that inset fluting fields terminating at major moraines were the product of palaeo-surges on the Canadian shield. Eyles *et al.* (1994) interpreted till diapirs as ridges deformed upward below masses of dead ice in shallow water following onshore surging of an ice margin in east Yorkshire, England. Because such studies have focused on a limited selection of landforms and sediments, they are too restricted to use 'globally' for the identification of palaeo-surging. Diagnostic landforms from contemporary surging glaciers have been previously identified by Sharp (1985a, b; thrust moraines, crevasse-fill ridges, flutings and hummocky moraine), Croot (1988a, b; thrust moraines) and Knudsen (1995; concertina eskers). However, individually these models are not comprehensive and, therefore, a surging glacier landsystems model is presented here, comprising a suite of landform-sediment associations believed to be diagnostic of glacier surging (cf. Evans and Rea, 1999).

11.2 GEOMORPHOLOGY AND SEDIMENTOLOGY OF CONTEMPORARY SURGING GLACIERS

This section provides details of the landform-sediment assemblages observed at contemporary surging glacier margins in Iceland, Svalbard, USA and Canada. Additionally, extensive reference is made to the 1982–83 surge of Variegated Glacier, Alaska, because it is the best documented and instrumented surge event. The time development of surging glacier geomorphology is best represented by the Icelandic examples due to the fact that the surges are historically documented by the local population over the last two centuries, and aerial photography since 1945 (Thorarinsson, 1964, 1969; Evans *et al.*, 1999a).

11.2.1 Thrust-Block Moraines and Push Moraines

Present day thrust-block moraines (including composite ridges and hill-hole pairs; Aber *et al.*, 1989) are found in two ice-marginal settings: at the margins of surging glaciers, (e.g. Sharp, 1985b; Croot, 1988a, b; Bennett *et al.*, 1999; Evans and Rea, 1999; Evans *et al.*, 1999b), and associated with sub-polar glacier margins in permafrost terrains (e.g. Kalin, 1971; Evans and England, 1991; Fitzsimons, 1996a). Proglacial thrusting at the margins of surging glaciers is due to rapid ice advance into proglacial sediments, which may be seasonally frozen, unfrozen or contain discontinuous permafrost. A surging glacier advancing into proglacial frozen sediments is most conducive to the failure and stacking of large contorted and faulted blocks, best exemplified by the wide belts of arcuate thrust ridges of glacimarine and glacifluvial materials at the margins of the polythermal surging glaciers of Svalbard (Fig. 11.1). However, rapid ice advance into unfrozen sands and gravels will still produce high proglacial and sub-marginal compressive stresses. High pore water pressures may be developed in silt and clay layers within the succession, which will act as décollement leading to shearing and stacking of the sand and gravel units.

Excellent examples of proglacially thrust unfrozen materials occur at the surge margins of the Icelandic glaciers Brúarjökull and Eyjabakkajökull, where pre-surge peat layers have been vertically stacked in a series of thrust overfolds (Fig. 11.2) within which individual beds often display slickensides. Thrust-block moraines may be produced also at the margins of polar and sub-polar glaciers where the proglacial sediments contain permafrost and high proglacial stresses can be produced (Evans and England, 1991; Ó Cofaigh *et al.*, Chapter 3). Because thrust-block moraines can be constructed by surging and non-surging glaciers, their existence in the ancient landform record cannot be ascribed unequivocally to glacier surging. However, former surging activity by sub-polar glaciers that have produced thrust-block moraines cannot be dismissed, particularly as surging has been reported in such environments (Hattersley-Smith, 1969a; Jeffries, 1984).

Although thrust-block moraines are the most spectacular constructional features produced by surging glacier margins, they can be produced only in areas where sufficient sediment is available for glacitectonic thrusting, folding and stacking. The wide surging margin of Brúarjökull has produced thrust blocks only in areas that lie down flow from braided outwash plains where large accumulations of fluvial and/or lacustrine material collected during the

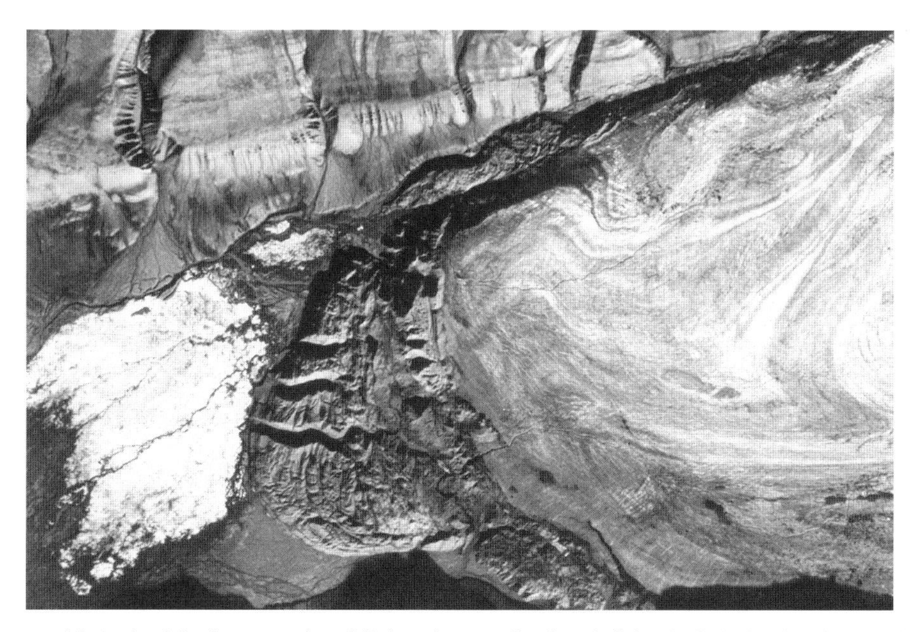

Figure 11.1 Aerial photograph of Rabotsbreen, Svalbard (Norsk Polarinstitutt) showing a thrust-block moraine produced during a recent surge.

Figure 11.2 Thrust overfolds in peat produced at the margin of Brúarjökull during the AD 1890 surge.

quiescent phase. Elsewhere, the glacier constructs low-amplitude push moraines by bulldozing and/or sub-marginally squeezing the veneer of till or peat that drapes the proglacial land surface.

11.2.2 Overridden Thrust-Block Moraines

A number of conspicuous ice-moulded hills occur in the proglacial forelands of Brúarjökull and Eyjabakkajökull. They occur down-ice of topographic depressions from which the hills were originally displaced by thrusting. The surfaces of these features appear extensively fluted and/or drumlinized (Fig. 11.3). Internal structures comprise glacitectonized outwash or lake sediments, the tops of which are usually modified into glacitectonite (Benn and Evans, 1996, 1998) and truncated by subglacial till (Fig. 11.4). These ice-moulded hills are interpreted as overridden thrust-block moraines. The sediments displaced by thrust-block construction are deposited during the surge quiescent phase when proglacial lakes and meltwater streams extensively modify the foreland. Each thrust block demarcates the former glacier margin during a surge, overriding taking place either during the same surge or during a later, more extensive surge. After prolonged periods of modification by overriding ice the former thrust blocks resemble the cupola hills of Aber *et al.* (1989).

11.2.3 Concertina Eskers

On aerial photographs of the margins of Brúarjökull and Eyjabakkajökull (Knudsen, 1995; Fig. 11.5), sinuous eskers and 'concertina' plan-form eskers are found. Extensive exposures of glacier ice at the base of the eskers (Fig. 11.5c) indicate they were produced englacially or supraglacially (Evans and Rea, 1999; Evans *et al.*, 1999b). Referring to a bulge observed in the snout of Brúarjökull during the 1964 surge, Knudsen (1995) suggests that the concertina

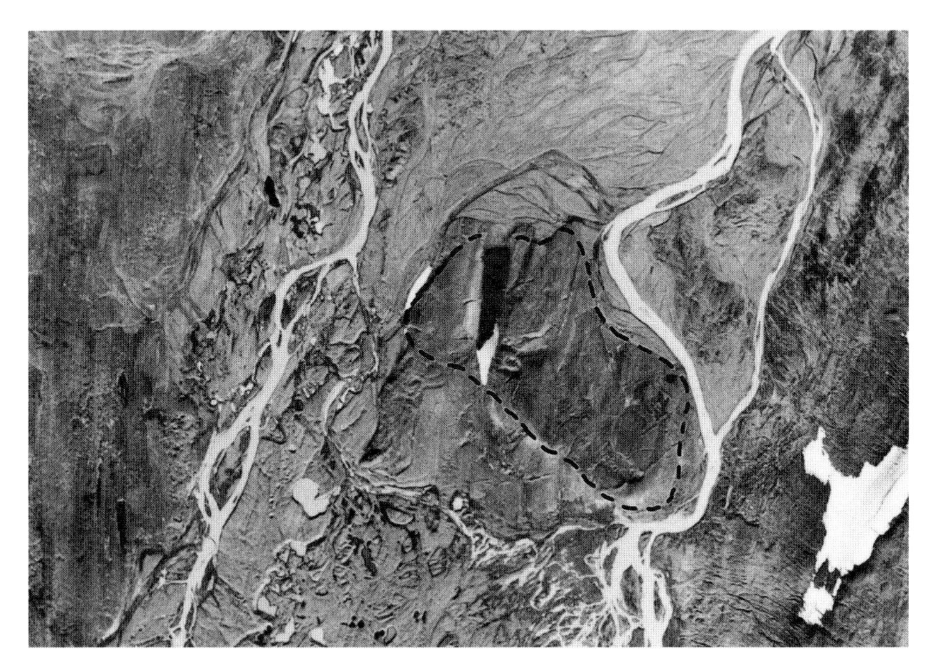

Figure 11.3 Part of an aerial photograph (Landmaelingar Islands, 1993) of an overridden thrust block on the foreland of east Brúarjökull (outlined by broken line). Note the occurrence of flutings and crevasse-squeeze ridges on the surface.

Figure 11.4 Exposure through an overridden thrust block near the AD 1890 surge limit of west Brúarjökull. Contorted interbeds of peat and stratified sediments are overlain by a till carapace that possesses a fluted surface expression.

(A)

(B)

(C)

Figure 11.5 A) Part of an aerial photograph (Landmaelingar Islands, 1993) of a concertina esker on the foreland of west Brúarjökull, Iceland, produced during the 1963 surge. B) Ground photograph of a concertina esker on the foreland of Brúarjökull, showing stratification and steep cliffs produced by the melt-out of buried glacier ice. C) Glacier ice exposed in a concertina esker on the Eyjabakkajökull foreland.

eskers were formed during surging similar in style to the Variegated Glacier surge of 1982–83. Knudsen proposes that the concertina eskers are produced by the shortening of pre-surge sinuous eskers by compression in the glacier snout during surging, the Brúarjökull examples having been compressed during the 1964 surge. It seems unlikely that a situation involving extreme tectonic activity and vertical thickening (>30 m during the surge) could deform an initially sinuous esker to produce the concertina plan-form. Moreover, substantial transverse extension of the snout is required in order for a concertina esker to form in the manner suggested by Knudsen (1995). Initially as the surge wave passed down through the snout region (and the sinuous eskers) there was no room for lateral extension and indeed the snout may have undergone some slight lateral shortening as it passed between topographic highs to the west (Kverkarnes) and east (high point between east Brúarjökull and Eyjabakkajökull). The principal extension at this time would then have been vertical, associated with extensive folding and thrusting. Once beyond these high points, the snout began to splay laterally, resulting in the formation of numerous conjugate shears with the principal axis of extension aligned approximately flow-transverse. The sinuous eskers have undergone flow-parallel compression, with initial vertical extension, followed by flow-transverse extension. The 'limbs' of the concertina eskers are sub-linear, the contact between the sediments and the ice is planar, and the sediments have undergone very little post deposition deformation, the most significant tectonics being normal faulting resulting from meltout of the underlying ice. Thus, it would seem unlikely that concertina eskers are formed by the shortening of pre-surge sinuous eskers in the manner proposed by Knudsen (1995).

It seems more reasonable to assume that the surge destroyed the pre-surge sinuous eskers, and the concertina eskers formed during the surge and/or the immediate post-surge period. The ice-cored nature of the concertina eskers (Fig. 11.5c) demonstrates that they were deposited either in englacial conduits or in supraglacial channels. During the 1982–83 Variegated Glacier surge, substantial quantities of water were stored behind the surge front and these were discharged periodically during, and, at the termination of the surge (Humphrey and Raymond, 1994). Basal water pressures are elevated and so water may drain via an englacial and/or supraglacial drainage system, which would rapidly exploit the extensive network of crevasses created during the surge. Such an event would be short lived, and indeed the high-angled apexes of the concertina eskers suggest the period of water flow along them was not of great longevity. We conclude that the concertina eskers formed during a short-lived, high-discharge event, occurring prior to or very shortly after surge termination. Drainage occurred either englacially or supraglacially and the channel was then rapidly abandoned.

Like some sinuous eskers (e.g. Price, 1969), the gradual melt-out of ice cores in concertina eskers either destroys or heavily disturbs their internal sedimentary structures. Additionally, because the majority of the initial relief of concertina eskers is due to their ice core, they may appear in ancient glacial landform assemblages as discontinuous chains of gravel and sand mounds and therefore be mis-identified as 'kame' forms.

11.2.4 Crevasse-Squeeze Ridges

Crevasse-squeeze ridges are best known from Brúarjökull and Eyjabakkajökull, Iceland and have been used as diagnostic criteria of surging (Sharp, 1985a,b); although Sharp referred to them as crevasse-fill ridges). Prominent crevasse-squeeze ridge networks can be seen on the forelands of both Brúarjökull and Eyjabakkajökull (Fig. 11.6), comprising cross-cutting diamicton ridges that can be traced from the foreland and into crevasse systems in the snouts (Sharp, 1985a, b; Evans and Rea, 1999; Evans *et al.*, 1999b). Crevasse-squeeze ridges have been reported also from Trapridge Glacier (Clarke, *et al.*, 1984) and Donjek Glacier (Johnson, 1972, 1975) in Yukon Territory and from Svalbard (Clapperton, 1975; Boulton *et al.*, 1996; Evans and Rea, 1999) where they are associated with surging glaciers (Fig. 11.7). The extreme tectonics experienced during a surge leave the glacier highly fractured (e.g. Kamb *et al.*, 1985; Raymond *et al.*, 1987; Herzfeld and Mayer, 1997), and many crevasses may extend to the glacier bed. At surge termination as the basal water pressures are reduced and effective pressure at the bed increases, water-saturated sediment rises into open basal crevasses. Currently these are seen as prominent cross-cutting diamict ridges melting out on the glacier surface (Fig. 11.6). Due to the inactivity of the quiescent phase the glacier ice melts away preserving the cross-cutting ridge network. If the glacier returns to slow-flow dynamics then the crevasse-squeeze ridges will be deformed into a typical glacier strain profile.

In the terminus region of non-surging glaciers, crevasses aligned perpendicular to the glacier margin are often found (e.g. Evans and Twigg, 2002). With a bed of deformable sediment such crevasses could fill with sediment and form crevasse-squeeze ridges. In this situation however, active retreat of the glacier margin would destroy or substantially modify the squeeze ridges. Only if the glacier stagnated and wasted away *in situ*, would preservation be favoured, and even then the radial crevasse pattern typical of such glacier snouts would indicate a non-surging origin. From the evidence presented above crevasse-squeeze ridges are a landform highly suggestive of surging glacier activity, but they cannot be regarded independently as diagnostic

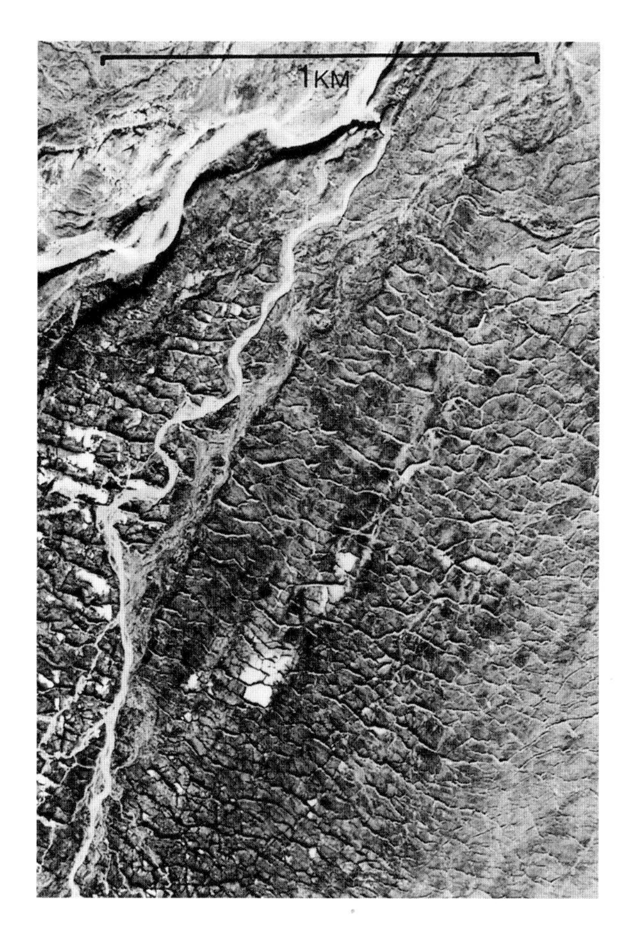

Figure 11.6a Part of an aerial photograph (Landmaelingar Islands 1993) of crevasse-squeeze ridges at the margin of west Brúarjökull, Iceland, produced during the 1963 surge.

Figure 11.6b Ground photograph of the crevasse-squeeze ridges and flutings at the margin of Eyjabakkajökull, Iceland.

Figure 11.7 A crevasse-squeeze ridge at the base of the surging glacier Osbornebreen, St Jonsfjorden, Svalbard.

features of palaeo-glacier surging even though widespread development of crevasse-squeeze networks clearly requires extensive fracturing of the glacier, normally associated with surging.

11.2.5 Flutings

Flutings occur on the forelands of many glaciers and are certainly not diagnostic of glacier surging. However, fluting length may provide important evidence for rapid advances over substantial distances. Excellent examples of this exist on the foreland of Brúarjökull where regularly spaced parallel-sided flutings (Benn and Evans, 1996, 1998) are continuous for more than 1 km inside the 1964 surge moraine (Fig. 11.8), and display remarkable uniformity in long- and cross-section. Another prominent feature of the fluting fields are numerous boulders with short sediment prows/flutes on their down-flow sides, which are interpreted as ploughs/incipient flutes produced by boulders embedded in the glacier sole or just lodged into the till at the surge termination. Clast fabrics were measured from a number of the Brúarjökull flutes and ploughs/incipient flutes. Low fabric strengths are recorded in both the elongate flutes

Figure 11.8 Long flutings formed during the 1964 surge of Brúarjökull.

and the short ploughs/incipient flutes (Fig. 11.9). Other clast fabric measurements made on flutes from non-surging glaciers suggest that either herring-bone or flow-parallel fabrics are the norm, with flow-parallel clustering being expected for parallel-sided flutings of the type described here (Rose, 1989; Benn, 1994, 1995; Eklund and Hart, 1996). The very low fabric strengths are taken to indicate that there was little coupling between the glacier and the bed. The elongation of these flutes suggests that they were formed during a single flow event when basal water pressures and thus the degree of ice-bed coupling and sediment strength remained 'constant'. Flutes formed by non-surging Icelandic glacier margins tend to be substantially shorter and much less uniform in long section, due probably to fluctuations in ice-bed interface conditions, driven by seasonal or annual cycles.

Additionally, the association of flutings and crevasse-squeeze ridges is an important aspect of the subglacial geomorphology of surging glaciers. Sharp (1985a, b) notes that fluting crests at Eyjabakkajökull rise to intersect the crevasse-squeeze ridge crests, indicating that subglacially deformed till was squeezed into basal crevasses as the glacier settled onto its bed at the end of the surge. The contemporaneous production of the flutings and crevasse-squeeze ridges at Eyjabakkajökull renders them diagnostic of glacier surging in the landform record. Due to the range of glaciers and glacial conditions under which flutings are produced they cannot be used independently as diagnostic of glacier surging.

11.2.6 Thrusting/Squeezing

Sharp (1985a), in his model of sedimentation for Eyjabakkajökull, suggests a zone of thrusting exists in the snout where sediment is lifted from the bed. Thrusting can occur in surging glaciers where the surge front propagates into thin ice (Raymond *et al.*, 1987). In such settings supraglacial sediment on the thin ice is entrained along the thrust. Similar features, dipping in an up-glacier direction, have been cited as the products of debris thrusting from basal to

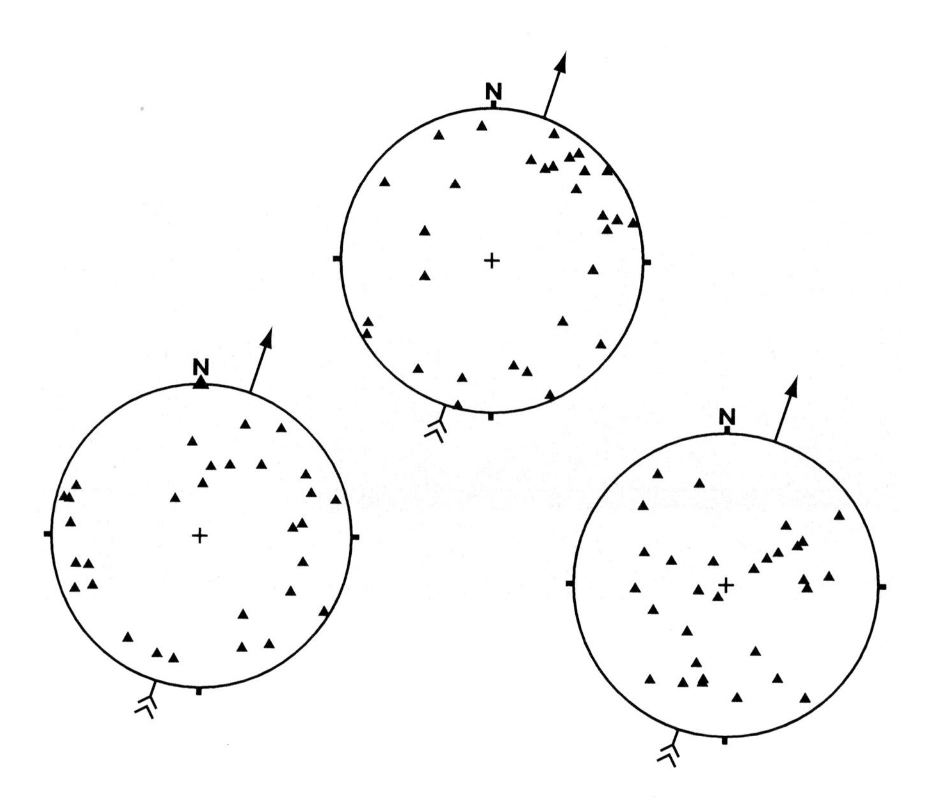

Figure 11.9 Typical low strength clast fabric plots measured in an elongate parallel sided flute on the Bruarjökull foreland. Arrows indicate flute long axis/ice flow direction and location of plot represents position of sample on the fluting (i.e. left side, crest and right side).

englacial/supraglacial positions in surging glacier snouts on Svalbard (Bennett *et al.*, 1996b; Hambrey *et al.*, 1996; Murray *et al.*, 1997; Porter *et al.*, 1997; Glasser *et al.*, 1998b). However, close inspection of a number of englacial diamict bands, which look temptingly like thrusts, exposed along the margins of Brúarjökull and Eyjabakkajökull, show no evidence of basal ice thrust over firnification ice, which would be the case if the diamict was emplaced by thrusting . It is believed that many of the features reported from Svalbard may be tilted crevasse-squeeze ridges rather than sediment emplaced by thrusting. Crevasse-squeeze ridges are predominantly formed vertically to sub-vertically. Subsequently, if normal glacier flow resumes or a small amount of forward momentum remains after ridge construction, crevasse fills will be compressed and tilted into a down-glacier direction (Fig. 11.10). Excavation of a crevasse-squeeze ridge melting out of the snout of Brúarjökull exposed slickensided, fine-grained sediment, indicating a small amount of post-emplacement shearing through the sediment in the ridge. If crevasse-squeeze ridge tilting occurs, the preservation potential of the squeeze ridge form is very poor. The melt-out of tilted ridges will produce a landform-sediment signature similar to that envisaged for englacial thrusting (see below). Specifically, this includes low-relief hummocky moraine comprising interbedded sediment gravity flows and crudely bedded stratified sediments with the possible preservation of small ridges where the thrust intersected the bed. Based upon observations of crevasse-squeeze ridges melting out from Brúarjökull, it is

Figure 11.10 Crevasse-squeeze ridge emerging at the surface of Brúarjökull. Note that the ridge is dipping up-glacier (towards the right) and mass flowage of the down-glacier side of the ridge is producing an extensive spread of bouldery rubble to the left.

clear that mass flowage of ridge sediment down the glacier surface results in the production of boulder lags and thin debris flow diamictons resting on the subglacial till surface (Fig. 11.10).

11.2.7 Hummocky Moraine

Production of prominent belts of hummocky moraine, particularly at the margins of lowland glaciers, is reliant upon widespread and effective transportation of large volumes of material to (eventual) supraglacial positions. In lowland surging glaciers, thrusting has been cited as the dominant process in transporting large volumes of debris into englacial and supraglacial positions. Subsequent ice stagnation leads to the production of hummocky moraine (Sharp, 1985b; Wright, 1980). Substantial amounts of sediment may also be intruded into the glacier by squeezing of sediment up into crevasses, which eventually melts out on the glacier surface during quiescence. Thick debris sequences on the glacier surface may preserve underlying stagnant ice for long periods, creating a feedback loop. Thus, successive surges may involve overriding, overthrusting and incorporation of debris-rich stagnant ice preserved from a previous surge, producing thick sequences of debris-rich and debris-covered ice in surging snouts (Johnson, 1972). The resulting landform assemblage after ice melt-out would comprise aligned hummocky moraine ridges and kame and kettle topography. The widespread distribution of hummocky moraine has been cited by Clayton *et al.* (1985) and Drozdowski (1986) as evidence of palaeo-surging at the margins of the Laurentide and Scandinavian Ice Sheets based upon observations of contemporary, debris-rich surge snouts by Clapperton (1975) and Wright (1980). However, such a landform assemblage can be produced by non-surging glaciers, particularly where debris provision rates are high, and so cannot be used solely as a diagnostic criterion.

Conspicuous mounds of hummocky topography occur on the down-ice sides of topographic depressions on the forelands of Brúarjökull and Eyjabakkajökull. These can be differentiated from overridden thrust-block moraines by the fact that they are characterized by extensive evidence of on-going melt-out of buried ice. This melt-out has disturbed the faint fluting patterns that occur on the hummock surfaces (Fig. 11.11). Stratigraphic exposures are rare but indicate that the hummocks comprise intensely glacitectonized, fine-grained stratified sediments and diamictons or poorly sorted gravels. Pockets of stratified sediments interbedded with diamictons occur in small depressions on the hummocky topography. These sediments have been contorted into low-amplitude folds by the melt-out of underlying ice. The interpretation of this hummocky topography is that it evolves from surging by the thrusting, squeezing and bulldozing of proglacial lake sediments and outwash over pre-existing stagnant ice dating to a previous surge. The faint flutings on the hummock surfaces indicate that the sediment and stagnant ice were overridden by the surging snout. Supraglacially reworked sediments are locally deposited over the bulldozed sediments as they emerge from beneath the melting ice, leading to the deposition of the small pockets of stratified sediments. The post-surge melt-out of the older buried ice results in the production of chaotic hummocky terrain upon which flutings may still be observed as discontinuous linear ridges, at least during the early stages of melt-out. Mapping

(B)

(C)

Figure 11.11 Examples of hummocky moraine at the margins of Icelandic surging glaciers. A) Part of an aerial photograph (Landmaelingar Islands, 1993) of ice-cored hummocky moraine tracts (H) located on the down-ice sides of topographic depressions at west Brúarjökull. The AD 1890 surge moraine is marked (S). B) View across ice-cored hummocky moraine located on the proximal side of the AD 1890 surge moraine at west Brúarjökull. Note surface pond produced by ice melt-out. C) Example of low-amplitude hummocky moraine produced after completion of ice melt-out, west Eyjabakkajökull.

of the hummocky moraine at Brúarjökull demonstrates that it occurs in discrete pockets on the foreland. These are located immediately down-glacier from extensive depressions that have been partially filled with proglacial outwash and glacilacustrine sediments since glacier recession. This distribution pattern, together with the sediment textures and structures and evidence of buried ice, strongly supports the contention outlined above that the material comprising the hummocky moraine originated as a drape over parts of a stagnant glacier snout occupying topographic depressions, and that more recent surging displaced both older ice and its sediment drape (Fig. 11.12). The production of high-relief hummocky moraine is more likely to be the result of the displacement of ice-cored outwash and lake sediment rather than the melt-out of englacially thrust or squeezed sediment, the land-forming limitations of which were mentioned above.

11.2.8 Ice-Cored Outwash and Glacilacustrine Sediments

Observations at the margins of Brúarjökull and Eyjabakkajökull indicate that proglacial outwash tracts and glacilacustrine depo-centres have been developed over the shallow stagnant margins of the snouts during quiescence (Evans and Rea, 1999). The outwash occurs as ice-contact fans fed by subglacial and englacial meltwater portals. During and at the termination of the 1982–83 surge of Variegated Glacier, outbursts of water were observed supraglacially (Kamb *et al.*, 1985) and so substantial quantities of sediment may be deposited onto the glacier surface at this time. However, this drainage network is transient and a drainage pattern controlled by the bed topography becomes re-established post-surge. During quiescence Brúarjökull is drained by four major rivers (Kverká, Kringilsá, Jökulsá á Brú and Jökulkvisl) from approximately seven ice-marginal outlets situated in topographic lows. Those parts of the glacier snout that occupy these topographic hollows are prone to burial by large outwash fans and, in more distal locations, glacilacustrine sediments.

Ice-cored outwash fans and glacilacustrine sediment bodies (including ice-contact deltas) are common along the margins of Brúarjökull (Fig. 11.13). An extensive outwash fan deposited over ice on the east side of Brúarjökull (which did not surge in 1963/64) has been modified gradually by melting of the underlying stagnant snout (Figs. 11.13b, 11.13c). This has resulted in the formation of kettles, followed by the collapse of tunnels in the stagnant ice to form ice-walled channels. The final stage in the evolution of the ice-cored outwash involves the production of chaotic sand and gravel hummocks within which sinuous eskers may be recognizable. Subsequent localized reworking of the outwash into terraces by proglacial streams and its draping by proglacial lake sediments may occur. Due to the extensive nature of the underlying ice, such outwash fans will be represented in the ancient landform record as a landscape of chaotic gravel mounds locally modified by fluvial activity. As such they may be difficult to differentiate from the hummocky moraine outlined above, although they will be located in topographic depressions as opposed to down-ice sides of topographic depressions. They may exhibit a fan shape when viewed as a landform assemblage, and will consist of hummocks with largely accordant summits. Additionally, the internal disturbance of the sedimentary structures of fans and lake sediments will be characterized by simple folds and normal faults rather than the compression structures seen in the hummocky moraine described above. A subsequent surge over these glacifluvial and glacilacustrine sediment bodies will result in the production of either thrust-block moraines where the underlying ice has melted out, or hummocky moraine where a substantial amount of buried ice remained.

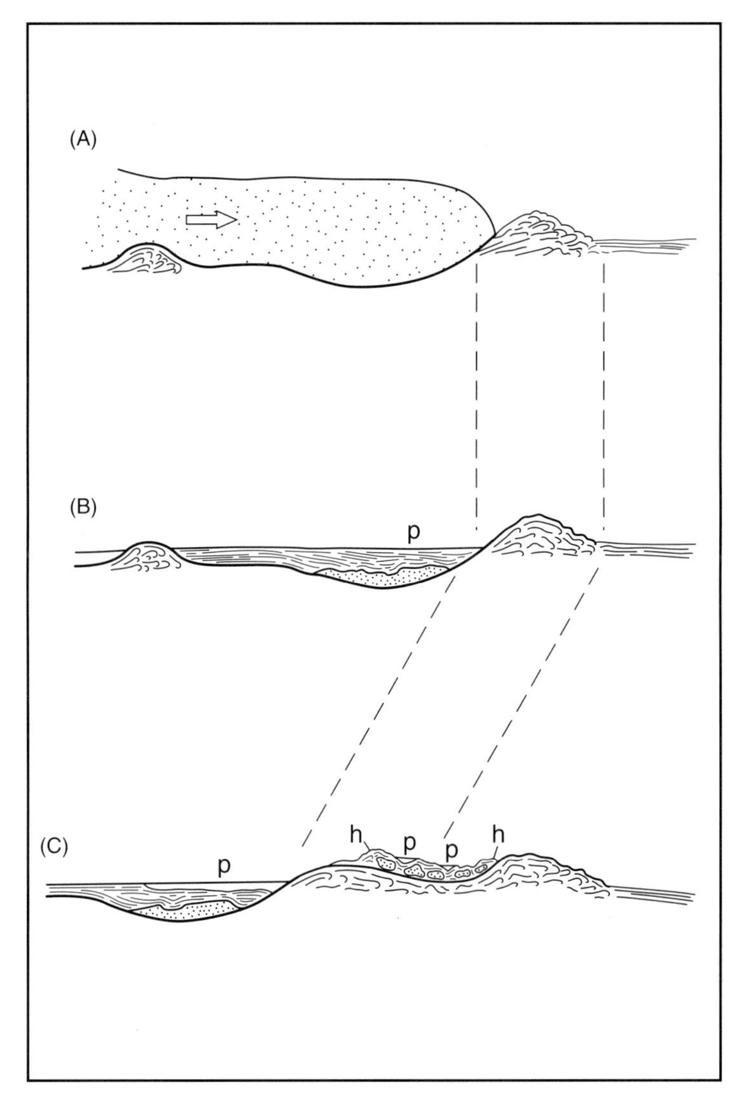

Figure 11.12 Schematic diagram showing the evolution of hummocky moraine at the margin of Brúarjökull, Iceland. A) Glacier surge overrides old thrust-block and constructs new thrust-block moraine from pre-existing sediments. B) Glacier stagnates and outwash, delta and lake sediments fill in the erosional basin, covering a large remnant of stagnating ice. C) Situation after a further surge and stagnation period (note change in position of the cross-section), showing the construction of another thrust-block moraine beyond the existing example and the transport of ice blocks and contorted lake and outwash sediments from the erosional basin to the top of the overridden thrust blocks (p = pond, h = hummocky moraine).

11.2.9 Complex Till Stratigraphies

Observations from Trapridge Glacier have prompted Clarke *et al.* (1984) and Clarke (1987) to suggest that a deformable substrate may play a significant role in cyclical surging activity, although exact causal mechanisms for surging are still unclear (Raymond, 1987). Similar to the flow of non-surging glaciers over deformable beds, the advance of surging glacier snouts into areas of soft sediment produces glacitectonic structures and the thickening of stacked glacitectonites, deformation tills and intervening deposits at the glacier margin (e.g. Boulton, 1996a, b; Alley *et al.*, 1997). This has been observed at the margin of Sefstrombreen, a surging glacier on Svalbard (Lamplugh, 1911; Boulton *et al.*, 1996; Fig. 11.14a), a contemporary setting that has been used as an analogue for till deposition in eastern England. The latter location is illustrative of the use of stratigraphic sequences comprising several tills and associated stratified sediments as diagnostic criteria for palaeo-surging (e.g. Eyles *et al.*, 1994). A stratigraphic section exposed by river erosion at the margin of Eyjabakkajökull, Iceland in 2000 is reproduced in Fig. 11.14b. The sediments and structures in this exposure record the most recent surge of part of the glacier margin in 1972. Glacifluvial gravels, including a stranded iceberg, have been glacitectonized by glacier advance (Fig. 11.14c) and capped by a surge till. The till is a massive diamicton that forms crevasse-squeeze ridges on the present day ground surface. In ancient settings, short periods of till deposition, identified by radiocarbon dating of organics lying between individual till layers in multiple till sequences, has been presented as compelling evidence for surging of the southern margin of the Laurentide Ice Sheet (e.g.

(A)

(B)

(C)

Figure 11.13 Ice-cored outwash fans and glacilacustrine sediments at the margin of Brúarjökull, Iceland. A) Proximal glacilacustrine sediments displaying evidence of extensive collapse due to melt-out of the 1964 surge snout. B) Ice-contact fan developed over the 1964 surge snout (photographed in 1995 with glacier towards the left), showing evidence of kettle production due to melt-out of buried glacier ice. C) The same ice-contact fan viewed from the glacier snout in 2000. Note that the pitted outwash has developed into a landform characterized by large ice-walled channels and chaotic mounds of gravel.

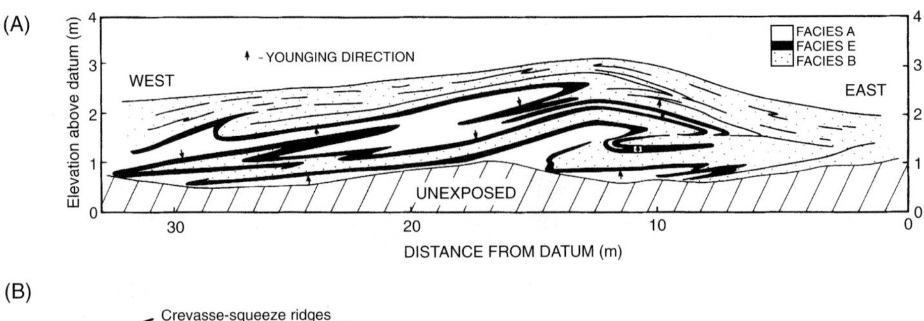

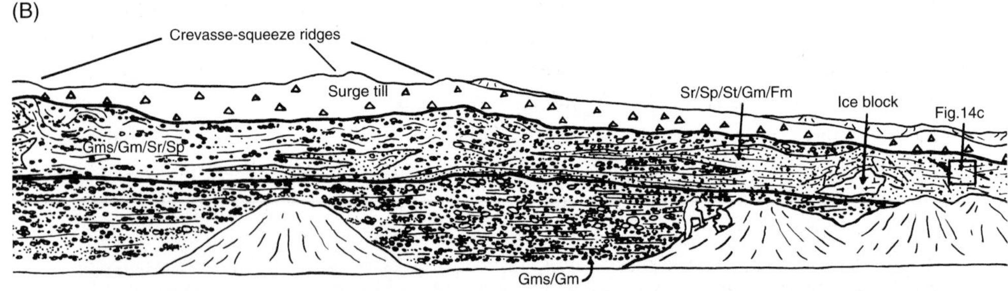

Figure 11.14 Examples of tills and associated stratified sediments from the forelands of contemporary surging glaciers. A) Glacitectonically folded and thrust sediments on Coralholmen, Svalbard, produced by a surge of Sefstrombreen. Facies A is red silt and clay-rich diamicton, Facies B is red silty-sandy diamicton and Facies E is green sand and gravel with large quantities of *Lithothamnium* molluscs. Modified from Boulton *et al.* (1996). B) Stratigraphic section at the margin of Eyjabakkajökull showing a sequence of poorly sorted gravels overlain by glacitectonized gravels and sands, containing a former grounded iceberg, and then massive diamicton (till) that forms the surface crevasse-squeeze ridges. Till deposition and glacitectonic disturbance was initiated by a surge in 1972 when the glacier overrode proximal proglacial outwash containing a stranded iceberg. C) Glacitectonically disturbed sand and mud in the fine-grained outwash to the right of the section in Figure 11.14 B.

Clayton *et al.*, 1985; Dredge and Cowan, 1989b). However, a sequence of glacier advances and retreats could feasibly produce a similar stratigraphy, so this alone is not diagnostic of palaeo-surging.

11.3 SUMMARY OF THE LANDSYSTEMS MODEL FOR SURGING GLACIERS

It has been demonstrated above that no single landform can be used to identify palaeo-glacier surging in the landform record. However, landform-sediment assemblages on the forelands of contemporary surging glaciers provide powerful diagnostic criteria for the recognition of surge imprints on the landscape (Figs. 11.15 and 11.16). Based upon a combination of observations from contemporary surging glacier margins, specifically Brúarjökull and Eyjabakkajökull, Iceland (Evans *et al.*, 1999b; Evans and Rea, 1999), and the published literature, a landsystems model is presented which includes the geomorphic and sedimentological signature of glacier surging (Fig. 11.16). The geomorphology is arranged in three overlapping zones: an outer zone (zone A) of thrust-block and push moraines, grading up-flow into patchy hummocky moraine (zone B), and then into flutings, crevasse-squeeze ridges and concertina eskers with areas of pitted, channelled and/or hummocky outwash and occasional overridden thrust and push moraines (zone C). Cyclic surging by a single glacier snout will often result in the overprinting of one surge event signature by another (Figs. 11.15 to 11.17). Further detail of the landsystems model based upon the signature of a single surge is presented below, although production of the hummocky moraine requires the pre-existence of stagnating ice from earlier surges.

The outer zone (zone A) represents the limit of the surge and is composed of weakly consolidated pre-surge sediments, proglacially thrust or pushed by rapid ice advance. Structurally and sedimentologically these thrust-block and push moraines comprise interbedded thrust slices or folded and sheared proglacial and older glacigenic deposits (glacitectonites; Benn and Evans, 1996). There may be organic layers found within these stacked sequences which represent pre-surge ground surfaces and so can be used to date individual surge events and/or sequences of multiple surge events (Clayton *et al.*, 1985; Dredge and Cowan, 1989b). The development of major thrust-block moraines (e.g. composite ridges and hill-hole pairs) is restricted to topographic depressions that are large enough to collect sufficient sediment during the quiescent phases. Specifically, in the case of Brúarjökull and Eyjabakkajökull it appears that they are confined to braided outwash deposits. Hence, topographically constrained surging glacier snouts like those on Svalbard commonly produce thrust-block moraines that stretch across most of the snout area (Fig. 11.1). In contrast, the topographic hollows along the wider glacier snouts such as Brúarjökull dictate the location of thrust-blocks versus push moraines during surges. In the event that more than one surge terminates in the same place, the overprinted tectonic signatures will produce a complex kinetostratigraphy (cf. Berthelsen, 1978).

The intermediate zone (zone B) consists of patchy hummocky moraine located on the down-glacier sides of topographic depressions and often draped on the ice-proximal slopes of the thrust-block and push moraines. Although hummocks may be the products of supraglacial melt-out and flowage of debris derived from the incorporation of stagnant ice into the glacier and material transported from the glacier bed along shear planes and via crevasses during and immediately after the surge, the relief of such forms is likely to be subdued. High-relief

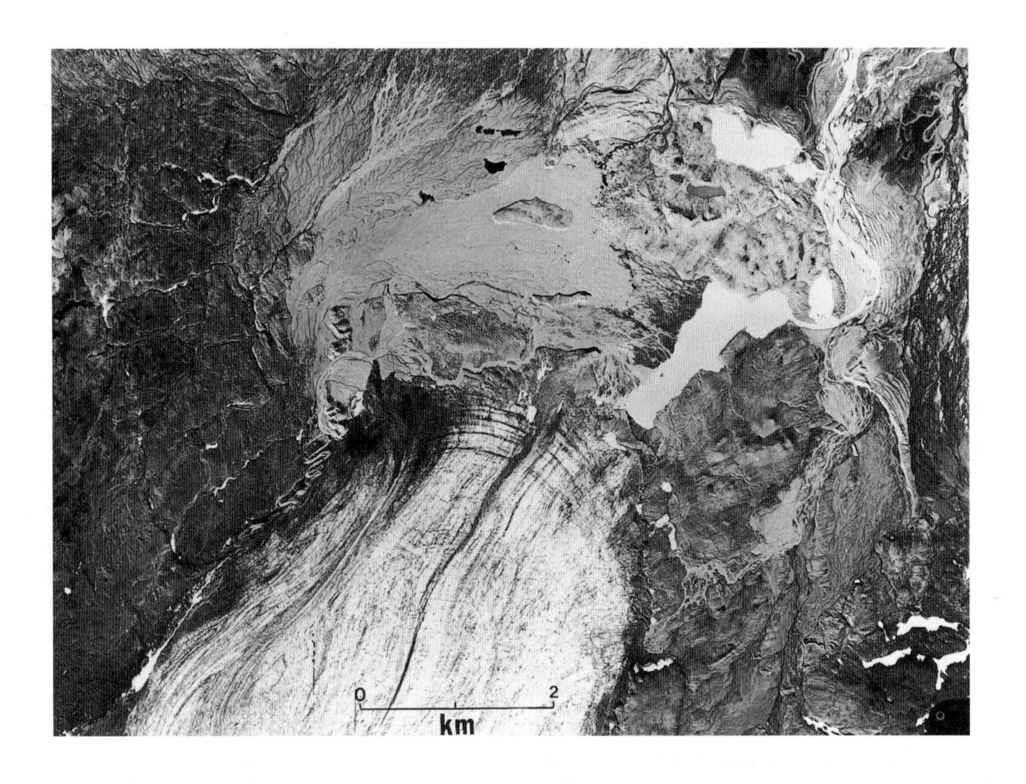
0
2
km

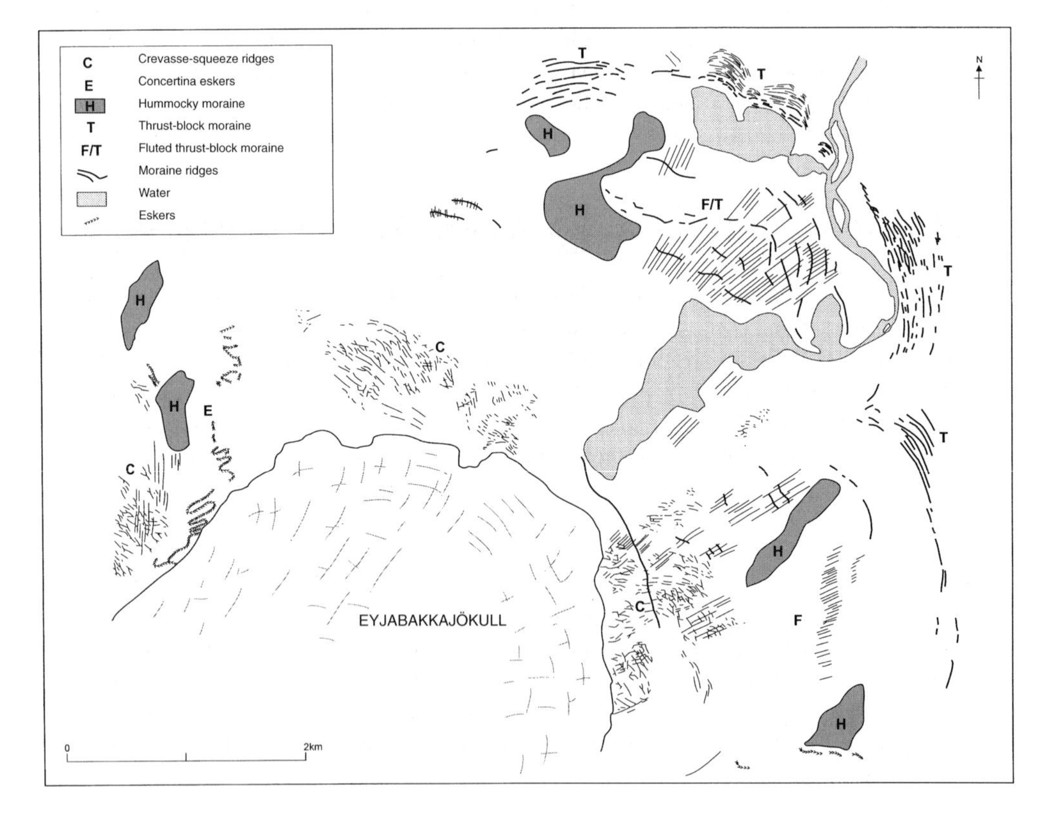
C
Crevasse-squeeze ridges
E
Concertina eskers
H
Hummocky moraine
T
Thrust-block moraine
F/T
Fluted thrust-block moraine
Moraine ridges
Water
Eskers
N
T
T
H
H
F/T
T
H
C
H
E
C
T
H
EYJABAKKAJÖKULL
C
F
H
0
2km

hummocky moraine composed of intensely glacitectonized, fine-grained stratified sediments and diamictons or poorly sorted gravels are the product of thrusting, squeezing and bulldozing of proglacial lake sediments and outwash over pre-existing stagnant ice. Evidence of overriding by the surging snout in the form of faint flutings precludes a supraglacial origin for such hummocks. Small pockets of interbedded mass flow diamictons and crudely bedded stratified sediments, disrupted by normal faulting and low-amplitude folds and occupying small depressions on the hummocky topography, have been produced by the most recent surge snout of Brúarjökull.

The inner zone (zone C) consists of subglacial deformation tills and long, low-amplitude flutings, produced by subsole deformation during the surge, and crevasse-squeeze ridges, documenting the filling of basal crevasses at surge termination. Concertina eskers can also occur in this zone where they are draped over the flutings and crevasse-squeeze ridges. The preservation potential of concertina eskers is likely to be poor although discontinuous gravel spreads and mounds normally referred to as kames or 'moulin kames' in the ancient landform record may represent concertina eskers. Evidence of the production of patchy boulder spreads and thin mass-flow diamictons from the melt-out of up-glacier dipping crevasse-squeeze ridges has been observed (Fig. 11.10). This process may help to explain some areas of low-amplitude hummocky moraine and boulder spreads preserved in the historical and ancient landform record.

Some diagnostic forms of surging are intrazonal, because either they are palimpsests of older surges (e.g. overridden moraines), or they relate to the location of proglacial outwash fans and streams (ice-cored, collapsed outwash), or they occur in ponded topographic depressions on the foreland (collapsed lake plains). Specifically, collapsed glacilacustrine sediment bodies and ice-contact fans may occur within topographic depressions where the stagnating glacier snout became buried during the quiescent phase. The locations and complexities of overridden moraines and multiple till sequences are dictated by the extents of subsequent surges in a single glacier basin.

11.4 APPLICATION OF THE SURGING GLACIER LANDSYSTEM

The surging glacier landsystem outlined above encompasses the landform-sediment associations typical of contemporary surging glacier margins and, because reconstructions of palaeo-ice dynamics rely on suitable analogues from glacierized landscapes where form is necessarily linked to process, it can be utilized by glacial geomorphologists/geologists when identifying possible surges in the Pleistocene glacial record. We now provide an example of the application of this model to the glacial geomorphology of part of the former southwest Laurentide Ice Sheet in western Canada.

Figure 11.15 Aerial photograph (upper: Landmaelingar Islands, 1993) and general geomorphology zonation map (lower) associated with the recent surging of Eyjabakkajökull, Iceland, displaying the juxtaposition/overlapping of thrust moraines, flutings, hummocky moraine, crevasse-squeeze ridges and concertina eskers (after Evans *et al.*, 1999b). Note that the concertina esker and associated crevasse-squeeze ridges on the west side of the snout were produced during the 1972 surge, whereas the outer zone of landforms were produced during the 1890 surge.

Some lobate margins of the Laurentide Ice Sheet have been explained in palaeo-glaciological models as the products of ice streams, possibly subject to surging over deformable substrates, based upon theoretical reasoning or selected geomorphological and sedimentological criteria (e.g. Boulton *et al.*, 1985; Clayton *et al.*, 1985; Fisher *et al.*, 1985; Clark, 1994b; Marshall *et al.*, 1996). However, the palaeo-geographical settings are also favourable for deforming bed development and the accretion of glacitectonites and deformation tills under non-surging conditions (e.g. Hicock, 1992; Boulton, 1996a, b; Evans, 2000a). Furthermore, a variety of genetic explanations have been provided for individual landform assemblages in this region that are not surge-related (e.g. hummocky terrain, thrust moraines, fluting and drumlin swarms). The surging glacier landsystem model presented above, is now applied to an area of east-central Alberta, Canada, that has previously been reported as having a strong palaeo-surge signature (Evans *et al.*, 1999b; Evans and Rea, 1999).

The glacial geomorphology of part of east-central Alberta, specifically part of National Topographic Survey (NTS) map area 73E, is summarized in Fig. 11.18. In the northern half of the map, mega-flutings document the passage of the Lac La Biche ice stream during recession of the Laurentide Ice Sheet. These have traditionally been related to ice streaming (Jones, 1982),

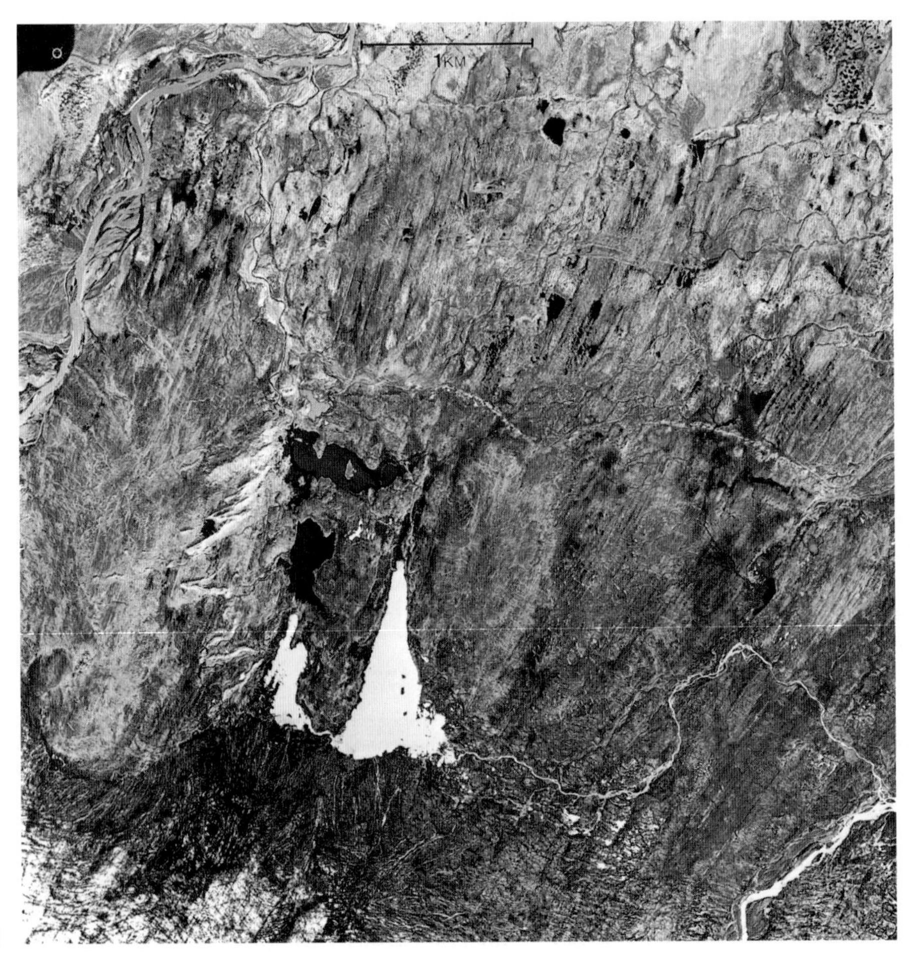

(A)

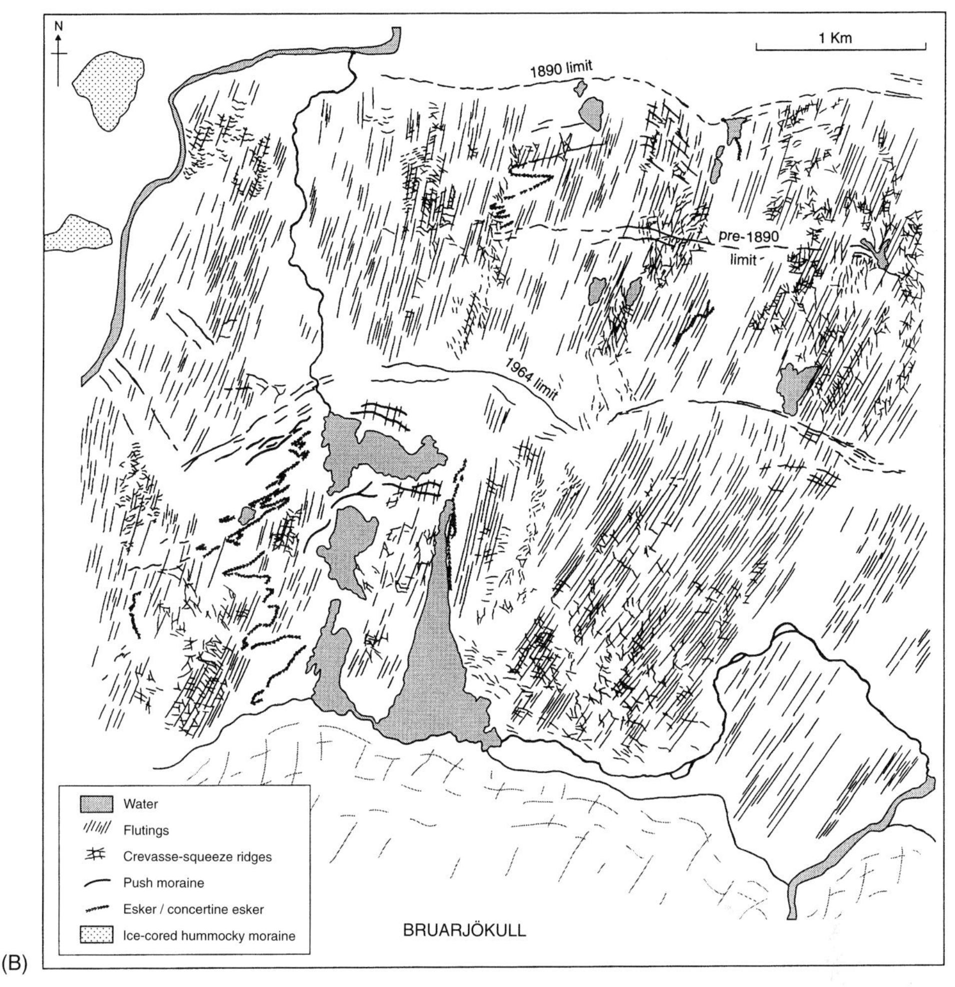

Figure 11.16 A) Aerial photograph (Landmaelingar Islands, 1993) and B) geomorphology map of part of the margin of Brúarjökull, Iceland (after Evans *et al.*, 1999b). The prominent features include flutings, crevasse-squeeze ridges, concertina eskers, small patches of hummocky moraine and extensive thrust-block and push moraines. The outermost moraine dates to the AD 1890 surge and the innermost moraine dates to the 1964 surge. An overridden moraine dating to a pre-1890 glacier margin lies between the two prominent moraines. Buried glacier ice from this pre-1890 surge is most likely responsible for the melt-out features in the hummocky moraine developed after the 1890 surge.

however, because of their association with other elements of the surging glacier landsystem, are linked here with former surge activity. The higher topography in the western half of the map is composed of large thrust-block moraine ridges (TM). Most of these ridges were formed at the margin of ice moving from the north-northwest, but the most recent features (marked by the southernmost TM on Fig. 11.18 and aligned NW-SE) were constructed during the surge of the Lac La Biche ice stream as it flowed from the northwest towards Lloydminster (L) and then from the northeast towards the south of the map area. All of the thrust-block moraines are

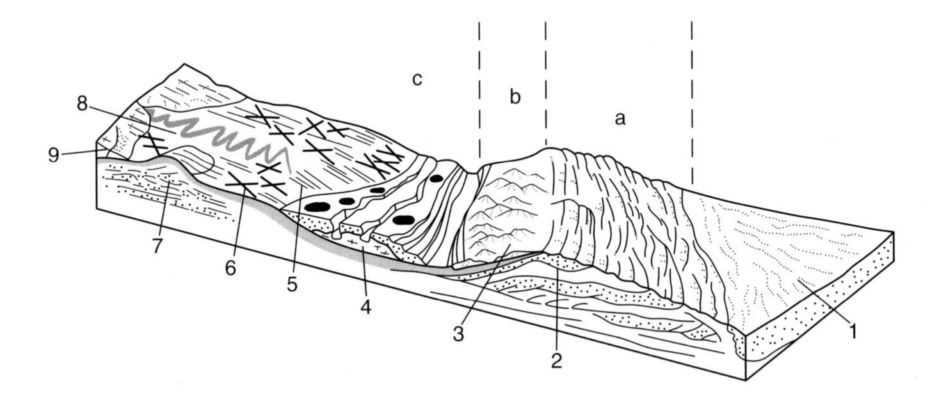

Figure 11.17 A landsystems model for surging glacier margins (after Evans *et al.*, 1999b; Evans and Rea, 1999): a = outer zone of proglacially thrust pre-surge sediment which may grade into small push moraines in areas of thin sediment cover, b = zone of weakly developed chaotic hummocky moraine located on the down-ice sides of topographic depressions, c = zone of flutings, crevasse-squeeze ridges and concertina eskers; 1 = proglacial outwash fan, 2 = thrust-block moraine, 3 = hummocky moraine, 4 = stagnating surge snout covered by pitted and channelled outwash, 5 = flutings, 6 = crevasse-squeeze ridge, 7 = overridden and fluted thrust-block moraine, 8 = concertina esker, 9 = glacier with crevasse-squeeze ridges emerging at surface.

truncated or overprinted by the surge geomorphology to the east. The eastern half of the map is dominated by a dense network of crevasse-squeeze ridges, which overwhelm the mega-flutings in a down-ice direction towards the centre of Fig. 11.18, and document a surge of the Lac La Biche lobe as it breached the large thrust-block moraines in the south. A small area of hummocky moraine and kame and kettle topography occurs amongst the crevasse-squeeze ridges in the southernmost part of the map. Stratigraphic sequences in the map area include severely glacitectonized deposits in the thrust-block moraines and multiple tills with interbedded stratified sediments at the centre of the former Lac La Biche lobe (Andriashek and Fenton, 1989; Mougeot, 1995).

It is evident that the landform-sediment assemblages shown in Fig. 11.18 exhibit all the characteristics (with the exception of concertina eskers) of the surging glacier landsystem model (Fig. 11.17). The juxtaposition of these landforms and sediments at the margin of the surging palaeo-ice stream demarcated by the Lac La Biche fluting field strongly suggests that surging affected this margin of the Laurentide Ice Sheet during its recession from western Canada.

11.5 CONCLUSION

The ability to identify former surging glaciers is crucial for reconstruction and interpretation of the dynamics of former ice sheets. However, no individual landform, except probably concertina eskers which are difficult to identify in the ancient landform record, can be unequivocally labelled as the product of glacier surging. The surging glacier landsystem presented in this chapter combines observations on the geomorphology, sedimentology and glaciology of surging glaciers in Iceland, Svalbard, the USA and Canada, integrating the suite of landforms and sediments known to be produced by surging glacier margins. The model is designed to assess the dynamics of palaeo-glaciers and ice sheets by identification of an integrated landform-sediment assemblage. For example, across large parts of the southwest margin of the Laurentide Ice Sheet the occurrence of numerous large thrust-block moraine arcs and associated mega-flutings, hummocky moraine tracts, areas of transverse (crevasse-squeeze?) ridges, and stratigraphic sequences comprising glacitectonized multiple tills/glacitectonites have been used to suggest that surging behaviour was characteristic during ice recession (e.g. Boulton *et al.*, 1985; Clayton *et al.*, 1985; Fisher *et al.*, 1985; Clark, 1994b). Application of a comprehensive landsystems model to an area of east-central Alberta corroborates the proposed palaeo-surging activity and indicates that further assessments of regional glacial geomorphology are warranted in order to identify large scale ice mass dynamics.

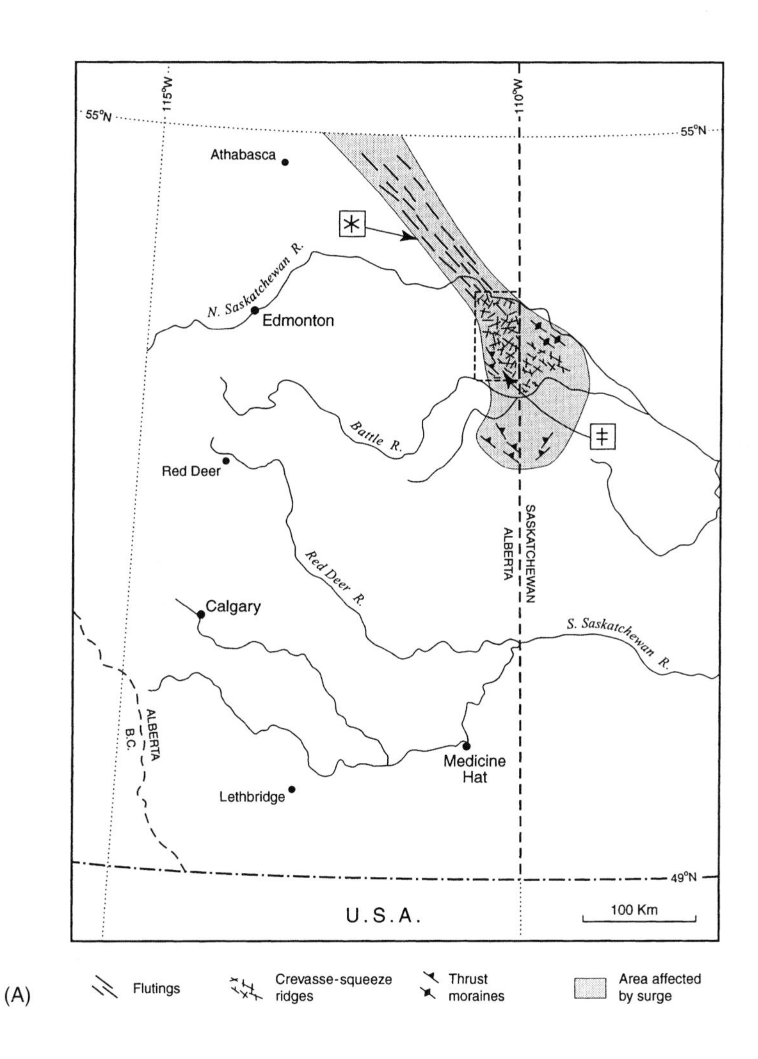

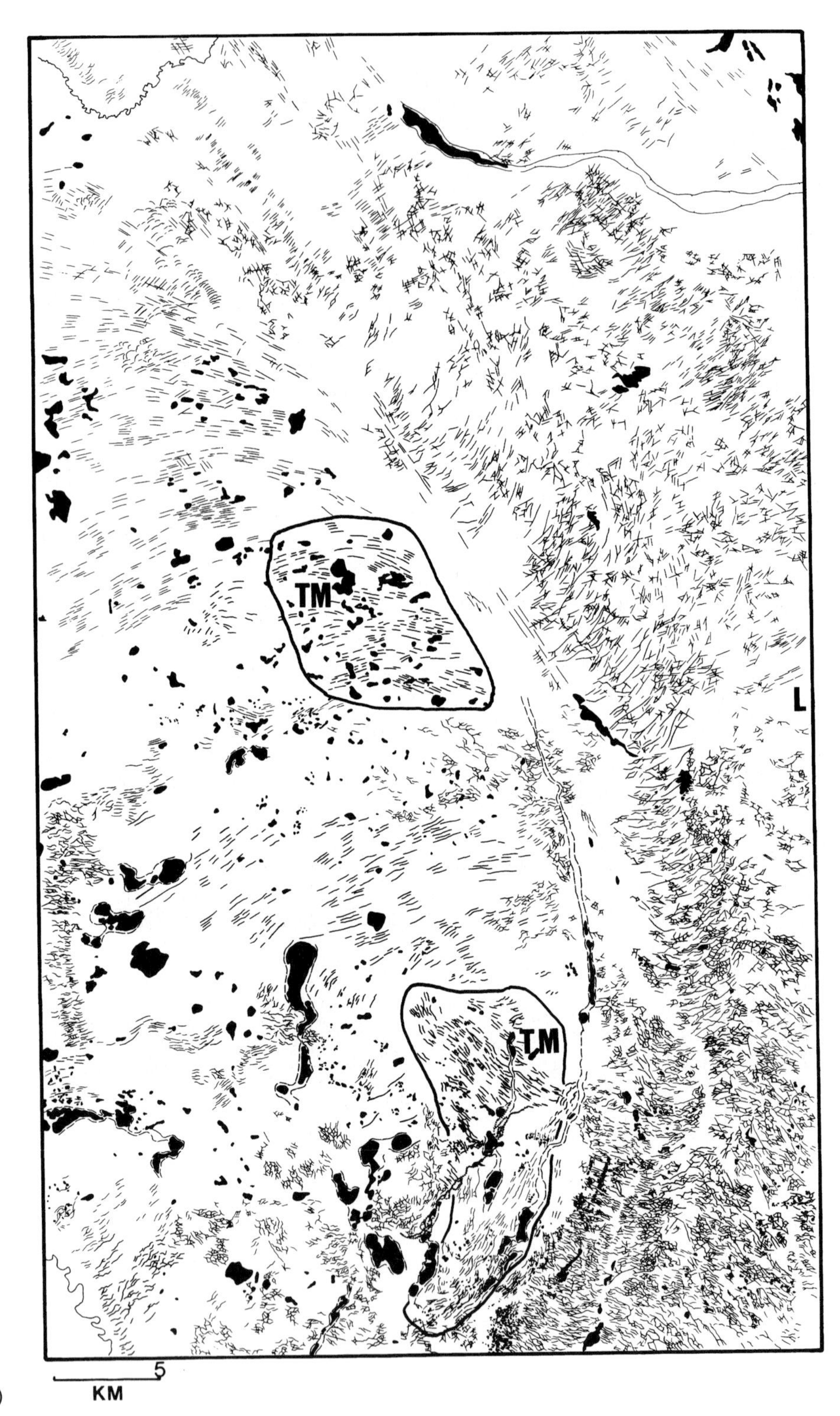

(B)

(C)

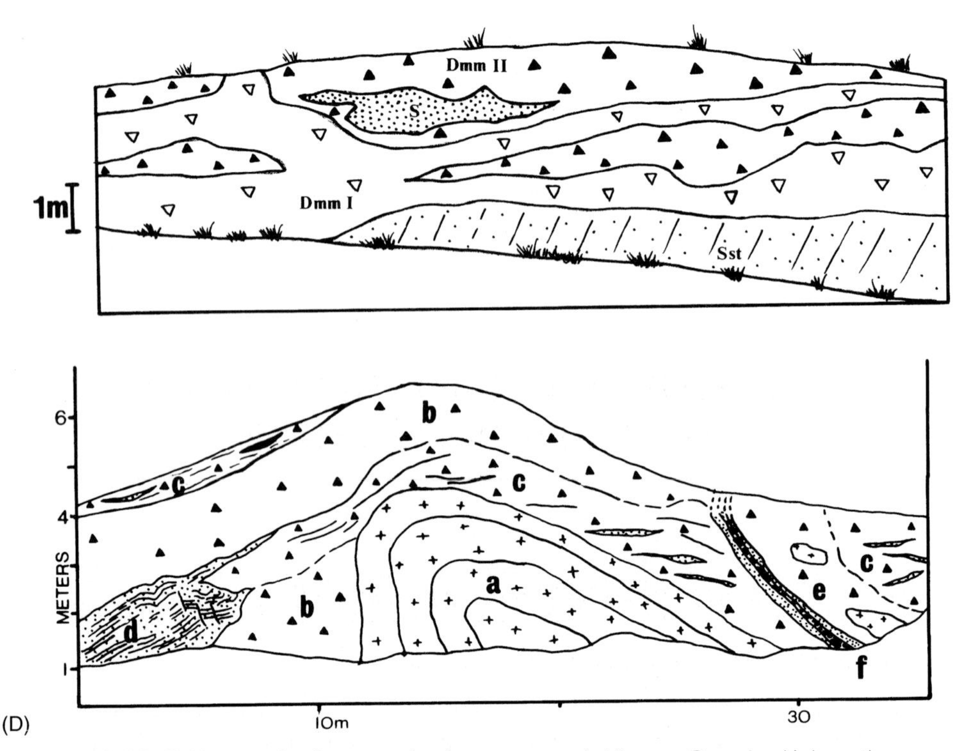

Figure 11.18 Evidence of palaeo-surging in east-central Alberta, Canada. A) Location map of the Lac la Biche palaeo-ice stream (only the geomorphic features relating to the ice stream are presented). Sections used in Fig. 11.18d are marked by asterisk and double cross symbols. B) The geomorphology of the east-central part of NTS map area 73E, east central Alberta, Canada (boxed area on Fig. 11.18a), showing all ice flow-parallel and flow-transverse ridges. TM = major thrust-block moraines breached by most recent surge. The last surge in the area produced the dense network of crevasse-squeeze ridges in the eastern half of the map, which grade into the southernmost flutings of the former Lac La Biche ice stream in the northern half of the map. Hummocky moraine and kame and kettle topography occurs at the southernmost edge of the map. L = the town of Lloydminster. C) Aerial photograph mosaic of the area represented in Figure 11.18 B, illustrating the major geomorphic features. D) selected, representative stratigraphic exposures from the region, showing (upper) interbedded tills and rafted sands overlying sandstone bedrock (after Andriashek and Fenton, 1989; see asterisk symbol on Fig. 11.18 A), and (lower) glacitectonized bedrock, faulted and folded stratified sediments and associated diamictons (tills) at the core of a thrust-block moraine (a = folded bedrock, b = massive diamicton, c = laminated diamicton with sand lenses, d = faulted and folded sands, e = massive diamicton with bedrock rafts, f = faulted and folded interbeds of laminated fines and gravels; after Mougeot, 1995; see double cross symbol on Fig. 11.18 A).

SUBAQUATIC LANDSYSTEMS: CONTINENTAL MARGINS

Tore O. Vorren

12.1 THE CONTINENTAL SHELF SYSTEM

12.1.1 Diamicton Formation on Continental Shelves

The most widespread sediments on glaciated continental shelves (Fig. 12.1) are diamictons. There has been, and still is, much discussion on the origin of these diamictons. The correct

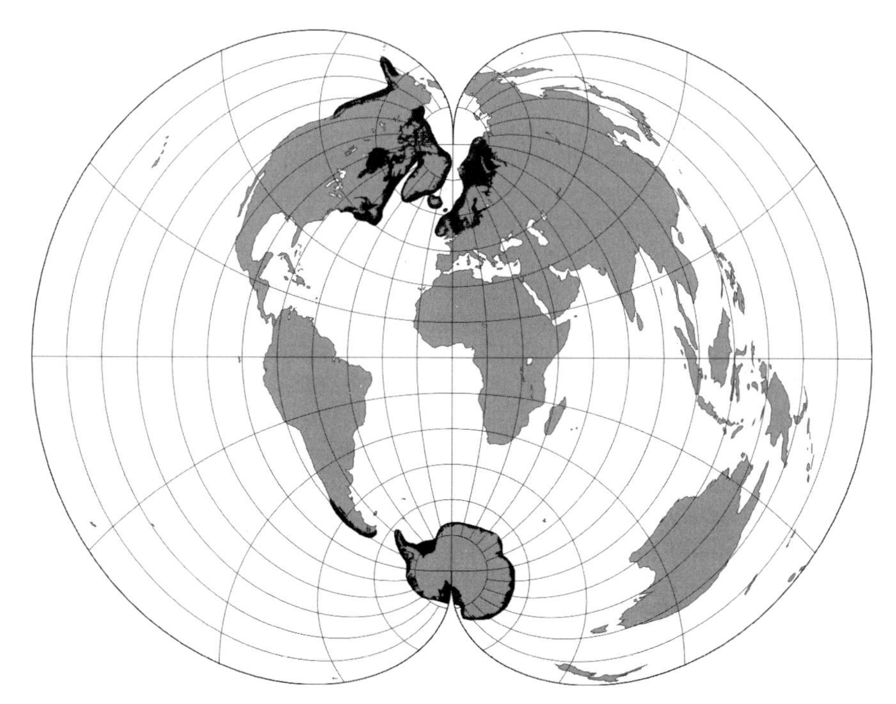

Figure 12.1 World map showing continental margins that were glaciated once or several times during the Late Cenozoic.

interpretation of the continental shelf diamictons is vital for understanding the glacial history of the shelf in question. In this chapter the possible glacial and glacimarine origins of the diamictons will be elucidated. In addition to the glacial and glacimarine diamicton-forming processes it should be kept in mind that other processes, particularly debris flows, might generate diamictons on shelves.

12.1.1.1 Grounded Glaciers

The same type of glacial basal deposits are to be expected in the marine as in the terrestrial environment, for example different types of basal tills (lodgement tills, melt-out tills, deformation tills; Fig. 12.2A). Discoveries of a deforming till layer beneath Ice Stream B in the Ross Sea region (e.g. Blankenship *et al.*, 1987; Alley *et al.* 1987) demonstrate that sediment deformation offers a means for efficient transportation of debris at the base of an ice sheet towards the grounding line.

In addition to these, a group of deposits called waterlain tills may occur (Fig. 12.2A). Dreimanis (1979) suggested three mechanisms for the deposition of waterlain tills; by grounded icebergs, as subaquatic flow tills and as waterlain basal melt-out tills. Basal melt-out of glacial debris can occur below floating glacier termini, ice shelves or in subglacial basins. Evidently for floating ice it is a matter of definition where to draw the line between waterlain till and glacimarine sediments. Dreimanis (1979) described waterlain till as being deposited either in direct contact with glacier ice or from glacier debris without substantial disaggregation or sorting. *A priori*, however, it is hard to see how the melt-out process will not disaggregate the sediments, unless they occur as a frozen or compacted block, and then it is an ice-rafted clast.

Subaquatic flow till was first described from a lacustrine environment (Evenson *et al.*, 1977), but was later identified in marine environments (e.g. Hicock *et al.*, 1981; Lønne, 1995). The sediment in this type of facies comprises proglacial cones of interbedded diamictons ('subaquatic flow tills') and outwash with glacimarine fine sediments occurring as discontinuous lenses between cones (Fig. 12.2B). Related to the subaquatic flow tills, are the 'till tongues' found on the Canadian and Norwegian continental shelves (King *et al.*, 1991), which are wedge-shaped deposits of sediments interbedded with stratified glacimarine sediment, and they constitute discrete stratigraphic units laid down near margins of marine ice sheets. According to the model, the till tongues "are formed through the accumulation of glacial debris by subglacial processes proximal to the grounding line, together with a penecontemporaneous, proglacial contribution from sediment gravity flows" (King *et al.*, 1991). Stravers and Powell (1997) interpret till-tongue-like deposits on the Baffin Island shelf originating from sediment failure and debris flows at the terminus of a temperate ice sheet. A better distinction regarding processes as well as terminology between 'subaquatic flow tills', 'till tongues' and 'debris flows' on glaciated shelves needs further work.

12.1.1.2 Ice Shelves

Deposition beneath ice shelves may take place by meltwater deposition near the grounding line, basal melting with deposition of basal and englacially transported debris, or basal melting with deposition of 'frozen-on sediments'. In addition to the siliclastic sediments, there may also be biogenic components from advection of plankton underneath the shelf.

In cases where the ice sheet proximal to the grounding line is at the pressure melting point, the sediments may be transported by subglacial meltwater. *A priori*, a shift in the grounding line

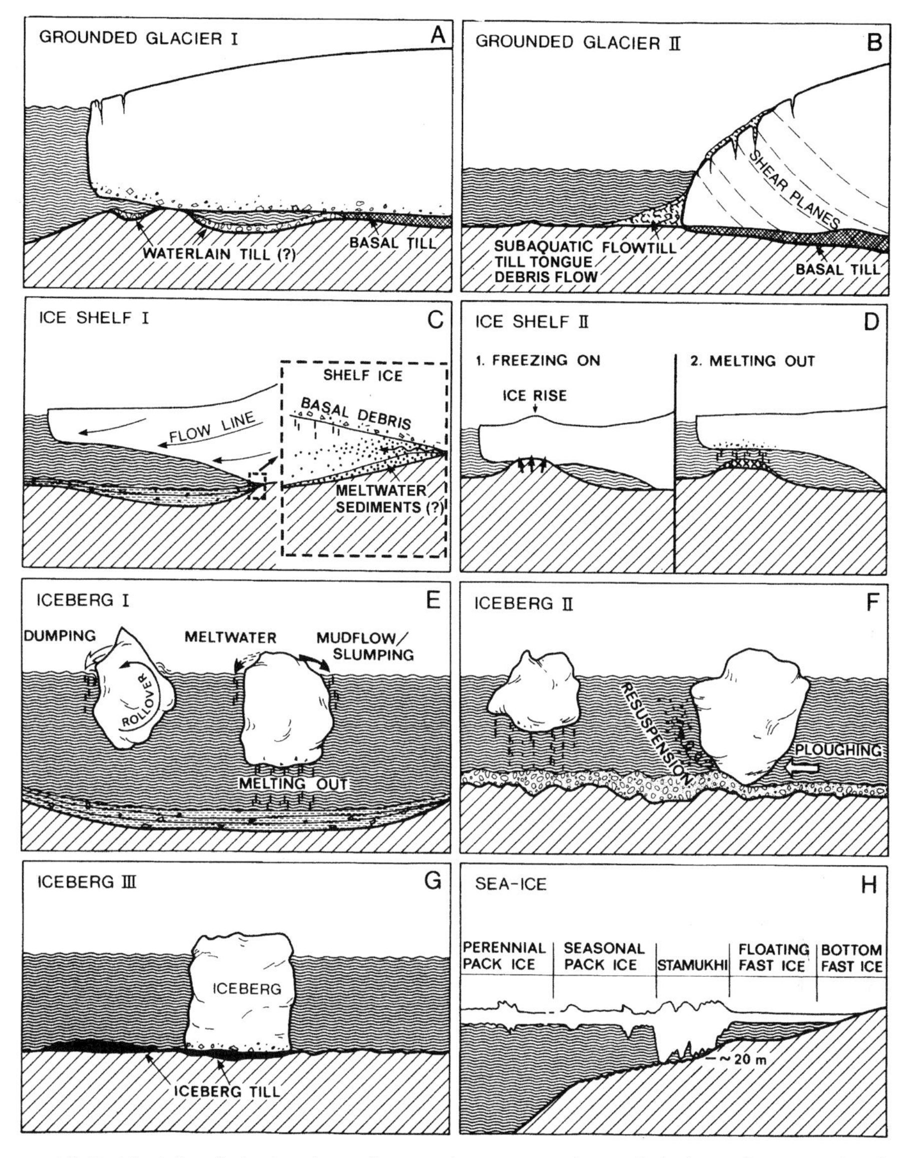

Figure 12.2 Models of glacigenic sedimentation on continental shelves. See text for further explanation. (Adapted from Vorren *et al.*, 1983).

could give rise to a complex stratigraphy of interbedded glacimarine diamictons, subglacial meltwater deposits and basal tills mixed with ice-rafted debris. However, no such processes have yet been described at modern ice-shelf grounding lines.

Orheim and Elverhøi (1981) suggested a process by which ice shelves can cause redeposition. If an ice shelf grounds (due to lower sea level and/or thicker ice) it may freeze to the bed. When the shelf then refloats it will contain frozen-on debris. Later melting releases this material (Fig. 12.2D). This process has actually been observed at a pinning point on a glacier tongue floating in Antarctica (Powell *et al.*, 1996).

12.1.1.3 Icebergs

Deposition from icebergs can occur by release from floating as well as grounded bergs. The extensive Heinrich layers in the North Atlantic are well known examples of deposition from floating icebergs (e.g. Heinrich, 1988; Bond *et al.*, 1992). Floating icebergs can release debris in several ways:

- by melting-out
- by dumping when the iceberg overturns, fragments or tilts
- by small mudflows
- by meltwater rivulets (Fig. 12.2E)

The textural and petrographic composition of the iceberg debris is almost identical to the till deposited by the source glacier of the iceberg (e.g. Anderson *et al.*, 1980 a). However, modification to this till composition will occur, depending on current activity (which causes winnowing), and the amount of fines added from suspension. The resulting sediment may be coarser or finer ('residual glacimarine' and 'compound glacimarine sediments, respectively; Anderson *et al.*, 1980b). The geometry of these sediment facies would be sheet drape and/or basin fill, and the stratification mostly massive or crudely stratified.

Redeposition by icebergs may be caused by iceberg ploughing and by adfreezing. The magnitude of iceberg-ploughing depends on the size of the iceberg, water depth, drifting velocity, bottom relief and shear strength of the bottom sediments. Iceberg ploughing causes deformation, reworking and some sorting/winnowing of the bottom sediments. Belderson *et al.* (1973) and King (1976) indicate that the berm of the furrow is composed of coarser sediments than the base of the furrow. Deformation structures beneath iceberg scours and sub-horizontal thrust faults related to lateral displacement and stacking of clay slabs are observed within the berms (Woodworth-Lynas and Guigne, 1990). The stirring action must cause turbulence and resuspension. Thus in the long run the sediments as a whole are probably depleted in fines; how much depends on the bottom current regime.

The composite geometry of this type of iceberg-redeposited sediment is sheet drape. The resulting sediment is a diamicton that is mostly coarser than the normal glacimarine sediments. Fossils in life position seldom exist. Vorren *et al.* (1983) suggested naming this sediment type 'iceberg turbate' (Fig. 12.2F). Barnes *et al.* (1988) suggested the term 'ice-keel turbates' to include sediments disturbed by sea ice as well.

Iceberg till (Dreimanis, 1979) is a type of waterlain till deposited by grounded icebergs (Fig. 12.2G). The most favourable situation for deposition of iceberg till would be a sudden decrease in water depth in the iceberg environment. The textural and petrographic composition should be similar to that of the till from the source glacier, but differ geometrically from the till being deposited as lenses.

12.1.1.4 Sea Ice

Several observations confirm sea ice as a debris-transporting medium (e.g. Carlson, 1975; Pfirman *et al.*, 1989; Nürnberg *et al.*, 1994; Stein and Korolev, 1994). Sea ice may acquire debris by adfreezing in the shore zone, by infreezing of suspended sediments, by settling of aeolian sand and silt, and from rivers flowing onto the sea-ice in spring.

Sediments may be deposited from sea ice in much the same way as from icebergs. The sediments may differ from iceberg-rafted debris with regard to clast roundness; most of the sea-ice clasts are derived from rounded littoral gravel (Lisitzyn, 1972). Another factor to consider is the subduing effect of sea ice on wind- and wave-generated currents, the consequence of which is reduced winnowing of the sediments. The nearshore sea-ice environment has been described in detail by Reimnitz *et al.* (1978), and Fig. 12.2H shows sea-ice zonation as found in the Alaskan Beaufort Sea. Between the seasonal pack ice and stamukhi zone, Kovac and Sondhi (1979) have described a zone with a floating fast ice extension. It should be noted that the ice gouging on the underlying sediments may be heavy (e.g. Rearic *et al.*, 1990), particularly in the stamukhi zone. The draft of the sea ice in pressure ridges may be as much as 47 m (Reimnitz *et al.*, 1972). Thus sea ice may cause bulldozing, resuspension and adfreezing at depths shallower than about 50 m.

12.1.2 First-Order Morphological Elements

The first-order morphological elements on a glaciated shelf are banks and large depressions (troughs/channels). Most banks on glaciated shelves reflect the bedrock morphology, but some are accentuated by shelf diamictons. Depressions are ubiquitous on glaciated continental shelves. In general there are two types of depressions, transverse troughs and longitudinal channels (Figs 12.3, 12.4 and 12.5). Examples of longitudinal channels are found off Norway (Holtedahl, 1958), Labrador (Josenhans, 1997), Alaska (Carlson *et al.*, 1982) and Antarctica (Anderson, (1999). Off Labrador a longitudinal channel runs parallel to the coast for more than 400 km, with depths of 600–800 m (Holtedahl, 1958). Another example is the Norwegian channel that encircles the southern part of Norway. It runs parallel to the coast, and the largest depths in the east are more than 700 m whereas it shallows to about 220 m in the middle before deepening to about 400 m at its mouth (Holtedahl, 1993). The longitudinal channels generally follow boundaries between sedimentary rocks on the shelf and the older crystalline troughs towards the coast, or faults and other zones of weakness (Holtedahl, 1958; Vogt, 1986; Josenhans, 1997; Anderson, 1999; Fig. 12.5).

Transverse troughs are normally overdeepened in their inner reaches. Off Norway the depths reach about 400–500 m (Fig. 12.5), but in Antarctica they may be deeper than 1000 m (Fig. 12.4). Most often the transverse troughs represent seaward extensions of fjord/glacial valleys. There is a tendency for troughs to be less than 20 km wide where the continental shelf is narrow. On broader shelves and in epicontinental seas, transverse troughs are much wider. Bear Island Trough in the epicontinental Barents Sea, is 170 km wide and 600 km long. The very deep inner parts of the troughs on the Antarctic shelves have been attributed partly to proglacial isostatic depression. However, this can only account for a minor part, and the great depth must primarily be due to glacial erosion (Anderson, 1999).

In recent years channels and troughs have often been demonstrated to be the drainage routes for ice streams (e.g. Anderson, 1999; Sejrup *et al.*, 1996; Vorren and Laberg, 1997; Ottesen *et al.*, 2001; Figs 12.6 and 12.7). At the front of many of these channels and troughs, trough mouth fans are deposited (see below).

12.1.3 Second-Order Morphological Elements

The main glacigenic morphological elements and lithofacies on a glaciated continental margin are shown in Fig. 12.8. Often glacigenic sediments are bounded below by a pronounced unconformity (e.g. Dekko, 1975; Solheim and Kristoffersen, 1984; Vorren *et al.*, 1986, 1990; Josenhans, 1997). Glacigenic sediments on a shelf are characterized by various types of diamictons

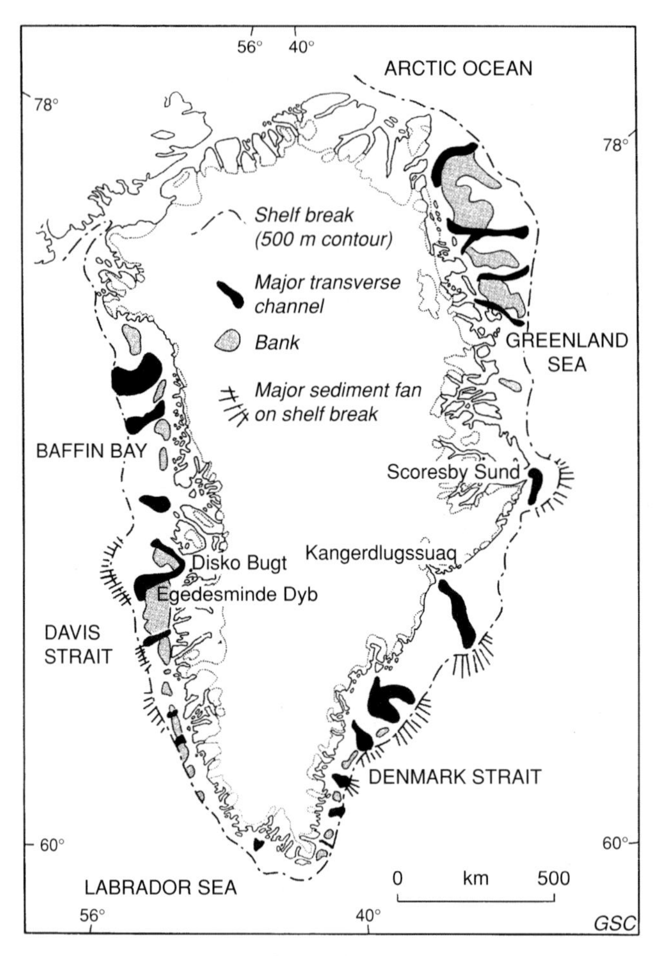

Figure 12.3 Morphological features of the Greenland continental margin. (After Funder, 1989).

with internal sub-horizontal conformable boundaries. In the Barents Sea, buried glacigenic conformable subsurfaces often have glacial lineations on them (Rafaelsen *et al.*, 2002). Angular unconformities and irregular boundaries may also occur, in particular related to subglacial glacifluvial drainage. The thicknesses of stratiform glacigenic sequences on continental shelves are normally less than a couple of hundred metres. On the Norwegian shelf proper, thickness varies between 0 and ~ 300 m, but at the shelf edge thicknesses often increase substantially (Vorren *et al.*, 1992).

The banks are often covered by palimpsest sediments (i.e. a mixture of the underlying diamictons and Holocene skeletal remains). The trough fills comprise tills and various types of glacimarine sediments (e.g. Vorren *et al.*, 1984; Sejrup *et al.*, 1996).

12.1.3.1 Moraine Ridges and Grounding Line Features

On many continental shelves, end-moraine systems or morainal banks as they are called by some (e.g. Powell, 1983; Powell and Molnia, 1989) are identified by acoustic methods. West of

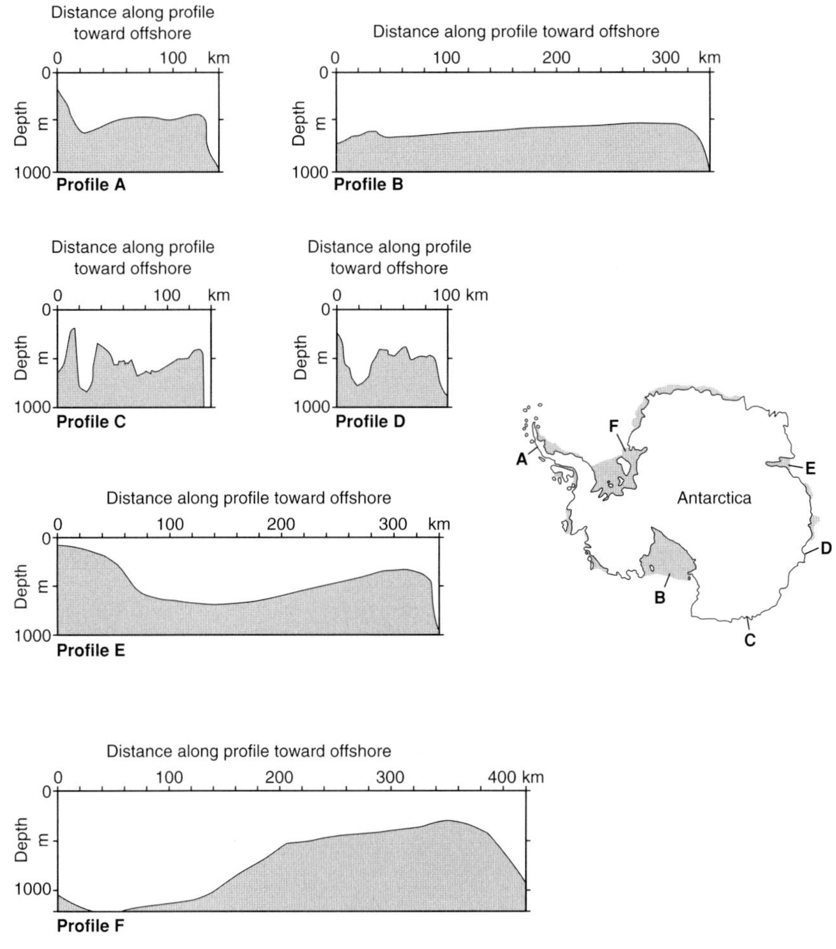

Figure 12.4 Representative bathymetric profiles from the Antarctic continental shelf illustrating the great depth and landward sloping profile of the shelf. (After Anderson, 1999).

Shetland are large end moraines up to 50 m high, 8 km wide and which can be traced for up to 60 km (Stoker, 1997). The moraine ridges are arranged in a series, and each ridge probably marks a stillstand during eastward glacial retreat. The proximal flank of each ridge was systematically overlain by a veneer of successively younger acoustic layers (Fig. 12.9A). Vorren and Kristoffersen (1986) and Gataullin and Polyak (1997) described moraine ridges in the southwestern and eastern Barents Sea, respectively, being 1.5–4 km broad and up to 40 m high. Slowly advancing or halting ice sheets probably deposited these moraines.

On the Antarctic continental shelf several so-called grounding zone wedges are observed (Bart and Anderson, 1997; Anderson, 1997; Shipp and Anderson, 1997b; Shipp *et al.*, 1999). The wedges average 35–75 m in thickness and extend for several tens of kilometres in length and width (Bart and Anderson, 1997; Fig. 12.9B). These features are what Alley *et al.* (1989) termed 'till deltas'. They suggested that unconsolidated, water-saturated sediment is transported at the base of the ice stream and deposited at the grounding line as a subglacial delta, with

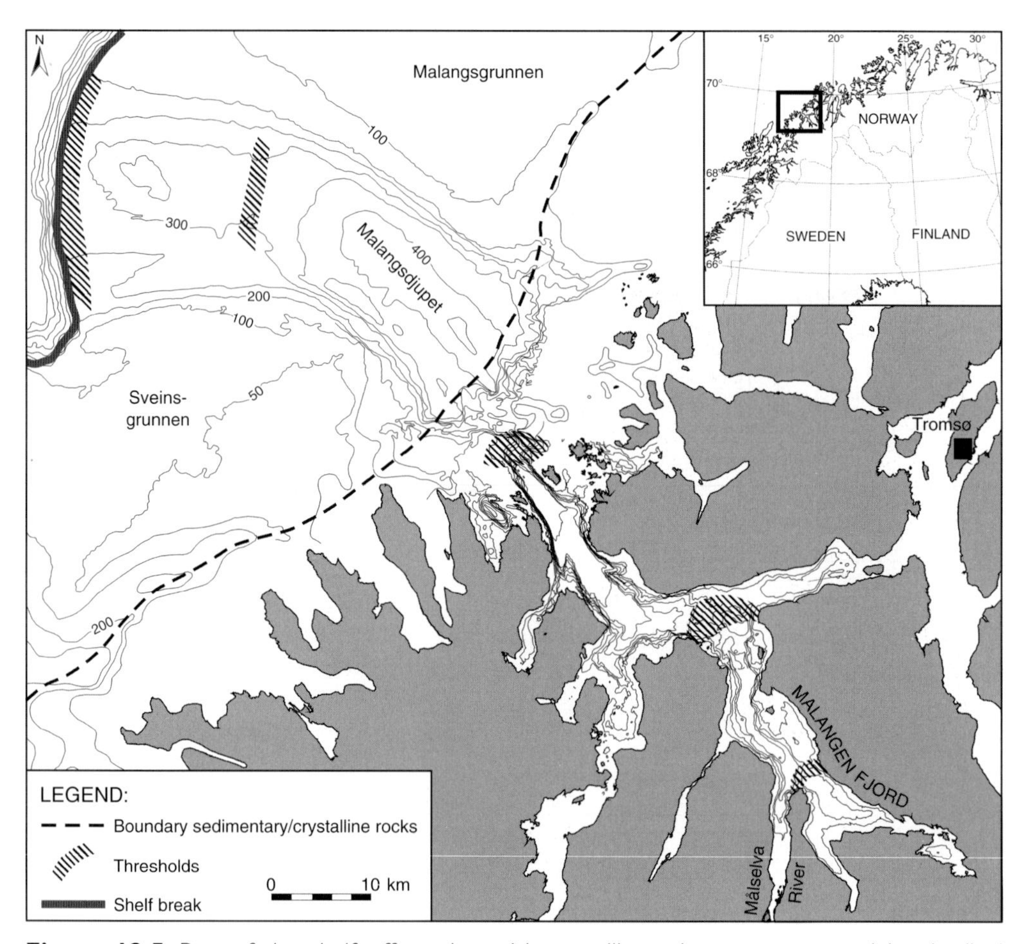

Figure 12.5 Part of the shelf off northern Norway illustrating transverse and longitudinal troughs and banks. Note the landward slope of the troughs and the steep headwall aligned along the boundary between crystalline and sedimentary bedrock. Malangsdjupet trough represents a seaward continuation of the Malangenfjord system and Målselv Valley.

topset beds composed of till and foreset and bottomset beds composed of sediment gravity flow deposits.

Marginal moraines deposited from surging glaciers are found off the modern ice cap Austfonna in the Svalbard archipelago. Typically the cross-sectional shape of a surge moraine is asymmetrical with a smooth distal slope of 1–3°, a proximal slope of 3–6°, and relief varying between 5 and 20 m. The width may exceed 1 km. The distal part of the ridge is associated with acoustically transparent slump lobes that drape the iceberg-ploughed seafloor. The morphology on the proximal side of the ridge is dominated by linear sediment ridges forming a rhombohedral cross pattern (Solheim, 1991; Fig. 12.9C; see also Chapter 11).

MacLean (1997) described an interesting lateral moraine from Hudson Strait (Fig. 12.9D). The moraine was probably deposited at the lateral margin of an ice sheet, the position of which was at

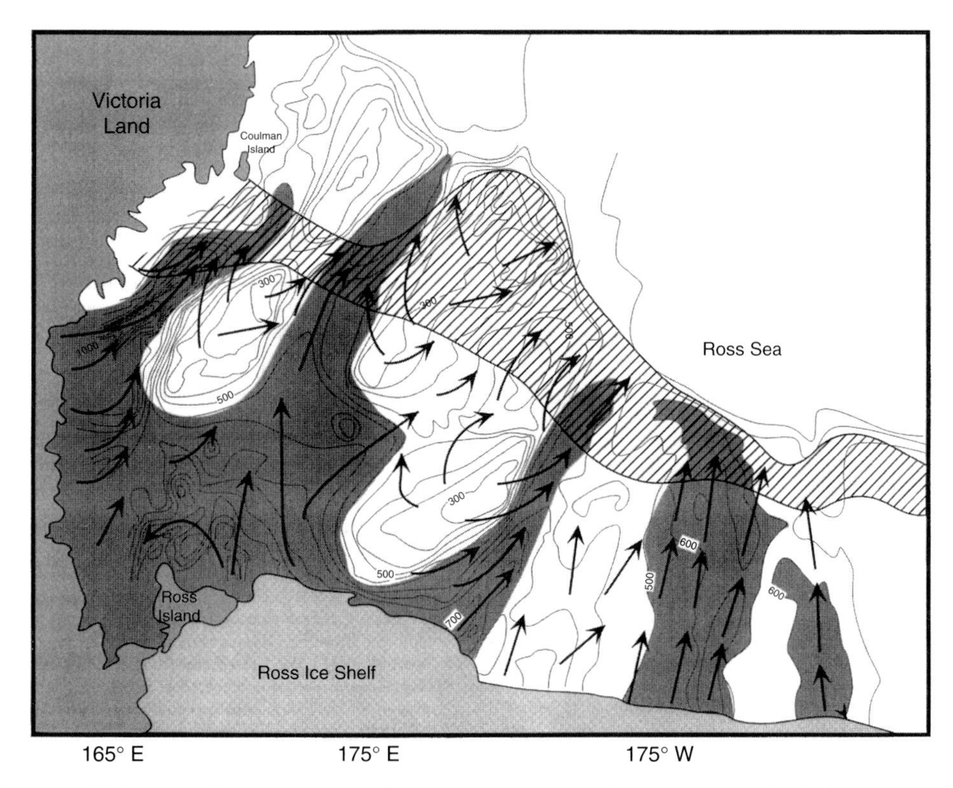

Figure 12.6 Late Pleistocene glacial reconstruction for the Ross Sea. Shaded areas in the troughs are where major ice streams existed, and the cross-hatched areas designate the location of the Last Glacial Maximum grounding zone. (After Anderson, 1999).

least in part controlled by water depth. The preservation of overridden sediments and the grounding depth relationships suggest that the ice sheet responsible for the ice-contact sediments was only lightly bearing on the seabed at the time of deposition (MacLean, 1997).

12.1.3.2 Linear Forms

Submarine drumlin fields are described from the inner Scotian shelf, Canada (Fader *et al.*, 1997), where the drumlins extend up to 35 m high and 800 m long, and range from 39 to 300 m wide. These drumlins typically have a flat upper surface. Drumlins from the Ross Sea in Antarctica are described by Shipp and Anderson (1997a). These average 2 km wide, range from 2 to 5 km long, and are tens of metres high. Some have well-developed hairpin-shaped scours rimming them. Down-glacier they merge with features identified as megaflutes by Shipp and Anderson (1997b).

A detailed swath bathymetry of the inner Norwegian channel (Longva and Thorsnes, 1997; Fig. 12.10) shows many linear forms. On the upstream part, crag and tails and drumlins occur. Flutes and drumlins from different ice movement directions are preserved. In the middle part of the channel, flutes 2–7 m high reflect the last ice movement in the area. The same succession of morphological elements is observed in the Ross Sea (i.e. a transition from striated basement rocks to drumlins (and crag and tails) to megascale lineations). Anderson (1999) interprets this

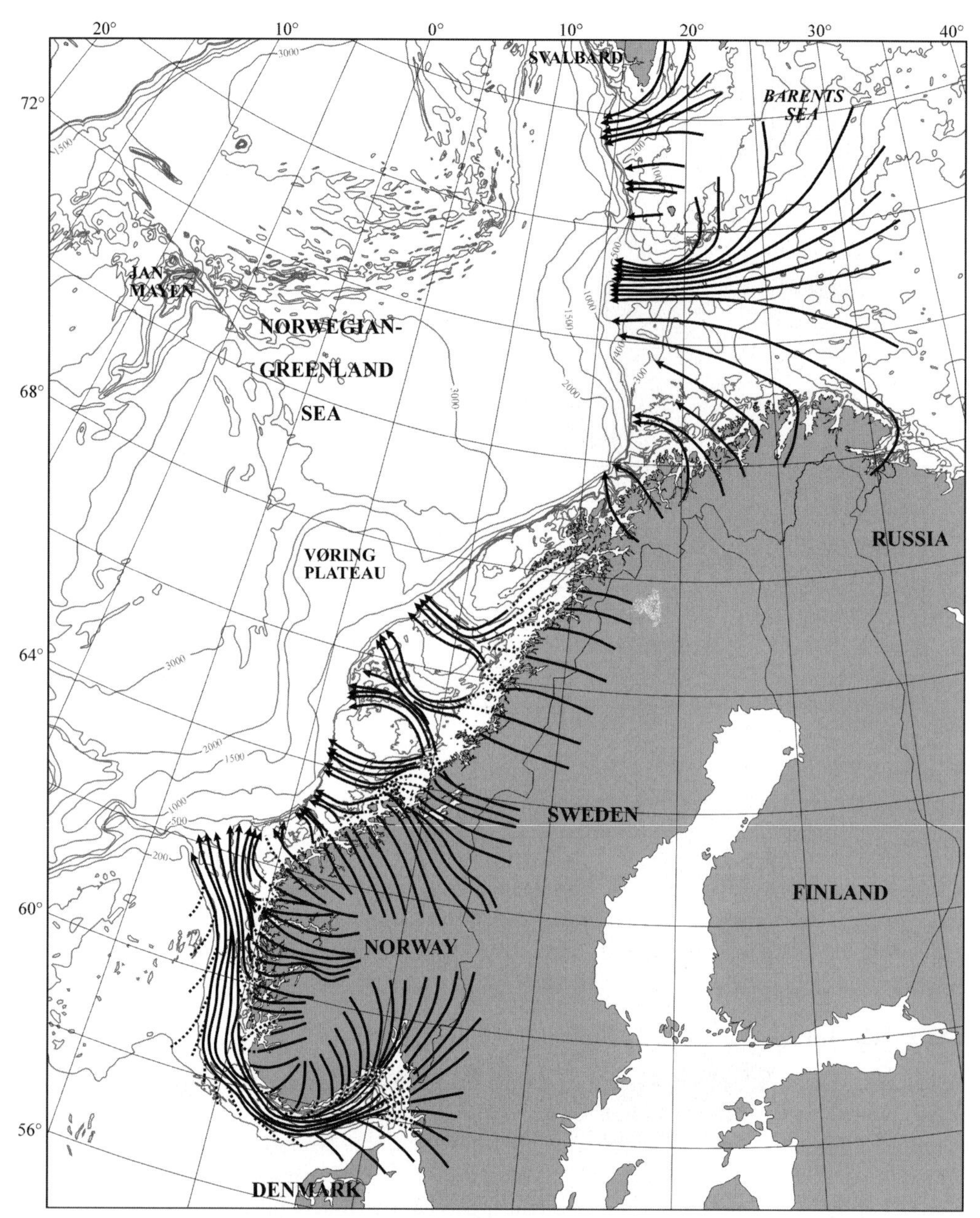

Figure 12.7 Inferred ice flow pattern of the Fennoscandian and Barents Sea Ice Sheets during Last Glacial Maximum on the Norwegian continental shelf. (Redrawn from Ottesent *et al.* (2001), south of Lofoten; Vorren and Laberg 1996, north of Lofoten).

as reflecting a transition from a 'sticky bed', where the ice sheet is coupled to the seafloor and basal meltwater is confined to the ice-bed interface, to a deforming bed, where meltwater is incorporated in the bed and the sediments are moulded into megaflutes and megascale lineations.

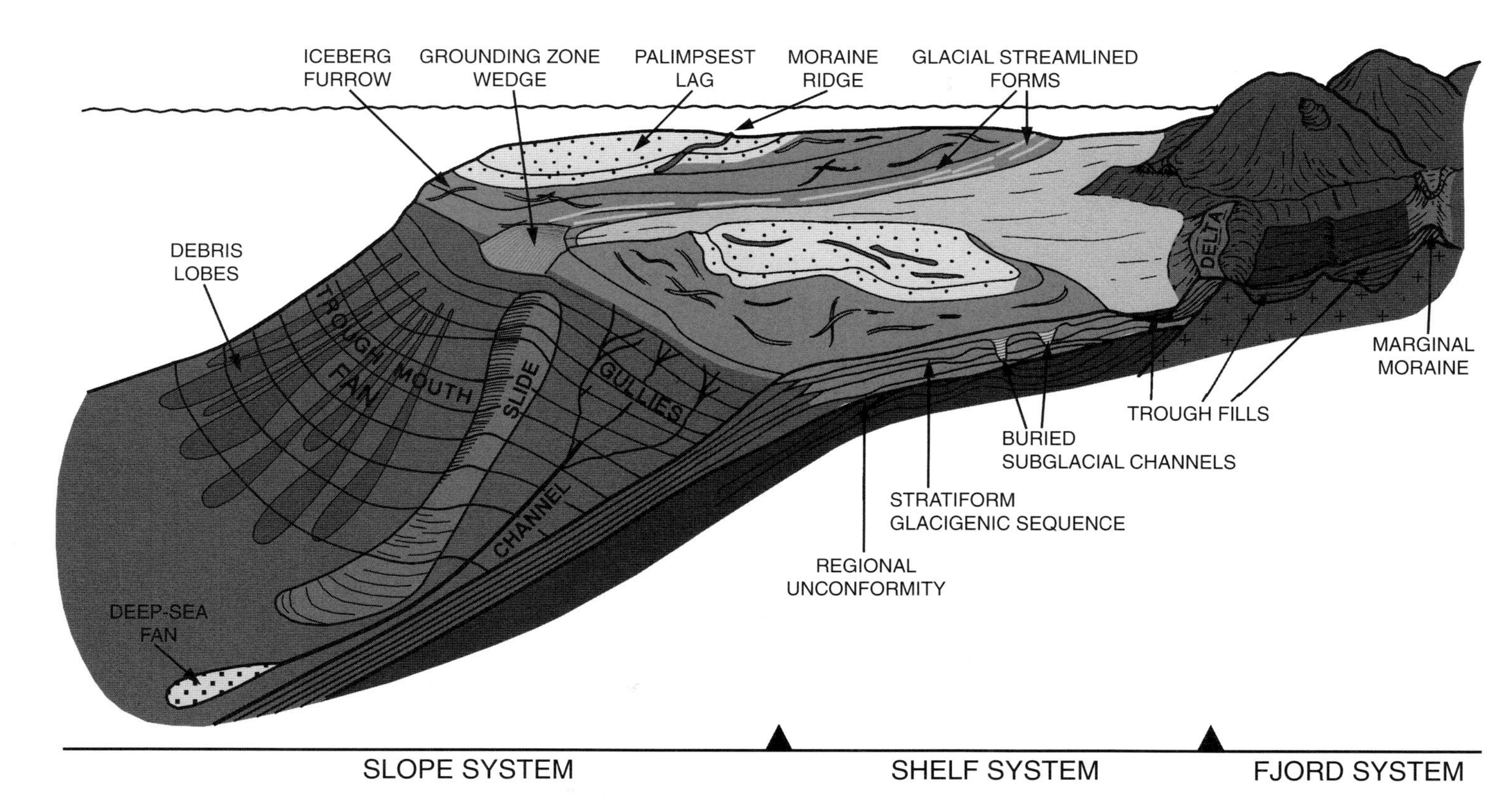

Figure 12.8 Model showing the main glacigenic morphological elements and lithofacies of a passive continental margin, exemplified by the margin off northern Norway.

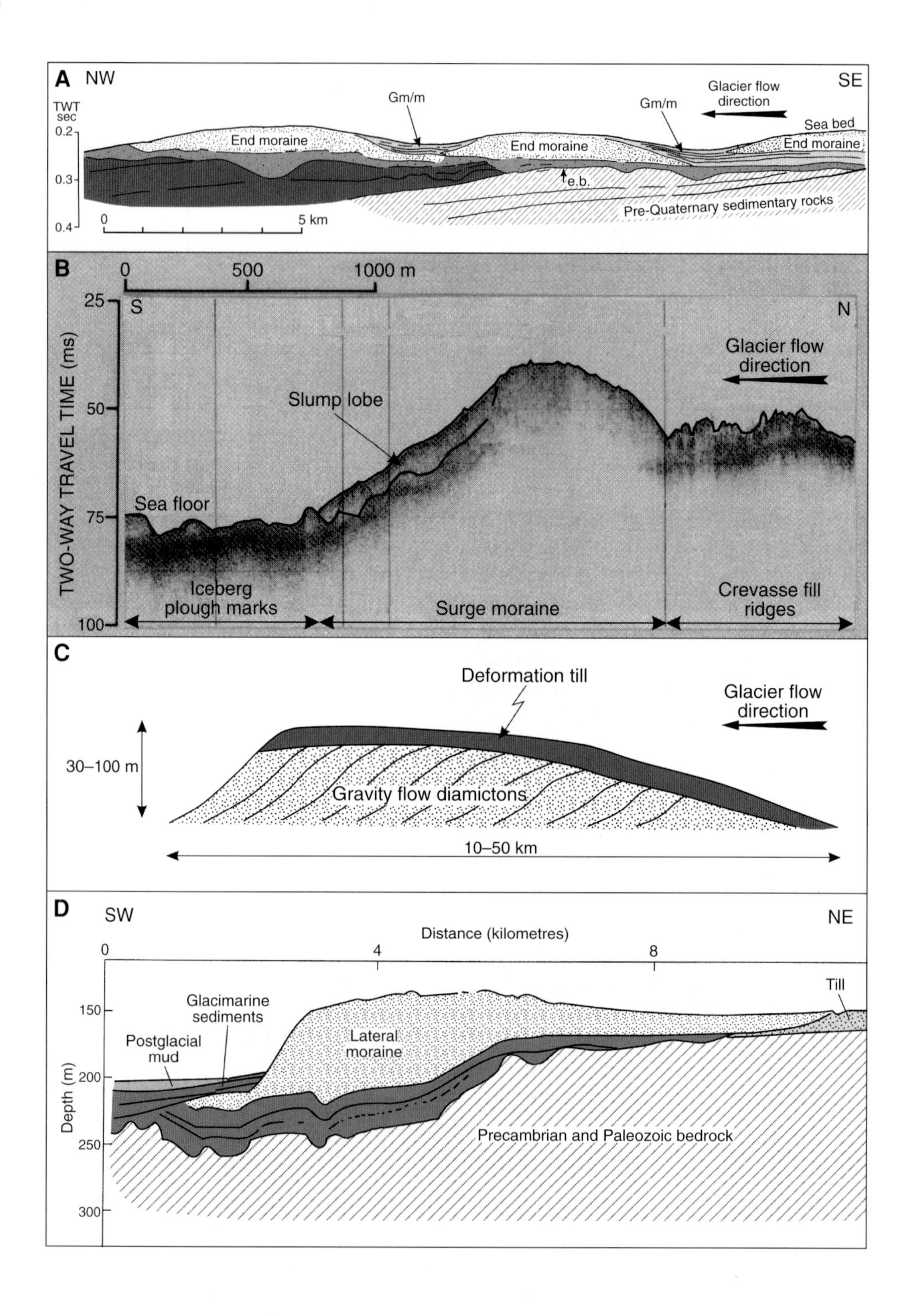
A NW
SE
TWT sec
0.2
0.3
0.4
Gm/m
Gm/m
Glacier flow direction
Sea bed
End moraine
End moraine
End moraine
e.b.
Pre-Quaternary sedimentary rocks
0
5 km
B
0
500
1000 m
25
50
75
100
S
N
TWO-WAY TRAVEL TIME (ms)
Glacier flow direction
Slump lobe
Sea floor
Iceberg plough marks
Surge moraine
Crevasse fill ridges
C
Deformation till
Glacier flow direction
30–100 m
Gravity flow diamictons
10–50 km
D
SW
NE
Distance (kilometres)
0
4
8
Till
150
200
250
300
Depth (m)
Glacimarine sediments
Postglacial mud
Lateral moraine
Precambrian and Paleozoic bedrock

12.1.3.4 Plough Marks

Iceberg plough marks (Fig. 12.11) form when keels of icebergs exceed the water depth and are therefore able to erode sea floor sediments. As an iceberg moves it ploughs aside sediments to create berms, it scrapes striae inside the furrow channels, and it may create a pile of sediments when the iceberg becomes grounded (Syvitski *et al.*, 2001). Iceberg plough marks have been revealed in many shelf seas by side-scan sonar and other acoustic studies during the last decades. Lien (1983), who studied the Norwegian shelf, found that the iceberg plough marks occurred in current depth ranges of 120–400/500 m. The maximum width and depth of the marks found were 250 and 25 m, respectively. Barrie (1980) observed maximum furrow depths of 17 m on the Labrador Bank; Syvitski *et al.* (2001) observed a maximum furrow depth of 28 m and a width of 274 m on the east Greenland margin; Rafaelsen *et al.* (2002) observed a maximum furrow depth of 25 m and a width of 500 m in the Barents Sea (Fig. 12.11); and Polyak *et al.* (2001) observed up to 30 m deep furrows on the Chucki borderland and adjacent continental margin. Clearly the draft of the iceberg limits the water depth at which this process can operate. On the Greenland and Antarctic margins iceberg scouring down to 550 m is observed (Barnes and Lien, 1988; Dowdeswell *et al.*, 1993). The largest sea depth (850 m) and width (1 km) of iceberg-ploughed sea floor is reported from the Yermak Plateau (Crane *et al.*, 1997).

12.1.3.5 Subglacial Channels

Tunnel valleys are characteristic elements of glaciated shelves that record meltwater drainage beneath ice sheets. They are not particular to marine settings but do occur on and beneath the present continental shelves. The southern North Sea contains large examples, locally up to 500 m in relief, 12 km wide and tens of kilometres long (Cameron *et al.*, 1987; Ehlers and Wingfield, 1991; Praeg, 1997). On the Grand Banks, Moran and Fader (1997) describe channels having a depth of about 25 m and widths of some hundreds of metres (Fig. 12.12). Various mechanisms for the formation of these valleys are commonly evaluated, but most researchers agree that they are formed by subglacial meltwater drainage.

Figure 12.9 Four examples of submarine moraine ridges. A) Geoseismic profile across submarine moraine ridges off northwest Britain showing stratigraphical relationships of the end moraines (dense stipples) and associated acoustically layered, ponded glacimarine/marine (Gm/m.) deposits. The underlying erosive sheetform unit (shaded and dotted) also belongs to the same seismostratigraphical sequence as the morainal complex; the erosional base (e.b.) is attributed to the expansive phase of the ice sheet. As the morainal complex formed during the subsequent gradual decay, this Late Devensian section appears to preserve a well-defined glacial advance-retreat cycle (redrawn from Stoker and Holmes, 1991; Stoker, 1997). B) Geoseismic profile (based on 3.5 kHz records) across the central part of the surge moraine of Bråsvellbreen in the northern Barents Sea. Inside the moraine is a rhombohedral pattern of ridges having a relief in the order of 5 m and spacing between individual ridges of 20–70 m; the ridges are interpreted as crevasse fills through squeeze-up of deformable sediments during an early phase of post-surge stagnation (see Chapter 11; redrawn from Solheim, 1997). C) Schematic cross profile of a grounding zone wedge (till delta/diamicton apron). D) Geoseismic profile (based on a single channel seismic reflection profile) from south-central Hudson Strait illustrating a lateral moraine up to 70 m thick and its relationship with underlying relatively undeformed acoustically statified sediment and younger glacimarine and postglacial sequences (modified from MacLean, 1997).

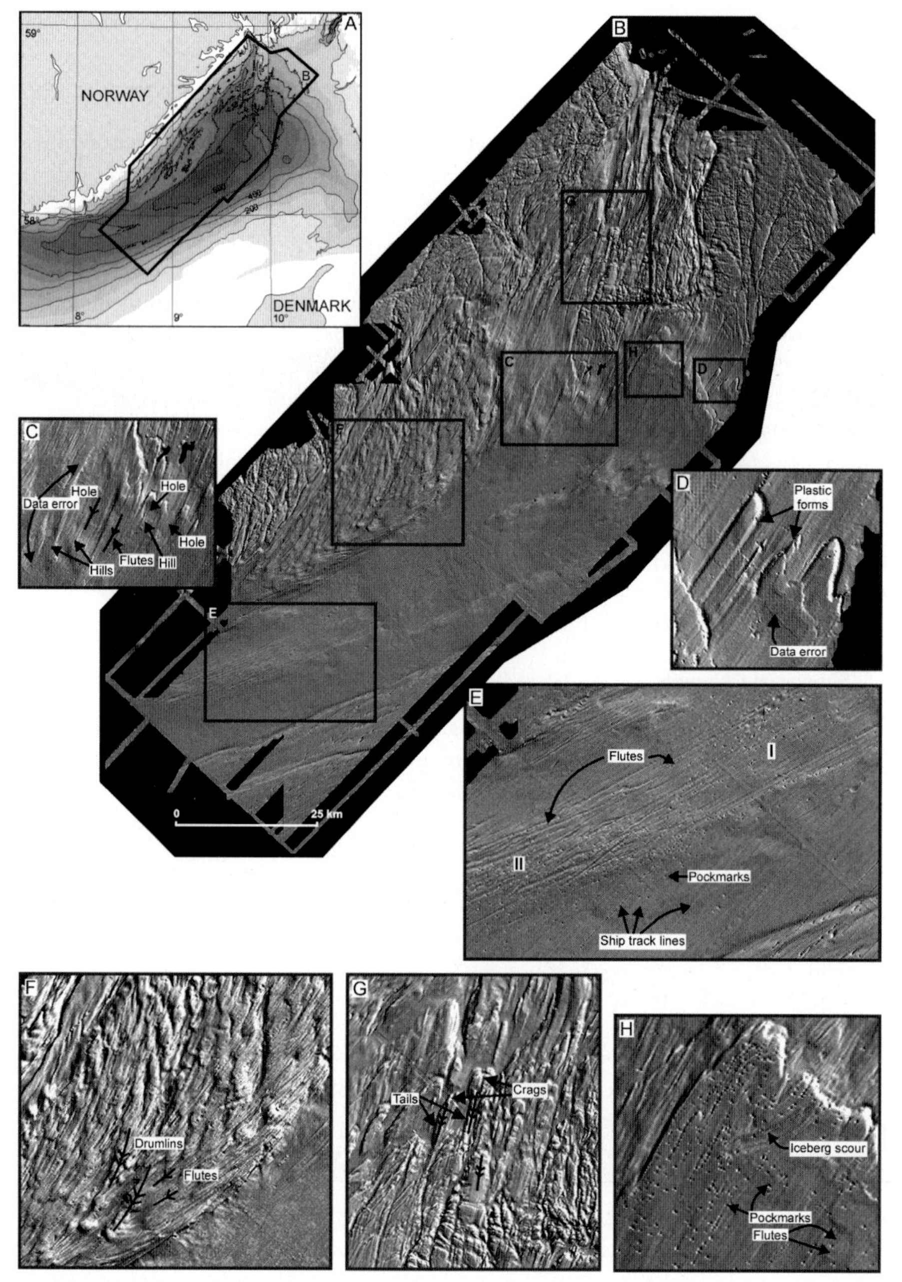

Figure 12.10 The central figure (B) shows a shaded image of processed multibeam echosounding data from the Norwegian Channel between Norway and Denmark (For location, see map (A) at upper left). The locations of the smaller images are shown by boxes. C) Glacitectonic hill-hole features. Slabs of bedrock or sediments were tectonically disrupted and dislocated. This erosion can be identified from the holes where the rock was removed and the hills where it was deposited. D) Plastic forms (p-forms) or grooves with a maximum depth of 20 m eroded into bedrock. E) Flutes and pockmarks. The flutes have a relief of 2–7 m and reflect the ice flow along the axis of the Norwegian Channel. Flutes from two different ice flow phases can be seen. F) Flutes and drumlins from different ice movement directions situated on top of the terrace landward of the Norwegian Channel. G) Crags and tails formed by glaciers moving toward the southwest. H). Flutes, iceberg scour and pockmarks. The iceberg scour marks cut the fluted surface. (Adapted from Longva and Thorsnes, 1997).

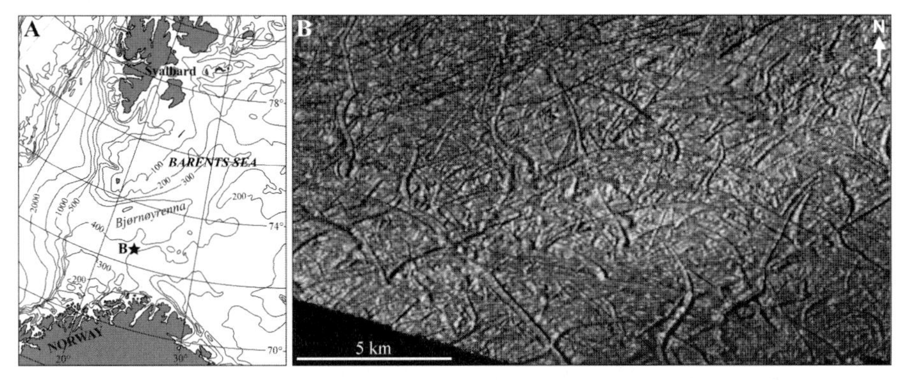

Figure 12.11 Heavily iceberg-ploughed sea floor in the Barents Sea at about 72° 30′ N and 23° 27′ E. The plough marks in this area are 30–500 m wide, 1–20 km long and 2.5–25 m deep. The figure is an illuminated time-structure map based on 3D seismic data where the light source is located east of the horizon. (Adapted from Rafaelsen *et al.*, in press).

12.1.3.6 Glacitectonic Forms

Sættem (1990, 1991) observed several types of glacitectonic phenomena on the Norwegian continental shelf. He concluded that glacitectonic deformation of bedrock strata occurs widely. In particular, he described glacitectonic hill-hole pairs (Fig. 12.13), earlier observed in the terrestrial environment. Hill-hole pairs are also observed in the Norwegian Channel (Longva and Thorsnes, 1997; Fig. 12.10). The inferred glacitectonic elements include both *in situ* deformation and large-scale displacement. Sættem (1990) concluded that glacitectonism has played an important role in the Cenozoic erosion, first by displacement of bedrock bodies or floes, resulting in instantaneous erosion of an amount depending on the size of displaced bodies, and second, by the transformation of displaced or *in situ* bedrock to glacitectonite and ultimately a till. It is theoretically elaborated that fluid overpressure and expulsion from the underlying bedrock may have played a role in the glacitectonic bedrock deformation.

12.2 THE CONTINENTAL SLOPE AND ADJOINING DEEP-SEA SYSTEM

The main large-scale morphological elements on continental slopes are trough mouth fans, gullies and channels, and large submarine slide scars and corresponding accumulations (Fig. 12.8). Some gullies and channels may be formed directly in association with continental ice sheets (e.g. Vorren *et al.*, 1998), but many of these and the slides do not. Thus, they will not be treated here. Here we will discuss the main glacigenic depo-centres on the continental slope, the trough mouth fans, and deep-sea drifts and fans directly related to glacigenic sediment input.

12.2.1 Trough Mouth Fans

12.2.1.1 Distribution of Trough Mouth Fans

On glaciated continental margins, fan or delta-like protrusions occur in front of many glacial troughs or channels crossing the continental shelf and ending on the shelf break (Figs 12.5, 12.6, 12.7 and 12.8). Many of these protrusions surrounding the Norwegian-Greenland Sea

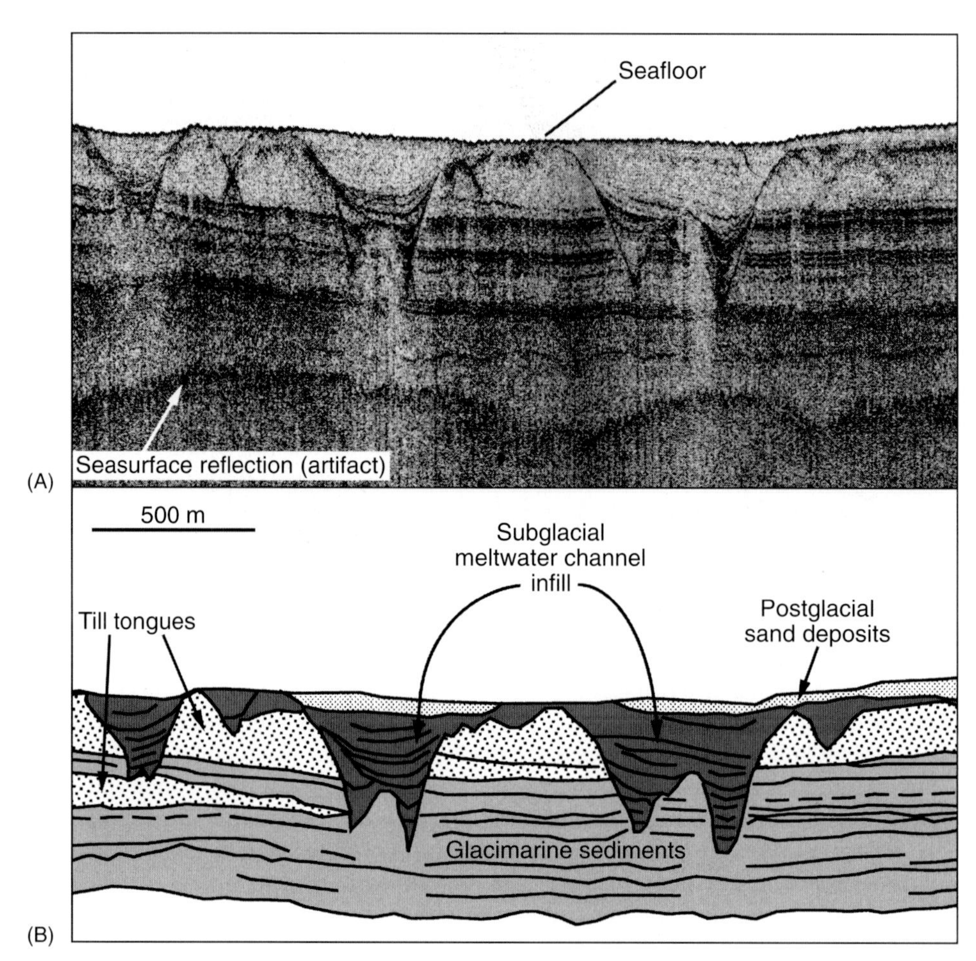

Figure 12.12 A) Huntec DTS high-resolution seismic reflection profile from the Halibut Channel, Grand Banks of Newfoundland. B) Geoseismic interpretation of A. Till tongues (wedge-shaped units of incoherent reflections) are interbedded with glacimarine sediments. The upper till tongue has been cut by a series of interpreted subglacial meltwater channels which have fragmented the till tongue into a series of erosional remnants. (Adapted from Moran and Fader (1997).)

were noted by Nansen (1904). Vogt and Perry (1978) pointed out that these protrusions are probably prograded deltas and attached fans. Vorren *et al.* (1988, 1989) proposed naming these features, which also occur on other glaciated margins, 'trough mouth fans' (TMF). Not all troughs ending on the shelf break have fans at their mouths, but on the eastern margin of the Norwegian-Greenland Sea they are particularly numerous (Fig. 12.14). On the Greenland continental margin the Scoresby Sund TMF is well developed (Dowdeswell *et al.*, 1997) and other fans around Greenland are identified by Funder (1989; Fig. 12.3). Although not originally denoted as TMFs, similar accumulations are described from the eastern continental margin of Canada (Aksu and Hiscott, 1992; Hiscott and Aksu, 1994, 1996) and from the continental margin off northwest Britain (Stoker, 1995). Bathymetric features indicate two TMFs in the eastern Arctic Ocean, namely in front of the St Anna Trough and Franz Victoria

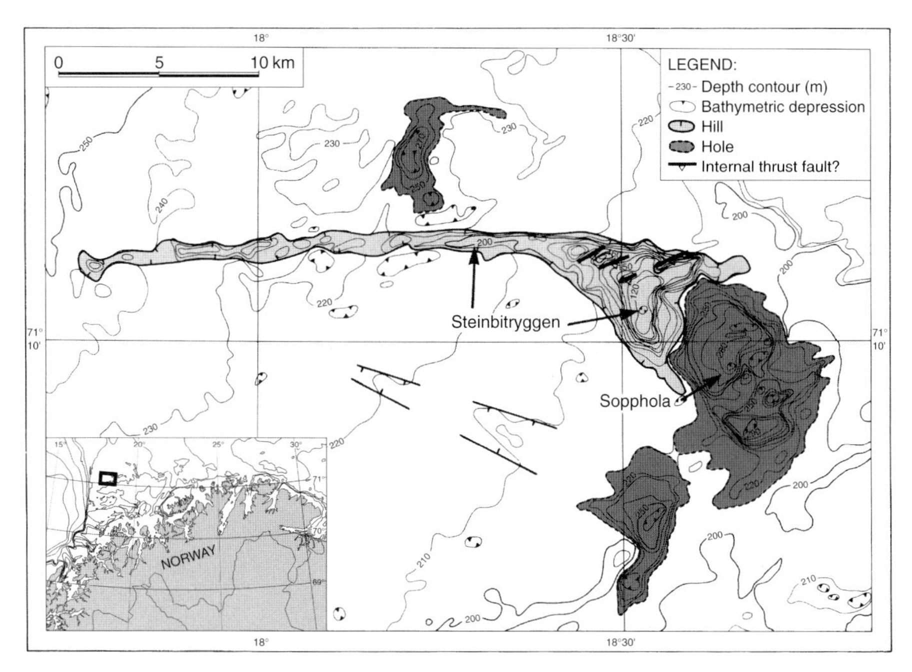

Figure 12.13 Detailed bathymetric map of an area of the southwestern Barents Sea showing a hill-hole pair. The hill (Steinbitryggen) is interpreted to consist of sediments tectonically displaced from the depression (Sopphola). The volume of Sopphola is equal to that of the ridge, which is partly shaped into a westerly trending streamlined form. Possible bathymetric expressions of boundaries between individual thrust masses is shown. Isolated depressions and a possible remnant of a glacitectonic hill southwest of Sopphola are also indicated. (Redrawn from Sættem, 1990).

Trough. TMFs in Antarctica include the Crary Fan in the Wedell Sea (Kuvaas and Kristoffersen, 1991; Batist *et al.*, 1997), the Prydz Bay fan (Hambrey, 1991) in the western Ross Sea, in the Bransfield Basin, and on the Wilkes Land continental margin (Anderson, 1999).

12.2.1.2 Morphology and Architecture of Trough Mouth Fans

The TMFs vary in size and shape. In the North Atlantic-Nordic Seas the smallest are found in the north off the archipelago of Svalbard and in the south off the British Isles. These are several orders of magnitude smaller than the largest, the Bear Island TMF (Table 12.1). The smallest TMFs have in general the steepest slopes (Table 12.1).

At least during the Late Quaternary, TMFs have been the sites of intense debris flow activity. Damuth (1978) was the first to indicate their presence. Vorren *et al.* (1988, 1989) found that the debris flows, seen in cross section, are bundled in sets of lenses separated by high-amplitude reflections. The middle part of the fans is dominated by a mounded seismic signature in cross section (Fig. 12.15), representing sections through debris flows. The debris flows are deposited in bathymetric lows between older deposits. The flows end on the lower fan. The high-amplitude reflections between each lens-set probably represent periods of low sediment

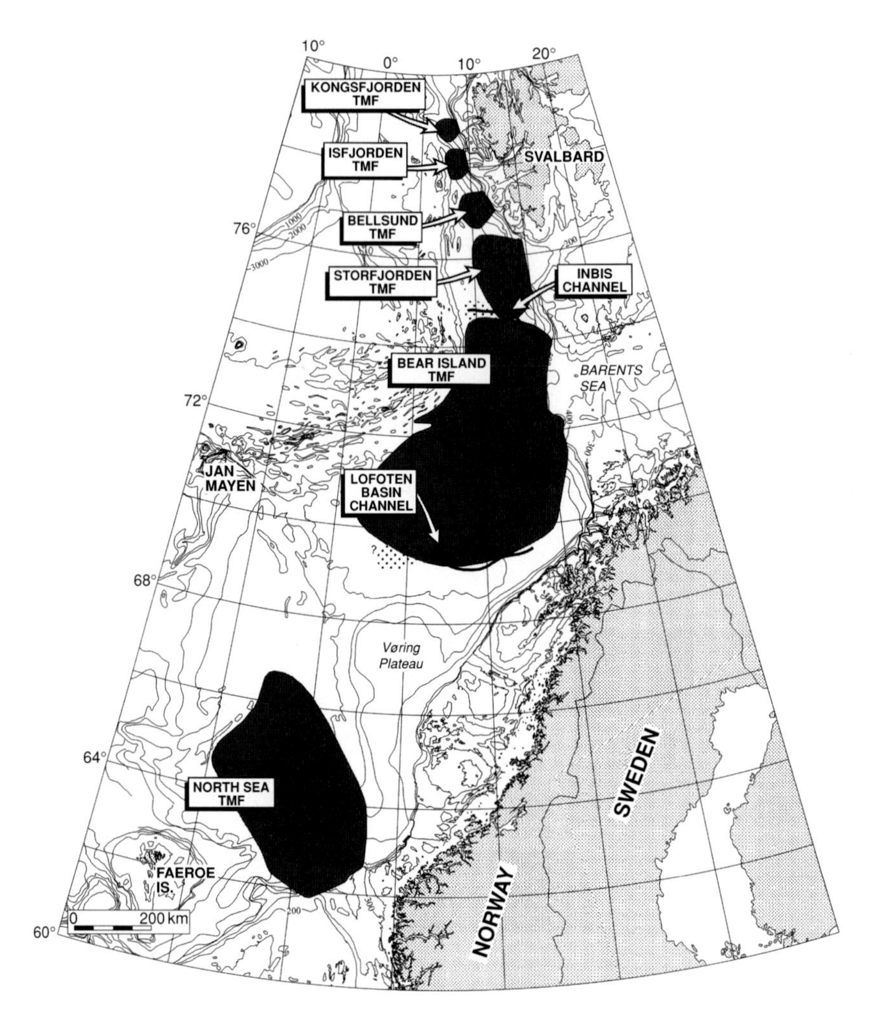

Figure 12.14 Bathymetric map showing the location and extent of trough mouth fans in the Norwegian Sea. (Adapted from Vorren *et al.*, 1978).

input/erosion during interstadials or interglacials. Later mapping by seismic and by side-scan sonars has confirmed that the debris flows are the main building blocks of the younger part of the TMFs in the Norwegian-Greenland Sea (Vogt *et al.*, 1993; Laberg and Vorren, 1995, 1996a, b; King *et al.*, 1996; Sejrup *et al.*, 1996; Dowdeswell *et al.*, 1996, 1997). Similar flows have been reported from the continental margin off northwest Britain (Stoker, 1995) and the eastern Canadian continental margin (Aksu and Hiscott, 1992; Hiscott and Aksu, 1994).

Vanneste (1995) suggested three basic types of TMF:

1. mostly stable TMFs characterized by the absence of large-scale mass-wasting deposits (e.g. Scoresby Sund TMF)
2. unstable TMFs characterized by the presence of large-scale mass-wasting deposits (e.g. the Bear Island TMF)
3. TMFs associated with deep sea-fan systems in their distal parts (e.g. Crary TMF).

	TMF						
	Kongsfjorden	**Isfjorden**	**Bellsund**	**Storfjorden**	**Bear Island**	**North Sea**	**Sula Sgeir**
Radius (km)	55	50	70	190	590	560	50
Width upper (km)	40	45	55	130	250	165	70
Width lower (km)	60	75	85	210	550	300	85
Depth upper (km)	0.2	0.25	0.15	0.4	0.5	0.4	0.2
Depth lower prox. (km)				2.4	3.0	2.7	1.3
Depth lower distal (km)	2.0	3.0	2.3	2.7	3.2	3.5	1.3
Area (km^2)	2,700	3,700	6,000	35,000	215,000	142,000	3,100
Gradient (upper)				1.8°	0.8°	0.6°	
Gradient (middle)	1.9°	3.2°	1.8	1.0°	0.4°	0.8°	1.3°
Gradient (lower)				0.2°	0.2°	0.3°	

Radius = radius along the longest axis; width upper = width at shelfbreak; width lower = maximum width of the lower fan; depth upper = depth at the shelf break; depth lower prox. = depth at the base of the proximal part of the fan; depth lower distal = depth of the base of the distal part of the fan; area = total fan area.
Gradients of the upper, middle and lower slope along the longest fan axis of the Storfjorden and Bear Island TMF according to Laberg and Vorren (1996b) and according to King et al. (1996) for the North Sea TMF. Average fan gradient is given for the Kongsfjord TMF, the Isfjord TMF, the Bellsund TMF and the Sula Sgeir TMF. (After Vorren and Laberg, 1997).

Table 12.1 Size and shape of most of the northwest European trough mouth fans (TMFs).

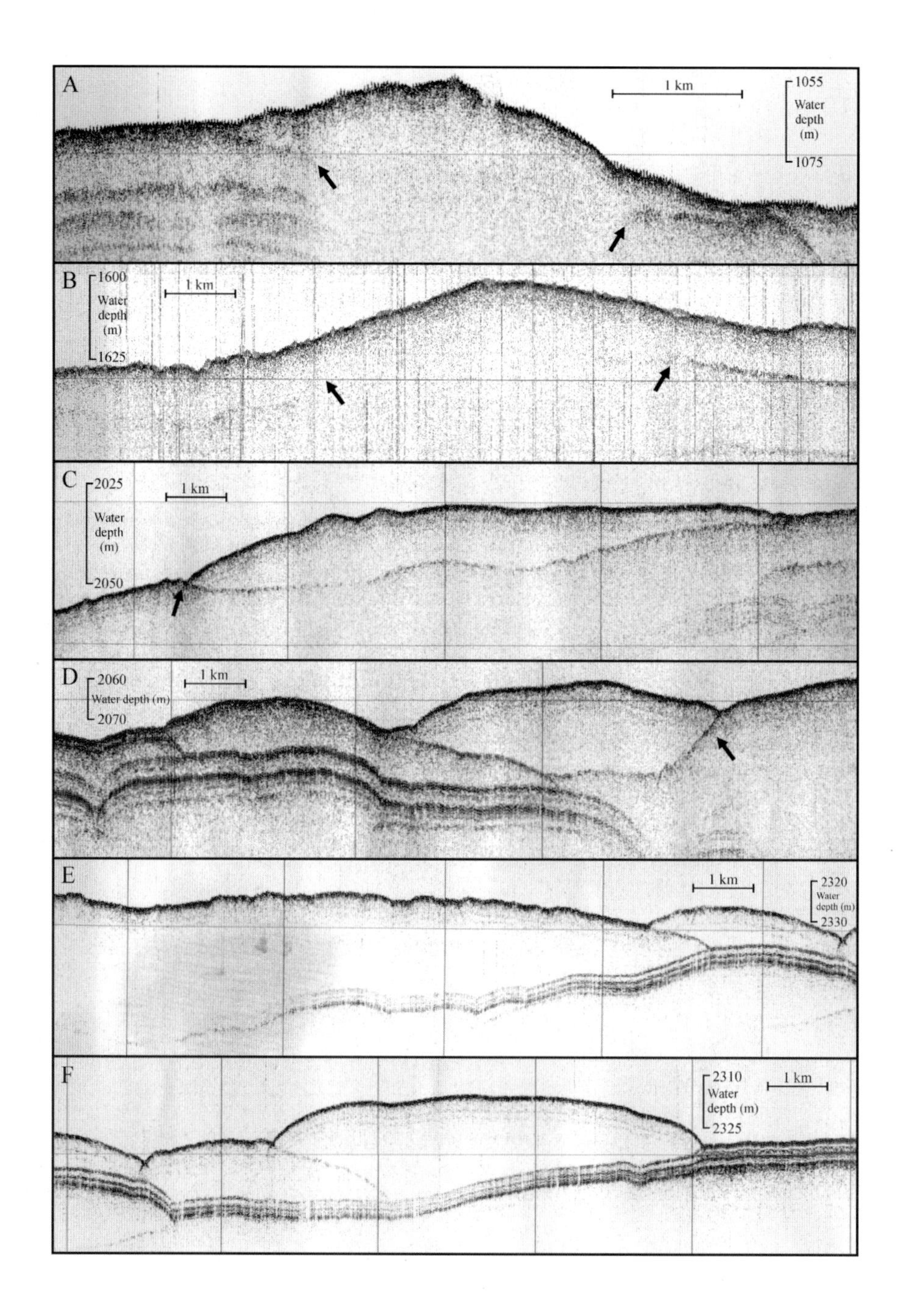
A
1 km
1055
Water
depth
(m)
1075
B
1600
Water
depth
(m)
1625
1 km
C
2025
Water
depth
(m)
2050
1 km
D
2060
Water depth (m)
2070
1 km
E
1 km
2320
Water
depth (m)
2330
F
2310
Water
depth (m)
2325
1 km

12.2.1.3 Debris Flows

King *et al.* (1998) suggested that the debris flows constituting the TMFs should be termed 'glacigenic debris flows' and Vogt *et al.* (1999) use the term 'glacigenic mudflows'. Here we will use glacimarine debris flows to make a distinction between glacigenic debris flows and mudflows that occur in the subaerial environment. The dimensions of the glacimarine debris flows vary: between 0.5 km and 40 km wide, between 5 and 60 m thick, from less than 10 km up to 200 km long, covering areas up to 1880 km^2, and with volumes from 0.5 to 50 km^3. There is a clear tendency towards larger fans having the larger and more voluminous debris flows.

The debris flow sediment is a homogeneous diamicton (Laberg and Vorren, 1995; King *et al.*, 1996, 1998; Elverhøi *et al.*, 1997; Vorren *et al.*, 1998). It contains 30–55 per cent clay, 30–50 per cent silt, 10–30 per cent sand and usually less than 10 per cent gravel. The grain-size distribution and water content are compatible with the youngest till units on the outer shelf (e.g. King *et al.*, 1998; Vorren *et al.*, 1998).

Transport to and accumulation at the shelf break: A generally accepted model (Fig. 12.16) is that glacigenic sediments were brought to the grounding line as a deforming till layer (Boulton, 1979; Alley *et al.*, 1989). This probably resulted in a build up either of 'till-deltas', according to the model of Alley *et al.* (1989), or 'diamict aprons' (Hambrey *et al.*, 1992), or 'grounding line or zone wedges' (e.g. Powell and Alley, 1997; Anderson, 1997) along the glacier terminus. The glacigenic sediments could also have continued directly downslope. Sediments deposited in the till deltas are probably inherently unstable, and not well preserved on a sloping subsurface (Dimakis *et al.*, 2000). However, there are examples on the seismic records across the shelf break, which could be interpreted as till deltas (Vorren and Laberg, 1997).

Release factors: The morphology, showing slide scars on the present uppermost slope surface and a chaotic seismic facies on the upper fan, indicates that several sediment slides were released near the shelf break (Laberg and Vorren, 1995). The slides may have been triggered by:

1. build up of excess pore pressure due to high sediment input (Dimakis *et al.*, 2000)
2. earthquakes
3. oversteepening

Figure 12.15 Segments of 3.5 kHz profiles across debris flow deposits on the Bear Island TMF. A) Profile across a debris flow deposit on the upper fan. Here the debris flows are identified from their surface relief. The reflection defining the base disappears immediately beneath the flow deposit (arrows). B) Profile across a debris flow deposit on the middle fan. In this area the debris flows are identified from their surface relief. The reflection defining the base disappears immediately beneath the flow deposit (arrows). C) Profile from about 2000 m water depth. Here the base reflection is clearly seen and the debris flow surface sometimes mirrors highs at the base. D) Profile from the middle fan. The deposit is characterized by a positive relief and the underlying deposits seem to be unaffected by the younger flow. E) Profile from the lower fan illustrating a slightly irregular surface relief. The underlying acoustic parallel unit seems to have been left unaffected by the deposition of the flow. F) Profile from the lower fan illustrating a relatively smooth surface relief. The underlying acoustic parallel unit seems to have been left unaffected by the deposition of the flow. (After Laberg and Vorren, 2000).

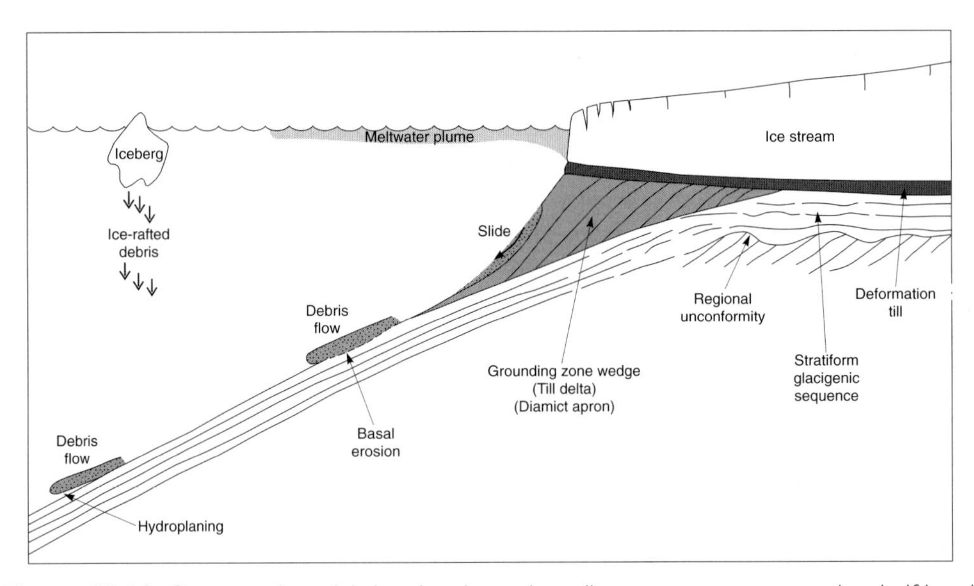

Figure 12.16 Conceptual model showing the main sedimentary processes on the shelf break and upper slope during the presence of an ice stream at the shelf break.

4. ice loading (e.g. Mulder and Moran, 1995), and/or
5. seepage of shallow gas.

The repeated release of slides on a regional scale suggests a trigger mechanism common to these particular settings. Thus, most likely the slides were released by build up of excess pore pressure and oversteepening. Dimakis *et al.* (2000) has calculated that after 95–170 years of high sedimentation rates, failures will take place that will remove the top 10–30 m of the deposited sediments.

Flow behaviour: The debris flows move downslope following bathymetric lows between older deposits (Aksu and Hiscott, 1992; Laberg and Vorren, 1995). Flows that move further downslope than their forerunners spread out laterally, resulting in a width and thickness increase on the lower fan. Generally, the debris flows containing the largest sediment volume have the longest run out distance (Laberg and Vorren, 1995).

Many of the observed debris flows have a large run out distance on low-gradient slopes, particularly on the Bear Island and North Sea TMFs. This indicates low viscosity behaviour. The mobility of debris flows involves an important contribution from excess pore fluid in allowing long run out distances.

Laberg and Vorren (2000) have shown that from at least 1600 m water depth debris flows erode and probably incorporate substrate debris, but further downslope they move passively over substrate sediments. The theory of hydroplaning of the debris flow front may explain why the debris flows moved across the lower fan without affecting the underlying sediments. When hydroplaning is established, the moving debris flow head is substantially decoupled from its bed and, as shown in experiments, runout distance and head velocity become independent of debris flow rheology (Mohrig *et al.*, 1998; Elverhøi *et al.*, 2000).

12.2.2 Deep-Sea Drifts and Fans

The Pacific continental shelf of the Antarctic Peninsula has typical glacial morphology, the main morphological elements being represented by banks separated by troughs. The continental slope is steep. On the continental rise are elongated sedimentary mounds bounded by erosional channels. The sedimentary mounds are sediment drifts. In contrast to trough mouths, elsewhere there is no morphological expression of fan growth on the Antarctic Peninsula slope. Rebesco *et al.* (1998) explain this by sediments supplied to the trough mouth having low shear strength. They therefore slide down the relatively steep continental slope and feed the turbidity current channels and drift systems (Fig. 12.17). The primary source of drift sediments is, therefore, thought to be from turbidity currents travelling in the channels that traverse the continental rise in deep areas between drifts. The fine-grained components of turbidity currents are entrained in a nepheloid layer within the ambient bottom currents and are then redeposited by those currents in response to sea floor topographic control. Thus, much of the glacial sediment transported to the shelf break accumulates in drifts and deep-sea fans rather than in TMFs in this setting. Also in other steep continental slope areas, glacigenic sediments transported through turbidity channels have been shown to accumulate in deep-sea fans (e.g. Vorren *et al.*, 1998; Anderson, 1999).

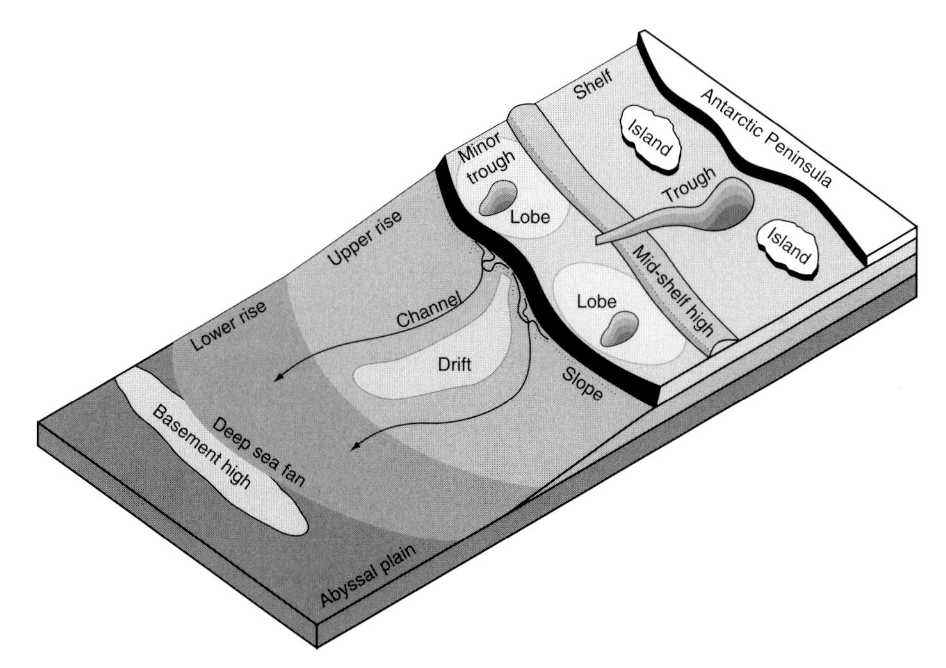

Figure 12.17 Synthetic schematic model showing the spatial relationship of the main physiographic elements on the continental margin west of the Antarctic Peninsula. Large glacial troughs traverse the continental shelf from the areas between the main islands, structurally guided towards the topographic lows of the mid-shelf high. Large prograding wedges develop on the outer shelf beyond the topographic highs of the mid-shelf high, next to the outward continuation of the major glacial troughs. Giant sediment drifts separated by large channel systems are present on the upper continental rise in between the prograding wedges. A large deep-sea fan, ponded against an oceanic basement high, is present on the lower continental rise beyond the lowermost reaches of the drifts. (After Rebesco *et al.*, 1998).

12.3 SUMMARY

- The first-order morphological elements on glaciated continental shelves are banks and longitudinal and transverse troughs. Many longitudinal troughs are located on bedrock boundaries or fracture zones, whereas transverse troughs often represent seaward continuations of fjords and glacial valleys on the hinterland. Most troughs have served as drainage routes for ice streams.
- The continental shelves are normally covered by less than 300 m of stratiform diamictons. The greatest thickness of glacigenic sediments occurs at the trough mouths.
- Many second-order morphological elements (e.g. morainal ridges/banks, streamlined forms, glacitectonic forms) are similar to the forms found in the terrestrial environment. Iceberg plough marks (normally at depths of less than 500 m, but locally up to 850 m depth) and iceberg turbate is typical for glaciated continental margins.
- TMFs are submarine fans at the mouths of troughs on presently or formerly glaciated continental shelves. TMFs are depo-centres containing kilometre-thick packages of sediments accumulated in front of ice streams draining ice sheets and ice caps. There is clearly reason to believe that many of the ancient diamicton records were deposited in a continental margin setting, particularly in TMFs.
- The main building blocks of TMFs are glacimarine debris flow deposits that might be as large as 2000 km^3 and have a runout distance of up to 200 km. The debris flows were remobilized sediments from till deltas/grounding-zone wedges deposited by ice streams at the shelf break. The long runout distance might be due to hydroplaning.
- On steep continental slopes the glacigenic sediments might be transported directly to the deep sea by turbidity currents through channels and accumulate in deep-sea fans and/or drifts.

CHAPTER

SUBAQUATIC LANDSYSTEMS: FJORDS

Ross D. Powell

13.1 INTRODUCTION

Fjords occur along many high-latitude coastlines today including those of Alaska, Antarctica (the Peninsula and along mainland coasts), Canada (British Columbia, Labrador, Newfoundland, Quebec, Baffin Island and Queen Elizabeth Islands), Chile, Greenland, Iceland, New Zealand, Norway (the mainland and Svalbard), Russia (the arctic and Kamchatka) and Scotland. Although the term 'fjord' is a generic Nordic name for a marine inlet, they are most commonly attributed to having a glacial erosional origin, being formed as land-based glaciers expand through fault-controlled or fluvially eroded troughs. Some fjords are thought to have existed for tens of millions of years since at least the Miocene; for example, Ferrar Valley (Barrett and Hambrey, 1992), Taylor Valley (McKelvey, 1981; Powell, 1981b; Hambrey *et al.*, 1989), Mackay Valley (Powell *et al.*, 1998; 2000), Beardmore Paleofjord (Webb *et al.*, 1996), Reedy Valley (Wilson *et al.*, 1998) and Lambert Graben (Hambrey and McKelvey, 2000) in Antarctica, and Ellesmere Island fjords in the Arctic (England, 2000; Ó Cofaigh *et al.*, 2000). Given either these long periods of time or very efficient erosion, fjords range from a few kilometres to several tens of kilometres wide and can be hundreds of kilometres long. Often there are entrance sills or thresholds of bedrock, which also may occur along the fjord axis. Basins separated by these sills may reach depths several hundred metres below sea level. Additionally, fjords are often within tectonically active mountain systems that can reach 5000–6000 m above sea level; therefore, fjord coastlines commonly exhibit some of the greatest relief in the world. Oceanographically, fjords are termed estuaries having some unique characteristics in terms of oceanic processes and characteristics. Understanding these oceanic processes, in addition to the glacial and fluvial processes, is important for interpreting fjord landsystems, and although space does not allow for detailed discussions here, a comprehensive review has been provided by Syvitski *et al.* (1987), with a more recent short update by Syvtiski and Shaw (1995). As can be discerned from their locations listed above, modern fjords occur under different climatic regimes ranging from temperate through sub-polar to polar (Table 13.1). Different landsystem models have been proposed for these differing conditions (Fig. 13.1), the details of which will be elaborated on through this chapter.

Climatic zone	Glacial flow velocity	Internal ice condition	Bed condition	Subglacial water free	Glacier terminus	Sediment contributions*										Modern examples
						Glacial		Glacifluvial		Marine			Terrestrial			
						Sub-	En-/supra-	Sub-	En-/supra-	Icebergs	Sea ice	Biogenic	Fluvial	Mass flow	Wind	
Temperate	Fast	Temperate	Deforming till, local	Conduit flow	Tidewater cliff	2	2	5	1	2		1	3	3	1	Alaska, British Columbia, Chile
Sub-polar	Fast	Slightly cold	Deforming till	Conduit flow	Tidewater cliff	3	1	3	1	1	2	1	3	1	1	Svalbard, Canadian and Russian arctic
	Moderate	Cold	Mostly frozen, local till	None to minor conduit flow	Short floating tongue or tidewater cliff	2		2		1		2	1		1	Antarctic Peninsula
Polar	Fast	Cold	Deforming till	Local conduit thin film	Floating tongue	3	1	3		3	1	1	2	1	1	Greenland, Ellesmere Island, Baffin Island
	Moderate	Very cold	Deforming till	None	Floating tongue	3	1			1	1	1	1	1	3	Antarctica (Mackay)
	Slow	Very cold	Mostly frozen, some till	None	Floating tongue or tidewater cliff	1	1			1	1	1		1	3	Antarctica (Ferrar and Blue)

*Relative scale estimating importance of sediment source contributions to fjords: 5=high to 1=low

Table 13.1 Current data on major local controls on modern fjord landsystems (modified from Syvitski *et al.* (1987) and Hambrey (1994)).

13.2 SEDIMENTARY PROCESSES AND EXTERNAL FORCES

Fjords are complex systems influenced by a wide range of processes, those processes interacting to produce positive and negative feedbacks on each other and controlling depositional systems. The major processes include glacial, fluvial, oceanic, iceberg, sea ice, aeolian, subaerial mass flow and

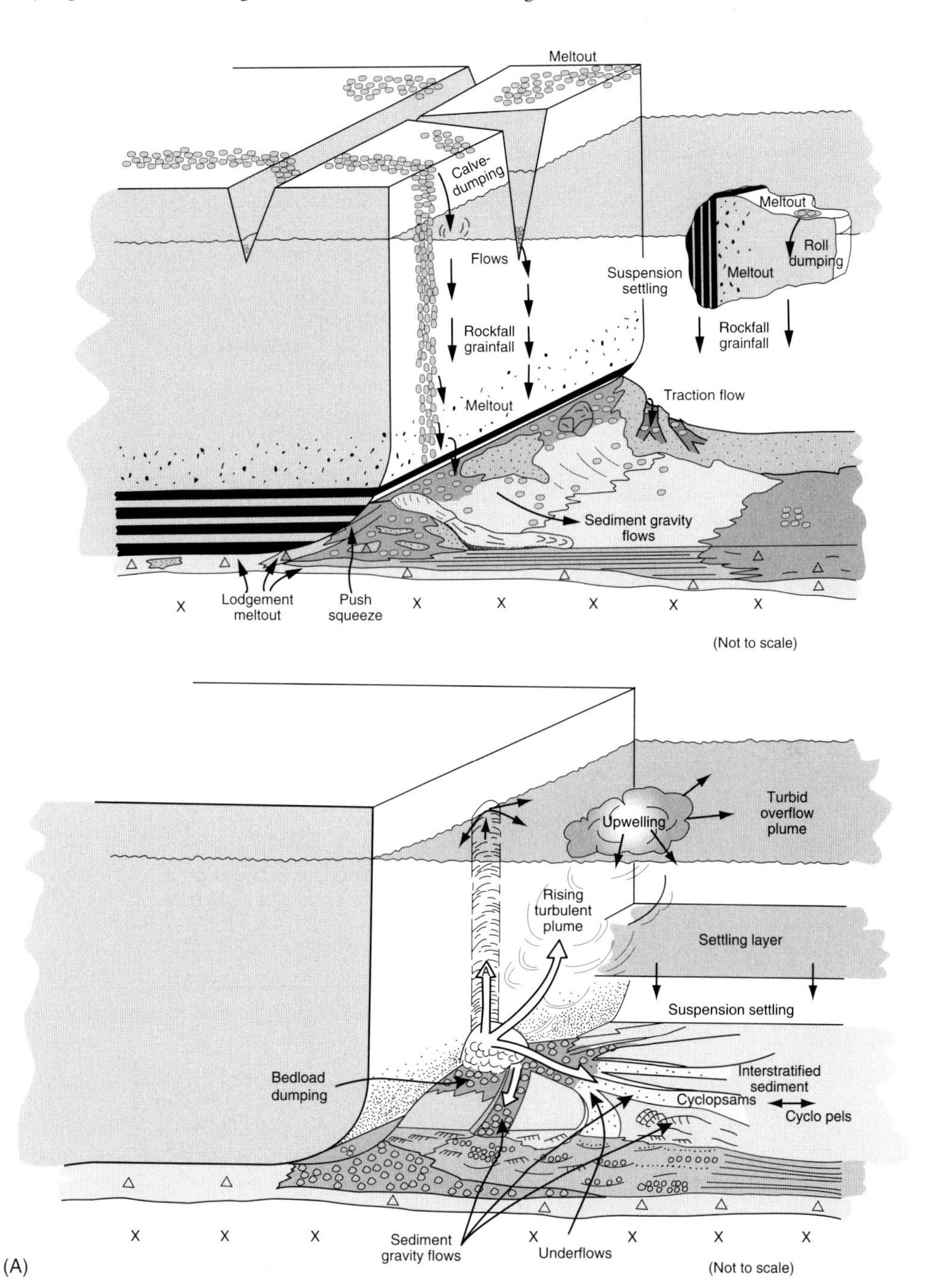

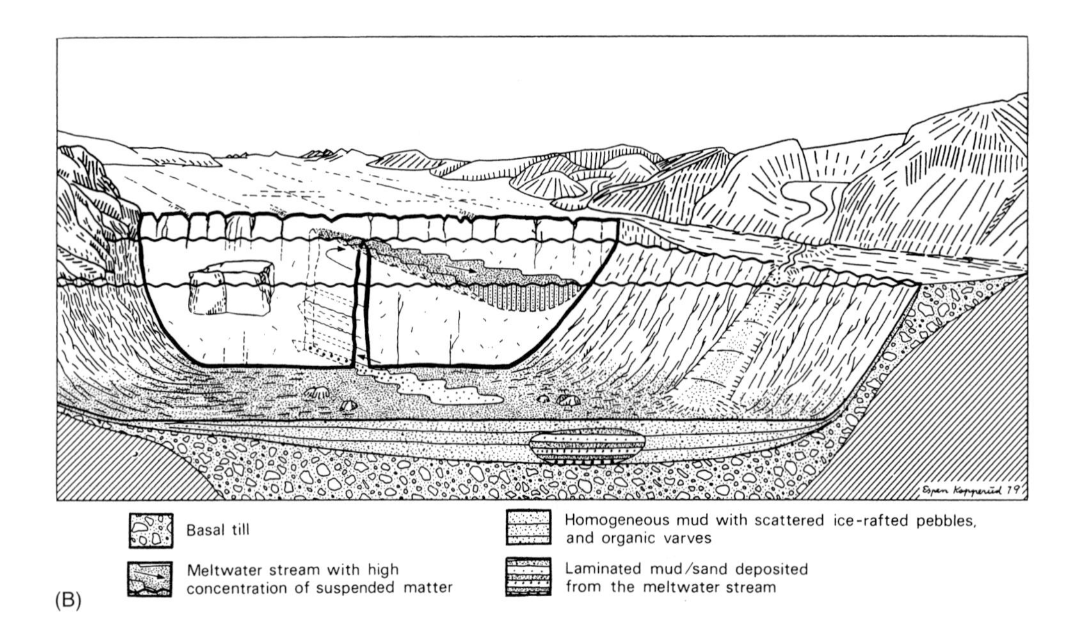
Basal till
Meltwater stream with high concentration of suspended matter
Homogeneous mud with scattered ice-rafted pebbles, and organic varves
Laminated mud/sand deposited from the meltwater stream

(B)

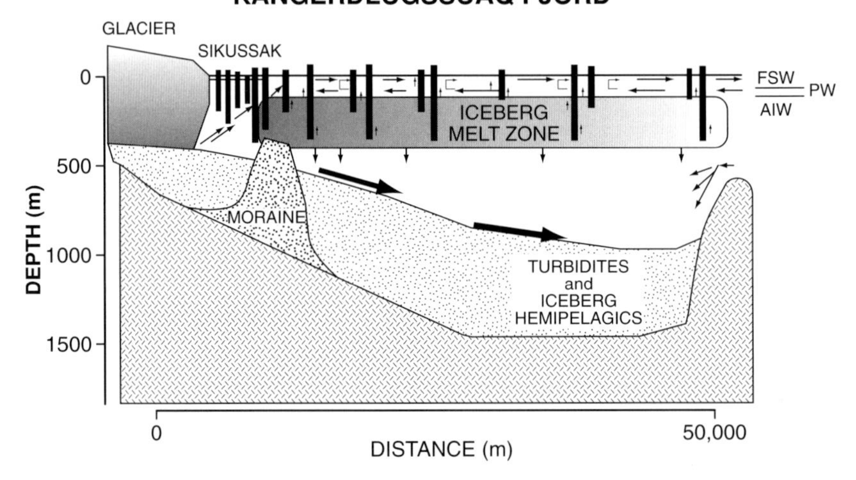
KANGERDLUGSSUAQ FJORD
GLACIER
SIKUSSAK
FSW
PW
AIW
ICEBERG MELT ZONE
MORAINE
TURBIDITES and ICEBERG HEMIPELAGICS
DEPTH (m)
0
500
1000
1500
DISTANCE (m)
0
50,000

1

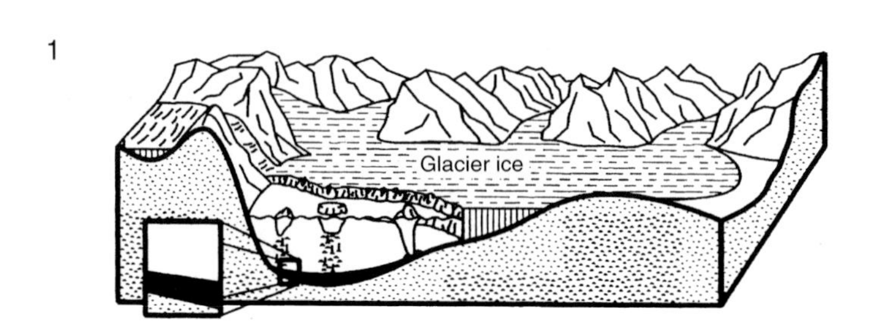
Glacier ice

(C)

(C) – *cont.* 2

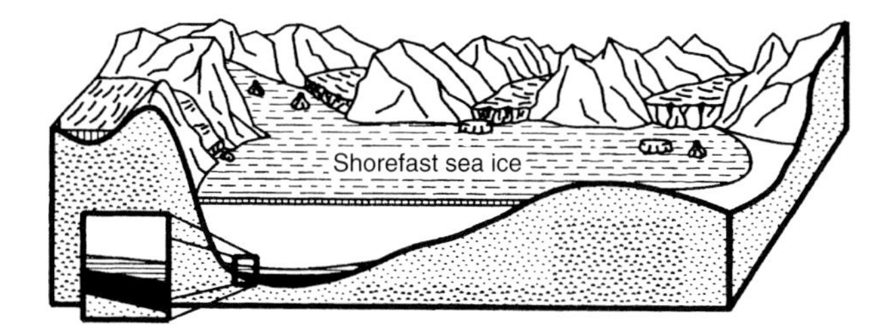

3

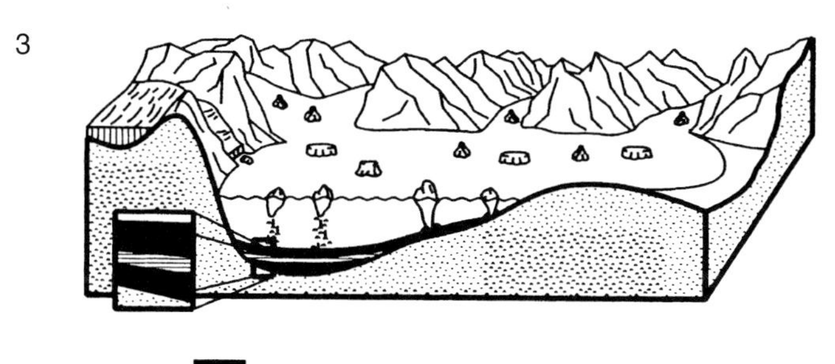

Massive, IRD-rich diamict

Laminated, IRD-poor mud

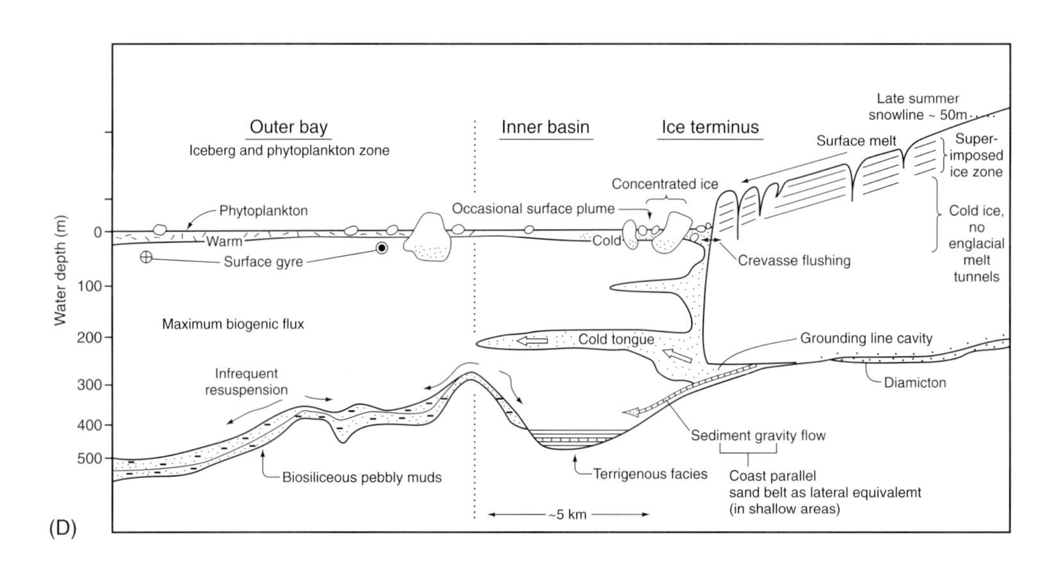

(D)

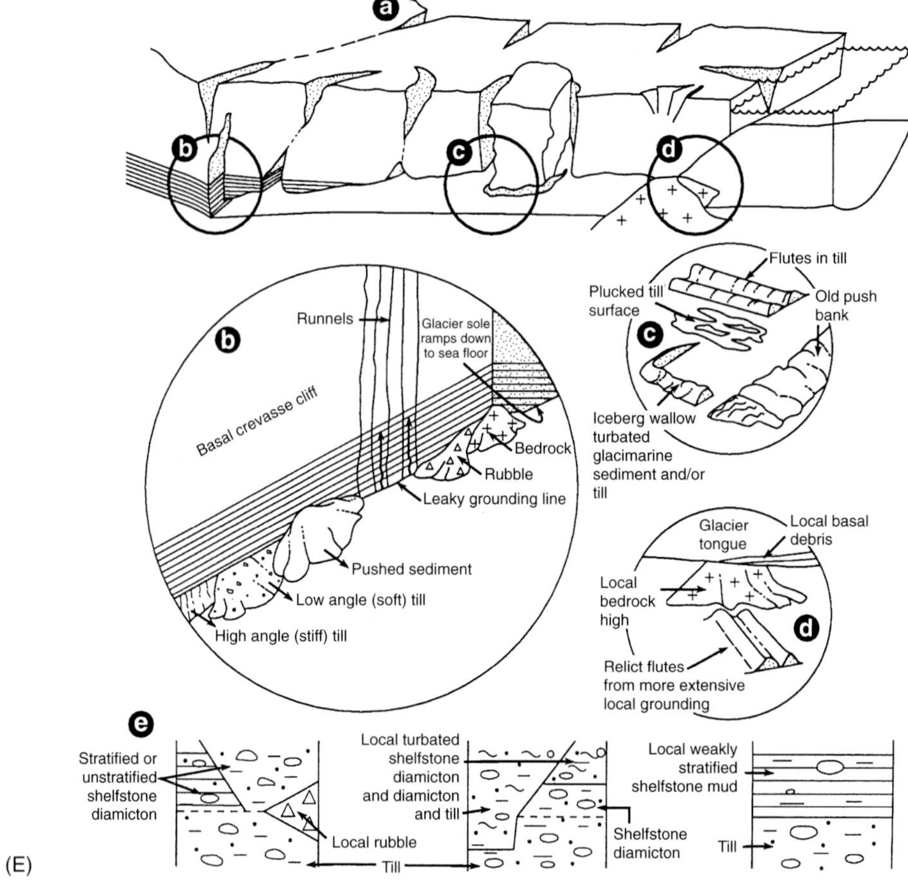

Figure 13.1 Sketches of processes and deposits of fjord landsystems under different climatic regimes; terms are defined and concepts are developed more fully throughout the chapter. A) Temperate glacimarine setting based on Alaskan examples (from Powell and Molnia, 1989). B) The warmer end of the sub-polar setting based on Svalbard examples. (After Elverhøi *et al.*, 1980). C) The cooler end of the sub-polar setting based on Greenland examples. (After Syvitski *et al.*, 1996; Ó Cofaigh and Dowdeswell, 2000). D) The warmer end of the polar setting based on Antarctic Peninsula examples. (After Domack and Ishman, 1993). E) The fully polar setting based on an East Antarctic example. (After Powell *et al.*, 1996).

biological (Fig. 13.2; Table 13.2). Before dealing with these processes and systems in detail we need to discuss controls external to the local fjord regime that influence sedimentary processes, facies geometries and stratigraphies (Fig. 13.2).

Climate, tectonics and sea level exert strong forces on the local processes that create and maintain fjord landsystems. Climate, primarily in the form of annual ranges of precipitation and temperature, as well as their rates of change, has a major influence on glacial regime (mass balance, dynamics, bed condition), bedrock weathering and vegetation all of which feed into erosion rates and sediment fluxes, sediment composition and chemical and carbon content of freshwater

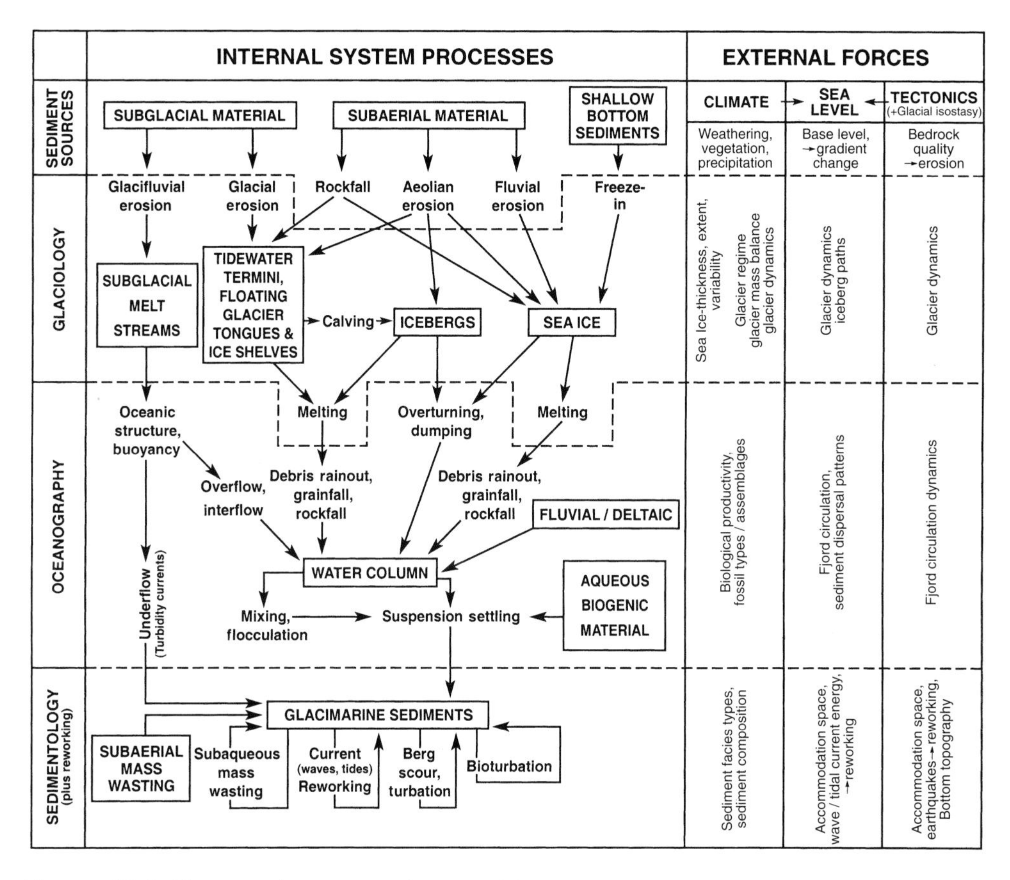

Figure 13.2 The complex system of external forces, sedimentary sources and transportation and depositional process in glacimarine systems. (Modified from Dowdeswell, 1987).

Depositional system	Fjord infilling phase
Bedrock walls and floor	Glacial
Subglacial systems	Glacial
Grounding-line systems	Glacial
Floating glacier-tongue systems	Glacial
Iceberg zone systems	Glacial
Open water systems	Glacial, paraglacial, non-glacial
Sidewall systems	All
Fjord-head systems	Paraglacial, non-glacial
see Glaciated Valley Landsystems	Fjord valley

Table 13.2 Major processes of depositional systems.

discharges into fjords. The thickness and extent of sea ice and its seasonal and inter-annual variability are important here, as well as the marine biology in terms of species, assemblages, productivity and their spatial and temporal variability. Ultimately, all of these determine the types, composition, geometries and sizes of depositional systems within fjords.

Fjords occur in the full range of tectonic settings from convergent to transform to divergent continental margins. Tectonics has a primary place in sediment production by influencing the bedrock quality relative to its glacial erodability. In active tectonic settings, tectonic uplift and depression are important in terms of snow accumulation and glacial dynamics, of fjord circulation in changing basinal geometries, and by providing the space in which the sediment can be accommodated.

Although internal forces such as sediment accumulation, drive changes in local water depth, changes in relative sea level and eustatic sea level primarily are the results of external forces: tectonics and global climate change, respectively. An exception to the external driving of relative sea level change is glacial isostasy, which plays an important role on facies sequences of shallow water depositional systems (e.g. Andrews, 1974; Bednarski, 1988; Boulton, 1990; Fig. 13.3). Such influences are primarily due to changing wave and tidal current energies that instigate sediment reworking. Another important consequence of fluctuating sea level is its changing of base level and gradients on land in terms of the sediment delivery systems to the fjord. Water

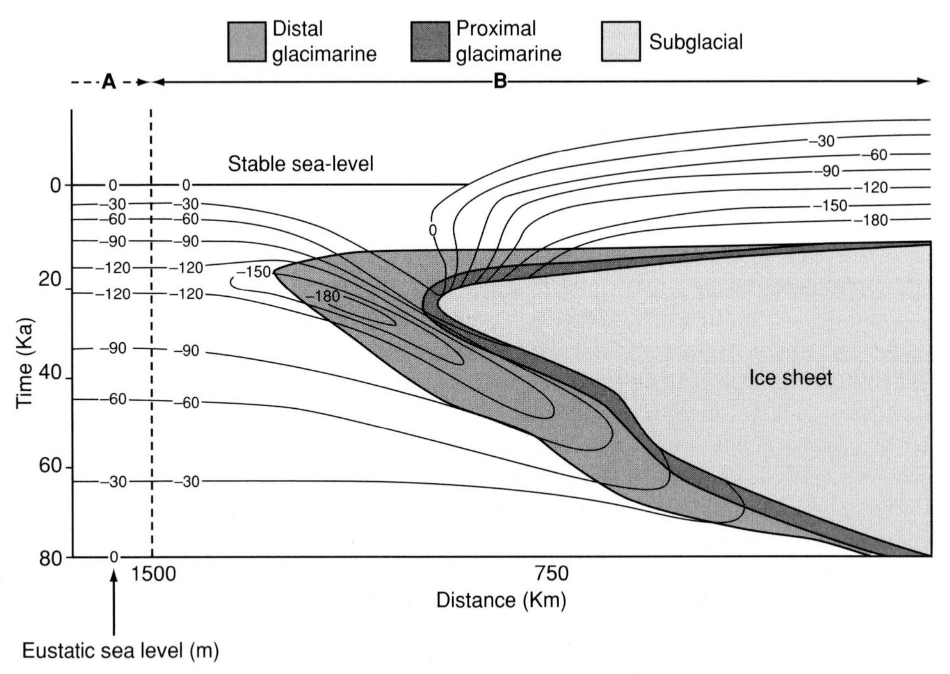

Figure 13.3 Model to show the effect of combining glacial eustasy and glacial isostasy relative to sea level changes around a glaciated continental margin shown in a spatial and temporal (0–80 ka) space. Variations in relative sea level and broad glacimarine landsystems at the margin of a continental ice sheet are shown with the ice sheet advancing and retreating from and to the right. (After Eyles and Eyles, 1992; who adapted the figure from Boulton, 1990).

depth has been found to be important in controlling calving speed of grounded glacial tidewater cliffs (Brown *et al.*, 1982; Pelto and Warren, 1991; Warren *et al.*, 1995), and although presently debated (van der Veen, 1996) has been inferred to influence the terminus position of glaciers ending in the sea. In fact, glaciers themselves may supply enough sediment to their terminus to decrease water depth there and help create their own stability (e.g. Powell, 1991; Fischer and Powell, 1998). Changes in water depth in a fjord also alter sediment accommodation space and change water masses and circulation patterns, which in turn can modify iceberg paths and sediment dispersal patterns.

13.3 GEOMORPHOLOGY AND DEPOSITIONAL SYSTEMS

The external forces on fjord landsystems described above tend to act over periods of several thousand to tens of thousands of years, but they remain significant when trying to understand the system over full glacial periods. Over the shorter term, local processes are the significant factors in generating a fjord landsystem record. Here we will discuss the major processes that go to producing different depositional systems, geomorphological units and allostratigraphic units within fjords, and describe each of the systems, their geometry and make-up relative to these processes. We will also deal with different climatic regimes in which fjords occur and compare and contrast their systems. First, however, we need to briefly mention erosion, as fjord systems are at least initially erosional at the large scale of the trough itself, commonly being formed by a combination of glacial, mass wasting and fluvial erosional processes.

13.3.1 Bedrock Walls and Floor

Bedrock composition and tectonic and climatic regimes influence degree of weathering and hence erodability. Some have argued that deep weathering is required for significant glacial erosion but that factor must be insignificant in the case of fjord overdeepening. However, soft bedrock does appear to be significant such as for the Lambert Graben, Antarctica, the world's largest (over 800 km long, 50 km wide and 3 km deep) glacierized fjord system (Hambrey, 1994, p. 90). Evidence of glacial erosion is seen in smaller-scale structures within fjord basins (e.g. hanging valleys, riegels, whalebacks, P-forms), but the specifics of subglacial erosional processes are poorly constrained simply due to the difficulty in observation and sampling. Erosion by abrasion, plucking and subglacial streams all occur and indeed must be active during the advanced glacial phases due to the common overdeepened profiles of fjords. Subglacial water pressure variations are probably critical for the overdeepening process (e.g. Hooke 1991; Iverson 1991; Hallet *et al.* 1996) even though erosion may be concentrated part-way up valley sides rather than at the bed (Harbor *et al.* 1988; Harbor 1992). Although bedrock erodability is a factor, overdeepenings which commonly end in bedrock sills appear to occur where ice velocity is high and convergent, such as at narrowings or trough confluences, and sills occur where ice flow becomes divergent where the glacier spreads from confined flow. Furthermore, if glacial erosion rates are on the same scale as sliding velocity, and sliding velocity reaches a maximum at the equilibrium line altitude (ELA; Andrews, 1972; Humphrey and Raymond, 1994; Paterson, 1994), Meigs and Sauber (2000) argue that the locus of primary bedrock erosion by glaciers should oscillate in synchrony with altitudinal shifts of the ELA during climatic fluctuations (Fig. 13.4). Fjord troughs may well be areas where the mean ELA oscillates for long periods of time to assist erosion there. Furthermore, Meigs and Sauber (2000) have argued that glacial erosion rates may well match rock influx/uplift rates and that erosion acts to

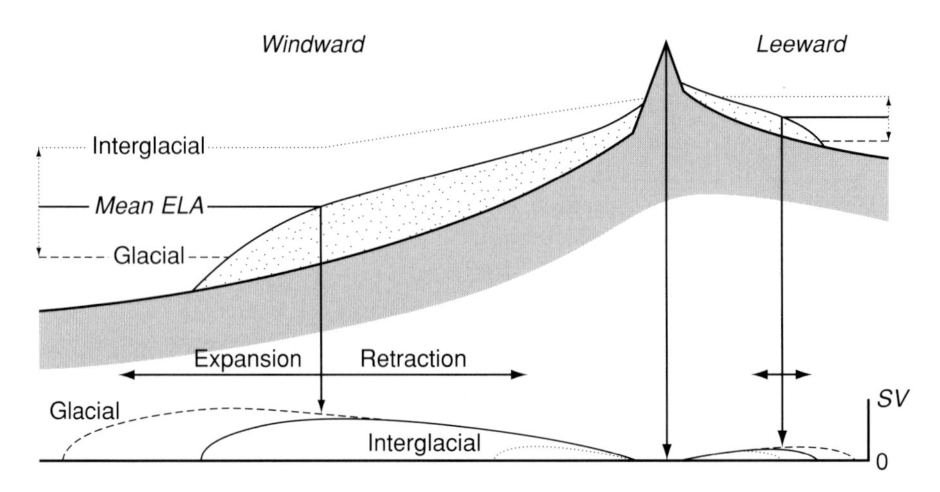

Figure 13.4 Model illustrating the relationship between sliding velocity (SV and small arrows at the base of the windward glacier) and the equilibrium line altitude (ELA) across an orogenic belt. An orographically induced rise in the ELA from the windward to the leeward side of the range leads to a higher mean ELA (solid black line). Smaller amplitude fluctuations in the ELA between glacial (dashed line) and interglacial (grey line) is shown schematically (the leeward range is ~33% of the windward range). Assuming bedrock erosion rate scales with basal sliding velocity, the model suggests concentration of erosion in a topographic band whose height is dictated by glacial/interglacial altitude limits to the ELA and whose width is a function of the concomitant glacial expansion/retraction in the landscape. The windward band width and height are likely to be greater than those of the leeward flank. Note that the range crest is defined by a topographic peak that corresponds spatially with a zone of low erosion rate by glaciers emphasized by the vertical downward arrow. (After Meigs and Sauber, 2000).

modulate the locus of uplift, at least in southern Alaska. They argue that positive feedback between erosion and uplift is manifest in the local topography where mean elevation reaches a maximum within the area where they infer high values of uplift and exhumation rates.

During glacial maxima, fjords are dominantly under erosional or non-depositional regimes, and sediment deposited during the glacial advance through the fjord as well as any deposits from previous cycles may be eroded out, perhaps down to bedrock (e.g. Andrews, 1987, 1990; Boulton, 1990; Powell, 1991). The degree of erosion, however, may depend on the glacial regime; sediment appears to be preserved from several glacial episodes in sub-polar (Andrews *et al.*, 1987, 1990) and polar fjords (e.g. McKelvey, 1981; Powell, 1981b; Barrett and Hambrey, 1992). When glaciers do erode almost all previously deposited sediment down to bedrock, large volumes of that sediment may be excavated. For example, in Glacier Bay, Alaska, much of the fjord system appears to have been filled with sediment, including outwash with buried trees and lacustrine deposits, up to about 200 m above present sea level after retreat from the Last Glacial Maximum (McKenzie and Goldthwait, 1971; Goodwin, 1988). During an advance of over 70 km during the Little Ice Age all of that sediment was eroded out down to bedrock which locally is over 300 m below sea level (Seramur *et al.*, 1997). The volume of sediment removed would probably have been in excess of about 10^{10} m^3 over approximately 850 years.

13.3.2 Subglacial Systems

The best ways to investigate subglacial systems of fjords is to look at overturned icebergs, interpret seismic reflection profiles, use ice radar to investigate bed characteristics, use submersible remotely operated vehicles to investigate the margins of glaciers where they end on the sea floor and also newly exposed sea floor from glacial retreat, and to study raised Quaternary deposits. Each method has its problems in terms of inferring former processes, but, although the glacier bed is below sea level in most areas of a fjord, processes and resulting sediments appear to be the same as for terrestrial glaciers (e.g. Glasser and Hambrey, 2001a). Subglacial till deformation occurs as inferred from overturned icebergs in temperate systems (cf. Powell and Molnia, 1989; Fig. 13.5) and from overturned icebergs (Glasser and Hambrey, 2001b) and terrestrially exposed deposits (Boulton *et al.*, 1999; Bennett *et al.*, 1996c) in sub-polar systems, and patterned ground, cored sediment strength (Solheim, 1991) and fluted till on recently exposed sea floor (Fig. 13.6; Powell *et al.*, 1996) in polar systems. Lodgement of till is also inferred to occur. However, subglacial melt-out must be rare because marine-ending glaciers rarely stagnate (Powell, 1984). Furthermore, the common terrestrial sequence of subglacial to englacial and/or supraglacial till facies does not occur for the same reason, and subglacial deposits are overlain by marine sediment. The till occurs in sheet or quasi-sheet geometries but its distribution appears to vary from temperate systems where it is patchy (Cai *et al.*, 1997) to being more continuous and extensive for sub-polar and polar systems (Elverhøi *et al.*, 1980, 1983; Powell *et al.*, 1996; Anderson, 1999; Domack *et al.*, 1999; Shipp *et al.*, 1999). These different styles of till distribution may well be a factor of subglacial water conditions in terms of volume and flow geometries. In glacial systems of temperate fjords conduit flow becomes well established during a melt season, and even though flow appears to decrease or stop during winters (Cowan and Powell, 1991a and b) it is commonly re-established in the same location the following season (Powell, 1990). This is not inconsistent with models which indicate that deforming till and low-pressure channels can co-exist (Alley, 1992), and perhaps the channels are not fully closed by ice or till deformation during winter as a result of being flushed with sea water. Whether marine-ending glaciers have similar debris-incorporation processes to the supercooling freeze-on of terrestrial glaciers (Lawson *et al.*, 1998; Alley *et al.*, 1998) is uncertain; however, the required overdeepened bed geometry is present in fjords and is a possible factor to consider in discharges at marine glacial termini (Powell, 1990; Dowdeswell *et al.*, 2001). In these overdeepened scenarios, a stream is unlikely to be sufficiently competent to transport coarse bedload up the back slope of the grounding-line fan that it's building. This forces coarse sediment under the glacier and on the up-glacier side of the fan.

13.3.3 Grounding-Line Systems

Sedimentary systems that are deposited in ice contact at grounding lines of marine-ending glaciers have been defined as grounding-line systems and may extend a few kilometres away from a grounding line. Geometries of the deposits differ according to whether a glacier is advancing or standing still, where sediment accumulates in bank, fan or wedge geometries, as opposed to retreating rapidly where sediment is spread in more of a sheet geometry. In colder glacial regimes where debris fluxes are low, the section may even be very condensed or absent in an hiatus. A wide variety of processes are commonly involved in producing these systems and include glacial, fluvial, marine, iceberg and mass-flow processes. Common end-member glacial processes have been defined by Powell and Alley (1997) as being (Fig. 13.7):

1. subglacial deforming bed and debris flowage beyond the grounding line
2. subglacial fluvial transport and rapid sedimentation from the consequent jet and plume as a stream issues from the grounding line
3. pushing and thrusting of glacial, fluvial and marine sediment at the grounding line
4. a group of processes that commonly produce lesser sediment, depending on glacial regime, including: lodgement of till on the stoss side of the grounding-line system; iceberg rafting both by dumping of supraglacial debris as ice is calved and as icebergs roll, and from rain-out by sea water gradually melting an iceberg; melting of glacial ice to produce grain- and rockfall from a grounded tidewater cliff; rain-out from the base of a floating glacier-tongue from undermelting by sea water; and squeezing of subglacial sediment out beyond the grounding line.

Figure 13.5 A) A submarine iceberg rising to sea level after calving from the basal ice area of a glacier in Alaska. B) After these bergs stabilize they provide a window of the basal debris stratigraphy of the glacier, including basal till and subglacial till. These may be sampled as is shown in (C) where subglacial till lies above basal till that is barely 1 per cent ice; the stratigraphy is inverted because the berg has rolled over to expose this basal section. (Photos A and B are from Powell and Molnia, 1989).

A unique process occurs at a grounding line when icebergs calve from the base of the glacier or when a floating glacier tongue lifts off its bed. Subglacial till is frozen onto what was the sole of the glacier and is carried with the ice into the water column. As this till drops to the sea floor it produces till clasts of varying sizes in an area near the grounding line before the till block is completely melted from the old basal ice. These till clasts help define grounding-line systems in combination with their distinctive facies associations and architecture.

Grounding-line systems are important depo-centres, receiving much of the sediment flux from a glacier where sediment accumulation is often very rapid; one case has been documented as accumulating over 80 m vertically against a temperate glacier from a point-source subglacial stream discharge in 1 year (Powell, 1990, Fig. 9). Sedimentation rates in these systems have been measured at tens of centimetres per year and they decay as a logarithmic function over the first few kilometres away from the glacier (Cai *et al.*, 1995). In other glacial regimes, even with slower sediment accumulation rates, they remain the major loci of sediment accumulation. Redepositional events by mass and gravity flow processes are common from these systems due to:

1. high sediment accumulation rates
2. icebergs calving from above sea level to disturb sea floor sediment after they penetrate through the water column
3. basal ice calving from the grounding line below sea level to disrupt sediment piled against the submarine cliff, and
4. active glaciers pushing sediment piled against a cliff.

Grounding-line deposits have been given various names in the literature but in general they have been divided into grounding-line wedges, morainal banks and grounding-line fans based upon their geometries (Powell and Domack, 1995; Powell and Alley, 1997). Their geometries primarily depend on the dominant processes by which they are made, but the terminus type at the grounding line is also important. A tidewater cliff places no restriction on sediment accommodation space and sediment may aggrade to sea level, whereas a floating glacier tongue constrains the vertical aggradation of grounding-line systems. As grounding-line systems grow under floating tongues they may actually create an advance of the grounding line because sediment comes in contact with the glacier sole, where it is likely to be deformed and sheared flat from glacier flow.

Formational processes of the main geometrical groups of grounding-line systems are thought to be as follows.

Figure 13.6 A) Aerial view of Mackay Glacier, an outlet glacier of the East Antarctic Ice Sheet in southern Victoria Land that ends in Granite Harbour along the coast of the Ross Sea at 77°S. The last 4 km of the glacier is a floating glacier tongue. B) and C) Flutes in subglacial till, observed using a submersible remotely operated vehicle, exposed on the sea floor following recent grounding-line retreat. C) A boulder has dropped from the floating glacier tongue above this site (B) and bergstone diamicton drapes flutes that have been exposed longer on the sea floor. (Photos (B) and (C) are after Powell *et al.*, 1996).

1. A line source along a grounding line being fed by a deforming till bed which, may accumulate up-glacier as a sheet, but when released from its subglacial position, moves away from the grounding line as diamictic debris flows to produce wedge geometries. The resulting debrites thin distally to form the wedge geometry and some may be tens of kilometres long (e.g. King *et al.*, 1991; Laberg and Vorren, 1995; Dowdeswell *et al.*, 1996, 1998; Mohrig *et al.*, 1999).
2. Morainal banks are also produced from line sources but as opposed to wedges, they are composed of a mixture of sediment originating from the wide range of glacial debris-supply processes described above. They also have the competency and accommodation space to stand in bank form (Fig. 13.7) and some are hundreds of metres high (e.g. Stravers and Syvitski, 1991). Sediment is redistributed from these banks as slides and slumps, debris flows and turbidity currents (e.g. Powell, 1981a, 1983; Syvitski and Hein, 1991; Syvitski, 1993; Syvitski and Shaw, 1995; Dowdeswell *et al.*, 1996, 1998; Cai *et al.*, 1997), and some may be steep enough for grain flows. Some evidence indicates that as glaciers surge into the sea, they form morainal banks with large debrites in the forebank areas (cf. Solheim, 1991; Boulton *et al.*, 1999), which may be useful for characterising these types of fjord glaciers.
3. Fan geometries are produced from point-sources at a grounding line as subglacial streams issue from a conduit and sorted sediment is transported away from the grounding line. The transport occurs both in a jet that produces marine outwash (Powell, 1990), and in a turbid plume transported within hypopycnal flows into which the jet transforms, and from which hemipelagic sedimentation occurs. Rarely, hyperpycnal flows may form at the efflux point to

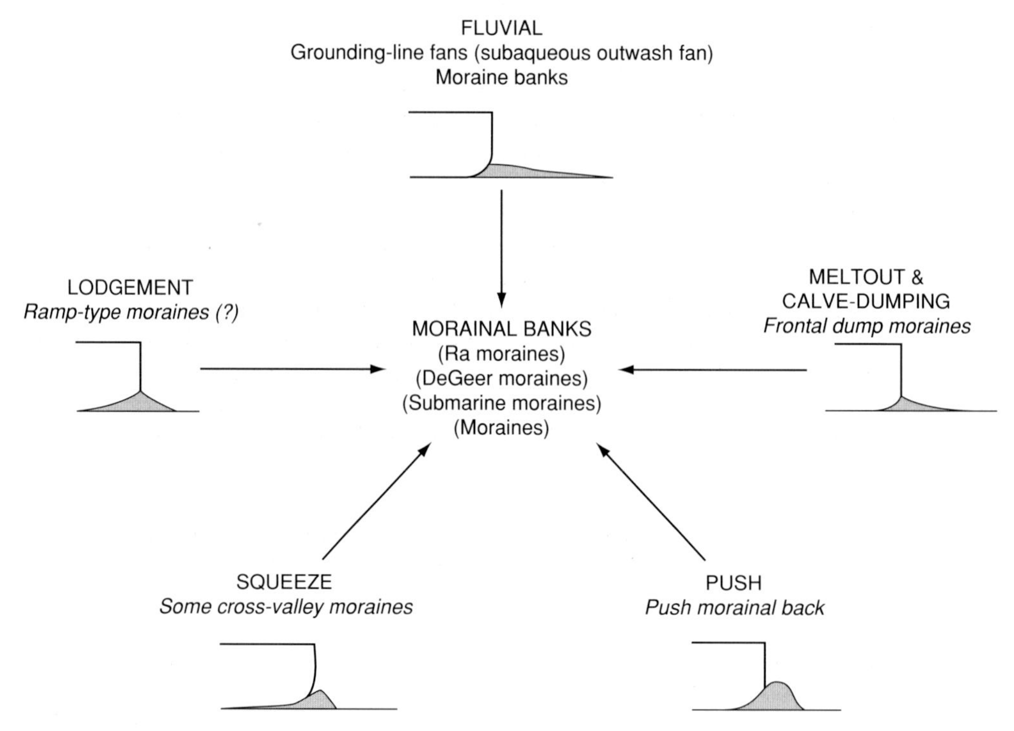

Figure 13.7 Major process components that contribute sediment to a morainal bank. Geometry of the banks varies depending on the end-member process and respective terminology used in the literature is presented with each one. Other terms used for morainal banks in the literature are also shown in parentheses under the term. The morainal bank is the preferred term here because it distinguished the deposits from moraines which are formed subaerially, and 'bank' is a standard allostratigraphic and maritime term used for such geometric bodies. (After Powell and Domack, 1995).

produce continuous underflows down the fan (Powell, 1990) in a similar way to those found in other environments (Mulder and Syvitski, 1995; Mulder *et al.*, 1998; Kassem and Imran, 2001). More commonly, rapid sedimentation at that site produces turbidity currents by repeated failing of sediment as it accumulates rapidly with a high water content in the zone where the jet detaches from the fan and rises through the water column. Often sediment within the turbid overflow plume is released episodically as the hypopycnal flow interacts with changes in tidal current velocities. That interaction produces deep-water tidalites via hemipelagic suspension settling processes. Laminae are produced in couplets of sand and mud proximal to the efflux (within about 1 km), and silt and mud distally in a radial sense, out to about 15 km from the efflux (Fig. 13.8). These facies have been termed cyclopsams and cyclopels, respectively (Mackiewicz *et al.*, 1984; Cowan and Powell, 1991a; Cowan *et al.*, 1998, 1999; cf. Ó Cofaigh and Dowdeswell, 2001). Local failures also may occur to redeposit sediment by mass flows within the fan complex.

Grounding-line systems in modern and Quaternary temperate settings are quite well documented and include a combination of morainal bank and grounding-line fan forms. Although less well known, sub-polar systems appear to be similar to temperate systems. Grounding-line systems in polar fjords are not well studied. We presume they are similar to those on continental shelves where they are dominantly grounding-line wedges (cf. King *et al.*, 1991; Anderson, 1999; Shipp *et al.*, 1999) with little subglacial conduit water flow. However, in fjords, glaciers may have more channelized subglacial water flow if the glaciers are thick enough for pressure melting at the bed. The grounding line of one polar outlet glacier has actually been observed by using a submersible, remotely operated vehicle. There, subglacial sediment was deforming and no channelized water flow was observed (Fig. 13.9). Grounding-line sediment was diamict occurring either as a wedge, or locally where basal crevasses at the grounding line allowed sufficient accommodation space, as a bank (Powell *et al.*, 1996). Beyond the grounding line, shelfstone diamicton (*sensu* Powell, 1984) was being formed by rain-out of basal debris from the sole of the tongue as it moved away from the grounding line. Under conditions close to being true polar in Antarctic Peninsula fjords, tongues of cold water have been documented extending away from glaciers. They are interpreted as originating from melting/freezing processes under floating or partially floating glacier tongues (Domack and Ishman, 1993; Ashley and Smith, 2000) and are discussed further below (see *Iceberg Zone Systems*).

Facies models based upon these processes have been verified and elaborated by detailed analyses of older uplifted Quaternary examples (e.g. Retelle and Bither, 1989; Stewart, 1991; McCabe and Ó Cofaigh, 1995; Lønne 1995, 2001; Lønne and Lauritsen, 1996; Hunter *et al.*, 1996b; Plink-Björklund and Ronnert, 1999) and high-resolution seismic reflection profiles (e.g. Elverhøi *et al.*, 1983; Gilbert, 1984; Belknap and Shipp, 1991; Stravers and Syvitski, 1991; Sexton *et al.*, 1992; Cai *et al.*, 1997; Lønne and Syvitski, 1997; Seramur *et al.*, 1997) from various locations. The advantage of using Quaternary deposits is that they allow a full documentation of lithofacies that in modern settings are depicted primarily as seismic facies with small local samples from sediment cores. In combination with seismic reflection profiles of fjord bottom sediment, they also allow verification of lateral facies changes and geometries, of large-scale deformation structures and of detailed facies successions, particularly in coarser-grained deposits which are difficult to sample in modern environments.

In general, the facies and geometries of Quaternary examples are the same as those of modern settings, but details of some facies and the complexities of interfacies geometries are better defined (Fig. 13.10). For example, small- and large-scale deformation structures associated with grounding-line oscillations and glacial over-riding are better defined (e.g. McCabe *et al.*, 1984; McCabe, 1986; Lønne, 1995, 2001; Hunter *et al.*, 1996b; Lønne and Lauritsen, 1996; Lønne and Syvitski, 1997; Bennett *et al.*, 1999; Boulton *et al.*, 1999; Plink-Björklund and Ronnert, 1999). Varieties of gravel and rubble facies associated with grounding-line systems, particularly grounding-line fans, are well documented (e.g. Powell, 1990; Plink-Björklund and Ronnert, 1999), as are structures generated by iceberg rafting (cf. Thomas and Connell, 1985).

13.3.4 Floating Glacier-Tongue Systems

In polar and some sub-polar fjords the area between the grounding-line system and the iceberg zone system discussed below may be under a floating glacier tongue. In this case, the calving line, where icebergs are produced, may be several tens of kilometres beyond the grounding line versus the situation where glaciers end with grounded cliffs, and the calving line and grounding line coincide. Due to extreme problems in accessibility, very little is known about these tongue systems but one study of an Antarctic system (Powell *et al.*, 1996, Dawber and Powell, 1998) shows it to

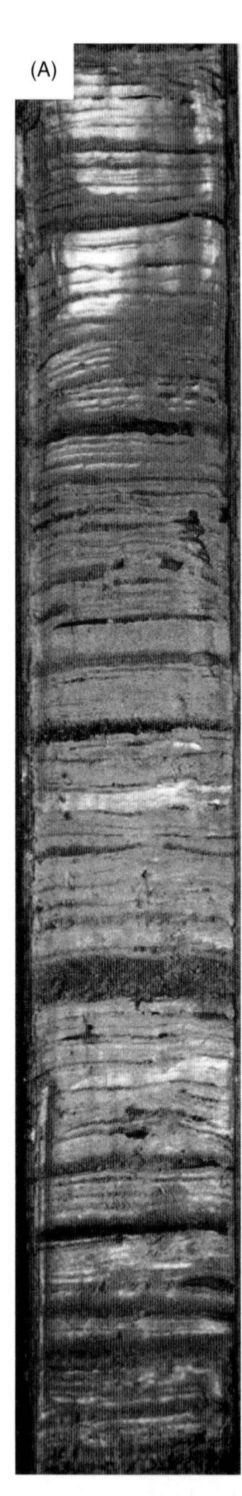

Figure 13.8 Laminites of cyclopels and cyclopsams deposited by suspension settling from turbid meltwater overflow plumes, Muir Inlet, southeast Alaska. A) A 7.5 cm wide core of ice-proximal laminites of cyclopsam sands and local turbidites with B) associated photomicrograph (~40 mm across) of a cyclopsam. C) A 7.5 cm wide core of more distal laminites of cyclopel silt laminae in suspension muds, with D) an associated photomicrograph (~100 mm across) of a cyclopel. (After Mackiewicz *et al.,* 1984).

(C)

(D)

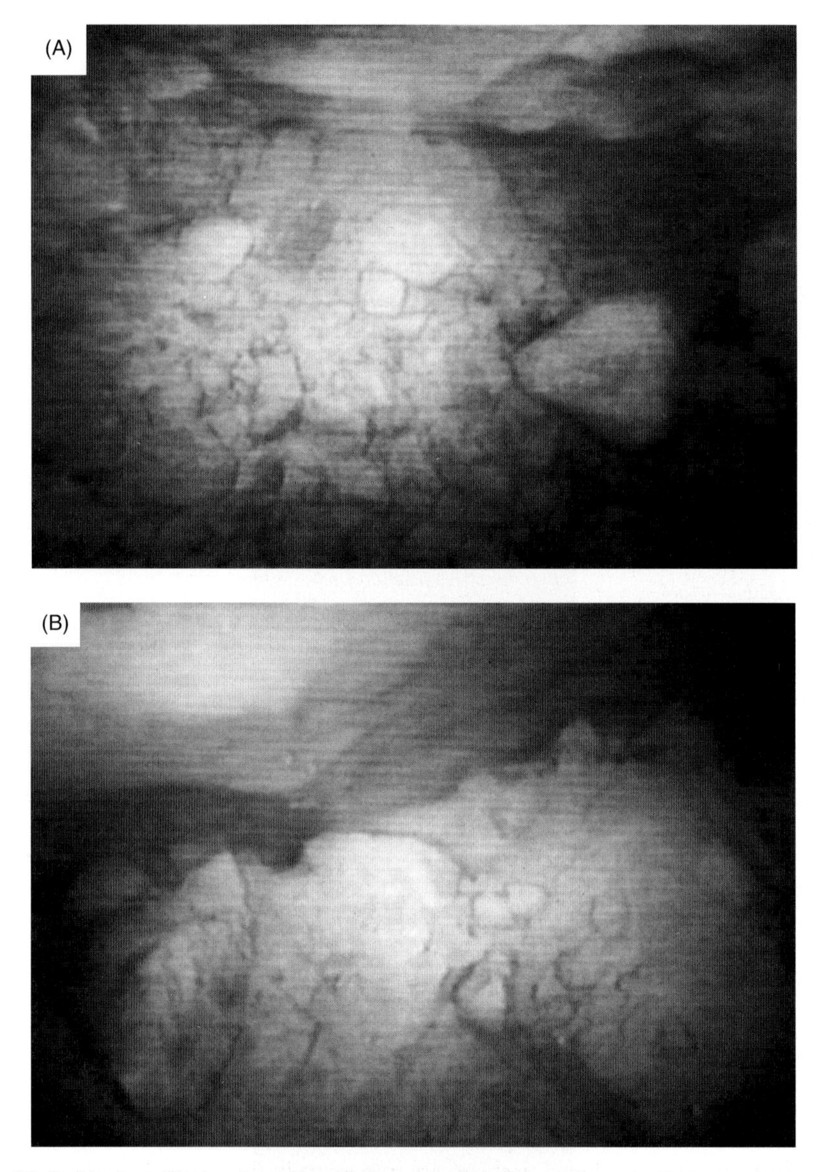

Figure 13.9 Mackay Glacier (see Fig. 13.6) grounding-line sediments that vary in texture and the degree of consolidation within tens of metres along the grounding line. Here stiff clast-rich till (a) contrasts with weak muddy till (b). (Photos after Powell *et al.*, 1996).

be sediment-starved and thus is represented by a depo-system that is a condensed section. As opposed to ice shelves where the distance between the grounding and calving lines may be hundreds of kilometres, distances here may be short enough that continental shelf waters rich in nutrients can circulate under the tongue. This circulation can bring in phytoplankton such as diatoms to produce a fossil record as well as providing a food source for quite diverse epibenthic and shallow infaunal communities.

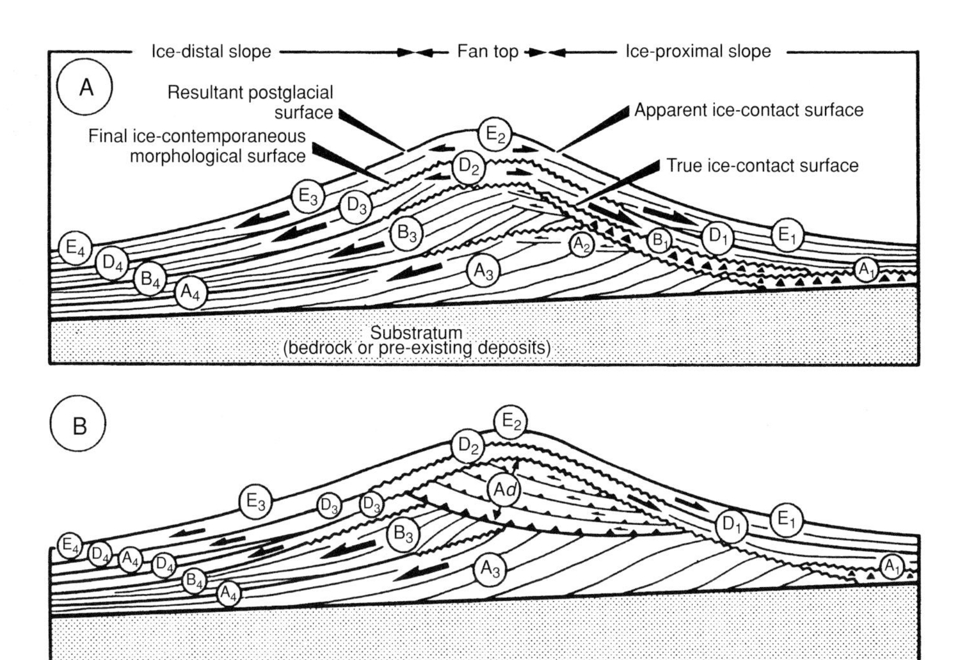

Figure 13.10 Conceptual model for the development of a grounding-line fan (schematic, not to scale). A) The basic depositional architecture comprises four allostratigraphic units, whose boundaries are erosional discontinuities ranging from discordant (thick wavy lines) to concordant (thick smooth lines); the arrows indicate dominant transport directions, and glacial tectonic deformation. Letter designations with sub-numerals indicate relative timing and location of each unit and their associations. Small-scale syn-sedimentary glacial tectonic deformation may be included in units (A) and (B). B) More compressive deformation, suggesting a pronounced glacier advance, is distinguished as 'Ad' (A-deformed). (After Lønne, 1995).

13.3.5 Iceberg Zone Systems

As with the systems already described, iceberg zone systems are absent in paraglacial and later phases of fjord evolution. Here, iceberg zones are taken to extend from the down-fjord end of a grounding-line system (even though icebergs are important in that system) to where icebergs are very rare, say covering about 1–2 per cent or less of the water surface. Sediment dispersal patterns relate mainly to water mass structure, fjord circulation including effects of Coriolis deflection, and wind forcing to a minor degree (Syvitski *et al.*, 1987; Gilbert, 1984). Iceberg zone depo-systems are dominated by sediment that originates from two or more sources and becomes mixed as it is being deposited. The main sources of sediment are:

- hemipelagic glacial rock flour coming from glacifluvial discharges
- mass failures from grounding-line systems
- iceberg-rafted debris
- sea-ice rafted debris
- aeolian sediment, and
- biogenic sediment.

Which source(s) dominates is primarily a function of climatic regime. For example, under polar conditions sea ice, aeolian and biogenic components are important, icebergs and hemipelagic mud sedimentation are significant in sub-polar regimes, and hemipelagic mud sedimentation and turbidity currents dominate all others in temperate systems. The periodicity of changes is also mainly climatically controlled (e.g. millennial periodicity or Milankovitch forcing; sub-millennial to decadal periodicity like the El Niño Southern Oscillation and Pacific Decadal Oscillation; annual periodicity (varves) and sub-annual periodicity (cyclopels)).

Icebergs impart their character to the sediment record primarily by rafting glacial debris. Because this debris ultimately becomes incorporated in fjord floor sediment, the deposits are termed genetically either bergstone muds or bergstone diamictons (Powell, 1984). Icebergs introduce their glacial debris load into the sea by rolling and dumping surface ablation debris and by progressive melt-out below the water line of individual englacial particles, a process termed 'rain-out'. It is important to remember that iceberg-rafted debris includes all particle sizes, thus, depending on iceberg concentration and the relative contribution of hemipelagic mud, sediments of a gravelly mud or a diamict texture are produced. Based upon these depositional processes iceberg zone systems are basin fill, sheet and sheet-draped deposits.

Iceberg-rafted debris produces lonestones from sand- and gravel-sized particles, and if they show evidence of having been dropped (e.g. Thomas and Connell, 1985) they may be termed 'dropstones'. Even with a lack of dropstone evidence it is still possible to decide lonestones have been rafted due to their out-sized character; that is, clasts could not have been hydrodynamically transported with their surrounding matrix based either upon particle-size distribution or thickness of stratification relative to clast size. However, care must be taken that this rafted sediment indeed originates from icebergs because other agents can raft sediment such as sea ice, plant roots and mammal gizzards (e.g. Gilbert, 1990b; Bennett *et al.*, 1996). Other features indicative of iceberg rafting are till pellets and clast nests, the latter being indicative of iceberg dumping and may serve as an indicator of an iceberg zone rather than a floating glacier-tongue system (cf. Powell, 1984). When icebergs are large they have large drafts and their keels can scour the fjord. As they do so they leave iceberg furrows (e.g. Syvitski *et al.*, 1983, 1996; Woodworth-Lynas and Guigné, 1990; Dowdeswell *et al.*, 1993) on the sediment surface (Fig. 13.11) and internally turbate the sediment in a process termed 'ice keel turbation' (Vorren *et al.*, 1983; Barnes and Lien, 1988). In fjords these processes most commonly occur along fjord walls and on fjord shoals where water depths are shallow, but in polar settings where icebergs are large they may be trapped behind sills and actively scour deeper basin sediment (e.g. Dowdeswell *et al.*, 1993).

In temperate and sub-polar settings such as Alaska and Norway (Powell, 1990; Cowan and Powell, 1991a and b; Plassen and Vorren, 2002; Vorren and Plassen, 2002). Baffin Island (Gilbert, 1982; Gilbert *et al.*, 1990; Syvitski and Hein, 1991; Winters and Syvitski, 1992) and west Spitsbergen (Elverhøi *et al.*, 1980, 1983; Görlich, 1986; Sexton *et al.*, 1992; Svendsen *et al.*, 2002), hemipelagic mud sedimentation occurs from baroclinic overflows that are turbid and charged with glacial rock flour from stream discharges at grounding-line fans and deltas. Sedimentation rates are lower than in the grounding-line system, being several centimetres per year in temperate systems and a factor or two lower in sub-polar systems. Instead of a logarithmic decay down-fjord as described above for more proximal locations, rates now decrease exponentially. Tides interact with the stream discharges to form deep-water tidalites termed 'cyclopels'. Commonly cyclopels are intimately inter-stratified with turbidites. If sedimentation rates are high enough this stratification is retained due to a lack of in-fauna and consequent absence of bioturbation. More distally (e.g.

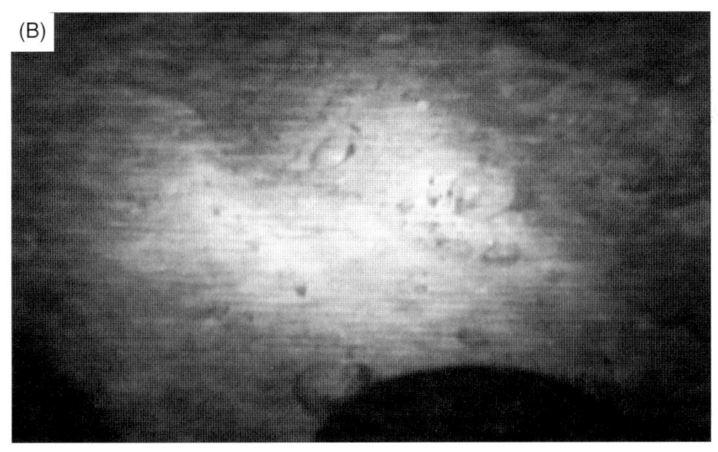

Figure 13.11 The corner of an iceberg calved from the edge of Mackay Glacier Tongue (see Fig. 13.6) that has impacted the sea floor and disturbed the epibenthic community: (A) after a berg detaches from the glacier tongue it rotates to its own centre of buoyancy, which may mean it impacts the sea floor causing iceberg wallows and scours, and turbating the sediment (B) as it is moved by wind and oceanic currents.

~15 km from a glacier terminus), where cyclopels are not formed because particle size variations in the plumes are minor, the mud may accumulate without internal structure and have a massive appearance even with an absence of bioturbation. This characteristic is probably due to processes of particle aggregation in the water column by agglomeration, flocculation and pelletization (e.g. Syvitski, 1980; Domack *et al.*, 1994; Hill *et al.*, 1998; Fig. 13.12). Mud may also accumulate like this in winter when stream discharges are low or absent.

In temperate systems mud dominates deposition during summers unless icebergs are concentrated for some reason and diamicton accumulates (e.g. Gottler and Powell, 1980). As mud accumulation rates decrease during winter, diamicton results, and if that occurs regularly each year, varves result; the summer laminae being bergstone muds (Cowan *et al.*, 1997). In sub-polar

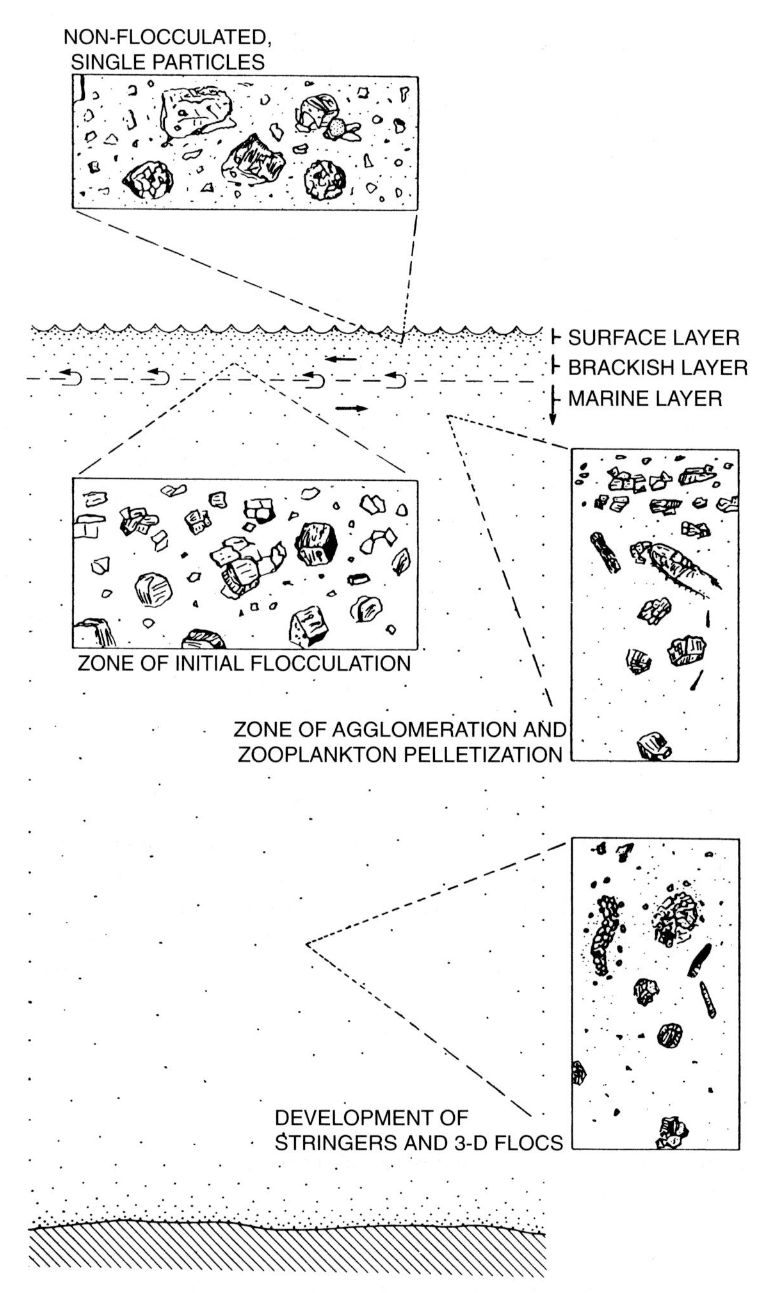

Figure 13.12 Schematic of the change in microtexture of suspended sediment particles as they settle through the fjord water column. Initially the particles are as non-flocculated single particles in the surface freshwater. As they move through the brackish layer they begin to flocculate, only to be consumed by zooplankton in marine waters. With the growth of organic matter, deep water flocs are large and three dimensional. (After Syvitski *et al.*, 1987).

to polar settings in Greenland, the dominant texture of iceberg zone sediment is diamicton beyond the limits of the turbid overflow plumes (Dowdeswell *et al.*, 1994), whereas closer to the glacier laminated bergstone muds accumulate similar to those in temperate systems (Ó Cofaigh *et al.*, 2001) although, that may vary among fjords too (Stein *et al.*, 1993; Syvitski *et al.*, 1996; Smith and Andrews, 2000). These facies are also influenced by permanent and annual sea ice, which traps icebergs in a formation termed 'sikussak', and inhibits their ability to widely disperse their debris; iceberg rafting is then concentrated to perhaps almost double sediment accumulation rates (Syvitski *et al.*, 1996). The distal diamicton facies in sub-polar settings differs from temperate conditions because no icebergs raft sediment the equivalent distance, as they are smaller and melt faster with a result that only mud accumulates. Small brash ice is the exception because it melts relatively fast in all regimes (cf. Smith and Ashley, 1996).

In polar settings siliciclastic sediment flux from subglacial streams appears to be minimal with a consequent very low sedimentation rate of 15 mm/a in polar/sub-polar conditions to 1 mm/a in true polar settings (Griffith and Anderson, 1989; Domack and Williams, 1990; Domack, 1990; Lemmen, 1990; Gilbert *et al.*, 1990b, 1993, 1998; Domack *et al.*, 1993b; Andrews *et al.*, 1994; Domack and Ishman, 1993; Domack and McClennen, 1996; Shevenelle *et al.*, 1996; Ashley and Smith, 2000). Tongues of cold water occur within the water column of some fjords (e.g. Domack and Ishman, 1993), although they are not found in all (Azetsu-Scott and Tan, 1994; Azetsu-Scott and Syvitski, 1999). The cold tongues carry fine-grained suspended sediment in very low concentrations and are thought to originate from a glacier melting directly in sea water (Fig. 13.1). The meltwater mixes with the ambient sea water and, enhanced by tidal currents, the mixed water then flows down-fjord as interflows. The depth of each interflow is controlled by its density resulting from a combination of its salinity, temperature and sediment concentration.

Turbidites and debrites originating from grounding-line systems are common facies of the iceberg zone nearest a glacier (see *Grounding-Line Systems* above). Mostly the turbidity currents and debris flows appear to be unconfined because no sea floor channels occur on most fjord floors although some incipient channels have been described (Cai *et al.*, 1997). It appears that slopes where they are initiated are not sufficiently steep to generate flows of erosive velocities, but the required steepness is reached by the time sediment has accumulated to, or near to, sea level (Carlson *et al.*, 1989). These deposits go to make up a significant portion of the fjord basin fills within the first 10–20 km proximal to grounding lines (Powell, 1991).

Other processes in iceberg zone systems include aeolian, sea ice and biogenic activity, which have their action recorded primarily in polar and some sub-polar settings. Aeolian sediment primarily originates from outwash plains, although some may be from bare fjord walls or supraglacial debris. These sources are available in the full range of climates, although they are active over longer periods of time in polar settings where even after deglaciation, exposed areas are poorly, if ever, vegetated. In regimes other than polar, the aeolian contribution to the depositional landsystems is relatively diluted by sediment originating from other sources. In polar regimes however, aeolian sediment may comprise a significant part of the siliciclastic sediment in a fjord because other processes provide such small volumes of sediment (e.g. Barrett *et al.*, 1983; Gilbert, 1984, 1990b; Dunbar *et al.*, 1989).

Sea ice can locally raft sediment less than 4 m across (Gilbert, 1990b), but generally sizes are less than boulder size and often smaller than that because of a lack of competence of the ice. Sea-ice-rafted debris cannot be carried as far as by icebergs due to the relative thinness of the ice and its speed of melting. Sediment may originate in sea ice by (Dayton *et al.*, 1969; Barnes *et al.*, 1982; Dionne, 1984; Reimnitz *et al.*, 1987; Reimnitz and Kempema, 1987; Gilbert, 1990b):

1. incorporation as agglomerates in frazil ice platelets during formation as seawater freezes
2. meltwater streams flowing out over shorefast sea ice before break up
3. aeolian sediment being blown on to sea ice, often during winter
4. freezing-on to the base of the sea ice grounding in shallow water, and
5. anchor ice attaching to sea floor material and eventually rising buoyantly to become frozen-in to the base of the sea ice floating above.

Sea ice may erode in shallow water environments, particularly when it is blown by wind into pressure ridges, the keels of which can reach several metres depth to cause scour and turbation (Hansom, 1983; Reimnitz *et al.*, 1984; Dionne, 1985). Sea ice can also influence the sedimentary record indirectly by controlling the movement of icebergs as described above for Greenland (e.g. Dowdeswell *et al.*, 1994; Syvitski *et al.*, 1996; Smith and Andrews, 2000; Ó Cofaigh *et al.*, 2001).

Biological activity and processes in fjords are primarily a function of temperature, salinity, turbidity, and the presence of ice. In temperate settings turbidity appears to be the dominant control particularly near glacial and riverine inputs. In sub-polar settings salinity and temperature can play a more significant role, but turbidity near high sediment fluxes from land can be locally significant. In polar conditions and in some sub-polar settings, sea ice plays a significant role in biological productivity as well as the abundance of different types of organisms. Floating glacier tongues are the most extreme form of ice influence where biological activity relies on ocean circulation to carry organisms under the thick floating ice to provide a biological record in themselves as well as being the food source for others living under the ice. In addition, preservation in the sediment record is also a function of silica and carbonate solubility and of subsequent erosional processes from glacial readvances, icebergs and sea ice turbation and from isostatic uplift resulting in wave or subaerial erosion.

Although only a few areas have been thoroughly studied (one of the more thorough overviews is from Kongsfjorden, Spitsbergen (Hop *et al.*, 2002)), records of microflora appear to be dominated by diatoms in the polar regions where they play an important role in biostratigraphic and palaeo-climatic research (e.g. Dunbar *et al.*, 1989; Leventer *et al.*, 1993). In subpolar and temperate settings Foraminifera become important beyond high siliciclastic sediment fluxes (e.g. Armentrout, 1980; Echols and Armentrout, 1980; Lagoe, 1980; Quinterno *et al.*, 1980; Schafer and Cole, 1986; Eyles *et al.*, 1991; Domack and Ishman, 1993; Jennings and Helgadottir, 1996). An important function of planktonic microfauna is in producing faecal pellets and enhancing the sedimentation rates of mud (Syvitski, 1980). Macrofauna and -flora can become established away from turbidity and high sediment accumulation rates, and in areas of sufficient productivity in the water column. Motile forms may occur in some turbid locations and soup-ground areas, or in areas of low water-column productivity (e.g. Syvitski *et al.*, 1989; Aitken and Bell, 1997; Dawber and Powell, 1998). Epibenthic communities dominate hard grounds, which are commonly bedrock walls or gravel lags such as relict morainal banks or consolidated till, whereas in-faunal communities are common in soft grounds (Plafker and Addicott, 1976; Armentrout, 1980; Hickman and Nesbitt, 1980; Dale *et al.*, 1989; Syvitski *et al.*, 1989; Aitken, 1990; Aitken and Bell, 1997; Carney *et al.*, 1999). The in-faunal communities also determine the preservation potential of finely laminated structures such as cyclopels and cyclopsams, which are better preserved in high sedimentation rate areas where biological productivity is low (Gilbert, 1984; Aitken *et al.*, 1988; Eyles *et al.*, 1992; Syvitski *et al.*, 1996; Cai *et al.*, 1997; Ashley and Smith, 2000; Ó Cofaigh and Dowdeswell, 2001).

13.3.6 Open Water Systems

Open water fjord systems occur in the distal reaches of glacial fjords and during the paraglacial and deglacial phases of fjord evolution. Although these are important systems, they are dealt with here cursorily because the emphasis is on glacial systems. Deep water estuarine processes dominate the system. Evidence for glacial action is slight and in temperate to sub-polar regimes sedimentation is predominantly hemipelagic having a fluvial source and a high organic content. In polar and some sub-polar regimes, sea ice, aeolian and biogenic processes are important. Once abandoned by glaciers, sea water circulation patterns change dramatically, and if fjord basins are deep relative to the sill height and outside water does not have the density to penetrate over the sill and descend into the basin, anoxia occurs forming chemically reduced, dark mud (e.g. Syvitski *et al.*, 1987).

Often the only evidence of cryogenic processes is from sea ice, and that is mainly in polar regimes. It is under these conditions that most systems experience little deposition; condensed sections are produced because fjords have very low sediment accumulation rates, even if they do not go anoxic (e.g. Pickerill, 1993). Under such conditions, diagenetic processes, particularly in the deeper water systems, can lead to characteristic chemistries and unique mineral formation, such as ikaite, which is thought to form in very cold water systems (Buchardt *et al.*, 2001) and preserved as what have been defined as 'glendonite crystals' in the stratigraphic record (e.g. Domack *et al.*, 1993a). If the fjords experience some riverine input, turnover and productivity at least in the upper waters, thick varved sediments may accumulate during full interglacial periods comprising diatomaceous spring-summer laminae and terrigenous winter laminae, such as has been recently recovered in Saanich Inlet, British Columbia (Blais-Stevens *et al.*, 2001).

13.3.7 Sidewall Systems

The walls of fjords are commonly very steep and often have hanging valleys left where tributary glaciers entered as a trunk glacier. Walls are not always straight and they often have bends or bays in which sediment can accumulate in the form of pocket beaches and tidal flats. The steep walls are most active sedimentologically during retreat of a glacier where glacial sediment had been ice-dammed along the margin of the glacier and then it loses its support during glacial downwasting and eventual terminus retreat. Immediately after a terminus retreats the highly water-charged sidewall sediment fails rapidly and is redeposited to the fjord floor by rock fall and mass flowage (Cowan and Powell, 1991a). The failed sediment includes a mix of rockfall and supraglacial ablation debris, till, glacifluvial and glacilacustrine deposits (see Chapter 7).

After this initial retreat phase, the most active sediment accumulation occurs in talus and alluvial fans and eventually fan deltas. Large-scale mass failures now mainly occur due to earthquake shocks or exceptional rainstorm events. Talus and cone-shaped fans are established early along fjord coasts even as the glacier is retreating. The cones are fed from residual glacial ice as it melted on the steep valley and may become abandoned after all of the remnant ice has melted and stops supplying sediment. Others fed by lateral streams from snow melt or hanging tributary glaciers may continue to build rapidly into the sea and form alluvial fan/delta systems (Postma, 1990; Prior and Bornhold, 1990; Nemec, 1990). In the initial stages of the development conical underwater deltas lack a subaerial distributary plain but with time they may develop into

Gilbert-type deltas. In polar climates it is the talus cones that form and develop no further due to a lack of running water and sea ice protection from waves.

If fjords are wide, oriented favourably to dominant wind directions and are long enough for sufficient fetch, then wave action on shorelines may develop beaches. They are often gravelly due to reworking of glacial deposits. These pocket beaches tend to form in small coastline embayments often associated with tributary valleys, which are the main sediment source. Close to glaciers, calving waves may also be important by eroding the beach face or carrying icebergs well above high tide (Syvitski and Shaw, 1995). In the intertidal zone icebergs can wallow and scour beaches.

Where tides are large such as in macrotidal or some mesotidal fjords, tidal flats, which are primarily mudflats, can form and become quite extensive over time due to an abundant supply of glacial rock flour (Bartsch-Winkler and Ovenshine, 1984; Bartsch-Winkler and Schmoll, 1984). A wide variety of structures have been observed as being formed on tidal flats by both icebergs and sea ice in association with more common structures such as lenticular bedding and mudcracks. Ice-related structures include evidence of:

- ice abrasion such as grooves, furrows and other scour marks, striae and intertidal boulder pavements
- ice wallowing and pushing, such as circular or irregular depressions, chaotic microrelief and bedding plane irregularities, ice-push ridges
- soft sediment deformation by ice such as ball-and-pillow structures, convoluted beds and microfaulting
- frost action including polygonal forms, and
- ice rafting

(e.g. Reimnitz and Kempema, 1982; Dionne, 1983, 1988, 1998; Hansom, 1983; Powell and Molnia, 1989).

In the longer term, these systems are very much influenced by glacial isostatic effects. Perhaps obviously, it is the rebound after glaciation that is critical for preservation of the systems, as those that are formed prior to and during advances are eroded. Rebound changes base level and alters facies types and geometries and ultimately can bring these systems into zones of erosion as they are raised to within wave-base or even above sea level.

13.3.8 Fjord Head Systems

During the last stage of the fjord deglacial phase the head of a fjord is the locus of most of the terrestrial sediment being supplied. This situation lasts through the paraglacial phase, and probably the fjord valley phase until the next glacial advance phase. In the initial stages grounding-line systems may accumulate as the glacier is retreating from the sea. In temperate and sub-polar regimes the grounding-line systems are likely to be dominated by grounding-line fans, which eventually aggrade to sea level until the fan apex becomes intertidal (Powell, 1990). Over time the landward end of the fan apex aggrades above sea level to form a delta plain and a new fan delta system becomes established (Fig. 13.13). During the intertidal stage many redepositional episodes occur, mainly as turbidity currents as the outwash stream channels migrate readily, often over tidal cycles as they are repeatedly raised from the intertidal portion of the delta plain at high tide and then lowered during ebb tides to erode the plain and any new sediment just accumulated.

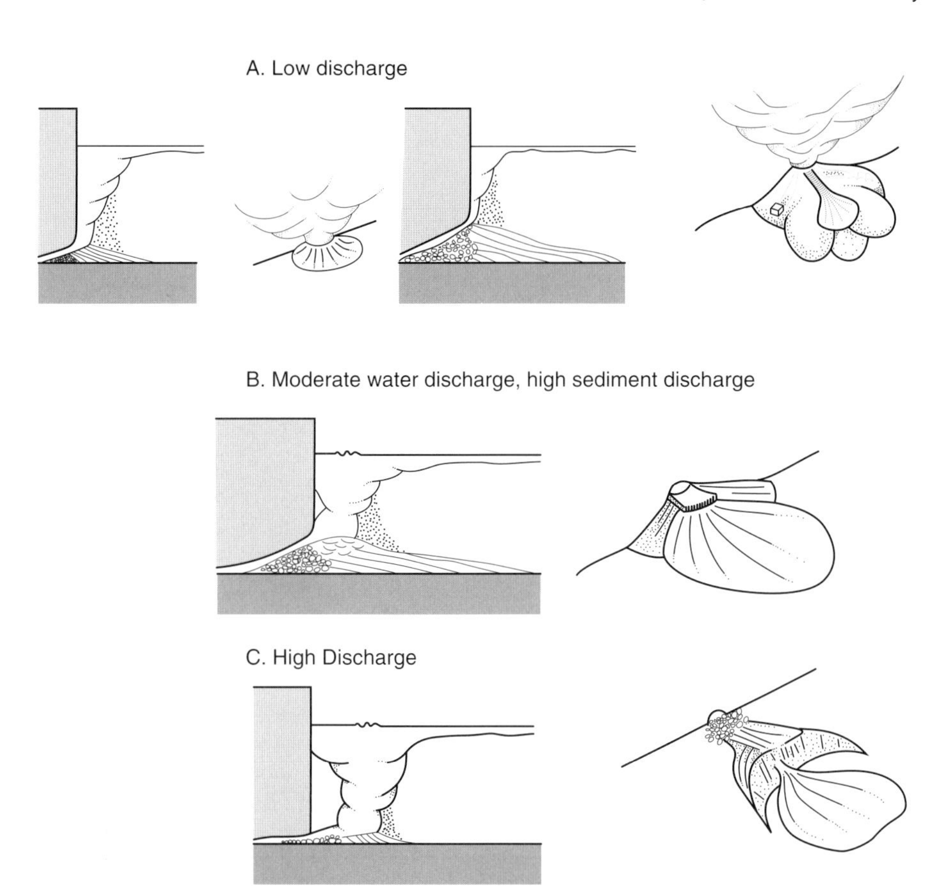

Figure 13.13 Sketches of the modes of formation of grounding-line fans during low, moderate and high discharges. If a terminus remains at one location for sufficient time (a longer time for lower discharges) then these fans grow to sea level where the fan apex becomes intertidal to eventually becomes a delta plain, having sea level as its base level. (After Powell, 1990).

As a delta becomes established it progrades and the delta plain expands under a braided stream system due to cyclic discharge and high sediment loads. Even as the delta plain expands, tides appear to influence re-sedimentation from the plain, at least in macrotidal areas. A process of tidal draw-down has been documented where during neap tides sediment may get stored on the outer reaches of a delta plain and then that sediment is redeposited during spring tidal periods when sea level falls below the lip of the delta to dramatically change local base level creating incision on the delta plain (Smith *et al.*, 1990; Phillips *et al.*, 1991). Delta fronts are sites of rapid sediment accumulation, which drive the progradation but also are areas of large mass failures and slides, and sediment by-pass in the form of turbidity currents (e.g. Prior *et al.*, 1981, 1987; Syvitski and Farrow, 1983; Prior and Bornhold, 1988, 1989; Syvitski, 1989, 1993; Carlson *et al.*, 1989, 1992, 1999; Fig. 13.14). Re-sedimentation appears to start in upper delta front areas from numerous small-scale gullies; eventually flows coalesce downslope and become channelized in turbidity current channels. These channels can run out along the fjord floor for tens of kilometres and some may end in small submarine fans. There are also good examples of these systems in uplifted Quaternary deposits (e.g. Plink-Björklund and Ronnert, 1999).

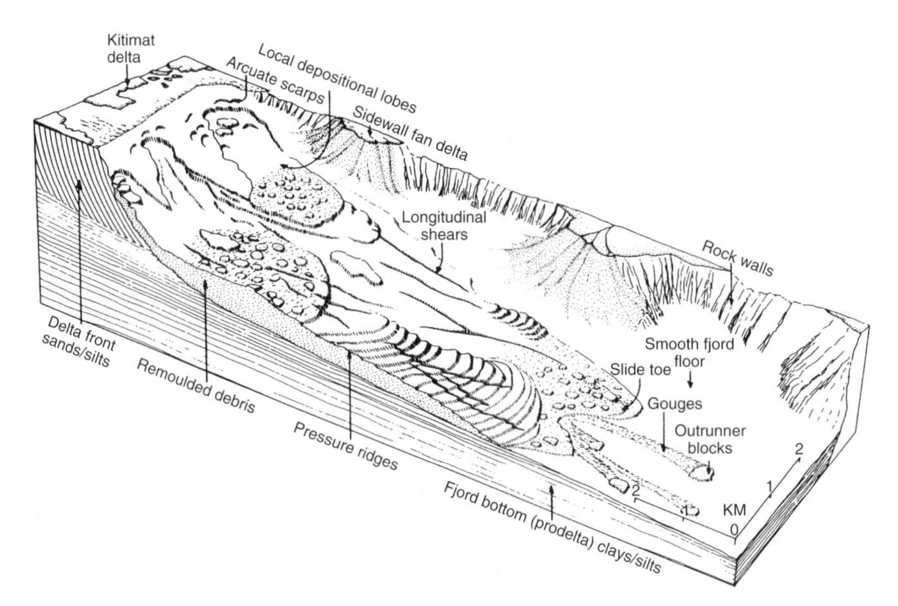

Figure 13.14 Schematic diagram of the Kitimat delta, British Columbia, Canada. The diverse mass flow processes and surface morphology are based upon high-resolution side-scan sonar data. (After Prior *et al.*, 1983).

In contrast, full polar deltas are sites of little activity where small ephemeral streams may be active for a few months over summer but carry low sediment loads (Chinn, 1993; Doran *et al.*, 2002). Delta discharges are ephemeral and the delta plains are protected from wave action by extensive sea ice cover (Powell, 1981b; Gilbert, 1984); open coastal systems on the other hand may be more active (e.g. Hill *et al.*, 2001). Major deposits are biogenic and similar to those described above (see *Open Marine Systems*).

In a similar way to sidewall systems, these fjord head systems are significantly influenced by isostatic rebound and may eventually be raised above sea level with concomitant changes in facies patterns and geometries, erosion and lag surfaces produced. The lowering of base level during rebound decreases the accommodation space for sediment on the delta plain creating stream incision and an increase in progradation of the delta.

13.3.9 Fjord Valley Systems

Fjord valley systems are created after deglaciation following the glacial and paraglacial fjord phases, when a fjord becomes filled with sediment to above sea level (Fig. 13.15). This stage can occur in temperate, sub-polar and polar settings and once reached, terrestrial glacial and non-glacial sedimentary systems are established.

13.4 MODELLING FJORD LANDSYSTEMS

Static conceptual models have been established for some fjord landsystems and those have been presented above. However, we are on the threshold of new developments where processes-driven numerical models can be used to build the landsystems in a predictive way. The past records

Figure 13.15 The cold desert environment of a polar fjord valley, McMurdo Sound Dry Valley system, southern Victoria Land, East Antarctica. The aerial view looks up Taylor Valley from its mouth in the sea-ice-covered Explorer's Cove on the Ross Sea coast. Drilling by the Dry Valley Drilling Project in the 1970s on the delta (foreground) and on the ridge between it and the first piedmont glacier to the right (Canada Glacier), showed more than 300 m of mainly glacimarine sediment filling a fjord dating back to at least Miocene time (McKelvey, 1981; Powell, 1981b).

contained in the landsystems can then act both as a basis to test and constrain the numerical models as well as providing the basis for simulation modelling of past processes that have combined to produce the record. The ideal is to achieve a climatically driven glacial model that integrates glacial dynamics, including:

- glacial erosion (e.g. Harbor *et al.*, 1988; Hooke, 1991; Iverson, 1991; Harbor, 1992; Hallet *et al.*, 1996)
- subglacial bed deformation (e.g. Iverson *et al.*, 1998, 1999; Tulaczyk, 1999; Tulaczyk *et al.*, 2000)
- glacial debris transport (Hunter *et al.*, 1996a)
- subglacial fluvial sediment transport (sheet, conduit and open-channel flow).

These data can provide erosion rates, sediment fluxes and yields that are critical for building the landsystems and also define the geometry of sediment input into the fjord (e.g. line source, point-source). In addition, modelling calving dynamics must be integrated into any time-dependent dynamic model because it may well be an important factor in influencing rates of debris supply to the terminus and beyond, as well as in determining geometries of deposits relative to terminus advance, stability and fast or slow retreat. Quantitative fjord sediment deposition and dispersal transport models, including those for each of the landsystems, need to be integrated to determine lateral and temporal extent and geometries of landsystems. Thus far, models for local cases have been determined for the following:

- grounding-line wedges (ten Brink and Schneider, 1995; ten Brink *et al.*, 1995)
- morainal banks (Alley, 1991b; Fischer and Powell, 1998)
- grounding-line fans (Powell, 1990)
- sediment dispersal in overflow plumes (Syvitski *et al.*, 1998; Morehead and Syvitski, 1999)
- suspension settling (Domack *et al.*, 1994; Hill *et al.*, 1998)
- turbidity currents (Zeng and Lowe, 1997a, b; Salaheldin *et al.*, 2000; Pratson *et al.*, 2001)
- debris flows (Elverhøi *et al.*, 1997; Mohrig *et al.*, 1999)
- iceberg rafting (Dowdeswell and Murray, 1990), and
- deltas (Syvitski and Daughny, 1992).

However, many of these individual glacimarine sediment models remain in their infancy and still require further development and testing. Furthermore, many dispersal processes are dependent on fjord circulation and thus tidal and estuarine circulation models must be considered. Partially independent parameters, biological productivity and ecosystem variability largely influence the sedimentary record preserved in the landsystems and provide important dating material. Thus factors such as the rates of community establishment and rates of bioturbation should also be quantified.

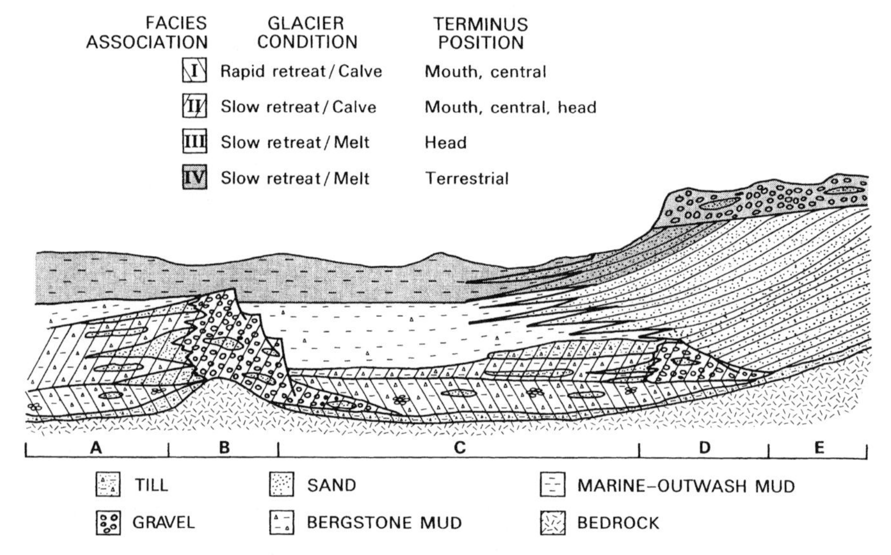

Figure 13.16 Hypothetical cross-section showing glacimarine allostratigraphic units deposited during the retreat of a temperate glacier with a tidewater terminus. During the time when the terminus was in sections A and C it was retreating rapidly by calving in deep water, and sediment (mainly bergstone mud) was deposited as a sheet above till. During the time the terminus was in section B, calving continued but recession was slowed by a channel constriction; a morainal bank was formed, beyond which mass flows interfingered with bergstone mud. When the terminus was in section D at the fjord head, calving and recession slowed and a morainal bank formed initially; eventually calving was superseded by surface melting and a grounding-line fan developed into a delta, with bergstone mud accumulating distally. Lastly, when the terminus was fully terrestrial in section E, the outwash delta prograded over previously deposited sediment with paraglacial marine outwash mud accumulating distally. (After Powell, 1981a).

The ultimate model needs to be designed such that time-steps through a full glacial cycle can be resolved, as has been done conceptually (e.g. Powell, 1981a; Powell *et al.*, 2000; Fig. 13.16). As quantitative models evolve and can be integrated into glaciological and oceanic process models, then predictions of sediment geometries given particular input parameters can be made. Then in addition, there needs to be an accounting for isostatic rebound effects (Andrews, 1974; Bednarski, 1988; Boulton, 1990). Timescales on a glacial/interglacial scale also require integration of external forces such as tectonics and eustatic sea level changes. Some conceptual models of sequence stratigraphic architecture for high latitude glacimarine landsystems have been established (Fig. 13.17) but they are mainly for continental shelves (e.g. Boulton, 1990; Bartek *et al.*, 1991; Cooper

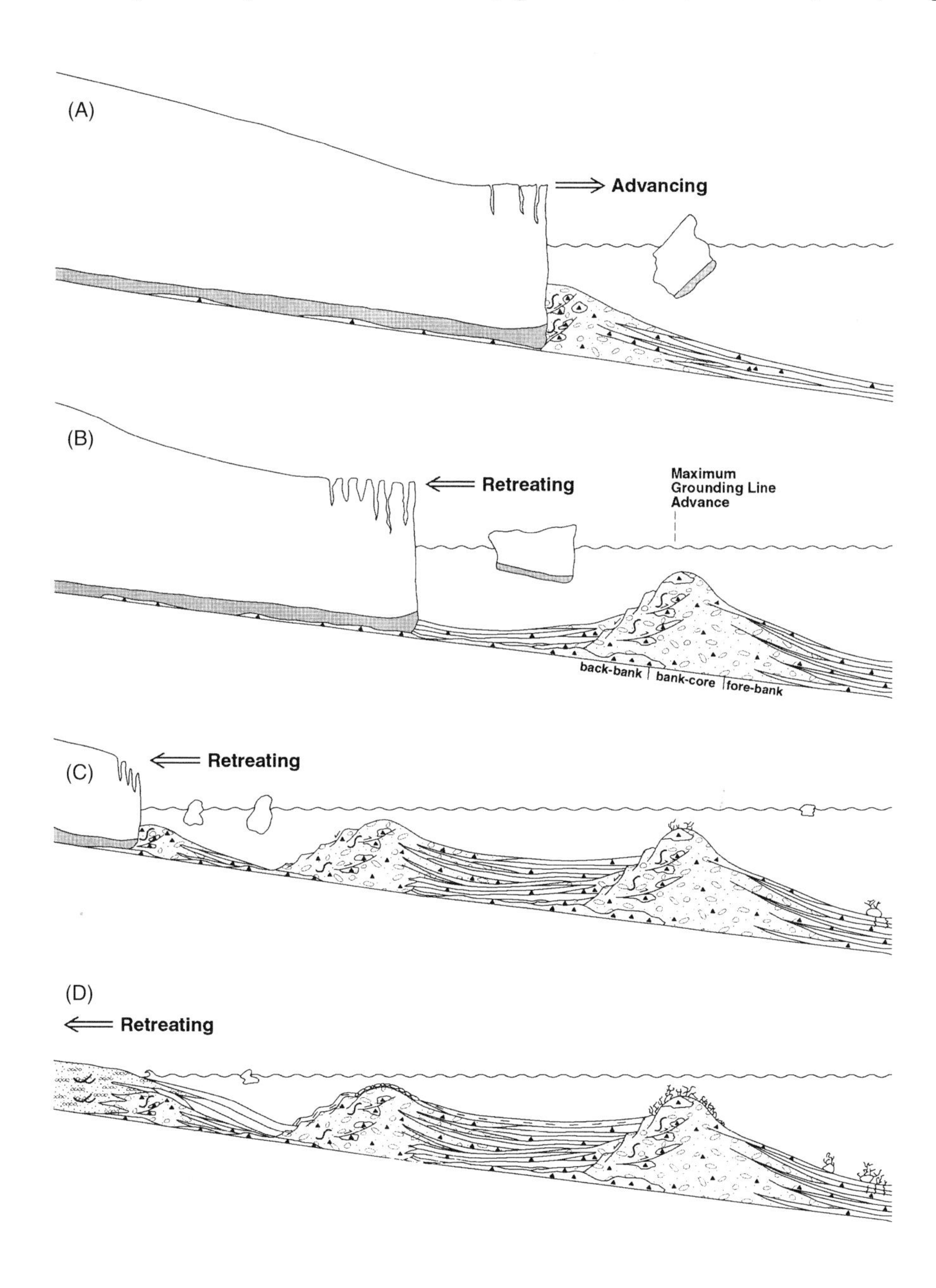

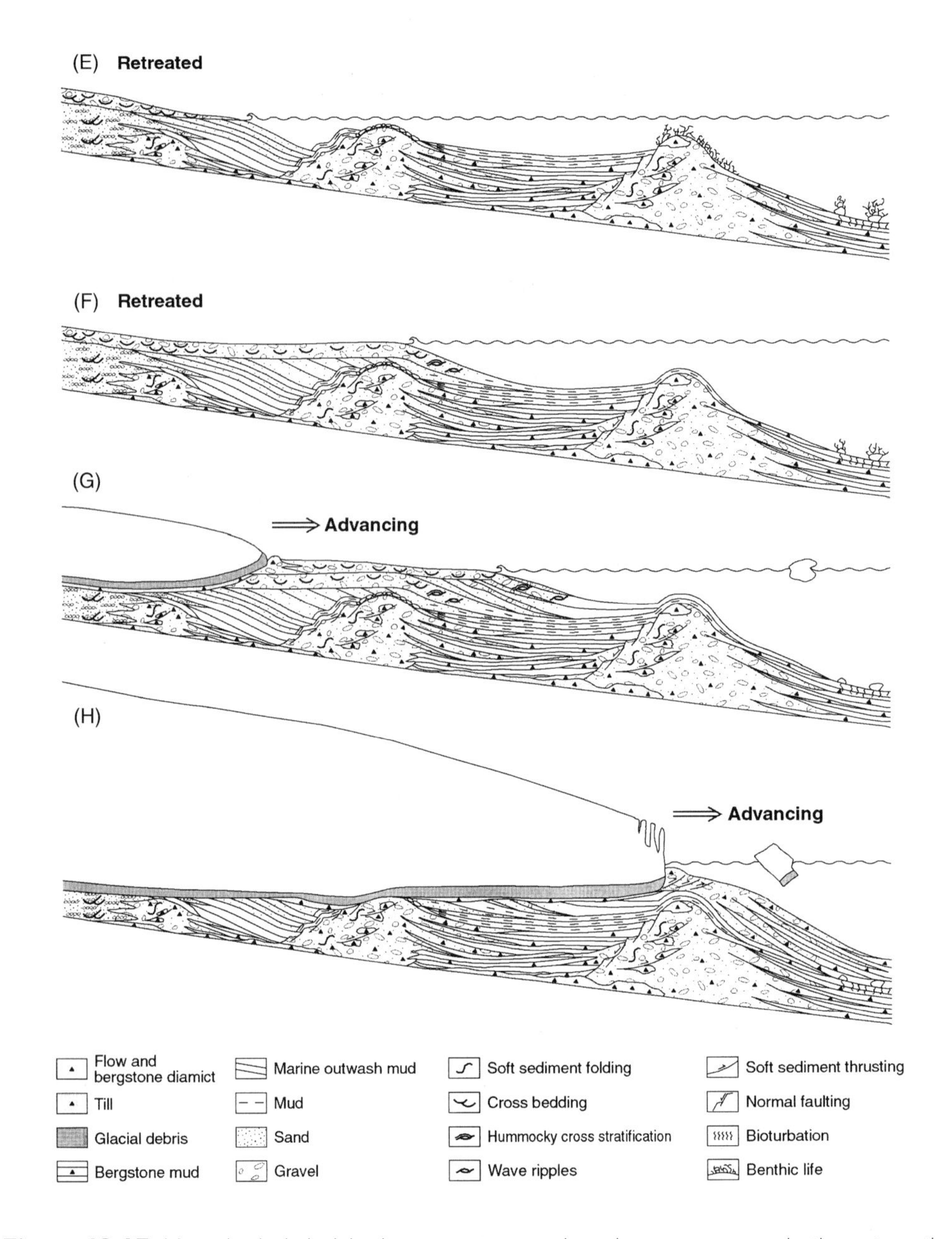

Figure 13.17 Hypothetical glacial advance, retreat and readvance sequence in time-steps that depict facies associations and sequences. Scale varies in order to allow pertinent features to be shown: vertical scale is ~1:15,000 and the horizontal scale varies between morainal banks (~1:140,000) and non-bank areas (>~1:40,000). The facies sequences are constructed to emphasize variations relative to glacial activity, so water depth does not vary; that is, effects of sea level, isostasy and tectonics are not considered. However, a particular preserved stratigraphic record will vary depending on relative lengths of time a glacier spends within different stages of its advance and retreat modes, the amount of erosion (glacial mainly) and the other variables not considered in relative water depth changes and their rates and their timing relative to the glacial advance and retreat cycle. (After Powell *et al.*, 2000).

et al., 1991, 1993; Bart and Anderson, 1995; De Santis *et al.*, 1995; Dowdeswell *et al.*, 1996; 1998; Elverhøi *et al.*, 1998; Solheim *et al.*, 1998; Vorren *et al.*, 1998; Anderson, 1999; Powell and Cooper, in press; see also Chapter 12), and fjordal systems need to be integrated and then used as tests of the numerical models. We perhaps have more information for these models for temperate and sub-polar systems than we do for polar systems.

13.5 CONCLUSION

Fjords have unique and some of the most dynamic landsystems. There is a large contrast among climatic settings and glacial regime and, to a degree, there still remains some lack of resolution among regimes. Temperate and sub-polar systems have similarities, but more data on the types of grounding-line systems, sediment accumulation rates, distribution and proportions of bergstone diamicton and bergstone mud, proportions of biogenic and organic carbon material and assemblages, and sequence architecture should pave the way to clarifying distinctions. These fjord landsystems are valuable repositories of scientific information for the earth sciences. We need to increase our efforts in understanding the processes in fjords that go to make the landsystems, as well as recover longer records of sediment within the landsystems for global change studies.

CHAPTER

SUBAQUATIC LANDSYSTEMS: LARGE PROGLACIAL LAKES

James T. Teller

14.1 Introduction

Whenever glaciers impede drainage, topographic closure may occur, resulting in the ponding of runoff. The term 'proglacial lake' has been used for lakes that owe their existence to the presence of a confining glacier margin (ice-marginal lakes), and for lakes that were strongly influenced by glacial meltwater, but which lay in a closed depression not directly in contact with the ice. In some ways this distinction is academic, and it is also arbitrary as to when a meltwater-fed lake not bounded by a glacier evolved from proglacial to non-glacial. However, ice-marginal lake basins commonly have a distinctive morphology and contain sediments with unique characteristics that reflect both their close proximity to the ice and the rapid changes that typically occur in that environment. As ice advanced and retreated, the distribution of these lakes changed; many eventually drained as the ice barrier disappeared, as isostatic rebound altered basin closure, and as outlets eroded.

During the last period of continental deglaciation, some large lakes remained ice marginal throughout their history. Other lakes, such as the glacial Great Lakes in North America, which occupied previously existing basins, were ice-marginal for only part of their history, progressively evolving to more distal meltwater-dominated proglacial lakes, and then to residual non-glacial bodies of water.

14.2 Controls on Formation and Extent of Proglacial Lakes

The main factors that controlled the extent and depth of Pleistocene ice-marginal lakes and their sedimentary and geomorphic landsystems were (Teller, 1987):

1. location of the glacier margin
2. elevation and topography of the newly emergent glacial landscape
3. location and elevation of the overflow channel
4. differential isostatic rebound, and
5. volume and nature of sediment supply.

When the glacier margin lay in the drainage basin of a lake, the hydrology, sediment load, and chemistry of inflow were strongly influenced by the proximity of the glacial source. In general, when the margin was close, lake turbidity (suspended sediment in the water column) and sedimentation rates were higher, and light penetration and temperatures were lower. As a result, animal and plant populations in the lake tended to be lower, as did taxonomic diversity (e.g. Risberg *et al.*, 1999; Björck, 1995; cf. Warner, 1990). When there was a glacial advance into the lake, some or all of the older lacustrine sediments and landforms were destroyed or buried. Glacier retreat in an ice-marginal lake was expedited by calving, and the typical convex outline of the ice margin on land probably became concave in deeper water, as waters promoted its retreat (see discussion in Benn and Evans, 1998). Both rapid retreat and advance (surging) of the ice margin is known to have occurred in large ice-marginal basins (e.g. Dredge and Cowan, 1989b; Clayton *et al.*, 1985; Evans and Rea, Chapter 11).

Of course the elevation of the lake's outlet controlled the level of the lake. Outbursts of overflow occurred when new, lower outlets were opened, which led to a draw down in the level of the lake (e.g. Teller *et al.*, 2002). The initial overflow through an outlet is the most likely time for channel deepening and further lake level lowering, mainly because discharge and flow energy were greater at this time, and because there usually was a cover of erodable glacial sediment over the bedrock.

Of particular importance to the formation of large Pleistocene proglacial lakes across North America, Asia and northeastern Europe is the fact that the continental land surface slopes northward toward the Arctic Ocean. This slope was accentuated by glacio-isostatic depression. As high latitude ice sheets expanded, drainage into the northern oceans was disrupted, resulting in large ice-marginal lakes and an extensive re-arrangement of northward drainage (Teller, 1987).

Differential isostatic loading of the northern (downslope) part of basins increased the potential storage capacity of proglacial basins. In North America and Scandinavia, the differential depression from south to north, as measured by now-deformed beaches, amounted to 200 m in some areas (e.g. Teller and Thorleifson, 1983; Andrews and Peltier, 1989; Dredge and Cowan, 1989b; Lemmen *et al.*, 1994b), although because beaches only measure rebound since ice retreated from a region, the total difference must have been greater. Measurements of modern isostatic rebound in the northern part of these basins, such as in the former Lake Agassiz and Baltic Ice Lake basins, indicate that uplift continues at more than 0.6 m per century (Barnett, 1970; Hunter, 1970; Andrews and Peltier, 1989; Eronen, 1983; Sjörberg, 1991), shifting residual lakes on their ancient floors southward through time.

When the ice margin retreated, the extent of the lake expanded. As new lower outlets opened, lake level dropped and the beach that had formed was abandoned. Subsequent differential isostatic rebound deformed the old water planes (beaches), raising those closest to the ice load more than those beyond (Fig. 14.1). And, because the rate of differential rebound decreased through time, younger (lower) strandlines developed gentler slopes.

After a draw-down in lake level, differential isostatic rebound resulted in the lake margin transgressing in the basin, south of the isobase line (line of equal isostatic rebound) that extended through the outlet, until a new outlet was opened. North of that isobase, the lake margin regressed (Larsen, 1987; Teller, 2001). A simple bathtub model in Fig. 14.2 shows a series of snapshots of changing lake levels related to overflow through three different outlets. Overflow through the southern outlet results in a gradual regression of the shoreline throughout the basin, shown in

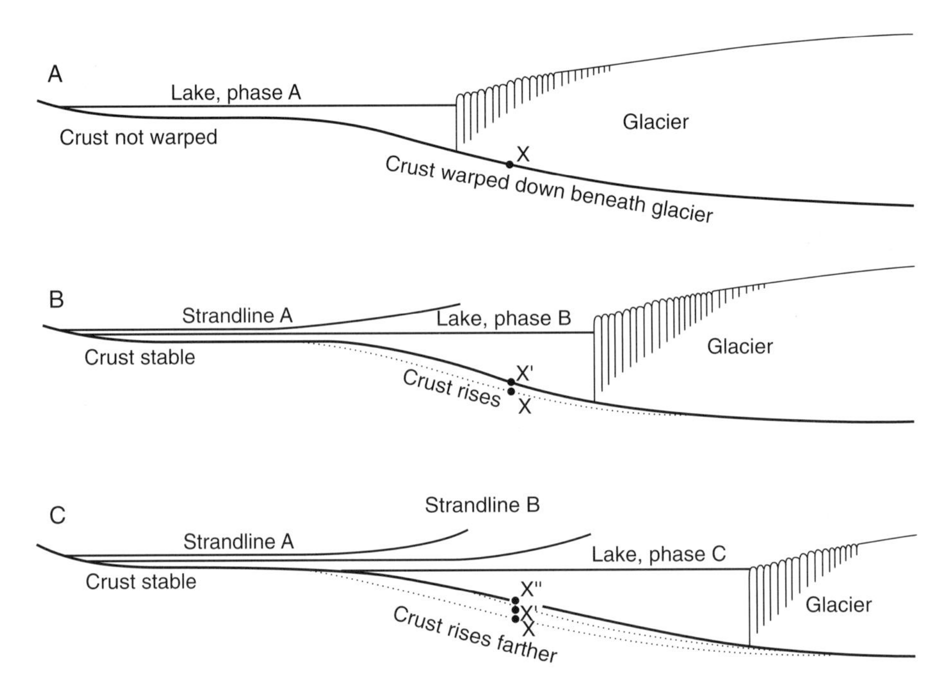

Figure 14.1 Ideal sequence of strandline formation resulting from ice retreat and overflow through an outlet at the southern end of the basin. Three different lake levels are shown (after Flint, 1971, fig. 13–11). Note different curvatures of the three ages of strandlines.

Fig. 14.2A by three lake levels. If overflow changes to the middle of the basin, differential isostatic rebound results in regression north of the isobase through the outlet, but transgression to the south (Fig. 14.2B). Overflow through an outlet in the northern part of the basin results in transgressing waters everywhere (Fig. 14.2C). During the life of most of North America's great proglacial lakes, two or three of these outlet scenarios occurred. Figure 14.3 shows how a change in outlets from south (S) to two different outlets in the middle of the basin (M1 and M2) could result in a complex series of beach levels. Superposed on all of this was outlet erosion (not shown in Fig. 14.3), which lowered lake level and was followed by continuing transgression and/or regression in the basin.

Thus, the elevation of the outlet controlled water level, and the geographic location of the overflow outlet controlled where transgression and regression of the lake margin occurred. As outlets changed during the life of a proglacial lake, mainly as a result of ice-marginal retreat (or advance), new strandlines formed. Because isostatic rebound varied exponentially from south to north, resultant strandlines are curved, not straight (Fig. 14.1).

14.3 THE SEDIMENTARY RECORD OF LARGE PROGLACIAL LAKES

14.3.1 Introduction

As glaciers melted, they relinquished their load of sediment. The hydrology and sediment transport in glacial rivers and lakes have been studied by many (e.g. Church and Gilbert, 1975;

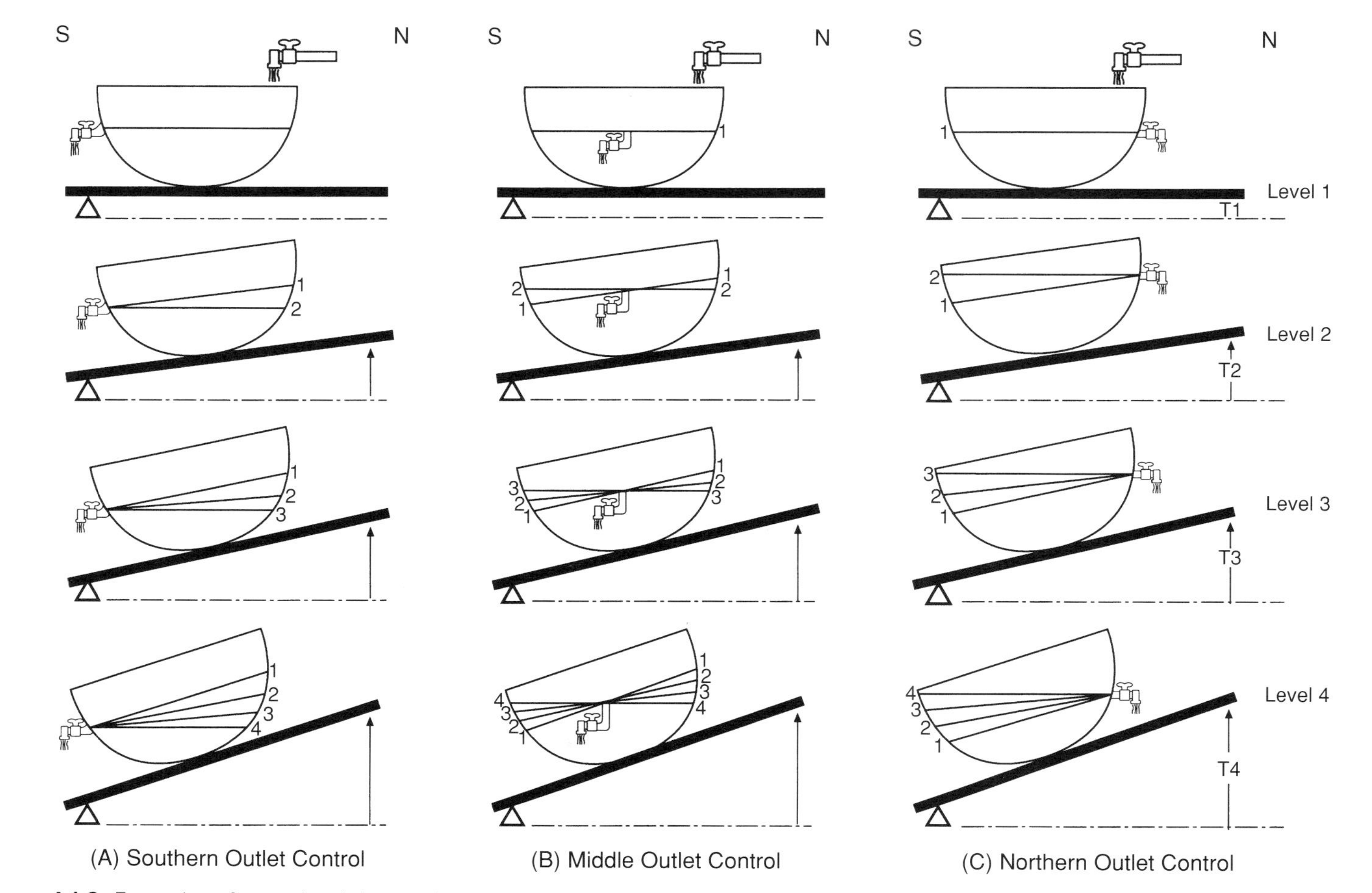

Figure 14.2 Examples of water level changes in a basin (levels 1, 2, 3, and 4 in bathtubs) undergoing differential isostatic rebound, where the outlet is in the southern end (S), northern end (N), and in the middle of the basin. A) Outlet at southern end. Lake regresses throughout basin and blanket of sand may be deposited. B) Outlet in middle of basin. Lake regresses to north of outlet and transgresses to south. Blanket of sand deposited in north, but lake and beach transgress upslope in southern end; eventually beach stranded at maximum (level 4) when lake level drops. C) Outlet in northern end of basin; lake level transgresses everywhere, moving beach upslope to maximum. (After Larsen, 1987; Teller, 2001).

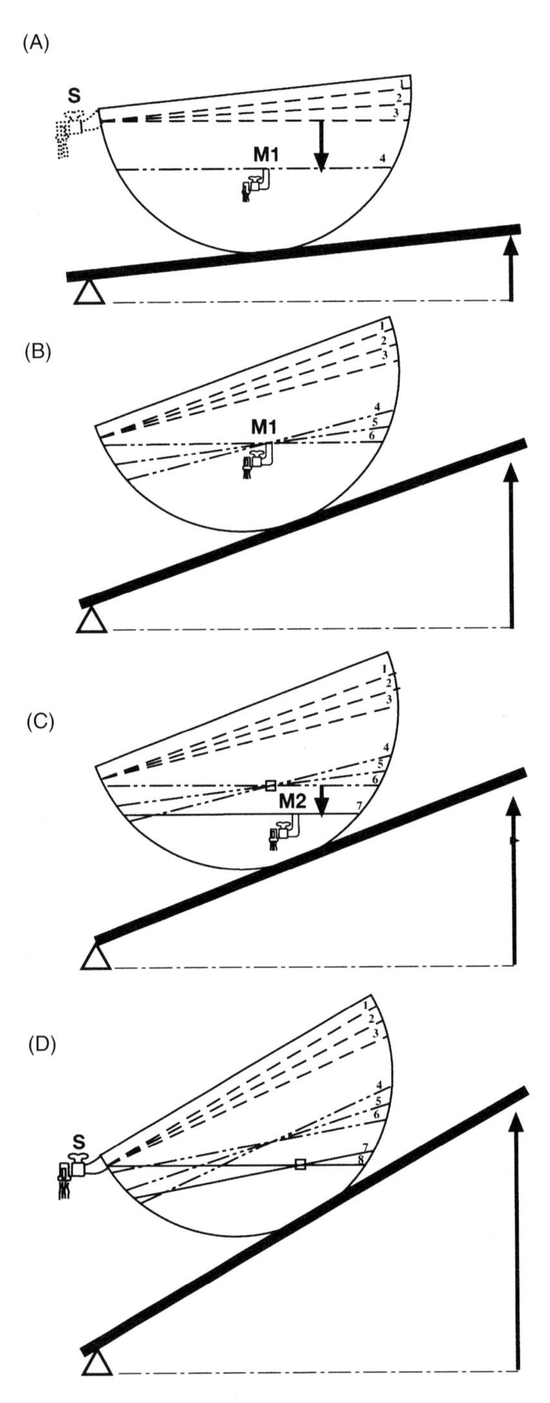

Figure 14.3 Example of more complex sequence of lake level development after southern outlet shown in Figure 14.2 A is abandoned (after Teller, 2003). A) Lake abruptly drops to level 4 after new outlet in middle of basin (M1) becomes ice-free; no large beaches formed. B) Overflow through outlet M1 produces transgression south of outlet, regression to north; beach moves upslope in front of transgression. C) New outlet in middle of basin opens (M2) and lake abruptly drops from level 6 to 7, stranding beach at the transgressive maximum of level 6. D) Lake transgresses from level 7 to 8, where it begins to overflow again through outlet at southern end of basin, stranding beach at transgressive maximum of level 8.

Gustavson, 1975; Ostrem, 1976), as has subaqueous deposition at the glacier margin and in deeper parts of proglacial lakes (e.g. Jopling and McDonald, 1975). Studies indicate that sediment loads are high in modern meltwater rivers (e.g. Maizels, 1995) and, in fact, Friedman and Sanders (1978, p. 25) show that the highest rate of sediment discharge to the oceans from any continent is from Antarctica. In the glacimarine environment of Alaskan fjords, Powell (1983) and Cowan and

Powell (1991b) measured sediment accumulation rates at an astonishing 9–13 m year^{-1}, and high sedimentation rates occur in lakes near the mouths of glacial rivers where subaqueous density currents and slumps deliver large amounts of sediment (Smith and Ashley, 1985). Away from the point of sediment influx, accumulation rates are much less, and the varved sedimentary record of many large Pleistocene proglacial basins indicates that offshore (distal) accumulation was rarely more than a few centimetres per year (e.g. Antevs, 1925, 1951; Teller and Mahnic, 1988; Ringberg, 1991; Hang, 1997). Overall, glacilacustrine sediment thins away from the ice margin and the mouths of rivers.

The rate of sediment released from a glacier is a function of several factors, among them rate of melting, type of glacier and concentration of debris in the ice. The latter is partly a function of the availability and erodibility of bedrock and the overlying unconsolidated material, so lakes in areas of resistant lithologies such as granite, which are not as easily weathered and eroded as are sedimentary strata, tend to receive relatively small influxes of sediment that contain low percentages of clay-sized material.

Ice-marginal lakes have higher sedimentation rates than do their downstream proglacial counterparts, and the nature of the sedimentary record is different. Abrupt jumps in sedimentation rate or style, grain size, mineralogy, and biological composition will occur when the ice retreats from (or advances into) a lake basin. Progressive changes in these parameters will occur as an ice margin advances or retreats across the lake basin, and these changes can be used to interpret the location of the margin in the basin through time.

Morphologically, most proglacial lake deposits tend to form flat landscapes of clay and silt that drape over (and may ultimately obscure) previous features in the offshore region of the lake. The flat offshore areas of lake sediment typically are outlined by wave-cut cliffs or by coarser sediment deposited in shallow water, such as beaches and lags of wave-washed sediment.

The focus of the following sections will be on those sediments and landforms in large ice-marginal lakes dammed by continental ice sheets. In many ways, processes in these lakes are the same as those in non-glacial environments. However, there are some distinctive (even unique) aspects about the sediments that relate to the presence of the bounding (or nearby) ice margin and its highly dynamic nature, the abruptness of water inflow and outflow, iceberg calving, and isostatic rebound. Unlike many other lake settings, transgressions and regressions are common during the life of proglacial lakes because of the frequent and often large changes in ice margin, outlet depth and location, runoff into the lake, and differential isostatic rebound. As a result of erosion during transgression and subaerial exposure, as well as glacial readvance, unconformities commonly are present in the sequence, particularly in shallower parts of the basin.

14.3.2 Surface Appearance

Commonly, the surface of a large proglacial lake basin is flat, partly because continental ice has smoothed the preglacial surface and partly because wave action and sediment accumulation have further reduced the relief. Relatively impermeable sediments and low relief in many large palaeo-lake basins have resulted in poor drainage; up to 4 m of peat has accumulated in places on the floor of these ancient lakes (e.g. Dredge and Cowan, 1989). In Canada, many of these areas today are covered by muskeg, and the Hudson Bay Lowland and adjacent Canadian Shield support the largest area of organic terrain in the world (Dredge and Cowan, 1989).

The keels or edges of icebergs, and even winter ice, sculpted long linear and curved grooves in the mud on the floor of many lakes, with associated ridges immediately adjacent to them (Fig. 14.4) (see Woodworth-Lynas, 1996 for summary and discriminating characteristics), and these features are found over extensive areas of proglacial lake basins in North America as well as in modern lakes (e.g. Clayton *et al.*, 1965; Dredge, 1982; Woodworth-Lynas and Guigné, 1990; Gilbert *et al.*, 1992). These straight-to-curved features criss-cross each other. In oceans, iceberg scours have been found in water depths of 500–950 m (Barnes and Lien, 1988; Polyak *et al.*, 2001). Sediment homogenization and faulting are commonly associated with these scour and plough marks. Although they are very subtle landforms, rarely having more than a metre or two of relief today, they are striking features from the air on an otherwise featureless offshore lake plain (Fig. 14.4). Tonal contrast results from the difference in organic accumulation in the soil between the better-drained ridges and adjacent lower areas. These ridges commonly lie adjacent to furrows and were interpreted by Clayton *et al.* (1965) as iceberg scour grooves and ploughed ridges, which are today more subdued than when originally formed. Woodworth-Lynas and Guigné (1990) argue that

Figure 14.4 Aerial photo of iceberg scour marks in the silty clay of the Lake Agassiz plain east of Winnipeg. Light linear tones are slightly elevated ridges. Main roads are spaced 1 mile (1.6 km) apart, and west to east distance across photo is about 5.5 miles; north is toward top.

these ridges were once the iceberg grooves which have been topographically inverted due to greater compaction of sediment on either side.

In addition to these extensive curvilinear scour features, Mollard (1983, 2000) describes and illustrates a variety of other irregular patterns that developed on lake floors in glacial regimes (Fig. 14.5), proposing various origins for them. The clayey and silty sediments in these features commonly exhibit convolutions, diapiric structures, and deformation (Mollard, 1983). All have very low relief. For example, as lake levels decline in an ice-marginal lake, icebergs may 'settle into the soft lake-bottom mud', resulting in folded and faulted silts and clays and a shallow depression that ranges from a few tens of metres to a few hundred metres in diameter (see Fig. 14.6). These may be rimmed by coarser drift that slipped off the ice mass, or by detritus that ablated into the depression (Thomas, 1984b; Mollard, 2000). Some of the low-relief 'doughnut' and brain-like patterns seen on aerial photographs in glacial terrains may be the result of this process (Fig. 14.5), although similar features are produced outside of lake basins in stagnant ice regions (Clayton and Freers, 1967; Mollard, 1983, 2000). Shallowing lake waters may also have resulted in the differential freezing (expansion) and thawing of lake muds, producing small pingo-like mounds that occasionally have central craters (Mollard, 2000). Even groundwater 'piping' from artesian

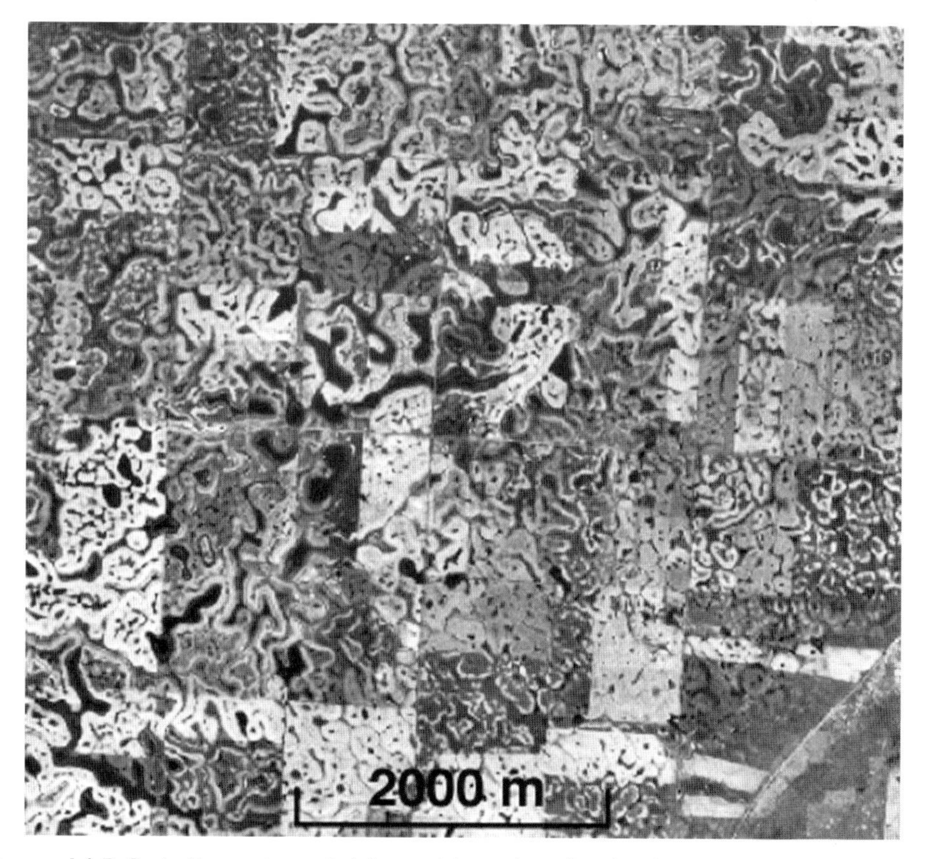

Figure 14.5 Brain-like pattern of deformed lacustrine silty clay, interpreted to have resulted from floating ice sinking into the soft mud (Mollard, 2000, fig. 6). The change to a less-distinct pattern with smaller rings in the lower right is interpreted to reflect deeper water.

flow from aquifers below the lacustrine muds may develop discharge pits and surrounding raised rims (Mollard, 1983).

Of course the morphology of a lake basin also reflects the underlying topography of the basin floor, which may be controlled by bedrock or an older erosional or depositional surface. Ice-marginal landforms deposited on the lake, such as end moraines, deltas and fans, may remain as distinct features on the lake floor, although they are vulnerable to wave erosion. A drape of younger lacustrine sediment may further diminish the relief, leaving relict (palimpsest) landforms.

14.3.3 Offshore Sediments

14.3.3.1 Iceberg Contribution

Icebergs may raft detritus of all sizes into ice-marginal lakes, releasing their load as they melt, tip or break up, and adding it in varying proportions to sediment on the floor of the lake (Fig. 14.6); this rain of detritus commonly deforms or even obliterates pre-existing laminae (Fig. 14.7), and consists of rock fragments, individual mineral grains, and fragments of unconsolidated sediment such as till (Ovenshine, 1970). The concentration of this ice-rafted detritus may be so great that the resulting sediment may resemble till (e.g. Dreimanis, 1979; see also Benn and Evans, 1998). In general, higher concentrations of this detritus are found close to the ice margin, beneath floating ice shelves, or where icebergs become grounded in shallow water and decay. In addition, sediment already deposited may be highly disturbed by icebergs ploughing into the soft sediment, producing an ice-keel turbate that may be similar in appearance to till (Vorren *et al.*, 1983; Barnes and Lien, 1988; Woodworth-Lynas and Guigné, 1990).

14.3.3.2 Fine-Grained Offshore Accumulation

Most sediment entering a lake from rivers beyond the ice margin will initially be reworked by waves. Coarser sediment will be deposited in shallower waters around the lake margins (see section

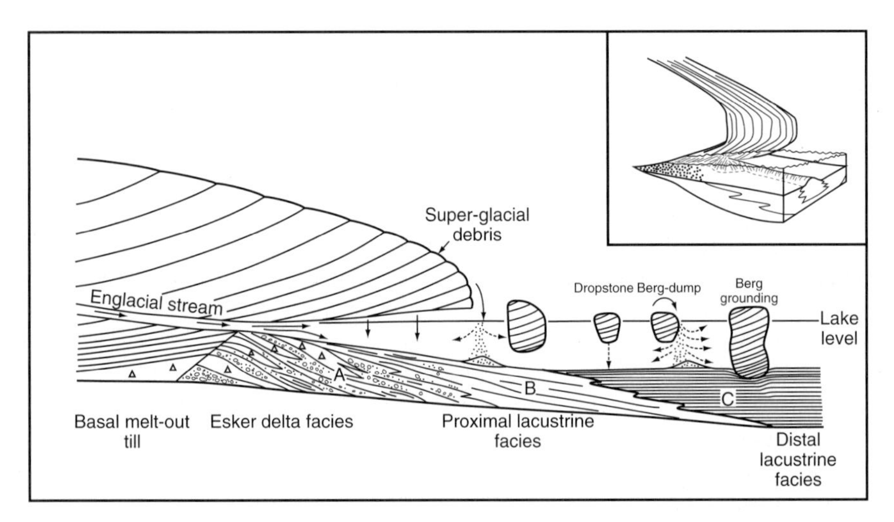

Figure 14.6 Construction of a fan delta (A) into a lake at the mouth of an englacial tunnel (commonly referred to as a grounding-line fan), showing more distal lacustrine facies (B and C) and additions of unsorted detritus from glacier overhang and icebergs. Note grounded iceberg and resultant depression. (From Thomas, 1984b, fig. 5).

Figure 14.7 Photo of laminated glacilacustrine sediment containing ice-rafted detritus. In the upper part a till-like sediment (diamicton) has been produced.

14.3.4 below), and accumulations may occur at the ice margin (see section 14.3.3.3, below). Where wave energy is greater, turbulently suspended finer sediment may be transported offshore in the surface layer of the lake (epilimnion), where it eventually is deposited during periods of low wave activity or after the lake surface has frozen and ends all wind stress. Of course, biogenic materials and even clays agglomerated by organisms into faecal pellets (Smith and Syvitski, 1982) may be added to these sediments. Where the load of a river entering a lake is fine grained, wave turbulence may keep some of the sediment suspended for long periods, resulting in some of these clays and silts overflowing from the basin through the outlet during the ice-free season (Smith *et al.*, 1982).

In deeper waters, particularly away from the ice margin, the clays and fine silts that settle out of the epilimnion are commonly laminated, with fine-coarse couplets resulting from the seasonal variation in runoff to the lake and the seasonal influence of wind that keeps some sediment suspended until the lake freezes. These annual sediment couplets, called varves (Fig. 14.8), form extensive blankets in many proglacial lake basins. Although the thickness and nature of the couplets vary spatially and through time – being related to sediment supply, proximity to the source (ice margin, river mouth), and depth below wave base – they commonly remain monotonously similar through time (Fig. 14.8). In fact, this repetitive aspect argues for control by an annual lake rhythm. Where couplets can be identified with confidence as annual increments, they have been effectively used to establish a glacial and glacilacustrine chronology, (e.g. DeGeer, 1912; Antevs, 1922; O'Sullivan, 1983; Ringberg, 1991; Wohlfarth *et al.*, 1993). Smith and Ashley (1985) discuss the difficulty of distinguishing true annual couplets from other rhythmic processes in lakes such as pulsing density underflow currents, repeated slumping, and

Figure 14.8 Typical light-dark (summer-winter) couplets (varves) of relatively coarse and fine sediment. (After Teller (1987, fig 3)

varying wave input. In fact, individual varves themselves commonly display seasonal grain-size variations that relate to wave energy and changes in sediment input due to meltwater runoff, precipitation, and slumping of unconsolidated sediment along the lake margin (Fig. 14.9). Ringberg (1984) has even identified a diurnal rhythm of accumulation within thick proximal varves. As described by Smith and Ashley (1985), true varves (versus surge rhythmites) may contain a lake-floor fauna, a seasonal progression of pollen and other organisms, internal interruptions in sedimentation, and grain sizes in the coarser part of the couplet that displays either an upward coarsening or no trend (versus an upward fining in gravitationally implaced sediments).

Interestingly, in some parts of the largest proglacial lakes, such as Lake Agassiz and the glacial Great Lakes, sediments are not varved, except in more protected areas (e.g. Antevs, 1951; Teller, 1976, 1987; Colman *et al.*, 1994). In part this may be because wind mixing of the water column re-suspended several years of previously-deposited sediment, homogenizing the fine-grained winter accumulation with the coarser summer increment. Continuous density underflow throughout the year may have played a role in inhibiting varving. Additionally, iceberg and winter ice scouring of bottom sediments in tens, if not hundreds, of metres of water may have resulted in homogenization of materials to a depth of several metres (Woodworth-Lynas and Guigné, 1990). Low lake levels may be reflected in the deeper-water sedimentary sequence by erosional unconformities, alluvial units, coarser lacustrine sediment, or by accumulation of units enriched in either *in situ* or reworked organic matter.

14.3.3.3 Accumulation at the Mouths of the Rivers and Near the Ice Margin

High sediment loads are common in rivers entering proglacial lakes directly from the ice margin, as well as from newly deglaciated terrestrial areas; rapid sedimentation is common in

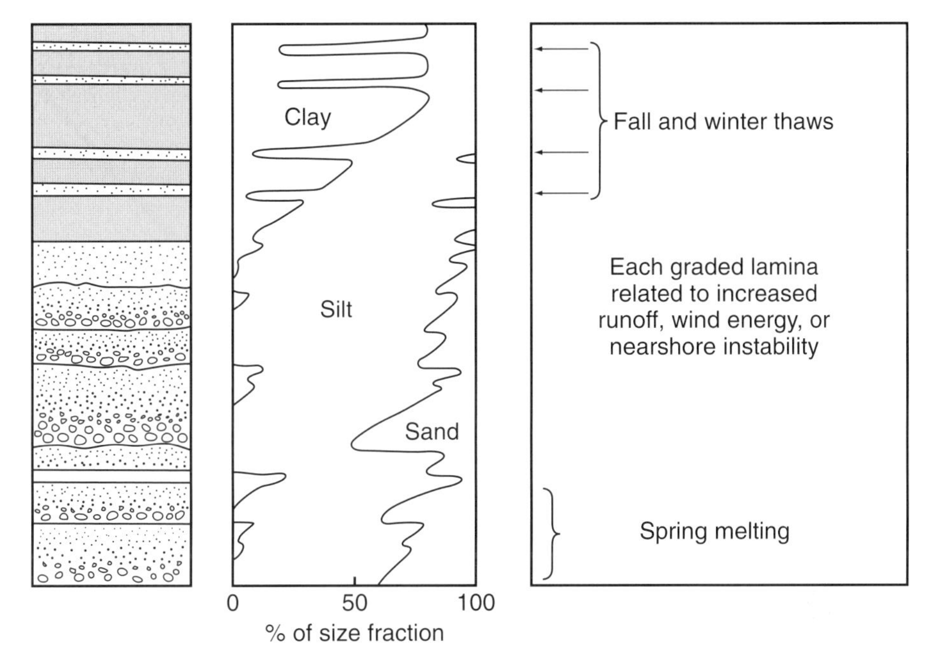

Figure 14.9 Schematic diagram of grain size variation in a single varve, resulting from changes in runoff, wind, or gravitational transfer. Note that the typical overall coarse to fine (summer-winter) nature still is present. (After Teller (1987, fig. 7).)

proximal areas. Under some circumstances, traditional Gilbert-type deltas with steep foreset beds and associated topset (subaerial) and bottomset (deeper water) beds may be deposited (Fig. 14.10A) (Fyfe, 1990; Postma, 1990). These commonly form at the mouths of rivers where the velocity of the inflowing water is abruptly checked, and the positive imbalance between sediment supply and water depth leads to sediment storage on the delta slope that exceeds removal and downslope transfer (Nemec, 1990). In an ice-marginal environment, the delta surface is graded to the supplying river and to lake level (Fig. 14.10A). If the ice margin remains stable for a relatively long period of time, a subaerial distributary river system develops, depositing the alluvial sediments of the topset portion of the delta. Foreset thicknesses are controlled by the depth of water, and range from a few metres to over 100 m. These beds prograde into the lake in a complex way, mainly by avalanching, with gentler (24–29°) slopes in sandy sediment and steeper (30–35°) slopes in gravels (Jopling, 1965; Nemec, 1990); gentler concave-up slopes and more tangential transitions to bottomset beds are found in finer sediment because more sediment remains in suspension past the crest of the foreset slope (Jopling, 1965). Foreset beds commonly parallel the delta face and may extend over the entire downslope length of the delta face. However, variable avalanching and sediment supply may lead to discontinuous beds that wedge-out upslope or downslope, or that assume lobe or tongue shapes (Nemec, 1990; Fig. 14.10). Because water levels in proglacial lakes frequently change, deltas may become 'stacked' (Thomas, 1984a), drowned, or eroded during transgressive phases and be incised by rivers when lake level drops; new deltas or fans then form at the new lake level.

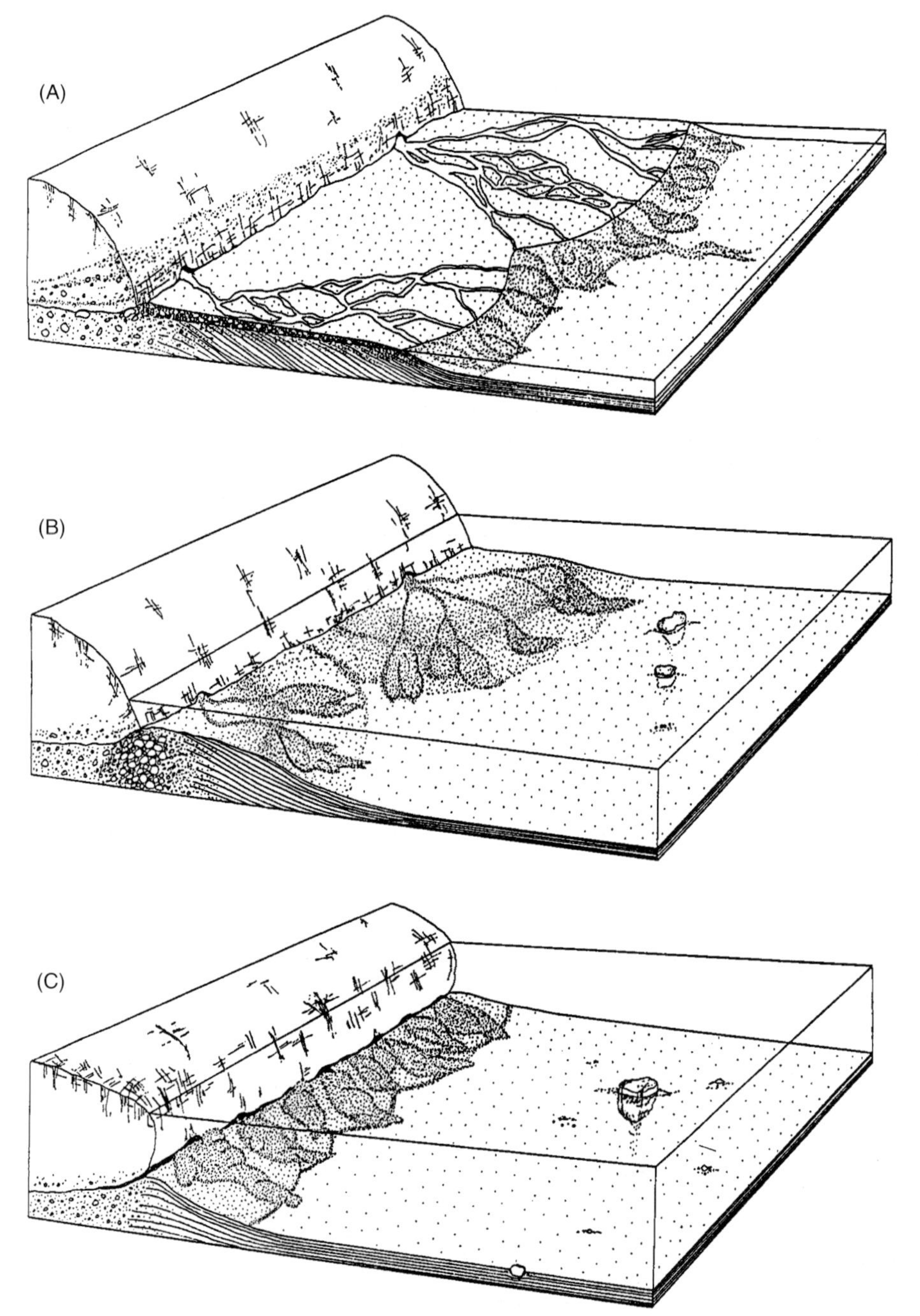

Figure 14.10 Three possible types of sedimentation in a lake at the margin of an ice sheet (Fyfe, 1990, fig. 7). A) Gilbert-type delta complex fed by two subglacial conduits, showing subaerial alluvial topsets and steeply-dipping foreset beds that merge with bottomset beds in deeper water. B) Subaqueous fan delta lobes fed by subglacial conduits, which may eventually aggrade enough to initiate a Gilbert delta; avalanching and mass movement dominate; sedimentation away from the ice margin is mainly from density underflow currents. C) Subaqueous fan apron in deep water fed by numerous conduits linked by a subglacial cavity system; the complex of rapidly-deposited fluvial sediment and diamicton from the ice commonly is called a grounding-line fan (see also Fig. 14.6).

Sediment entering ice-marginal lakes through meltwater rivers at or near the base of a glacier may be deposited as subaqueous fans, the beds of which slope more gently into deeper water (e.g. Gustavson *et al.*, 1975; Powell, 1990; Eyles and Eyles, 1992; Ashley, 1995; Figs. 14.10B, 14.10C, 14.11A). Benn and Evans (1998) discuss this environment in detail. High sediment concentrations in the inflowing river favour gentler slopes and fan (versus Gilbert delta) construction by density underflow currents (Jopling, 1965). As the ice margin retreats, a broad esker-like ridge may develop by overlapping fans if the tunnel remains active (Fig. 14.11B). These fan-shaped deposits may merge with other fans deposited at the mouths of subglacial conduits along the ice margin, forming a variably continuous ridge of subaqueous fluvial sediment (Fig. 14.10B and 14.10C), such as those of the extensive Salpausselka moraines across southern Finland that were deposited as ice retreated into shallower water along the northern side of the Baltic Ice Lake (e.g. Fyfe, 1990; Raino *et al.*, 1995). These are commonly called grounding-line fans (Fig. 14.6).

Sediment in the resultant asymmetrical bedform has a steep ice-contact side, fines downslope into the basin and typically consists of gravels, sands and occasional diamicton units (Fig. 14.12); some have considered these ridges DeGeer moraines. Because sediment transport across ice-marginal subaqueous fans is mainly in highly turbulent flows and by gravitational transfer, sorting and bedding may be poor (Nemec, 1990; Benn and Evans, 1998). A drape or fringe of diamicton on these fans is common where there was an overhanging ice margin (Fig. 14.10), and Thomas (1984b) reports conical mounds of diamicton within the fan sequence that have

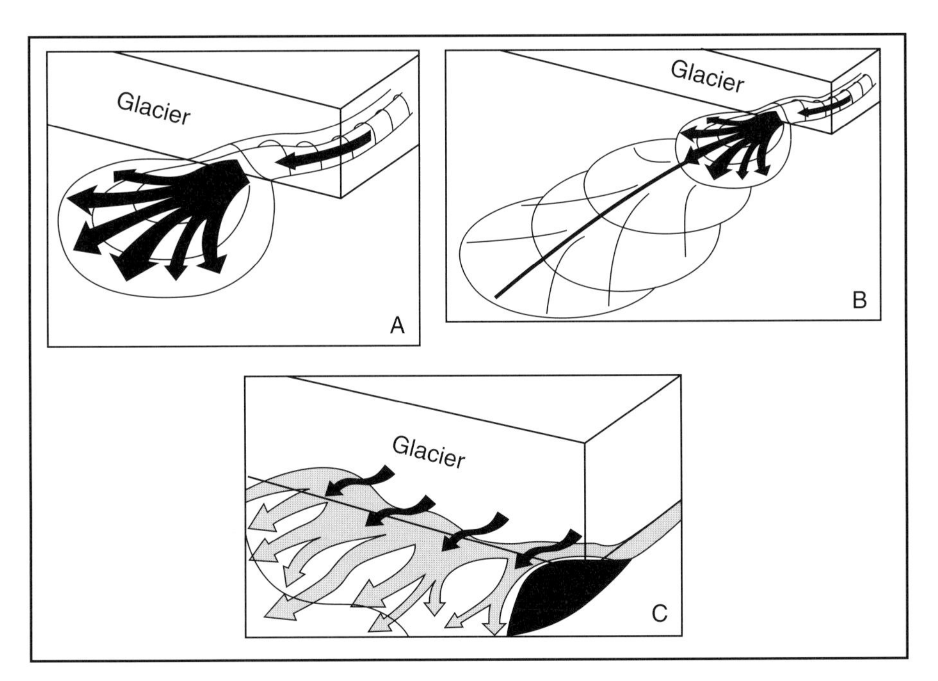

Figure 14.11 Schematic of meltwater deposition in proglacial lake at glacier margin. (After Sharpe *et al.*, 1992). A) Subaqueous fan at mouth of subglacial tunnel. B) Series of overlapping subaqueous fans producing a broad esker-like ridge as ice retreats. C) Subglacial meltwater outburst that produces a broad moraine or extended ice-marginal fan below lake level.

Figure 14.12 Moderately well-sorted and bedded ice-contact gravel in fan delta near Winnipeg that is overlain by 1 m thick diamicton (flow till) that thins toward the left and is capped by laminated glacial Lake Agassiz silts and clays.

up to 60 cm of relief (Fig. 14.6). Faulting and deformation in these units is common where deposition was on or against the ice (Fig. 14.13). Readvances of the margin may result in considerable deformation and thrusting of these ice-marginal sediments (e.g. Boulton, 1986); this may occur on an annual basis as a result of the decline in winter ablation of the ice margin (Krüger, 1993).

On the steep slopes of subaqueous fans deposited in deep water, debris avalanching is common across the surface, resulting in deposition of coarse lobes and anastomosing gravel- and sand-filled chutes like those found in subaerial alluvial fans (Fig. 14.10C) (e.g. Fyfe, 1990; Nemec, 1990; Prior and Bornhold, 1990). Debris-flow diamictons are commonly deposited, reflecting the proximity to the ice margin, and slumping of the proximal side of ice-contact deltas may occur as a result of the melting of buried ice and removal of ice-marginal support. These episodic, subaqueous mass movements may evolve into dense turbulent underflows or turbidity currents, which erode channels in pre-existing substrate and rapidly deposit finer sediment in deeper water (e.g. Prior and Bornhold, 1990; Benn and Evans, 1998).

Abrupt releases of subglacial meltwater may occur during glacial retreat (e.g. Shaw *et al.*, 1989; Shoemaker, 1992; Shaw, 1996) and during rapid advances such as surges (Sharpe and Cowan, 1990). Short-distance surges (see Dredge and Cowan, 1989, Fig. 8.23), as well as extensive surges (Clayton *et al.*, 1985; Clark, 1994a), appear to have been common in ice-marginal lakes and may have coincided with outbursts of subglacial water. Although the processes and magnitudes of such outbursts are controversial, they would have delivered large volumes of stored water and debris to the glacier margin in a short time. A study by Sharpe and Cowan (1990) in the Lake Agassiz basin led to the conclusion that some subaqueous moraines with a

Figure 14.13 Highly faulted bedded delta sands that were deposited over ice along retreating ice sheet in southern Sweden; note coarse ice-marginal gravels at right.

steep proximal and gentle distal form might have been deposited along the ice margin during such outbursts (Fig. 14.11C); these moraines consist of stratified-to-massive sediment that grades upward from gravel to sand and silt. They suggest that this outburst may have resulted from a rapid lowering of lake level, which produced a large hydrological differential (Fig. 14.14). This type of moraine may be similar in nature to subaqueous grounding-line fans that accumulated by 'normal' meltwater discharge from various subglacial tunnels. Beyond the ice margin, sedimentation rates, grain size and sedimentation style reflect such outbursts. In addition, when the confining margins of ice-marginal lakes fail, large slugs of water are normally released. This may have a domino effect on other lakes downstream, as well as on fluvial systems, and subaqueous fans may be very rapidly deposited in these downstream lakes (e.g. Kehew and Clayton, 1983; Kehew and Lord, 1987; Kehew and Teller, 1994). Turbulent high discharges of sediment from subglacial tunnels (or proglacial rivers), as well as those related to mass sediment transfer such as slumps and avalanches from delta faces and ice margins, commonly evolved into density underflows that moved along the lake floor until their turbulence and energy dissipated (Fig. 14.15). These underflows commonly were episodic, although some glacial rivers may have delivered a continuing, albeit varying, high concentration of sediment to proglacial lakes for long periods, which retained their identity over long distances and over gentle bottom slopes (e.g. Smith and Ashley, 1985). Much of this sediment was focused into relatively low topographic areas on the basin floor. In fact, most sediment in the deepest part of large proglacial lakes was probably transported there by these subaqueous gravity currents (see Smith and Ashley, 1985), and the so-called bottomset units of Gilbert-type deltas are partly composed of sediment from these flows and partly of material settling out from higher in the water column. Where there is a high influx of suspended fine sediment (usually fine sands and coarse silts) and where the sediment accumulation rate is high, a climbing-ripple

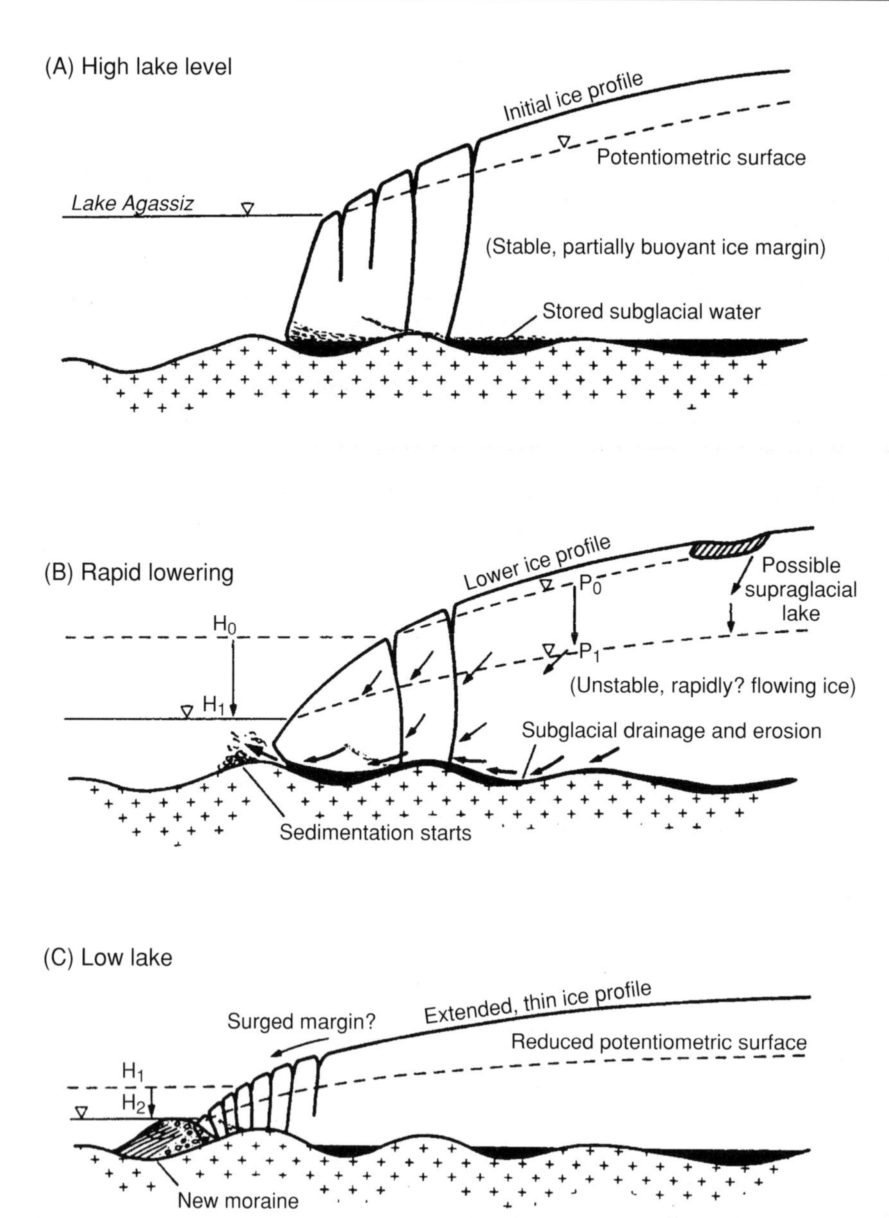

Figure 14.14 Schematic of subaqueous moraine formation by subglacial outbursts into an ice-marginal lake. A) Initial high lake stage, showing stored subglacial water. B) Abrupt lowering of lake that increases pressure head and promotes outward flow of subglacial water and sediment; glacier surging may also occur at this stage. C) New moraine, expanded and thinner ice margin, and new stable condition following subglacial outburst. (Sharpe and Cowan, 1990, fig. 10).

sequence may be deposited (Fig. 14.16), which consists of a long succession of ripples whose crests are incrementally offset from one another (in-drift climbing ripples) or lie directly above each other (in-phase climbing ripples) (e.g. McKee, 1965; Jopling and Walker, 1968; Gustavson *et al.*, 1975).

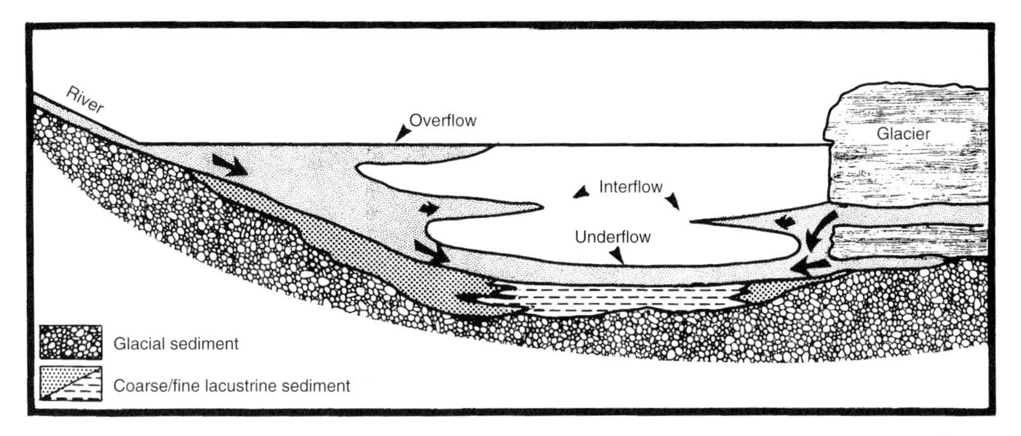

Figure 14.15 Ice-marginal lake showing distribution of suspended sediment (light grey) arriving from glacier and river as underflow, interflow, and overflow in water. All but finest grains move across lake floor as relatively dense underflow suspensions into centre of basin, progressively depositing finer and finer sediment. Finer grains may be transported by wave energy as overflows and intermediate sizes may move at the epilimnion-hypolimnion density interface. (After Teller, 1987, fig. 4).

14.3.4 Nearshore Sediments

Wave action and longshore drift commonly reworked sediment delivered to a lake, depositing grains around the margin that were too large to be transported into deeper water. In simplest situations, each beach is an indicator of lake level, and is a chrono-morphological and chrono-stratigraphic marker (Fig. 14.17). However, subaqueous bars, just offshore from the main beach, which commonly result from wave return flow, may form contemporaneously with the main beach (Figs. 14.18) (e.g. Walker and Plint, 1992). Around large lakes, where wind fetch is high, storm beaches may introduce further interpretive difficulties, because their elevation may extend well above the normal level of the lake (cf. Otvos, 2000). Thus, several levels of 'beaches' may form at the same time. Furthermore, wind activity during and after beach formation may add a cap of dunes on these lake-level landforms, adding to the difficulty of establishing exactly what the water level was in a lake at a given time.

As waves eroded their margins, helping to provide sediment for beaches, wave-trimmed scarps developed on headlands and exposed shorelines. Less distinct 'washing limits' of waves around the lake margin can also be correlated, and provide evidence for palaeo-lake level (Veillette, 1994).

When lake levels declined, nearshore deposits regressed downslope at the lake margin (e.g. Fig. 14.2A). Unless there were interruptions to this decline in lake level, or unless storm beaches were constructed, there was little morphology associated with deposition, and a blanket of sandy sediment was deposited, generally with only low-relief ridges on it (e.g. Posamentier *et al.*, 1988; Walker and Plint, 1992; Thompson and Baedke, 1997).

As lake levels rose, beaches migrated upslope (e.g. Fig. 14.2C). This transgression resulted in waves reworking shoreline deposits so that, except when rapid drowning of the shoreline occurred, older

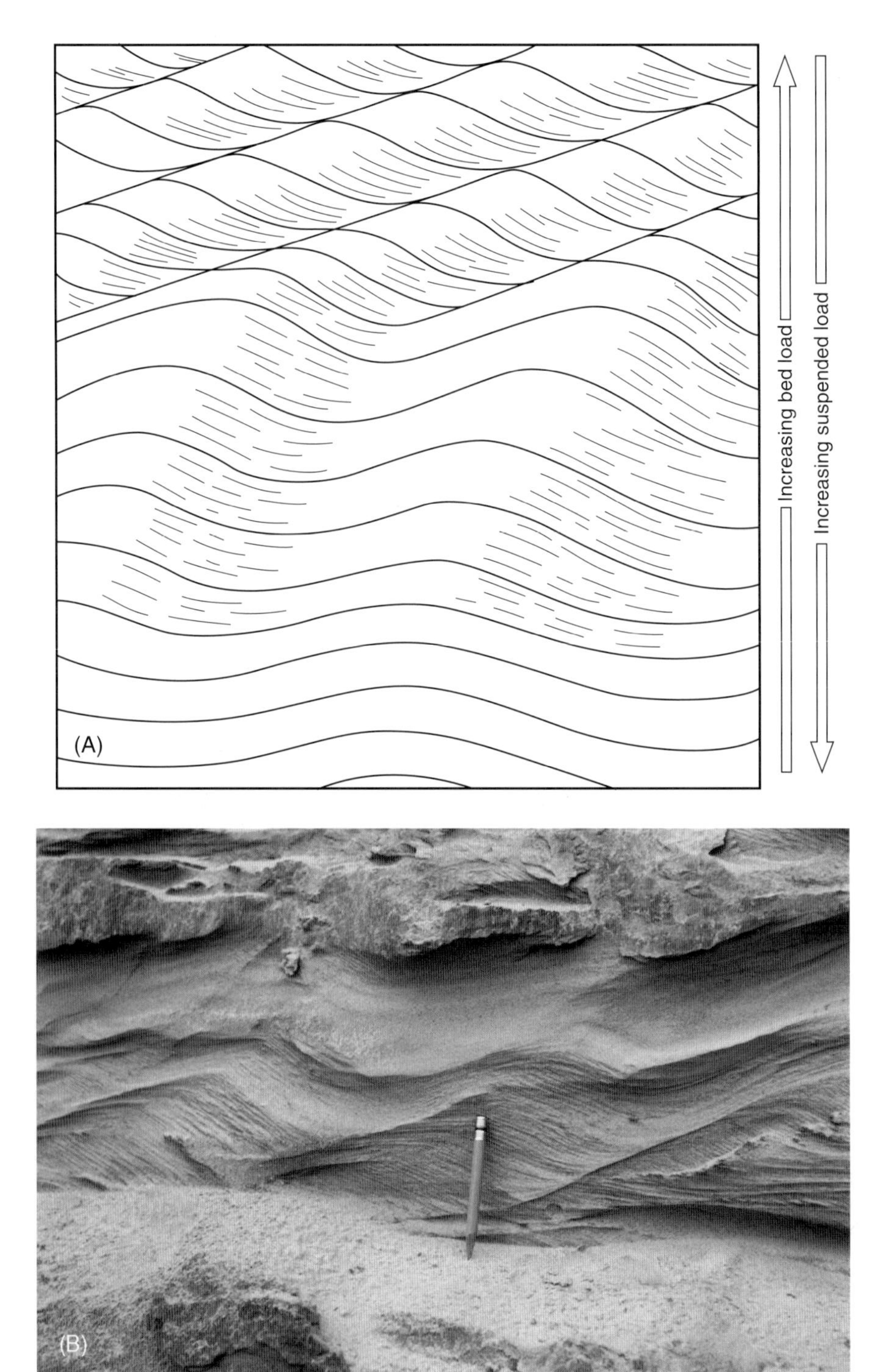

Figure 14.16 A) Schematic of climbing ripple lamination, showing evolution of in-phase ripples to in-drift ripples and the decrease in angle of climb as the ratio of fallout from suspension to bedload decreases (after Jopling and Walker, 1968). Flow left to right. Preservation of laminae on stoss side of ripple decreases upward. If suspended load/bedload decreases to where there is little sediment in suspension, ripples will only migrate and will not grow upward at the same time. B) In-drift climbing ripples.

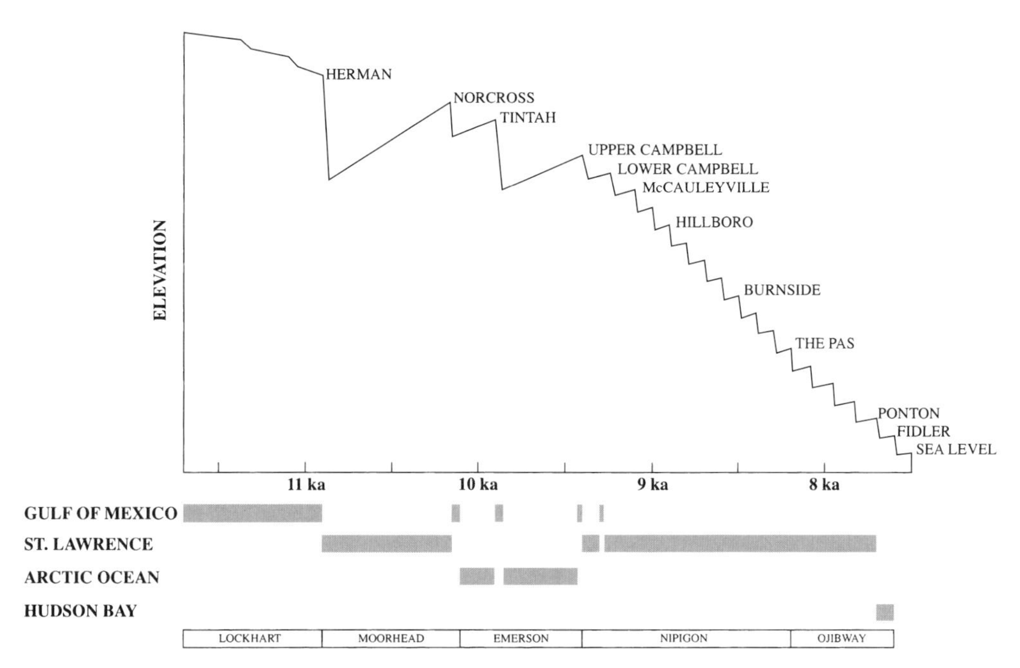

Figure 14.17 Schematic of relative changes in level of Lake Agassiz from 11.7 to 7.7 ka, showing rising (transgressing) stages resulting from differential isostatic rebound after the opening of a new lower outlet caused an abrupt drop in lake level. The occurrence of transgressions indicates that these shorelines were south of the outlet. (After Teller, 2003).

beach morphology was destroyed. As discussed previously and illustrated in Fig. 14.2, in proglacial lakes around continental ice sheets, differential isostatic rebound resulted in lake level transgression south of the isobase through the overflow outlet, and regression north of the isobase. This means that large beaches did not form north of the isobase that extends through the outlet, and that beaches to the south reflect only the maximum level of transgression for any given phase of a large lake (Teller, 2001).

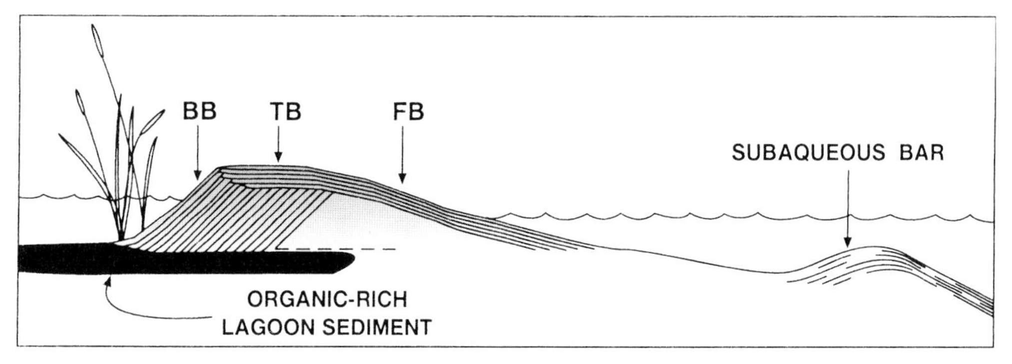

Figure 14.18 Cross-section sketch of barrier beach between lagoon and lake, showing gently dipping (planar) forebeach laminae (FB), nearly flat top-beach laminae (TB), and angle of repose backbeach laminae (BB) that prograde into lagoon during periods when waves overtop the barrier. Subaqueous bar may be tip of growing spit or ridge accumulated by wave return flow.

As with beaches around non-glacial lakes, bedding gently slopes lakeward on the foreshore (Fig. 14.18). Straight, narrowly spaced, symmetrical ripples, typically with rounded crests, form in the nearshore zone where there is an overall balance between wave uprush and return flow, whereas asymmetrical ripples form where unidirectional flow dominates; close to shore, megaripples may form and these become planar laminae on the beach face (e.g. Clifton *et al.*, 1971). Although grain size is related to wave energy, which is partly a function of wind fetch, it is limited by the supply of grains from shoreline erosion and longshore drift. Headlands commonly have no beaches, or only coarse-grained lags because wave energy is focused on them, in contrast to the finer-grained beaches in protected bays and in shallow water. Although beach sediments are typically well sorted and bedded, this is not always the case in ice-marginal lakes because of the shorter ice-free season for wave action and the potential for disturbance by floating ice. Ice-rafted coarse material may be added to these beaches.

Morphologically, beaches may form berms attached to the mainland or offshore barrier islands, separated from the mainland by a lagoon. Barrier beaches commonly form by the growth of a spit and may have an additional steeply dipping set of cross beds (backset beds) that form when waves overtop the barrier and flow into the lagoon, burying these organic-rich sediments (Fig. 14.18); beds on the lakeward-facing side may be eroded if the barrier is transgressing into the lagoon. In large proglacial lakes, such as Lake Agassiz, these barrier beaches can be hundreds of metres wide and stand tens of metres above the adjacent floor of the lake. In contrast to the generally sparse organic content in sediment of large ice-marginal lakes, these lagoons commonly provided an important palaeo-ecological record because of their warmer and less-turbid environment and may lie buried beneath barrier beaches of proglacial lakes (e.g. Teller *et al.*, 2000; Fig. 14.18).

In general, most proglacial lakes developed an irregular stair-step of beaches (Fig. 14.17) around the basin south of the isobase that extended through the outlet. Most beaches are discontinuous and some have associated wave-trimmed shorelines. In differentially rebounding areas these shorelines can be used to define the outline of the lake at a given transgressive maximum. Today, these once-horizontal beaches have been deformed by differential isostatic rebound, with elevations rising toward the centre of maximum rebound. In the Lake Agassiz basin, most beaches in northern regions are now over 150 m above their contemporaneously formed equivalents 500–700 km to the south (Fig. 14.19) (Teller and Thorleifson, 1983).

Rebounded ancient beaches can be used, in concert with ancient marine beaches, to reconstruct the rates of isostatic rebound as well as total postglacial rebound of a continent (e.g. Andrews and Peltier, 1989; Dyke, 1996). In addition, by comparing the curvature (slope) of upwarped shorelines in the same region, their relative age can be determined. Specifically, older beaches in a sequence will have undergone more differential rebound than younger beaches, so will have steeper gradients which diverge from those of younger beaches (Fig. 14.1).

14.4 SUMMARY OF DEVELOPMENT OF LARGE PROGLACIAL LAKES

The advance of continental ice sheets led to the formation of large proglacial lakes because glaciers blocked river systems. As these ice sheets expanded, the fringe of ice-marginal lakes expanded upslope, forcing them to seek new outlets; eventually some lakes were completely displaced by the ice. The morphological and sedimentary record of these lakes is incomplete.

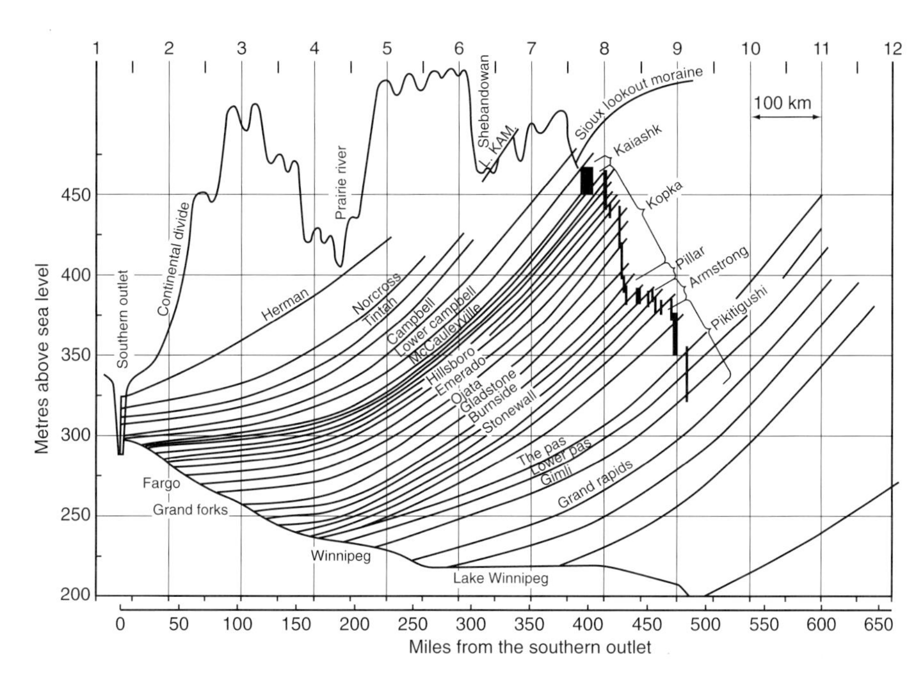

Figure 14.19 Isostatically deformed strandlines of Lake Agassiz. Each line represents a water plane that was once horizontal. Each lower strandline is younger and relates to a new transgressive level of water reached by the lake following the opening of a lower outlet (outlet channels represented by short vertical lines). (After Teller and Thorleifson, 1983, fig. 2).

As ice retreated, ice-marginal lakes redeveloped within the ice-dammed lowlands. Smaller lakes (e.g. Fig. 14.20, lakes A, B, and C) generally left only a thin record of lacustrine sediment without distinct shorelines (cf. Bluemle, 1974; Fenton *et al.*, 1983; Hobbs, 1983; Benn and Evans, 1998), and many of them soon drained and re-formed in new locations along the retreating ice margin (Kehew and Teller, 1994). Others expanded along the ice margin and, during each period of global deglaciation, a vast and interconnected system of proglacial lakes developed across Canada and the adjacent USA, and in northern Europe and Asia (e.g. Vincent and Hardy, 1979; Grosswald, 1980; Teller and Clayton, 1983; Karrow and Calkin, 1985; Dyke and Prest, 1987; Teller, 1987, 2003; Teller and Kehew, 1994; Mangerud *et al.*, 1999, 2002; in the Baltic Sea basin, both the Scandinavian ice margin and isostasy helped create a large proglacial lake, which was eventually replaced by the ocean (e.g. Björck, 1995).

Glacier margins in large and deep proglacial lakes were less stable than when they were on land (cf. Benn and Evans, 1998; Eyles and McCabe, 1991), and surging and iceberg calving made it difficult for an ice sheet to establish a fixed boundary of equilibrium. This, plus the high potential for subsequent wave erosion, means that there are few end moraines in large proglacial lake basins except where the water was shallow, so the details of the history of ice retreat across these basins is not well established. However, inferences about the glacial boundary are possible using strandlines (Fig. 14.1), the history of outlet use during retreat (Fig. 14.19), and lithostratigraphy (see Teller, 2001).

In deeper parts of proglacial lake basins, fine sediments were deposited, partly from density underflows and partly from suspension. In some areas, varves composed of clays and silts

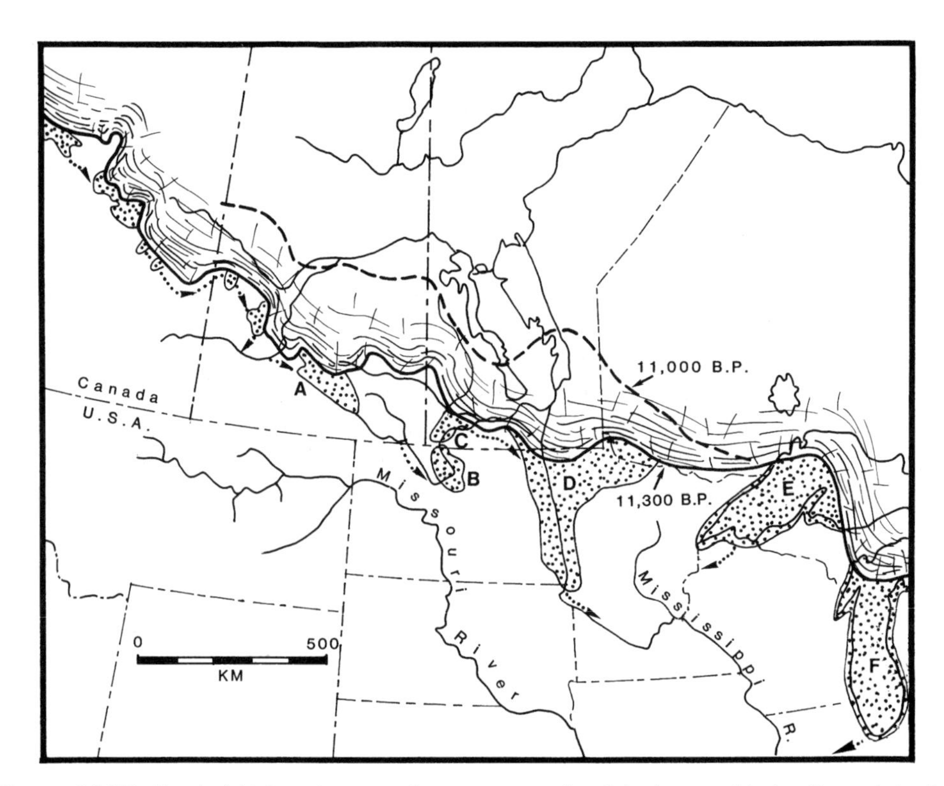

Figure 14.20 Proglacial lakes along southwestern margin of the Laurentide Ice Sheet (stippled) about 11.3 ka; dashed line is ice margin several hundred years later (after Teller, 1987, fig. 15). Overflow linked many of these lake basins, and retreat of the confining ice margin led to northward expansion and periodic catastrophic releases of waters that impacted on rivers and lakes downstream, as well as to a dramatic change in their size and extent (e.g. Kehew and Teller, 1994; Teller, 1987; Teller *et al.*, 2002). Lake Agassiz (D) eventually expanded more than a thousand kilometres north and east but smaller lakes to the west drained and re-formed along the retreating ice front.

accumulated, although massive to poorly laminated sediment is common in some basins. Closer to the margin of the ice sheet and the mouths of rivers, the clay-rich sequence may be interrupted by sand and coarse silt units deposited from sediment-rich subaqueous density underflows; these may display climbing ripples as well as graded bedding. Close to the retreating ice margin, iceberg-rafted detritus is abundant, but the concentration varies temporally at any one site, reflecting the proximity of the ice margin. Along the ice margin, fans of gravel and sand were deposited in places where subglacial conduits delivered sediment to the lake. Beds in these fans dip lakeward and coalesced into asymmetric ridges in some areas that are commonly called grounding-line fans (Figs. 14.6, 14.10B and C). Diamicton was commonly deposited on the up-ice side of this ridge and may also be interbedded with the fan complex. Traditional Gilbert-type deltas with topset, foreset and bottomset beds may also form at the mouth of a subglacial tunnel (Fig. 14.10A).

The floors of glacilacustrine basins are generally very flat, as original relief was masked by sediment accumulation and as elevated areas on the floor were eroded by waves. Distinct large, but very low relief, linear and curvilinear scour marks and plough ridges criss-cross the surface of many proglacial lake floors, reflecting the influence of icebergs as they dragged across the lake floor (Fig. 14.4); a variety of other low-relief patterns formed in some areas (Fig. 14.5).

Toward the shallower parts of the basin, glacilacustrine sediments are more silty and sandy. The absence of lacustrine sediment over large regions of some basins, such as in the Lake Agassiz basin (see Dredge and Cowan, 1989, Fig. 3.18), suggests that, in some cases, current strength on the lake floor prevented sediment from being deposited or that previously deposited sediments were eroded by waves or perhaps by subaqueous outbursts of water.

Around most proglacial lakes, sand and gravel beaches and wave-cut cliffs outline old lake levels, with younger beaches assuming progressively lower positions on the landscape (Figs. 14.17 and 14.19). Although small beaches may reflect storms or short episodes in the life of the lake, large beaches were formed by long-term transgressive events that allowed wave action to accumulate sediment over time and move it upslope. Because of differential isostatic rebound, large beaches only formed south of the isobase through the outlet carrying overflow during a particular phase in the lake's history, and reflect the maximum extent to which waters rose before the beach was stranded (fixed in place) when a new outlet began to carry the overflow (Fig. 14.3D); the new outlet may have been one opened by ice retreat or simply have been the new low point on the divide reached by transgressing waters. As shown in Fig. 14.2, north of the overflow outlet, the shoreline regressed, and a relatively thin and smooth blanket of shoreline sediment was deposited, interrupted in places by low beach ridges related to storms. Lake levels in most proglacial lakes rose and fell repeatedly because outlet elevation changed through time as a result of glacial advance and retreat, differential isostatic rebound, and outlet erosion. Each time the level of a lake declined, waters immediately began to rise throughout the region south of the overflow channel (Figs. 14.3 and 14.17). Today, because of isostatic rebound, all of these beaches rise in elevation toward the former centres of maximum ice thickness (Fig. 14.19).

Glacilacustrine sediments extend across a large part of continents in high northern latitudes, and their stratigraphic and morphological records are integral in our interpretation of continental glacial history. Additionally, these lakes played a major role in landsystem development far beyond their own boundaries, with their overflow influencing rivers, lakes, and oceans downstream. The potential impact of overflow from these lakes on oceans and climate has been discussed by Björck *et al.* (1996), Clark *et al.* (2001), Teller *et al.* (2002), and others.

CHAPTER

GLACIATED VALLEY LANDSYSTEMS

Douglas I. Benn, Martin P. Kirkbride, Lewis A. Owen and Vanessa Brazier

15.1 INTRODUCTION

The concept of the glaciated valley landsystem was introduced by Boulton and Eyles (1979) and Eyles (1983b), to describe the characteristic sediments and landforms associated with valley glaciers in upland and mountain environments. By focusing on the scale of the whole depositional basin, the glaciated valley landsystem has a broader compass than most of the other landsystems explored in this book, which are specific to particular depositional environments. Indeed, glaciated valley landsystems may incorporate ice-marginal, supraglacial, subglacial, proglacial, periglacial and paraglacial landsystems, recording the juxtaposition and migration of very different depositional environments. Additionally, because glaciated valleys occur in every latitudinal environment from equatorial to polar regions, the dimensions of climate and glacial thermal regime add even more variability. Thus the 'glaciated valley landsystem' should be regarded as a family of landsystems, which exhibits considerably more variety than suggested by the original Boulton and Eyles model (Fig. 15.1).

Despite this variability, landsystems in glaciated valleys tend to have certain recurrent features, as a result of two main factors:

1. the strong influence of topography on glacier morphology, sediment transport paths and depositional basins
2. the importance of debris from supraglacial sources in the glacial sediment budget.

In this chapter, we emphasize the contrasts between glaciers with limited supraglacial debris ('clean glaciers') and glaciers with substantial debris covers in their ablation zones ('debris-covered glaciers'), although it should be recognized that intermediate forms occur between these end members. Before examining the landsystems of glaciated valleys, we begin by considering debris sources and transport pathways through valley glaciers, and the ways in which debris cover influences glacier dynamics.

15.2 SEDIMENT SOURCES

Processes delivering debris to glacier surfaces include debris flows, snow avalanches, rockfalls, and rock avalanches (Gordon and Birnie, 1986). In tectonically active regions, earthquake-generated rock avalanches are significant (Post, 1964; Hewitt, 1988), but many rock avalanches have no obvious trigger (Gordon *et al.*, 1978; McSaveney, 1992). The input of debris to glacier surfaces by mass movement processes is partly controlled by catchment topography, which determines the gradient, area and distribution of debris source areas. Additionally, bedrock lithology can introduce contrasts in supraglacial debris supply between otherwise similar catchments. For example, granite batholiths overlooking the Baltoro Glacier (Karakoram Mountains) yield little debris, whereas fissile sedimentary rocks in the same region are associated with high debris supply. The importance of the relative rates of debris and snow/ice delivery to the glacier surface is often overlooked. Glaciers with high debris concentrations tend to occur where rockfall rates are high and/or snowfall is low. For example, on the main Himalayan chain in Nepal, where precipitation is low (~500 mm year^{-1}) and the rapid uplift of young sedimentary rocks and extreme topography encourages very high rates of rockfall delivery to glacier surfaces, the majority of glaciers have debris-mantled ablation zones, and rock glaciers are widespread (Müller, 1980). In contrast, in western Norway, where crystalline rocks yield low rockfall rates, and precipitation is in the range 2,000–3,000 mm year^{-1}, glaciers have little or no surface debris and rock glaciers are absent.

15.3 SEDIMENT TRANSPORT PATHWAYS

Transport routeways through valley glaciers are varied (Fig. 15.2). Boulton (1978) distinguished two main sediment transport pathways:

1. active subglacial transport, and
2. passive supraglacial or englacial transport.

Subglacial debris transport was termed active because sediment in the basal shear zone of glaciers is subjected to high inter-particle contact forces and consequently undergoes significant abrasion, fracture and comminution. Boulton (1978) argued that, in contrast, sediment in higher-level transport undergoes little modification and thus retains the characteristics of the parent debris. While this distinction is a useful one (e.g. Vere and Benn, 1989; Benn and Ballantyne, 1994), it is an oversimplification because supraglacial transport is not always 'passive'. Boulders may undergo edge-rounding as debris is redistributed by ablation of the underlying ice (Benn and Evans, 1998; Owen *et al.*, 2002).

Glacifluvial processes were overlooked in Boulton's classification, though they may transport large amounts of sediment over, beneath and through many valley glaciers. Glacifluvial transport is particularly important on low-gradient glaciers with extensive debris cover, where reservoirs of sediment can be accessed by meltwater (Kirkbride and Spedding, 1996; Spedding, 2000). Subglacial conduits can entrain sediment from the glacier bed, and englacial conduits collect debris by wall melting. Glacifluvial sediment can then be delivered to the supraglacial transport zone where conduits emerge at the surface, or following conduit closure or freezing and debris melt-out. Fluvially rounded cobbles and sorted sand occur in the debris covers of New Zealand valley glaciers, Ngozumpa Glacier (Nepal) and other glaciers. Such facies probably represent a tiny

(A)

(B)

(C)

Figure 15.1 Three examples of mountain glaciers, showing the diversity of glaciated valley landsystems. A) Chola Glacier, a debris-covered glacier in the Khumbu Himal. Note avalanche cones below the headwall, and large lateral moraines. B) Slettmarkbreen, a cirque glacier with little supraglacial debris, Jotunheimen, Norway. C) Un-named hanging glacier and reconstituted lower tongue, Lahul Himalaya. Note avalanche track leading to the lower glacier, and dissected moraines at lower right.

fraction of the debris flushed out of such glaciers during the ablation season, which is supplied directly to the proglacial outwash system (Kirkbride, 2002). In terms of sediment discharge, the apparent dominance of supraglacial debris is misleading, because the supraglacial load represents an inefficient pathway. Englacial and supraglacial fluvial pathways, although rarely observed, may dominate landsystem development at the termini of large debris-covered glaciers, as shown by the volume of Holocene outwash valley fills, many orders of magnitude greater than the volume of Holocene ice-marginal moraines.

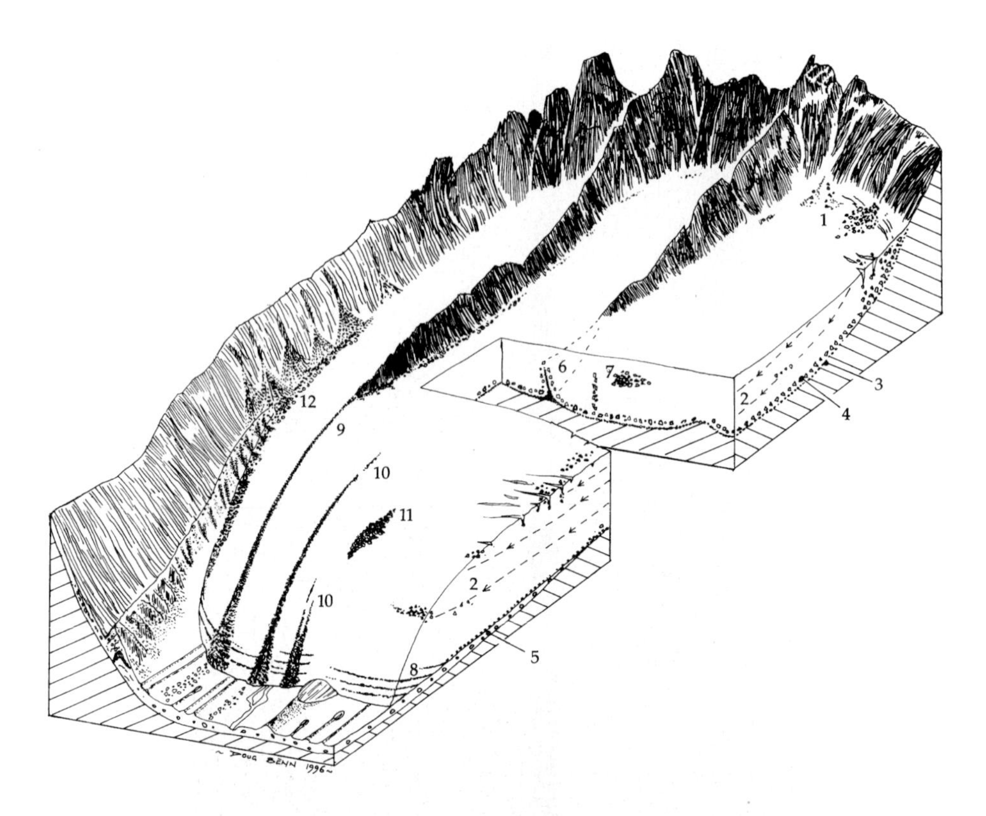

Figure 15.2 Debris transport paths in a valley glacier. 1 = burial of rockfall debris in accumulation area, 2 = englacial transport and melt-out in ablation area, 3 = basal traction zone, 4 = suspension zone, 5 = basal till (may undergo deformation), 6 = elevated debris septum below glacier confluence, 7 = diffuse cluster of rockfall debris, 8 = debris elevated from the bed by compressive flow and shear near the margin, 9 = ice-stream interaction medial moraine, 10 = ablation-dominant medial moraine, 11 = avalanche-type medial moraine, 12 = supraglacial lateral moraine. (From Benn and Evans (1998).)

Debris passes between transport pathways by several processes, including melt-out, burial by snow, and ingestion by crevasses. In high-relief terrain, steep icefalls above low-gradient ablation zones commonly elevate debris from basal transport to high-level transport by avalanching and glacier reconstitution, supplying large volumes of debris to supraglacial covers (Fig. 15.3).

Medial moraines are distinctive features of many valley and cirque glaciers. Eyles and Rogerson (1978) proposed a comprehensive classification based upon the relationship between debris supply and the morphological development of the moraine. Three main types were recognized:

1. ablation-dominant (AD) moraines, which emerge at the surface as the result of the melt-out of englacial debris

Figure 15.3 Avalanche-reconstitution of a glacier tongue transferring englacial and basal debris septa into a supraglacial debris cover. Kaufmann Glacier, Mt Haidinger, New Zealand.

2. ice-stream interaction (ISI) moraines, which find immediate surface expression downstream from glacier confluences, often by the merging of two supraglacial lateral moraines, and
3. avalanche-type (AT) moraines, which are transient features formed by exceptional rockfall events onto a glacier surface (Fig. 15.2).

Although there are shortcomings with this classification (Small *et al.*, 1979; Vere and Benn, 1989), no satisfactory alternative has been proposed, and it remains in common use (Benn and Evans, 1998).

Where supraglacial sediment is high relative to snow inputs, continuous debris covers typically form in glacier ablation zones. Such debris-covered glaciers are distinctively different from clean glaciers (Higuchi *et al.*, 1980), both in terms of their dynamics and their associated depositional landsystems.

15.4 DYNAMICS OF DEBRIS-COVERED GLACIERS

Thin debris cover (<~5 cm) enhances ablation due to reduced albedo and increased absorption of short and longwave radiation, whereas thicker debris insulates the underlying ice and reduces ablation, because of its low thermal conductivity (Nakawo and Young, 1981). On debris-covered glaciers, debris thickness generally increases towards the glacier terminus, reversing the ablation gradient and causing ablation rates to be very small on the lower part of the glacier. The reduced ablation causes ablation zones to enlarge to offset mass gains in the accumulation zone. As a consequence, debris-covered glaciers in equilibrium have accumulation-area ratios (AARs) of 0.2–0.4, compared with values of 0.6–0.7 for clean glaciers (Benn and Evans, 1998).

Glacier response to climate fluctuations is strongly influenced by the degree of supraglacial debris cover. For clean glaciers, ice volume changes are reflected in oscillations of the glacier terminus. The response of debris-covered glaciers to climatic warming is dampened by the insulating effect of debris. However, if warming is sustained, such glaciers can enter a phase of very rapid ablation if ice-contact lakes expand by calving (Kirkbride, 1993; Reynolds, 2000; Benn *et al.*, 2001). Retreat of debris-covered glaciers and the cessation of sediment delivery to terminal moraines may thus significantly lag climate changes (Benn and Owen, 2002).

During periods of glacier stability or thickening, the termini of heavily debris-loaded glaciers are foci of dramatic sediment aggradation, forming some of the most impressive glacial depositional landforms in the world (Owen and Derbyshire, 1993; Kirkbride, 2000). Considerable variation in landsystem development occurs between glaciers, ranging from steep fronted lateral-frontal moraines (sometimes referred to as latero-frontal or lateral-terminal) to lower gradient ice-contact debris fans and outwash heads. This variation largely depends on the relative supply of ice and debris to the terminal area, and the efficiency of its removal by meltwater. Shroder *et al.* (2000) contrast the terminus environments of three glaciers in the Nanga Parbat massif, Pakistan, and identify three primary controls on landform development:

1. overall sediment supply to the glacier by rockfall and avalanching, which determines the amount of debris available for ice-marginal deposition
2. the velocity of ice in the ablation zone, which controls whether debris accumulates supraglacially or is transferred to the ice-margin for deposition, and
3. the ability of fluvial processes to remove sediment from the ice margin, which determines whether sediment accumulation is focused in the ice-marginal or proglacial zones.

The common view that debris-covered glaciers are unresponsive to climate is not strictly true. Large lateral-terminal moraines can act as significant barriers to glacier advance, particularly if depositional rates are high, so mass balance variation on debris-covered glaciers is commonly manifest as thickening and thinning instead of advance and retreat. Research on such glaciers emphasizes negative mass balance conditions where ice is increasingly insulated under thickening supraglacial debris (Kirkbride and Warren, 1999; Nakawo *et al.*, 1999; Naito *et al.*, 2000). Under positive balance, when gradients and velocities are increased, debris covers may accentuate the effects of kinematic waves (Thomson *et al.*, 2000), enhancing an expansionary tendency over multiple mass balance cycles.

Kirkbride (2000) suggested that supraglacial load increases over several mass balance cycles, which complicates the response of covered glaciers to climate variation. Indeed, it is debatable whether a true equilibrium between glacier volume and climate can ever be achieved. If mean specific ablation rates decline due to supraglacial loading, continued expansion of the ablation zone is a necessary consequence. Thus, under constant climate, a debris-covered glacier will have to advance to maintain equilibrium mass balance. The slow, sustained advances of glacier ice-cored rock glaciers similarly reflect glaci-dynamic influences and an expansionary tendency.

In summary, the terminus positions of debris-covered glaciers tend to be stable for long periods. Debris delivered to the ice margin from glacial transport is concentrated into large landforms with high preservation potential. Because high supraglacial loads are largely offset by low terminus velocities, it is the long-term stability of the ice-contact zone that is a key determinant on landsystem evolution.

15.5 LANDSYSTEMS OF GLACIAL DEPOSITION

While clean and debris-covered glaciers show many similar features within their landsystems, they differ in the dominant processes operating at the ice margin and in the relative development of landforms (Figs 15.4 and 15.5).

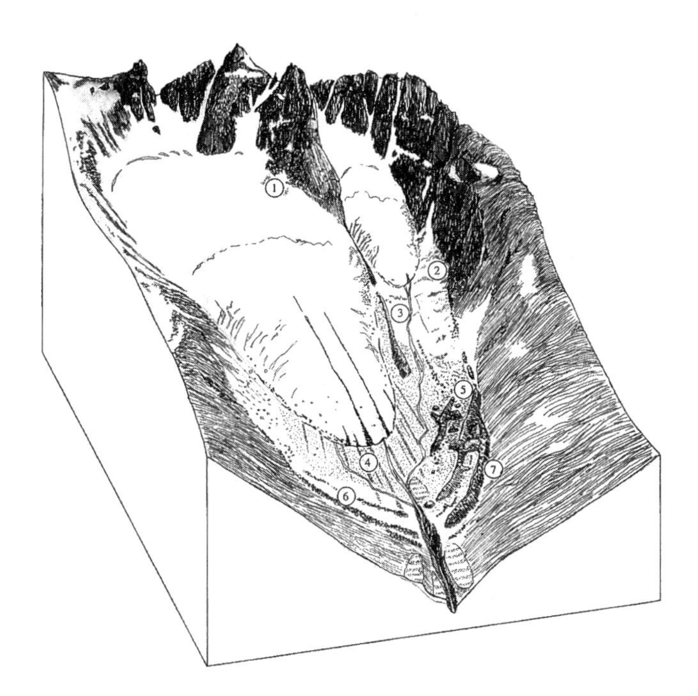

Figure. 15.4 A landsystem model for valley glaciers with relatively low supraglacial debris inputs. (From Benn and Evans, 1998). 1= Supraglacially entrained debris; 2 = Periglacial trimline above ice-scoured bedrock; 3 = Medial moraine; 4 = Fluted till surface; 5 = Paraglacial reworking of glacigenic deposits; 6 and 7 = Lateral moraines, showing within-valley asymmetry.

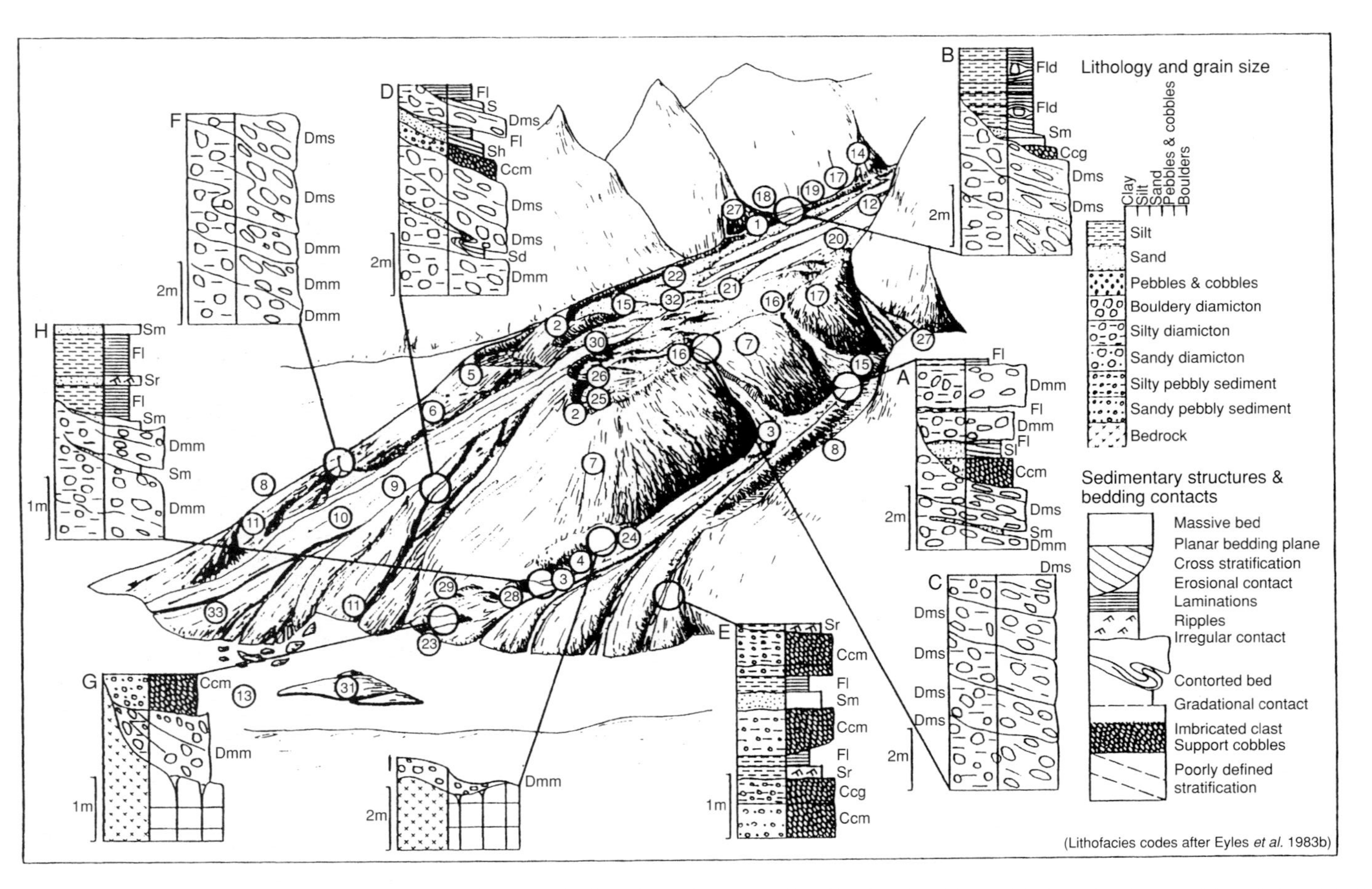

Figure 15.5 A landsystem model for debris-mantled valley glaciers, based on the Ghulkin Glacier, Hunza Valley, Karakoram Mountains, Northern Pakistan. (From Owen (1994).)

15.5.1 Ice-Marginal Moraines and Related Landforms

15.5.1.1 Processes of Moraine Formation

Moraine formation at glacier margins with limited supraglacial debris involves one or more processes, including pushing (particularly where the margin is buried by glacifluvial deposits or debris flows), dumping of supraglacial debris, and, where fine-grained saturated sediment is present at the margin, squeezing (Benn and Evans, 1998). Thrust moraines may form where glaciers come into contact with thick, unconsolidated sediments such as glacimarine clays and silts (e.g. Gray and Brooks, 1972; Benn and Evans, 1993). Moraines are typically less than 10 m high, but size is strongly influenced by catchment lithology, debris availability and glacier dynamics. In valleys underlain by resistant crystalline rocks, lateral moraines may be little more than boulder lines.

15.5.1.2 Lateral-Terminal Moraine Complexes

Giant lateral moraines are common along the margins of glaciers with large discharges of debris in high-level transport and poor linkage between the glacial and proglacial transport systems. Such moraines represent major sediment sinks, with distal slopes 100–300 m high near the terminus. Lateral moraines extend from the contemporary equilibrium line as continuous, sharp-crested ridges that increase down-glacier in cross-sectional area. In the upper ablation zone, moraines may have little or no distal slope, being effectively a debris veneer accreted onto the valley side. Down-valley, moraines become increasingly separated from the valley wall, forming lateral morainic troughs (Hewitt, 1993; see below) which act as traps for glacifluvial sediment and slope debris derived from the valley side.

Debris-covered glaciers may be perched far above the valley floor as the result of positive feedbacks between moraine deposition and the dynamics of the glacier terminus (Fig. 15.6). If sediment

Figure 15.6 Hatunraju Glacier, Cordillera Blanca, Peru – a debris-covered cirque glacier on a raised bed terminating at a large moraine dam. (Photo: C.M. Clapperton).

supply is sufficiently large, accumulated debris around the glacier margin constitutes a major barrier to glacier flow, preventing forward advance of the glacier during periods of positive mass balance and forcing the margin to thicken *in situ*. In turn, this focuses more deposition on the moraine, creating a greater impediment to glacier advance.

Large lateral-frontal moraines are formed as debris falls, slumps, slides or flows down the ice edge and accumulates around the glacier margin (Humlum, 1978; Small, 1983; Owen and Derbyshire, 1989, 1993; Owen, 1994; Benn and Owen, 2002). A depositional model has been presented by Owen and Derbyshire (1989, 1993) and Owen (1994), who termed it the Ghulkin-type association after the glacier of that name in the Karakoram Mountains (Fig. 15.5). If the glacier margin remains in a stable position, the accumulation of dumped material produces a wedge-shaped moraine with crude internal bedding dipping away from the glacier at angles between 10° and 40°. Facies consist of stacked diamictons with variable clast content interbedded with thin sand and gravel layers reflecting intermittent glacifluvial deposition and reworking (Figs. 15.7 and 15.8).

In general terms, the constituent debris reflects the mix of supraglacial debris types on the glacier, with the addition of thin basal tills by lodgement and/or basal melt-out. Debris covers

Figure 15.7 Diamicts interbedded with gravels and sands exposed within a large lateral moraine complex, recording deposition from mass movements and meltwater, and within ephemeral ponds in sections in the Hunza Valley, Karakoram. The largest boulders are ~ 1m in diameter.

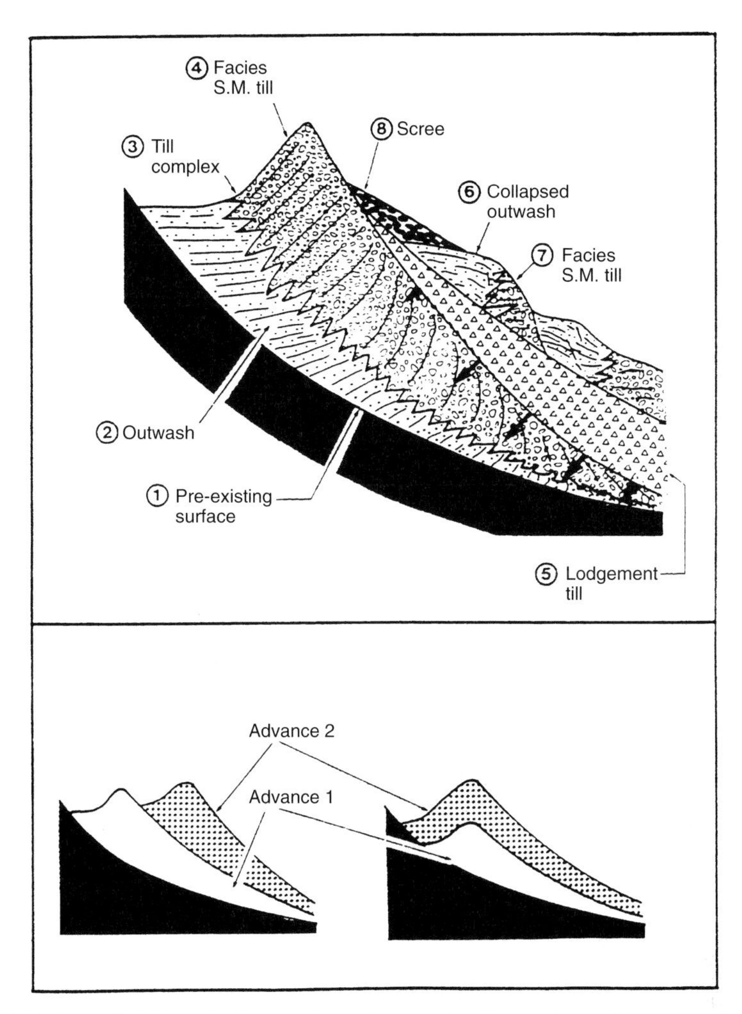

Figure 15.8 Schematic internal stratigraphy of large lateral moraines. (From Boulton and Eyles, 1979).

that are heterogeneous mixes of lithological types will produce lateral moraine complexes containing identifiable compositional modes over quite short distances. Coarse, bouldery layers within the moraine may be derived from supraglacially-transported rock avalanche material (Humlum, 1978). The majority of diamict facies within large lateral moraines are sandy boulder gravels containing predominantly angular debris from passive transport of rockfall material. There is sometimes a significant component of finer debris and a subangular to subrounded clast mode. Various explanations account for the varied quantities of more rounded clasts within lateral moraines:

1. delivery of a higher proportion of basal transport zone debris to the ice margin towards the terminus, giving a clast shape gradient along the moraine (Matthews and Petch, 1982)

2. entrainment of proglacial sediment by the glacier during advance (Slatt, 1971)
3. corrosion of debris in englacial conduits before the debris is returned to passive transport (Kirkbride and Spedding, 1996; Kruger and Aber, 1999; Spedding, 2000), and
4. rounding and comminution of clasts in shear zones within the glacier (Glasser *et al.*, 1999).

The ice-proximal parts of lateral-frontal moraines tend to be structurally complex because of widespread collapse and reworking following the removal of ice support. Bedding is commonly contorted as a result of the melt-out of buried ice and gravitational reworking. Individual facies may be hard to distinguish but subglacial tills, including melt-out tills, may be recognized. Diamicts may be interbedded with supraglacial lacustrine and glacifluvial sediments, which typically form deformed channel fills.

As noted above, the termini of debris-covered glaciers may repeatedly occupy similar positions in a valley. Consequently, moraines may undergo several aggradational episodes separated by periods of erosion or non-deposition. The old moraine crest may be completely buried, or a new inset moraine may be formed within the older one, forming multi-crested lateral moraine complexes. Complex depositional histories may be preserved in the internal structure, in the form of multiple depositional sequences bounded by erosion surfaces (Boulton and Eyles, 1979; Small, 1983; Richards *et al.*, 2000; Benn and Owen, 2002; Fig. 15.8). Periods of non-deposition may be recorded by palaeosols or even buried trees, providing a valuable source of palaeoclimatic data (Röthlisberger *et al.*, 1980; Röthlisberger, 1986; Grove, 1988). In many mountain areas, moraine chronologies demonstrate repeated superposition by successive advances over millennia, often over the entire Neoglacial period since *c.* 5 ka BP. It is probable that at some glaciers, giant moraines store the entire Holocene yield of coarse sediment from the catchment (Lliboutry, 1986). Following glacier thinning and retreat, lateral-frontal moraines are abandoned and their inner faces subject to collapse and paraglacial reworking (Blair, 1994; Ballantyne, 2002b, Chapter 17). Inset moraines will therefore be unstable and short lived, particularly if they are deposited on top of dead ice masses.

The presence of thick supraglacial and ice-marginal sediments means that subglacial landsystems are rarely exposed on the forelands of debris-covered glaciers. The existence of deeply incised glacial troughs and the presence of striated, actively transported clasts in lateral-frontal moraines, indicate that subglacial erosion and transport must be effective, although it is likely to be volumetrically less significant than supraglacial debris entrainment and transport in most basins (Small, 1987a, b; Gardner and Jones, 1993).

Lateral-frontal moraines in Ghulkin-type landsystems form continuous ramparts around the glacier margin. Such landsystems tend to develop where meltwater discharges are low relative to debris fluxes and where the moraines are formed predominantly by mass-movement processes. Where meltwater discharges are higher (such as in the monsoon-dominated environments along the southern slope of the Himalaya), glacifluvial processes keep open a central corridor, preventing a continuous moraine loop from forming (Fig. 15.9). Glacifluvial deposition in the proximal proglacial zone forms fans of gravel and sand, sometimes interbedded with diamictons (debris flows and hyperconcentrated flood-flow deposits).

15.5.1.3 Breach-Lobe Moraines

Breach-lobe moraines have not previously been recognized as a distinct landform. They form as inset loops or lobate moraines where a glacier exploits a gap in a major moraine ridge, either in lateral or lateral-frontal positions, and they may evolve either into large superposed moraine complexes or into accreted suites of individual ridges. Sedimentologically, they are identical to the moraines adjacent to the breach.

Multi-lobed glaciers such as Miage Glacier, Italy (Deline, 1999a) and Sachen Glacier, Pakistan (Shroder *et al.*, 2000) owe their terminus morphology to a long-term expansionary tendency and repeated breaching of the main moraine barrier. An explanation of the formation of secondary lobate termini is provided by the moraine-breaching model of Lliboutry (1977), which explains the unusual morphology of Glaciar Hatunraju in Peru (Fig. 15.6). At Miage Glacier (Italy; Fig. 15.10), three phases of breaching have punctuated phases of lateral moraine construction (Deline, 1999a, b). The earliest breach dates from *c.* 2300–2900 BP, when seven moraine ridges formed before the main lateral moraine sealed the breach by 1250 BP (AD 700). A second set of breach-lobe moraines represent glacier expansion in the *c.* 8–11th centuries AD, again followed by lateral moraine construction, before final breaching beginning in the early 17th century deposited at least ten breach-lobe moraines before AD 1930.

At Tasman Glacier (New Zealand), breaches in the lateral moraines formed much earlier and have evolved into broad embayments in the lateral moraines (Kirkbride, 2000). One embayment (Fig. 15.11) has evidence of four expansionary phases since *c.* 3700 years BP, the latest representing the 'Little Ice Age' and culminating at *c.* AD 1860. Sediments forming the breach are complex, with superposed moraines comprising couplets of bouldery basal and supraglacial melt-out tills, each couplet representing a glacier thinning/stagnation cycle (Fig. 15.12).

15.5.1.4 Infills of Lateral Morainic Troughs ('ablation valleys')

Lateral morainic troughs were originally termed ablationsschlucht ('ablation valleys') by Oestreich (1906), and are features of many large valley glaciers. They separate lateral moraines from valley side slopes and act as gutters trapping slope debris transported by processes including rock fall, debris flow, snow avalanche and fluvial transport (Hewitt, 1993), and may contain ponds between debris cones. When ice levels are low, retreat of the proximal moraine slope may breach the moraine to capture the valley side drainage (Fig. 15.13). Sediments deposited in lateral morainic troughs have low preservation potential in areas of tectonic uplift, where they are rapidly lost from active slope systems.

In the Karakoram Mountains, infills of lateral morainic troughs tend to be more complete further up-glacier, where lateral moraines are smaller and slope processes more effective at supplying material to the trough (Hewitt, 1993). An altitudinal zonation of slope processes, conditioned by decreasing temperature and increasing precipitation with altitude, is manifest in the depositional processes along very long lateral moraines which intersect more than one altitudinal geomorphological zone. At the altitudes of the lateral moraines of Europe and New Zealand, such zonation is less well expressed and fluvial processes are more important than in the greater ranges.

15.5.1.5 Within-Valley Asymmetry of Lateral Moraines

'Within-valley asymmetry' refers to larger moraine volumes on one side of the valley than the other (Matthews and Petch, 1982; Fig. 15.4). There are several causes of within-valley asymmetry (Matthews and Petch, 1982; Benn, 1989a):

1. larger moraines occur on valley sides with extensive rock walls, which increase debris supply to that side of the glacier. Debris may be delivered to the glacier via rockfalls, or indirectly by the subglacial incorporation of preglacial screes
2. where lateral moraines are formed by pushing or thrusting of pre-existing materials, within-valley asymmetry can result from differences in the thickness and type of sediment on the foreland
3. cross-valley differences in lithology or structure can influence debris supply, either to the surface or bed of the glacier, and
4. asymmetry may occur due to differences in glacier dynamics on either side of a valley. For example, a stable ice margin can build a large moraine, but if the other margin retreats, a series of smaller moraines will form.

Total moraine volume may be the same on each valley side, but an impression of asymmetry arises because deposition on one side is more focused.

(A)

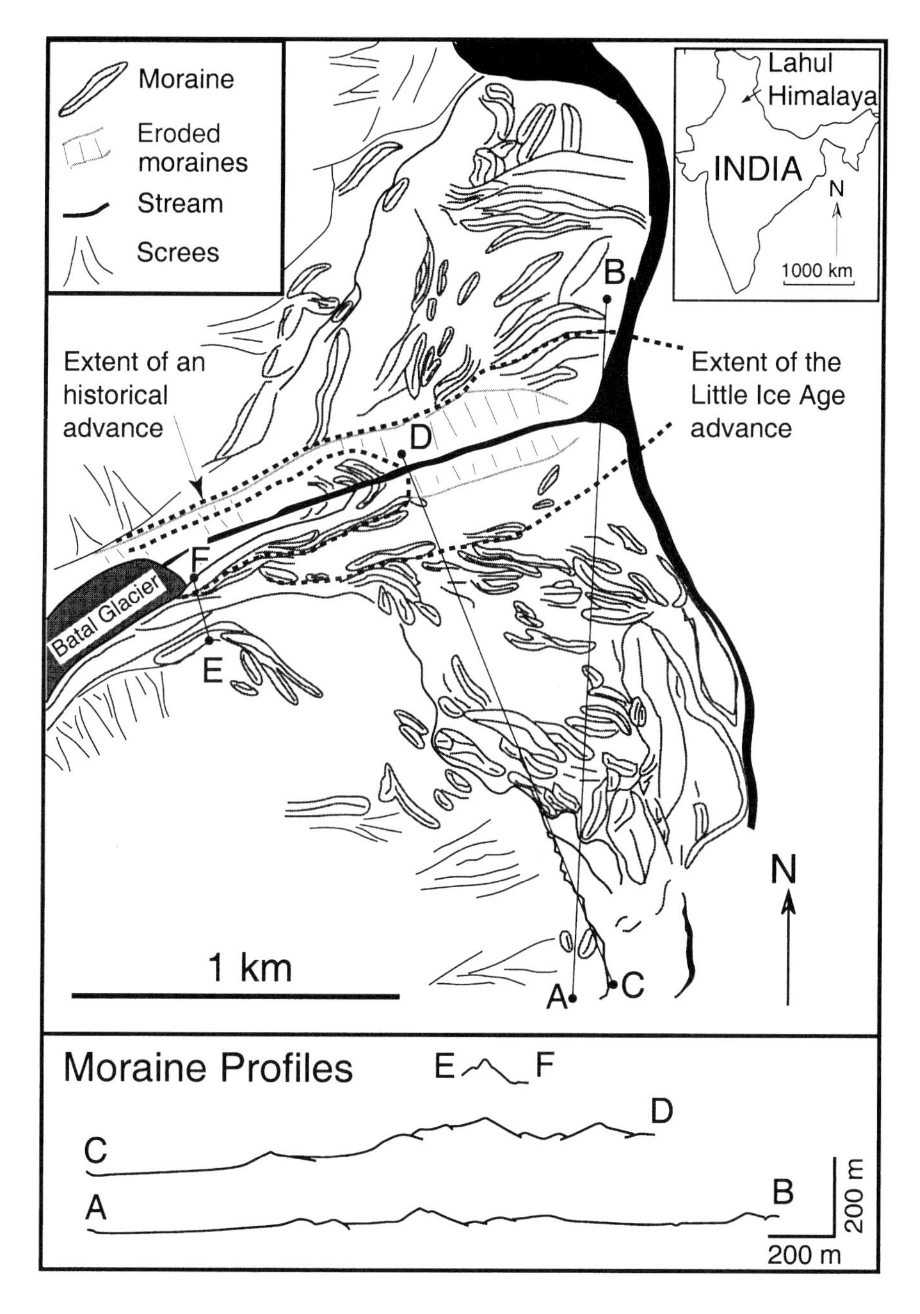

(B)

Figure 15.9 The Batal Glacier in the Lahul Himalaya. A) View looking southwest at two large lateral moraines. B) Geomorphic map and profiles showing the main landform elements associated with this debris-covered glacier (Adapted after Benn & Owen, 2002).

15.5.2 Subglacial Sediments and Landforms

Where valley glaciers carry relatively small amounts of supraglacial debris, subglacial landsystems may be well exposed on the depositional surface (Figs. 15.4 and 15.14). Indeed, the beds of former valley glaciers may provide the best laboratories for studying the geomorphological and sedimentological products of subglacial processes, because landforms can be clearly placed within the spatial and temporal context of the glacier system as a whole, unlike areas of former ice sheet beds, where the wider context is often rather ill-defined (e.g. Hallet and Anderson, 1982; Sharp *et al.*, 1989; Benn, 1994).

Extensive areas of ice-moulded bedrock, recording net erosion of the bed, typically occupy the upper parts of the beds of former valley glaciers. Striated roches moutonnées, whalebacks and overdeepened rock-basins document abrasion and quarrying of the bed by sliding, debris-charged basal ice. Further downvalley, localized subglacial deposits occur in the swales between rock outcrops and downstream of roches moutonnées (lee-side cavity fills; Levson and Rutter,

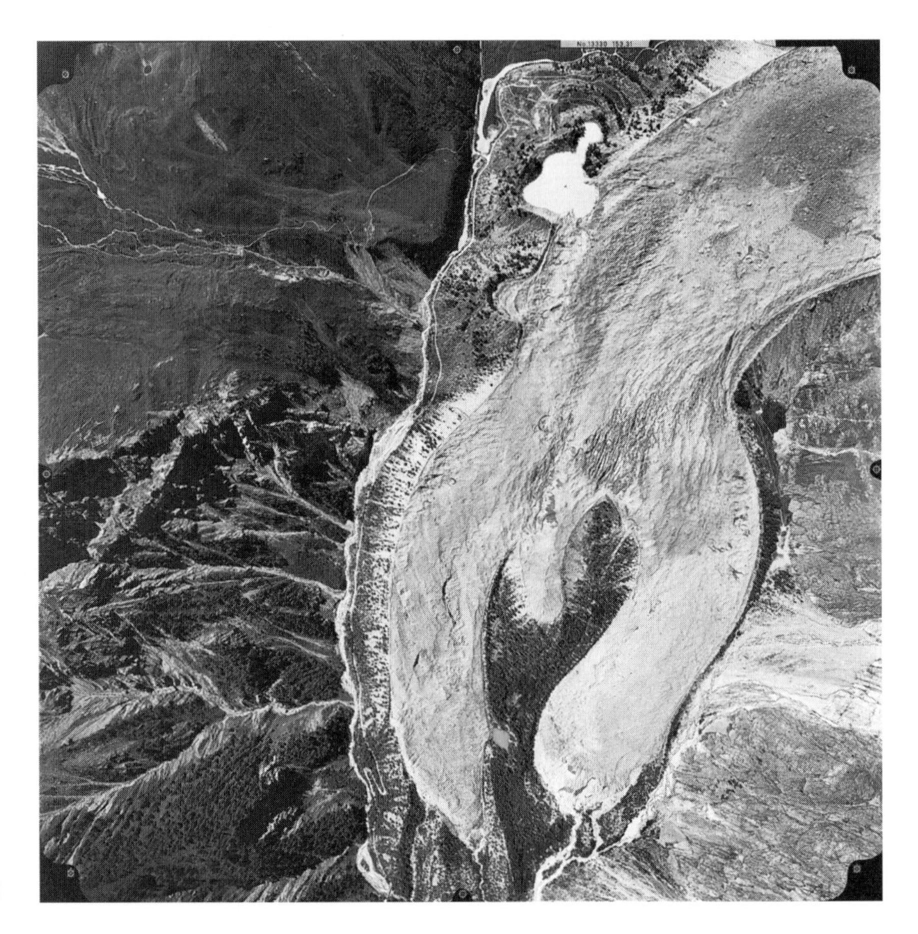

(A)

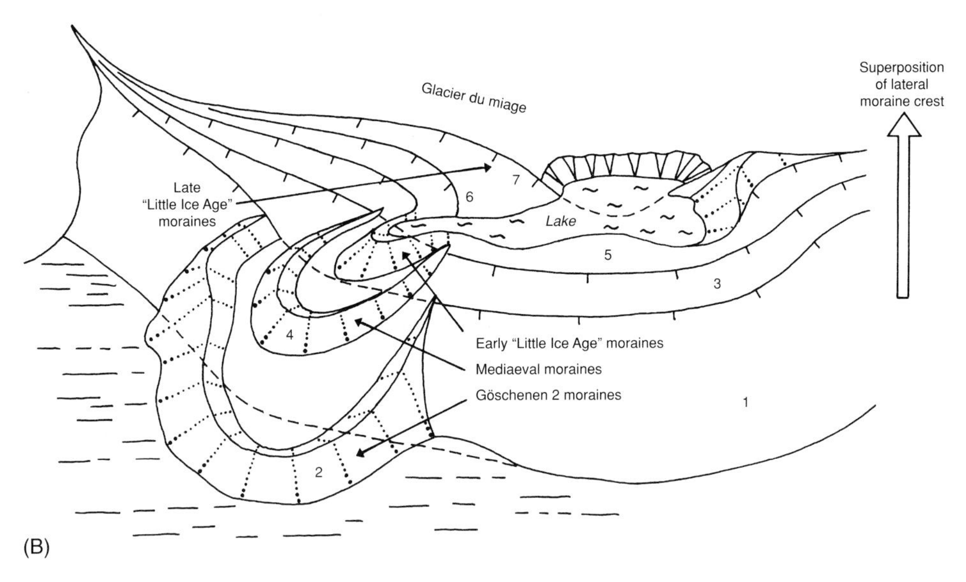

Figure 15.10 Breach-lobes of the Miage Glacier, Monte Bianco massif, Italy. A) Vertical aerial photograph of the debris-covered tongues of Miage Glacier. The breach-lobe moraines of the Amphitheatre date from >1700 years BP. Other breaches have been exploited to form the three extant lobes of debris-covered ice. (Reproduced with permission of the Compagnia Generale Ripreseaeree, Parma, Italy.) B) Diagrammatic representation of the sequential development of phases of upbuilding and outbuilding of the breach-lobe moraines of Miage Amphitheatre (after Deline, 1999a).

1989). Down-glacier, more of the bed is occupied by subglacial till. The most abundant subglacial tills of mountain glaciers are overconsolidated lodgement- or high-strength deformation till, with matrix-support, a fissile structure, and abundant faceted, striated clasts (Benn, 1994; Benn and Evans, 1996). Fluted moraines commonly occur down-glacier of boulders and other obstructions (Boulton, 1976; Rose, 1989; Benn, 1994; Benn and Evans, 1996). The preservation potential of fluted moraines is low, and they may not survive as prominent landforms for more than a few decades. In older deglaciated terrain, fluted moraines and other subglacial till surfaces may be reduced to scattered flow-parallel stoss-and-lee boulders (Rose, 1992).

Where glaciers extended from the confines of valleys onto fringing lowlands, the depositional zone can resemble the soft-bed subglacial landsystem described in Chapter 2 with drumlinized surfaces underlain by deformation tills and glacitectonites (Rose, 1987; Benn and Evans, 1996). This type of landsystem occurs at the margins of the Scottish Highlands, where piedmont glaciers flowed into lowlands underlain by glacimarine and glacilacustrine sediments during the Loch Lomond (Younger Dryas) Stade (Thorp, 1991; Benn and Evans, 1996).

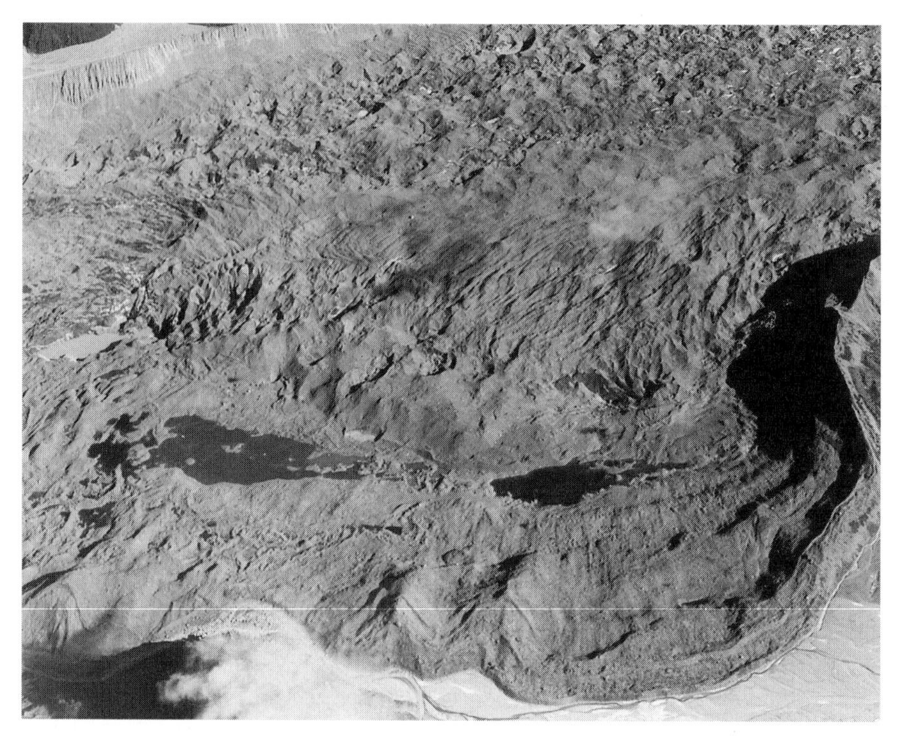

Figure 15.11 Oblique aerial photograph of the eastern margin of Tasman Glacier, New Zealand, where ice has expanded to form a lateral embayment in the mouth of Murchison Valley. Ice flow from top right.

Figure 15.12 Bouldery supraglacial melt-out till overlying basal melt-out till, forming a couplet deposited since the mid-1960s by stagnating ice in the Murchison embayment (Tasman Glacier, New Zealand). (Figure shown (arrow) for scale).

Figure 15.13 Infills of lateral morainic troughs in the Mount Cook region, New Zealand. Exposure of interfingering deposits of bouldery supraglacial till forming the eastern lateral moraine of Hooker Glacier, and colluvium derived from valley-side debris flow and fluvial reworking in the marginal trough. (Figure for scale.)

15.5.3 Facies of Glacier Retreat

15.5.3.1 Recessional and 'Hummocky' Moraine

Recessional moraines mark the positions of annual (winter) readvances or more significant longer-term advances of the margin, and are common in low-relief mountains. Moraines on the forelands of climatically sensitive clean glaciers can form a detailed archive of glacier oscillations and decadal climate change (Nesje and Dahl, 2000).

In Scotland, complex assemblages of moraine mounds and ridges (so-called 'hummocky moraine') blanket the floors of many valleys. Sissons (1974, 1977) interpreted these landforms as the products of widespread glacier stagnation at the termination of the Loch Lomond (Younger Dryas) Stade. Recently research shows hummocky moraine to be polygenetic, consisting of up to three superimposed landsystems:

1. recessional moraines forming converging cross-valley pairs
2. flow-parallel drumlins and flutings, and
3. non-aligned mounds and ridges recording uncontrolled ice-marginal deposition (Benn, 1992; Bennett, 1994; Wilson and Evans, 2000; Fig. 15.15).

Recessional moraines, recording oscillating ice margins during overall retreat, are common within Scottish hummocky moraine and show that widespread glacier stagnation did not occur at the end of the Younger Dryas Stade (Bennett and Glasser, 1991; Benn, 1992, 1993; Bennett

Figure 15.14 Recessional moraines superimposed on fluted moraines, Maradalsbreen, Norway.

and Boulton, 1993; Bennett, 1994). Moraine morphology and sedimentology indicate that most are dump and push moraines, with varying amounts of proglacial tectonics. Bennett *et al.* (1998) argue that transverse elements within Scottish hummocky moraine result from proglacial and englacial thrusting, similar to the contemporary margins of polythermal Arctic glaciers. Aspects of this model are appealing, as it explains anomalously large amounts of debris within hummocky moraine in terms of glacial tectonics in a permafrost environment (Graham and Midgely, 2000). However, an alternative depositional model invoking multiple episodes of moraine deposition and reworking appears more plausible (Wilson and Evans, 2000). Ballantyne (2002b, Chapter 17) argues that inheritance of large volumes of glacial and paraglacial sediment is a characteristic of periods of renewed glaciation in mountain environments.

15.5.3.2 Till Sheets

Debris-covered glaciers undergoing frontal retreat or progressive downwasting deposit their debris load as a veneer of variable thickness, superimposed on the underlying topography. Where

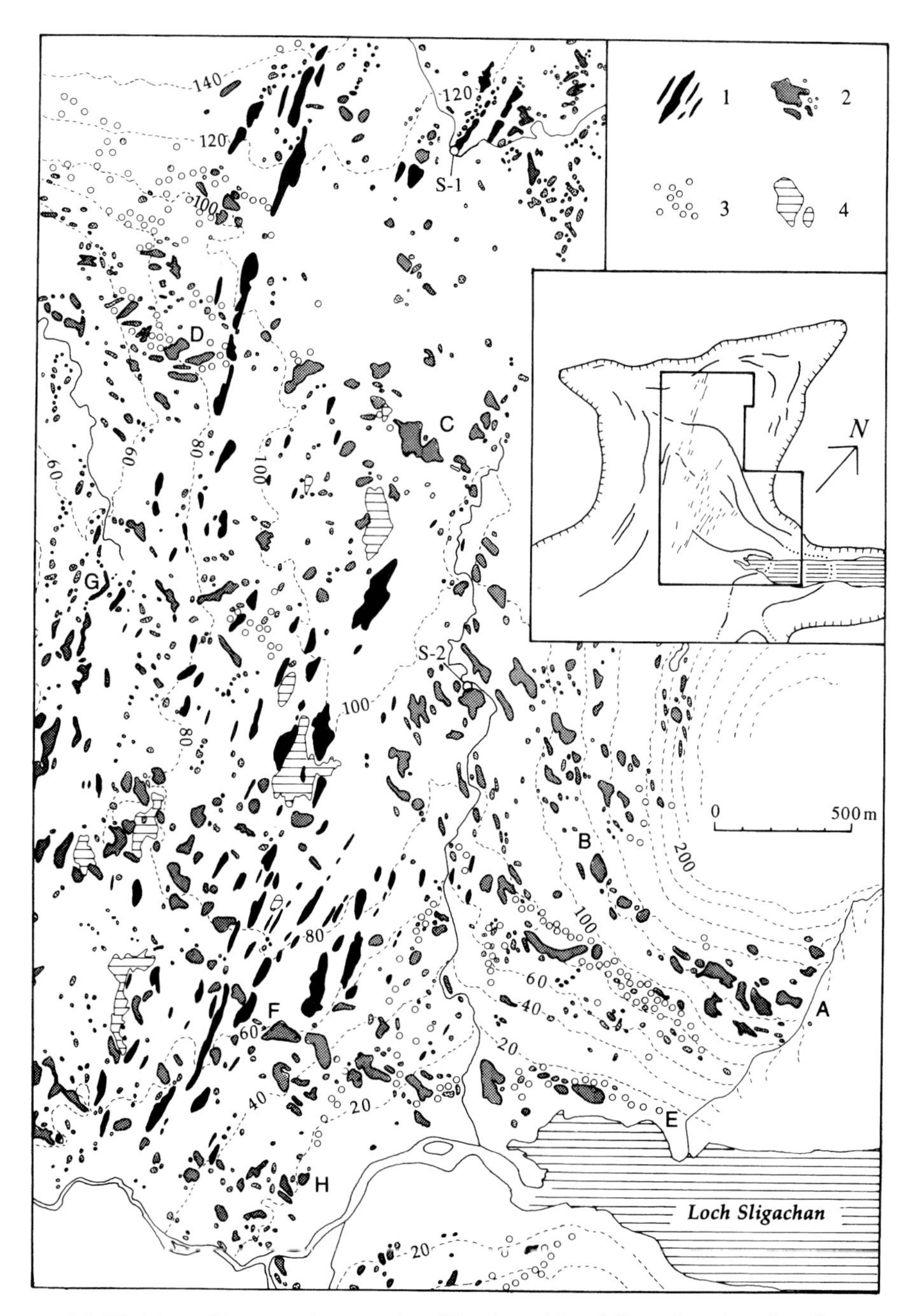

Figure 15.15 Map of hummocky moraine, Sligachan, Isle of Skye. 1 = drumlins, 2 = moraine mounds and ridges. Note transverse chains of moraines marking ice-margin positions (e.g. A–D, E–F), a chain of eskers (G), and chaotic hummocky moraine (H); 3 = boulders, 4 = water bodies. Inset: Younger Dryas limits and main recessional moraines around Sligachan. (From Benn, 1992).

foreland relief is low, the deposits of retreating glaciers in high-relief environments show close similarities to those of debris-rich temperate lowland ice lobes (e.g. Krüger, 1994a).

Three facies of deposition have been defined by Eyles (1979) based upon associations between activity of the terminus, thickness of supraglacial cover and reworking by meltwater.

Facies 1 consists of thick reworked accumulations of supraglacial till deposited by backwasting and decay of melting ice cores buried beneath thick debris cover. Where ice at the terminus is stationary (or nearly so), a predominance of melt-out processes results in a chaotic distintegration topography which Eyles termed 'uncontrolled', because the final product does not reflect the geometry of the ice margin or structures within the ice.

Facies 2 is laid down as a dispersed bouldery veneer by dumping from a retreating terminus, which can display down-glacier lineated patterns, reflecting deposition focused by structures such as gullies in the ice front. Debris cover is thin, and no relief inversions are associated with Facies 2 deposition. Seasonal dump moraines may contain internal bedding due to gravity-sorting after release of the debris from the ice margin.

Facies 3 describes a supraglacial morainic till complex comprising interfingering lensate horizons of supraglacial melt-out till and glacifluvial sediment. The areal extent of facies 3-type successions is greater at inactive, low-gradient termini where meltwater streams and ponds occur within ice-cored terrain. The distribution and relative development of these facies may aid in the reconstruction of ice-margin dynamics during glacier retreat.

The evolution of downwasting, debris-covered glacier termini in Iceland has been documented by Krüger (1994a). His model emphasizes the importance of debris bands within the ice in supplying sediment to the ablating ice surface, and of the interaction between various gravitational, glacifluvial and glacilacustrine reworking processes. The final depositional assemblage consists of low-relief, hummocky topography underlain by variably deformed diamicts and sorted sediments.

15.5.3.3 Medial Moraines

Supraglacial debris on many valley glaciers is delivered to the terminus as medial moraines. Though conspicuous features on many valley glaciers, they are seldom preserved as prominent landforms following deglaciation. This is because medial moraines generally contain relatively small amounts of debris and tend to undergo considerable reworking during glacier ablation. Where medial moraines consist of distinctive lithologies they may be clearly preserved, particularly where they consist of unusually large boulders. Deposition of medial moraines may form longitudinal bands of facies 1 or 2, superimposed on the underlying lodgement till plain (Eyles 1979; Fig. 15.16).

15.6 ROCK GLACIERS

Rock glaciers are tongue-like or lobate masses of ice and coarse debris that flow downslope by internal deformation. They commonly have ridges, furrows and sometimes lobes on their surfaces, and have steep fronts down which debris collapses and is then over-ridden by the advancing mass (Washburn, 1979; Ballantyne and Harris, 1994). A wide range of models has been proposed to

Figure 15.16 Deposition of the medial moraine of the Haut Glacier d'Arolla by slow terminus retreat, giving juxtaposition of Eyles (1979) facies 1 and 2 in the foreland. (Photograph by Ben Brock).

explain the genesis of rock glaciers. Some researchers use the term 'rock glacier' broadly, to include features with cores of glacier ice or ground ice (e.g. Humlum, 1982): others reserve the term exclusively for 'periglacial' phenomena (e.g. Haeberli, 1985; Barsch, 1987). A twofold genetic classification is used in some texts, consisting of periglacial rock glaciers, which involve the slow deformation of ground ice below talus slopes (e.g. Kirkbride and Brazier, 1995), and glacial rock glaciers, which form by the progressive burial and deformation of a core of glacier ice by a thick, bouldery debris mantle (e.g. Whalley *et al.*, 1995b). This classification can be difficult to apply in practice, and rock glaciers probably form a genetic continuum with no clear division between periglacial and glacial rock glaciers.

In mountain environments, rock, snow and ice are delivered to the base of slopes by avalanches and other mass movement processes, in varying proportions over space and time. Where the rock component is negligible, clean glaciers will form where snow and ice can survive ablation over the balance year. Where the snow and ice component is zero, talus slopes will result. Between these end-members exists a continuum of forms. Debris-covered glaciers form where the rock component is relatively high, and debris accumulates as a lag on the ablation zone of the dirty ice mass. Where the rock component is much higher, avalanche snow and ice will occur as isolated but deformable lenses within a talus, and the resulting form will be a rock glacier. It is probable that many rock glaciers in high mountain environments such as the Khumbu Himal, the Karakoram Mountains and Lahul Himalaya form by this mechanism (Barsch and Jakob, 1998; Owen and England, 1998; Fig. 15.17). In the Khumbu region, rock glaciers commonly occur in relatively low-lying catchments (5,000–5,600 m) where snow input occurs only during the drier winter months. Debris-covered glaciers typically occupy higher catchments, where temperatures are low enough for summer monsoon precipitation to fall as snow. Other origins of rock glaciers,

Figure 15.17 View over the debris-covered Ngozumpa Glacier to an avalanche-fed rock glacier below the peak of Cholo, Khumbu Himal, Nepal.

such as the formation of interstitial ice within a talus by the freezing of groundwater, can also be interpreted within this continuum model.

The relative proportions of rock and snow/ice delivered to the base of a slope will change with climate. A decrease in precipitation or an increase in temperature (more precipitation falling as rain rather than snow) will increase the relative importance of the rock component, producing conditions less favourable for glaciers but more favourable for rock glacier formation (Brazier *et al.*, 1998; Nicholson, 2000). During periods of glacier retreat, active rock glaciers may develop at the heads of former debris-covered glaciers, while the remnants of the ablating glacier tongue also evolve into rock glacier forms. Remnant ice-cored moraines sometimes develop into rock glaciers as the protected ice core begins to flow internally under the stresses imposed by the debris overburden and distal slope (Vere and Matthews, 1985; Owen and England, 1998). Such features have been referred to as rock-glacierized moraines in the Canadian arctic (Dyke *et al.*, 1982; Evans, 1993).

15.7 LANDSYSTEMS OF PROGLACIAL DEPOSITION

15.7.1 Glacial-Proglacial Linkage

The proglacial landsystem comprises landform-sediment associations constructed by fluvial, glacilacustrine mass movement and aeolian processes, which redistribute glacigenic sediment. Landforms include outwash fans, sandar, terraces formed by fluvial incision into valley fills, and drapes of wind-blown sand and silt. Volumetrically, the proglacial deposits dominate at large glaciers, particularly in maritime ranges, where most of the coarse sediment from glaciers is redistributed as fluvial bedload in proglacial valley trains. At small glaciers and in arid ranges, where proglacial fluvial deposition may be negligible, a proglacial river may be a bedrock channel whose load is primarily fine-grained glacial sediment in suspension.

15.7.2 Outwash Fans and Sandar

15.7.2.1 Aggrading Outwash Fans and Sandur (Valley Trains)

During periods of advance and extended stillstands of maritime glaciers, linkage between the ice-contact and proglacial zones is strong. Efficient fluvial redistribution of sediment aggrades the whole valley width by braided rivers, aided by switching of the loci of the outwash portal at wide glacier termini. Aggradation occurs when sediment is supplied to a proglacial river either directly from glaciers or during periods of paraglacial activity (Ballantyne, 2002b). The valley fill usually forms a sharp break of slope with the valley walls, except where tributaries build fans interfingering with the aggrading sandur surface (Fig. 15.18). During periods of glacier retreat, ice-contact lakes

Figure 15.18 Aggrading braided valley fill in the Godley Valley, New Zealand, a major sediment sink in the glaciated valley landsystem. Note the lack of valley-side fan development and great width of the active outwash plain.

influence downstream sedimentation by acting as sediment traps for the coarse sediment delivered from the glacier.

Glacifluvial valley fills attain thicknesses of several hundred metres and extend tens to hundreds of kilometres downstream from glacier termini. They form large sediment sinks on timescales of 10^4–10^5 years, even in tectonically active regions where they may be the last remaining depositional evidence for former glacial advances in some valleys. Proximal to the glacier, the valley fill takes the form of an alluvial fan with a broad apex at the ice margin. Downstream, decreases in gradient and sediment size are associated with the gradation of the proglacial fan into the braided river plain (sandur, or valley train). The detailed facies architecture of sandur in relation to fluvial processes is described elsewhere (e.g. Boothroyd and Ashley, 1975; Boothroyd and Nummedal, 1978; Maizels, 2002).

15.7.2.2 Incised Outwash Streams

During glacier retreat, sediment supply to the proglacial zone may be reduced for three reasons:

1. opening of a terminal ice-contact proglacial lake trapping coarse sediment
2. exhaustion of glacigenic slope mantles, and
3. stabilization of slopes by vegetation.

Reduced sediment supply can lead to incision of sandur surfaces, to give major paired river terraces, within which degradational (unpaired) terraces are inset. Multiple flights of paired terraces are associated with complex glacial histories (eg. Maizels, 1989). The transition is an important threshold in the sediment transfer system. Incision propagates downstream from the glacier terminus, initially forming a narrow inset floodplain (Fig. 15.19). The incised reach extends diachronously downstream, and may mark the first phase of a period of complex fluvial response triggered by glacier retreat and/or slope stability.

15.7.2.3 Ice-Contact Adverse Slopes (Outwash Heads)

The term outwash head describes the up-valley or adverse ice-contact slope bounding a proglacial sandur or fan. Outwash heads are associated with well-connected glacial and fluvial transport systems, in which little debris is incorporated into terminal moraines. If moraines form they have low preservation potential due to destruction by powerful, migratory outwash rivers. Though common landforms along the southern margins of the Laurentide Ice Sheet (e.g. Koteff, 1974), outwash heads are under-represented in research on valley glaciers in humid alpine regions. They are a major ice-marginal form in areas where debris-rich glaciers terminate in wide, gentle valleys (e.g. Alaska, New Zealand). The formation of an outwash head is not dependent on the presence of a debris-covered glacier, though such glaciers in humid regions invariably terminate in outwash heads.

The ice-marginal and outwash head environments at Tasman Glacier (Kirkbride, 2000) provide evidence of how Holocene glacier fluctuations reflect the dynamics of the debris-covered ablation zone. Proglacial fluvial aggradation during the Holocene created the outwash head, which now constrains a growing ice-contact lake (Fig. 15.20). Neoglacial terminal moraines are clustered in lateral-frontal positions and represent a tiny proportion of the debris discharge from the glacier. The vast majority has been transferred directly into the proglacial zone by glacial dumping and syndepositional redistribution in the proglacial fan.

Figure 15.19 Oblique aerial view of the terminus region of Maud Glacier in the Godley Valley, New Zealand. Ice retreat has opened up an ice-contact lake leading to incision of the outwash stream. Note the abandoned braided channels, and the late 19th century trimline and drift limit of the glacier. (Light autumn snow cover.)

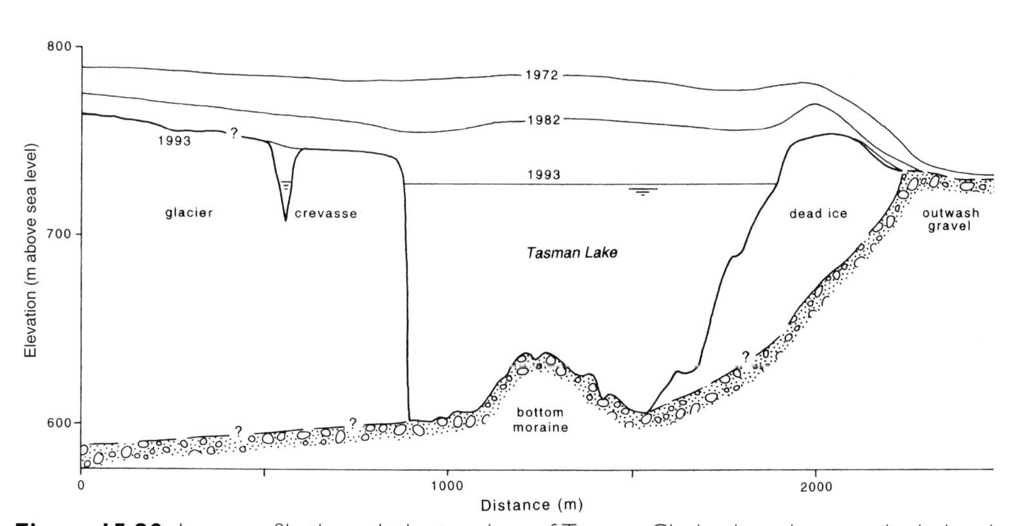

Figure 15.20 Long profile through the terminus of Tasman Glacier, based on geophysical and bathymetric surveys. The outwash head (the former ice-contact adverse slope) ponds the growing proglacial lake. A block of separated dead ice decaying on the adverse slope will eventually form the irregular hummocky topography typical of such landforms. (Adapted from Hochstein *et al.*, 1995).

15.8 PROGLACIAL AEOLIAN LANDSYSTEMS

Supraglacial debris or morainic, glacilacustrine and glacifluvial sediments may be rich in silt and fine-grained sand. Aeolian processes easily deflate the fine-grained material within these deposits and complex suites of glacioaeolian landforms are present within many valley glacial environments (Derbyshire and Owen, 1996). Patterns of sediment transport are strongly influenced by local wind systems, including katabatic, anabatic and föhn winds. In some regions such as the Himalaya in northern Pakistan and Tibet, thick deposits of loess are present, although these are frequently subjected to colluviation (Owen *et al.*, 1995; Lehmkuhl, 1997; Rose *et al.*, 1998).

15.9 ICE- AND MORAINE-DAMMED LAKES

Temporary lakes dammed by either glacier ice or moraines are common features of mountain environments, and are formed in four main situations (Yamada, 1998; Clague and Evans, 2000):

1. where a glacier emanating from a side-valley blocks the drainage of the trunk valley
2. where a glacier in a trunk valley blocks drainage from side valleys
3. at the junction between two valley glaciers, and
4. behind lateral-frontal moraines and outwash heads.

Cases (1), (2) and (3) form during glacier advance, and case (4) during glacier retreat. In the case of ice-dammed lakes, water level may be controlled either by low bedrock or sediment-floored cols in the lake catchment or by the glacier dam itself. An example of the former case is recorded in Glen Roy, Scotland, where ice advance during the Loch Lomond (Younger Dryas) Stade blocked the drainage of a major valley system (Sissons, 1981). Cols on the watershed controlled water level, and the lake rose and fell through three distinct levels as successive cols were blocked by glacier advance or exposed by retreat. The former water levels are recorded by very prominent shorelines known as the 'Parallel Roads', which remain strikingly clear after approximately 11,000 years of weathering and erosion (Fig. 15.21). Subaqueous fans mark the former glacier terminus, and drapes of laminated sediments cover much of the former lake floor.

In situations where the ice dam controls lake level, lakes are inherently unstable, as high lake levels will tend to destabilize the dam, thus precipitating catastrophic lake drainage (Clarke, 1982). Such lakes will tend to undergo multiple drainage and filling cycles during a single glacial cycle (Benn, 1989b).

The transition from small supraglacial ponds into a large moraine-dammed lake may be quite rapid, occurring within 2–3 decades in some cases (Ageta *et al.*, 2000). Glacier lake outburst floods (GLOFs) from moraine-dammed lakes are currently a significant environmental hazard in high mountain environments in the Himalaya, Andes and North American Cordillera as a result of recent rapid climatic warming and deglaciation (Lliboutry, 1977; Richardson and Reynolds, 2000).

Although short lived, ice- and moraine-dammed lakes can have profound effects on glaciated valleys. High sediment fluxes mean that they infill rapidly with sediment. Different combinations of sedimentary facies are deposited according to locally dominant processes. In

Figure 15.21 The Parallel Roads of Glen Roy: erosional shorelines formed along ice-dammed lakes during the Loch Lomond (Younger Dryas) Stade.

supraglacial moraine-dammed lakes, dumping of the debris cover into the lake produces an ice-cored lake floor where relief inversions and buoyant berg release will occur against a background of fine sedimentation from suspension and iceberg dumping. Backwasting of the ice shorelines and coalescence of neighbouring ponds opens up larger lakes until a calving terminus develops in deep water (Kirkbride, 1993). Retreat of the calving margin causes a transition from an ice-bounded to a predominantly moraine-bounded lake during growth, causing a shift in depositional processes to more distal, fine-grained sedimentation and mass movements from moraine walls. Loss of the subaqueous ice floor produces syndepositional deformation of lake-floor sediments.

Depositional facies in Lateglacial valley lakes in Canada and New Zealand (Shaw, 1977c; Pickrill and Irwin, 1983; Eyles *et al.*, 1987; Ryder *et al.*, 1991; Ashley, 2002; Fig. 15.22) record deposition dominated by subaqueous mass flow deposits affected by numerous deformation structures caused by the melt-out of buried ice. In some areas, valley-side delta and kame terraces are preserved, recording former lake levels or positions of the ice surface. During the wastage of the Cordilleran Ice Sheet (British Columbia), stagnant glacier ice was isolated in deep valley bottoms and broke up into lake basins. Clague and Evans (1994b) have argued that environmental conditions were similar to those occurring today in the St Elias Mountains, where major debris-covered glacier tongues have thinned dramatically during the last hundred years. Useful sedimentological studies have also been conducted at small, younger lakes elsewhere (e.g. Gilbert and Desloges, 1987; Liverman, 1987; Hicks *et al.*, 1990; Bennett *et al.*, 2000).

GLOFs can erode and rework large volumes of sediment. Flood tracks may be preserved in the landscape in the form of channels and boulder fans extending down-valley from the moraine breach (Clague and Evans, 1994a; Coxon *et al.*, 1996).

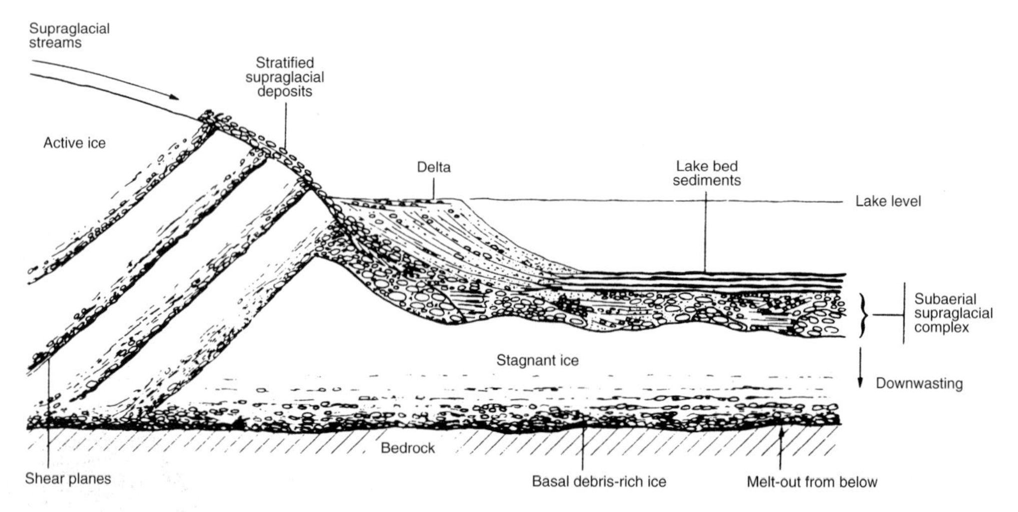

Figure 15.22 Relationships between ice-margin structures, sediment delivery, and supraglacial lacustrine sedimentation in Lateglacial Okanagan Valley, B.C., Canada. (Shaw and Archer, 1977).

15.10 GLACIATED VALLEY LANDSYSTEMS: SYNTHESIS

The wide variety of landforms and landform associations found in glaciated valleys reflects several interrelated controls. The most important identified in this Chapter are:

1. topography, itself a function of tectonic and denudational history
2. debris supply to glacier surfaces, in particular the relative supply of debris and snow/ice, which determines where glaciers will lie on the clean glacier–debris-covered glacier–rock glacier continuum, and
3. efficiency of sediment transport from the glacier to the proglacial environment by the glacifluvial system.

Changes to one or more of these variables will influence the character of glacial landsystems. The influence of factors (1) and (2) on glacier morphology and dynamics has already been discussed. Here we focus on the divergence of glacier margins into distinct landsystem associations due to variations in the degree of coupling between glacial and proglacial environments by the glacifluvial system (factor 3). Water discharges from glacier termini are a function of catchment area, climatically determined changes in water storage in the catchment, and mean precipitation. There is thus a climatic significance attached to the morphological outcomes, but one that is modulated by catchment size and other topographic characteristics. For example, in dry climates, the discharge needed to redistribute material supplied from supraglacial transport might require a large glacier, whereas in a more humid climate the same discharge will be issued from a smaller catchment. Similar morphologies may therefore occur under different climates but at glaciers of different sizes. Fig. 15.23 summarizes the landsystem associations typical of coupled and decoupled ice margins, that is, systems with efficient and inefficient glacifluvial transport between glacial and proglacial systems.

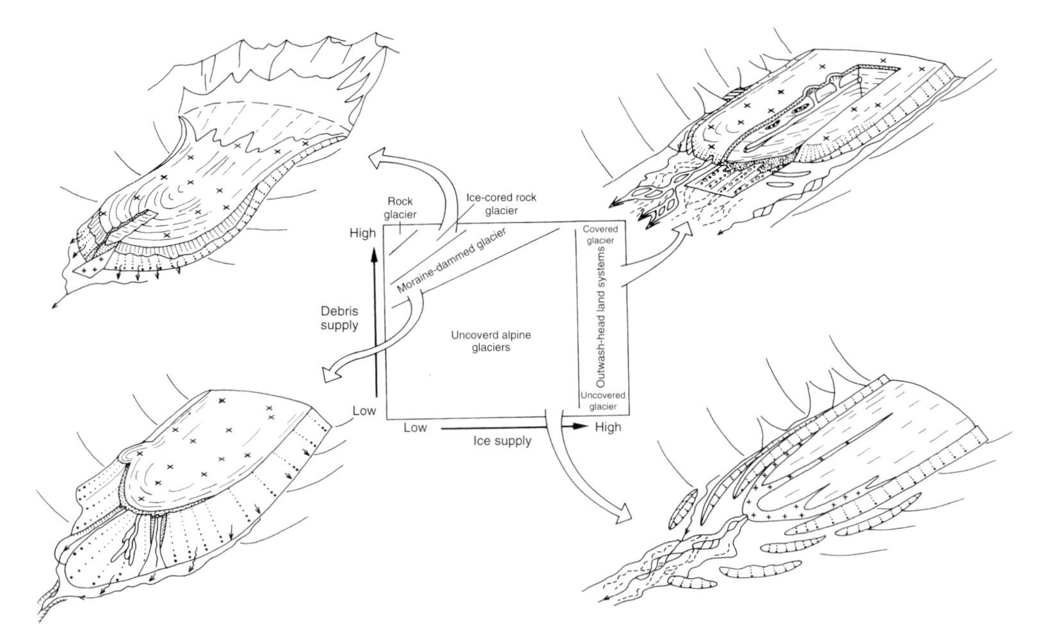

Figure 15.23 Conceptual relationships between constraints on landsystem development at valley glacier termini and developmental pathways into four landsystem associations. 'Covered' and 'uncovered glaciers' refer to debris-covered and clean glaciers, respectively.

15.10.1 Coupled Ice Margins

At glaciers where transfer of sediment between the glacier and the proglacial fluvial system is efficient, a coupled landsystem association exists. The efficiency of sediment transfer from the glacial to the fluvial zones means that moraine development is limited while large amounts of sediment pass into and through the proglacial zone. Most sediment leaves the glacier as fluvial load, and switching of meltwater streams across the glacier front ensures that if terminal moraines do form, they are usually rapidly destroyed. The dominant landsystem is the outwash head/aggrading sandur.

Coupled ice margins occur where powerful migratory outwash rivers leave the glacier, and the limiting factor on their rate of development is the competence and capacity of proglacial streams. Coupled landsystem associations are well displayed in the humid mountain ranges of Alaska and New Zealand. After deglaciation, the preservation potential of the ice-marginal landsystems is very high. Beautifully preserved examples date from the last glacial/interglacial transition in New Zealand and southern South America.

15.10.2 Decoupled Ice Margins

At smaller glaciers and in arid mountain areas, insufficient outwash discharge is available to transfer sediment away from the glacier margin. Coupling of the glacial and fluvial transport systems is then lost and debris is dumped around the glacier perimeter to form moraines. Over millennia, repeated superposition of moraines around the margins of debris-covered glaciers constructs giant bounding moraines, constraining small, steep glaciers on elevated beds. Breach-lobe moraine formation is commonly associated with the continued thickening of such glaciers.

'Decoupled' ice margins are characteristic of many high-altitude, semi-arid to arid mountain ranges such as the central and northern Andes and High Asia. High firn lines prevent the formation of large compound glaciers, favouring instead cirque and small valley glaciers with debris concentration ratios conducive for sediment-choked, decoupled margins. The limiting factor governing landsystem development is low ice supply due to aridity, while sediment supply to the glacier surface remains very high. These smaller, drier glaciers have enhanced morainic landsystems at the expense of the proglacial landsystem. During recession, conditions favour moraine-dammed lake development. Well-documented examples include Ngozumpa Glacier (Nepal), Hatunraju Glacier (Peru), and Miage Glacier (Italy) Benn *et al.* 2001; Lliboutry, 1977; Deline, 1999a, b). Rock glacier fronts fed by debris-covered glaciers are a variant on decoupled ice margins. Rock glaciers represent stores of sediment, which again reflect the inability of ice-marginal processes to evacuate sediment.

15.11 CHANGES IN LANDSYSTEM DISTRIBUTION OVER SPACE AND TIME

Glacial landsystems change over space and time due to climate change. The Ben Ohau Range (New Zealand) demonstrates the response of a range of glaciated valley landsystems as glacier equilibrium altitudes (ELAs) have risen since Late Glacial time (Brazier *et al.*, 1998). A steep precipitation gradient allows a variety of forms associated with humid and arid climates to have developed in close proximity. It is possible to see how landsystems within individual catchments have evolved as debris supply and transport have altered under changing Holocene climates.

Spatial and temporal patterns are revealed by mapping and rock weathering-rind dating of landsystems within single catchments (Fig. 15.24) and along the range. At present, outwash-dominated landsystems of the major Mount Cook valley glaciers occupy large catchments to the north of the range. Relatively clean alpine glaciers occupy the humid north of the range itself, debris-covered cirque glaciers with moraine dams in central catchments, and active talus rock glaciers towards the arid south (Birkeland, 1982; Brazier *et al.*, 1998). The spatial pattern occurs over only ~30 km when plotted normal to the regional isohyets, due to the steepness of the climatic gradient. Variability occurs within single catchments, where forms occupy local topoclimatic 'niches' defined by particular mixes of ice and debris in the transport system (cf. Morris, 1981).

The modern spatial succession demonstrates how limiting factors for landform development change with increasing aridity. At the humid extreme, fluvial competence and capacity limit sediment transfer from debris-rich glaciers to their proglacial zones. With reduced precipitation, 'alpine' glaciers fail to build larger moraines because of more limited sediment supply and dynamically fluctuating ice-margins. At the arid extreme, abundant rockfall supply but precipitation starvation limit the transport of debris largely to gravitational processes, so that uncoupled landsystems dominated by rock glaciers and talus predominate.

The temporal dimension of landsystem development has been combined with the regional spatial pattern in Fig. 15.25. It can be seen how landsystems that occupy particular process-form domains have shifted in space and time as ELAs have risen since the Last Glacial Maximum. Rising ELAs have caused a shift along the spectrum from 'coupled' to 'uncoupled' landsystems, as the ice-covered area and importance of glacial transport and runoff within the catchment has reduced.

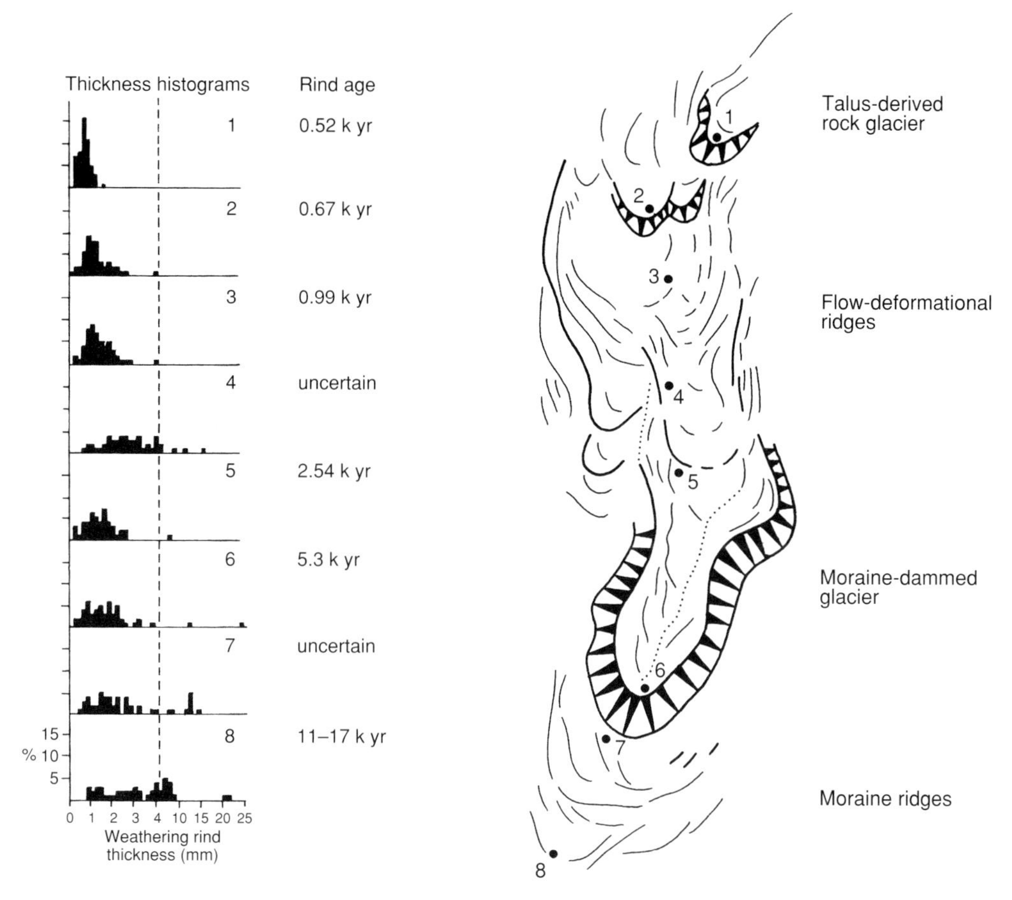

Figure 15.24 Relict moraine dams, supraglacial melt-out till, and rock glacier landsystems in MacMillan Creek, Ben Ohau Range (New Zealand). The rock weathering-rind histograms give calibrated ages in ^{14}C years BP. The landform sequence shows a trend to smaller, more debris-rich, and 'decoupled' landsystems through the Neoglacial period. The proportion of ice to debris in transport has declined as equilibrium line altitudes have risen.

Preserved landsystem sequences reflect both the spatial and temporal climate controls on catchment ice and debris fluxes.

15.12 CONCLUSIONS

The process-form associations outlined in this chapter allow interpretations to be made of relict forms belonging to past glacial advances. While no single landform is necessarily diagnostic of glacier type, assemblages of landforms (landsystems) allow former glacier systems to be reconstructed with some confidence. We have emphasized the wide variety of landforms and landform associations found in glaciated valleys and that these landforms reflect the local topography, lithology, climate, debris supply and the efficiency of sediment transport from the glacier to the proglacial environment. We have also presented many of the numerous landsystems

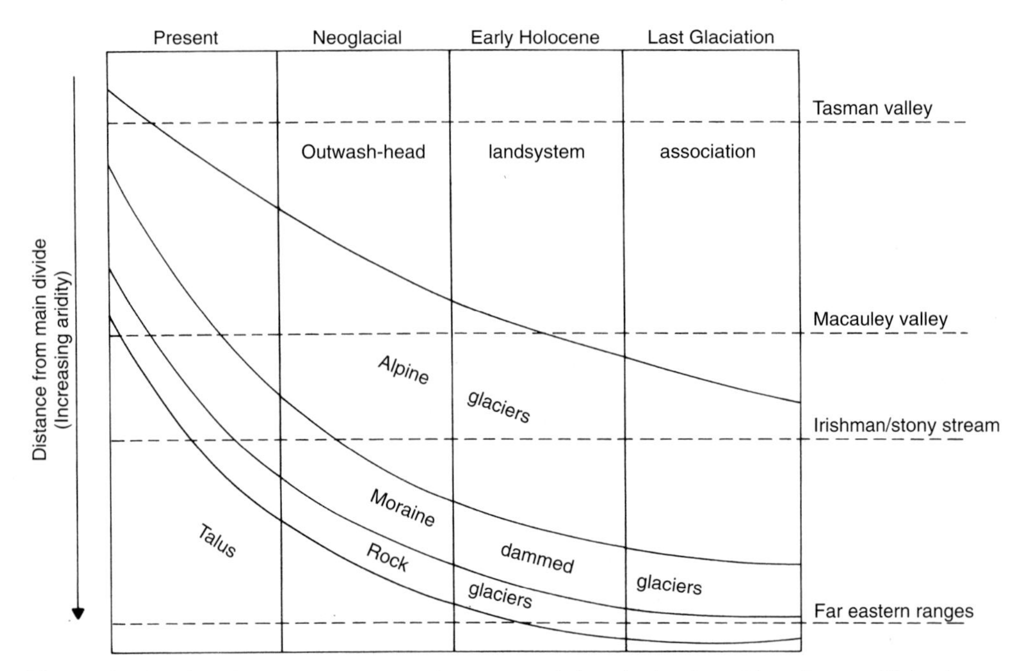

Figure 15.25 Landsystem occurrence in space and time in the Ben Ohau Range. Older, larger landsystems represent more open/'coupled' sediment-transfer systems. As later glacier advances have been smaller, more closed/'uncoupled' landsystems have developed. Thus, sediment transfer through catchments has become more interrupted from the Last Glacial Maximum through to the late Holocene.

and lithofacies models that exist to help examine the dynamics, landforms and sediments that characterize particular, and often unique, glacial valley landsystems. This plethora of models illustrates the complexity of glacial valley landsystems, but they help develop a framework for interpreting and reconstructing ancient landform and sediment associations in glacial valleys. Such reconstructions are important for palaeo-climatic inferences (Benn and Lehmkuhl, 2000) and for developing models of how mountain landscapes evolve. Furthermore, an understanding of the dynamics of erosion, sediment transfer, deposition and landform evolution is essential for efficient and effective sampling and interpretation of numerical dating such as cosmogenic radionuclide surface exposure and optical stimulated luminescence dating (Benn and Owen, 2002; Owen *et al.* 2002).

CHAPTER

PLATEAU ICEFIELD LANDSYSTEMS

Brice R. Rea and David J.A. Evans

16.1 INTRODUCTION AND RATIONALE

Plateaux are topographical features typical of many mountain regions and are perhaps best developed in passive margin settings, that have experienced significant periods of tectonic stability. Such conditions allow the preservation of palaeo-landsurfaces, which may have subsequently been uplifted and/or incised (Goddard, 1989). Good examples of such landscapes are found on the west and east sides of the North Atlantic, for example in Norway, the British Isles, USA, Greenland and in the island archipelago of the Canadian Arctic. Areas of Iceland also contain plateaux (Brown and Ward, 1996), where the plateaux (stapis or tuyas) are often the products of geologically recent subglacial volcanic eruptions (van Bemmelen and Rutten, 1955; Jones 1969). This chapter compiles a diagnostic assemblage of landforms and criteria that can be used to identify or infer the existence of former plateau icefields. Ice masses reconstructed from glacial geology are often used to reconstruct regional palaeo-climate, but evidence of glacierization on plateaux is often subtle and/or missing. It is therefore essential that plateau-style glaciation is recognized and the glaciers are reconstructed with realistic hypsometry (e.g. Furbish and Andrews, 1984; Benn and Lehmkuhl, 2000) as this can have a major impact on equilibrium line altitudes (ELAs) and on the climate thus inferred. A lack of appreciation of valley and associated plateau glaciers has often led to the misinterpretation of the geometry of ice coverage and thus the reconstruction of erroneous palaeo-ELAs.

Plateau and valley landscapes are synonymous with landscapes of selective linear erosion (Sugden, 1968, 1974; Sugden and John, 1976), and it is the juxtaposition of glacier erosion and glacier protection that proved one of the major problems in understanding the geomorphic evolution of these landscapes. The preserved palaeo-landsurfaces containing remnant landforms, sediments and weathering profiles which in some instances are found adjacent to the deeply incised, glacially eroded troughs led to the misinterpretation that these areas had acted as refugia during multiple Pleistocene glaciations. Contrary to this opinion, plateaux have since been viewed as the loci of glacier initiation and expansion during cooling events and as a haven for glaciers during warming in interglacials and interstadials, depending upon the latitude and altitude of the plateau and the global climate characteristics at the time. In most instances it is the valleys that contain the bulk of the geological information (e.g. deltas, shorelines, moraines.) used in reconstructing the geometry of former glacier cover. The thermal and dynamic properties of the plateau ice in such landscapes dictate the geomorphic signature that is produced. In some instances this signature may be very subtle or non-existent.

The plateau icefield landsystem model presented in this chapter is based upon contemporary examples from Ellesmere Island in arctic Canada, Troms and Finnmark regions in North Norway, and Iceland. These systems provide geomorphological evidence from both cold-based and warm-based plateau icefields and associated valley glaciers. Potential errors in ELA calculation associated with the misinterpretation of glaciation style will be highlighted using examples of plateau icefields dating to the Little Ice Age (LIA) and Younger Dryas, in North Norway.

16.2 AREA ALTITUDE RELATIONSHIP FOR PLATEAU ICEFIELDS

Many reconstructions of glaciation in mountain regions have ignored the potential impact of plateaux on glacier mass balance, probably due to the fact that early research focused largely on alpine regions. However, the general controls on plateau icefield existence were established by Manley (1955, 1959). He suggested that the larger the breadth of a glacier summit, perpendicular to the prevailing accumulation season wind, the closer to the regional snow/firn line an ice cover can be sustained. Fig. 16.1 shows data taken from Manley's original publication (Manley, 1955)

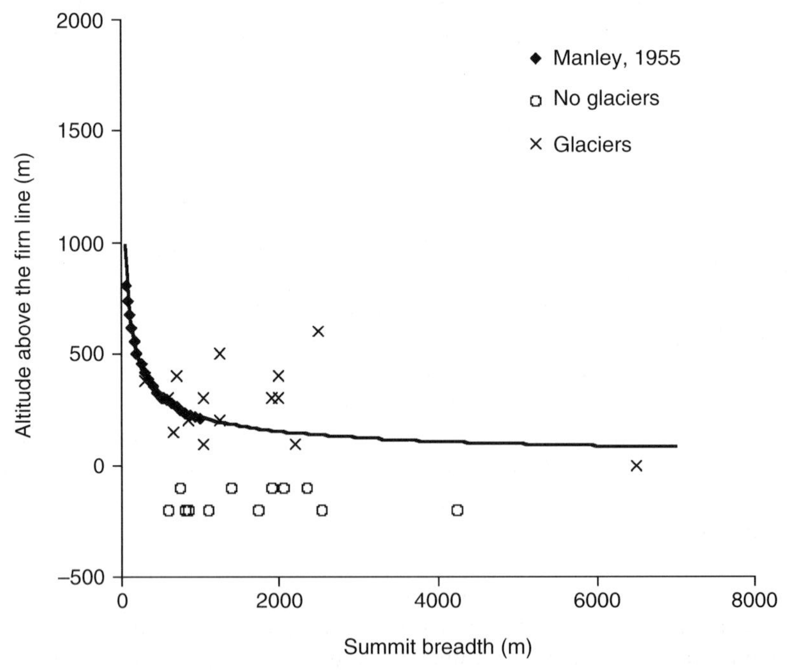

Figure 16.1 Graphic plot showing Manley's original data (with a power law fitted), and additional data for summits in North Norway. (From Rea *et al.* 1998).

with additional data from north Norway added by Rea *et al.* (1998). Manley's curve on Fig. 16.1 can be best approximated by the power law:

$$s_h = as_b^{\ c}$$

Where s_h is the summit altitude above the firn line, a and c are empirical constants and s_b is the summit breadth perpendicular to the predominant accumulation wind direction. This equation can be used to evaluate icefield existence if firn line altitude can be approximated from other sources (e.g. lateral moraine elevations), where ELA and firn line are assumed to be one and the same (Porter, 2001). For decreasing plateau size, plateau altitude must increase and so correspondingly a reduction in accumulation temperature will occur, so the ice that forms is more likely to be cold-based and non-erosive, thereby reducing the likelihood of major geomorphic impact. In such situations it is important to use the evidence found in the valleys radiating from plateaux to constrain glacier geometries and to establish the local firnline/ELA using other techniques (e.g. maximum elevation of lateral moraines, accumulation area ratio (AAR) from corrie and alpine style valley glaciers). The potential presence of an icefield can be then determined by employing Fig. 16.1 and the equation above.

16.3 CONTEMPORARY PLATEAU ICEFIELDS

16.3.1 Ellesmere Island, Arctic Canada

Plateau icefields drained by narrow outlet glaciers, terminating either as piedmont lobes or calving snouts in troughs and fjords, respectively, characterize glaciation of the Clements Markham Fold Belt of northwest Ellesmere Island, (Evans 1990a, b; Rea *et al.*, 1998; Fig. 16.2). Typical landform assemblages produced by the plateau icefields and associated outlet glaciers are associated with three depositional settings:

1. plateau surfaces
2. fjord/trough ice marginal depo-centres, and
3. undulating bedrock lowlands.

The plateau summits of northwest Ellesmere Island are characterized by a thin, often patchy residuum or weathered bedrock surface. In some areas the lithological properties of the residuum mimic exactly the underlying bedrock, indicating an autochthonous weathering origin. A few highly weathered erratics are scattered across the plateaux and document regional ice flow of unknown age over the area. Preservation of the residuum indicates that glacial coverage was cold-based and protective. Other features diagnostic of cold-based ice coverage are evident on and around the plateaux (cf. Ó Cofaigh *et al.*, 2003). The summits are devoid of obvious glacial erosional features and are often characterized by well-developed patterned ground features and tors. Retreat of the plateau ice margins is often recorded by meltwater incision into residuum and/or bedrock, indicating marginal rather than subglacial drainage (Fig. 16.3).

In fjord/trough ice marginal settings, there may be considerable thicknesses of pre-existing sediments that are accessible for glacigenic erosion, transport and re-deposition. Valley systems surrounding individual plateaux receive sediment directly from lateral/proglacial streams resulting in the accumulation of thick valley-bottom sequences and alluvial fans. On Ellesmere Island, sequences of aggradation and incision are predominantly controlled by glacio-isostatically

Figure 16.2 Part of oblique aerial photograph (1950) of plateau icefields on northwest Ellesmere Island, Canadian high arctic. (T408R-222, Energy, Mines and Resources, Canada.)

influenced sea level changes (England, 1983; Evans, 1990a). Commonly, piedmont lobes debouching from plateau icefields dam the regional drainage creating extensive ice-dammed lakes. An aerial photograph (Fig. 16.4) of a plateau icefield south of Phillips Inlet on northwest Ellesmere Island provides a clear illustration of well-developed river terraces and incised alluvial fans around the margins of the northerly draining piedmont lobes. Extensive stratigraphic sections in the middle reaches of the main river include delta foresets and other glacilacustrine sediments, documenting the damming of the drainage by advance of the northern piedmont lobes. The development of thrust-block moraine complexes, composed of old raised glacimarine, glacilacustrine or glacifluvial sediment (Evans, 1989b; Evans and England, 1991, 1992; Fig. 16.5) is common during such ice advances. Overriding, entrainment and subsequent release of this reworked valley bottom sediment during periods of glacio-isostatically higher sea level results in the deposition of subaqueous or grounding-line fans (Fig. 16.6). The ice contact deposits are used to demarcate the former marginal positions of outlets emanating from the surrounding plateaux (Evans, 1990a). The lateral margins of former outlet glaciers are often marked by rock glaciers, which originate as accumulations of talus over stagnant glacier ice or as supraglacial lateral moraines (Evans, 1993; Ó Cofaigh *et al.*, 2003).

Figure 16.3 A) Part of aerial photograph (1959) of plateau icefield near Phillips Inlet, northwest Ellesmere Island, Canadian high arctic, showing lateral meltwater channels cut during the recession of the largest outlet glacier lobe (A16760-99, Energy, Mines and Resources, Canada). B) Contemporary ice-marginal meltwater channels forming at the snout of a plateau outlet lobe on eastern Ellesmere Island, Canadian high arctic.

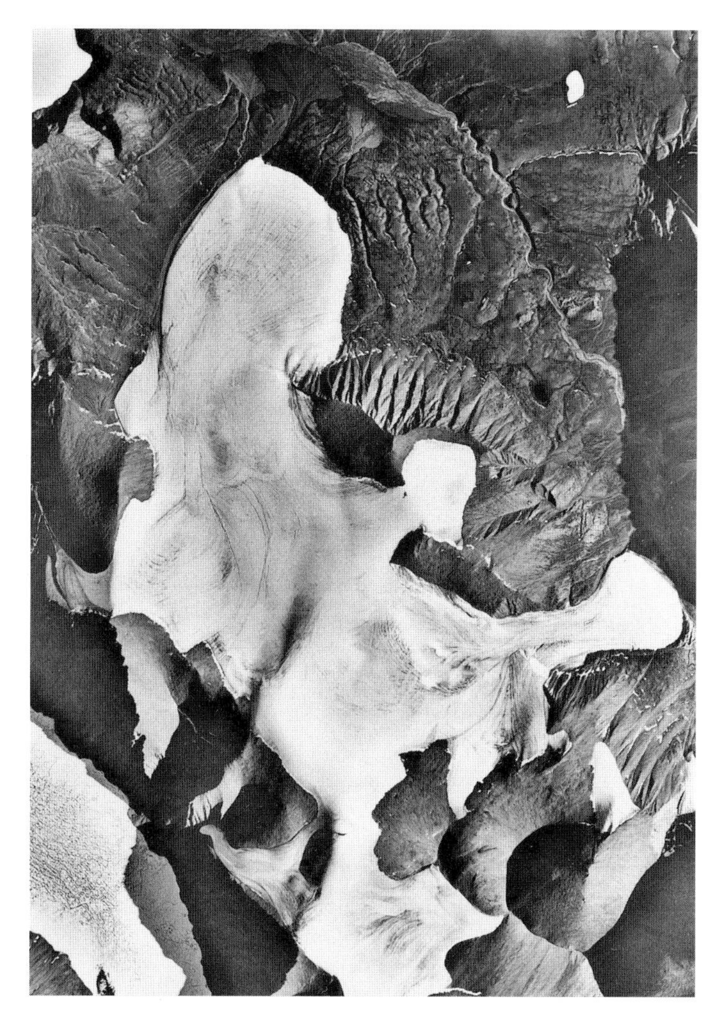

(A)

(B)

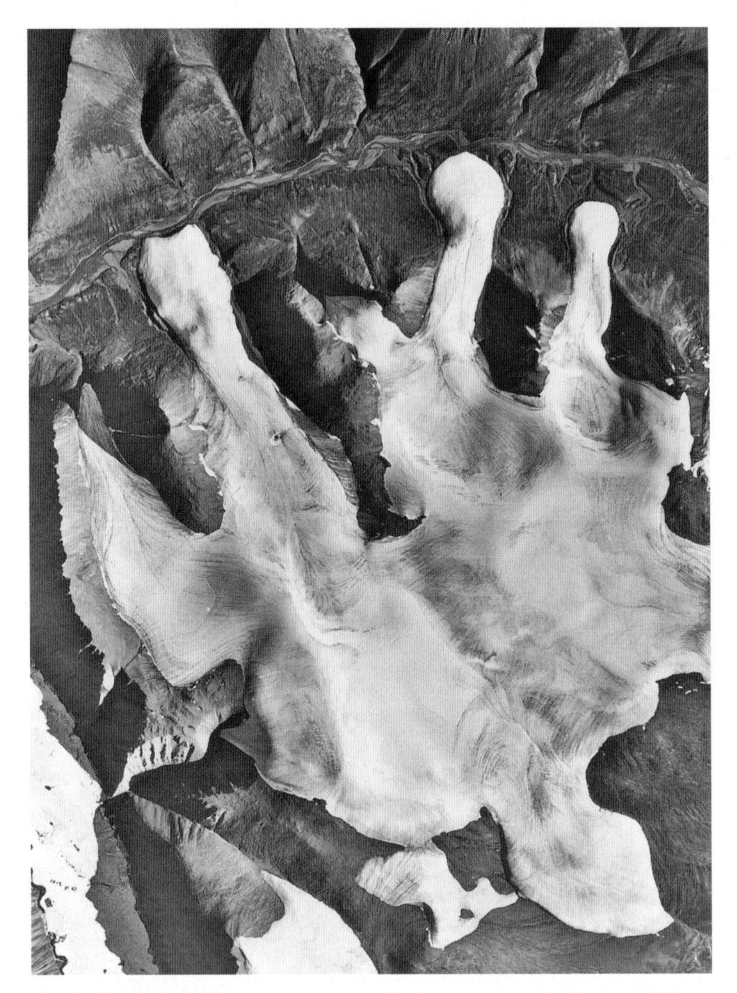

Figure 16.4 Part of aerial photograph (A16760-101, Energy, Mines and Resources, Canada 1959) of a plateau icefield south of Phillips Inlet, northwest Ellesmere Island, Canadian high arctic. Note the incised and terraced valley floor sediments. These sediments are of glacilacustrine, glacifluvial and alluvial fan origin and document former damming of the valley by the outlet glaciers. Incision resulted from valley drainage and glacio-isostatic rebound/relative sea level fall.

Undulating bedrock lowlands rather than fjord/trough systems border some plateaux. Here the lack of thick sedimentary sequences restricts the development of till blankets and ice-contact glacimarine deposits. A typical bedrock lowland formerly covered by an expanded plateau ice cap is found to the south of Cape Armstrong, Phillips Inlet (Fig. 16.7). This area is characterized by thin, discontinuous till veneers, extensive bedrock exposures and residuum. The presence and retreat of glaciers in such areas is mapped using abandoned lateral meltwater channels, making them similar geomorphologically to plateau summits.

The evidence presented thus far has been related to the more restricted ice coverage when plateaux act as individual accumulation centres. As glacierization proceeds, plateau icefields will

Figure 16.5 Thrust-block moraine composed of glacifluvial outwash and glacilacustrine sediments, formed by proglacial thrusting at the margins of a plateau icefield outlet glacier lobe on northwest Ellesmere Island, Canadian high arctic.

Figure 16.6 Subaqueous grounding-line fan at the head of a fjord on northwest Ellesmere Island, Canadian high arctic. Main picture shows the location of the section face in inset picture. The fan was constructed by meltwater emanating from a glacier outlet lobe that was nourished by plateau icefields surrounding the fjord. The glacier lobe flowed from left to right and occupied the fjord head, thereby damming, at least initially, the main valley in the foreground. Gravel mounds located upslope of the fan demarcate the glacier margin.

Figure 16.7 Undulating bedrock terrain to the south of Cape Armstrong, Phillips Inlet, northwest Ellesmere Island. A river valley in the middle distance was cut by meltwater from an expanded plateau outlet lobe that advanced during the last glaciation into the bedrock lowland. This view shows the till veneer, residuum and bedrock exposures typical of plateau outlet advance into bedrock terrains.

begin to coalesce, eventually becoming overwhelmed by a regional ice sheet that may imprint its own geomorphological signature. Similarly, some linearly eroded landscapes containing plateau icefields may be dominated by outlet glaciers fed from mountainous terrain located further inland, as is the case on eastern Ellesmere Island (e.g. Rea *et al.*, 1998; England *et al.*, 2000). The response time of the local ice was shorter, resulting in an early thinning of the plateau ice during deglaciation and allowing the development of regional geomorphological imprints by the less responsive trunk glaciers fed from further inland. Fig. 16.8 shows the low profile lateral meltwater channels cut along the southern wall of Hayes Fjord, eastern Ellesmere Island. These channels document the recession of the Hayes Fjord trunk glacier, which drained the expanded Prince of Wales Icefield during the last glaciation. The more recently regenerated plateau ice cap on the Thorvald Peninsula plateau summit to the south of the fjord had obviously receded sufficiently to allow the incision of the regional ice-configured meltwater channels during deglaciation. In such physiographic/glaci-dynamic settings, the differentiation of plateau icefield and regional trunk glacier geomorphology is clearly essential for accurate reconstructions of palaeo-glaciation.

16.3.2 North Norway

16.3.2.1 Lyngen

Glaciers centred on the peak of Jiek'kevárri (1833 m) in the southern Lyngen Peninsula provide an excellent example of plateau icefields and valley outlet glaciers (Fig. 16.9). The landscape is highly fretted with ice supply from plateaux to valleys dominated by ice avalanching. Presently

Figure 16.8 The Thorvald Peninsula plateau icefield draining into Hayes Fjord, eastern Ellesmere Island, Canadian high arctic. Lateral meltwater channels occur along the southern wall of Hayes Fjord and document the recession of a trunk glacier in the fjord at the end of the last glaciation. The outlet glaciers from the plateau icefield on the Thorvald Peninsula have advanced across the meltwater channels since their construction.

glaciated plateaux tend to be reasonably small with Jiek'kevárri being the largest in the region at 3.70 km^2, and lie well above the regional firn line. Around the margins of the icefields and on lower unglaciated plateaux, autochthonous blockfield cover is ubiquitous; in some places banding can be observed in the blockfield reflecting banding in the layered gabbros below. Despite the existence of easily removable blockfield material on the plateaux, the cold-based nature of the ice cover (Whalley *et al.*, 1981; Gellatly *et al.*, 1988; Gordon *et al.*, 1988) and insignificant supraglacial debris sources restrict bed erosion and moraine formation, respectively. Some localized erosion may occur at the plateau edge where an outlet glacier exists, for example on the north side of Bálgesvárri (Fig. 16.9). Gordon *et al.* (1988) and Gellatly *et al.* (1988) highlight the absence of meltwater channels, although meltwater was observed around the margins of the remnant ice cover on Bredalsfjellet (Fig. 16.9). It appears that the low bed slope angles produce meltwater ponding but this was insufficient to form significant channelized drainage. Thus, on plateaux where ice is cold-based, evidence of former glacier cover is very subtle.

In valleys below the plateaux the geomorphological signature is similar to what is traditionally expected of any temperate valley glacier in a mountain environment. After recession of the main fjord glaciers, where ice contacted the sea, ice-contact deltas delimit former glacier margins. Up-valley from these deltas significant quantities of glacifluvial sediment accumulated in lower gradient and overdeepened valley sections. Sequences of bouldery, frontal and lateral moraines document ice margins in the valleys. In areas where the till cover forms a thin veneer over bedrock, striae and roches moutonnées dominate the geomorphic signature of glaciation.

Figure 16.9 Aerial photograph (1978) of the fretted plateau and valley landscape centred on Jiek'kevárri at the head of Lyngsdalen (Fjellanger Wideroe 7802, 33-8 – 5820). For glacier names see Figure 16.17).

16.3.2.2 Troms-Finnmark

Plateaux in this region are lower and larger than those found in Lyngen (where plateaux are dissected), and as a result tend to act as accumulation centres for outlet glaciers in a similar style to the examples on northwest Ellesmere Island. Outlets exit steeply into valley heads as icefalls that remain connected to the main icefield above (Fig. 16.10). Generally the altitudinal range of a plateau is at most in the order of 100–200 m. Many of the surfaces are relatively flat, with slope angles generally less than 10°, only steepening in areas where they rise upward to meet nunataks/tors (Rea *et al.*, 1996a). These larger, lower icefields are in places at the pressure melting point and sliding over their beds (Rea and Whalley, 1994). Some ice margins terminate behind moraines produced during the LIA (Gellatly *et al.*, 1988; Fig. 16.11). In places around the ice

Figure 16.10 The main outlet of Langfjordjøkelen, which drops steeply from the bedrock plateau (800–1000 m), down to the snout at just above 300 m.

Figure 16.11 Bouldery moraine at the eastern margin of Øksfjordjøkelen. The middle ground shows ice plunging directly over the plateau edge, thereby prohibiting moraine formation. Note also the subglacial drainage.

margins, exposed by retreat since the LIA, the results of bedrock quarrying and abrasion are evident (Rea and Whalley, 1994, 1996). Measurements in subglacial cavities and observations of subglacial, channelized outflow streams indicate that ice is at the pressure melting point (PMP) around at least parts of the icefield margin (Fig. 16.11).

Non-erosive ice is also present, as indicated by the preservation of extensive blockfields containing patterned ground (Fig. 16.12), heavily weathered nunataks and bedrock, and a lack of moraines. Some bedrock areas show signs of 'older' subglacial erosion, but the extent of the subsequent weathering suggests that this is most likely to have taken place during a glaciation of at least pre-Weichselian age. In places, the blockfields are in excess of 1 m thick and are frost-sorted, although no permafrost has been found in the blockfields around Øksfjordjøkelen (Rea *et al.*, 1996a, b), unlike the higher plateaux in Lyngen (Gellatly *et al.*, 1988). In both regions studies have suggested that the blockfields represent remnants of a weathering sequence, perhaps pre-Pleistocene in age (Rea *et al.*, 1996a, b; Whalley *et al.*, 1997).

In a thorough assessment of the geomorphic impact of plateau glaciation centred on Øksfjordjókelen by Evans *et al.* (2002), it has been suggested that the largest accumulations of glacially derived materials occur as lateral and latero-frontal moraines in the valleys. The bouldery nature of these moraines and the angularity of the individual boulders suggest that rock avalanches and rock falls from the extensive bedrock cliffs are the main source of the debris. Moraine asymmetry attests to the variability of cliff exposures within some basins (eg. Matthews and Petch, 1982; Benn, 1989a; Evans, 1999). Rock glaciers have developed below some precipitous bedrock walls, reflecting the locally high rates of debris provision. The occurrence of some subglacially derived material in end moraines, in addition to roches-moutonnées and patchy till covers, attest to basal sliding, bed erosion and sediment deposition. Much of this was likely initiated by strain heating in basal ice passing through the steep icefalls that linked plateau summit ice to valley outlet glaciers. Where plateau outlet glaciers terminated in the surrounding fjord heads they sometimes deposited ice-shelf moraines, indicative of cold-based snouts. Recession of snouts is recorded in the shallow marine waters of some fjord heads by De Geer moraines and ice-contact, Gilbert-type deltas.

Figure 16.12 An area of weathered bedrock and blockfield showing patterned ground beyond the margin of Øksfjordjøkelen.

16.3.3 Iceland

The temperate glacier snouts of Iceland are mainly wet-based for at least part of the year with a narrow frozen zone developing at most margins during the winter due to the penetration of the seasonal atmospheric cold wave (see Evans, 2003). With the exception of glaciers that may 'freeze-on' large quantities of debris due to supercooling in overdeepenings (Spedding and Evans, 2002), debris-rich basal ice sequences are typically thin or absent. The stapis or tuyas are ideal physiographic features for the accumulation of small plateau icefields in areas located at the limit of glacierization, good examples being þórisjökull, Eiriksjökull and Hrútfell, which surround the larger Langjökull icecap, Drangajökull in the northwest and Torfajökull in the south. Glaciers nourished on plateau surfaces are unlikely to accumulate large volumes of supraglacial debris due to the lack of extraglacial debris sources (Rea *et al.*, 1998; Evans, in press). However, substantial accumulations of rockfall debris characterize the lateral margins of some outlet glaciers where they descend through precipitous cliffs at the plateau edge (Fig. 16.13). Consequently, lateral moraines are developed only at those margins lying beneath steep cliffs and are therefore discontinuous and asymmetrical on individual outlet lobes. Under some highly active rock walls, debris supply is sufficient to bury glacier ice during recession from the valley head, thereby producing ice-cored lateral moraines and talus-foot rock glaciers (Fig. 16.13). Beyond the influence of active rockwalls, plateau outlet glaciers deposit inset sequences of latero-frontal moraines during recession from valley heads. However, annual push moraines are not evident around the margins of many of the Icelandic plateau icefields. Instead, the valley floors are covered by a boulder veneer, which thickens at individual latero-frontal moraines (Fig. 16.13). This type of till cover suggests that the glaciers are transporting regolith from the plateau summit only short distances into nearby valley heads and that ice-marginal still-stands only occur in response to climate signals of lower frequency than the annual cycle.

Substantial rubbly moraines have been constructed on plateau summits, for example at the southwest margin of þórisjökull, although the extent of the underlying ice core is unknown (Fig. 16.14). The moraines demonstrate that at least parts of the glacier bed are warm-based and capable of eroding regolith and/or bedrock. Some areas cleared of regolith indicate glacier ice descended from a plateau into surrounding valleys (Rea *et al.*, 1998). Elsewhere, plateau geomorphology is dominated by or even restricted to meltwater channels, indicating predominantly cold-based ice and/or ineffective subglacial erosion.

16.4 DYNAMICS OF PLATEAU ICEFIELDS

In order to compile a plateau icefield landsystem it is important to understand the dynamics and style of glacierization of plateau summits. Broadly speaking, the glacierization style can be divided into two types, each representing a distinct phase of glaciation:

- large-scale ice cover (ice sheet) usually experienced under full glacial conditions where the topography is submerged and exerts less control on ice flow directions
- smaller scale, regional- to local-scale glaciation with ice sources centred on plateaux, the pattern of ice distribution being dictated by the altitude and latitude of the region (in highly dissected landscapes with small plateaux e.g. Lyngsdalen, valley glaciers may or may not exist during parts of this phase of glacierization)

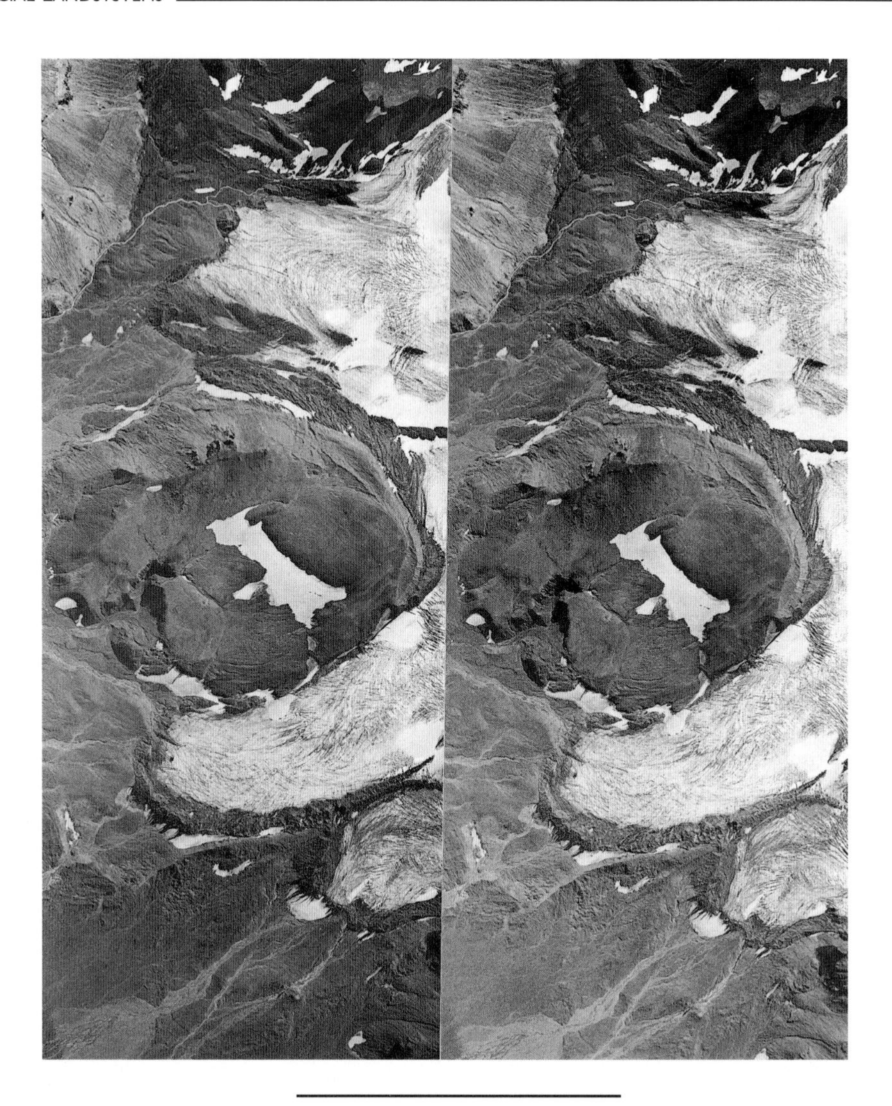

Figure 16.13 Aerial photograph stereopair (Isgraf/Loftmyndir and University of Glasgow, 1999) of the outlet glaciers on the northwest margin of þorisjökull. Visible are asymmetrically developed supraglacial lateral moraines and rubbly latero-frontal moraines whose distribution is controlled by the location of bedrock cliffs relative to the ice margins. Scale bar represents 1 km.

The onset of full glacial conditions will generally induce a decrease in annual air temperatures ensuring that newly forming or expanding plateau icefields are likely to be/become cold-based. Such cold-based ice acts as a protective cover over the plateau surface. Gradually, as the larger regional and continental ice masses grow, thick valley glaciers and eventually ice streams may form, which reach the PMP (periodically or for long periods) enabling glacial erosion and overdeepening of valleys. This is classic 'selective linear erosion' (Sugden, 1968; 1974), which is believed to be responsible for the formation of through valleys and the overdeepening of favourably orientated valleys. It is during this phase of glaciation that erratics may be transported onto plateaux (see Ellesmere Island above; Sugden, 1968; Sugden and Watts, 1977; Rea *et al.*, 1998).

Figure 16.14 Aerial photograph (Isgraf/Loftmyndir and University of Glasgow, 1999) of the southwest corner of Þorisjökull. A substantial rubbly, ice-cored moraine has developed at the Little Ice Age maximum limit, indicating that the glacier margin has eroded and transported the underlying blockfield. Scale bar represents 0.5 km.

Under conditions of local, topographically controlled glacier coverage, ice on plateaux is thinner but will thicken with increasing plateau size. It is during this phase of glaciation that the thermal regime may become more complex. In order for an icefield to form or expand on a plateau there must clearly be a positive mass balance. Manley (1955), and Fig. 16.1 demonstrates that there is an important area/altitude relationship that dictates plateau icefield accumulation. Simplistically, below a critical summit size, the smaller the plateau the higher it must be above the firn line to support an icefield, and thus, the lower will be the temperature of the ice. The warmest ice will therefore be found on the lowest plateaux, which Fig. 16.1 suggests will tend to have large icefields. Thus, ice which forms on plateaux that are well above the firn line (e.g. Lyngsdalen) is most likely to be cold-based and non-erosive. In some situations the ice may reach the PMP (e.g. Øksfjordjøkelen). In polar regions, where accumulation temperatures are lower, even the lowest altitude, large icefields will be unlikely to reach the PMP at their base. However, the larger the icefield the greater the potential for producing a clear geomorphic signature.

Due to the shallow slope angles, basal shear stresses and thus strain rates will be low in the basal layers of ice on plateaux. In areas where the bed steepens (i.e. towards outlets), the opposite occurs. If the basal ice temperature is close to the PMP the increase in strain heating may be sufficient to induce basal melting and thus sliding (e.g. Bálgesvárri mentioned earlier). At this point the glacier will begin to erode its bed. Basal ice may also reach the PMP around parts of an icefield margin. For example, at least two parts of the plateau-terminating margin of Øksfjordjøkelen are warm-based (Gellatly *et al.*, 1988; Rea and Whalley, 1994) and have eroded the bed to produce moraines; a similar situation has arisen at the southwest margin of þórisjökull in Iceland (see above). The ice in these locations may reach the PMP due to:

1. the percolation of meltwater towards the bed
2. the penetration of a summer warm wave through the thin ice
3. increases in strain heating as bed slope angles increase, or
4. some combination of these factors.

16.5 THE PLATEAU ICEFIELD LANDSYSTEM

This section integrates information from contemporary environments in the Canadian Arctic, north Norway and Iceland and provides the main criteria for identifying the former existence of plateau icefields in deglaciated landscapes. As mentioned above, glacierization may be experienced at both a smaller, local scale and at a larger, continental scale. If affected by both, the plateau icefield landsystem may comprise a composite signature. It is climate reconstructions of the smaller, local scale glaciation that are most sensitive to the non-identification of former plateau glaciers.

During successive glaciations the combination of glacial overdeepening and interglacial fluvial and slope activity results in the dissection of the landscape into isolated, remnant high altitude surfaces and incised valleys and fjords. In suitable tectonic settings, regional faulting may have been instrumental in the initial production of relief through the development of grabens (e.g. England, 1987a). In Iceland it is the juxtaposition of active volcanism and ice coverage that combined to produce plateaux. Once isolated, each plateau is an ideal collection area for permanent snowfields at the onset of glacial conditions ('instantaneous glacierization'; Ives *et al.*, 1975). In areas where the plateau margins have precipitous drops into fjords or troughs, 'fall glaciers' (cf. 'reconstituted glaciers'; Benn and Lehmkuhl, 2000) are nourished below the regional ELA by dry calving at the plateau cliff edge (Gellatly *et al.*, 1986). This is representative of most of the valley glaciers surrounding the Lyngen plateau icefields. During the more advanced stages of local scale glaciation, glaciers develop in the valley bottoms between plateaux and it is at this stage that the most significant landforms are likely to be produced.

Under conditions of pervasive ice sheet cover associated with full glacial conditions, the role of the topography in controlling ice discharge routeways can be reduced or overcome, and through-valleys may be excavated by transection glaciers. Conversely, valleys cut during plateau-centred glaciations will possess a radial drainage pattern (see Figs 16.17 and 16.19 later). Plateau tops may show no evidence of subglacial erosion if they hosted cold-based ice (Dyke, 1993). If there is evidence of erosion, indicating that the ice crossing the plateau was warm-based, then the erosional forms will again be aligned parallel to sub-parallel across the area or will radiate out from the plateau centre, depending upon the pattern of ice cover. Regional ice sheet imprints may be recorded as major landscape features such as through-valleys and fjords separating plateaux. Such troughs often contain adornments such as lateral meltwater channels and moraines associated with their occupation by regional ice (e.g. the fjords of southern Ellesmere Island; see above). Erratics transported onto plateaux from outside the local area often further demonstrate regional ice sheet coverage.

More localized styles of glacierization are characterized by the growth of plateau-centred icefields. As mentioned above, the climate reconstructions based upon this stage of ice coverage are most sensitive to erroneous interpretations of the valley glacier – plateau icefield configuration. The various landform units diagnostic of plateau icefield glaciation can now be isolated and reviewed (Fig. 16.15).

16.5.1 Moraines

Provided some parts of the ice are warm-based, and thus erosive, moraines may be found on top of the plateau or leading on to the plateau from the surrounding valley-head outlets (Whalley *et al.*, 1995a). A number of good examples occur around Øksfjordjøkelen (Fig. 16.11). If the ice is cold-based throughout, lateral and frontal moraines will be found in the valleys only. Valley moraines tend to be dominated by large, cobble to boulder size, angular material even where there is evidence of basal sliding, indicating the dominant debris source is rock fall (Fig. 16.16). Many latero-frontal moraines in the valley heads that surround the plateaux are ice-cored and display within-valley asymmetry (Benn, 1989a). This reflects the distribution of bedrock free faces and concomitant variability in rockfall/avalanche and debris flow activity. Active free faces may provide sufficient debris to produce supraglacial lateral moraines. In some circumstances these may develop into rock glacierized ice-cored moraines that may persist after deglaciation of the valley heads. Where considerable thicknesses of valley floor sediments are available, plateau outlet glaciers may construct large end moraine sequences, particularly where proglacial thrusting takes place (Figs 16.5 and 16.15). This appears to be most effective where glacio-isostatic uplift has resulted in the recent aggradation of permafrost and the production of a shallow décollement surface within the valley floor sediments (e.g. northwest Ellesmere Island; see Ó Cofaigh *et al.*, 2003).

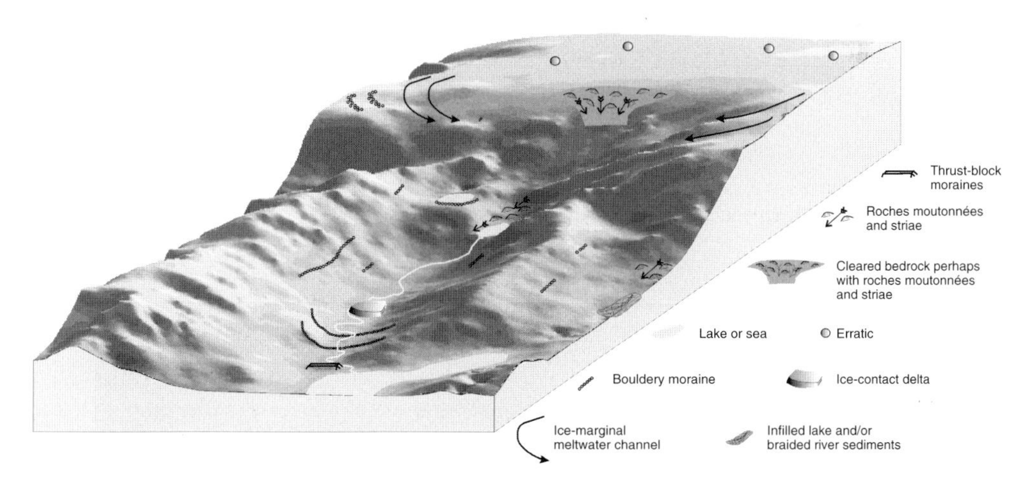

Figure 16.15 A schematic diagrammatic representation of the various elements produced by the plateau icefield landsystem. It is assumed that the undulating plateau surface has a blockfield cover. Higher, smaller summits (not depicted) that show no signs of glacierization must be evaluated using other means.

16.5.2 Ice-Contact Deltas

In areas where the glacier terminated in contact with standing water, ice-contact deltas and/or grounding-line fans are used as evidence for ice-marginal locations (Fig. 16.6). In marine environments where sea level history is known, associations between the deltas and associated shorelines can be used as a proxy dating technique for reconstructing the deglaciation chronology (Evans 1990a; Evans *et al.*, 2002).

16.5.3 Sediments

Sediments found on the plateau are most likely to be allochthonous or autochthonous weathering products described variously as blockfields, felsenmeer, weathering residuum and/or highly weathered bedrock exposures (Fig. 16.12). The depths, features and mineralogy of the weathering have been used to infer significant ages for these sediments and associated landforms (e.g. Sugden and Watts, 1977; Rea *et al.*, 1996a and b). As mentioned above, moraines may be found on plateaux that contain material indicative of active subglacial transport. Significant till cover does not form, due to the commonly short transport distances to plateau outlets and the dynamics of the plateau ice cover.

The valley floors are characterized by thin, generally patchy tills, which locally thicken in depressions. Proglacial reworking of sediments occurs to some degree, and in places large braided river networks produce extensive spreads of reworked material. Lake sediments may be deposited in ice or moraine dammed lakes and in overdeepened basins.

Figure 16.16 The bouldery latero-frontal moraine found at the mouth of Storelvdalen (see Figure 16.19) produced during the Younger Dryas. Note the steep valley head outlets exiting the icefield in the background.

16.5.4 Meltwater Channels

Lateral meltwater channels produced by the retreat of outlet glaciers on to plateaux (Figs 16.3 and 16.4) are typical of large areas of deglaciated upland terrain throughout the Canadian and Greenland high arctic, where they are interpreted as the products of recession by cold-based ice (e.g. Maag, 1969, 1972; Dyke, 1978, 1993; Edlund, 1985; Ó Cofaigh *et al.*, 2003). Meltwater channels have not been found around Lyngsdalen where plateaux are small and high and ice is cold-based. Melting and downwasting has been observed to produce meltwater that ponds around the ice margins (Gellatly *et al.*, 1988) but the quantities appear to be insufficient to produce channelized drainage. The larger, lower icefields in north Norway (e.g. Troms-Finnmark) tend to have warm-based outlets with basal ice at the PMP, producing subglacial meltwater and allowing surface meltwater to percolate to the bed. The subglacial bedrock topography and ice surface slope act to drain the meltwater towards the outlets, invariably in channelized subglacial meltwater streams (Fig. 16.11). Meltwater at the margins of the Icelandic plateau icefields is directed by both subglacial and ice marginal drainage pathways and ultimately feeds into proglacial channels that connect to steep alluvial fans at the base of plateau edge gullies.

16.5.5 Glacial Erosion

On plateaux the extent and duration of the coverage by warm-based ice will influence the quantity of erosion and thus the landforms produced. If only limited erosion occurs, there may only be a clearing out of pre-existing regolith or blockfield, forming a moraine if the ice terminates on top of the plateau. However, if the ice margin avalanches off the plateau edge then moraines may not form (Fig. 16.11), but the presence of bedrock surfaces cleared of regolith or blockfield with striae and roches moutonnées formed when the erosion is more extensive, indicates the former presence of warm-based ice. If extensive evidence of erosion is available then ice flow directional indicators can be investigated. If the erosion is the product of a local ice cover, erosional forms should indicate that ice flowed out radially from a central ice accumulation centre. Alternatively, they may occur as parallel suites of forms cutting obliquely across the plateau if produced under a regional glacier cover. Striated bedrock and roches moutonnées may be exposed in the surrounding valleys where the sediment cover was thin (Fig. 16.15).

16.5.6 Bedrock Weathering Zones

During periods when the ice does not cover the whole of the plateau, exposed bedrock is subjected to subaerial weathering. Providing sufficient time elapses between successive exposure periods (the time required being site-specific and dependent upon the characteristics of the rock exposed), then it may be possible to identify a progressively 'younging' zonation towards the centre of the plateau. Gellatly *et al.* (1988) reported finding two distinct bedrock weathering zones around parts of the margins of Øksfjordjøkelen. Cosmogenic exposure history studies could potentially identify younging zonation, provided the difference between zones is greater than the resolution of the dating.

16.5.7 Erratics

The presence of erratics on plateaux (Fig. 16.15) provides evidence of former glacier coverage and may be used to identify palaeo-iceflow directions. They are likely to represent large-scale ice cover when the topography is submerged and ice surface slopes can force ice to flow over the summits.

It must be remembered that as full glacial conditions give way to regional ice cover, plateau glaciers may become erosive and so local erratic transport may occur. Erratics, which are sourced from beyond the plateau, can be used to infer ice flow directions during large-scale ice inundation, and local erratic transport may be used to reconstruct former ice flow patterns of the plateau icefield.

16.5.8 Plateaux With No Evidence of Former Glacier Cover

Such plateaux provide the most problems in reconstructing regional glaciation style but are obviously very important in climate reconstructions if they were supplying ice to valleys where evidence of glaciation can be found. Providing that evidence of glaciation can be found on surrounding summits, a relationship of the form shown in Fig. 16.1 can be used to assess whether or not the plateau is broad enough to have sustained ice. If there is no indication of glaciation on any plateaux, evidence from the valleys must be employed to constrain glaciation style. The regional palaeo-ELA/firn line can be reconstructed from alpine style valley glaciers and cirques that would have had no plateau ice contributions, and the equation above (see section 16.2) can again be employed for icefield detection. Also, if valley glaciers below plateaux, when reconstructed without accounting for any accumulation input from the plateau above, show lower ELAs than for the alpine and cirque glaciers, then additional accumulation from surrounding plateaux is most likely. Conversely, when reconstructed without accounting for any accumulation input from the plateau above they show similar ELAs to those calculated for the alpine-type valley glaciers; then additional accumulation from above was insignificant. However, a plateau ice cover with no connections to valley glaciers at lower altitudes is still possible.

16.6 IMPACT OF PLATEAU ICE ON RECONSTRUCTED EQUILIBRIUM LINE ALTITUDES

Once the presence of glaciers has been confirmed by the application of a landsystems approach, palaeo-climate reconstructions can be undertaken based upon the limits and interpreted dynamics of the glacier coverage. Where plateau icefields existed but left little or no evidence of their presence they have often been ignored and reconstructions have been confined to valley glaciers only (e.g. Sissons, 1980; Ballantyne, 1990). In the following subsections the potential impact of overlooking plateau icefield contributions on climate interpretations (reconstructed ELAs) is assessed. One example is based on the LIA maximum glaciers in the southern Lyngen Peninsula centred on the peak of Jiek'kevárri, where ice supply from icefield to valley was mainly by avalanching. The other is the Younger Dryas glacier coverage of Øksfjordjøkelen where the icefield was physically connected to the valley glaciers via steep valley-head ice falls.

16.6.1 Lyngen

The reconstructed LIA (1700–1750) valley glaciers and plateau icefields centred on Jiek'kevárri in Lyngen are presented in Fig. 16.17. Rea *et al.* (1998) calculated ELAs for the area using the AAR method (Meier and Post, 1962; Porter, 1975; Kuhle, 1988) and these are presented in Table 16.1. The results highlight the potential impact on ELA calculations and thus inferences of palaeo-climate in such landscapes if contributions from plateau icefields are ignored.

Figure 16.18 shows the icefield area as a percentage of the combined glacier (the valley glacier plus the contributing plateau icefield area) plotted against the measured ELA shifts. This supports the intuitive prediction that the larger the plateau contributing area, the greater the

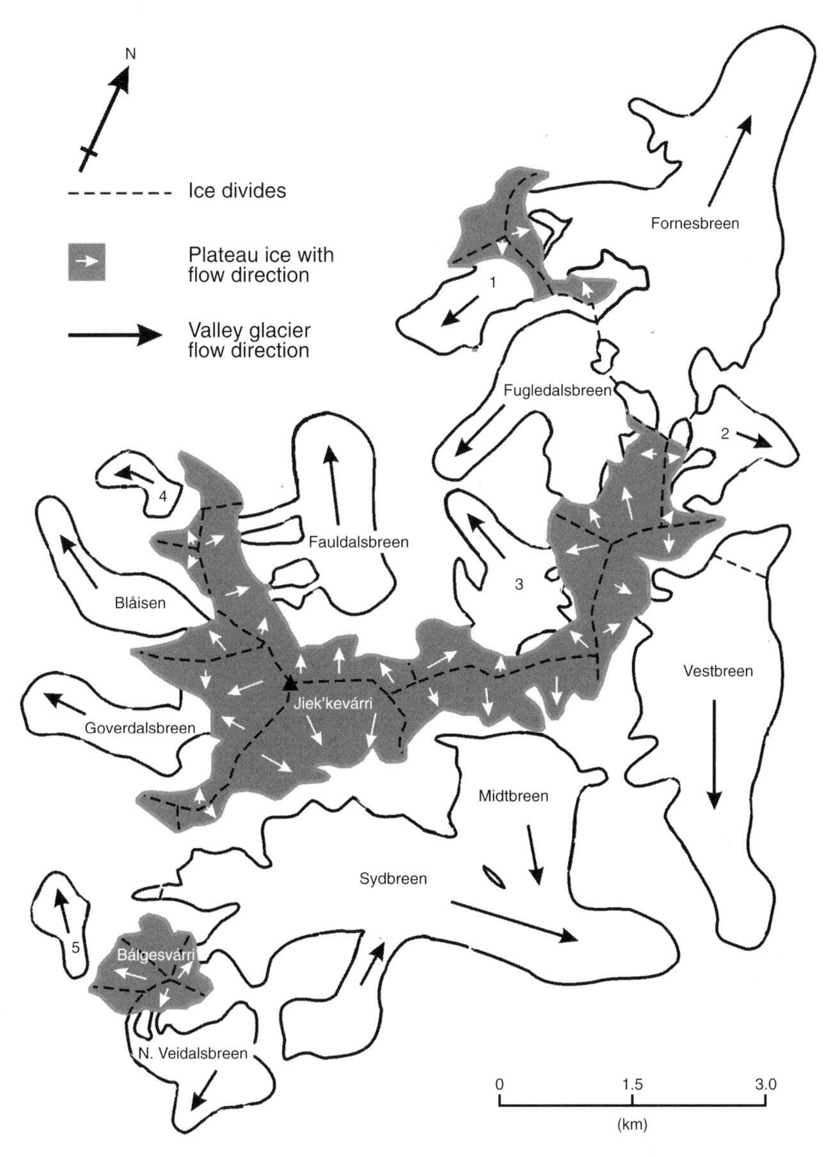

Figure 16.17 Little Ice Age maximum reconstruction of glaciers and plateau ice centred on Jiek'kevárri. Note the radial drainage pattern of the valleys. (Adapted from Rea *et al.* (1998).)

resulting vertical shift in the ELA. Excessive shifts in ELA occur when the ELA moves above the main valley glacier body and up a narrow, steep gully connection. Rea *et al.* (1998) compared the plateau-corrected ELAs with previous climate reconstructions (Ballantyne, 1990) and found that, for the minimum interpreted temperature depression during the LIA maximum (1700–1750), winter accumulation rates up to 20 per cent lower than those calculated by Ballantyne (1990) were possible.

Glacier	Glacier area (km^2)	Plateau area (km^2)	Total area (km^2)	Per cent area plateau
Fornesbreen	7.97	0.39	8.36	4.67
1	1.02	0.19	1.21	15.69
Fugledalsbreen	2.75	1.00	3.75	26.54
2	1.13	0.18	1.31	13.98
3	1.68	1.29	2.97	43.49
Vestbreen	5.64	0.96	6.60	14.48
Fauldalsbreen	2.03	2.05	4.08	50.23
Sydbreen-Midtbreen	10.56	3.44	14.00	24.54
4	0.40	0.11	0.51	20.75
Blåisen	1.38	0.54	1.92	28.03
Goverdalsbreen	1.52	1.37	2.89	47.44
5	0.52	0.31	0.84	37.47
N. Veidalsbreen	0.66	0.27	0.93	29.27

Table 16.1 Little Ice Age maximum reconstructed plateau icefield and valley glacier area distributions for 13 glaciers in the southern Lyngen Alps.

16.6.2 Øksfjordjøkelen

Evans *et al.* (2002) reconstructed the deglaciation chronology of the Bergsfjord Peninsula, and the Younger Dryas extent of Øksfjordjøkelen. Figure 16.19 shows the Younger Dryas icefield, and Table 16.2 presents the statistics on each outlet glacier of the icefield. All but one glacier has a significant proportion of its area above the 800 m contour (chosen as a representative altitude for the plateau edge), and so the implications for reconstructing ELAs using the AAR method are obvious. The data are plotted on Fig. 16.20 and show again the same trend as Lyngen. In comparison with the Lyngen data, there are more glaciers with plateau-contributing areas in excess of 40 per cent of the combined glacier area, resulting in greater ELA shifts. It is obvious from Figs 16.18 and 16.20 that any climate reconstructions based on ELAs that are calculated without accounting for the contributions of plateau ice will be erroneous.

16.7 DISCUSSION

In many reconstructions of formerly glaciated landscapes the role of plateau icefield inputs to the mass balance of disconnected and connected valley glaciers has been ignored. Indeed in some areas this misinterpretation of the style of glaciation may be the cause of large variations in reconstructed ELAs and appeals to large gradients in winter precipitation and excessive snow blow effects.

Sissons (1980) reconstructed the Younger Dryas glacier cover in the Lake District, England and needed to invoke three separate snowfall intensity zones, in combination with the effects of

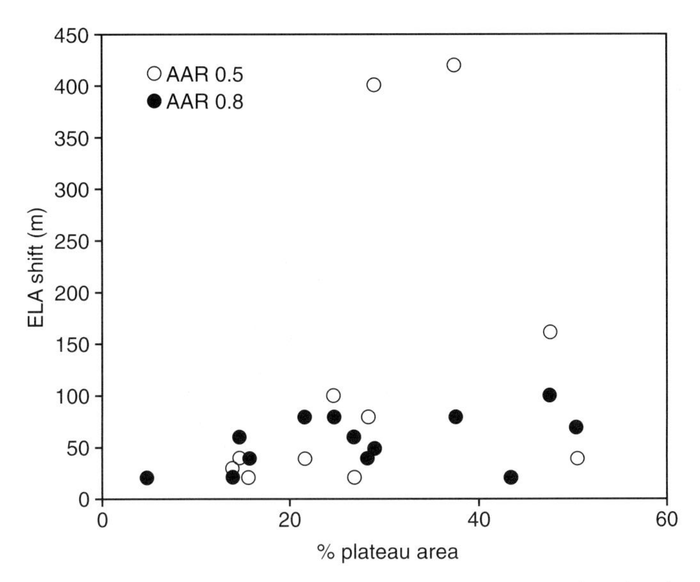

Figure 16.18 Data for the Little Ice Age maximum Lyngen glaciers showing the altitudinal shift in equilibrium line altitude (ELA) as a function of the percentage of the plateau icefield area to the combined glacier area, in accumulation area ratio (AAR) calculations. (The two outliers represent the ELAs shifting up very narrow icefalls.)

Glacier	Area above/below 800 m contour (km^2)		Glacier area (km^2)	Per cent area plateau
	Above 800 m	Below 800 m		
Sörfjorddalen – 1	3.28	4.32	7.6	43.2
Fjorddalen – 2	4.71	7.39	12.1	38.9
Tverrfjorddalen – 3	1.00	2.80	3.8	26.3
Storelvdalen – 4	4.65	2.85	7.5	62
Bac'cavuonvag'gi – 5	1.40	2.20	3.6	39
Isfjorden – 6	11.67	2.43	14.1	82.8
Skalsavatnet – 7	5.39	4.41	9.8	55
Isdalen – 8	11.31	4.99	16.3	69.4
Skognesdalen – 9	0.28	2.52	2.8	10
n-Tverrfjorddalen – 10	2.82	6.58	9.4	30

Table 16.2 Plateau and valley glacier area distributions of Øksfjordjøkelen reconstructed for the Younger Dryas. (Number after name refers to glacier location shown on Fig. 16.19)

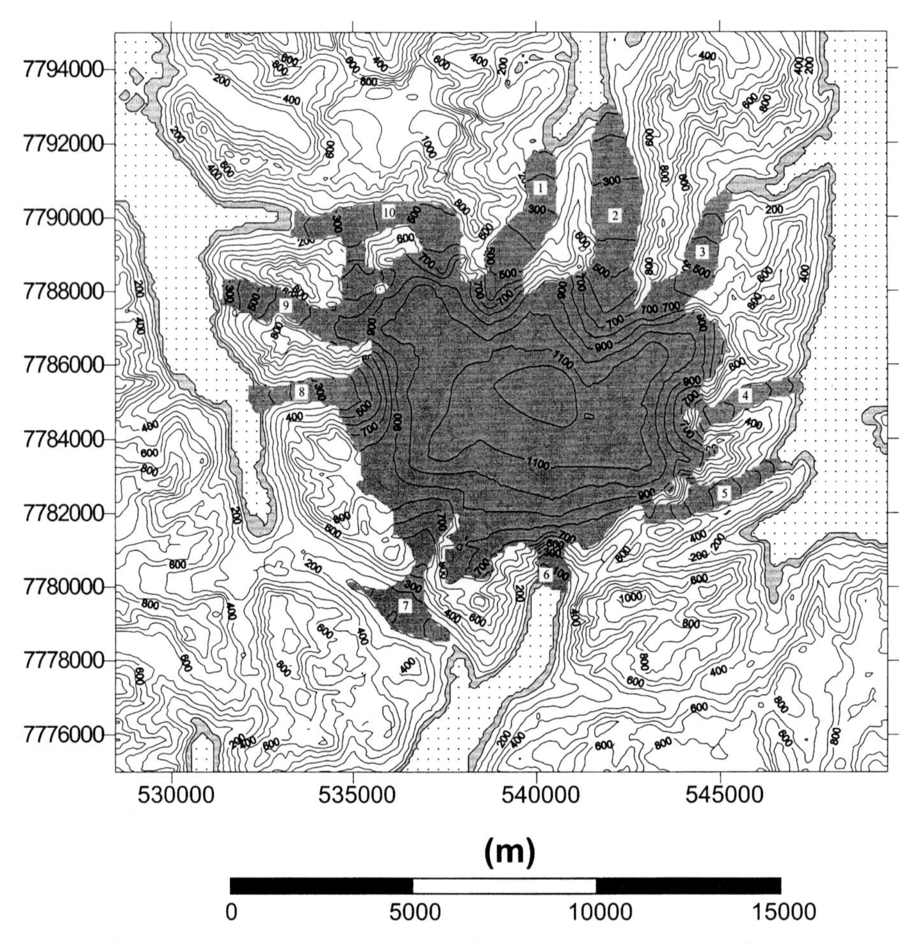

Figure 16.19 Modelled reconstruction of the Younger Dryas extent of Øksfjordjøkelen (dark grey). The sea is white and the land submerged by glacio-isostatically controlled higher sea level is shown in light grey. The boxed numbers on the icefield refer to the glacier names given in Table 16.2. Note the radial drainage pattern shown by the outlet glaciers. (After Evans *et al.*, 2002).

snowblow and aspect, to account for considerable variations in glacier extent and ELAs (the latter varying by 500 m). However, while Sissons envisaged that variations in glacier extent and ELA were primarily climatically driven, recent research in the Lake District suggests a likelihood that the majority of these variations reflect topo-climatic controls (McDougall, 1998, 2001). Specifically, the more extensive 'valley' glaciers mapped by Sissons were actually outlet glaciers of small plateau icefields. Similar to the contemporary examples cited above, the Lake District valley heads contained substantial volumes of glacial sediment in latero-frontal and linear hummocky moraines that extend in many cases onto plateau margins. Evidence of plateau ice is patchy and often restricted to scoured bedrock at plateau edges. McDougall's (1998) palaeo-glacier reconstructions for the English Lake District, incorporating plateau icefields, suggest that ELAs during the Younger Dryas were up to 90 m higher than those calculated by Sissons (1980) based upon an alpine style of glaciation with no plateau ice.

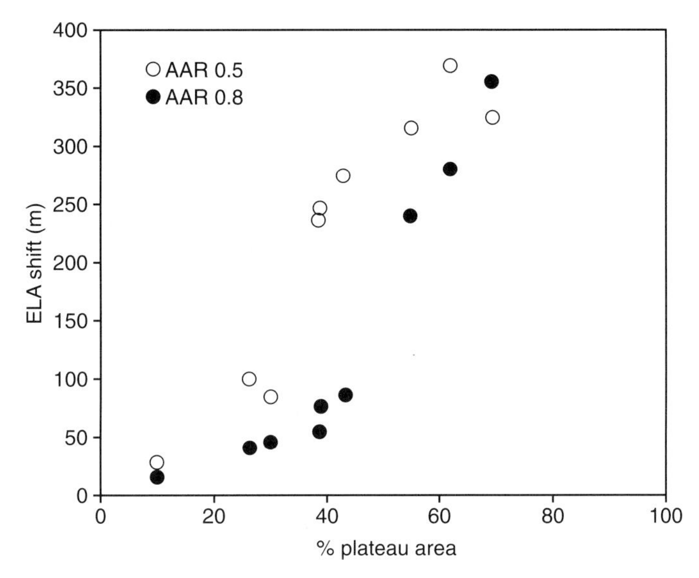

Figure 16.20 Equilibrium line altitude (ELA) shifts plotted as a function of plateau area for each Øksfjordjøkelen Younger Dryas outlet glacier, for the upper and lower accumulation area ratio (AAR) limits. The plateau summit was arbitrarily chosen as bedrock areas above 800 m.

16.8 CONCLUSIONS

The evidence presented above provides a diagnostic assemblage of landforms (landsystem) for use in the identification of former plateau icefields (Fig. 16.15). Contemporary plateau icefields in both sub-polar and temperate environments have developed similar geomorphic signatures, which have been identified in ancient landform assemblages and associated with a plateau icefield style of glaciation rather than a more traditional alpine reconstruction (e.g. English Lake District). The landsystem can be employed in conjunction with the plateau altitude/breadth relationship depicted in Fig. 16.1 and the empirical relationship shown in the equation to identify the presence of former glaciers and the possible existence of plateau icefields. The importance of identifying plateau icefields (beyond the need for correct geomorphological and glaciological interpretations of former glaciation) is highlighted by the data presented in Figs 16.18 and 16.20. The ELA for a region could be significantly miscalculated if plateau icefield mass contributions to valley glaciers are ignored. The error increases as the proportion of the contributing area increases. As with all geomorphological research, it is apparent that knowledge of modern day analogues is central to an understanding of ancient glaciations. The landsystem model presented here provides the tool for reassessing regional/local glaciation in areas where plateaux are found. In some instances this may provide the key to explaining previously anomalous ELA calculations.

CHAPTER

PARAGLACIAL LANDSYSTEMS

Colin K. Ballantyne

17.1 INTRODUCTION

The retreat of glacier ice exposes land surfaces to processes that progressively modify glacial landforms, landscapes and landsystems. Such modification is often described as *paraglacial*, a term first defined by Church and Ryder (1972, p. 3059) as referring to 'non-glacial processes that are directly conditioned by glaciation'. Use of 'paraglacial' has, however, subsequently widened to include description of landforms and sediments as well as geomorphic processes. In this chapter the term 'paraglacial' is therefore redefined as describing 'non-glacial earth-surface processes, sediment accumulations, landforms, landsystems and landscapes that are directly conditioned by glacation and deglaciation'. This revised definition retains the essence of the original, but recognizes the more eclectic use of the term now current.

The paraglacial concept is one of adjustment of glacial landsystems to non-glacial conditions, and involves the progressive relaxation of unstable or metastable elements of the glaciated landscape to a new stable state. The concept cannot be defined by process, as all of the geomorphological processes identified as components of paraglacial adjustment operate outside glaciated areas (Eyles and Kocsis, 1989). Nor can relaxation time be considered a unifying factor, as different components of deglaciated landscapes equilibriate to non-glacial conditions over vastly different timescales (Ballantyne, 2002a, b). Paraglacial landform adjustment and the period over which this operates may, however, be conceptualized in terms of glacially conditioned sediment release. All forms of paraglacial adjustment share a common rudiment, namely that deglaciation has resulted in the exposure of unstable or metastable sediment stores that are subsequently released and reworked by a wide variety of processes over a wide range of timescales. The 'paraglacial period', defined by Church and Ryder (1972) as 'the time during which paraglacial processes occur', may thus be redefined as the timescale over which a glacially conditioned sediment source either becomes exhausted or attains stability in relation to particular reworking processes. Once this has occurred, sediment release may be envisaged as having relaxed to an 'equilibrium' or 'non-glacial' state, quantitatively indistinguishable from that which would result from primary denudation of the land surface.

For primary paraglacial systems in which glacigenic sediment sources are not replenished, the rate of sediment reworking (and thus the duration of the paraglacial period) may be approximated by an exhaustion model (Ballantyne, 2002a), in which sediment yield is related to the amount of remaining available sediment by a negative exponential function:

$$S_t = S_0 \, e^{-\lambda t}$$

where t is time elapsed since deglaciation, S_t is the proportion of available sediment remaining for reworking at time t, S_0, is the total available sediment at $t = 0$, and λ is the rate of change in the loss of available sediment by release and/or stabilization (Fig. 17.1).

Secondary paraglacial systems (primarily fluvial systems) in which sediment inputs include both *in situ* glacigenic sediment and reworked paraglacial sediment from upstream sources may behave in an intrinsically more complex fashion. Church and Slaymaker (1989) interpreted a downstream increase in specific sediment yield in rivers draining glaciated terrain as implying a delayed peak in paraglacial sediment yield in large catchments. This concept has been developed by Harbor and Warburton (1993), who suggested that the temporal pattern of fluvial paraglacial sediment transport can be described by a family of curves, with those for the smallest (upland) catchments peaking immediately after deglaciation and those for larger basins peaking progressively later as catchment size increases (Fig. 17.2A). Ballantyne (2002a) has argued that a downstream increase in specific sediment yield is equally consistent with an exhaustion model (Fig. 17.2B), provided that initial sediment availability (S_0) is greater in steep tributary basins and that the rate of change in sediment removal (λ) declines with increasing catchment size, because whereas glacigenic

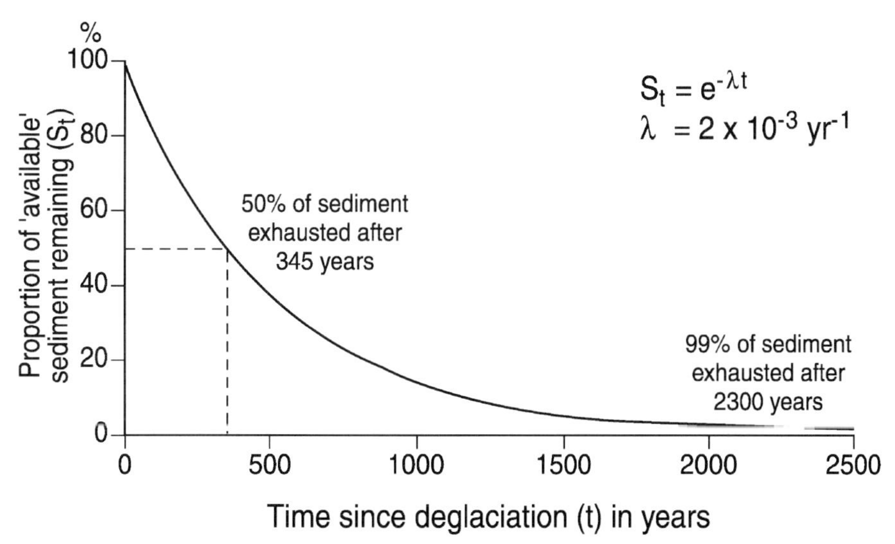

Figure 17.1 Exhaustion model of paraglacial sediment release, in which rate of decline in sediment release (λ) is related to the proportion of 'available' sediment (S_t) at time (t) since deglaciation as $\lambda = \ln(S_t) / -t$. In this example $\lambda = 0.002$ year^{-1} (i.e. 0.002% of remaining 'available' sediment is released per year), 50% of initial 'available' sediment is removed in the first 345 years and 99% of 'available' sediment has been removed after 2300 years, defining the approximate length of the paraglacial period. (From Ballantyne (2002a).)

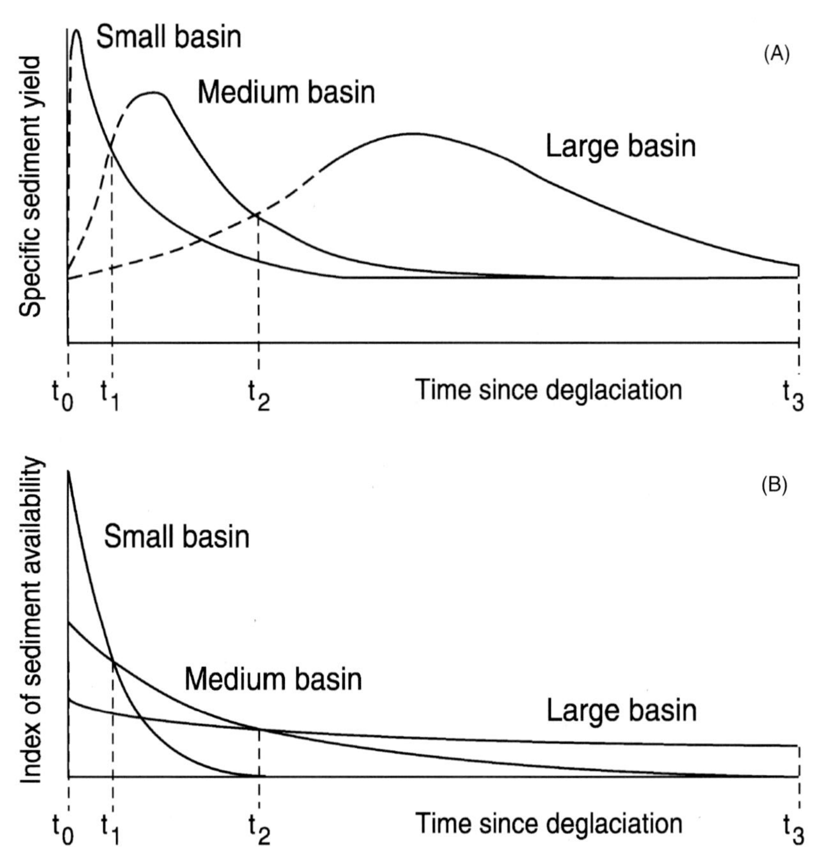

Figure 17.2 Two explanations for the downstream increase in specific sediment yield characteristic of glaciated terrain. A) The model proposed by Harbor and Warburton (1993), in which sediment yield in the smallest catchments peaks immediately after deglaciation, but the peak of paraglacial sediment transport is progressively lagged as catchment size increases. This model predicates limited sediment release in large basins immediately after deglaciation, which appears unlikely. B) The exhaustion model proposed by Ballantyne (2002a), which assumes (1) that initial sediment availability is greatest in small upland catchments and decreases with catchment size and (2) that the rate of decline of sediment reworking is highest in small upland basins but decreases as catchment size increases. Both models predict that sediment yield is initially greatest in small basins during period t_0–t_1, then in medium-sized basins during t_1–t_2, and then in large basins during t_2–t_3, thus accounting for a present-day increase in specific sediment yield with increasing catchment area.

sediments removed from upland tributaries are not replenished, sediment recruitment in larger catchments is sustained not only by release of *in situ* glacigenic deposits but also by reworked sediment supplied by tributary streams (Church and Ryder, 1972; Brooks, 1994). It seems likely that these two competing models (Figs 17.2A and 17.2B) represent end-members of a continuum of possibilities determined by initial catchment relief and particularly the initial abundance and distribution of glacigenic sediment sources.

The models of paraglacial sediment release illustrated in Figs 17.1 and 17.2 assume steady-state conditions in which there is no systematic change in process-generating mechanisms (e.g. in the magnitude and frequency of extreme rainstorm events) or other boundary conditions. This is rarely the case over millennial timescales, and consequently the actual course of paraglacial landscape response often departs from the monotonic decline in sediment release predicted by the exhaustion model (Fig. 17.3). Climatic change may affect rates of sediment release, for example through the regeneration or readvance of glaciers in mountain catchments (Brooks, 1994), accelerated retreat of mountain glaciers (Leonard, 1997) or changes in runoff regime, all of which may produce secondary peaks in sediment yield. Extreme climatic events may also trigger renewed reworking of glacigenic sediment, sometimes millennia after termination of the initial period of paraglacial adjustment (e.g. Ballantyne and Benn, 1996; Ballantyne and Whittington, 1999). In paraglacial fluvial systems, glacio-isostatic uplift may cause a fall in base levels, rejuvenating sediment release as rivers cut down into glacigenic valley-fills and paraglacial sediment stores such as debris cones and alluvial fans (Church and Ryder, 1972). Similar effects are evident on paraglacial barrier coasts where rising seas permit coastal erosion to tap pristine sources of glacigenic sediment, causing renewed pulses of reworked sediment to enter the nearshore sediment budget (Boyd *et al.*, 1987; Forbes and Taylor, 1987; Forbes *et al.*, 1995a, b). At all scales, therefore, the exhaustion model of paraglacial sediment release is vulnerable to extrinsic perturbation, which may produce secondary peaks of sediment transfer as sources of glacigenic sediment are mobilized or remobilized. Such effects tend to prolong or rejuvenate paraglacial sediment release. Thus, although it is possible in many contexts to identify a discrete duration for the operation of paraglacial sediment release and reworking (the paraglacial period), external perturbation may remobilize glacigenic sediment long after the period of initial adjustment has ended.

17.2 PARAGLACIAL LANDSYSTEMS

The term 'landsystem' is generally employed to decribe a large-scale grouping of landforms and related sediment facies that characterize a particular geomorphic context or environment (Eyles,

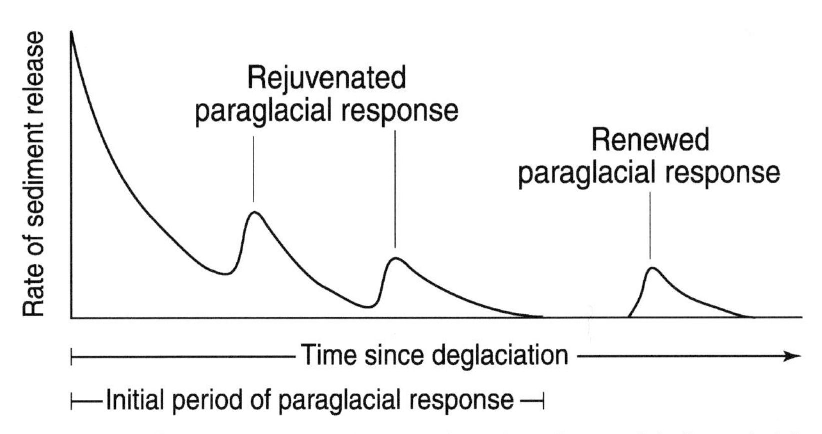

Figure 17.3 Effects of extrinsic perturbations on the exhaustion model of paraglacial sediment release. Factors that may trigger a rejuvenated paraglacial response (during the initial period of paraglacial activity) or a renewed paraglacial response (after the initial period of paraglacial activity) include climate change, extreme climatic events, neotectonic activity, anthropogenic activity and, in the coastal zone, sea level change.

1983a; Benn and Evans, 1998). As previous chapters have demonstrated, a landsystems approach is particularly useful for identifying, classifying and explaining sediment-landform associations in different glacial environments. A similar approach can be applied to paraglacial landform and sediment assemblages. Ballantyne (2002b) has proposed a sixfold subdivision of paraglacial landsystems based on locational context (rock slopes, drift-mantled slopes, glacier forelands and alluvial, lacustrine and coastal settings). In the review that follows, the salient characteristics of these landsystems are summarized, with particular emphasis on process-landform-sediment relationships. The chapter concludes by:

1. identifying the sediment transport pathways that link individual landsystems
2. exploring the implications for the interpretation of Holocene landscape evolution in glaciated areas, and
3. discussing the general implications of paraglacial landscape modification for the interpretation of ancient glacial landsystems.

17.3 PARAGLACIAL ROCK SLOPE LANDSYSTEMS

Glaciation and deglaciation affect rock slope stability in two ways. First, glacial erosion may steepen and lengthen rock slopes, thereby increasing the self-weight (overburden) shear stresses acting within the rock mass. Second, during glacial periods the weight of overlying or adjacent glacier ice increases stress levels within rock masses. Part of the resulting ice-load deformation is elastic and stored within the rock as strain energy. During deglaciation and consequent unloading of glacially stressed rock, release of strain energy causes 'rebound' or stress release within the rock. Stress release causes extension of the internal joint network, together with reduced cohesion along joint planes and a reduction of internal locking stresses (Wyrwoll, 1977). These changes may initiate three types of response, namely catastrophic rock-slope failure during or after deglaciation, rock-slope readjustment through slow rock-mass deformation, or progressive adjustment through numerous small-scale rockfalls. The type of response is determined primarily by joint density and the orientation and inclination of fractures relative to the newly exposed rock face (Augustinus, 1995a).

17.3.1 Catastrophic Rock-Slope Failure

Examples of catastrophic rockslides or rock avalanches occurring during or shortly after recent deglaciation are legion (Evans and Clague, 1994). Such failures have been widely attributed to a combination of stress release and debuttressing of glacially steepened rockwalls by the thinning of glacier ice (Fig. 17.4). There is also evidence for widespread paraglacial rock-slope failure following Late Pleistocene deglaciation. Caine (1982), for example, has shown that scarp-edge toppling in Tasmania represents a paraglacial response to Late Pleistocene deglaciation, and in Great Britain paraglacial rock-slope failure was widespread in mountain areas in Lateglacial and Early Holocene times (Ballantyne, 1986, 1997; Shakesby and Matthews, 1996), though some failures did not occur until several millennia after deglaciation (Ballantyne *et al.*, 1998).

Cruden and Hu (1993) have proposed that the temporal pattern of paraglacial rock-slope failure following Late Pleistocene deglaciation can be approximated by an exhaustion model similar to that outlined above (see equation). Their model assumes that there are a finite number of potential failure sites following deglaciation, that each site fails only once, and that the probability of occurrence of individual rockslides remains constant. Under these

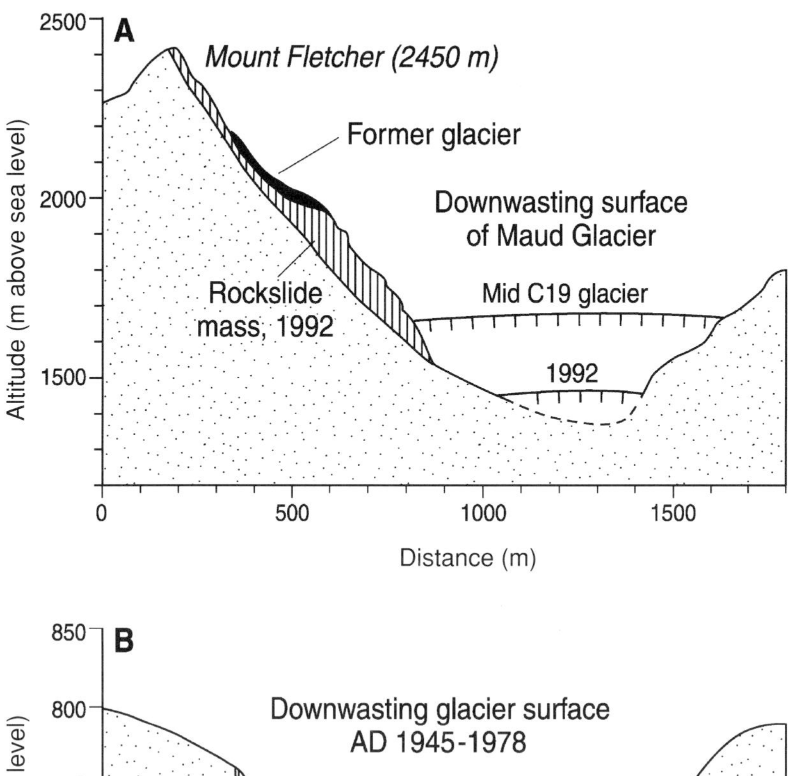

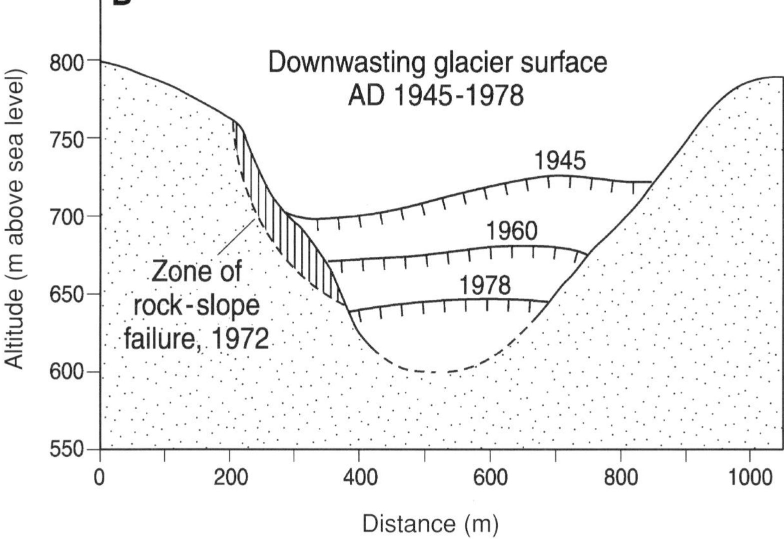

Figure 17.4 A) Cross-section through a rockslide initiated by recent downwastage of the Maud glacier in New Zealand (adapted from McSaveney, 1993). B) Cross-section through a rock slope failure initiated by downwastage of an outlet glacier of the Myrdalsjökull ice cap, southern Iceland (adapted from Sigurdsson and Williams, 1991).

assumptions the overall probability of failure within a given area diminishes exponentially with time elapsed since deglaciation as the number of potential (i.e. unfailed) failure sites is progressively reduced. For glacially steepened overdip slopes in the Canadian Rockies, they calculated that the average pre-failure lifespan of individual sites is *c.* 5,700 years, implying that glacially conditioned rock-slope failure may occur many millennia after deglaciation in response to progressive stress release and consequent ramification of the joint network causing reduction of rock-mass strength. Some support for their proposal comes from Yosemite Valley, California, where the volume of rock-slope debris deposited at the foot of granite rockwalls over the period AD 1851–1992 implies an average accumulation rate that is less than half the postglacial average (Wieczorek and Jäger, 1996), consistent with a progressive slowing in the rate of rock-slope failure as predicted by the exhaustion model.

17.3.2 Rock-Slope Deformation

Stress release due to deglacial unloading and debuttressing may also result in slow paraglacial rock-slope deformation, also referred to as rock-mass creep. Such deformation represents wholesale failure of large rock masses, but occurs (at least initially) without catastrophic runout of debris. Landforms characteristic of rock-slope deformation include ridge-top graben, crevasse-like tension cracks, antiscarps and convex 'bulging' slopes (Fig. 17.5). Such deformation may reduce slopes to a state of conditional stability, and sometimes precedes catastrophic failure (Chigira, 1992).

A close relationship between debuttressing of rock slopes during deglaciation and initiation of rock slope deformation has been observed by several authors (Tabor, 1971; Radbruch-Hall, 1978). In New Zealand, thinning of the Tasman Glacier has resulted in widespread deformation of the adjacent bedrock slope, producing tension cracks, flexural topples and antiscarps (Blair, 1994). Similarly, the slopes overlooking the downwasting Affliction Glacier in British Columbia exhibit widespread evidence of recent rock-slope deformation, including fractures, antiscarps, elongate graben and collapse pits, collectively indicative of tensional deformation in the near-surface zone. A survey of rock-slope movement at this site by Bovis (1990) indicates that glacier thinning has

Figure 17.5 Antiscarps and bulging slopes due to paraglacial rock mass deformation on Beinn Fhada, Scotland.

triggered gravitational movement of ~3 × 107 m^3 of rock, with surface velocities of a few millimetres to a few centimetres per year (Fig. 17.6). The widespread occurrence of similar indications of rock-slope deformation in mountainous areas that now lack glacier ice strongly suggests that paraglacial rock-slope deformation commonly accompanied deglaciation in Late Pleistocene or Early Holocene times (e.g. Tabor, 1971; Radbruch-Hall *et al.*, 1976; Mahr, 1977; Radbruch-Hall, 1978; Jarman and Ballantyne, 2002).

17.3.3 Paraglacial Rockfall and Talus Accumulation

The third type of response of glacially steepened rockwalls to deglaciation is through enhanced rockfall activity that results in the development of paraglacial talus accumulations (Augustinus, 1995a). Several authors have noted that the large volumes of talus at the foot of rockwalls deglaciated at the end of the Pleistocene are inconsistent with presently modest rockfall activity, and have concluded that the rockfall inputs were very much greater immediately after deglaciation (e.g. Luckman, 1981; Gardner, 1982; Johnson, 1984, 1995; Marion *et al.*, 1995). Luckman and Fiske (1995) found that rockfall input over the past 300 years at a site in the Canadian Rockies has been roughly an order of magnitude too low to have produced the volume of talus now present. Similarly, Hinchliffe and Ballantyne (1999) found that at a site in the Scottish Highlands, roughly 80 per cent of talus accumulation took place within the first 6,000 years after deglaciation, and only about 20 per cent in the ensuing 11,000 years, implying a marked reduction in rockfall rate through time.

Though enhanced rockfall immediately after deglaciation may partly reflect freeze-thaw activity under former periglacial conditions, there is evidence that paraglacial stress-release and associated

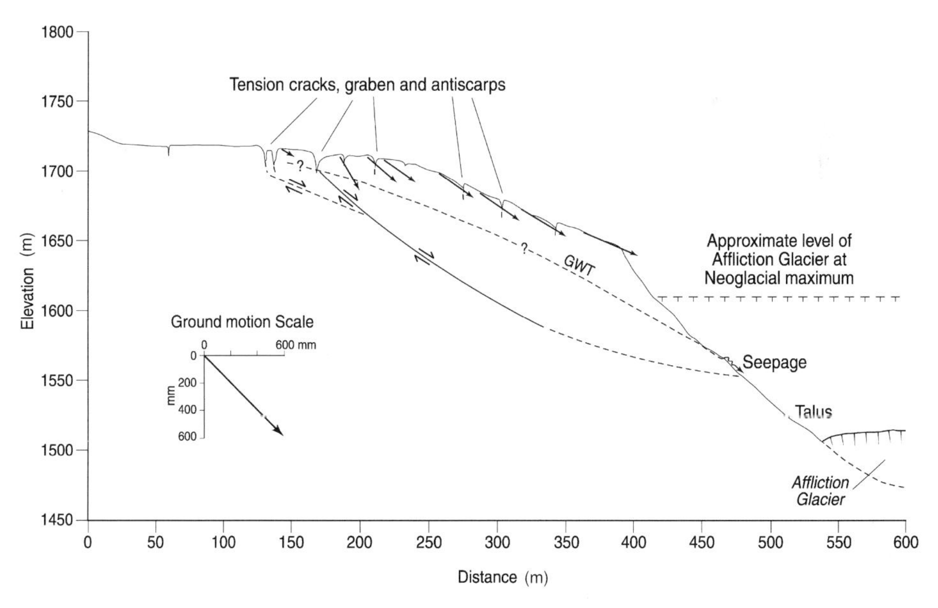

Figure 17.6 Cross-section depicting vectors of paraglacial rock slope deformation triggered by downwastage of Affliction Glacier, British Columbia. The vectors represent 4 years of movement. The position of the groundwater table (GWT) and the underlying shear plane are schematic. (Adapted from Bovis (1990).)

rock-slope instability has exerted the dominant control on postglacial rockfall rates. High rockfall inputs from densely jointed rockwalls in New Zealand have resulted in rapid talus accumulation near the snouts of retreating glaciers (Augustinus, 1995a), and the recent retreat of glaciers on Mexican stratovolcanoes has similarly been accompanied by rapid rockwall degradation and talus accumulation (Palacios and de Marcos, 1998), yet neither location currently experiences severe periglacial conditions. André (1997) has shown that on Svalbard, rates of rockwall retreat due to paraglacial stress release are roughly an order of magnitude higher that those due to freeze-thaw effects. If such findings are representative, they imply that intrinsic paraglacial effects have been much more important than freeze-thaw cycling in promoting rockfall and talus accumulation in formerly glaciated areas, and that many supposedly 'periglacial' talus accumulations reflect a strong element of paraglacial inheritence.

17.3.4 Landforms and Deposits

Paraglacial rock-slope adjustment produces a wide range of landforms and deposits. Mountain slopes and summits are characterized by tension cracks and crevasse-like fractures, split summit ridges, toppled blocks, antiscarps and bulging slopes seamed with nested antiscarp arrays (Bovis, 1990; Chigira, 1992; Jarman and Ballantyne, 2002; Figs 17.5 and 17.6). Sliding failure along slope-parallel stress-release joints is evident in the formation of crescentic overhangs, and large-scale catastrophic rock-slope failures (rock avalanches and deep-seated rotational slides) produce deep arcuate failure scars. Large-scale catastrophic failures may accumulate as valley-side cones of large angular boulders, or extend well beyond the slope-foot as an excess-runout flowslide (Sturzstrom) deposit. Such excess-runout failures may extend several kilometres down valley floors (Evans and Clague, 1994), and are characterized by steep, well-defined lateral margins or levées of very coarse debris and concentric flow ridges. Paraglacial rockfall produces talus cones, coalescing talus cones and talus sheets along the lower slopes of cirque headwalls and glacial troughs. Such talus accumulations generally consist of smaller boulders than the products of large-scale catastrophic failure, and consist of clast-supported diamicts overlain by a surface layer of openwork boulders that exhibit a general increase in size downslope (fall-sorting). They often overlie valley-side morainic deposits and exhibit evidence for surface reworking by debris flows and snow avalanches. In permafrost environments, rockfall talus accumulations may terminate downslope in protalus (lobate) rock glaciers (Ballantyne, 2002b).

17.3.5 Wider Significance

The evidence outlined above suggests that paraglacial rock-slope adjustment represents an important if rarely acknowledged component of postglacial landscape evolution, particularly in alpine environments. It is worth noting two important implications. First, not only do paraglacial rock-slope failure, rock-slope deformation and rockfall alter the form of rockwalls during interglacials, but also rock-mass weakening due to paraglacial stress release may determine the foci of glacial erosion during later periods of glacial advance, raising the possibility that archetypal glacial landforms such as cirques and glacial troughs owe their present form as much to successive episodes of paraglacial adjustment as to successive periods of glacial quarrying and abrasion (Augustinus, 1995b). Second, because paraglacial rock-slope failures, slope deformations and talus accumulations provide an abundant source of readily entrainable debris, much of the sediment transported during the initial stages of a later glacial advance may ultimately be of paraglacial origin, rather than eroded from intact bedrock. If so, it is likely that the initial stages of renewed glaciation are marked by enhanced rates of glacial sediment transport; when glacial recycling of paraglacial debris is complete, glaciers transport only sediment supplied by 'normal' entrainment processes such as subglacial erosion and direct debris delivery from valley-side slopes. A full understanding of the sources of glacigenic sediment therefore requires not only appreciation of the mechanics of glacial erosion, but also the role of earlier paraglacial rock-slope adjustment in providing an abundant source of readily-entrainable debris (Ballantyne, 2002b). Just as paraglacial processes are by definition glacially-conditioned, so some aspects of glacial processes may to some extent reflect paraglacial conditioning or inheritance.

17.4 PARAGLACIAL DRIFT-MANTLED SLOPE LANDSYSTEMS

Retreat of glacier ice in mountain areas often exposes valley-side slopes mantled by thick glacigenic deposits, usually composed of stacked lateral moraines (Mattson and Gardner, 1991; Ballantyne and Benn, 1994, 1996). Such deposits are susceptible to erosion by translational slope failure (slumping), debris flow, snow avalanches, streamflow and surface wash. These processes may completely rework drift-mantled slopes within a few decades or centuries, forming a paraglacial landsystem of intersecting gullies, coalescing debris cones and valley-floor deposits of reworked sediment (Fig. 17.7).

Figure 17.7 Gullying of drift-mantled slopes near the snout of Fåbergstølsbreen, Jostedalen, Norway. Photographic evidence shows that as late as AD 1943 this slope was ungullied and supported a prominent Little Ice Age lateral moraine. By 1988 the slope was extensively gullied, the lateral moraine had been completely removed, and extensive areas of fresh bedrock were exposed as a result of annually recurrent debris-flow activity. Most reworked sediment has accumulated in coalescing debris cones at the slope foot.

17.4.1 Drift-Mantled Slopes: Processes

The dominant agent of sediment reworking on recently deglaciated drift-mantled slopes is debris flow, the rapid downslope movement of a poorly sorted mixture of boulders, fine sediment and water (Zimmermann and Haeberli, 1992; Evans and Clague, 1994; Owen, 1994; Solomina *et al.*, 1994). Snow avalanches and streams generally play a secondary role in redistributing glacigenic sediment downslope. On the foreland of Fåbergstølsbreen in Norway, for example, Ballantyne and Benn (1994) recorded an average of five debris flows per year per kilometre of slope, many of which had been triggered by rapid snowmelt at gully heads. Numerous flow tracks marked by parallel levées of debris descend from gullies, often cross-cutting earlier flows to produce a complex hummocky microtopography of dissected levées. By no means all recently deglaciated drift-mantled slopes experience extensive paraglacial modification, however. Research by Curry (2000a) suggests that initial gradients over 30° are essential for extensive slope erosion by debris flows, and that on such slopes a high density of gullies (>20 km^{-1}) is associated with thick drift cover and sediments with high void ratios. Local hydrological controls (particularly focusing of runoff by rock gullies upslope) may be critical in initiating widespread reworking of drift-mantled slopes by debris flow activity.

In valleys where recent ice downwastage has exposed steep-sided lateral moraines, proximal moraine slopes may be extensively modified by slumping, debris falls and debris flows. Widespread failure of moraine walls has occurred along the flanks of Tasman Glacier in New Zealand at sites where moraine relief exceeds 120 m (Blair, 1994). Ice-cored lateral moraines are particularly susceptible to failure as the underlying ice melts, reducing the strength of the overlying sediment so that debris is released through a combination of slumping and flow (Fitzsimons, 1996b; Bennett *et al.*, 2000a; Etzelmüller, 2000). At Boundary Glacier in Alaska, Mattson and Gardner (1991) recorded 25 slope failures incorporating ~35,000 m^3 of debris from ice-cored moraines over two summers. Most involved failure at the ice-sediment boundary, and the majority occurred near the glacier snout, indicating rapid modification of moraine slopes following deglaciation. The distal slopes of steep terminal moraines may also be affected by slumping and debris flow (Palacios *et al.*, 1999).

17.4.2 Drift-Mantled Slopes: Landforms

The morphological consequences of recent drift-slope modification have been intensively studied on the foreland of Fåbergstølsbreen (Norway), where steep drift-mantled slopes have been so extensively modified by debris flows that little of the original slope remains. The modified slope comprises two zones. The upper comprises broad gullies up to 25 m deep and 80 m wide that are incised into valley-side drift, locally exposing areas of underlying bedrock, and separated by 'arêtes' of drift (Fig. 17.7). The lower consists mainly of reworked sediment, mainly in the form of coalescing debris cones that overlie bedrock or till. Ballantyne and Benn (1994) found that within 50 years, gullying of upper drift slopes resulted in a reduction in slope gradient from ~35° to ~30°, the latter probably representing the minimum gradient for debris-flow initiation. Curry (1999) has shown that gully incision occurs rapidly after deglaciation, and that gullies thereafter undergo progressive widening until sidewall slopes have declined to a gradient of ~25°, after which parallel retreat of gully sides predominates until inter-gully arêtes are consumed or gully-side slopes attain stability. The final form of the drift-slope landsystem comprises an upper, bedrock-floored source area, a midslope zone of broad gullies with sidewalls resting at stable, moderate gradients, and a lower zone of coalescing debris cones and fans, a landform assemblage common in many upland valleys that were deglaciated in Late Pleistocene times (Miller *et al.*, 1993; Ballantyne and Benn, 1996). This assemblage achieves spectacular dimensions in the Karakoram Mountains and Lahul Himalaya, where paraglacial debris-flow deposits form sediment sequences up to 90 m thick (Owen and Derbyshire, 1989; Owen, 1991; Owen *et al.*, 1995).

17.4.3 Drift-Mantled Slopes: Sediments

Differentiation of *in situ* glacigenic deposits from those reworked by debris flow is often problematic, particularly in the case of glacigenic deposits that have experienced flow during deposition (Lawson, 1988; Owen and Derbyshire, 1989; Zielinski and van Loon, 1996). Comparative studies of recent paraglacial debris-flow deposits and their parent tills have demonstrated that the former retain most of the characteristics of the latter, being indistinguishable in terms of macrofabric strength or type, clast imbrication, angularity, shape and texture, matrix granulometry or void ratio (Owen, 1991, 1994; Ballantyne and Benn, 1994; Curry and Ballantyne, 1999). Significant differences occur, however, in terms of the alignment of discontinuities, stratification, shear structures and bedding, which in paraglacial debris-flow deposits tends to be parallel or sub-parallel to valley-side slope, and in terms of the aggregate preferred orientation of elongate clasts, which tends to be aligned downslope in reworked deposits but down-valley in *in situ* basal tills. Micromorphological analyses also show promise for distinguishing reworked from *in situ* glacigenic deposits. Owen (1991, 1994) detected differences in the characteristics of microshears, and Harris (1998) found that till deposits reworked by debris flows exhibited a range of diagnostic characteristics, including preferred downslope grain orientations, shear-induced birefringence, evidence for clast rotation and sheared wavy textural domains, together with the presence of wash layers and well-sorted sand and gravel lenses.

17.4.4 Drift-Mantled Slopes: Rates of Paraglacial Modification

Extensive paraglacial reworking of drift-mantled slopes may occur within a few decades or centuries of deglaciation. The drift-mantled slope exposed by retreat of Fåbergstølsbreen was transformed into a badland of deep gullies within 50 years (Ballantyne and Benn, 1994; Curry, 1999), and in neighbouring Bergsetdalen paraglacial debris cones that began to accumulate between AD 1750 and AD 1908 had completely stabilized by 1965 as a result of sediment exhaustion (Ballantyne, 1995). Such rapid changes imply minimum gully erosion rates of 19–169 mm $year^{-1}$. From the volume of paraglacial debris cones in the Nepal Himalaya, Watanabe *et al.* (1998) inferred mean catchment denudation rates of 0.4–8.0 mm $year^{-1}$ over the past 550 years, but acknowledged that denudation rates were probably much higher immediately after deglaciation. Such rates imply that paraglacial drift-slope reworking is likely to be completed within a few centuries of deglaciation, a conclusion that is supported by evidence of rapid attainment of drift-slope stability following Late Pleistocene deglaciation. Miller *et al.* (1993) have shown that paraglacial gully erosion and concomitant debris cone formation in the Andes of northern Peru commenced during deglaciation at *c.* 12–10 ka BP but was complete before *c.* 8 ka BP. Similarly, Jackson *et al.* (1982) estimated that 80 per cent of postglacial debris flow activity in the Bow River valley of the Canadian Rockies took place between deglaciation at *c.* 13–12 ka BP and establishment of spruce forest at *c.* 10.4–10.0 ka BP. Drift-mantled slopes, however, may experience renewed or delayed paraglacial reworking long after the end of the initial period of paraglacial activity has ended, particularly in response to extreme storm events (Ballantyne and Benn, 1996; Curry, 2000b).

17.4.5 Glacial-Paraglacial Sediment Recycling

It was noted above that paraglacial rock-slope debris and rock weakening may provide important sources of readily entrainable sediment. This principle also applies to paraglacial sediment accumulations reworked from valley-side drift mantles. Sections exposed in the sidewalls of gullies incised in valley-side drifts exposed by retreating outlet glaciers draining Jostedalsbreen in southern Norway exhibit two distinct sediment associations (Ballantyne and Benn, 1994; Curry and Ballantyne, 1999). The upper consists of a massive diamicton that represents glacigenic deposits emplaced during recent (Little Ice Age) glacier advance. The lower exhibits crude slope-parallel stratification and preferred downslope clast orientation, and represents paraglacial reworking of much earlier (Preboreal) glacigenic deposits by debris flows. As the contact between the two is usually erosional (Fig. 17.8), it implies that Preboreal paraglacial sediments were re-entrained by the outlet glaciers during the Little Ice Age advance. This sequence therefore implies a cycle of alternating glacial and paraglacial sediment transfer, the former being dominant during glacier advance and the latter during glacier retreat. It suggests that many glacigenic deposits in mountain areas contain sediments that have undergone at least one previous cycle of glacial/paraglacial reworking.

17.5 PARAGLACIAL MODIFICATION OF GLACIER FORELANDS

Glacier retreat exposes unvegetated valley-floor deposits that undergo paraglacial modification by mass movement, frost sorting, wind and running water, resulting in changes to both the

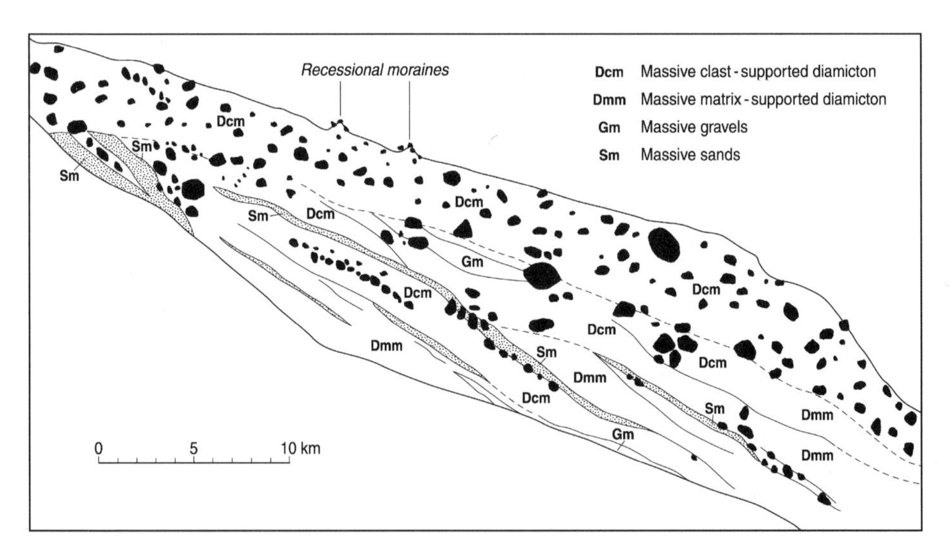

Figure 17.8 Section through drift-mantled slopes, Fåbergstølsbreen, Norway. The upper massive clast-supported diamicton represents deposition by the glacier during the Little Ice Age. The underlying crudely-stratified diamicton consists of early Holocene paraglacial debris-flow deposits. The erosional contact between the two demonstrates that the Little Ice Age advance of the glacier reworked the earlier paraglacial slope deposits. (From Ballantyne and Benn, 1994).

morphology and the near-surface sedimentological characteristics of proglacial landforms and deposits. In high-latitude permafrost environments, paraglacial modification of recently deglaciated forelands is dominated by slumping and flow of soil from ice-cored moraines (Fitzsimons, 1996a,b; Bennett *et al.*, 2000a) and thermo-erosional incision into ice-cored terrain by meltwater streams. Etzelmüller (2000) has estimated that, on glacier forelands on Svalbard, thermo-erosion of large ice-cored moraines and slumping of sediments into the resulting ice-walled channels generates a specific sediment yield of 400–960 t km^{-2} $year^{-1}$, and thus forms a major component of sediment loss. Solifluction is also of widespread importance in modifying foreland topography in both arctic and alpine environments, particularly at the margins of retreating glaciers where sediments often contain abundant water, and may significantly modify the form of both moraines and subglacial bedforms. Slope measurements on recently deposited annual or recessional moraines show that initially steep gradients are reduced to stable angles within a decade or so (Welch, 1970; Sharp, 1984; Fig. 17.9), implying that many recessional moraines and other depositional landforms rapidly lose both their original form and near-surface structure (Rose, 1991).

Near-surface sediment structure may also be altered by frost-action processes. Lateral sorting of clasts into miniature sorted nets has been observed on several forelands (Ballantyne and Matthews, 1983; Krüger, 1994b), and the formation of large-scale sorted circles up to 3 m in diameter on recently deglaciated terrain (Ballantyne and Matthews, 1982) may imply convective soil movement within a saturated dilatant layer of till exposed by glacier retreat (Boulton and Dent, 1974). On the foreland of Styggedalsbreen in Jotunheimen, Norway, Matthews *et al.* (1998) recorded a range of frost-action phenomena, including sorted nets, sorted stripes, surface cracks, boulder-cored frost boils produced by the upfreezing of large clasts, miniature solifluction lobes and ploughing boulders, all of which imply widespread disruption of near-surface glacigenic deposits.

Wind erosion may also modify unvegetated recently deglaciated terrain. In Iceland, Boulton and Dent (1974) observed rapid loss of surface fines due to deflation by strong katabatic winds, forming a stony lag deposit overlying till. On Svalbard, aeolian erosion of push moraines composed of silts and fine sand deposits has planed off the original relief, causing significant surface lowering (Riezebos *et al.*, 1986). Aeolian deposits may also accumulate in hollows on recently deglaciated forelands (Derbyshire and Owen, 1996), and particularly on outwash plains where unvegetated overbank deposits are exposed during periods of low streamflow (McKenna-Neumann and Gilbert, 1986).

Slopewash and rainsplash also play an important role in redistributing sediment within sparsely vegetated glacier forelands (Rose, 1991; Krüger, 1994b; Fitzsimons, 1996b; Matthews *et al.*, 1998), sometimes forming miniature fans of reworked silt and sand (Theakstone, 1982). A more subtle effect of water movement on recently deglaciated terrain is downwash (eluviation) of silt and clay from surficial sediments, particularly where the uppermost sediment comprises dilatant till of high permeability. Boulton and Dent (1974) showed that the uppermost layers of such deposits experience rapid loss of silt and clay particles, which initially accumulate at the base of the dilatant layer and are then translocated more slowly through the underlying compact till.

Bank erosion and slopewash also rapidly enrich the sediment load of proglacial meltwater streams flowing across recently deglaciated terrain (Warburton, 1990). At Hilda Glacier in Alberta,

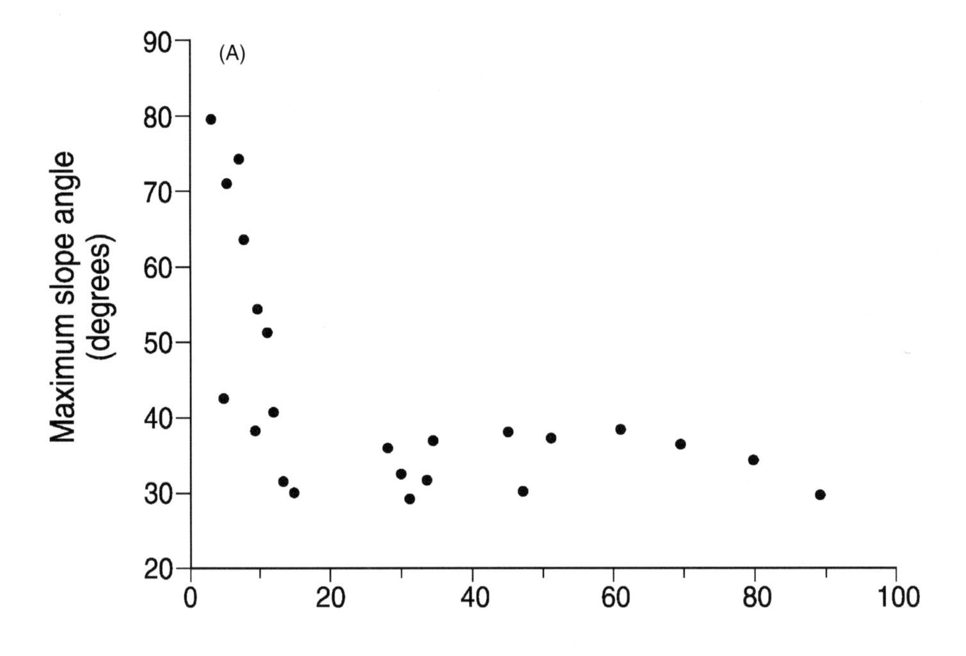

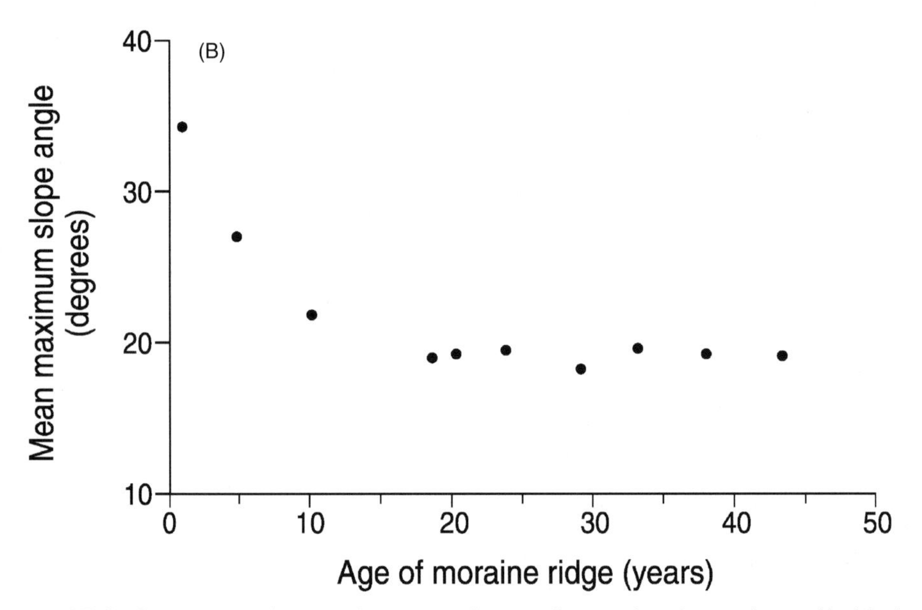

Figure 17.9 Changes in the maximum gradients of recessional moraines. A) Maximum gradients of recessional moraines on the foreland of the Athabasca Glacier, Canada (from Welch, 1970). B) Mean maximum gradients of recessional moraines on the foreland of Skálafellsjökull, Iceland (from Sharp, 1984).

Hammer and Smith (1983) compared the sediment load in the meltwater stream at the glacier snout with that 1 km downstream. Their data imply an average enrichment of 80 per cent of suspended load and about 37 per cent of bedload between the two points, with much greater sediment increases during episodes of bank collapse. They concluded that erosion of channel banks cut into recently exposed till provides a major portion of total sediment load at the downstream sampling point, and that sediment yields are at or close to their peak rates due to paraglacial reworking of glacigenic deposits. As banks stabilize and vegetation colonization reduces sediment input due to slopewash, sediment yields decline significantly. Leonard (1985, 1986) has shown that maximum sediment input into a distal proglacial lake coincides with the initial phases of ice retreat from readvance maxima, and attributed this to the exposure of unstable glacigenic deposits to fluvial erosion.

The most significant effects of paraglacial modification of glacier forelands are sediment redistribution, relief modification and changes in sediment characteristics. Sediment redistribution and relief modification are accomplished by mass movement, slopewash and streamflow, although wind erosion may also be locally significant. The characteristics of surficial sediments may be altered by slumping, solifluction, slopewash, lateral and vertical frost sorting, downwash of fines and deflation. Most paraglacial processes operating on glacier forelands, however, are effective only over a decadal timescale (Matthews *et al.*, 1998). Their effects nonetheless have potentially important consequences, as they imply rapid and significant modification of both the form and near-surface sedimentological characteristics of proglacial landsystems (Boulton and Dent, 1974; Rose, 1991).

17.6 PARAGLACIAL ALLUVIAL LANDSYSTEMS

Paraglacial alluvial landsystems comprise three main categories of landform, namely debris cones, alluvial fans and valley fills. All three may be regarded as paraglacial sediment stores that form within a few centuries or millennia after deglaciation but which frequently experience later fluvial erosion due to decreased sediment supply and/or base-level lowering. Research on recently deglaciated terrain has shown that small and intermediate-sized debris cones and alluvial fans may form, stabilize and decay within a few decades or centuries (Broscoe and Thompson, 1969; Ballantyne, 1995; Harrison and Winchester, 1997). In the Garwhal Himalaya, for example, glacier retreat over the past 200 years was accompanied by the development of paraglacial fans composed of reworked morainic debris, but most have now ceased to accumulate and exhibit fan-head entrenchment and fluvial erosion (Owen and Sharma, 1998). Most research on paraglacial alluvial landsystems, however, has been carried out in the context of fans and valley fills that accumulated after Late Pleistocene deglaciation, particularly in British Columbia and Alberta.

17.6.1 Paraglacial Fans of Late Pleistocene and Early Holocene Age

Ryder (1971a) showed that relict, vegetated alluvial fans in British Columbia owe their origins to reworking of glacigenic sediments (till, glacifluvial and glacilacustrine deposits) by streams and debris flows in tributary valleys. Stratigraphic evidence suggests that fan accumulation commenced soon after deglaciation and continued until shortly after the deposition of a near-surface tephra layer at *c.* 6.6 ka BP. Many fans were subsequently dissected by fan-head trenching or incision due to lowering of local base level, although in locations where fan accumulation continued during base-level lowering, nested multi-level fans were formed. The paraglacial fans described by Ryder are composed of fluvial gravels and debris-flow diamictons with occasional intercalated lacustrine or aeolian sediments. Debris-flow-dominated fans tend to have higher gradients and to occur at the outlets of small, steep tributary catchments (Ryder, 1971a, b). The volumes of most fans imply tributary denudation rates of the order of 0.25–2.0 m ka^{-1} (Church and Ryder, 1972).

Subsequent work in Alberta (Roed and Waslyk, 1973; Kostaschuk *et al.*, 1986; Beaudoin and King, 1994) has confirmed the generality of Ryder's findings, although in some instances at least, fan accumulation was episodic. Radiocarbon dating of fan deposits in the Lower Seymour Valley of British Columbia shows that paraglacial sedimentation commenced prior to *c.* 11.4 ka BP and was largely complete by *c.* 9.0 ka BP, but within this period renewed sediment accumulation occurred at *c.* 10 ka BP and accompanied climatic warming and a reduction in precipitation; charcoal-rich beds within fan sediments suggest that later depositional events may reflect slope instability triggered by fire in tributary catchments (Lian and Hickin, 1996). The lower Cheekye fan, which occupies an area of 8.3 km^2 in southwest British Columbia, reached roughly its present dimensions before *c.* 6.0 ka BP, but deposition continued intermittently until *c.* 1.3 ka BP. Friele *et al.* (1999) calculated that approximately 14×10^8 m^3 of sediment accumulated on this fan prior to *c.* 6. 0 ka BP, but an order of magnitude less ($\sim 1.4 \times 10^8$ m^3) since then.

Particularly impressive paraglacial fans occur in the valleys of the Karakoram and Himalayan Mountains. Since deglaciation there has been massive reworking of glacial deposits by debris flows and rivers and consequent burial of till beneath thick fan sediments. In the Hunza Valley of the Karakoram, 44 per cent of the valley floor is covered by paraglacial fan deposits compared with only 14 per cent mantled by intact glacigenic deposits (Li Jijun *et al.*, 1984), and the floor of the Gilgit Valley is similarly dominated by Late Pleistocene paraglacial fans (Owen, 1989; Derbyshire and Owen, 1990; Fig. 17.10). Fan deposits are commonly several tens of metres thick, implying paraglacial re-sedimentation on a truly grand scale.

The lithofacies architecture of large paraglacial fans is often complex. Eyles and Kocsis (1988) showed that the sediments in a fan that accumulated between *c.* 11 ka BP and *c.* 7 ka BP in British Columbia are dominated by diamict facies deposited by debris flows (48 per cent of fan volume) and sheetflood gravels (37 per cent), intercalated with occasional beds of aeolian silt and sheetwash deposits. Crude bedding within diamict facies represents superimposition of multiple debris-flow units 0.2–3.0 m thick, and alluvial gravel facies are characteristically massive, crudely bedded and poorly sorted, and thus similar to those in the proximal reaches of shallow braided rivers. Within the large paraglacial fans of the Karakoram, individual debris flow units are characterized by discrete shears and pronounced fabric anisotropy near their bases, and their upper surfaces are often draped in fine silt reflecting post-depositional slopewash (Derbyshire and Owen, 1990). Some fans in this area consist of a few thick debris-flow units interbedded with fluvial and glacifluvial sediments, the latter demonstrating fan accumulation very soon after glacial retreat. The main debris-flow deposits consist of diamict sheets 5–20 m thick that cover areas of up to 30 km^2. The vast size of these units implies flow of exceptional volumes of fluidized debris on a catastrophic scale.

17.6.2 Paraglacial Valley Fills

The term 'valley fill' describes unconsolidated deposits that overlie bedrock in valley-floor locations. Valley fills are often compositionally complex, but in glaciated areas it is often possible

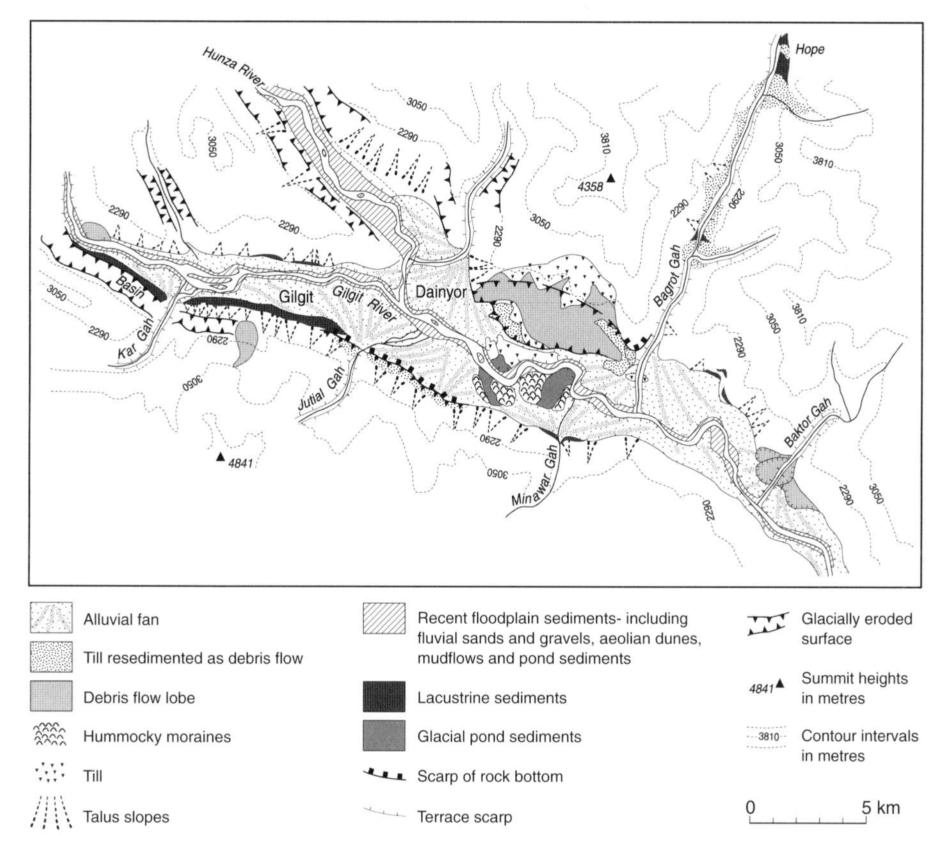

Figure 17.10 Landforms and surficial deposits of the Gilgit Valley, Karakoram Mountains. The surface of the valley fill is dominated by large paraglacial fans. The outcrop of *in situ* glacigenic deposits is extremely limited. (Adapted from Owen (1989).)

to identify a lower sequence of glacigenic deposits and, overlying these, a paraglacial sequence of debris flow, floodplain, alluvial fan, lacustrine and/or aeolian deposits in some combination (Owen, 1989; Figs 17.10 and 17.11).

An interesting sequence of paraglacial valley-fill aggradation and later incision occurs in the valley of the Bow River, which drains the Rocky Mountains in Alberta. In the upper valley a basal valley-fill deposit of proximal outwash sediments is overlain by multiple beds of massive diamict facies with a total maximum thickness of ~30 m. The diamicts have a sheet-like configuration with largely conformable bedding contacts, dip downvalley at 5–10° and were interpreted by Eyles *et al.* (1988) as the product of successive massive paraglacial debris flows that reworked glacifluvial and glacilacustrine sediments from up-valley and adjacent slopes. Farther down-valley, the Bow River is flanked by terraces cut in gravel fill, which diminishes in thickness from ~30 m near the mountain edge to ~10 m near Calgary, 100 km downstream. Radiocarbon dating of the terrace sediments implies that gravel aggradation occurred mainly within the period 11.5–10.0 ka BP, two millennia after the last glacial readvance reached the mountain edge. Jackson *et al.* (1982) therefore interpreted the gravel fill in the lower valley as the product of fluvial reworking of the debris-flow deposits in the upper valley, and thus as 'second-generation' paraglacial fluvial

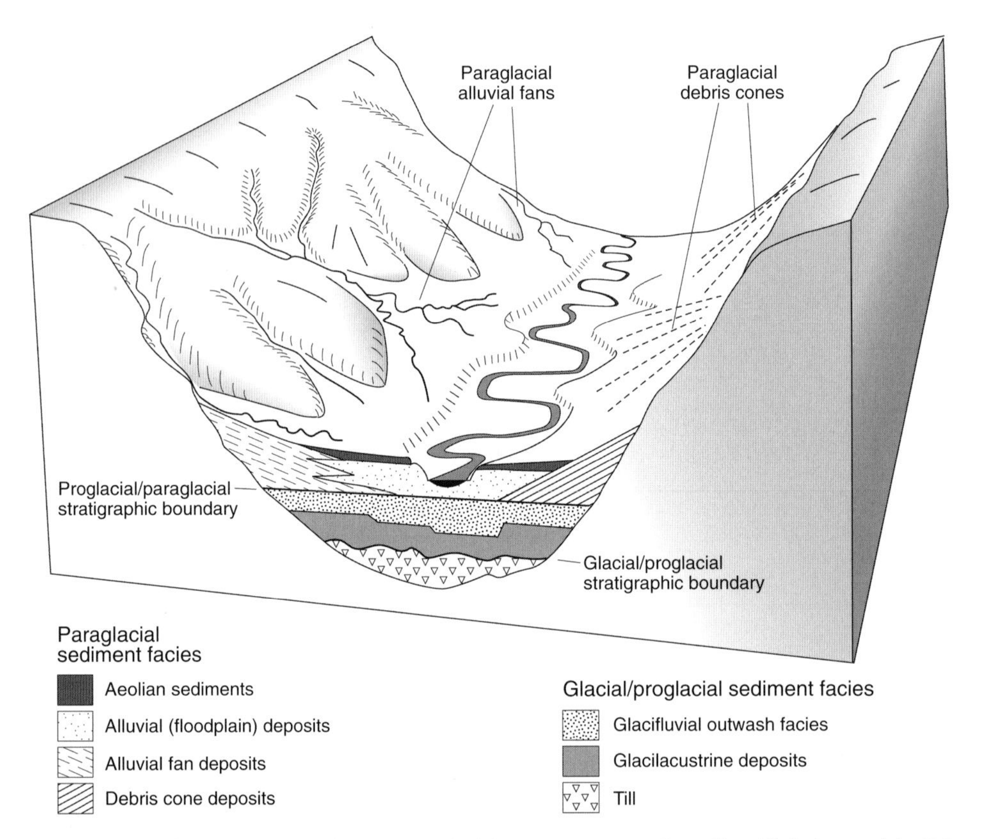

Figure 17.11 Glacial, proglacial and paraglacial components of a valley fill (schematic). Often fewer sediment units are present. Interfingering of paraglacial units (particularly debris flow or alluvial fan deposits) and glacigenic sediments implies rapid paraglacial resedimentation during and immediately after deglaciation.

deposits. Since *c.* 4.6 ka BP, the Bow River has incised into the gravel fill in its lower reaches in response to diminished sediment supply from upstream, cutting the terraces that now flank its course.

The Bow River model of valley-fill aggradation (Fig. 17.12) is by no means the only possible sequence of paraglacial valley-fill accumulation (Owen, 1989). Lacustrine and distal glacilacustrine sedimentation may also constitute a major component of paraglacial valley fills (Clague, 1986). In the South Thompson Valley of British Columbia, for example, a distal glacilacustrine fill up to 150 m thick accumulated in only 100–200 years in the Late Pleistocene. Near the valley sides, paraglacial debris flow deposits interrupt lacustrine

Figure 17.12 Model of paraglacial valley-fill development based on the sequence in the Bow River valley, Alberta. A) Emplacement of thick accumulations of paraglacial debris-flow deposits in the upper part of a mountain catchment. B) Fluvial erosion of valley-head debris-flow deposits and deposition of a paraglacial alluvial fill farther downvalley. C) Fluvial incision and terracing of the alluvial fill as sediment input from upstream is reduced.

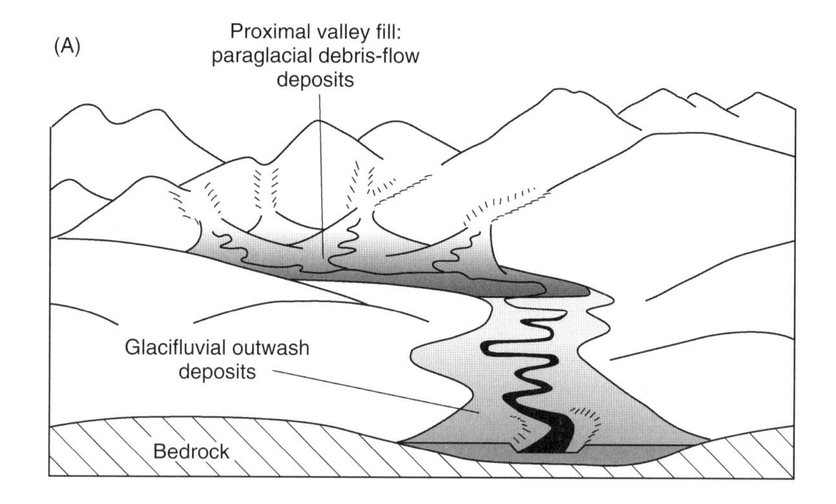
(A)
Proximal valley fill:
paraglacial debris-flow
deposits
Glacifluvial outwash
deposits
Bedrock

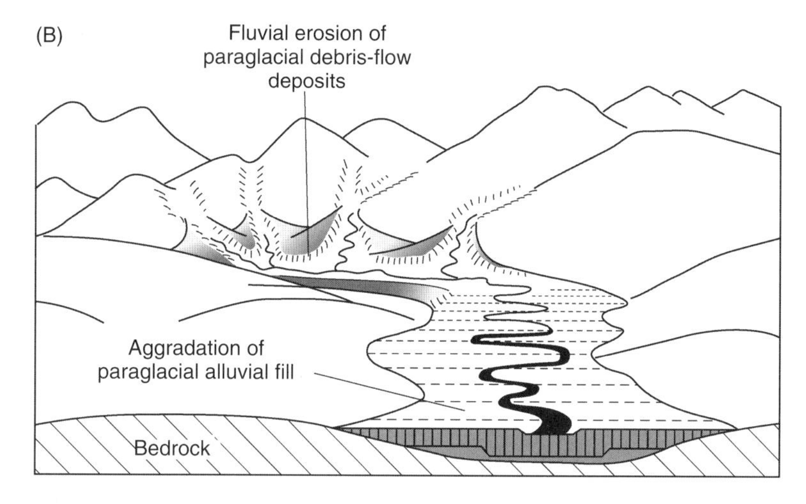
(B)
Fluvial erosion of
paraglacial debris-flow
deposits
Aggradation of
paraglacial alluvial fill
Bedrock

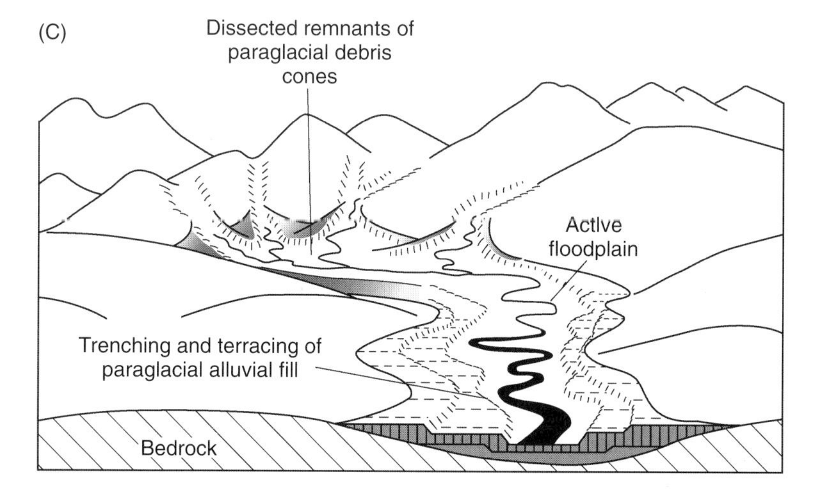
(C)
Dissected remnants of
paraglacial debris
cones
Actlve
floodplain
Trenching and terracing of
paraglacial alluvial fill
Bedrock

rhythmites, and sand bodies were deposited where tributary streams entered the lake. After lake drainage, aeolian reworking of lake deposits produced a paraglacial loess deposit that caps terrace fragments, and river incision cut a floodplain through the lake sediments (Roberts and Cunningham, 1992). In the Walensee Valley in Switzerland, the valley fill comprises patchy till deposits overlain in turn by glacilacustrine sediments then postglacial lacustrine sediments, the latter being interrupted laterally by deltaic deposits overlain by alluvial gravels deposited below two major paraglacial fans. Müller (1999) calculated that the sediments underlying the two fans accumulated at an average rate of 70–100 mm $year^{-1}$ prior to *c.* 12.3 ka BP, but thereafter at an average rate of only 3–4 mm $year^{-1}$, attributing the rapid sedimentation prior to *c.* 12 ka BP to the reworking of glacigenic sediments in steep tributary valleys by powerful debris flows immediately after deglaciation.

17.7 PARAGLACIAL SEDIMENTATION IN LAKES

During deglaciation, many lakes experience an upwards depositional succession from proximal, ice-contact sediment accumulation, to ice-distal sedimentation, to paraglacial sedimentation (e.g. van Husen, 1979; Eberle, 1987; Müller, 1999). Proximal glacilacustrine sedimentation is characterized by the formation of ice-contact deltas, subaqueous fans and moraines, submerged ramps produced by subaqueous mass movement and rhythmic laminated bottom sediments (Ashley, 1995). Distal glacilacustrine sedimentation and paraglacial lacustrine sedimentation are both characterized by delta progradation, often of fine-grained, gently dipping foreset sediments, and the accumulation of bottom deposits (sometimes rhythmic) of sand, silt and clay. It is thus often difficult to distinguish distal glacilacustrine deposition from paraglacial lacustrine deposition on the basis of the sediment record alone. In mountainous areas, slumping of valley-side glacigenic deposits may generate large subaqueous debris flows (Eyles, 1987) that may interrupt the depositional sequence.

The influence of paraglacial sediment input is particularly evident in small lake basins. Studies of Late Holocene varve thickness in a distal glacial lake in Alberta led Leonard (1986, p. 211) to note that the '...highest sedimentation rates occur during the initial stages of glacial recession, rather than at the time of the glacial maximum, indicating the existence of a "paraglacial" peak in the sedimentation rate'. This conclusion is supported by later studies of Holocene sedimentation in distal lakes (e.g. Desloges, 1994; Dirszowsky and Desloges, 1997), and has important implications for the interpretation of periods of enhanced minerogenic sedimentation in Holocene lake sequences. Over a longer timescale, Smith (1975) showed that sediment influx in a shallow lake in the Canadian Rockies has diminished during the Holocene, attributing the trend to a declining input of reworked Pleistocene glacigenic sediment. Conversely, an increase in the sedimentation rate in Nicolay Lake in the Canadian arctic over the past five centuries has been identified as representing the delayed arrival of a wave of paraglacial sediment driven by base-level lowering (Lamoureux, 1999).

The pattern of lacustrine sedimentation during and after the retreat of Late Pleistocene glaciers has been investigated in several of the large, deep inland 'fjord lakes' of British Columbia (e.g. Eyles *et al.*, 1990, 1991b; Mullins *et al.*, 1990; Desloges and Gilbert, 1991; Gilbert and Desloges, 1992). These studies suggest that vast quantities of sediment accumulated on lake floors within decades or a few centuries during and immediately after glacier retreat from lake basins. Seismic studies suggest that the largest of these, Okanagan Lake, is occupied by about 90 km^3 of sediment up to 792 m thick. The great majority of this infill was interpreted by Eyles *et al.* (1990, 1991b) as reflecting rapid proglacial accumulation and associated subaqueous mass movement, but a discontinuous upper stratified unit typically less than 25 m thick appears to represent a deglacial/paraglacial varve sequence. Similar seismostratigraphic sequences are evident in Harrison Lake (Desloges and Gilbert, 1991) and Stave Lake (Gilbert and Desloges, 1992), where accoustically stratified units interpreted as the products of rapid proximal glacilacustrine sedimentation are overlain by an upper sediment unit up to 28 m thick that appears to reflect paraglacial/postglacial sedimentation. A drawback of these studies is the lack of deep borehole records to confirm interpretations of the seismostratigraphic record, or of dated horizons to verify the inferred rapidity of glacilacustrine sedimentation or to establish variations in the rate of postglacial infill. The consensus of interpretation nonetheless suggests very rapid sedimentation during glacier retreat, and that though depositional rates slowed after deglaciation, substantial volumes of sediment continued to accumulate through fluvial influx of reworked glacigenic sediment.

Finally, it is worth noting that glacilacustrine deposits constitute a major source of fluvially reworked sediment. Gordon (1979) attributed the high sediment yield of New England rivers over the past 8 ka to continuing erosion of Late Pleistocene glacilacustrine deposits, and the removal of huge volumes of glacilacustrine sediment from the South Thompson Valley (Clague, 1986; Roberts and Cunningham, 1992) and from the Ain Basin in France (Campy *et al.*, 1998) highlights the importance of glacilacustrine deposits as paraglacial sediment sources. Such deposits occur as valley fills and are thus readily accessible to fluvial reworking, and being predominantly fine-grained are readily erodible. The outcrop of glacilacustrine deposits is extensive (Teller, 1987), and it is likely that in many areas glacilacustrine sediments have provided the most important source of fluvially reworked sediment during the Holocene (Church and Slaymaker, 1989; Ashmore, 1993).

17.8 PARAGLACIAL COASTAL LANDSYSTEMS

The concept of paraglacial coasts is dominated by the progressive release of a vast store of *in situ* and fluvially reworked glacigenic sediment into the nearshore and offshore sediment budget. This happens in two ways. First, the supply of reworked glacigenic sediment by rivers enhances sediment influx in estuarine and fjord-head locations. Second, where glacigenic sediments crop out in the littoral zone they are eroded directly by waves and tidal currents, introducing reworked sediment that strongly influences coastal landforms and nearshore sediment transfer. Forbes and Syvitski (1994) pointed out that the timing and duration of sediment delivery or availability is largely determined by the disposition of sources of glacigenic sediment relative to the coastline and particularly by changes in relative sea level. On coasts where glacigenic sediments are reworked at the shorefront, for example, rising seas allow wave action to tap pristine sources of sediment, prolonging sediment reworking; conversely, falling seas may isolate glacigenic sediment sources from wave action, terminating sediment reworking. In the account below the dynamics of paraglacial coasts are described in terms of three subsystems: fjords, barrier coasts and glaciated shelves.

17.8.1 Fjords

Fjords have immense sediment storage capacity, and because the mouths of many fjords are crossed by a shallow sill of bedrock or glacigenic deposits, they are often effectively closed basins from which sediment escape is limited. Many fjords therefore contain a record of sediment accumulation that spans the last glacial-interglacial cycle (Gilbert, 1985; McCann and Kostaschuk, 1987; Syvitski and Lee, 1997). Syvitski and Shaw (1995) identified five stages in the deglacial sedimentation sequence:

1. occupance by glacier ice
2. proximal or ice-contact sedimentation associated with retreating tidewater or floating glacier termini
3. distal proglacial sedimentation after glacier termini have become land-based
4. postglacial sedimentation following disappearance of glacier ice, and
5. complete infill of fjord basins.

Paraglacial sedimentation effectively commences when retreating ice becomes land-based (Powell and Molnia, 1989; Forbes and Syvitski, 1994) and has four main elements (Fig. 17.13):

- progradation of a fjord-head delta
- reworking of sediment by subaqueous mass movement and turbidity currents
- settling of suspended sediment on the fjord floor, and
- in some fjords, reworking of glacigenic or older marine sediment around the fjord margins.

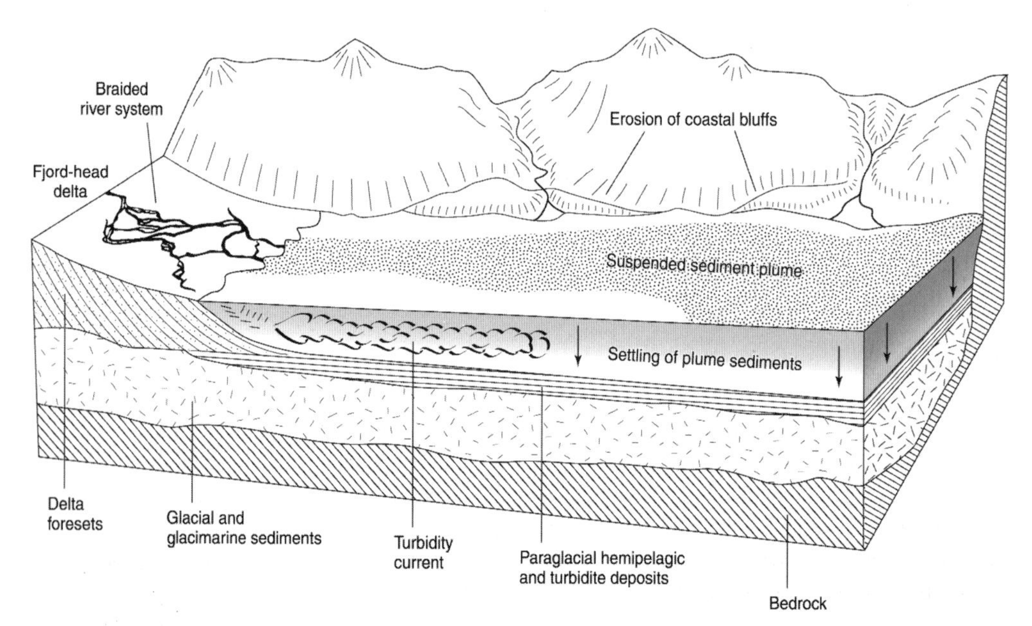

Figure 17.13 Paraglacial sedimentation in fjord basins (schematic). Sedimentation is dominated by: (1) deposition of sand and gravel foreset beds at the front of a prograding fjord-head delta; (2) localized failure of the fjord-head delta front, which generates turbidity currents and results in deposition of turbidite layers on the fjord floor; (3) settlement of fine-grained suspended particles from a surface sediment plume, and associated accumulation of hemipelagic bottom sediments; (4) reworking of shoreface bluffs by wave action. Submarine failure of sidewall sediments may also generate turbidity currents and turbidite deposition.

Aeolian reworking of outwash sediments may also supply silt and fine sand to some fjords, but is quantitatively significant only in arid high-arctic environments (Gilbert, 1983).

In most fjords, paraglacial sedimentation has been dominated by the influx of reworked glacigenic sediment carried by the trunk stream entering the fjord head. Such rivers may reach the fjord as a single or braided channel, or tidal channels cut through mudflats. The sediment load separates into two components at the river mouth: suspended sediment is carried out into the fjord, but bedload settles on to the front of a prograding fjord-head delta (Kostaschuk, 1985). The foreset beds of fjord-head deltas typically dip seawards at angles of 5–30° to depths of 10–50 m, then prograde along bottomset beds at much gentler gradients. Slope failures are common along steep delta fronts, forming chutes 10–30 m wide (Kostaschuk and McCann, 1983; Syvitski *et al.*, 1988). The failed sediments liquefy as they move, forming turbidity currents that sweep down channels in the prodelta zone before depositing sediment on the fjord floor. Similar sediment slumping and turbidity current formation may also occur on sidewall slopes, side-entry deltas and fjord-mouth sills (Syvitski and Shaw, 1995).

Suspended sediment influx moves seawards at the water surface in a buoyant plume (Kostaschuk and McCann, 1983), and sedimentation rates tend to decrease exponentially down-fjord. Most paraglacial fjord-bottom sediments thus comprise dominantly fine-grained hemipelagic deposits, often fining down-fjord, intercalated with turbidite deposits derived from subaqueous sediment failures. Reworking of Pleistocene deposits by wave action may also contribute to fjord infill, but tends to be significant only in shallow fjords where sills are absent or deep. The Holocene sediment infill of such 'wave-dominated' fjords in eastern Canada consists largely of such directly reworked sediment (Syvitski and Shaw, 1995), which currently accumulates on the fjord floor at up to 3 mm year^{-1} (Piper *et al.*, 1983).

The effectiveness and duration of the above processes may be significantly influenced by changes in relative sea level. A fall in sea level may reduce the accessibility of glacigenic deposits to wave attack, but the associated lowering in base level causes incision of fjord-head deltas and outwash deposits, thus prolonging the period of terrestrial paraglacial sediment influx. A rise in sea level may result in the submergence of paraglacial deltas and shoreface platforms. In fjords along the south coast of Newfoundland, for example, paraglacial deltas formed during the Early Holocene have been drowned by rising seas (Shaw and Forbes, 1992, 1995).

Data on fjord sedimentation rates suggest a radical decline since deglaciation. Stravers *et al.* (1991) calculated that in Cambridge Fjord on Baffin Island, paraglacial sedimentation during and after deglaciation (*c.* 10–6 ka BP) was an order of magnitude greater than current sedimentation rates. In Sanguenay Fjord in eastern Canada, present rates of sediment delivery account for only 7 per cent of overall sediment accumulation (Syvitsky and Praeg, 1989). If representative, these data suggest that in many fjords deglaciated at the end of the Pleistocene, the effects of paraglacial re-sedimentation are now much diminished.

17.8.2 Barrier Coasts

Barrier coasts are those characterized by nearshore sediment accumulations such as beaches, baymouth bars, spits and barrier islands. On most paraglacial barrier coasts the main source of sediment in barrier structures is derived from reworking of glacigenic sediment by coastal erosion, such as occurs on the southern Baltic coast and the Beaufort Sea coast of Alaska. Paraglacial barrier coastlines exhibit great variations in the size, configuration, sedimentary characteristics and stability of barrier structures, reflecting the complex interaction of several controls, including antecedent coastal configuration and relief, coastal compartmentalization (interruption by bedrock headlands), sediment availability and texture, wave and tidal energy, and relative sea level change. A classification of paraglacial barrier systems by FitzGerald and van Heteren (1999) is based primarily upon compartmentalization and recognizes a progressive transition from small barriers in isolated rocky inlets to large barriers fed by abundant sources of glacigenic sediment. In the latter situation, sediment release may keep pace with sea level rise, producing progradational barriers, but both progradational and retrogradational behaviour is possible on all paraglacial barrier systems. Within their general typology, barriers may be characterized by morphology (spit, baymouth bar, etc), texture (sand, mixed sand-gravel, gravel- and boulder-dominated), and stratigraphy (progradational, aggradational, retrogradational or complex). Most paraglacial barriers are coarse-grained, except where the parent glacigenic sediment sources are themselves fine-grained, and stratigraphic complexity is typical (Forbes and Taylor, 1987; van Heteren *et al.*, 1998).

The evolution of paraglacial barrier coasts is often dominated by the balance between rates of sediment input and rates of sea level change (Boyd *et al.*, 1987; Forbes and Taylor, 1987; Carter *et al.*, 1989; Shaw *et al.*, 1990; Forbes *et al.*, 1995a, b). Where the net sediment balance is positive, barrier systems may continue to prograde seaward. Under conditions of diminishing sediment supply, however, barrier retreat, erosion and destruction occurs, and a new barrier system may develop landward of the original (Fig. 17.14), nourished not only by new onshore glacigenic sediment sources, but also by sediments reworked from the original barrier and associated back-barrier sediments. Barrier evolution is also influenced by inherited submarine morphology and sediment supply. At St George's Bay in Nova Scotia, for example, paraglacial barrier structures rest on large subaqueous sandy platforms that are partly composed of sediment reworked from glacigenic sources during an early Holocene lowstand (Shaw and Forbes, 1992). Under steady-state conditions, paraglacial barrier systems exhibit gradual evolution involving littoral cell development, beachface realignment, crest build-up and progressive sediment sorting. Changes in relative sea level, sediment supply, storm intensity or wave regime, however, may trigger rapid destabilization, barrier breakdown, sediment remobilization and a cascade of secondary effects both alongshore and in back-barrier embayments (Forbes *et al.*, 1995a, b). On paraglacial coasts where erosion of boulder-rich till forms the main sediment source, boulder lag deposits may accumulate, forming boulder barricades, boulder pavements and boulder-strewn beaches, tidal flats and rock platforms (Lauriol and Gray, 1980; Dionne, 1981; Hansom, 1983; Forbes, 1984). Boulder barricades may armour an eroding shoreface, reducing the sediment supply to barrier structures, and thus contributing to their eventual segmentation and destruction.

Drumlin coasts represent a subset of paraglacial barrier coasts, with individual drumlins acting as discrete sediment sources. In areas of Holocene marine transgression, such as eastern Nova Scotia and western Ireland, paraglacial barriers fed by coastal erosion of drumlins tend to experience a distinctive cycle of growth then destruction as sediment supply fails to keep pace with rising sea level, leading to shoreward movement of sediment and re-establishment of barriers (Boyd *et al.*, 1987; Forbes and Taylor, 1987; Carter and Orford, 1988; Fig. 17.14). Individual drumlins may be eroded down to sea level then submerged by transgressive seas, forming offshore shoals of lag boulders (Piper *et al.*, 1986; Carter *et al.*, 1990). Where drumlin islands occur in broad, shallow bays, cuspate spits form at sites of longshore drift convergence, often forming tombolos that link drumlin headlands

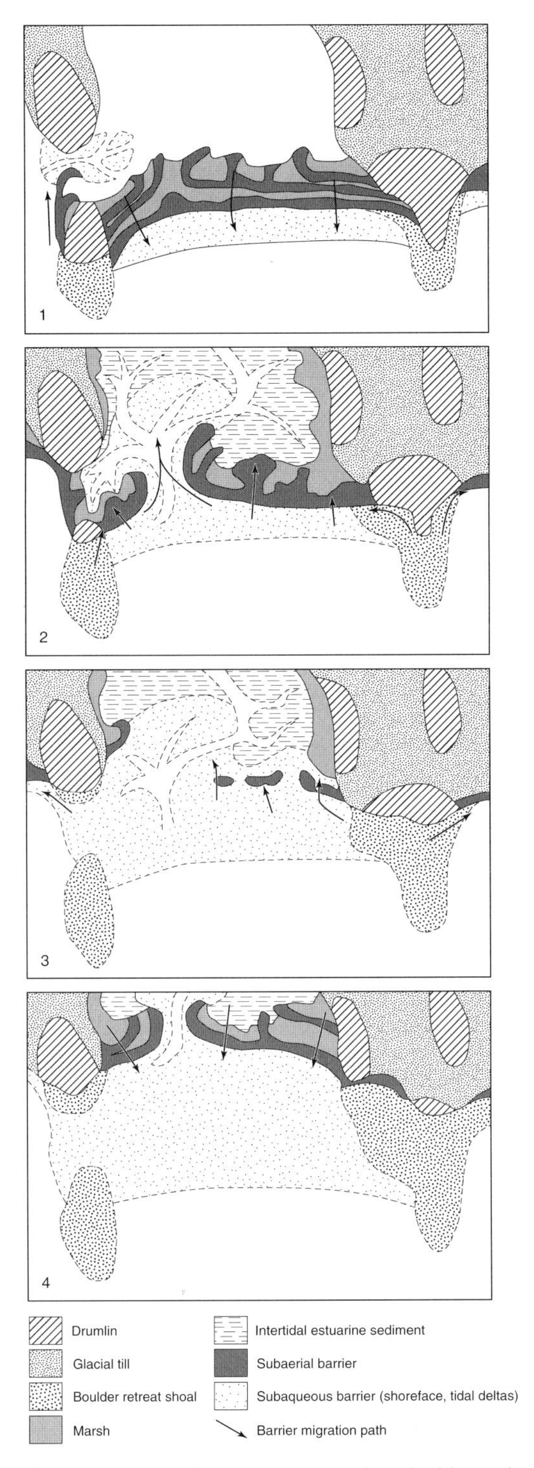

Figure 17.14 Evolution of paraglacial barrier systems associated with marine transgression across drumlin headlands. 1) Coastal reworking of glacigenic deposits into a prograding barrier system. 2) Erosion of barriers as sediment supply diminishes. 3) Destruction of the original barrier. 4) Growth of a new barrier system, fed by new sources of glacigenic sediments and by sediments reworked from the earlier barrier and associated back-barrier deposits. (Adapted from Boyd *et al.* (1987).)

(Rosen and Leach, 1987). Where inter-drumlin valleys have been invaded by the sea, barriers evolve at the mouths of individual inlets (Carter *et al.*, 1990, 1992). Back-barrier sedimentation in larger enclosed basins is dominated by flood-tidal sand sheets and basin muds, but open estuaries are characterized by incipient beach and barrier formation, lateral flood-delta expansion and saltmarsh accretion (Carter *et al.*, 1989, 1992; Shaw *et al.*, 1993).

Active outwash coasts form a further subset of paraglacial barrier coasts, and occur where unconfined outwash (sandur) plains fronting present-day glaciers meet the sea. Examples occur in northwest Svalbard, the Gulf of Alaska and the coast of southeast Iceland. In such locations, coastal configuration and barrier formation reflects the delivery of sediment by braided meltwater streams and the reworking of such sediment by waves, tidal current and longshore drift. In southeast Iceland, for example, outwash sediments reworked by longshore drift form a predominantly sandy barrier complex of beaches, spits and islands that increases westwards in width from about 200 m to 700 m. Under conditions of stable relative sea level, the shoreline has locally prograded up to 1 km during the past century (Nummedal *et al.*, 1987).

17.8.3 Glaciated Shelves

The stratigraphy of Quaternary deposits on glaciated shelves comprises five major elements:

1. till or ice-contact sediments, overlain by
2. glacimarine sediment, sometimes subdivided into a lower ice-proximal unit and
3 an upper ice-distal unit
4. a paraglacial sequence, and
5. an uppermost postglacial unit, comprising lag deposits, basin fill, estuarine muds and pelagic oozes.

Syvitski (1991) suggested that deposition of paraglacial sediments typically occurred during a period of falling relative sea level, and reflects high rates of fluvial sediment influx associated with widespread reworking of terrestrial glacigenic sediment. He characterized the paraglacial sequence as rapidly prograding deltaic wedges comprising '... a mixture of sediment gravity flow and hemipelagic deposits with syndepositional mass flow deposits, buried channels and shear planes related to submarine landslides' (Syvitski, 1991, p. 910). The paraglacial sediments of the eastern Canadian shelf (Syvistski and Praeg, 1989) and Alaskan shelf (Carlson, 1989; Powell and Molnia, 1989) are typical examples. On the latter, where paraglacial re-sedimentation is still active, recent sediment fines seawards from littoral sand to clayey silts 75 km offshore, and the total thickness of Holocene sediment accumulation on the shelf averages ~55 m (Molnia *et al.*, 1978). On many glaciated shelves, the distinction between 'paraglacial' and 'postglacial' sediment accumulation (*sensu* Syvitski, 1991) is rather arbitrary, as the latter may comprise a substantial component of reworked glacigenic sediment carried seaward from estuaries or the littoral zone; present sediment accumulation on many glaciated shelves is thus essentially paraglacial in the sense that the greater part of the sediment ultimately derives from glacigenic deposits.

17.9 PARAGLACIAL LANDSYSTEMS AND POSTGLACIAL LANDSCAPE CHANGE

Subdivision of paraglacial landscape response into six landsystems (rock-slope, drift-slope, glacier foreland, alluvial, lacustrine and coastal) provides a useful context for analysis of process-landform-sediment relationships. This approach, however, fails to highlight the links between individual landsystems. Ballantyne (2002b) has suggested that this may be achieved by representing paraglacial response as an interrupted sediment cascade with four primary sediment sources (rockwalls, drift-mantled slopes, valley-floor glacigenic deposits and coastal glacigenic deposits), four terminal sediment sinks (valley fills, lake basins, coastal/nearshore settings and shelf/offshore settings) and numerous intervening sediment stores and transport pathways (Fig. 17.15). In the initial stages of the cascade, sediment inputs such as rockfall and reworking of glacigenic sediments by debris flows and rivers produces a range of primary sediment stores, such as talus accumulations, debris cones and alluvial fans, while in littoral settings reworking of glacial drift initiates the formation of coastal barriers such as spits and

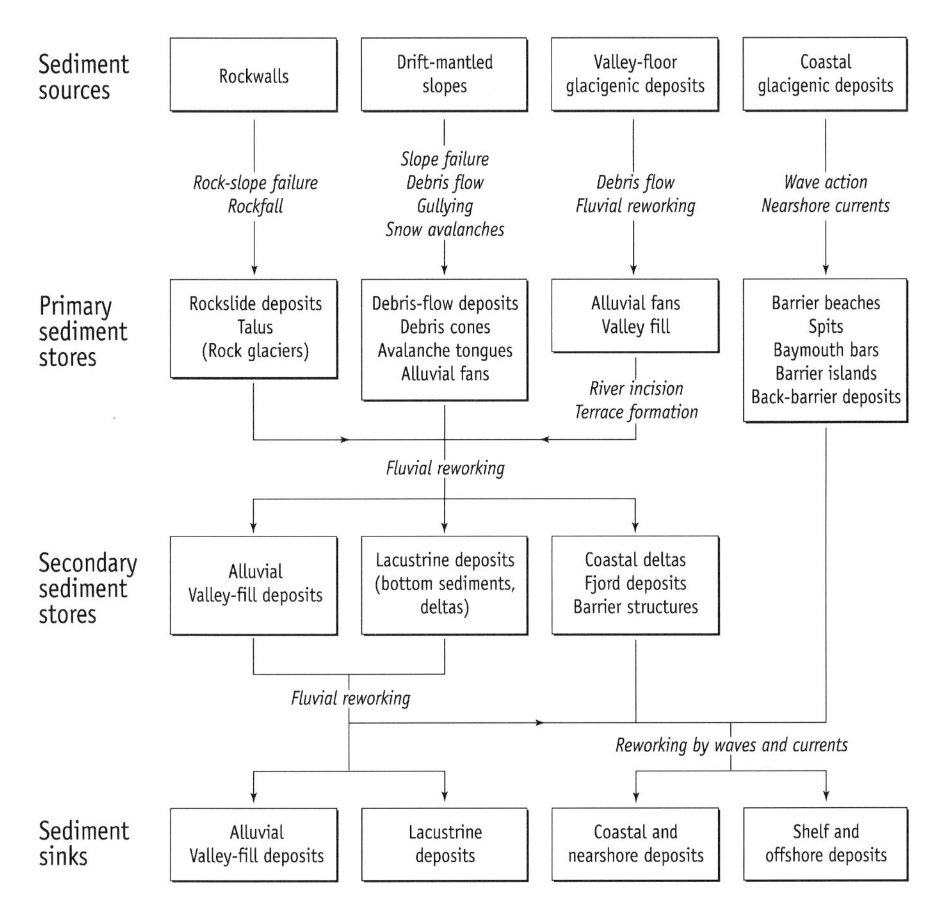

Figure 17.15 Simplified paraglacial sediment cascade, showing the principal primary and secondary sediment stores and the main sediment transfer routeways. Sediment is likely to enter storage on numerous occasions during source-to-sink transport, for example through multiple episodes of floodplain deposition or storage in barrier structures. All terrestrial sediment stores may survive throughout the Holocene (i.e. over a timescale of 10^4 years), but most experience a transition from net accumulation to net erosion within this timescale.

barrier islands. As sediment input to primary sediment stores slows, it is succeeded by net erosion, primarily by rivers in terrestrial settings and by waves and currents in coastal settings. The subsequent transport history of reworked sediment varies in complexity. Some may be transported directly to its final destination within valley fills, lake basins or nearshore or offshore depo-centres. Other sediments may experience several cycles of deposition and re-entrainment, entering secondary sediment stores (Fig. 17.15) that later experience net erosion. The timing and duration of sediment storage is determined not only by sediment budget and the capacity of transporting agents, but also by a wide range of extrinsic influences such as climate change, extreme climatic events, neotectonic uplift and tilting, sea level change and anthropogenic activity.

The sediment cascade model of paraglacial landscape response provides a useful framework for reconstruction of postglacial landscape evolution. It is possible to distinguish an initial, often rapid, period of paraglacial adjustment that immediately follows deglaciation and is characterized by reorganization of landforms and sediments to a more stable state, sometimes over a few decades (modification of glacier forelands), centuries (adjustment of drift-mantled slopes) millennia (accumulation of talus, large alluvial fans or valley fills) or tens of millennia (reworking of glacigenic sediments by rivers or the sea). Superimposed on this pattern are episodes of rejuvenation triggered by external perturbations (Fig. 17.3). The study of Holocene geomorphic change in glaciated environments can thus be conceptualized entirely within a context of initial, prolonged, delayed and/or renewed paraglacial sediment reworking, in which any part of the sediment cascade may be remobilized long after the initial period of paraglacial adjustment.

17.10 Implications for Glacial Landsystems

The study of ancient (Pleistocene) glacial landsystems often assumes, albeit implicitly, that component landforms and sediments are in a pristine state. As this chapter has emphasized, this is not always the case, as both may have experienced significant paraglacial modification by non-glacial processes. Depending on the processes involved, the original sediments may be only slightly modified, or altered beyond recognition by selective entrainment or sorting during deposition, potentially leading to misinterpretation of the significance of Quaternary landform assemblages or stratigraphic sequences. Diamictons previously interpreted as *in situ* till or gelifluctate have been shown to represent paraglacial debris-flow deposits (e.g. Holmes and Street-Perrott, 1989; Wright, 1991; Harris, 1998; Bennett, 1999), and valley-fill gravels originally interpreted as outwash have been shown to have a paraglacial alluvial origin (Jackson *et al.*, 1982). The form of cirques and glacial troughs may be extensively altered after deglaciation by rock-slope failure, rockfall and reworking of drift-mantled slopes (e.g. Bovis, 1990; Ballantyne and Benn, 1994, 1996; Augustinus, 1995a, b; Curry, 1999, 2000a). Individual depositional landforms such as moraines or subglacial bedforms may have experienced paraglacial modification that has altered their surface form and sediment characteristics (e.g. Boulton and Dent, 1974; Rose, 1991; Dardis *et al.*, 1994; Matthews *et al.*, 1998; Etzelmüller, 2000). Few attributes of glacial landsystems are thus immune to significant paraglacial modification. One implication is that the investigation of ancient glacial landsystems requires an understanding not only of the environment and mechanics of orginal sediment deposition by glacier ice or glacial meltwater, but also of the paraglacial processes that may have affected, sometimes substantially, their subsequent modification.

Acknowledgements

The author thanks the following for permission to reproduce published material: the Regents of the University of Colorado (Fig. 17.8); Professor Martin Sharp and the International Glaciological Society (Fig. 17.9); E. Schweizerbart'sche Verlagsbuchhandlung (Fig. 17.10); and Academic Press (Fig. 17.14).

REFERENCES

Aario, R., 1977. Classification and terminology of morainic landforms in Finland. *Boreas*, 6, 87–100.

Aario, R., 1992. Puljo moraines and Sevetti moraines. In A.-M. Robertson, B. Ringberg, U. Miller and L. Brunnberg (eds), *Quaternary stratigraphy, glacial morphology and environmental changes.* Sveriges Geologiska Undersökning, Series Ca 81, pp. 7–14.

Aartolahti, T., 1974. Ring ridge hummocky moraines in northern Finland. *Fennia*, 134, 22.

Aartolahti, T., 1975. Two glacial mound fields in northern Savo, Finland. *Fennia*, 139, 23.

Aartolahti, T., 1995. Glacial morphology in Finland. In J. Ehlers, S. Kozarski and P. Gibbard (eds), *Glacial deposits in north-east Europe*. Balkema, Rotterdam, pp. 37–50.

Aber, J.S., 1988., Ice-shoved hills of Saskatchewan compared with Mississippi mudlumps – implications for glaciotectonic models. In D.G. Croot, D.G. (ed.), *Glaciotectonics – forms and processes.* Balkema, Rotterdam, pp. 1–9.

Aber, J.S., Croot, D.G. and Fenton, M.M., 1989. *Glaciotectonic landforms and structures.* Kluwer Academic, Boston, 201 pp.

Acomb, L.J., Mickelson, D.M. and Evenson, E.B., 1982 Till stratigraphy and late glacial events in the Lake Michigan lobe of eastern Wisconsin. *Geological Society of America Bulletin*, 93, 289–296.

Ageta, Y., Iwata, S., Yabuki, H. *et al.*, 2000. Expansion of glacier lakes in recent decades in the Bhutan Himalayas. In *Debris-covered glaciers*. Proceedings of a workshop held in Seattle, Washington, USA, September 2000. International Association of Hydrological Sciences, Publication 264, pp. 165–175.

Aitken, A.E., 1990. Fossilization potential of Arctic fjord and continental shelf macrofaunas. In J.A. Dowdeswell and J.D. Scourse (eds), *Glacimarine environments: processes and sediments.* Geological Society London, Special Publication, 53, pp. 155–176.

Aitken, A.E. and Bell, T.J., 1997. Holocene glacimarine sedimentation and macrofossil palaeoecology in the Canadian high Arctic: environmental controls. *Marine Geology*, 145, 151–171.

Aitken, A.E., Risk, M.J. and Howard, J.D., 1988. Animal-sediment relationships on a subarctic intertidal flat, Pangnirtung Fiord, Baffin Island, Canada. *Journal of Sedimentary Petrology*, 58, 969–978.

Aksu, A.E. and Hiscott, R.N., 1992. Shingled Quaternary debris flow lenses on the north-east Newfoundland Slope. *Sedimentology* 39, 193–206.

Alden, W.C., 1905. Drumlins of Southeastern Wisconsin. *US Geological Survey Bulletin*, 273, 111.

Alexandersson, H., Adrielsson, L., Hjort, C. *et al.*, 2002. Depositional history of the North Taymyr ice-marginal zone, Siberia – a landsystem approach. *Journal of Quaternary Science*, 17, 361–382.

Alley, R.B., 1991a. Deforming-bed origin for southern Laurentide till sheets? *Journal of Glaciology*, 37, 67–77.

Alley, R.B., 1991b. Sedimentary processes may cause fluctuations of tidewater glaciers. *Annals of Glaciology*, 15, 119–124.

Alley, R.B., 1992. How can low-pressure channels and deforming tills coexist subglacially? *Journal of Glaciology*, 38, 200–207.

Alley, R.B., Blankenship, D.D., Bentley, C.R. and Rooney, S.T., 1987. Till beneath ice stream B, 4. A coupled ice-till flow model. *Journal of Geophysical Research* 92, 8931–8940.

Alley, R.B., Blankenship, D.D., Rooney, S.T. and Bentley, C.R., 1989. Sedimentation beneath ice shelves – the view from ice stream B. *Marine Geology*, 85, 101–120.

Alley, R.B., Cuffey, K.M., Evenson, E.B., Strasser, J.C., Lawson, D.E. and Larson, G.J., 1997. How glaciers entrain and transport basal sediment: physical constraints. *Quaternary Science Reviews*, 16, 1017–1038.

Alley, R.B., Lawson, D.E., Evenson, E.B., Strasser, J.C. and Larson, G.J., 1998. Glaciohydraulic supercooling: a freeze-on mechanism to create stratified, debris-rich basal ice. II. Theory. *Journal of Glaciology*, 44, 563–569.

Alley, R.B., Strasser, J.C., Lawson, D.E., Evenson, E.B. and Larson, G.J., 1999. Glaciological and geological implications of basal-ice accretion in overdeepenings. In D.M. Mickelson and J.W. Attig (eds), *Glacial processes: past and present*. Geological Society of America, Special Paper 337, pp. 1–9.

Åmark, M., 1986. Clastic dikes formed beneath an active glacier. *Geologiska Foereningen i Stockholm Foerhandlingar*, 108, 13–20.

Anderson, J.B., 1997. Grounding zone wedges on the Antarctic continental shelf, Weddell Sea. In T.A. Davies, T. Bell, A.K. Cooper, *et al.* (eds), *Glaciated continental margins: an atlas of acoustic images*. Chapman and Hall, London, pp. 98–99.

Anderson, J.B., 1999. *Antarctic marine geology*. Cambridge University Press, Cambridge, 289 pp.

Anderson, J.B., Domack, E.W. and Kurtz, D.D., 1980a. Observation on sediment laden icebergs in Antarctic waters: implication to glacial erosion and transport. *Journal of Glaciology*, 25, 387–396.

Anderson, J.B., Kurtz, D.D., Domack, E.W. and Balshaw, K.M., 1980b. Glacial and glacial marine sediments of the Antarctic continental shelf. *Journal of Geology* 88, 399–414.

Andersson, G., 1998. Genesis of hummocky moraine in the Bolmen area, southwestern Sweden. *Boreas*, 27, 55–67.

André, M.-F., 1997. Holocene rockwall retreat in Svalbard: a triple-rate evolution. *Earth Surface Processes and Landforms*, 22, 423–440.

Andrews, J.A. and Peltier, W.R., 1989. Quaternary geodynamics in Canada. In R.J. Fulton (ed.), *Quaternary geology of Canada and Greenland*. Geological Society of America, The Geology of North America, K-1, pp. 543–572.

Andrews, J.T., 1972. Glacier power, mass balances, velocities and erosion potential. *Zeitschrift für Geomorphologie Suppl. Bd.*, 13, 1–17.

Andrews, J.T., 1973. The Wisconsinan Laurentide ice sheet: dispersal centers and climatic implications. *Arctic and Alpine Research*, 5, 185–199.

Andrews, J.T., 1974. Cainozoic glaciations and crustal movements of the Arctic. In J.D. Ives and R.G. Barry (eds), *Arctic and Alpine environments*. Methuen, London, pp. 277–317.

Andrews, J.T., 1975. *Glacial systems: an approach to glaciers and their environments*. Duxbury Press, North Scituate, Mass.

Andrews, J.T., Jull, A.J.T., Donahue, D.J., Short, S.K., Osterman, L.E., 1985. Sedimentation rates in Baffin Island fiord cores from comparative radiocarbon dates. *Canadian Journal of Earth Sciences*, 22, 1827–1834.

Andrews, J.T., 1987. Late Quaternary marine sediment accumulation in fiord-shelf-deep transects, Baffin Island to Baffin Bay. *Quaternary Science Reviews*, 6, 231–243.

Andrews, J.T., 1989. Quaternary geology of the northeastern Canadian Shield. In R.J. Fulton (ed.), *Quaternary geology of Canada and Greenland*. Geological Survey of Canada, Geology of Canada, No. 1, pp. 276–301.

Andrews, J.T., 1990. Fiord to deep sea sediment transfers along the north-eastern Canadian continental margin: models and data. *Géographie Physique et Quaternaire*, 44, 55–70.

Andrews, J.T., 1998. Abrupt changes (Heinrich events) in late Quaternary North Atlantic marine environments. *Journal of Quaternary Science*, 13, 3–16.

Andrews, J.T. and Sim, V.W., 1964. Examination of the carbonate content of drift in the area of Foxe Basin, N.W.T. *Geographical Bulletin*, 8, 174–193.

Andrews, J.T. and Tedesco, K., 1992. Detrital carbonate-rich sediments, northwestern Labrador Sea: implications for ice-sheet dynamics and iceberg rafting (Heinrich) events in the North Atlantic. *Geology*, 20, 1087–1090.

Andrews, J.T., Milliman, J.D., Jennings, A.E., Rynes, N. and Dwyer, J., 1994. Sediment thickness and Holocene glacial marine sedimentation rates in three Greenland fjords (ca. 68 °N). *Journal of Geology*, 102, 669–683.

Andriashek, L.D. and Fenton, M.M., 1989. *Quaternary stratigraphy and surficial geology of the Sand River Area 73L*. Alberta Research Council, Bulletin 57.

Antevs, E., 1922. The recession of the last ice sheet in new England. *American Geographical Society Research Series* 11, 120 pp.

Antevs, E., 1925. *Retreat of the last ice sheet in eastern Canada*. Geological Survey of Canada, Memoir 146, 142 pp.

Antevs, E., 1951. Glacial clay in Steep Rock Lake, Ontario, Canada. *Geological Society of America Bulletin* 62, 1223–1262.

Armentrout, J.M., 1980. Surface sediments and associated faunas of upper slope, shelf and bay environments, Yakataga-Yakutat area, northern Gulf of Alaska. In: M.E. Field, A.H. Bouma, I.P. Colburn, R.G. Douglas and J.C. Ingle (Eds), *Proceedings of the Quaternary depositional environments of the Pacific Coast. Pacific Coast Paleogeography Symposium*, 4, pp. 241–255.

Arnold, N. and Sharp, M., 1992 Influence of glacier hydrology on the dynamics of a large Quaternary ice sheet. *Journal of Quaternary Science*, 7, 109–124.

Ashley, G.M., 1975 Rythmic sedimentation in glacial Lake Hitchcock, Massachusetts-Connecticutt. In A.V. Jopling and B.C. McDonald (eds), *Glaciofluvial and glaciolacustine sedimentation*. SEPM Special Publication No. 23, pp. 304–320.

Ashley, G.M., 1995. Glaciolacustrine environments. In J. Menzies (ed.), *Modern glacial environments: processes, dynamics and sediments*. Butterworth-Heinemann, Oxford, pp. 417–444.

Ashley, G.M., 2002. Glaciolacustrine environments. In J. Menzies (ed.), *Modern and past glacial environments*. Butterworth-Heinemann, Oxford, pp. 335–359.

Ashley, G.M. and Smith, N.D., 2000. Marine sedimentation at a calving glacier margin. *Geological Society of America Bulletin*, 112, 657–667.

Ashmore, P., 1993. Contemporary erosion of the Canadian landscape. *Progress in Physical Geography*, 17, 190–204.

Attig, J.W., 1993. *Pleistocene geology of Taylor County*. Wisconsin Geological and Natural History Survey Bulletin 90, 25 pp.

Attig, J.W. and Clayton, L., 1993. Stratigraphy and origin of an area of hummocky glacial topography, northern Wisconsin. *Quaternary International*, 18, 61–67.

Attig, J.W., Clayton, L. and Mickelson, D.M., 1985 Correlation of late Wisconsin glacial phases in the western Great Lakes area. *Geological Society of America Bulletin*, 96, 1585–1593.

Attig, J.W., Mickelson, D.M. and Clayton, L., 1989. Late Wisconsin landform distribution and glacier-bed conditions in Wisconsin. *Sedimentary Geology*, 62, 399–405.

Augustinus, P.C., 1995a. Rock mass strength and the stability of some glacial valley slopes. *Zeitschrift für Geomorphologie*, 39, 55–68.

Augustinus, P.C., 1995b. Glacial valley cross-profile development: the influence of in situ rock stress and rock mass strength, with examples from the southern Alps, New Zealand. *Geomorphology*, 14, 87–97.

Aylsworth, J.M. and Shilts W.W., 1989. *Glacial features around the Keewatin ice divide: districts of Mackenzie and Keewatin*. Geological Survey of Canada, Paper 88-4.

Azetsu-Scott, K. and Syvitski, J.M.P., 1999. Influence of melting icebergs on distribution, characteristics and transport of marine particles in an East Greenland fjord. *Journal of Geophysical Research*, 104(C3), 5321–5328.

Azetsu-Scott, K. and Tan, F.C., 1994. Oxygen isotope studies from Iceland to an East Greenland fjord; behaviour of glacial meltwater plume. *Marine Chemistry*, 56, 239–251.

Ballantyne, C.K., 1986. Landslides and slope failures in Scotland: a review. *Scottish Geographical Magazine*, 102, 134–150.
Ballantyne, C.K., 1990. The Holocene glacial history of Lyngshalvöya, northern Norway: chronology and climatic implications. *Boreas*, 19, 93–117.
Ballantyne, C.K., 1995. Paraglacial debris cone formation on recently-deglaciated terrain. *The Holocene*, 5, 25–33.
Ballantyne, C.K., 1997. Holocene rock slope failures in the Scottish Highlands. *Paläoklimaforschung*, 19, 197–205.
Ballantyne, C.K., 2002a. A general model of paraglacial landscape response. *The Holocene*, 12, 371–376.
Ballantyne, C.K., 2002b. Paraglacial geomorphology. *Quaternary Science Reviews*, 18/19, 1935–2017.
Ballantyne, C.K. and Benn, D.I., 1994. Paraglacial slope adjustment and resedimentation following glacier retreat, Fåbergstølsdalen, Norway. *Arctic and Alpine Research*, 26, 255–269.
Ballantyne, C.K. and Benn, D.I., 1996. Paraglacial slope adjustment during recent deglaciation and its implications for slope evolution in formerly glaciated environments. In M.G. Anderson and S. Brooks (eds), *Advances in hillslope processes, Volume 2*. John Wiley and Sons, Chichester, pp. 1173–1195.
Ballantyne, C.K. and Harris, C., 1994. The periglaciation of Great Britain. Cambridge University Press, Cambridge, 330 pp.
Ballantyne, C.K. and Matthews, J.A., 1982. The development of sorted circles on recently-deglaciated terrain, Jotunheimen, Norway. *Arctic and Alpine Research*, 14, 341–354.
Ballantyne, C.K. and Matthews, J.A., 1983. Desiccation cracking and sorted polygon development, Jotunheimen, Norway. *Arctic and Alpine Research*, 15, 339–349.
Ballantyne, C.K. and Whittington, G., 1999. Late Holocene floodplain incision and alluvial fan formation in the central Grampian Highlands, Scotland: chronology, environment and implications. *Journal of Quaternary Science*, 14, 651–671.
Ballantyne, C.K., Stone, J.O. and Fifield, L.K., 1998. Cosmogenic Cl-36 dating of postglacial landsliding at The Storr, Isle of Skye, Scotland. *The Holocene*, 8, 347–351.
Bamber, J.L., 1989. Ice/bed interface and englacial properties of Svalbard ice masses deduced from airborne radio echo-sounding. *Journal of Glaciology*, 35, 30–37.
Bamber, J.L, Vaughan, D.G. and Joughin, I., 2001. Widespread complex flow in the interior of the Antarctic Ice Sheet. *Science*, 287, 1248–1250.
Bammens, E., 1986. *Elster stuwingen Bruine Bank, Noordzee*. Unpublished MSc thesis, University of Amsterdam.
Banerjee, I. and McDonald, B.C., 1975. Nature of esker sedimentation. In A.V. Jopling and B.C. McDonald (eds), *Glaciofluvial and glaciolacustrine sedimentation*. SEPM Special Publication 23, pp. 133–154.
Banham, P.H., 1975. Glacitectonic structures: a general discussion with particular reference to the Contorted Drift of Norfolk. In A.E. Wright and F. Moseley (eds), *Ice ages: ancient and modern*. Seel House Press, Liverpool, pp. 69–94.
Banham, P.H., 1977. Glacitectonites in till stratigraphy. *Boreas*, 6, 101–106.
Barnes, P.W. and Lien, R., 1988. Icebergs rework shelf sediments up to 500 m off Antarctica. *Geology*, 16, 1130–1133.
Barnes, P.W., Reimnitz, E. and Fox, D., 1982. Ice rafting of fine-grained sediment: a sorting and transport mechanism. *Journal of Sedimentary Petrology*, 52, 493–502.
Barnes, P.W., Rawlinson, S.E. and Reimnitz, E., 1988. Coastal geomorphology of Arctic Alaska. In A. Chen and C. Leidersdorf (eds), *Arctic coastal processes and slope projection*. American Society of Civil Engineers, New York, pp. 3–30.
Barnett, D.M., 1970. A re-examination and re-interpretation of tide-gauge data for Churchill, Manitoba. *Canadian Journal of Earth Sciences*, 3, 77–88.
Barnett, P.J., Sharpe, D.R., Russell, H.A.J. *et al.*, 1998. On the origin of the Oak Ridges Moraine. *Canadian Journal of Earth Sciences*, 35, 1152–1167.
Barrett, P.J. and Hambrey, M.J., 1992. Plio-Pleistocene sedimentation in Ferrar Fiord, Antarctica. *Sedimentology*, 39, 109–123.

Barrett, P.J., Pyne A.R. and Ward, B.L., 1983. Modern sedimentation in McMurdo Sound, Antarctic. In R.L. Oliver, P.R. James and J.B. Jago (eds), *Antarctic earth science*. Cambridge University Press, Cambridge, pp. 550–555.

Barrie, J.V. 1980. Iceberg–seabed interaction (northern Labrador Sea). *Annals of Glaciology* 1, 71–76.

Barsch, D., 1987. The problem of the ice-cored rock glacier. In J.R. Giardino, J.F. Shroder and J.D. Vitek (eds), *Rock glaciers*. Allen and Unwin, London, pp. 45–53.

Barsch, D. and Jakob, M., 1998. Mass transport by active rockglaciers in the Khumbu Himalaya. *Geomorphology*, 26, 215–222.

Bart, P.J. and Anderson, J.B., 1995. Seismic record of glacial events affecting the Pacific margin of the northwestern Antarctic Peninsula. *AGU, Antarctic Research Series*, 68: 74–95.

Bart, P.J. and Anderson, J.B., 1997. Grounding zone wedges on the Antarctic continental shelf, Antarctic Peninsula. In T.A. Davies, T. Bell, A.K. Cooper *et al.* (eds), *Glaciated continental margins: an atlas of acoustic images*. Chapman and Hall, London, pp. 96–97.

Bartek, L.R., Vail, P.R., Anderson, J.B., Emmet, P.A. and Wu, S., 1991. Effect of Cenozoic ice sheet fluctuations in Antarctica on the stratigraphic signature of the Neogene. *Journal of Geophysical Research*, 96B, 6753–6778.

Bartsch-Winkler, S. and Ovenshine, A.T., 1984. Macrotidal subarctic environment of Turnagain and Knik Arms, upper Cook Inlet, Alaska: sedimentology of the intertidal zone. *Journal of Sedimentary Petrology*, 54, 1221–1238.

Bartsch-Winkler, S. and Schmoll, H.R., 1984. Bedding types in Holocene tidal channel sequences, Knik Arm, upper Cook Inlet, Alaska. *Journal of Sedimentary Petrology*, 54, 1239–1250.

Batist, M.De, Bart, P.J. and Miller, H., 1997. Trough-mouth fans: Crary Fan, Eastern Weddell Sea, Antarctica. In T.A. Davies, T. Bell, A.K. Cooper *et al.* (eds), *Glaciated continental margins: an atlas of acoustic images*. Chapman and Hall, London, pp. 276–279

Beaudoin, A.B. and King, R.H., 1994. Holocene palaeoenvironmental record preserved in a paraglacial alluvial fan, Sunwapta Pass, Jasper National Park, Alberta, Canada. *Catena*, 22, 227–248.

Bednarski, J., 1988. The geomorphology of glaciomarine sediments in a High arctic fiord. *Géographie physique et Quaternaire*, 42, 65–74.

Bednarski, J., 1998. Quaternary history of Axel Heiberg Island bordering Nansen Sound, NWT, emphasizing the last glacial maximum. *Canadian Journal of Earth Sciences*, 35, 520–533.

Bednarski, J., 2003. *Deglaciation of Bathurst Island, Nunavut*. Geological Survey of Canada, Map 2020a, scale 1:250 000.

Beget, J.E., 1986. Modelling the influence of till rheology on the flow and profile of the Lake Michigan lobe, southern Laurentide ice sheet, U.S.A.. *Journal of Glaciology*, 32, 235–241.

Belderson, R.H., Kenyon, N.H. and Wilson, J.B., 1973. Iceberg plough marks in the northeast Atlantic. *Paleogeography, Paleoclimatology, Paleoecology*, 13, 215–224.

Belknap, D.F. and Shipp, R.C., 1991. Seismic stratigraphy of glacial marine units, Maine inner shelf. In J.B. Anderson and G.M. Ashley (eds), *Glacial marine sedimentation – paleoclimatic significance*. Geological Society America., Special Paper, 261, pp. 137–158.

Bell, R.E., Blankenship, D.D., Finn, C.A. *et al.*, 1998. Influence of subglacial geology on the onset of a West Antarctic ice stream from aerogeophysical observations. *Nature*, 394, 58–62.

Benn, D.I., 1989a. Debris transport by Loch Lomond Readvance glaciers in northern Scotland, basin form and the within-valley asymmetry of lateral moraines. *Journal of Quaternary Science*, 4, 243–254.

Benn, D.I., 1989b. Controls on sedimentation in a Late Devensian ice-dammed lake, Achnasheen, Scotland. *Boreas*, 18, 31–42.

Benn, D.I., 1992. The genesis and significance of 'hummocky moraine': evidence from the Isle of Skye, Scotland. *Quaternary Science Reviews*, 11, 781–799.

Benn, D.I., 1993. Scottish Landform Examples – 9: moraines in Coire na Creiche, Isle of Skye. *Scottish Geographical Magazine*, 109, 187–191.

Benn, D.I., 1994. Fluted moraine formation and till genesis below a temperate glacier: Slettmarkbreen, Jotunheimen, Norway. *Sedimentology*, 41, 279–292.

Benn, D.I., 1995. Fabric signature of subglacial till deformation, Breiðamerkurjökull, Iceland. *Sedimentology*, 42, 735–747.

Benn, D.I. and Evans, D.J.A., 1993. Glaciomarine deltaic deposition and ice-marginal tectonics: the 'Loch Don Sand Moraine', Isle of Mull, Scotland. *Journal of Quaternary Science*, 8, 279–291.

Benn, D.I. and Evans, D.J.A., 1996. The interpretation and classification of subglacially-deformed materials. *Quaternary Science Reviews* 15, 23–52.

Benn, D.I. and Evans, D.J.A., 1998. *Glaciers and glaciation*. Edward Arnold, London, 734 pp.

Benn, D.I. and Ballantyne, C.K., 1994. Reconstructing the transport history of glacigenic sediments: a new approach based on the co-variance of clast form indices. *Sedimentary Geology*, 91, 215–227.

Benn, D.I. and Lehmkuhl, F., 2000. Mass balance and equilibrium-line altitudes of glaciers in high mountain environments. *Quaternary International*, 65/66, 15–29.

Benn, D.I. and Owen, L.A., 2002. Himalayan glacial sedimentary environments: a framework for reconstructing and dating former glacial extents in high mountain regions. *Quaternary International*, 97/98, 3–26.

Benn, D.I., Wiseman, S. and Warren, C.R., 2000. Rapid growth of a supraglacial lake, Ngozumpa Glacier, Khumbu Himal, Nepal. In M. Nakawo, C.F. Raymond and A. Fountain (eds), *Debris-covered glaciers*. IAHS Publication No. 264, IAHS Press, Wallingford, pp. 177–185.

Benn, D.I., Wiseman, S. and Hands, K.A., 2001. Growth and drainage of supraglacial lakes on the debris-mantled Ngozumpa Glacier, Khumbu Himal, Nepal. *Journal of Glaciology*, 47, 626–638.

Bennett, M.R., 1994. Morphological evidence as a guide to deglaciation following the Loch Lomond Readvance: a review of research approaches and models. *Scottish Geographical Magazine*, 110, 24–32.

Bennett, M.R., 1999. Paraglacial and periglacial slope adjustment of a degraded lateral moraine in Glen Torridon. *Scottish Journal of Geology*, 35, 79–83.

Bennett, M.R., 2001. The morphology, structural evolution and significance of push moraines. *Earth Science Reviews*, 53, 197–236.

Bennett, M.R. and Boulton, G.S., 1993. A reinterpretation of Scottish 'hummocky moraine' and its significance for the deglaciation of the Scottish highlands during the Younger Dryas or Loch Lomond Stadial. *Geological Magazine*, 130, 301–318.

Bennett, M.R. and Glasser, M.F., 1991. The glacial landforms of Glen Geusachan, Cairngorms: a reinterpretation. *Scottish Geographical Magazine*, 107, 116–123.

Bennett, M.R., Hambrey, M.J., Huddart, D. and Ghienne, J.F., 1996a. Moraine development at the high-arctic valley glacier Pedersenbreen, Svalbard. *Geografiska Annaler*, 78A, 209–222.

Bennett, M.R., Hambrey, M.J., Huddart, D. and Ghienne, J.F., 1996b. The formation of a geometrical ridge network by the surge-type glacier Kongsvegen, Svalbard. *Journal of Quaternary Science*, 11, 437–449.

Bennett, M.R., Doyle, P. and Mather, A.E., 1996c. Dropstones; their origin and significance. *Palaeogeography, Palaeoclimatology, Palaeoecology*, 121, 331–339.

Bennett, M.R., Hambrey, M.J., Huddart, D. and Glasser, N.F., 1998. Glacial thrusting and moraine-mound formation in Svalbard and Britain: the example of Coire a' Cheud-chnoic (Valley of a Hundred Hills). *Quaternary Proceedings*, 6, 17–34.

Bennett, M.R., Hambrey, M.J., Huddart, D., Glasser, N.F. and Crawford, K.R., 1999. The landform and sediment assemblage produced by a tidewater glacier surge in Kongsfjorden, Svalbard. *Quaternary Science Reviews*, 18, 1213–1246.

Bennett, M.R., Huddart, D., Glasser, N.F. and Hambrey, M.J., 2000a. Resedimentation of debris on an ice-cored lateral moraine in the High-Arctic (Kongsvegen, Svalbard). *Geomorphology*, 35, 21–40.

Bennett, M.R., Huddart, D. and McCormick, T., 2000b. The glaciolacustrine landform-sediment assemblage at Heinabergsjökull, Iceland. *Geografiska Annaler*, 82A, 1–16.

Bennett, M.R., Huddart, D. and Waller, R.I., 2000c. Glaciofluvial crevasse and conduit fills as indicators of supraglacial dewatering during a surge, Skeiðarárjökull, Iceland. *Journal of Glaciology*, 46, 25–34.

Bentley, C.R., 1987. Antarctic ice streams: a review. *Journal of Geophysical Research*, 92 (B9), 8843–8858.

Bergsma, B.M., Svoboda, J. and Freedman, B., 1984. Entombed plant communities released by a retreating glacier at central Ellesmere Island, Canada. *Arctic*, 37, 49–52.

Berthelsen, A., 1978. The methodology of kineto-stratigraphy as applied to glacial geology. *Bulletin of the Geological Society of Denmark*, 27, 25–38.

Berthelsen, A., 1979. Recumbent folds and boudinage structures formed by subglacial shear: an example of gravity tectonics. *Geologie en Mijnbouw*, 58, 253–260.

Bik, M.J.J., 1968. Morphoclimatic observations on prairie mounds. *Zeitschrift für Geomorphologie*, 4, 409–469.

Bird, J.B., 1967. *The physiography of Arctic Canada*. Johns Hopkins Press, Baltimore.

Birkeland, P.W., 1982. Subdivision of Holocene glacial deposits, Ben Ohau Range, New Zealand, using relative-dating methods. *Geological Society of America Bulletin*, 93, 433–449.

Bishop, B.C., 1957. *Shear moraines in the Thule area, northwest Greenland*. US Snow, Ice and Permafrost Research Establishment, Research Report 17.

Bitinas, A., 1992. Peculiarities of formation of flat glaciolacustrine hills. *Proceedings of the 29th International Geological Congress*, Part B, 193–199.

Björck, S., 1995. A review of the history of the Baltic Sea, 13.0–8.0 ka BP. *Quaternary International*, 27, 19–40.

Björck, S., Kromer, B., Johnsen, S. *et al.*, 1996. Synchronized terrestrial-atmospheric deglacial records around the North Atlantic. *Science*, 274, 1155–1160.

Björnsson, H., 1975. Subglacial water reservoirs, jokulhlaups and volcanic eruptions. *Jökull*, 25, 1–12.

Björnsson, H., 1976. Marginal and supraglacial lakes in Iceland. *Jökull*, 26, 40–50.

Björnsson, H., 1992. Jokulhlaups in Iceland: prediction, characteristics and simulation. *Annals of Glaciology*, 16, 95–106.

Björnsson, H., 1996. Scales and rates of glacial sediment removal: a 20km long, 300m deep trench created beneath Breiðamerkurjökull during the Little Ice Age. *Annals of Glaciology*, 22, 141–146.

Björnsson, H., 1998. Hydrological characteristics of the drainage system beneath a surging glacier. *Nature*, 395, 771–774.

Björnsson, H., Gjessing, Y., Hamran, S-E. *et al.*, 1996. The thermal regime of sub-polar glaciers mapped by multi-frequency radio-echo sounding. *Journal of Glaciology*, 42, 23–32.

Black, R.F., 1976a. Periglacial features indicative of permafrost: ice and soil wedges. *Quaternary Research*, 6, 3–26.

Black, R.F., 1976b. Quaternary geology of Wisconsin and contiguous Upper Michigan. In W.H. Mahaney (ed.), *Quaternary stratigraphy of North America*. Dowden, Hutchinson, and Ross, Stroudesburg, pp. 93–117.

Blackadar, R.G., 1958. Patterns resulting from glacier movements north of Foxe Basin, N.W.T. *Arctic*, 11, 157–165.

Blair, R.W., 1994. Moraine and valley wall collapse due to rapid deglaciation in Mount Cook National Park, New Zealand. *Mountain Research and Development*, 14, 347–358.

Blais-Stevens, A., Bornhold, B.D., Kemp, A.E.S., Dean, J.M. and Vaan, A.A., 2001. Overview of Late Quaternary stratigraphy in Saanich Inlet, British Columbia: results of Ocean Drilling Program Leg 169S. *Marine Geology*, 174, 3–20.

Blake, W., Jr., 1963. *Notes on glacial geology, northeastern district of Mackenzie*. Geological Survey of Canada, Paper 63-28.

Blake, W., Jr., 1970. Studies of glacial histories in Arctic Canada I: pumice, radiocarbon dates, and differential postglacial uplift in the eastern Queen Elizabeth Islands. *Canadian Journal of Earth Sciences*, 7, 634–664.

Blake, W., Jr., 1981. Neoglacial fluctuations of glaciers, southeastern Ellesmere Island, Canadian Arctic Archipelago. *Geografiska Annaler*, 63A, 201–218.

Blake, W., Jr., 1989. Application of 14C AMS dating to the chronology of Holocene glacier fluctuations in the High Arctic, with special reference to Leffert Glacier, Ellesmere Island, Canada. *Radiocarbon*, 31, 570–578.

Blake, W., Jr., 1992. Shell-bearing till along Smith Sound, Ellesmere Island-Greenland: age and significance. *Sveriges Geologiska Undersokning*, Series Ca 81, 51–58.

Blankenship, D.D., Bentley, C.R., Rooney, S.T. and Alley, R.B., 1987. Till beneath ice stream B. 1. Properties derived from seismic travel times. *Journal of Geophysical Research*, 92, 8903–8911.

Blatter, H., 1987. On the thermal regime of an arctic valley glacier: a study of White Glacier, Axel Heiberg Island, N.W.T. Canada. *Journal of Glaciology*, 33, 200–211.

Blatter, H. and Hutter, K., 1991. Polythermal conditions in arctic glaciers. *Journal of Glaciology*, 37, 261–269.

Bluemle, J.P., 1974. Early history of Lake Agassiz in southeast North Dakota. *Geological Society of American Bulletin*, 85, 811–814.

Bluemle, J.P., 1993. Hydrodynamic blowouts in North Dakota. In J.S. Aber (ed.), *Glaciotectonics and mapping glacial deposits*. Canadian Plains Research Center, Regina, pp. 259–266.

Bluemle, J.P. and Clayton, L., 1984 Large-scale glacial thrusting and related processes in North Dakota. *Boreas*, 13, 279–299.

Bluemle, J.P., Lord, M.L. and Hunke, N.T., 1993. Exceptionally long, narrow drumlins formed in subglacial cavities, North Dakota. *Boreas*, 22, 15–24.

Bodéré, J.-C., 1977. Les kettles du sud-est de L'islande. *Revue de Géographie Physique et de Géologie Dynamique*, 19, 259–270.

Bond, G.C. and Lotti, R., 1990. Iceberg discharges into the North Atlantic on millennial time scales during the last glaciation. *Science*, 267, 1005.

Bond, G., Heinrich, H., Broecker, W. *et al.*, 1992. Evidence for massive discharges of icebergs into the glacial Northern Atlantic. *Nature*, 360, 245–249.

Boone, S.J. and Eyles, N., 2001. Geotechnical model for great plains hummocky moraine formed by till deformation below stagnant ice. *Geomorphology*, 38, 109–124.

Boothroyd, J.C. and Ashley, G.M., 1975. Processes, bar morphology and sedimentary structures on braided outwash fans, northeastern Gulf of Alaska. In A.V. Jopling and B.C. MacDonald (eds), *Glaciofluvial and glaciolacustrine sedimentation*. Society of Economic Mineralogists and Palaeontologists Special Publication 23, pp. 193–222.

Boothroyd, J.C. and Nummedal, D., 1978. Proglacial braided outwash: a model for humid alluvial-fan deposits. In A.D. Miall (ed.), *Fluvial sedimentology*. Canadian Society for Petroleum Geologists, Memoir 5, pp. 641–668.

Böse, M., 1995. Problems of dead ice and ground ice in the central part of the North European Plain. *Quaternary International*, 28, 123–125.

Bostock, H.S., 1970. Physiographic subdivisions of Canada. In R.J.W. Douglas (ed.), *Geology and economic minerals of Canada*. Geological Survey of Canada, Economic Geology Report No.1, 10-30.

Boulton, G.S., 1967. The development of a complex supraglacial moraine at the margin of Sørbreen, Ny Friesland, Vestspitsbergen. *Journal of Glaciology*, 6, 717–735.

Boulton, G.S., 1968. Flow tills and related deposits on some Vestspitsbergen glaciers. *Journal of Glaciology*, 7, 391–412.

Boulton, G.S., 1970. On the origin and transport of englacial debris in Svalbard glaciers. *Journal of Glaciology*, 9, 213–229.

Boulton, G.S., 1972a. Modern arctic glaciers as depositional models for former ice sheets. *Journal of the Geological Society of London*, 128, 361–393.

Boulton, G.S., 1972b. The role of the thermal regime in glacial sedimentation. In R.J. Price and D.E. Sugden (eds), *Polar geomorphology*. Institute of British Geographers, Special Publication 4, pp. 1–19.

Boulton, G.S., 1975. Processes and patterns of subglacial sedimentation: a theoretical approach. In A.E. Wright and F. Moseley (eds), *Ice ages: ancient and modern*. Seel House Press, Liverpool, pp. 7–42.

Boulton, G.S., 1976. The origin of glacially fluted surfaces: observations and theory. *Journal of Glaciology*, 17, 287–309.

Boulton, G.S., 1978. Boulder shapes and grain-size distributions as indicators of transport paths through a glacier and till genesis. *Sedimentology*, 25, 773–799.

Boulton, G.S., 1979. Processes of glacier erosion on different substrata. *Journal of Glaciology*, 23, 15–38.

Boulton, G.S., 1986. Push moraines and glacier contact fans in marine and terrestrial environments. *Sedimentology*, 33, 677–698.

Boulton, G.S., 1987. A theory of drumlin formation by subglacial deformation. In J. Menzies and J. Rose (eds), *Drumlin symposium*. Balkema, Rotterdam, pp. 25–80.

Boulton, G.S., 1990. Sedimentary and sea level changes during glacial cycles and their control on glacimarine facies architecture. In J.A. Dowdeswell and J.D. Scourse (eds), *Glacimarine environments: processes and sediments*. Geological Society London, Special Publication, 53, pp. 15–52.

Boulton, G.S., 1996a. Theory of glacial erosion, transport and deposition as a consequence of subglacial sediment deformation. *Journal of Glaciology*, 42, 43–62.

Boulton, G.S., 1996b. The origin of till sequences by subglacial sediment deformation beneath mid-latitude ice sheets. *Annals of Glaciology*, 22, 75–84.

Boulton, G.S. and Caban, P.E., 1995. Groundwater flow beneath ice sheets: Part II – its impact on glacier tectonic structures. *Quaternary Science Reviews*, 14, 563–588.

Boulton, G.S. and Dent D.L., 1974. The nature and rates of post-depositional changes in recently deposited till from south-east Iceland. *Geografiska Annaler*, 56A, 121–134.

Boulton, G.S. and Dobbie, K.E., 1998. Slow flow of granular aggregates: the deformation of sediments beneath glaciers. *Philosophical Transactions of the Royal Society of London*, A356, 2713–2745.

Boulton, G.S. and Eyles, N., 1979. Sedimentation by valley glaciers: a model and genetic classification. In C. Schluchter (ed.), *Moraines and varves*. Balkema, Rotterdam, pp. 11–23.

Boulton, G.S. and Hindmarsh, R.C.A., 1987. Sediment deformation beneath glaciers: rheology and sedimentological consequences. *Journal of Geophysical Research* 92, 9059–9082.

Boulton, G.S. and Jones, A.S., 1979. Stability of temperate ice sheets resting on beds of deformable sediments. *Journal of Glaciology*, 24, 29–43.

Boulton, G.S. and Paul, M.A., 1976. The influence of genetic processes on some geotechnical properties of till. *Journal of Engineering Geology*, 9, 159–194.

Boulton, G.S., Smith, G.D., Jones, A.S. and Newsome, J., 1985. Glacial geology and glaciology of the last mid-latitude ice sheets. *Journal of the Geological Society of London*, 142, 447–474.

Boulton, G.S., Caban, P.E. and van Gijssel, K., 1995. Groundwater flow beneath ice sheets: Part I – Large scale patterns. *Quaternary Science Reviews*, 14, 545–562.

Boulton, G.S., van der Meer, J.J.M., Hart, J. *et al.*, 1996. Till and moraine emplacement in a deforming bed surge – an example from a marine environment. *Quaternary Science Reviews*, 15, 961–987.

Boulton, G.S., Van der Meer, J.J.M., Beets, D.J., Hart, J.K. and Ruegg, G.H.J., 1999. The sedimentary and structural evolution of a recent push moraine complex: Holmstömbreen, Spitsbergen. *Quaternary Science Reviews*, 18, 339–371.

Boulton, G.S., Dobbie, K.E. and Zatsepin, S., 2001a. Sediment deformation beneath glaciers and its coupling to the subglacial hydraulic system. *Quaternary International*, 86, 3–28.

Boulton, G.S., Dongelmans, P., Punkari, M. and Broadgate, M., 2001b. Palaeoglaciology of an ice sheet through a glacial cycle: the European ice sheet through the Weichselian. *Quaternary Science Reviews*, 20, 591–625.

Bourne, R., 1931. Regional survey and its relation to stock-taking of the agricultural resources of the British Empire. *Oxford Forestry Memoirs*, 13.

Bovis, M.J., 1990. Rock-slope deformation at Affliction Creek, southern Coast Mountains, British Columbia. *Canadian Journal of Earth Sciences*, 27, 243–254.

Bovis, M.J. and Barry, R.G., 1974 A climatological analysis of north polar desert areas. In T.L. Smiley and J.H. Zumberge (eds), *Polar deserts and modern man*. University of Arizona Press, Tucson, pp. 23–31.

Bowen, D.Q., Richmond, G.M., Fullerton, D.S., Sibrava, V., Fulton, R.J. and Velichko, A.A., 1986. Correlation of Quaternary glaciations in the northern Hemisphere. *Quaternary Science Reviews*, 5, 509–510 and chart 1.

Boyce, J.I. and Eyles, N., 1991. Drumlins carved by deforming till streams below the Laurentide Ice Sheet. *Geology*, 19, 787–790.

Boyd, R., Bowen, A.J. and Hall, R.K., 1987. An evolutionary model for transgressive sedimentation on the eastern shore of Nova Scotia. In D.M. FitzGerald and P.S. Rosen (eds), *Glaciated coasts*. Academic Press, San Diego, pp. 87–114.

Brazier, V., Kirkbride, M.P. and Owens, I.F., 1998. The relationship between climate and rock glacier distribution in the Ben Ohau Range, New Zealand. *Geografiska Annaler*, 80A, 193–207.

Brehmer, A., 1990. En sedimentologisk og geokemisk undersøgelse samt magnetostratigrafisk datering af en sen Mellem-Weichsel issøaflejring I Isteberg lergrav, NV-Sjælland. *Dansk Geologisk Forening, Årsskrift for 1987-1989*, 49–54.

Brodzikowski, K., 1995. Pre-Vistulian glaciotectonic features in southwestern Poland. In J. Ehlers, S. Kozarski and P. Gibbard (eds), *Glacial deposits in north-east Europe*: Balkema, Rotterdam, pp. 339–359.

Brodzikowski, K. and van Loon, A.J., 1987. A systematic classification of glacial and periglacial environments, facies and deposits. *Earth Science Reviews*, 24, 297–381.

Brodzikowski, K. and van Loon, A.J., 1991. *Glacigenic sediments*. Elsevier, Amsterdam.

Broecker, W.S. and Hemming, S., 2001. Climate swings come into focus. *Science*, 294, 2308–2309.

Brooks, G.R., 1994. The fluvial reworking of Late Pleistocene drift, Squamish River drainage basin, southwest British Columbia. *Géographie Physique et Quaternaire*, 48, 51–68.

Broscoe, A.J. and Thompson, S., 1969. Observations on an alpine mudflow, Steele Creek, Yukon. *Canadian Journal of Earth Sciences*, 6, 219–229.

Brown, C.S., Meier, M.F. and Post, A., 1982. *The calving relation of Alaskan tidewater glaciers with application to Columbia Glacier*. US Geological Survey Professional Paper, 1258-C, 13 pp.

Brown, I. and Ward, R., 1996. The influence of topography on snowpatch distribution in southern Iceland: A new hypothesis for glacier formation? *Geografiska Annaler*, 78A, 197–207.

Buchardt, B., Israelson, C., Seaman, P. and Stockmann, G., 2001. Ikaite tuffa towers in Ikka Fjord, southwest Greenland: their formation by mixing of seawater and alkaline spring water. *Journal of Sedimentary Research*, 71, 176–189.

Budd, W.F., 1975. A first simple model of periodically self-surging glaciers. *Journal of Glaciology*, 14, 3–21.

Burn, C.R., 1997. Cryostratigraphy, paleogeography, and climate change during the early Holocene warm interval, western Arctic coast, Canada. *Canadian Journal of Earth Sciences*, 34, 912–925.

Burton, H.R. and Campbell, P.J., 1980. *The climate of the Vestfold Hills, Davis Station, Antarctica, with a note on its effect on the hydrology of hypersaline Deep Lake*. ANARE Scientific Reports, Series D, No. 129, 50 pp.

Byrne, T., 1994. Sediment deformation, dewatering and diagenesis: illustrations from selected mélange zones. In A.J. Maltman, B. Hubbard and M.J. Hambrey (eds), *Deformation of glacial materials*. Geological Society Special Publication no. 176, London, pp. 240–260.

Cai, J., Powell, R.D. and Cowan, E.A., 1995. Sediment accumulation rates estimated from ^{210}Pb dating in a glacimarine system of Disenchantment and Yakutat Bays, Alaska. *EOS*, 76(46), 290.

Cai, J., Powell, R.D., Cowan, E.A. and Carlson, P.R., 1997. Lithofacies and seismic reflection interpretations of temperate glacimarine sedimentation in Tarr Inlet, Glacier Bay, Alaska. *Marine Geology*, 145, 5–37.

Caine, N., 1982. Toppling failures from alpine cliffs on Ben Lomond, Tasmania. *Earth Surface Processes and Landforms*, 7, 133–152.

Cameron, T.D.J., Stoker, M.S. and Long, D., 1987. The history of Quaternary sedimentation in the UK sector of the North Sea Basin. *Journal of the Geological Society, London* 144, 43–58.

Cameron, T.D.J., Schüttenhelm, R.T.E. and Laban, C., 1989. Middle and Upper Pleistocene and Holocene stratigraphy in the southern North Sea between 52° and 54°N, 2° to 4°E. In J.P. Henriet and G. De Moor (eds), *The Quaternary and Tertiary geology of the Southern Bight, North Sea. International Colloquium*. Belgian Geological Survey, Gent, pp. 119–135.

Campy, M., Buoncristiani, J.F. and Bichet, V., 1998. Sediment yield from glacio-lacustrine calcareous deposits during the postglacial period in the Combe d'Ain (Jura, France). *Earth Surface Processes and Landforms*, 23, 429–444.

Canals, M., Urgeles, R. and Calafat, A.M., 2000. Deep sea-floor evidence of past ice streams off the Antarctic Peninsula. *Geology*, 28(1), 31–34.

Carlson, R.F., 1975. A theory of spring river discharge into the Arctic icepac. *Proceedings International Conference on Port and Ocean Engineering under Arctic Conditions*. University of Alaska, pp. 165–166.

Carlson, P.R., 1989. Seismic reflection characteristics of glacial and glacimarine sediment in the Gulf of Alaska and adjacent fjords. *Marine Geology*, 85, 391–416.

Carlson, P.R., Bruns, T.R., Molnia, B.F. and Schwab, W.C., 1982. Submarine valleys in the northeastern Gulf of Alaska: characteristics and probable origin. *Marine Geology*, 47, 217–242.

Carlson, P.R., Powell, R.D. and Rearic, D.M., 1989. Turbidity-current channels in Queen Inlet, Glacier Bay, Alaska. *Canadian Journal of Earth Science*, 26, 807–820.
Carlson, P.R., Powell, R.D. and Phillips, A.C., 1992. Submarine sedimentary features on a fjord delta front, Queen Inlet, Glacier Bay Alaska. *Canadian Journal of Earth Science*, 29, 565–573.
Carlson, P.R., Cowan, E.A., Powell, R.D. and Cai, J., 1999. Growth of a post Little-Ice-Age submarine fan, Glacier Bay, Alaska. *Geo-Marine Letters*, 19, 227–236.
Carney, D., Oliver, J.S. and Armstrong, C., 1999. Sedimentation and composition of wall communities in Alaskan fjords. *Polar Biology*, 22, 38–49.
Carter, R.W.G. and Orford, J.D., 1988. Conceptual model of coarse clastic barrier formation from multiple sediment sources. *Geographical Review*, 78, 221–239.
Carter, R.W.G., Forbes, D.L., Jennings, S.C., Orford, J.D., Shaw, J. and Taylor, R.B., 1989. Barrier and lagoon coast evolution under differing sea-level regimes: examples from Ireland and Nova Scotia. *Marine Geology*, 88, 221–242.
Carter, R.W.G., Orford, J.D., Forbes, D.L. and Taylor, R.B., 1990. Morpho-sedimentary development of drumlin-flank barriers with rapidly-rising sea-level, Story Head, Nova Scotia. *Sedimentary Geology*, 69, 117–128.
Carter, R.W.G., Orford, J.D., Jennings, S.C., Shaw, J. and Smith, J.P., 1992. Recent evolution of a paraglacial estuary under conditions of rapid sea level rise: Chezzetcook Inlet, Nova Scotia. *Proceedings of the Geologists' Association*, 103, 167–185.
Caspers, G., Jordan, H., Merkt, J., Meyer, K.-D., Müller, H. and Streif, H., 1995. III. Niedersachsen. In L. Benda (ed.), *Das Quartär Deutschlands*. Gebrüder Borntraeger, Berlin, pp. 23–58.
Chigira, M., 1992. Long-term gravitational deformation of rocks by mass creep. *Engineering Geology*, 32, 157–184.
Chinn, T.J.H., 1991. Polar margin debris features. *Memorie della Societe Geologica Italiana*, 46, 25–44.
Chinn, T.J.H, 1993. Physical hydrology of the dry valley lakes. *AGU Antarctic Research Series*, 59, 1–51.
Christian, C.S. and Stewart, G.A., 1952. *Summary of General Report on Survey of Katherine-Darwin Region, 1946*. CSIRO, Australia, Land Research Series 1.
Chrobok, S.N. and Nitz, B., 1995. A remarkable series of Late-glacial sediments in the hinterland of the Frankfurt end moraine, north of Berlin. In J. Ehlers, S. Kozarski and P. Gibbard (eds), *Glacial deposits in north-east Europe*. Balkema, Rotterdam, pp. 493–500.
Church, M. and Gilbert, R., 1975. Proglacial fluvial and lacustrine environments. In A.V. Jopling and B.C. McDonald (eds), *Glaciofluvial and glaciolacustrine sedimentation*. SEPM, Special Paper 23, pp. 22–100.
Church, M. and Ryder, J.M., 1972. Paraglacial sedimentation: a consideration of fluvial processes conditioned by glaciation. *Geological Society of America, Bulletin*, 83, 3059–3071.
Church, M. and Slaymaker, O., 1989. Disequilibrium of Holocene sediment yield in glaciated British Columbia. *Nature*, 337, 452–454.
Churski, Z., 1973. Hydrographic features of the proglacial area of Skeidararjökull. *Geographia Polonia*, 26, 209–254.
Clague, J.J., 1986. The Quaternary stratigraphic record of British Columbia – evidence for episodic sedimentation and erosion controlled by glaciation. *Canadian Journal of Earth Sciences*, 23, 885–894.
Clague, J.J. and Evans, S.G., 1994a. *Formation and failure of natural dams in the Canadian Cordillera*. Geological Survey of Canada, Bulletin, 464, 35 pp.
Clague, J.J. and Evans, S.G., 1994b. Historic retreat of Grand Pacific and Melbern Glaciers, Saint Elias Mountains, Canada: an analogue for decay of the Cordilleran ice sheet at the end of the Pleistocene? *Journal of Glaicology*, 39, 619–624.
Clague, J.J. and Evans, S.G., 2000. A review of catastrophic drainage of moraine-dammed lakes in British Columbia. *Quaternary Science Reviews*, 19, 1763–1783.
Clapperton, C.M., 1975. The debris content of surging glaciers in Svalbard and Iceland. *Journal of Glaciology*, 14, 395–406.
Clark, C.D., 1993. Mega-scale glacial lineations and cross-cutting ice-flow landforms. *Earth Surface Processes and Landforms*, 18, 1–29.

Clark, C.D., 1994. Large-scale ice-moulding: A discussion of genesis and glaciofluvial significance. *Sedimentary Geology*, 91, 253–268.
Clark, C.D., 1999. Glaciodynamic context of subglacial bedform generation and preservation. *Annals of Glaciology*, 28, 23–32.
Clark, C.D. and Stokes, C.R., 2001. Extent and basal characteristics of the M'Clintock Channel Ice Stream. *Quaternary International*, 86, 81–101.
Clark, P.U., 1991. Striated clast pavements: products of deforming subglacial sediment? *Geology*, 19, 530–533.
Clark, P.U., 1992. Surface form of the southern Laurentide Ice Sheet and its implications to ice-sheet dynamics. *Geological Society of America Bulletin*, 104, 595–605.
Clark, P.U., 1994. Unstable behavior of the Laurentide Ice Sheet over deforming sediment and its implications for climate change. *Quaternary Research*, 41, 19–25.
Clark, P.U. and Walder, J.S., 1994. Subglacial drainage, eskers, and deformable beds beneath the Laurentide and Eurasian ice sheets. *Geological Society of America Bulletin*, 106, 304–314.
Clark, P.U., Marshall, S., Clarke, G. Licciardi, J. and Teller, J., 2001. Freshwater forcing of abrupt climate change during the last glaciation. *Science*, 293, 283–287.
Clarke, G.K.C., 1982. Glacier outburst flood from 'Hazard lake', Yukon Territory, and the problem of flood magnitude prediction. *Journal of Glaciology*, 28, 3–21.
Clarke, G.K.C., 1987. Subglacial till: a physical framework for its properties and processes. *Journal of Geophysical Research*, 92, 9023–9036.
Clarke, G.K.C. and Blake, E.W., 1991. Geometric and thermal evolution of a surge-type glacier in its quiescent state, Trapridge Glacier, Yukon Territory, Canada. *Journal of Glaciology*, 37, 158–169.
Clarke, G.K.C., Collins S.G. and Thompson D.E., 1984. Flow, thermal structure and subglacial conditions of a surge-type glacier. *Canadian Journal of Earth Sciences*, 21, 232–240.
Clarke, G.K.C., Schmok, J.P., Ommaney, C.S.L. and Collins, S.G., 1986. Characteristics of surge-type glaciers. *Journal of Geophysical Research*, 91, 7165–7180.
Clarke, T.S. and Echelmeyer, K., 1996. Seismic reflection evidence for a deep subglacial trough beneath Jakobshavns Isbræ, West Greenland. *Journal of Glaciology*, 42, 219–232.
Clayton, K.M., 1965. Glacial erosion in the Finger Lakes region New York State, USA. *Zeitschrift für Geomorphologie*, 9, 50–62.
Clayton, L., 1967. *Stagnant-glacier features of the Missouri Coteau in North Dakota*. North Dakota Geological Survey, Miscellaneous Series 30, pp. 25–46.
Clayton, L., Attig, J.W. and Mickelson, D.M., 2001. Effects of late Pleistocene permafrost on the landscape of Wisconsin, USA. *Boreas*, 30, 173–188.
Clayton, L. and Moran, S.R., 1974. A glacial process-form model. In D.R. Coates (ed.), *Glacial geomorphology*. State University of New York, Binghamton, pp. 89–119.
Clayton, L., 1964. Karst topography on stagnant glaciers. *Journal of Glaciology*, 5, 107–112.
Clayton, L. and Attig, J.W., 1997. *Pleistocene Geology of Dane County, Wisconsin*. Wisconsin Natural History and Geological Survey, 64 pp.
Clayton, L. and Cherry, J.A., 1967. *Pleistocene superglacial and ice-walled lakes of west-central North America*. North Dakota Geological Survey, Miscellaneous Series 30, pp. 47–52.
Clayton, L., and Freers, T.F., 1967. *Glacial geology of the Missouri Coteau and adjacent area*. North Dakota Geological Survey, Miscellaneous Series 30, 170 pp.
Clayton, L. and Moran, S.R., 1982. Chronology of late Wisconsin glaciation in middle North America. *Quaternary Science Reviews*, 1, 55–82.
Clayton, L., Laird, W.M., Klassen, R.W. and Kupsch, W.O., 1965. Intersecting minor lineations on Lake Agassiz plain. *Journal of Geology*, 73, 652–656.
Clayton, L., Teller, J.T. and Attig, J.W., 1985. Surging of the southwestern part of the Laurentide Ice Sheet. *Boreas*, 14, 235–241.
Clayton, L., Mickelson, D.M. and Attig, J.W., 1989. Evidence against pervasively deformed bed material beneath rapidly moving lobes of the southern Laurentide Ice Sheet. *Sedimentary Geology*, 62, 203–208.

Clayton, L., Attig, J.W. and Mickelson, D.M., 1999. Tunnel channels formed in Wisconsin during the last glaciation. In D.M. Mickelson and J.W. Attig (eds), *Glaciers past and present.* Geological Society of America Special Paper 337, pp. 69–82.

Clifton, H.E., Hunter, R.E., and Phillips, R.L., 1971. Depositional structures and processes in the non-bared high-energy nearshore. *Journal of Sedimentary Petrology*, 41, 651–670.

Colgan, P.M., 1996. *The Green Bay and Des Moines lobes of the Laurentide ice sheet: evidence for stable and unstable glacier dynamics 18,000 to 12,000 BP.* Unpublished Ph.D. thesis. Madison, University of Wisconsin, 266 pp.

Colgan, P.M. and Mickelson, D.M., 1997. Genesis of streamlined landforms and flow history of the Green Bay lobe, Wisconsin, USA. *Sedimentary Geology*, 111, 7–25.

Colhoun, E.A., Mabin, M.C.G., Adamson, D.A. and Kirk, R.M., 1992. Antarctic ice volume and contribution to sea-level fall at 20,000 yr BP from raised beaches. *Nature*, 358, 316–319.

Colman, S.M., Clark, J.A., Clayton, L., Hansel, A.K., and Larsen, C.E., 1994. Deglaciation, lake levels, and meltwater discharge in the Lake Michigan basin. *Quaternary Science Reviews*, 13, 879–890.

Cooke, R.U. and Doornkamp J.C., 1990. *Geomorphology in Environmental Management.* Oxford University Press, Oxford.

Cooper, A.K., Barrett, P.J., Hinz, K., Traube, V., Leitchenkov, G. and Stagg, H.M.J., 1991. Cenozoic prograding sequences of the Antarctic continental margin: a record of glacio-eustatic and tectonic events. *Marine Geology*, 102, 175–213.

Cooper, A.K., Eittreim, S., ten Brink, U. and Zayatz, I., 1993. Cenozoic glacial sequences of the Antarctic continental margin as recorders of Antarctic ice sheet fluctuations. *AGU, Antarctic Research Series*, 60, 75–89.

Cowan, E.A. and Powell, R.D., 1991a. Suspended sediment transport and deposition of cyclically interlaminated sediment in a temperate glacial fiord. Alaska, USA. In J.A. Dowdeswell and J.D. Scourse (eds), *Glacimarine environments, processes and sediments.* Geological Society of London, Special Publication 53, pp. 75–89.

Cowan, E.A. and Powell, R.D., 1991b. Ice-proximal sediment accumulation rates in a temperate glacial fjord, Southeast Alaska. In J.B. Anderson and G.M. Ashley (eds), *Glacial marine sedimentation – paleoclimatic significance.* Geological Society America, Special Paper, 261, pp. 61–74.

Cowan, E.A., Cai, J., Powell, R.D., Clark, J.D. and Pitcher, J.N., 1997. Temperate glacimarine varves from Disenchantment Bay, Alaska. *Journal of Sedimentary Research*, 67, 536–549.

Cowan, E.A., Cai, J., Powell, R.D., Seramur, K.C. and Spurgeon, V.L., 1998. Modern tidal rhythmites deposited in deep water. *Geo-Marine Letters*, 18, 40–48.

Cowan, E.A., Cai, J., Powell, R.D. and Seramur, K.C., 1999. Tidal and meltwater controls on cyclic rhythmic sedimentation in an Alaskan fjord. *Sedimentology*, 46, 1109–1126.

Coxon, P., Owen, L.A. and Mitchell, W.A., 1996. A late Quaternary catastrophic flood in the Lahul Himalayas. *Journal of Quaternary Science*, 11, 495–510.

Crane, K., Vogt, P.R. and Sundvor, E., 1997. Deep Pleistocene iceberg plowmarks on the Yermak Plateau. In T.A. Davies, T. Bell, A.K. Cooper *et al.* (eds), *Glaciated continental margins: an atlas of acoustic images.* Chapman and Hall, London, pp. 140–141.

Crary, A.P., 1958. Arctic ice islands and ice shelf studies, part 1. *Arctic*, 11, 3–42.

Crary, A.P., 1960 Arctic ice islands and ice shelf studies, part 2. *Arctic*, 13, 32–50.

Croot, D.G., 1988a. Morphological, structural and mechanical analysis of neoglacial ice-pushed ridges in Iceland. In D.G. Croot (ed.), *Glaciotectonics: forms and processes.* Balkema, Rotterdam, pp. 33–47.

Croot, D.G., 1988b. Glaciotectonics and surging glaciers, a correlation based on Vestspitsbergen, Svalbard, Norway. In D.G. Croot (ed.), *Glaciotectonics: forms and processes.* Balkema, Rotterdam, pp. 49–61.

Cruden, D.M. and Hu, X.Q., 1993. Exhaustion and steady-state models for predicting landslide hazards in the Canadian Rocky Mountains. *Geomorphology*, 8, 279–285.

Curry, A.M. and Ballantyne, C.K., 1999. Paraglacial modification of glacigenic sediments. *Geografiska Annaler*, 81A, 409–419.

Curry, A.M., 1999. Paraglacial modification of slope form. *Earth Surface Processes and Landforms*, 24, 1213–1228.

Curry, A.M., 2000a. Observations on the distribution of paraglacial reworking of glacigenic drift in western Norway. *Norsk Geografisk Tidsskrift*, 54, 139–147.

Curry, A.M., 2000b. Holocene reworking of drift-mantled hillslopes in the Scottish Highlands. *Journal of Quaternary Science*, 15, 529–541.

Cutler, P.M., MacAyeal, D.R., Mickelson, D.M., Pariezek, B.R. and Colgan, P.M., 2000. An ice-flow–permafrost model to investigate the paleoglaciology of the southern Laurentide Ice Sheet. *Journal of Glaciology*, 46, 311–325.

Cutler, P.M., Mickelson, D.M., Colgan, P.M., MacAyeal, D.R. and Parizek, B.R., 2001. Influence of the Great Lakes on the dynamics of the southern Laurentide ice sheet: numerical experiments. *Geology*, 29, 1039–1042.

Cutler, P.M., Colgan, P.M. and Mickelson, D.M., 2002. Sedimentologic evidence for outburst floods from the Laurentide Ice Sheet margin in Wisconsin, USA: implications for tunnel-channel formation. *Quaternary International*, 90, 23–40.

Dale, J.E., Aitken, A.E., Gilbert, R. and Risk, M.J., 1989. Macrofauna of Canadian Arctic fjords. *Marine Geology*, 85, 331–358.

Damuth, J.E., 1978. Echo character of the Norwegian-Greenland Sea: relationship to Quaternary sedimentation. *Marine Geology*, 528, 1–36.

Dardis, G.F., Hanvey, P.M. and Coxon, P., 1994. Late-glacial resedimentation of drumlin till facies in Ireland. In W.P. Warren and D.G. Croot (eds), *Formation and deformation of glacial deposits*. Balkema, Rotterdam, pp. 127–137.

Dawber, M. and Powell, R.D., 1998. Epifaunal distributions at marine-ending glaciers: influences of ice dynamics and sedimentation. In C.A. Ricci (ed.), The Antarctic region: geological evolution and processes (Proceedings VII International Symposium on Antarctic Earth Sciences, Siena, Italy, 1995). *Terra Antartica, Siena*, 875–884.

Dawes, P.R., 1987. Topographical and geological maps of Hall Land, north Greenland. *Grønlands Geologiske Undersøgelse* Bulletin 155.

Dawson, A.G., 1979. A Devensian medial moraine in Jura. *Scottish Journal of Geology*, 15, 43–48.

Dayton, P.K., Robbilliard, G.A. and De Vries, A.L., 1969. Anchor ice formation in McMurdo Sound, Antarctica, and its biological effects. *Science*, 163, 273–274.

De Gans, W., Beets, D.J. and Centineo, M.C., 2000. Late Saalian and Eemian deposits in the Amsterdam glacial basin. *Geologie en Mijnbouw*, 79, 147–160.

De Gans, W., De Groot, T. and Zwaan, H., 1987. The Amsterdam basin, a case study of a glacial basin in The Netherlands. In J.J.M. van der Meer (ed.), *Tills and glaciotectonics*. Balkema, Rotterdam, pp. 205–216.

De Geer, G., 1912. A geochronology of the last 12,000 years. 11th International Geological Congress, Stockholm. *Compte Rendu*, 1, 241–258.

De Santis, L., Anderson, J.B., Brancolini, G. and Zayatz, I., 1995. Seismic record of late Oligocene through Miocene glaciation on the central and eastern continental shelf of the Ross Sea. *AGU, Antarctic Research Series*, 68, 235–260.

Dekko, T., 1975. Refleksjonsseismiske undersøkelser i Vestfjorden 1972. *Institutt for kontinentalsokkelundersøkelser*, 77, 47 pp.

Deline, P. 1999a. La mise en place de l'amphithéâtre morainique du Miage (Val Veny, Val d'Aoste). *Géomorphologie: Relief, Processus, Environnement*, 1, 59–72.

Deline, P., 1999b. Les variations Holocènes récentes du Glacier du Miage (Val Veny, Val d'Aoste). *Quaternaire*, 10, 5–13.

Denton, G.H. and Hughes, T.J., 1981. *The last great ice sheets*. John Wiley and Sons, New York.

Denton, G.H., Brockheim, J.C., Wilson, S.C. and Stuiver, M., 1989. Late Wisconsin and early Holocene glacial history, inner Ross Embayment, Antarctica. *Quaternary Research*, 31, 151–182.

Derbyshire, E. and Owen, L.A., 1990. Quaternary alluvial fans in the Karakoram Mountains. In A.H. Rachocki and M. Church (eds), *Alluvial fans: a field approach*. Wiley, Chichester, pp. 27–53.

Derbyshire, E. and Owen, L.A., 1996. Glacioaeolian processes, sediments and landforms. In J. Menzies (ed.), *Glacial environments, 2. Past glacial environments – sediments, forms and techniques*. Butterworth-Heinemann, Oxford, pp. 213–237.

Desloges, J.R., 1994. Varve deposition and sediment yield record at three small lakes of the southern Canadian Cordillera. *Arctic and Alpine Research*, 26, 130–140.

Desloges, J.R. and Gilbert, R.E., 1991. Sedimentary record of Harrison Lake: implications for deglaciation in southwestern British Columbia. *Canadian Journal of Earth Sciences*, 28, 800–815.

Dimakis, P., Elverhøi, A., Høeg, K. *et al.*, 2000. Submarine slope stability on high-latitude glaciated Svalbard-Barents Sea margin. *Marine Geology*, 162, 303–316.

Dionne, J-C., 1981. A boulder-strewn tidal flat, north shore of the Gulf of St. Lawrence, Québec. *Géographie Physique et Quaternaire*, 35, 261–267.

Dionne, J.-C., 1983. Forms, figures and glacial sedimentary facies of muddy tidal flats of cold regions. *Palaeogeography, Palaeoclimatology, Palaeoecology*, 51, 415–451.

Dionne, J.-C., 1984. An estimate of ice-drifted sediments based on the mud content of the ice cover at Montmagny, middle St. Lawrence estuary. *Marine Geology*, 31, 237–241.

Dionne, J.C., 1985. Drift-ice abrasion marks along rocky shores. *Journal of Glaciology*, 31, 237–241.

Dionne, J.-C., 1988. Characteristic features of modern tidal flats in cold regions. In P.L. de Boer, A. van Gelder and S.D. Nio (eds), *Tide-influenced sedimentary environments and facies*. Sedimentology and Petroleum Geology, Utrecht, The Netherlands, pp. 301–332.

Dionne, J.-C., 1998. Sedimentary structures made by shore ice in muddy tidal-flat deposits, St. Lawrence Estuary, Quebec. *Sedimentary Geology*, 116(3–4), 261–274.

Dirszowsky, R.W. and Desloges, J.R., 1997. Glaciolacustrine sediments and neoglacial history of the Chephren Lake Basin, Banff National Park, Alberta. *Géographie Physique et Quaternaire*, 51, 41–53.

Doake, C.S.M., Frolich, R.M., Mantripp, D.R., Smith, A.M. and Vaughan, D.G., 1987. Glaciological studies on Rutford Ice Stream, Antarctica. *Journal of Geophysical Research*, 92, 8951–8960.

Domack, E.W., 1990. Laminated terrigenous sediments from the Antarctic Peninsula: the role of subglacial and marine processes. In J.A. Dowdeswell and J.D. Scourse (eds), *Glacimarine environments: processes and sediments*. Geological Society London, Special Publication, 53, pp. 91–103.

Domack, E.W. and Ishman, S., 1993. Oceanographic and physiographic controls on modern sedimentation within Antarctic fjords. *Geological Society of America Bulletin*, 105, 1175–1189.

Domack, E.W. and McClennen, C.E., 1996. Accumulation of glacial marine sediments in fjords of the Antarctic Peninsula and their use as late Holocene paleoenvironmental indicators. *AGU, Antarctic Research Series*, 70, 135–154.

Domack, E.W. and Williams, C.R., 1990. Fine structure and suspended sediment transport in three Antarctic fjords. *AGU, Antarctic Research Series*, 50, 71–89.

Domack, E.W., Jull, A.J.T. and Nakao, S., 1991. Advance of east Antarctic outlet glaciers during the hypsithermal: implications for the volume state of the Antarctic ice sheet under global warming. *Geology*, 19, 1059–1062.

Domack, E.W., Burkley, L.A., Domack, C.R. and Banks, M.R., 1993a. Facies analysis of glacial marine pebbly mudstones in the Tasmania Basin; implications for regional paleoclimates during the late Paleozoic. In R.H. Findlay, R. Unrug, M.R. Banks and J.J. Veevers (eds), *Assembly, evolution and dispersal*, Proceedings International Gondwana Symposium, 8. Balkema, Rotterdam, pp. 471–484.

Domack, E.W., Mashiotta, T.A. and Burkley, L.A., 1993b. 300-year cyclicity in organic matter preservation in Antarctic fjord sediments. *AGU, Antarctic Research Series*, 60, 265–272.

Domack, E.W., Foss, D.J.P., Syvitski, J.P.M. and McClennen, C.E., 1994. Transport of suspended particulate matter in an Antarctic fjord. *Marine Geology*, 121, 161–170.

Domack, E.W., Jacobson, E.A., Shipp, S. and Anderson, J.B., 1999. Late Pleistocene-Holocene retreat of the West Antarctic ice-sheet system in the Ross Sea; Part 2, Sedimentologic and stratigraphic signature. *Geological Society of America Bulletin*, 111, 1517–1536.

Dongelmans, P., 1996. *Glacial dynamics of the Fennoscandian Ice Sheet: a remote sensing study*. Unpublished Ph.D. thesis, University of Edinburgh, Edinburgh, UK.

Doran, P.T., Priscu, J.C., Lyons, W.B., Walsh, J.E., Fountain, A.G., McKnight, D.M., Moorehead, D.L., Virginia, R.A., Wall, D.H., Clow, G.D., Fritsen, C.H., McKay, C.P. and Parsons, A.N., 2002. Antarctic climate cooling and terrestrial ecosystem response. Nature, 415: 517–520.

Dowdeswell, J.A. and Kenyon, N.H., 1997. Long-range side-scan sonar (GLORIA) imagery of the eastern continental margin of the glaciated Polar North Atlantic. In T.A. Davies, T. Bell, A.K. Cooper *et al.* (eds), *Glaciated continental margins: an atlas of acoustic images*. Chapman and Hall, London, pp. 260–263.

Dowdeswell, J.A. and Murray, T., 1990. Modelling rates of sedimentation from icebergs. In J.A. Dowdeswell and J.D. Scourse (eds), *Glacimarine environments: processes and sediments*. Geological Society of London, Special Publication 53, pp. 121–137.

Dowdeswell, J.A. and Sharp, M.J., 1986. Characterization of pebble fabrics in modern terrestrial glacigenic sediments. *Sedimentology*, 33, 699–710.

Dowdeswell, J.A., Hamilton, G.S. and Hagen, J.O., 1991. The duration of the active phase of surge-type glaciers: contrasts between Svalbard and other regions. *Journal of Glaciology*, 37, 388–400.

Dowdeswell, J.A., Villinger, H., Whittington, R. J. and Marienfeld, P., 1993. Iceberg scouring in Scoresby Sund and on the East Greenland continental shelf. *Marine Geology*, 111, 37–53.

Dowdeswell, J.A., Whittington, R.J. and Marienfeld, P., 1994. The origin of massive diamicton facies by iceberg rafting and scouring, Scoresby Sund, East Greenland. *Sedimentology*, 41, 21–35.

Dowdeswell, J.A., Kenyon, N.H., Elverhøi, A. *et al.*, 1996. Large-scale sedimentation on the glacier-influenced Polar North Atlantic margins: long-range side-scan sonar evidence. *Geophysical Research Letters*, 23, 3535–3538.

Dowdeswell, J.A., Kenyon, N.H. and Laberg, J.S., 1997. The glacier-influenced Scoresby Sund Fan, East Greenland continental margin: evidence from GLORIA and 3.5 kHz records. *Marine Geology*, 143, 207–221.

Dowdeswell, J.A., Elverhøi, A. and Spielhagen, R., 1998. Glacimarine sedimentary processes and facies on the polar North Atlantic margins. *Quaternary Science Reviews*, 17, 243–272.

Dowdeswell, J.A., Andrews, J.T. and Scourse, J.D., 2001. Workshop explores debris transported by icebergs and paleoenvironmental implications. *EOS*, 82, 382.

Dredge, L.A., 1982. Relict ice-scour marks and late phases of Lake Agassiz in northernmost Manitoba. *Canadian Journal of Earth Sciences*, 19, 1079–1087.

Dredge, L.A., 1990. The Melville Moraine: sea level change and response of the western margin of the Foxe Ice Dome, Melville Peninsula, Northwest Territories. *Canadian Journal of Earth Sciences*, 27, 1215–1224.

Dredge, L.A., 1995. *Quaternary Geology of Northern Melville Peninsula, District of Franklin, Northwest Territories*. Geological Survey of Canada, Bulletin 484.

Dredge, L.A., 2000. Carbonate dispersal trains, secondary till plumes, and ice streams in the west Foxe Sector, Laurentide Ice Sheet. *Boreas*, 29, 144–156.

Dredge, L.A., 2002. *Quaternary Geology of Southern Melville Peninsula, Nunavut*. Geological Survey of Canada, Bulletin 561.

Dredge, L.A. and Cowan, W.R., 1989a. Lithostratigraphic record on the Ontario Shield. In R.J. Fulton (ed.), *Quaternary geology of Canada and Greenland*. Geological Survey of Canada, pp. 235–249.

Dredge, L.A. and Cowan, W.R., 1989b. Quaternary geology of the southwestern Canadian Shield. In R.J. Fulton (ed.), *Quaternary geology of Canada and Greenland*. Geological Society of America, The Geology of North America, K-1, pp. 214–235.

Dreimanis, A., 1979. The problem of waterlain tills. In C. Schlüter (ed.), *Moraines and varves*. Balkema, Rotterdam, pp. 167–177.

Dreimanis, A., 1992. Downward injected till wedges and upward injected till dikes. *Sveriges Geologiska Undersoekning, Serie*, Ca 4, 91–96.

Dreimanis, A. and Rappol, M., 1997. Late Wisconsinan sub-glacial clastic intrusive sheets along Lake Erie bluffs, at Bradtville, Ontario, Canada. *Sedimentary Geology*, 111, 1–4.

Driscoll, F.G., Jr., 1980. Wastage of the Klutlan ice-cored moraines, Yukon Territory, Canada. *Quaternary Research*, 14, 31–49.

Drozdowski, E., 1986. Surge moraines. In V. Gardiner (ed.), *International geomorphology Part II*. Wiley, Chichester, pp. 675–692.

Dunbar, M., 1978. Petermann Gletscher: possible source of a tabular iceberg off the coast of Newfoundland. *Journal of Glaciology*, 20, 595–597.

Dunbar, R.B., Leventer, A.R. and Stockton, W.L., 1989. Biogenic sedimentation in McMurdo Sound, Antarctica. *Marine Geology*, 85, 155–180.

Dyke, A.S., 1974. *Deglacial chronology and uplift history of the northeastern sector of the Laurentide Ice Sheet.* Institute of Arctic and Alpine Research, University of Colorado, Occasional Paper 12.

Dyke, A.S., 1978. *Indications of neoglacierization of Somerset Island, District of Franklin.* Geological Survey of Canada Paper, 78-1B, pp. 215–217.

Dyke, A.S., 1979. Glacial and sea level history of southwestern Cumberland Peninsula, Baffin Island, NWT, Canada. *Arctic and Alpine Research*, 11, 179–202.

Dyke, A.S., 1984. *Quaternary geology of Boothia Peninsula and northern district of Keewatin, central Canadian Arctic.* Geological Survey of Canada, Memoir 407.

Dyke, A.S., 1987. A reinterpretation of glacial and marine limits around the northwestern Laurentide Ice Sheet. *Canadian Journal of Earth Sciences*, 24, 591–601.

Dyke, A.S., 1990. *Quaternary Geology of the Francis Lake Map Area, Yukon and Northwest Territories.* Geological Survey of Canada, Memoir 426.

Dyke, A.S., 1993. Landscapes of cold-centred Late Wisconsinan ice caps, Arctic Canada. *Progress in Physical Geography*, 17, 223–247.

Dyke, A.S., 1996. *Preliminary paleogeographic maps of glaciated North America.* Geological Society of Canada Open File 3296. Sheet 6. Isobase maps.

Dyke, A.S., 1999. Last glacial maximum and deglaciation of Devon Island, arctic Canada: Support for an Innuitian Ice Sheet. *Quaternary Science Reviews*, 18, 393–420.

Dyke, A.S., 2000. Holocene delevelling of Devon Island, Arctic Canada: implications for ice sheet geometry and crustal response. *Canadian Journal of Earth Sciences*, 35, 885–904.

Dyke, A.S. and Dredge, L.A., 1989. Quaternary geology of the northwestern Canadian Shield. In R.J. Fulton (ed.), *Quaternary geology of Canada and Greenland.* Geological Survey of Canada, Geology of Canada, No.1, pp. 189–214.

Dyke, A.S. and Hooper, M.J.G., 2001 *Deglaciation of northwest Baffin Island, Nunavut.* Geological Survey of Canada, Map 1999A, scale 1:500 000, with marginal notes, table and 14 figures.

Dyke, A.S. and Morris, T.F., 1988. Drumlin fields, dispersal trains and ice streams in Arctic Canada. *Canadian Geographer*, 32, 86–90.

Dyke, A.S. and Prest, V.K., 1987. Late Wisconsinan and Holocene history of the Laurentide Ice Sheet. *Geographie physique et Quaternaire*, 41, 237–263.

Dyke, A.S. and Savelle, J.M., 2000. Major end moraines of Younger Dryas age on Wollaston Peninsula, Victoria Island, Canadian arctic: implications for paleoclimate and for formation of hummocky moraine. *Canadian Journal of Earth Sciences*, 37, 601–619.

Dyke, A.S., Andrews, J.T. and Miller, G.H., 1982. *Quaternary Geology of Cumberland Peninsula, Baffin Island, District of Franklin.* Geological Survey of Canada, Memoir 403.

Dyke, A.S., Morris, T.F., Green, D.E.C. and England, J., 1992. *Quaternary Geology of Prince of Wales Island, Arctic Canada.* Geological Survey of Canada, Memoir 433, pp. 142.

Dyke, A.S., Andrews, J.T., Clark, P.U. *et al.*, 2002. The Laurentide and Innuitian ice sheets during the Last Glacial Maximum. *Quaternary Science Reviews*, 21, 9–31.

Dyke, A.S., St Onge, D.A. and Savelle, J.M., (in press) *Deglaciation of southwest Victoria Island and adjacent Canadian Arctic Mainland, Nunavut and Northwest Territories.* Geological Survey of Canada, Map, scale 1:500 000, with marginal notes, table and figures.

Eberle, M., 1987. Zur Lockergesteinfülling des St. Galler und Liechtensteiner Rheintales. *Eclogae Geologica Helvetica*, 80, 207–222.

Echelmeyer, K., Clarke, T.S. and Harrison, W.D., 1991. Surficial glaciology of Jakobshavns Isbræ, West Greenland: Part I. Surface morphology. *Journal of Glaciology*, 37(127), 368–382.

Echelmeyer, K.A. and Harrison, W.D., 1990. Jakobshavns Isbræ, West Greenland: seasonal variations in velocity – or lack thereof. *Journal of Glaciology*, 36(122), 82–88.

Echelmeyer, K.A., Harrison, W.D. and Mitchell, J.E., 1994. The role of the margins in the dynamics of an active ice stream. *Journal of Glaciology*, 40, 527–538.

Edlund, S.A. and Alt, B.T., 1989. Regional congruence of vegetation and summer climate patterns in the Queen Elizabeth Islands, Canada. *Arctic*, 42, 3–22.

Edlund, S.A., 1985. *Lichen-free zones as neoglacial indicators on Western Melville Island, District of Franklin.* Geological Survey of Canada, Paper 85-1, 709–712.

Ehlers, J., 1990. *Untersuchungen zur Morphodynamik der Vereisungen Norddeutschlands unter Berücksichtigung benachbarter Gebiete.* Universität Bremen, Bremen, 166 pp.

Ehlers, J., 1996. *Quaternary and Glacial Geology.* John Wiley and Sons, Chichester, 578 pp.

Ehlers, J. and Stephan, H.-J., 1983. Till fabric and ice movement. In J. Ehlers (ed.), *Glacial deposits in north-west Europe.* Balkema, Rotterdam, pp. 267–274.

Ehlers, J. and Wingfield, R., 1991. The extension of the Late Weichselian/Late Devensian ice sheets in the North Sea Basin. *Journal of Quaternary Science*, 6, 313–326

Ehlers, J., Kozarski, S. and Gibbard, P., 1995. Glacial deposits in North-East Europe: general overview. In J. Ehlers, S. Kozarski and P. Gibbard (eds.), *Glacial deposits in north-east Europe.* Balkema, Rotterdam, pp. 547–552.

Eissmann, L. 1994. Grundzüge der Quartärgeologie Mitteldeutschlands (Sachsen, Sachsen-Anhalt, Südbrandenburg, Thüringen). In, Eissmann, L. and Litt, T. (eds.), *Das Quartär Mitteldeutschlands* (The Quaternary in Central Germany), Volume 7. Altenburger Naturwissenschaftliche Forschungen. Altenburg, Naturkundliches Museum, 55-135.

Eissmann, L. 1995. The pre-Elsterian Quaternary deposits of central Germany. In, Ehlers, J., Kozarski, S. and Gibbard, P. (eds.), *Glacial Deposits in North-East Europe.* Balkema, Rotterdam, 423-437.

Eissmann, L., Litt, T. and Wansa, S., 1995. Elsterian and Saalian deposits in their type areas in central Germany. In J. Ehlers, S. Kozarski and P. Gibbard (eds.), *Glacial deposits in north-east Europe.* Balkema, Rotterdam, pp. 439–464.

Ekberg, M.P., Lowell, T.V. and Stuckenrath, R., 1993. Late Wisconsin glacial advance and retreat patterns in southwestern Ohio, USA. *Boreas*, 22, 189–204.

Eklund, A and Hart, J.K., 1996. Glaciotectonic deformation within a flute from the Isfallsglaciären, Sweden. *Journal of Quaternary Science*, 11, 299–310.

Ekman, I.M., Ilyin, V.A. and Lukashov, A.D., 1981. Degradation of the late glacial sheet on the territory of the Karelian ASSR. In *Glacial Deposits and Glacial History in Eastern Fennoscandia.* Academy of Sciences USSR, pp. 103–117.

Elson, J.A., 1957. Origin of washboard moraines. *Geological Society of America Bulletin*, 68, 1721.

Elson, J.A., 1981. Deformation till. INQUA Commission on genesis and lithology of Quaternary deposits, Work group 1, Genetic classification of tills and criteria for their recognition, Circular 20.

Elverhøi, A.., Liestøl, O. and Nagy, J., 1980. Glacial erosion, sedimentation and microfauna in the inner part of Kongsfjorden, Spitsbergen. *Norsk Polarinstitutt Skifter*, 172, 33–58.

Elverhøi, A., Lønne, Ø. and Seland, R., 1983. Glaciomarine sedimentation in a modern fjord environment, Spitsbergen. *Polar Research*, 1, 127–149.

Elverhøi, A., Norem, H., Andersen, E.S. *et al.*, 1997. On the origin and flow behaviour of submarine slides on deep sea fans along the Norwegian-Barents Sea continental margin. *Geo-Marine Letters*, 17, 119–125.

Elverhøi, A., Hooke, R.Le B. and Solheim, A., 1998. Late Cenozoic erosion and sediment yield from the Svalbard-Barents Sea region: implications for understanding erosion of glacierised basins. *Quaternary Science Reviews*, 17(1-3), 209–242.

Elverhøi, A., Harbitz, C.B., Dimakis, P., Mohrig, D. and Marr, J., 2000. On the dynamics of subaqueous debris flows. *Oceanography*, 13, 109–117.

Engelhardt, H.F. and Kamb, B., 1998 Basal sliding of ice stream B, West Antarctica. *Journal of Glaciology*, 44, 223–230.

England, J., 1978. The glacial geology of northeastern Ellesmere Island, NWT, Canada. *Canadian Journal of Earth Sciences*, 15, 603–617. England, J., 1983. Isostatic adjustments in a full glacial sea. *Canadian Journal of Earth Sciences*, 20, 895–917.

England, J., 1985. The late Quaternary history of Hall Land, northwest Greenland. *Canadian Journal of Earth Sciences*, 22, 1394–1408.

England, J., 1986. Glacial erosion of a high arctic valley. *Journal of Glaciology*, 32, 60–64.

England, J., 1987a. Glaciation and the evolution of the Canadian high arctic landscape. *Geology*, 15, 419–424.

England, J., 1987b. The late Quaternary history of Hall Land, northwest Greenland. Reply. *Canadian Journal of Earth Sciences*, 24, 374–380.

England, J., 1990. The late Quaternary history of Greely Fiord and its tributaries, west-central Ellesmere Island. *Canadian Journal of Earth Sciences*, 27, 255–270.

England, J., 1999. Coalescent Greenland and Innuitian ice during the last glacial maximum: Revising the Quaternary of the Canadian High Arctic. *Quaternary Science Reviews*, 18, 421–456.

England, J., Bradley, R.S. and Miller, G.H., 1978. Former ice shelves in the Canadian high arctic. *Journal of Glaciology*, 20, 393–404.

England, J., Smith, I.R. and Evans, D.J.A., 2000. The last glaciation of east-central Ellesmere Island, Nunavut: ice dynamics, deglacial chronology, and sea level change. *Canadian Journal of Earth Sciences*, 37, 1355–1371.

Ensminger, S.L., Alley R.B., Evenson E.B., Lawson D.E. and Larson, G.J., 2001. Basal crevasse fill origin of laminated debris bands at Matanuska Glacier, Alaska, USA. *Journal of Glaciology*, 47, 412–422.

Eronen, M., 1983. Late Weichselian and Holocene shore displacement in Finland. In D.E. Smith and A.G. Dawson (eds), *Shorelines and Isostasy*. Academic Press, London, pp. 183–208.

Etzelmüller, B., 2000. Quantification of thermo-erosion in pro-glacial areas – examples from Svalbard. *Zeitschrift für Geomorphologie*, 44, 343–361.

Etzelmüller, B., Hagen, J.O., Vatne, G., Odegard, R.S. and Sollid, J.L., 1996. Glacier debris accumulation and sediment deformation influenced by permafrost: examples from Svalbard. *Annals of Glaciology*, 22, 53–62.

Evans, D.J.A., 1989a. Apron entrainment at the margins of sub-polar glaciers, northwest Ellesmere Island, Canadian high arctic. *Journal of Glaciology*, 35, 317–324.

Evans, D.J.A., 1989b. The nature of glacitectonic structures and sediments at sub-polar glacier margins, northwest Ellesmere Island, Canada. *Geografiska Annaler*, 71A, 113–123.

Evans, D.J.A., 1990a. The effect of glacier morphology on surficial geology and glacial stratigraphy in a high arctic mountainous terrain. *Zeitschrift für Geomorphologie*, 34, 481–503.

Evans, D.J.A., 1990b. The last glaciation and relative sea level history of northwest Ellesmere Island, Canadian high arctic. *Journal of Quaternary Science*, 5, 67–82.

Evans, D.J.A., 1993. High latitude rock glaciers: a case study of forms and processes in the Canadian arctic. *Permafrost and Periglacial Processes*, 4, 17–35.

Evans, D.J.A., 1999: Glacial debris transport and moraine deposition: a case study of the Jardalen cirque complex, Sogn-og-Fjordane, western Norway. *Zeitschrift für Geomorphologie*, 43, 203–234.

Evans, D.J.A., 2000a. A gravel outwash/deformation till continuum, Skalafellsjökull, Iceland. *Geografiska Annaler*, 82A, 499–512.

Evans, D.J.A., 2000b. Quaternary geology and geomorphology of the Dinosaur Provincial Park area and surrounding plains, Alberta, Canada: the identification of former glacial lobes, drainage diversions and meltwater flood tracks. *Quaternary Science Reviews*, 19, 931–958.

Evans, D.J.A., 2003. Ice marginal terrestrial landsystems: active temperate glacier margins. In D.J.A. Evans (ed.), *Glacial landsystems*. Arnold, London.

Evans, D.J.A., (in press). The glacier-marginal landsystems of Iceland. In C.J. Caseldine, A.J. Russell, O. Knudsen and J. Hardardottir (eds), *Iceland: modern processes and past environments*. Elsevier, Amsterdam.

Evans, D.J.A. and England, J., 1991. Canadian Landform Examples 19: high arctic thrust block moraines. *Canadian Geographer*, 35, 93–97.

Evans, D.J.A. and England, J., 1992. Geomorphological evidence of Holocene climatic change from northwest Ellesmere Island, Canadian high arctic. *The Holocene*, 2, 148–158.

Evans, D.J.A. and Rea, B.R., 1999. Geomorphology and sedimentology of surging glaciers: a landsystems approach. *Annals of Glaciology*, 28, 75–82.

Evans, D.J.A. and Twigg, D.R., 2000. *Breiðamerkurjökull 1998*. 1:30,000 Scale Map. University of Glasgow and Loughborough University.

Evans, D.J.A. and Twigg, D.R., 2002. The active temperate glacial landsystem: a model based on Breiðamerkurjökull and Fjallsjökull, Iceland. *Quaternary Science Reviews*, 21, 2143–2177.

Evans, D.J.A., Owen, L.A. and Roberts, D., 1995. Stratigraphy and sedimentology of Devensian (Dimlington Stadial) glacial deposits, east Yorkshire, England. *Journal of Quaternary Science*, 10, 241–265.

Evans, D.J.A., Rea, B.R. and Benn, D.I., 1998. Subglacial deformation and bedrock plucking in areas of hard bedrock. *Glacial Geology and Geomorphology* rp04/1998 – http://ggg.qub.ac.uk/ggg/papers/full/1998/rp041998/rp04.html

Evans, D.J.A., Archer, S. and Wilson, D.J.H., 1999a. A comparison of the lichenometric and Schmidt hammer dating techniques based on data from the proglacial areas of some Icelandic glaciers. *Quaternary Science Reviews*, 18, 13–41.

Evans, D.J.A., Lemmen, D.S. and Rea, B.R., 1999b. Glacial landsystems of the southwest Laurentide Ice Sheet: modern Icelandic analogues. *Journal of Quaternary Science*, 14, 673–79.

Evans, D.J.A., Salt, K.E. and Allen, C.S., 1999c. Glacitectonized lake sediments, Barrier Lake, Kananaskis Country, Canadian Rocky Mountains. *Canadian Journal of Earth Sciences*, 36, 395–407.

Evans, D.J.A., Rea, B.R., Hansom, J.D. and Whalley, W.B., 2002. The geomorphology and style of plateau icefield glaciation in a fjord terrain, Troms-Finnmark, north Norway. *Journal of Quaternary Science*. 17, 221–239.

Evans, I.S., 1996. Abraded rock landforms (whalebacks) developed under ice streams in mountain areas. *Annals of Glaciology*, 22, 9–15.

Evans, S.G. and Clague, J.J., 1994. Recent climatic change and catastrophic geomorphic processes in mountain environments. *Geomorphology*, 10, 107–128.

Evenson, E. B., Dreimanis, A. and Newsome, J.W., 1977. Subaquatic flow tills: a new interpretation for the genesis of some laminated till deposits. *Boreas*, 6, 115–133.

Evenson, E.B., Lawson, D.E., Strasser, J.C. *et al.*, 1999. Field evidence for the recognition of glaciohydrologic supercooling. In D.M. Mickelson and J.W. Attig (eds), *Glacial processes past and present*. Geological Society of America Special Paper 337, pp. 23–35.

Eyles, C.H., Eyles N. and Miall, A.D., 1985. Models of glaciomarine sedimentation and their application to the interpretation of ancient glacial sequences. *Palaeogeography, Palaeoclimatology, Palaeoecology*, 51, 15–84.

Eyles, C.H., Eyles, N. and Lagoe, M.B., 1991a. The Yakataga Formation; A late Miocene to Pleistocene record of temperate glacial marine sedimentation in the Gulf of Alaska. In: J.B. Anderson and G.M. Ashley (eds), *Glacial marine sedimentation; paleoclimatic significance*. Geological Society America, Special Paper 261, pp. 159–180.

Eyles, N., 1979. Facies of supraglacial sedimentation on Icelandic and alpine temperate glaciers. *Canadian Journal of Earth Sciences*, 16, 1341–1361.

Eyles, N., 1983a. Glacial geology: a landsystems approach. In N. Eyles (ed.), *Glacial geology*. Pergamon, Oxford, pp. 1–18.

Eyles, N., 1983b. The glaciated valley landsystem. In N. Eyles (ed.), *Glacial geology*. Pergamon, Oxford, pp. 91–110.

Eyles, N., 1983c. Modern Icelandic glaciers as depositional models for 'hummocky moraine' in the Scottish Highlands. In E.B. Evenson (ed.), *Tills and related deposits*. Balkema, Rotterdam, pp. 47–59.

Eyles, N., 1987. Late Pleistocene debris flow deposits in large glacial lakes in British Columbia and Alaska. *Sedimentary Geology*, 53, 33–71.

Eyles, N., 1993. Earth's glacial record and its tectonic setting. *Earth Science Reviews*, 35, 1–248.

Eyles, N. and Dearman, W.R., 1981. A glacial terrain map of Britain for engineering purposes. *Bulletin of the International Association of Engineering Geology*, 24, 173–184.

Eyles, N., and Eyles, C.H., 1992. Glacial depositional systems. In R.G. Walter and N.P. James (eds), *Facies models, response to sea-level change*. Geological Association of Canada, Toronto, pp. 73–100.

Eyles, N. and Kocsis, S., 1988. Sedimentology and clast fabric of subaerial debris flow facies in a glacially-influenced alluvial fan. *Sedimentary Geology*, 59, 15–28.

Eyles, N. and Kocsis, S., 1989. Reply to M. Church and J.M. Ryder's discussion of 'Sedimentology and clast fabrics of subaerial debris flow facies in a glacially-influenced alluvial fan'. *Sedimentary Geology*, 59, 15–28.

Eyles, N. and McCabe, A.M., 1991. Glaciomarine deposits of the Irish Sea Basin: the role of glacioisostatic disequilibrium. In J. Ehlers, P.L. Gibbard and J. Rose (eds), *Glacial deposits in Great Britain and Ireland.* Balkema, Rotterdam, pp. 311–331.

Eyles, N. and Menzies, J., 1983. The subglacial landsystem. In N. Eyles (ed.), *Glacial geology.* Pergamon, Oxford, pp. 19–70.

Eyles, N. and Rogerson, R.J., 1978. A framework for the investigation of medial moraine formation: Austerdalsbreen, Norway, and Berendon Glacier, British Columbia, Canada. *Journal of Glaciology*, 20, 99–113.

Eyles, N., Sladen, H.A. and Gilroy, S., 1982. A depositional model for stratigraphic complexes and facies superimposition in lodgement till. *Boreas*, 11, 317–333.

Eyles, N., Dearman, W.R. and Douglas, T.D., 1983a. The distribution of glacial landsystems in Britain and North America. In N. Eyles (ed.), *Glacial geology.* Pergamon, Oxford, pp. 213–228.

Eyles, N., Eyles C.H. and A.D. Miall, 1983b. Lithofacies types and vertical profile models; an alternative approach to the description and environmental interpretation of glacial diamict and diamictite sequences. *Sedimentology*, 30, 393–410.

Eyles, N., Clark, B.M. and Clague, J.J., 1987. Coarse-grained sediment-gravity flow facies in a large supraglacial lake. *Sedimentology*, 34, 193–216.

Eyles, N., Eyles, C.H. and McCabe, A.M., 1988. Late Pleistocene subaerial debris-flow facies of the Bow Valley, near Banff, Canadian Rocky Mountains. *Sedimentology*, 35, 465–480.

Eyles, N., Mullins, H.T. and Hine, A.C., 1990. Thick and fast: sedimentation in a Pleistocene fiord lake of British Columbia, Canada. *Geology*, 18, 1153–1157.

Eyles, N., Mullins, H.T. and Hine, A.C., 1991b. The seismic stratigraphy of Okanagan Lake: a record of rapid deglaciation in a deep 'fiord lake' basin. *Sedimentary Geology*, 73, 13–41.

Eyles, N., Vossler, S.M. and Lagoe, M.B., 1992. Ichnology of a glacially influenced continental shelf and slope: the Late Cenozoic Gulf of Alaska (Yakataga Formation). *Palaeogeography, Palaeoecology, Palaeoclimatology*, 94, 193–221.

Eyles, N., McCabe, A.M. and Bowen, D.Q., 1994. The stratigraphic and sedimentological significance of Late Devensian ice-sheet surging in Holderness, Yorkshire, UK. *Quaternary Science Reviews*, 13, 727–759.

Eyles, N., Boyce, J.I. and Barendregt, R.W., 1999a. Hummocky moraine: sedimentary record of stagnant Laurentide Ice Sheet lobes resting on soft beds. *Sedimentary Geology*, 123, 163–174.

Eyles, N., Boyce, J.I. and Barendregt, R.W., 1999b. Hummocky moraine: sedimentary record of stagnant Laurentide Ice Sheet lobes resting on soft beds – reply. *Sedimentary Geology*, 129, 169–172.

Fader, B.J., Stea, R.R. and Courtney, R.C., 1997. A seabed drumlin field on the inner Scotian shelf, Canada. In T.A. Davies, T. Bell, A.K. Cooper *et al.* (eds), *Glaciated continental margins: an atlas of acoustic images.* Chapman and Hall, London, pp. 98–99.

Fairchild, H.L., 1907. Drumlin structure and origin. *Geological Society of America Bulletin*, 17, 702–706.

Falconer, G., Ives J.D., Loken O.H. and Andrews J.T., 1965. Major end moraines in eastern and central Arctic Canada. *Geographical Bulletin*, 7, 137–153.

Falconer, G., 1966. Preservation of vegetation and patterned ground under a thin ice body in Northern Baffin Island, N.W.T. *Geographical Bulletin*, 8, 194–200.

Fastook, J.L., 1987. Use of a new finite element continuity model to study the transient behaviour of Ice Stream C and causes of its present low velocity. *Journal of Geophysical Research*, 92, 8941–8949.

Fastook, J.L., Brecher, H.H. and Hughes, T.J., 1995. Derived bedrock elevations, strain rates and stresses from measured surface elevations and velocities: Jakobshavns Isbræ, Greenland. *Journal of Glaciology*, 41(137), 161–173.

Fay, H., (2002) Formation of ice block obstacle marks during the November 1996 glacier-outburst flood (jökulhlaup), Skeiðarársandur, southern Iceland. In I.P. Martini, V.R. Baker and G. Garzon (eds), *Flood and megaflood deposits: recent and ancient.* Special Publication of the International Association of Sedimentologists.

Fenton, M.M., Moran, S.R., Teller, J.T. and Clayton, L., 1983. Quaternary stratigraphy and history in the southern part of the Lake Agassiz basin. In J.T. Teller and L. Clayton (eds), *Glacial Lake Agassiz.* Geological Association of Canada, Special Paper 26, pp. 49–74.

Ferrigno, J.G., Williams, R.S., Rosanova, C.E., Lucchitta, B.K. and Swithibank, C., 1998. Analysis of coastal change in Marie Byrd Land and Ellsworth Land, West Antarctica, using Landsat imagery. *Annals of Glaciology*, 27, 33–40.

Fischer, M.P. and Powell, R.D., 1998. A simple model for the influence of push morainal banks on the calving and stability of glacial tidewater termini. *Journal of Glaciology*, 44, 31–41.

Fisher, D.A., Koerner R.M., Bourgeois J.C. *et al.*, 1998. Penny Ice Cap cores, Baffin Island, Canada, and the Wisconsinan Foxe Dome connection: two states of Hudson Bay ice cover. *Science*, 279, 692–695.

Fisher, D.A., Reeh, N. and Langley, K., 1985. Objective reconstructions of the Late Wisconsinan Ice Sheet. *Géographie Physique et Quaternaire*, 39, 229–238.

FitzGerald, D.M. and Van Heteren, S., 1999. Classification of paraglacial barrier systems: coastal New England, USA. *Sedimentology*, 46, 1083–1108.

Fitzsimons, S.J., 1990. Ice marginal depositional processes in a polar maritime environment, Vestfold Hills, Antarctica. *Journal of Glaciology*, 36, 279–286.

Fitzsimons, S.J., 1996a. Formation of thrust-block moraines at the margins of dry-based glaciers, south Victoria Land, Antarctica. *Annals of Glaciology*, 22, 68–74.

Fitzsimons, S.J., 1996b. Paraglacial redistribution of glacial sediments in the Vestfold Hills, East Antarctica. *Geomorphology*, 15, 93–108.

Fitzsimons, S.J., 1997a. Glaciotectonic deformation of glaciomarine sediments and formation of thrust-block moraines at the margin of an outlet glacier, Vestfold Hills, East Antarctica. *Earth Surface Processes and Landforms*, 22, 175–187.

Fitzsimons, S.J., 1997b. Depositional models for moraines in east Antarctic coastal oases. *Journal of Glaciology*, 43, 256–264.

Fitzsimons, S.J. and Domack, E.W., 1993. Evidence for early Holocene deglaciation of the Vestfold Hills, Antarctica. *Polar Record*, 29, 237–240.

Fitzsimons, S.J., McManus, K.J. and Lorrain, R., 1999. Structure and strength of basal ice and substrate of a dry based glacier: evidence for substrate deformation at subfreezing temperatures. *Annals of Glaciology*, 28, 236–240.

Flint, R.F., 1971. *Glacial and Quaternary geology*. John Wiley and Sons, New York, 892 pp.

Florin, B.-B. and Wright, H.E., 1969. Diatom evidence for the persistence of stagnant glacial ice in Minnesota. *Geological Society of America Bulletin*, 80, 695–704.

Fookes, P.G., Gordon D.L. and Higginbottom, I.E., 1978. Glacial landforms, their deposits and engineering characteristics. In *The engineering behaviour of glacial materials*. Proceedings of Symposium, University of Birmingham. Geoabstracts: 18-51.

Forbes, D.L., 1984. *Coastal geomorphology and sediments of Newfoundland*. Geological Survey of Canada, Paper 84-1B, pp. 11–24.

Forbes, D.L. and Syvitski, J.P.M., 1994. Paraglacial coasts. In R.W.G. Carter and C.D. Woodroffe (eds.), *Coastal evolution: Late Quaternary shoreline morphodynamics*. Cambridge University Press, Cambridge, pp. 373–424.

Forbes, D.L. and Taylor, R.B., 1987. Coarse-grained beach sedimentation under paraglacial conditions, Canadian Atlantic coast. In D.M. Fitzgerald and P.S. Rosen (eds.), *Glaciated coasts*. Academic Press, San Diego, pp. 51–86.

Forbes, D.L., Orford, J.D., Carter, R.W.G., Shaw, J. and Jennings, S.C., 1995a. Morphodynamic evolution, self-organization, and instability of coarse-clastic barriers on paraglacial coasts. *Marine Geology*, 126, 63–85.

Forbes, D.L., Shaw, J. and Taylor, R.B., 1995b. Differential preservation of coastal structures on paraglacial shelves: Holocene deposits of southeastern Canada. *Marine Geology*, 124, 187–201.

Foster, J.D. and Palmquist, R.C., 1969. Possible subglacial origin for 'minor moraine' topography. *Proceedings of the Iowa Academy of Science*, 76, 296–310.

Fountain, A.G., Dana, G.L., Lewis, K.J., Vaughn, B.H. and McKnight, D., 1998. Glaciers of the McMurdo dry valleys, southern Victoria Land, Antarctica. In J.C. Priscu (ed.), *Ecosystem dynamics in a polar desert*. American Geophysical Union, Antarctic Research Series 72, 65–75.

Fowler, A.C., 1987. A theory of glacier surges. *Journal of Geophysical Research*, 92, 9111–9120.

French, H.M. and Harry, D.G., 1988. Nature and origin of ground ice, Sandhills Moraine, southwest Banks Island, Western Canadian Arctic. *Journal of Quaternary Science*, 3, 19–30.

French, H.M. and Harry, D.G., 1990. Observations on buried glacier ice and massive segregated ice, western arctic coast, Canada. *Permafrost and Periglacial Processes*, 1, 31–43.

Friedman, G.M. and Sanders, J E., 1978. *Principles of Sedimentology*. John Wiley and Sons, New York.

Friele, P.A., Ekes, C. and Hicken, E.J., 1999. Evolution of Cheekye fan, Squamish, British Columbia: Holocene sedimentation and implications for hazard assessment. *Canadian Journal of Earth Sciences*, 36, 2023–2031.

Frolich, R.M. and Doake, C.S.M., 1988. Relative importance of lateral and vertical shear on Rutford Ice Stream, Antarctica. *Annals of Glaciology*, 11, 19–22.

Funder, S., 1989. Quaternary geology of the ice-free areas and adjacent shelves of Greenland. In R.J. Fulton (ed.), *Quaternary geology of Canada and Greenland*. Geological Survey of Canada, Geology of Canada no.1, pp. 743–792.

Furbish, D.J. and Andrews, J.T., 1984. The use of hypsometry to indicate long-term stability and response of valley glaciers to changes in mass transfer. *Journal of Glaciology*, 30, 199–211.

Fyfe, G.J., 1990. The effect of water depth on ice-proximal glaciolacustrine sedimentation: Salpausselkä I. southern Finland. *Boreas*, 19, 147–164.

Fyles, J.G., 1963. *Surfical Geology of Victoria and Stefansson Islands, District of Franklin*. Geological Survey of Canada, Bulletin 101.

Galon, R., 1973. Geomorphological and geological analysis of the proglacial area of Skeidararjökull: central section. *Geographia Polonica*, 26, 15–57.

Gardner, J.S. and Jones, N.K., 1993. Sediment transport and yield at the Raikot Glacier, Nanga Parbat, Punjab Himalaya. In J.F. Shroder (ed.), *Himalaya to the sea*. Routledge, London, pp. 184–197.

Gardner, J.S., 1982. Alpine mass wasting in contemporary time: some examples from the Canadian Rocky Mountains. In C.E. Thorn (ed.), *Space and time in geomorphology*. Allen and Unwin, London, pp. 171–192.

Gautallin, V., Polyak L. 1997 Morainic ridge complex, eastern Barents Sea. In: Davies, T.A., Bell, T., Cooper, A.K., Josenhans, H., Polyak, L., Solheim, A., Stoker, M.S., and Stravers, J.A. (eds): *Glaciated continental margins: an atlas of acoustic images*. Pp 82–83. Chapman & Hall.

Garetsky, R.G., Ludwig, A.O., Schwab, G. and Stackebrandt, W., 2001. *Neogeodynamics of the Baltic Sea depression and adjacent areas – results of IGCP Project 346, Brandenburgische Geowissenschaftliche Beiträge, Volume 8*. Kleinmachnow, Landesamt für Geowissenschaften und Rohstoffe Brandenburg, 47 pp., 8 maps.

Gellatly, A.F., Whalley, W.B. and Gordon, J.E., 1986. Topographic control over recent glacier changes in Southern Lyngen Peninsula, North Norway. *Norsk Geografisk Tidsskrift*, 41(1), 211–218.

Gellatly, A.F., Whalley, W.B., Gordon, J.E. and Hansom, J.D., 1988. Thermal regime and geomorphology of plateau ice caps of northern Norway: observations and implications. *Geology*, 16, 983–986.

Gilbert, R., 1982. Contemporary sedimentary environments on Baffin Island, N.W.T. Canada: glaciomarine processes in fiords of eastern Cumberland Peninsula. *Arctic and Alpine Research*, 14, 1–12.

Gilbert, R., 1983. Sedimentary processes of Canadian arctic fjords. *Sedimentary Geology*, 36, 147–175.

Gilbert, R., 1985. Quaternary glacimarine sedimentation interpreted from seismic surveys of fjords on Baffin Island, NWT. *Arctic*, 38, 271–280.

Gilbert, R., 1990a. Sedimentation in Expedition Fiord, Axel Heiberg Island, Northwest Territories. *Géographie physique et Quaternaire*, 44, 71–76.

Gilbert, R., 1990b. Rafting in glacimarine environments. In J.A. Dowdeswell and J.D. Scourse (eds.), *Glacimarine environments: processes and sediments*. Geological Society London, Special Publication, 53, pp. 105–120.

Gilbert, R. and Desloges, J.R., 1987. Sediments of ice-dammed, self-draining Ape Lake, British Columbia. *Canadian Journal of Earth Sciences*, 24, 1735–1747.

Gilbert, R. and Desloges, J.R., 1992. The late Quaternary sedimentary record of Stave Lake, southwestern British Columbia. *Canadian Journal of Earth Sciences*, 29, 1997–2006.

Gilbert, R., Naldrett, D.L. and Horvath, V., 1990. Holocene sedimentary environment of Cambridge Fiord, Baffin Island, Northwest Territories. *Canadian Journal of Earth Sciences*, 27, 271–280.

Gilbert, R., Handford, K.J. and Shaw, J., 1992. Ice scours in the sediments of glacial Lake Iroquois, Prince Edward County, eastern Ontario. *Géographie physique et Quaternaire* 46, 189–194.

Gilbert, R., Aitken, A.E. and Lemmen, D.S., 1993. The glacimarine sedimentary environment of Expedition Fiord, Canadian High Arctic. *Marine Geology*, 110, 257–273.

Gilbert, R., Nielsen, N., Desloges, J.R. and Rasch, M., 1998. Contrasting glacimarine sedimentary environments of two arctic fiords on Disko, West Greenland. *Marine Geology*, 147, 63–83.

Glasser, N.F. and Hambrey, M.J., 2001a. Styles of sedimentation beneath Svalbard valley glaciers under changing dynamic and thermal regimes. *Journal of the Geological Society of London*, 158, 697–707.

Glasser, N.F. and Hambrey, M.J., 2001b. Tidewater glacier beds: insights from iceberg debris in Kongsfjorden, Svalbard. *Journal of Glaciology*, 47, 295–302.

Glasser, N.F. and Hambrey, M.J., 2002. Sedimentary facies and landform genesis at a temperate outlet glacier: Soler Glacier, North Patagonian Icefield. *Sedimentology*, 49(1), 43–64.

Glasser, N.F., Hambrey, M.J., Crawford, K.R., Bennett, M.R. and Huddart, D., 1998a. The structural glaciology of Kongsvegen, Svalbard and its role in landform genesis. *Journal of Glaciology*, 44, 136–148.

Glasser, N.F., Bennett, M.R. and Huddart, D., 1998b. Ice-marginal characteristics of Fridtjovbreen (Svalbard) during its recent surge. *Polar Research*, 17, 93–100.

Glasser, N.F., Bennett, M.R. and Huddart, D., 1999. Distribution of glaciofluvial sediment within and on the surface of a high arctic valley glacier: Marthabreen, Svalbard. *Earth Surface Processes and Landforms*, 24, 303–318.

Goddard, A., 1989. Les vestiges des manteaux d'altération sur les socles des hautes latitudes: identification, signification. *Zeitschrift für Geomorphologie, Supplementband*, 72, 1–20.

Goldsmith, R., 1987. Ledyard recessional moraine, Glacial Park, Connecticut. In D.C. Roy (ed.), *Centennial Field Guide*. Geological Society of America. 5, pp. 175–180.

Goldthwait, R.P., 1951. Development of end moraines in east-central Baffin Island. *Journal of Geology*, 59, 567–577.

Goldthwait, R.P., 1960. *Study of ice cliff in Nunatarssuaq, Greenland*. Snow, Ice and Permafrost Research Establishment, Technical Report 39, 1–103.

Goldthwait, R.P., 1961. Regimen of an ice cliff on land in northwest Greenland. *Folia Geographica Danica*, 9, 107–115.

Goldthwait, R.P., 1971. Introduction to till, today. In R.P. Goldthwait (ed.), *Till, a symposium*. Ohio State University Press, pp. 3–26.

Goldthwait, R.P. and Mickelson, D.M., 1982. Glacier Bay: A model for the deglaciation of the White Mountains in New Hampshire. In G.J. Larson and B.D. Stone (eds.), *Late Wisconsinan glaciation of New England*. Kendall/Hunt, Dubuque IA, pp. 167–181.

Goodwin, R.G.,1988. Holocene glaciolacustrine sedimentation in Muir Inlet and ice advance in Glacier Bay, Alaska, USA. *Arctic and Alpine Research*, 20, 55–69.

Gordon, J.E. and Birnie, R.V., 1986. Production and transfer of subaerially-generated rock debris and resulting landforms on South Georgia: an introductory perspective. *British Antarctic Survey Bulletin*, 72, 25–46.

Gordon, J.E., Birnie, R.V. and Timmis, R., 1978. A major rockfall and debris slide on the Lyell Glacier, South Georgia. *Arctic and Alpine Research*, 10, 49–60.

Gordon, J.E., Darling, W.G., Whalley, W.B. and Gellatly, A.F., 1988. ?D-d18O relationships and the thermal history of basal ice near the margins of two glaciers in Lyngen, North Norway. *Journal of Glaciology*, 34, 265–268.

Gordon, R.B., 1979. Denudation rate of central New England determined from estuarine sedimentation. *American Journal of Science*, 279, 632–642.

Görlich, K., 1986. Glacimarine sedimentation of muds in Hornsund fjord, Spitsbergen. *Annales Société Géologique de Pologne*, 56, 433–477.

Gottler, P.F. and Powell, R.D., 1980. Processes and deposits of iceberg-rafted debris, Glacier Bay, Alaska. In A.M. Milner and J.D. Wood (eds), *Proceedings of the Second Glacier Bay Science Symposium*. US National Park Service, Anchorage, pp. 56–61.

Gozhik, P.F., 1995. Glacial history of the Ukraine. In J. Ehlers, S. Kozarski and P. Gibbard (eds), *Glacial deposits in north-east Europe*. Balkema, Rotterdam, pp. 213–215.

Graham, D.J and Midgley, N., 2000. Moraine-mound formation by englacial thrusting: the Younger Dryas moraines of Cwm Idwal, North Wales. In A.J. Maltman, B. Hubbard and M.J Hambrey (eds), *Deformation of glacial materials*. Geological Society of London Special Publication 176, pp. 321–336.

Gravenor, C.P. 1955. The origin and significance of prairie mounds. *American Journal of Science*, 253, 475–483.

Gravenor, C.P. and Kupsch, W.O., 1959. Ice-disintegration features in western Canada. *Journal of Geology*, 67, 48–64.

Gray, J.M. and Brooks, C.L., 1972. The Loch Lomond Readvance moraines of Mull and Meneith. *Scottish Journal of Geology*, 8, 95–103.

Griffith, T.W. and Anderson, J.B., 1989. Climatic control on sedimentation in bays and flords of the northern Antarctic Peninsula. *Marine Geology*, 85, 181–204.

Gripp, K., 1952. Inlandeis und Salzaufstieg. *Geologische Rundschau*, 40, 74–81.

Gripp, K., 1975. Hochsandur-Satzmorane-Endmoranenvertreter. *Zeitschrift für Geomorphologie*, 19, 490–496.

Grosswald, M.G., 1980. Late Weichselian ice sheet of northern Eurasia. *Quaternary Research*, 13, 1–32.

Grove, J.M., 1988. *The Little Ice Age*. Methuen, London, 498 pp.

Gry, H., 1942. Diskussion om vore dislocerede Klinters Dannelse. *Meddelelser fra dansk geologisk Forening*, 10, 39–51.

Gustavson, T.C., 1975. Sedimentation and physical limnology in proglacial Malaspina Lake, southeastern Alaska. In A.V. Jopling and B.C. McDonald (eds), *Glaciofluvial and glaciolacustrine sedimentation*. SEPM, Special Paper 23, pp. 249–263.

Gustavson, T.C. and Boothroyd, T.C., 1987. A depositional model for outwash, sediment sources, and hydrologic characteristics, Malaspina Glacier, Alaska: a modern analog of the southeastern margin of the Laurentide ice sheet. *Geological Society of America Bulletin*, 99, 187–200.

Gustavson, T.C., Ashley, G.M. and Boothroyd, J.C., 1975. Depositional sequences in glaciolacustrine deltas. In A.V. Jopling and B.C. McDonald (eds), *Glaciofluvial and glaciolacustrine sedimentation*. SEPM, Special Paper 23, pp. 264–280.

Gwynne, C.S., 1942. Swell and swale pattern of the Mankato lobe of the Wisconsin drift plain in Iowa. *Journal of Geology*, 50, 200–208.

Gwynne, C.S., 1951. Minor moraines in South Dakota and Minnesota. *Geological Society of America Bulletin*, 62, 223–250.

Haeberli, W., 1985. Creep of mountain permafrost: internal structure and flow of alpine rock glaciers. *Mitteilungen der Versuchsanstalt fur Wasserbau, Hydrologie und Glazialogie*, ETH, Zurich. No. 77, 142 pp.

Hagen, J.O., 1987. Glacier surge at Usherbreen, Svalbard. *Polar Research*, 5, 239–252.

Hagen, J.O., 1988. Glacier surges in Svalbard with examples from Usherbreen. *Norsk Geografisk Tidsskrift*, 42, 202–213.

Hagen, J.O. and Liestøl, O., 1990. Long term glacier mass balance investigations in Svalbard 1950-88. *Annals of Glaciology*, 14, 102–106.

Hagen, J.O. and Saetrang, A., 1991. Radio-echo soundings of sub-polar glaciers with low-frequency radar. *Polar Research*, 26, 15–57.

Hagen, J.O., Korsen, O.M. and Vatne, G., 1991. Drainage pattern in a subpolar glacier: Brøggerbreen, Svalbard. In Y. Gjessing, J.O. Hagen, K.A. Hassel, K. Sand and B. Wold (eds), *Arctic hydrology. Present and future tasks*. Oslo, Norwegian National Committee for Hydrology Report No. 23.

Hagen, J.O., Liestøl, O., Roland, E. and Jørgensen, T., 1993. *Glacier atlas of Svalbard and Jan Mayen*. Norsk Polarinstitutt Medd, Oslo, 129 pp.

Hald, M., Dahlgren, T., Olsen, T.-E., Lebesbye, E., 2001. Late Holocene palaeoceanography in Van Mijenfjorden, Svalbard. *Polar Research*, 20, 23–35.

Hald, M., Korsun, S., 1997. Distribution of modern benthic Foraminifera from fjords of Svalbard, European Arctic. *J Foraminiferal Research*, 27, 101–122.

Haldorsen, S., 1982. The genesis of tills from Åstadalen, southeastern Norway. *Norsk Geologisk Tidsskrift*, 62, 17–38

Hallberg, G.R. and Kemmis, T.J., 1986. Stratigraphy and correlation of the glacial deposits of the Des Moines and James Lobes and adjacent areas in North Dakota, South Dakota, Minnesota, and Iowa. *Quaternary Science Reviews*, 5, 65–68.

Hallet, B. and Anderson, R.S.l, 1982. Detailed glacial geomorphology of a proglacial bedrock area at Castleguard Glacier, Alberta, Canada. *Zeitschrift für Gletscherkunde und Glazialgeologie*, 16, 171–184.

Hallet, B., Hunter, L. and Bogen, J., 1996. Rates of erosion and sediment evacuation by glaciers: A review of field data and their implications. *Global and Planetary Change*, 12, 213–235.

Ham, N.R. and Attig, J.W., 1996. Ice wastage and landscape evolution along the southern margin of the Laurentide Ice Sheet, north-central Wisconsin. *Boreas*, 25, 171–186.

Ham, N.R. and Attig, J.W., 2001. Minor end moraines of the Wisconsin Valley Lobe, north-central Wisconsin, USA. *Boreas*, 30, 31–41.

Ham, N.R., 1994. *Glacial geomorphology and dynamics of the Wisconsin Valley lobe of the Laurentide ice sheet, Lincoln County, Wisconsin*. Unpublished Ph.D. thesis, University of Wisconsin, Madison., 235 pp.

Ham, N.R. and Attig, J.W., 1997. *Pleistocene geology of Lincoln County, Wisconsin*. Wisconsin Geological and Natural History Survey, Bulletin 93, 31 pp.

Hambrey, M., 1991. Structure and dynamics of the Lambert Glacier-Amery Ice Shelf system: implications for the origin of the Prydz Bay sediments. In J. Barron, B. Larsen *et al. Proceedings ODP Sci. Results, College Station TX (Ocean Drilling Program)*, 119, 61–75.

Hambrey, M.J., 1994. *Glacial environments*. University of British Columbia Press, Vancouver, 296 pp.

Hambrey, M.J. and Dowdeswell, J.A., 1997. Structural evolution of a surge-type polythermal glacier: Hessbreen, Svalbard. *Annals of Glaciology*, 24, 375–381.

Hambrey, M.J. and Huddart, D., 1995. Englacial and proglacial glaciotectonic processes at the snout of a thermally complex glacier in Svalbard. *Journal of Quaternary Science*, 10, 313–326.

Hambrey, M.J. and Lawson, W.J., 2000. Structural styles and deformation fields in glaciers: a review. In A.J. Maltman, B. Hubbard and M.J. Hambrey (eds), *Deformation of glacial materials*. Geological Society of London Special Publication 176, pp. 59–83.

Hambrey, M.J. and McKelvey, B.C., 2000. Neogene fjord sedimentation on the western margin of the Lambert Graben, East Antarctica. *Sedimentology*, 47, 577–608.

Hambrey, M.J. and Müller, F., 1978. Structures and ice deformation in White Glacier, Axel Heiberg Island, NWT, Canada. *Journal of Glaciology*, 20, 41–67.

Hambrey, M.J., Barrett, P.J. and Robinson, P.H., 1989. Stratigraphy and sedimentology. In P.J. Barrett (ed.), *Antarctic Cenozoic history from the CIROS-1 drillhole, McMurdo Sound*. Department of Scientific Industrial Research Bulletin, 245, pp. 19–47.

Hambrey, M.J., Barrett, P.J., Ehrmann, W.U. and Larsen, B., 1992. Cenozoic sedimentary processes on the Antarctic continental margin and the record from deep drilling. *Zeitschrift für Geomorphologie*, N.F. 86, 77–103.

Hambrey, M.J., Dowdeswell, J.A., Murray, T. and Porter, P.R., 1996. Thrusting and debris entrainment in a surging glacier: Bakaninbreen, Svalbard. *Annals of Glaciology*, 22, 241–248.

Hambrey, M.J., Huddart, D., Bennett, M.R. and Glasser, N.F., 1997. Genesis of 'hummocky moraine' by thrusting in glacier ice: evidence from Svalbard and Britain. *Journal of the Geological Society of London*, 154, 623–632.

Hambrey, M.J., Bennett, M.R., Dowdeswell, J.A., Glasser, N.F. and Huddart, D., 1999a. Debris-entrainment and transport in polythermal valley glaciers, Svalbard. *Journal of Glaciology*, 45, 69–86.

Hambrey, M.J., Bennett, M.R., Glasser, N.F., Huddart, D. and Crawford, K., 1999b. Facies and landforms associated with ice deformation in a tidewater glacier, Svalbard. *Glacial Geology and Geomorphology*, http:boris.qub.ac.uk/ggg/papers/full/1999/ rp071999/rp07.html.

Hamilton, G.S. and Dowdeswell, J.A., 1996. Controls on glacier surging in Svalbard. *Journal of Glaciology*, 42, 157–168.

Hammer, K.M. and Smith, N.D., 1983. Sediment production and transport in a proglacial stream: Hilda Glacier, Alberta, Canada. *Boreas*, 12, 91–106.

Hang, T., 1997. Clay varve chronology in the eastern Baltic area. *Geologiska Foreningens i Stockholm Forhandlingar*, 119.
Hansel, A.K. and Johnson, W.H., 1987. Ice marginal sedimentation in a late Wisconsinan end moraine complex, Northeastern Illinois, USA. In J.J.M. van der Meer (ed.), *Tills and glaciotectonics*. Balkema, Rotterdam, pp. 97–104.
Hansel, A.K. and Johnson, H.K., 1996. Wedron and Mason Groups: Lithostratigraphic reclassification of deposits of the Wisconsin Episode, Lake Michigan lobe area. *Illinois Geological Survey*, Bulletin 104, 116 pp.
Hansel, A. and Johnson, H.K., 1999. Wisconsin episode glacial landscape of central Illinois: a product of subglacial deformation processes? In D.M. Mickelson and J.W. Attig (eds), *Glacial processes past and present*. Geological Society of America Special Paper 337, pp. 121–135.
Hansen, S., 1940. Varvighed i danske og skånske senglaciale aflejringer. *Danmarks Geologiske Undersøgelse*, II, Række, Nr. 63., 473 pp.
Hansom, J.D., 1983. Ice-formed intertidal boulder pavements in the Sub-Antarctic. *Journal of Sedimentary Petrology*, 53, 135–145.
Harbor, J.M., 1992. Numerical modelling of the development of U-shaped valleys by glacial erosion. *Geological Society of America Bulletin*, 103, 1364–1375.
Harbor, J. and Warburton, J., 1993. Relative rates of glacial and nonglacial erosion in alpine environments. *Arctic and Alpine Research*, 25, 1–7.
Harbor, J.M., Hallet, B. and Raymond, C.F., 1988. A numerical model of landscape development by glacial erosion. *Nature*, 333, 347–349.
Harris, C., 1998. The micromorphology of paraglacial and periglacial slope deposits: a case study from Morfa Bychan, west Wales, UK. *Journal of Quaternary Science*, 13, 73–84.
Harris, C. and Bothamley, K., 1984. Englacial deltaic sediments as evidence for basal freezing and marginal shearing, Leirbreen, Norway. *Journal of Glaciology*, 30, 30–34.
Harrison, S. and Winchester, V., 1997. Age and nature of paraglacial debris cones along the margins of the San Rafael glacier, Chilean Patagonia. *The Holocene*, 7, 481–487.
Hart, J.K., 1994. Till fabric associated with deformable beds. *Earth Surface Processes and Landforms*, 19, 15–32.
Hart, J.K., 1995. Recent drumlins, flutes and lineations at Vestari-Hagafellsjökull, Iceland. *Journal of Glaciology*, 41, 596–606.
Hart, J.K., 1999. Identifying fast ice flow from landform assemblages in the geological record: a discussion. *Annals of Glaciology*, 28, 59–67.
Hart, J.K. and Boulton, G.S., 1991. The interrelation of glaciotectonic and glaciodepositional processes within the glacial environment. *Quaternary Science Reviews*, 10, 335–350.
Hart, J.K. and Roberts, D.H., 1994. Criteria to distinguish between subglacial glaciotectonic and glaciomarine sedimentation. I. Deformation styles and sedimentology. *Sedimentary Geology*, 91, 191–213.
Hart, J.K., Gane, F. and Watts, R.J., 1996. Deforming bed conditions on the Dänischer Wohld Peninsula, northern Germany. *Boreas*, 25, 101–113.
Hart, J.K., Gane, F. and Watts, R.J., 1997. Deforming bed conditions on the Dänischer Wohld Peninsula, northern Germany: Reply to comments. *Boreas*, 26, 79–80.
Hartshorn, J.H., 1958. Flow till in southeastern Massachusetts. *Geological Society of America Bulletin*, 69, 477–482.
Hattersley-Smith, G., 1957. The Ellesmere ice shelf and the ice islands. *Canadian Geographer*, 9, 65–70.
Hattersley-Smith, G., 1969a. Recent observations on the surging Otto Glacier, Ellesmere Island. *Canadian Journal of Earth Sciences*, 6, 883–889.
Hattersley-Smith, G., 1969b. Glacial features of Tanquary Fjord and adjoining areas of northern Ellesmere Island, NWT. *Journal of Glaciology*, 8, 23–50.
Hattestrand, C., 1997. Ribbed moraines in Sweden – distribution pattern and palaeoglaciological implications. *Sedimentary Geology*, 111, 41–56.
Heim, D., 1983. Glaziare Entwasserung und Sanderbildung am Kotlujökull, Südisland. *Polarforschung*, 53, 17–29.
Heim, D., 1992. Sandergenese und Gletscherentwasserung am Kotlujökull (Hofdabrekkujökull), Südisland. *Polarforschung*, 62, 95–128.

Heinrich, H., 1988. Origin and consequences of cyclic ice rafting in the north-east Atlantic during the past 130,000 years. *Quaternary Research*, 29, 143–152.
Hertzfeld, U.C. and Mayer, H., 1997. Surge of Bering Glacier and Bagley Ice Field, Alaska: an up-date to August 1995 and an interpretation of brittle-deformation patterns. *Journal of Glaciology*, 43, 427–734.
Hewitt, K., 1988. Catastrophic landslide deposits in the Himalaya. *Science*, 242, 64–77.
Hewitt, K., 1993. Altitudinal organisation of Karakoram geomorphic processes and depositional environments. In J.F. Shroder (ed.), *Himalaya to the sea*. Routledge, London, pp. 159–183.
Hickman, C.S. and Nesbitt, E.A., 1980. Holocene mollusk distribution patterns in the northern Gulf of Alaska. In M.E. Field, A.H. Bouma, I.P. Colburn, R.G. Douglas and J.C. Ingle (eds), *Proceedings of the Quaternary depositional environments of the Pacific Coast, Pacific Coast Paleogeography Symposium* 4, pp. 305–312.
Hicks, D.M., McSaveney, M.J. and Chinn, T.J.H., 1990. Sedimentation in proglacial Ivory Lake, Southern Alps, New Zealand. *Arctic and Alpine Research*, 22, 26–42.
Hicock, S.R., 1992. Lobal interactions and rheologic superposition in subglacial till near Bradtville, Ontario, Canada. *Boreas*, 21, 73–88.
Hicock, S.R. and Dreimanis, A., 1992. Sunnybrook drift in the Toronto area, Canada: re-investigation and re-interpretation. In P.U. Clark and D. Lea (eds.), *The last interglacial-glacial transition in North America*. Geological Society of America, Special Paper 270, pp. 139–161.
Hicock, S.R. and Fuller, E.A., 1995. Lobal interactions, rheologic superposition, and implications for a Pleistocene ice stream on the continental shelf of British Columbia. *Geomorphology*, 14, 167–184.
Hicock, S.R., Dreimanis, A. and Broster, B.E., 1981. Submarine flow tills at Victoria, British Columbia. *Canadian Journal of Earth Science*, 18, 71–80.
Higgins, A., 1989. North Greenland ice islands. *Polar Record*, 25, 207–212.
Higgins, A. and Weidick, A., 1988. The world's northernmost surging glacier? *Zeitschrift für Gletscherkunde and Glazialgeologie*, 24, 111–123.
Higuchi, K., Fushimi, H., Ohata, T. *et al.*, 1980. Glacier inventory in the Dudh Kosi region, East Nepal. *IASH Publication*, 126 World Glacier Inventory, 95–103.
Hill, P.R., Mudie, P.J., Moran, K. and Blasco S.M., 1985. A sea level curve for the Canadian Beaufort Shelf. *Canadian Journal of Earth Sciences*, 22, 1383–1393.
Hill, P.R., Lewis, C.P., Desmarais, S., Kauppaymuthoo, V. and Rais, H., 2001. The Mackenzie delta: sedimentary processes and facies of a high-latitude, fine-grained delta. *Sedimentology*, 48, 1047–1078.
Hill, P.S., Syvitski, J.P.M., Cowan, E.A. and Powell, R.D., 1998. Floc settling velocities under a buoyant discharge plume in Glacier Bay, Alaska. *Marine Geology*, 145, 85–94.
Hinchliffe, S. and Ballantyne, C.K., 1999. Talus accumulation and rockwall retreat, Trotternish, Isle of Skye, Scotland. *Scottish Geographical Journal*, 115, 53–70.
Hiscott, R.N. and Aksu, A.E., 1994. Submarine debris flows and continental slope evolution in front of Quaternary ice sheets, Baffin Bay, Canadian Arctic. *American Association Petroleum Geology Bulletin*, 78, 445–460.
Hiscott, R.N. and Aksu, A.E., 1996. Quaternary sedimentary processes and budgets in Orphan Basin, Southwestern Labrador Sea. *Quaternary Research*, 45, 160–175.
Hjelle, A, 1993. *Geology of Svalbard*. Norsk Polarinstitutt, Oslo.
Hobbs, H.C., 1983. Drainage relationship of glacial Lake Aikins and Upham and early Lake Agassiz in northeastern Minnesota. In J.T. Teller and L. Clayton (eds.), *Glacial Lake Agassiz*. Geological Association of Canada, Special Paper 26, pp. 245–259.
Hochstein, M.P., Claridge, D., Henrys, S.A., Pyne, A., Nobes, D.C. and Leary, S.F., 1995. Downwasting of the Tasman Glacier, South Island, New Zealand: changes in the terminus region between 1971 and 1993. *New Zealand Journal of Geology and Geophysics*, 38, 1–16.
Hodge, S.M. and Doppelhammer, S.K., 1996. Satellite imagery of the onset of streaming flow of Ice Streams C and D, West Antarctica. *Journal of Geophysical Research*, 101, 6669–6677.
Hodgkins, R., 1997. Glacier hydrology in Svalbard, Norwegian High Arctic. *Quaternary Science Reviews*, 16, 957–973.

Hodgson, D.A., 1985. The last glaciation of west-central Ellesmere Island, arctic archipelago, Canada. *Canadian Journal of Earth Sciences*, 22, 347–368.

Hodgson, D.A., 1989. Quaternary geology of the Queen Elizabeth Islands. In R.J. Fulton (ed.), *Quaternary Geology of Canada and Greenland.* Geological Survey of Canada, Geology of Canada, No. 1, pp. 441–477.

Hodgson, D.A., 1994. Episodic ice streams and ice shelves during retreat of the northwestern sector of the late Wisconsinan Laurentide Ice Sheet over the central Canadian Arctic Archipelago. *Boreas*, 23, 14–28.

Hodgson, D.A., Vincent, J.-S. and Fyles, J.G., 1984. Quaternary Geology of Central Melville Island, Northwest Territories. *Geological Survey of Canada*, Paper 83-16.

Hodgson, D.A. and Vincent, J.-S., 1984. A 10,000 yr BP extensive ice shelf over Viscount Melville Sound, arctic Canada. *Quaternary Research*, 22, 18–30.

Holmes, J.A. and Street-Perrott, F.A., 1989. The Quaternary history of Kashmir, north-west Himalaya: a revision of de Terra and Paterson's sequence. *Zeitschrift für Geomorphologie, Supplementband*, 76, 195–212.

Holtedahl, H., 1958. Some remarks on geomorphology of continental shelves off Norway, Labrador, and southeast Alaska. *Journal of Geology*, 66, 461–471.

Holtedahl, H., 1993. Marine geology of the Norwegian continental margin. *Norges Geologiske Undersøkelse, Special Publication* 6. 150 pp.

Homci, H., 1974. Jungpleistozäne Tunneltäler im Nordosten von Hamburg (Rahlstedt-Meiendorf). *Mitteilungen aus dem Geologisch-Paläontologischen Institut der Universität Hamburg*, 43, 99–126.

Hooke, R.LeB., 1968 Comments on 'The formation of shear moraines: an example from south Victoria Land, Antarctica'. *Journal of Glaciology*, 7, 351–352.

Hooke, R.LeB., 1970. Morphology of the ice sheet margin near Thule, Greenland. *Journal of Glaciology*, 9, 303–324.

Hooke, R.LeB., 1973a. Flow near the margin of the Barnes Ice Cap, and the development of ice-cored moraines. *Geological Society of America Bulletin*, 84, 3929–3948.

Hooke, R.LeB., 1973b. Structure and flow at the margin of the Barnes Ice Cap, Baffin Island, NWT, Canada. *Journal of Glaciology*, 12, 423–438.

Hooke, R.LeB., 1991. Positive feedbacks associated with erosion of glacial cirques and overdeepenings. *Geological Society of America Bulletin*, 103, 1104–1108.

Hop, H., Pearson, T., Hegseth, E.N., Kovacs, K.M., Wiencke, C., Kwasniewski, S., Ketil, E., *et al.*, 2002. The marine ecosystem of Kongsfjorden, Svalbard. *Polar Research*, 21, 167–208.

Hoppe, G., 1952. Hummocky moraine regions with special reference to the interior of Norrbotten. *Geografiska Annaler*, 34, 1–72.

Houmark-Nielsen, M., 1983. Glacial stratigraphy and morphology of the northern Bælthav region. In J. Ehlers (ed.), *Glacial deposits in north-west Europe.* Balkema, Rotterdam, pp. 211–218.

Houmark-Nielsen, M., 1987. Pleistocene stratigraphy and glacial history of the central part of Denmark. *Bulletin of the Geological Society of Denmark*, 36, 1–189.

Houmark-Nielsen, M., 1989. The last interglacial-glacial cycle in Denmark. *Quaternary International*, 3/4, 31–39.

Houmark-Nielsen, M., 1994. Late Pleistocene stratigraphy, glaciation chronology and middle Weichselian environmental history from Klintholm, Mon, Denmark. *Bulletin of the Geological Society of Denmark*, 41, 181–202.

Houmark-Nielsen, M. and Berthelsen, A., 1981. Kineto-stratigraphic evaluation and presentation of glacial-stratigraphic data, with examples from northern Samsö, Denmark. *Boreas*, 10, 411–422.

Howarth, P.J., 1968. *Geomorphological and Glaciological Studies, Eastern Breiðamerkurjökull, Iceland.* Unpublished PhD Thesis, University of Glasgow.

Howarth, P.J., 1971. Investigations of two eskers at eastern Breiðamerkurjökull, Iceland. *Arctic and Alpine Research*, 3, 305–318.

Howarth, P.J. and Price, R.J., 1969. The proglacial lakes of Breiðamerkurjökull and Fjallsjökull, Iceland. *Geographical Journal*, 135, 573–581.

Høydal, Ø.A., 1996. A force balance study of ice flow and basal conditions of Jutulstraumen, Antarctica. *Journal of Glaciology*, 42, 413–425.

Hubbard, B. and Sharp, M.J., 1989. Basal ice formation and deformation: a review. *Progress in Physical Geography*, 13, 529–558.

Huddart, D. and Hambrey, M.J., 1996. Sedimentary and tectonic development of a high-arctic, thrust-moraine complex: Comfortlessbreen, Svalbard. *Boreas*, 25, 227–243.

Hudleston, P.J., 1976. Recumbent folding in the base of the Barnes Ice Cap, Baffin Island, NWT, Canada. *Geological Society of America Bulletin*, 87, 1684–1692.

Hughes, O.L., Harington, C.R., Janssens, J.A. *et al.*, 1981. Upper Pleistocene stratigraphy, paleoecology, and archaeology of the northern Yukon interior, eastern Beringia, 1, Bonnet Plume Basin. *Arctic*, 34, 329–365.

Humlum, O., 1978. Genesis of layered lateral moraines: implications for palaeoclimatology and lichenometry. *Geografisk Tidsskrift*, 77, 65–72.

Humlum, O., 1982. Rock glacier types on Disko, central West Greenland. *Geografisk Tidsskrift*, 82, 59–66.

Humlum, O., 1985. Genesis of an imbricate push moraine, Hofdabrekkujökull, Iceland. *Journal of Geology*, 93, 185–195.

Humphrey, N.F. and Raymond, C.F., 1994. Hydrology, erosion and sediment production in a surging glacier: Variegated Glacier, Alaska, 1982–83. *Journal of Glaciology*, 40, 539–552.

Hunter, G.T., 1970. Postglacial uplift at Fort Albany, James Bay. *Canadian Journal of Earth Sciences*, 7, 547–548.

Hunter, L.E., Powell, R.D. and Lawson, D.E., 1996a. Flux of debris transported by ice at three Alaskan tidewater glaciers. *Journal of Glaciology*, 42, 123–135.

Hunter, L.E., Powell, R.D. and Smith, G.W., 1996b. Facies architecture and grounding-line fan processes of morainal banks during the deglaciation of coastal Maine. *Geological Society of America Bulletin*, 108, 1022–1038.

Huuse, M. and Lykke-Andersen, H., 2000. Overdeepened Quaternary valleys in the eastern Danish North Sea: morphology and origin. *Quaternary Science Reviews*, 19, 1233–1253.

Iken, A., 1972. Measurements of water pressure in moulins as part of a movement study of the White Glacier, Axel Heiberg Island, Northwest Territories. *Journal of Glaciology*, 11, 53–58.

Iken, A., 1974. *Velocity fluctuations of an arctic valley glacier – a study of the White Glacier, Axel Heiberg Island.* Axel Heiberg Island Research Reports, Glaciology no. 5. McGill University, Montreal, 116 pp.

Iverson, N.R., 1991. Potential effects of subglacial water pressure fluctuations on quarrying. *Journal of Glaciology*, 37, 27–36.

Iverson, N.R. and Iverson, R.M., 2001. Distributed shear of subglacial till due to Coulomb slip. *Journal of Glaciology*, 47, 481–488.

Iverson, N.R., Hooyer, T.S. and Baker, R.W., 1998. Ring shear studies of till deformation: coulomb-plastic behavior and distributed strain in glacier beds. *Journal of Glaciology*, 44, 634–642.

Iverson, N.R., Baker, R.W., Hooke, R.LeB., Hanson, B. and Jansson, P., 1999. Coupling between a glacier and a soft bed. I. A relation between effective pressure and local shear stress determined from till elasticity. *Journal of Glaciology*, 45, 31–40.

Ives, J.D., 1960. Glaciation and deglaciation of the Helluva Lake area, central Labrador-Ungava. *Geographical Bulletin*, 15, 46–64.

Ives, J.D. and Andrews, J.T., 1963. Studies in the physical geography of north central Baffin Island. *Geographical Bulletin*, 19, 5–48.

Ives, J.D., Andrews, J.T. and Barry, R.G., 1975. Growth and decay of the Laurentide Ice Sheet and comparisons with Fenno-Scandinavia. *Naturwissenschaften*, 62, 118–125.

Jackson, L.E., MacDonald, G.M. and Wilson, M.C., 1982. Paraglacial origin for terraced river sediments in Bow Valley, Alberta. *Canadian Journal of Earth Sciences*, 19, 2219–2231.

Jaritz, W., 1973. Zur Entstehung der Salzstrukturen Nordwestdeutschlands. *Geologisches Jahrbuch*, A10, 1–77.

Jarman, D. and Ballantyne, C.K., 2002. Beinn Fhada, Kintail: an example of large-scale paraglacial rock slope deformation. *Scottish Geographical Journal* 118, 59–68.

Jeffries, M.O., 1984. Milne Glacier, northern Ellesmere Island, NWT: a surging glacier? *Journal of Glaciology*, 30, 251–253.

Jeffries, M.O., 1987. The growth, structure and disintegration of arctic ice shelves. *Polar Record*, 23, 631–649.

Jenkins, A., Vaughan, D.G., Jacobs, S.S., Hellmer, H.H. and Keys, J.R., 1997. Glaciological and oceanographic evidence of high melt rates beneath Pine Island Glacier, West Antarctica. *Journal of Glaciology*, 43(143), 114–121.

Jiskoot, H, Boyle, P. and Murray, T., 1998. The incidence of glacier surging in Svalbard: evidence from multivariate statistics. *Computers and Geosciences*, 24, 387–399.

Jiskoot, H., Murray, T., Boyle, P., 2000. Controls on the distribution of surge-type glaciers in Svalbard. *Journal of Glaciology*, 46, 412–422.

Johnson, P.G., 1975. Recent crevasse fillings at the terminus of the Donjek Glacier, St Elias Mountains, Yukon Territory. *Quaestiones Geographicae*, 2, 53–59.

Johnson, W.H. and Hansel, A.K., 1999. Wisconsin episode glacial landscape of central Illinois: a product of subglacial deformation processes? In D.M. Mickelson and J.W. Attig (eds.), *Glacial processes: past and present*. Geological Society of America, Special Paper 337, pp. 121–135.

Johnson, M.D., 1986. *Pleistocene geology of Barron County, Wisconsin*. Wisconsin Geological and Natural History Survey, Information Circular 55, 42 pp.

Johnson, M.D., 2000. *Pleistocene geology of Polk County, Wisconsin*. Wisconsin Geological and Natural History Survey, Bulletin 92, 70 pp.

Johnson, M.D., Mickelson, D.M., Clayton, L. and Attig, J.W., 1995. Composition and genesis of glacial hummocks, western Wisconsin. *Boreas*, 24, 97–116.

Johnson, M.D., Addis, K.L., Ferber, L.R., Hempsted, C., Meyer, G.N. and Komai, L.T., 1999. Glacial Lake Lind, Wisconsin and Minnesota, USA. *Geological Society of America Bulletin*, 111, 1371–1386.

Johnson, P.G., 1972. The morphological effects of surges of the Donjek Glacier, St Elias Mountains, Yukon Territory, Canada. *Journal of Glaciology*, 11, 227–234.

Johnson, P.G., 1984. Paraglacial conditions of instability and mass movement: a discussion. *Zeitschrift für Geomorphologie*, 28, 235–250.

Johnson, P.G., 1995. Debris transfer and sedimentary environments: alpine glaciated areas. In O. Slaymaker (ed.), *Steepland geomorphology*. Wiley, Chichester, pp. 27–44.

Johnson, W.H., 1990. Ice-wedge casts and relict patterned ground in central Illinois and their environmental significance. *Quaternary Research*, 33, 51–72.

Johnson, W.H. and Hansel, A.K., 1990. Multiple Wisconsinan glacigenic sequences at Wedron, Illinois. *Journal of Sedimentary Petrology*, 60, 26–42.

Jones, J.G., 1969. Intraglacial volcanoes in the Laugarvatn region, south-west Iceland. *Quarterly Journal of the Geological Society of London*, 495, 197–211.

Jones, N.D., 1982. The formation of glacial flutings in east-central Alberta. In R. Davidson-Arnott, W. Nickling and B.D. Fahey (eds), *Research in glacial, glacio-fluvial and glacio-lacustrine systems*. Geobooks, Norwich, pp. 49–70.

Jopling, A.V., 1965. Hydraulic factors and the shape of laminae. *Journal of Sedimentary Petrology*, 35, 777–791.

Jopling, A.V. and MacDonald, B.C. (eds), 1975. *Glaciofluvial and glaciolacustrine sedimentation*. SEPM, Special Publication 23.

Jopling, A.V. and Walker, R.G., 1968. Morphology and origin of ripple drift cross lamination, with examples from the Pleistocene of Massachusetts. *Journal of Sedimentary Petrology*, 38, 971–984.

Josenhans, H., 1997. Glacially overdeepened troughs on the Labrador Shelf, Canada. In T.A. Davies, T. Bell, A.K. Cooper *et al.* (eds), *Glaciated continental margins: an atlas of acoustic images*. Chapman and Hall, London, pp. 248–249.

Joughin, I, Gray, L., Bindschadler, R.A. *et al.*, 1999. Tributaries of West Antarctic Ice streams revealed by RADARSAT interferometry. *Science*, 286(5438), 283–286.

Kalin, M., 1971. *The active push moraine of the Thompson Glacier, Axel Heiberg Island*. Research Report (Glaciology) No. 4, McGill University, Montreal, 68 pp.

Kamb, B., 1987. Glacier surge mechanism based on linked cavity configuration of the basal water conduit system. *Journal of Geophysical Research*, 92, 9083–9100.

Kamb, B., 2001. Basal zone of the West Antarctic ice streams and its role in lubrication of their rapid motion. In R.B. Alley and R.A. Bindschadler (eds), *The West Antarctic Ice Sheet: behavior and environment*. Antarctic Research Series, 77, American Geophysical Union, Washington, D.C, pp. 157–199.

Kamb, B., Raymond, C.F., Harrison, W.D. *et al.*, 1985. Glacier surge mechanism: 1982–1983 surge of Variegated Glacier, Alaska. *Science*, 327, 469–479.

Kaplan, M.R., 1999. Retreat of a tidewater margin of the Laurentide ice sheet in eastern coastal Maine ca. 14 to 13,000 14C yrs BP. *Geological Society of America Bulletin*, 111, 620–633.

Karczewski, A., 1987. Lithofacies variability of a drumlin in Pomerania, Poland. In J. Menzies and J. Rose (eds), *Drumlin symposium*. Balkema, Rotterdam, pp. 177–183.

Karrow, P.R. and Calkin, P.E. (eds), 1985. Quaternary Evolution of the Great Lakes. *Geological Association of Canada*, Special Paper 30.

Kasparek, L. and Kozarski, S., 1989. Ice-lobe contact sedimentary scarps in marginal zones of the major Vistulian ice-sheet positions, west-central Poland. *Quaestiones Geographicae*, Special Issue 2, 69–81.

Kassem, A. and Imran, J., 2001. Simulation of turbid underflows generated by the plunging of a river. *Geology*, 29, 655–658.

Kaufman, D.S., Miller, G.H., Stravers, J.A. and Andrews, J.T., 1993. Abrupt early Holocene (9.9–9.6 ka) ice stream advance at the mouth of Hudson Strait, Arctic Canada. *Geology*, 21, 1063–1066.

Kehew, A.E. and Lord, M.L., 1987. Glacial-lake outbursts along the mid-continent margins of the Laurentide ice sheet. In L. Mayer and D. Nash (eds), *Catastrophic flooding, Binghampton Symposium in Geomorphology*. Allen and Unwin, Boston, pp. 95–120.

Kehew, A.E. and Clayton, L., 1983. Late Wisconsinan floods and development of the Souris-Pembina spillway system in Saskatchewan, North Dakota, and Manitoba. In J.T. Teller and L. Clayton (eds), *Glacial Lake Agassiz*. Geological Association of Canada, Special Paper 26, 187–209.

Kehew, A. E. and Teller, J.T., 1994. History of late glacial runoff along the southwestern margin of the Laurentide Ice Sheet. *Quaternary Science Reviews*, 13, 859–877.

Kemmis, T.J., 1992. *Glacial landforms, sedimentology, and depositional environments of the Des Moines lobe, northern Iowa*. Unpublished Ph.D. thesis, University of Iowa, 393 pp.

Kemmis, T.J., Hallberg, G.R. and Lutenegger, A.J., 1981. *Depositional environment of glacial sediments and landforms on the Des Moines Lobe, Iowa*. Iowa Geological Survey Guidebook 6, 132 pp.

Kemmis, T.J., Bettis, E.A., III and Quade, D.J., 1994. The Des Moines Lobe in Iowa: a surging Wisconsinan glacier. *American Quaternary Association, Programs and Abstracts*, 13th biennial meeting, Minneapolis, Minnesota, 19–22 June, 1994, 112.

Kerr, J.W., 1980. Structural Framework of Lancaster Aulacogen, Arctic Canada. *Geological Survey of Canada*, Bulletin 319.

Khristophorova, T.F., 1995. Early Pleistocene glacial deposits in the Middle Dnieper region. In J. Ehlers, S. Kozarski and P. Gibbard (eds), *Glacial deposits in north-east Europe*. Balkema, Rotterdam, pp. 217–220.

King, E.L., Sejrup, H.P., Haflidason, H., Elverhøi, A. and Aarseth, I., 1996. Quaternary seismic stratigraphy of the North Sea Fan: glacially-fed gravity flow aprons, hemipelagic sedimentation, and submarine sliding. *Marine Geology*, 130, 293–315.

King, E.L., Haflidason, H., Sejrup, H.P. and Løvlie, R., 1998. Glacigenic debris flows on the North Sea Trough Mouth Fan during ice stream maxima. *Marine Geology*, 152, 217–246.

King, L.H., 1976. Relict iceberg furrows on the Laurentian Channel and western Grand Banks. *Canadian Journal of Earth Science*, 13, 1083–1092.

King, L.H., Rokoengen, K., Fader, G.B.J. and Gunleiksrud, T., 1991. Till-tongue stratigraphy. *Geological Society of America Bulletin*, 103, 637–659.

King, R.B., 1987. Review of geomorphic description and classification in land resource surveys. In V. Gardner (ed.), *International geomorphology 1986, Part II*. Wiley, Chichester, pp. 384–403.

Kirkbride, M.P., 1993. The temporal significance of transitions from melting to calving termini at glaciers in the central Southern Alps of New Zealand. *The Holocene*, 3, 232–240.

Kirkbride, M.P., 2000a. Ice marginal geomorphology and Holocene expansion of debris-covered Tasman Glacier, New Zealand. In M. Nakawo, C. Raymond and A. Fountain (eds), *Debris-covered glaciers*. IAHS Publication 264, pp. 211–217.

Kirkbride, M.P., 2002b. Processes of glacial transportation. In J. Menzies (ed.), *Modern and past glacial environments*. Butterworth-Heinemann, Oxford, pp. 147–169.

Kirkbride, M.P. and Brazier, V., 1995. On the sensitivity of Holocene talus-derived rock glaciers to climate change in the Ben Ohau Range, New Zealand. *Journal of Quaternary Science*, 10.

Kirkbride, M.P. and Spedding, N.F., 1996. The influence of englacial drainage on sediment-transport pathways and till texture of temperate valley glaciers. *Annals of Glaciology*, 22, 160–166.

Kirkbride, M.P. and Warren, C.R., 1999. Tasman Glacier, New Zealand: 20th-century thinning and predicted calving retreat. *Global and Planetary Change*, 22, 11–28.

Kirkham, R.V., 1995. *Generalized Geological Map of the World*. Geological Survey of Canada.

Kjær K.H. and Krüger, J., 2001. The final phase of dead-ice moraine development: processes and sediment architecture, Kötlujökull, Iceland. *Sedimentology*, 48, 935–952.

Klassen, R.A., 1982. *Glaciotectonic thrust plates, Bylot Island, District of Franklin*. Geological Survey of Canada, Paper 82-1A, pp. 369–373.

Klassen, R.A., 1993. *Quaternary geology and glacial history of Bylot Island, Northwest Territories*. Geological Survey of Canada, Memoir 429.

Klassen, R.W., 1971. *Surficial geology, Franklin Bay (97C) and Brock River (97D)*. Geological Survey of Canada, Open File 48, scale 1:250 000.

Klassen, R.W., 1993. Moraine plateaus: relicts of stagnant ice in southwestern Saskatchewan. In R.W. Barendregt, M.C. Wilson and F.J. Jakunis (eds), *The Paliser Triangle: a region in time and space*. Thompson Communications, Calgary, Alberta, pp. 77–87.

Kleman, J. and Borgström, I., 1994. Glacial landforms indicative of a partly frozen bed. *Journal of Glaciology*, 40, 255–264.

Kleman, J., Hättestrand, C. and Clarhäll, A., 1999. Zooming in on frozen-bed patches: scale-dependent controls on Fennoscandian ice sheet basal thermal zonation. *Annals of Glaciology*, 28, 189–194

Klimek, K., 1973. Geomorphological and geological analysis of the proglacial area of the Skeidararjökull. *Geographia Polonica*, 26, 89–113.

Klostermann, J., 1995. IV. Nordrhein-Westphalen. In L. Benda (ed.), *Das Quartär Deutschlands*. Gebrüder Borntraeger, Berlin, pp. 59–94.

Kluiving, S.J., 1994. Glaciotectonics of the Itterbeck-Uelsen push moraines, Germany. *Journal of Quaternary Research*, 9, 235–244.

Kluiving, S.J., Bartek, L.R. and Van der Wateren, F.M., 1999. Multi-scale analyses of subglacial and glaciomarine deposits from the Ross Sea continental shelf, Antarctica. *Annals of Glaciology*, 28, 90–96.

Kluiving, S.J., Rappol, M. and Van der Wateren, F.M., 1991. Till stratigraphy and ice movements in eastern Overijssel, The Netherlands. *Boreas*, 20, 193–205.

Knight, P.G., 1997. The basal ice layer of glaciers and ice sheets. *Quaternary Science Reviews*, 16, 975–993.

Knoth, W., 1995. VII. Sachsen-Anhalt. In L. Benda (ed.), *Das Quartär Deutschlands*. Gebrüder Borntraeger, Berlin, pp. 148–170.

Knudsen, O., 1995. Concertina eskers, Brúarjökull, Iceland: an indicator of surge-type glacier behaviour. *Quaternary Science Reviews*, 14, 487–493.

Koenig, L.S., Greenaway, K.R., Dunbar, M. and Hattersley-Smith, G., 1952. Arctic ice islands. *Arctic*, 5, 67–103.

Koerner, R.M., 1977. Ice thickness measurements and their implications with respect to past and present ice volumes in the Canadian high arctic ice caps. *Canadian Journal of Earth Sciences*, 14, 2697–2705.

Koerner, R.M., 1989. Queen Elizabeth Islands glaciers. In R.J. Fulton (ed.), *Quaternary geology of Canada and Greenland*. Geological Survey of Canada, Geology of Canada no.1, pp. 464–473.

Korsun, S. and Hald, M., 1998. Modern benthic Foraminifera off Novaya Zemlya tidewater glaciers, Russian Arctic. *Arctic and Alpine Research*, 30, 61–77.

Korsun, S. and Hald, M., 2000. Seasonal dynamics of benthic Foraminifera in a glacially fed fjord of Svalbard, European Arctic. *Journal of Foraminiferal Research*, 30, 251–271.

Kostaschuck, R.A., 1985. River mouth processes in a fjord-delta, British Columbia. *Marine Geology*, 69, 1–23.

Kostaschuk, R.A. and McCann, S.B., 1983. Observations on delta-forming processes in a fjordhead delta, British Columbia. *Sedimentary Geology*, 36, 269–288.

Kostaschuck, R.A., MacDonald, G.M. and Putnam, P.E., 1986. Depositional process and alluvial fan-drainage basin morphometric relationships near Banff, Alberta, Canada. *Earth Surface Processes and Landforms*, 11, 471–484.

Koteff, C., 1974. The morphosequence concept and and deglaciation of southern New England. In D.R. Coates (ed.), *Glacial geomorphology*. Binghampton, New York. State University of New York, pp. 121–144.

Koteff, C. and Pessl, F., Jr., 1981. *Systematic ice retreat in New England*. US Geological Survey Professional Paper 1179, 20 pp.

Kotler, E., Michel, F.A. and Hodgson, D.A., 1998. Gravimetric investigation of mounded till deposits, central Victoria Island, Northwest Territories, Canada. In A.G. Lewkowicz and M. Allard (eds), *Permafrost, Seventh International Conference Proceedings*. Centre d'études nordiques, Université Laval, pp. 607–610.

Kovac, A. and Sondhi, D.S., 1979. Ice pile-up and ride-up on Arctic and Subarctic beaches. *Proceedings 5. International Conference on Port and Ocean Engineering under Arctic Conditions*. Technical University, Trondheim, pp. 127–146.

Kozarski, S., 1981. Ablation end moraines in western Pomerania, NW Poland. *Geografiska Annaler*, 63A, 169–174.

Krüger, J., 1969. Landskabsformer i sydlige Sjælland. Studier over glaciallandskabets morfologi, opbygning og dannelse. *Geografisk Tidsskrift*, 68, 105–212.

Krüger, J., 1983. Glacial morphology and deposits in Denmark. In J. Ehlers (ed.), *Glacial deposits in north-west Europe*. Balkema, Rotterdam, pp. 181–191.

Krüger, J., 1985. Formation of a push moraine at the margin of Hofdabrekkujökull, south Iceland. *Geografiska Annaler*, 67A, 199–212.

Krüger, J., 1987. Traek af et glaciallandskabs udvikling ved nordranden af Myrdalsjökull, Iceland. *Dansk Geologisk Foreningens*, Arsskrift for 1986, 49–65.

Krüger, J., 1993. Moraine ridge formation along a stationary ice front in Iceland. *Boreas*, 22, 101–109.

Krüger, J., 1994a. Glacial processes, sediments, landforms and stratigraphy in the terminus region of Myrdalsjökull, Iceland. *Folia Geographica Danica*, 21, 1–233.

Krüger, J., 1994b. Sorted polygons on recently deglaciated terrain in the highland of Maelifellsandur, south Iceland. *Geografiska Annaler*, 76A, 49–55.

Krüger, J., 1995. Origin, chronology and climatological significance of annual-moraine ridges at Myrdalsjökull, Iceland. *The Holocene*, 5, 420–427.

Krüger, J., 1996. Moraine ridges formed from subglacial frozen-on sediment slabs and their differentiation from push moraines. *Boreas*, 25, 57–63.

Krüger, J., 1997. Development of minor outwash fans at Kotlujökull, Iceland. *Quaternary Science Reviews*, 16, 649–659.

Krüger, J. and Aber, J.S., 1999. Formation of supraglacial sediment accumulations on Kötlujökull, Iceland. *Journal of Glaciology*, 45, 400–402.

Krüger, J. and Kjær, K., 2000. De-icing progression of ice-cored moraines in a humid, subpolar climate, Kötlujökull, Iceland. *The Holocene*, 10, 737–747.

Krüger, J. and Thomsen, H.H., 1984. Morphology, stratigraphy and genesis of small drumlins in front of the glacier Myrdalsjökull, south Iceland. *Journal of Glaciology*, 30, 94–105.

Kuhle, M. 1988. Topography as a fundamental element of glacial systems. *Geojournal* 17, 545–568.

Kupetz, M., 1997. Geologischer Bau und Genese der Stauchendmoräne Muskauer Faltenbogen [Geological structure and genesis of the Muskau push moraine]. *Brandenburgische Geowissenschaftliche Beiträge*, 4, 1–20.

Kuster, H. and Meyer, K.-D., 1979. Glaziäre Rinnen im mittleren und nordöstlichen Niedersachsen. *Eiszeitalter und Gegenwart*, 29, 135–156.

Kutzbach, J.E., 1987. Model simulations of the climatic patterns during the deglaciation of North America. In W.F. Ruddiman and H.E. Wright, Jr. (eds), *North America and adjacent oceans during the last deglaciation*. Geological Society of America, The Geology of North America, v. K-3, pp. 425–446.

Kuvaas, B. and Kristoffersen, Y., 1991. The Crary Fan: a trough mouth fan on the Weddell Sea continental margin, Antarctica. *Marine Geology*, 97, 345–362.

Laban, C., 1995. *The Pleistocene glaciation in the Dutch sector of the North Sea: a synthesis of sedimentary and seismic data.* Unpublished PhD thesis, University of Amsterdam.

Laberg, J.S. and Vorren, T.O., 1995. Late Weichselian submarine debris flow deposits on the Bear Island Trough Mouth Fan. *Marine Geology*, 127, 45–72.

Laberg, J.S., Vorren, T.O., 1996a The Middle and Late Pleistocene evolution of the Bear Island Trough Mouth Fan. In A. Solheim, F. Riis, A. Elverhøi, J.I. Faleide, L.N. Jensen and S. Cloetingh (eds), *Impact of glaciations on basin evolution: data and models from the Norwegian margin and adjacent areas.* Global and Planetary Change 12, pp. 309–330

Laberg, J.S. and Vorren, T.O., 1996b. The glacier fed fan at the mouth of Storfjorden Trough, western Barents Sea: a comparative study. *Geologische Rundschau*, 85, 338–349.

Laberg, J.S. and Vorren, T.O., 2000. Flow behaviour of the submarine glacigenic debris flows on the Bear Island Trough Mouth Fan, western Barents Sea. *Sedimentology*, 47, 1105–1117.

Lagerbäck, R., 1988. The Veiki moraines in northern Sweden – widespread evidence of an Early Weichselian deglaciation. *Boreas*, 17, 469–486.

Lagoe, M.B., 1980. Recent arctic Foraminifera: an overview. In M.E. Field, A.H. Bouma, I.P. Colburn, R.G. Douglas and J.C. Ingle (eds), *Proceedings of the Quaternary depositional environments of the Pacific Coast, Pacific Coast Paleogeography Symposium*, 4, 33–42.

Lamoureux, S.F., 1999. Catchment and lake controls over the formation of varves in monomictic Nicolay Lake, Cornwall Island, Nunavat. *Canadian Journal of Earth Sciences*, 36, 1533–1546.

Lamplugh, G.W., 1911. On the shelly moraine of the Sefstromglacier and other Spitsbergen phenomena illustrative of the British glacial conditions. *Proceedings of the Yorkshire Geological Society*, 17, 216–241.

Larsen, C.E., 1987. Geological history of glacial Lake Algonquin and the Upper Great Lakes. *US Geological Survey Bulletin*, B1801.

Larsen, E. and Mangerud, J., 1992. Subglacially formed clastic dikes. *Sveriges Geologiska Undersoekning*, Serie Ca. 4, 163–170.

Lauriol, B. and Gray, J.T., 1980. *Processes responsible for the concentration of boulders in the intertidal zone in Leaf Basin, Ungava.* Geological Survey of Canada, Paper, 80-10, pp. 281–292.

Lawrance, C.J., 1972. *Terrain evaluation in west Malaysia: Part I: Terrain classification and survey methods.* Transport and Road Research Laboratory, Crowthorne, UK, Report LR 506.

Lawson, D.E., Strasser, J.C., Evenson, E.B., Alley, R.B., Larson, G.J. and Arcone, S.A., 1998. Glaciohydraulic supercooling: a freeze-on mechanism to create stratified, debris-rich basal ice. I. Field evidence. *Journal of Glaciology*, 44, 547–562.

Lawson, D.E., 1979. *Sedimentological analysis of the western terminus region of the Matanuska Glacier, Alaska.* US Army Cold Regions Research and Engineering Laboratory, 79-9, 112 pp.

Lawson, D.E., 1981. *Sedimentological characteristics and classification of depositional processes and deposits in the glacial environment.* US Army Cold Regions Research and Engineering Laboratory Report No. 81-27.

Lawson, D.E., 1988. Glacigenic resedimentation: classification concepts and application to mass-movement processes and deposits. In R.P. Goldthwait and C.L. Matsch (eds), *Genetic Classificaion of Glacial Deposits.* Balkema, Rotterdam, pp. 147–169.

Lawson, W., 1996. Structural evolution of Variegated Glacier, Alaska, USA, since 1948. *Journal of Glaciology*, 42, 261–270.

Lee, H.A., 1959. *Surficial Geology of Southern District of Keewatin and the Keewatin Ice Divide NWT.* Geological Survey of Canada, Bulletin 51.

Lefauconnier, B. and Hagen, J.O., 1990. Glaciers and climate in Svalbard: statistical analysis and reconstruction of the Brøgger Glacier mass balance for the last 77 years. *Annals of Glaciology*, 14, 148–152.

Lefauconnier, B. and Hagen, J.O., 1991. Surging and calving glaciers in eastern Svalbard. *Norsk Polarinstitutt Medd*, 116.

Lefauconnier, B., Hagen, J.O., Orbaek, J.B., Melvold, K. and Isaksson, E., 1999. Glacier balance trends in the Kongsfjorden area, western Spitsbergen, Svalbard, in relation to the climate. *Polar Research*, 18, 307–313.

Lehmann, R., 1992. Arctic push moraines, a case study of the Thompson Glacier moraine, Axel Heiberg Island, NWT, Canada. *Zeitschrift fur Geomorphologie* 86, 161–171.

Lehmkuhl, F. 1997. The spatial distribution of loess and loess-like sediments in the mountain areas of Central and High Asia. *Zeitschrift für Geomorphologie*, 111, 97–116.

Lemmen, D.S., 1989. The last glaciation of Marvin Peninsula, northern Ellesmere Island, high arctic, Canada. *Canadian Journal of Earth Sciences*, 26, 2578–2590.

Lemmen, D.S., 1990. Glaciomarine sedimentation in Disraeli Fiord, High Arctic Canada. *Marine Geology*, 94, 9–22.

Lemmen, D.S., Evans, D.J.A. and England, J., 1988. Canadian landform examples 10: ice shelves of northern Ellesmere Island, NWT. *Canadian Geographer*, 32, 363–367.

Lemmen, D.S., Gilbert, R. and Aitken, A.E., 1991. *Quaternary investigations in the Expedition Fiord area, west-central Axel Heiberg Island, NWT.* Geological Survey of Canada, Paper 91-1B, pp. 1–7.

Lemmen, D.S., Aitken, A.E. and Gilbert, R., 1994a. Early Holocene deglaciation of Expedition and Strand fiords, Canadian high arctic. *Canadian Journal of Earth Sciences*, 31, 943–958.

Lemmen, D.S., Duk-Rodkin, A. and Bednarski, J.M., 1994b. Late glacial drainage systems along the northwestern margin of the Laurentide Ice Sheet. *Quaternary Science Reviews*, 13, 805–825.

Leonard, E.M., 1997. The relationship between glacial activity and sediment production: evidence from a 4450-year varve record of neoglacial sedimentation in Hector Lake, Alberta, Canada. *Journal of Palaeolimnology*, 17, 319–330.

Leonard, E.M., 1985. Glaciological and climatic controls on lake sedimentation, Canadian Rocky Mountains. *Zeitschrift für Gletscherkunde und Glazialgeologie*, 21, 35–42.

Leonard, E.M., 1986. Varve studies at Hector Lake, Alberta, Canada, and the relationship between glacial activity and sedimentation. *Quaternary Research*, 25, 199–214.

Leventer, A., Domack, E.W., Ishman, S.E., Brachfeld, S., McClennen, C.E. and Manley, P., 1996. Productivity cycles of 200-300 years in the Antarctic Peninsula region: understanding linkages among sun, atmosphere, oceans, sea ice and biota. *Geological Society of America Bulletin*, 108, 1626–1644.

Leventer, A., Dunbar, R.B. and DeMaster, D.J., 1993. Diatom evidence for late Holocene climatic events in Granite Harbor, Antarctica. *Paleoceanography*, 8, 373–386.

Levson, V.M. and Rutter, N.W., 1989a. Late Quaternary stratigraphy, sedimentology and history of the Jasper townsite area, Alberta, Canada. *Canadian Journal of Earth Sciences*, 26, 1325–1342.

Levson, V.M. and Rutter, N.W., 1989b. A lithofacies analysis and interpretation of depositional environments of montane glacial diamictons, Jasper, Alberta, Canada. In R.P. Goldthwait and C.L. Matsch (eds), *Genetic classification of glacigenic deposits*. Balkema, Rotterdam, pp. 117–140.

Li Jijun, E., Derbyshire, E. and Xu S, 1984. Glacial and paraglacial sediments of the Hunza Valley, North-West Karakoram, Pakistan: a preliminary analysis. In K. Millar (ed.), *International Karakoram Project*. Cambridge University Press, Cambridge, pp. 496–535.

Lian, O.B. and Hickin, E.J., 1996. Early postglacial sedimentation of lower Seymour Valley, southwestern British Columbia. *Géographie Physique et Quaternaire*, 50, 95–102.

Liedtke, H., 1981. *Die nordischen Vereisungen in Mitteleuropa*. 307 pp.

Lien, R., 1983. *Pløyemerker etter isfjell på norsk kontinentalsokkel.* Institutt for Kontinentalsokkelundersøkelser, Trondheim, 109, 147 pp.

Liestøl, O., 1988. The glaciers in the Kongsfjorden area, Spitsbergen. *Norsk Geografisk Tidsskrift*, 42, 231–238.

Lineback, J.A., Bleuer, N.K., Mickelson, D.M., Farrand, W.R. and Goldthwait, R.P., 1983. *Quaternary Geologic Map of the Chicago 4° × 6° quadrangle, United States.* US Geological Survey Miscellaneous Investigations Series Map I-1420 (NK-16), scale 1:1,000,000.

Lingle, C.S.T., Hughes, T.J. and Kollmeyer, R.C., 1981. Tidal flexure of Jakobshavns Glacier, West Greenland. *Journal of Geophysical Research*, 86(B5), 3960–3968.

Lisitzyn, A.P., 1972. *Sedimentation in the World Ocean*. Society of Econ. Paleon. Min., Special Publication 17, 218 pp.

Liverman, D.G.E., 1987. Sedimentation in ice-dammed Hazard Lake, Yukon. *Canadian Journal of Earth Sciences*, 24, 1797–1806.

Lliboutry, L.A., 1977. Glaciological problems set by the control of dangerous lakes in the Cordillera Blanca, Peru. II. Movement of a covered glacier embedded within a rock glacier. *Journal of Glaciology*, 18, 255–273.

Lliboutry, L.A., 1986. Discharge of debris by glacier Hatunraju, Cordillera Blanca, Peru. *Journal of Glaciology*, 32, 133.

Long, D., Laban, C., Streif, H., Cameron, T.D.J. and Schüttenhelm, R.T.E., 1988. The sedimentary record of climatic variation in the southern North Sea. *Philosophical Transactions of the Royal Society of London*, B 318, 523–537.

Longva, O. and Thorsnes, T., 1997. Skagerrak in the past and at the present – an integrated study of geology, chemistry, hydrography and microfossil ecology. *Norges Geologiske Undersøkelse Special Publication*, 8, 98 pp.

Lønne, I., 1995. Sedimetary facies and depositional architecture of ice-contact glacimarine systems. *Sedimentary Geology*, 98, 13–43.

Lønne, I., 2001. Dynamics of marine glacier termini read from moraine architecture. *Geology*, 29, 199–202.

Lønne, I. and Lauritsen, T., 1996. The architecture of a modern push-moraine at Svalbard as inferred from ground-penetrating radar measurements. *Arctic and Alpine Research*, 28, 488–495.

Lønne, I. and Syvitski, J.P.M., 1997. Effects of the readvance of an ice margin on the seismic character of the underlying sediment. *Marine Geology*, 143, 81–102.

Lorrain, R.D. and Demeur, P., 1985. Isotopic evidence for relict Pleistocene glacier ice on Victoria Island, Canadian Arctic Archipelago. *Arctic and Alpine Research*, 17, 89–98.

Lorrain, R.D., Souchez, R.A. and Tison, J-L., 1981. *Characteristics of basal ice from two outlet glaciers in the Canadian arctic – implications for glacier erosion.* Geological Survey of Canada, Paper 81-1B, pp. 137–144.

Lorrain, R.D., Fitzsimons, S.J., Vandergoes, M.J. and Stievenard, M., 1999. Ice composition evidence for the formation of basal ice from lake water beneath a cold-based Antarctic glacier. *Annals of Glaciology*, 28, 277–281.

Lowe, D.R., 1975. Water escape structures in coarse-grained sediments. *Sedimentology*, 22, 157–204.

Lowell, T.V. and Kite, J.S., 1986. Glaciation style of northwestern Maine. In J.S. Kite, T.V. Lowell and W.B. Thompson (eds), *Contributions to the Quaternary geology of northern Maine and adjacent Canada.* Maine Geological Survey, Bulletin 37, 75–85.

Lucchitta, B.K., Rosanova, C.E. and Mullins, K.F., 1995. Velocities of Pine Island Glacier, West Antarctica from ERS-1 SAR images. *Annals of Glaciology*, 21, 277–283.

Luckman, B.H., 1981. The geomorphology of the Alberta Rocky Mountains – a review and commentary. *Zeitschrift für Geomorphologie Supplementband*, 37, 91–119.

Luckman, B.H. and Fiske, C.J., 1995. Estimating long-term rockfall accretion rates by lichenometry. In O. Slaymaker (ed.), *Steepland geomorphology.* Wiley, Chichester, pp. 233–255.

Lundqvist, J., 1981. Moraine morphology – terminological remarks and regional aspects. *Geografiska Annaler*, 63A, 127–138.

Lundqvist, J., Clayton, L. and Mickelson, D.M., 1993. Deposition of the late Wisconsin Johnstown Moraine, south-central Wisconsin. *Quaternary International*, 18, 53–59.

Lyså, A. and Ida Lønne, I., 2001. Moraine development at a small High-Arctic valley glacier: Rieperbreen, Svalbard. *Journal of Quaternary Science*, 16, 519–529.

Maag, H., 1969. *Ice-dammed lakes and marginal glacial drainage on Axel Heiberg Island.* Axel Heiberg Island Research Report, McGill University, Montreal.

Maag, H., 1972. *Ice-dammed lakes on Axel Heiberg Island, with special reference to the geomorphological effect of the outflowing lake water.* Axel Heiberg Island, Research Report, McGill University, Montreal.

Mabbutt, J.A., 1968. Review of concepts of land classification. In G.A. Stewart (ed.), *Land evaluation.* Macmillan, Melbourne, pp. 11–28.

Mackay, J.R., 1963. The Mackenzie Delta Area, N.W.T. Geographical Branch, Memoir 8.

Mackay, J.R., 1971. The origin of massive icy beds in permafrost, western Arctic coast, Canada. *Canadian Journal of Earth Sciences*, 8, 397–422.

Mackay J.R. 1979. Pingos of the Tuktoyaktuk Peninsula area, Northwest Territories. *Géographie physique et Quaternaire*, 33, 3–61.

Mackay, J.R., 1983. Oxygen isotope variations in permafrost, Tuktoyaktuk Peninsula area, Northwest Territories. In *Current Research, Part B*, Geological Survey of Canada, Paper 83-1B, pp. 67–74.

Mackay, J.R. and Mathews, W.H., 1964. The role of permafrost in ice thrusting. *Journal of Geology*, 72, 378–380.

Mackiewicz, N.E., Powell, R.D., Carlson, P.R. and Molnia, B.F., 1984. Interlaminated ice-proximal glacimarine sediments in Muir Inlet, Alaska. *Marine Geology*, 57, 113–147.

MacLean, B., 1997. Submarine lateral moraine in the south central region of Hudson Strait, Canada. In T.A. Davies, T. Bell, A.K. Cooper *et al.* (eds.), *Glaciated continental margins: an atlas of acoustic images*. Chapman and Hall, London, pp. 98–99.

Macoun, J. and Králík, F., 1995. Glacial history of the Czech Republic. In J. Ehlers, S. Kozarski and P. Gibbard (eds), *Glacial deposits in north-east Europe*. Balkema, Rotterdam, pp. 389–405.

Mahr, T., 1977. Deep-reaching gravitational deformations of high-mountain slopes. *Bulletin of the International Association of Engineering Geologists*, 16, 121–127.

Maizels, J., 1977. Experiments on the origin of kettle holes. *Journal of Glaciology*, 18, 291–303.

Maizels, J., 1989. Differentiation of late Pleistocene terrace outwash deposits using geomorphic criteria: Tekapo Valley, South Island, New Zealand. *New Zealand Journal of Geology and Geophysics*, 32, 225–241.

Maizels, J., 1992. Boulder ring structures produced during jokulhlaup flows: origin and hydraulic significance. *Geografiska Annaler*, 74A, 21–33.

Maizels, J.K. (1995) Sediments and landforms of modern proglacial terrestrial environments. In Menzies, J. (ed.). Modern Glacial Environments. Butterworth-Heinemann, Oxford, 365-416.

Maizels, J., 1997. Jokulhlaup deposits in proglacial areas. *Quaternary Science Reviews*, 16, 793–819.

Maizels, J., 2002. Sediments and landforms of modern proglacial terrestrial environments. In J. Menzies (ed.), *Modern and past glacial environments*. Butterworth-Heinemann, Oxford, pp. 279–316.

Malmberg Persson, K., 1991. Internal structures and depositional environment of a kame deposit at V. Ingelstad, Skåne, southern Sweden. *Geologiska Föreningens i Stockholm Förhandlingar*, 113, 163–170.

Maltman, A., 1987. Shear zones in argillaceous sediments – an experimental study. In M.E. Jones and R.M. Preston (eds), *Deformation of sediments and sedimentary rocks*. Geological Society, Special Publication 29, pp. 77–87.

Maltman, A., 1994. Deformation structures preserved in rocks. In A.J. Maltman, B. Hubbard and M.J. Hambrey (eds), *Deformation of glacial materials*. Geological Society Special Publication no. 176, London, 261–307.

Manabe, S. and Broccoli, A.J., 1984. Ice-age climate and continental ice sheets: some experiments with a general circulation model. *Annals of Glaciology*, 5, 100–105.

Mandl, G. and Harkness, R.M., 1987. Hydrocarbon migration by hydraulic fracturing. In M.E. Jones and R.M.F. Preston (eds.), *Deformation of sediments and sedimentary rocks*. Geological Society of London, Special Publication 29, pp. 39–53.

Mangerud, J., Svendsen, J.I. and Astakhov, V., 1999. Age and extent of the Barents and Kara ice sheets in northern Russia. *Boreas*, 28, 46–80.

Mangerud, J., Astakhov, V., Jakobsson, M. and Svendsen, J.I., 2002. Huge ice-age lakes in Russia. *Journal of Quaternary Science*, 16, 773–777.

Manley, G., 1955. On the occurrence of ice-domes and permanently snow-covered summits. *Journal of Glaciology*, 2, 453–456.

Manley, G., 1959. The late-glacial climate of north-west England. Liverpool and Manchester *Geological Journal*, 2, 188–215.

Marcussen, I., 1973. Studies of flow till in Denmark. *Boreas*, 2, 213–231.

Marion, J., Filion, L. and Hétu, B., 1995. The Holocene development of a debris slope in subarctic Québec, Canada. *The Holocene*, 5, 409–419.

Marks, L., 2002. Last glacial maximum in Poland. *Quaternary Science Reviews*, 21, 103–110.

Marshall, S.J., Clarke, G.K.C., Dyke, A.S. and Fisher, D.A., 1996. Geologic and topographic controls on fast flow in the Laurentide and Cordilleran Ice Sheets. *Journal of Geophysical Research*, 101(B8), 17827–17839.

Mathews, W.H., 1974. Surface profile of the Laurentide Ice Sheet in its marginal areas. *Journal of Glaciology*, 13, 37–43.

Mathews, W.H., 1991. Ice sheets and ice streams: Thoughts on the Cordilleran Ice Sheet Symposium. *Geographie Physique et Quaternaire*, 45, 263–267.
Mathews, W.H. and Mackay, J.R., 1960. Deformation of soils by glacier ice and the influence of pore pressures and permafrost. *Transactions of the Royal Society of Canada*, Section 3, 54, 27–36.
Matthews, J.A. and Petch, J.R., 1982. Within-valley asymmetry and related problems of Neoglacial lateral moraine development at certain Jotunheimen glaciers, southern Norway. *Boreas*, 11, 225–247.
Matthews, J.A., Cornish, R. and Shakesby, R.A., 1979. 'Saw-tooth' moraines in front of Bodalsbreen, southern Norway. *Journal of Glaciology*, 22, 535–546.
Matthews, J.A., McCarroll, D. and Shakesby, R.A., 1995. Contemporary terminal moraine ridge formation at a temperate glacier: Styggedalsbreen, Jotunheimen, southern Norway. *Boreas*, 24, 129–139.
Matthews, J.A., Shakesby, R.A., Berrisford, M.S. and McEwen, L.J., 1998. Periglacial patterned ground in the Styggedalsbreen glacier foreland, Jotunheimen, southern Norway: micro-topographical, paraglacial and geochronological controls. *Permafrost and Periglacial Processes*, 9, 147–166.
Mattson, L.E. and Gardner, J.S., 1991. Mass wasting on valley-side ice-cored moraines, Boundary Glacier, Alberta, Canada. *Geografiska Annaler*, 73A, 123–128.
Matveyev, A., 1995. Glacial history of Belarus. In J. Ehlers, S. Kozarski and P. Gibbard (eds), *Glacial deposits in north-east Europe*. Balkema, Rotterdam, pp. 267–276.
Matveyev, A. and Nechiporenko, L.A., 1995. The influence of tectonic activity on glacial deposition and landforms in Belarus. In J. Ehlers, S. Kozarski and P. Gibbard (eds), *Glacial deposits in north-east Europe*. Balkema, Rotterdam, pp. 277–284.
Maxwell, J.B., 1980. *The climate of the Canadian arctic islands and adjacent waters*. Canada Department of Environment, Atmospheric Environment Service, Climatological Studies no. 30, V. 1, 532pp.
McCabe, A.M., 1986. Glaciomarine facies deposited by retreating tidewater glaciers; an example from the late Pleistocene of Northern Ireland. *Journal of Sedimentary Petrology*, 56(6), 880–894.
McCabe, A.M. and Ó Cofaigh, C., 1995. Late Pleistocene morainal bank facies at greystones, eastern Ireland: an example of sedimentation during ice-marginal re-equilibration in an isostatically depressed basin. *Sedimentology*, 42, 647–664.
McCabe, A.M., Dardis, G.F. and Hanvey, P.M., 1984. Sedimentology of a late Pleistocene submarine-moraine complex, County Down, Northern Ireland. *Journal of Sedimentary Petrology*, 54(3), 716–730.
McCann, S.B. and Kostaschuck, R.A., 1987. Fiord sedimentation in northern British Columbia. In D.M. FitzGerald and P.S. Rosen (eds), *Glaciated coasts*. Academic Press, San Diego, pp. 33–49.
McDougall, D.A., 1998. *Loch Lomond Stadial Plateau Icefields in the Lake District, Northwest England.* Unpublished PhD Thesis, University of Glasgow.
McDougall, D.A., 2001. The geomorphological impact of Loch Lomond (Younger Dryas) Stadial plateau icefields in the central Lake District, northwest England. *Journal of Quaternary Science*, 16, 531–543.
McKee, E.D., 1965. Experiments on ripple lamination. In G.V. Middleton (ed.), *Primary sedimentary structures and their hydrodynamic interpretation*. SEPM, Special Publication 12, 66–83.
McKelvey, B.C., 1981. The lithologic logs of DVDP cores 10 and 11, eastern Taylor Valley. *AGU, Antarctic Research Series*, 33, 63–94.
McKenna-Neumann, C. and Gilbert, R., 1986. Aeolian processes and landforms in glaciofluvial environments of southeastern Baffin Island, N.W.T., Canada. In W.G. Nickling (ed.), *Aeolian geomorphology*. Allen and Unwin, Boston, pp. 213–235.
McKenzie, G.D. and Goldthwait, R.P., 1971. Glacial history of the last eleven thousand years in Adams Inlet, southeastern Alaska. *Geological Society of America Bulletin*, 82, 1767–1782.
McSaveney, M.J., 1992. The Mount Fletcher rock avalanche of May 2, and again on September 16. *New Zealand Alpine Journal*, 45, 99–103.
McSaveney, M.J., 1993. Rock avalanches of 2 May and 16 September 1992, Mount Fletcher, New Zealand. *Landslide News*, 7, 2–4.
Meier, M.F. and Post, A.S., 1962. Recent variations in mass net budgets of glaciers in western North American *IASH* 58, 63–77.

Meier, M.F. and Post, A.S., 1969. What are glacier surges? *Canadian Journal of Earth Sciences*, 6, 807–819.

Meigs, A. and Sauber, J., 2000. Southern Alaska as an example of the long-term consequences of mountain building under the influence of glaciers. *Quaternary Science Reviews*, 19, 1543–1562.

Melander, O., 1976. *Geomorfologiska kartbladet 29 J Kiruna–beskrivning och naturvärdsbedömning*. Statens Naturvårdsverk PM 741, 74 pp.

Melvold, K. and Hagen, J.O., 1998. Evolution of a surge-type glacier in its quiescent phase: Kongsvegen, Spitsbergen, 1964–95. *Journal of Glaciology*, 44, 394–404.

Menzies, J. and Shilts, W.W., 2002. Subglacial environments. In J. Menzies (ed.), *Modern and past glacial environments – revised student edition*. Butterworth-Heinemann, Oxford, pp. 183–278.

Merritt, J.W., Auton, C.A. and Firth, C.R., 1995. Ice-proximal glaciomarine sedimentation and sea-level change in the Inverness area, Scotland: a review of the deglaciation of a major ice stream of the British Late Devensian Ice Sheet. *Quaternary Science Reviews*, 14, 289–329.

Meyer, K.-D., 1983. Zur Anlage der Urstromtäler in Niedersachsen. *Zeitschrift für Geomorphologie*, N.F. 27, 147–160.

Meyer, K.-D., 1987. Ground and end moraines in Lower Saxony. In J.J.M. van der Meer (ed.), *Tills and glaciotectonics*. Balkema, Rotterdam, pp. 197–204.

Mickelson, D.M., 1986. *Glacial and related deposits of Langlade County, Wisconsin*. Wisconsin Geological and Natural History Survey Information Circular 52, 30 pp.

Mickelson, D.M., Clayton, L., Fullerton, D.S. and Borns, H.W., Jr., 1983. The Late Wisconsin glacial record of the Laurentide ice sheet in the United States. In H.E. Wright, Jr. (ed.), *Late Quaternary environments of the United States*. Vol. 1, University of Minnesota Press, Minneapolis, pp. 3–37.

Miller, D.C., Birkeland, P.W. and Rodbell, D.T., 1993. Evidence for Holocene stability of steep slopes, northern Peruvian Andes, based on soils and radiocarbon dates. *Catena*, 20, 1–12.

Miller, G.H., Bradley, R.S., Andrews, J.T. 1975. The glaciation level and lowest equilibrium line altitude in the high Canadian arctic: maps and climatic interpretation. *Arctic and Alpine Research* 7, 155–168.

Milthers, V., 1948. Det danske istidslandskabs terrænformer og deras opstaaen. *Danmarks Geologiske Undersøgelse*, III, Række. Nr. 28, 233 pp.

Minell, H., 1979. The genesis of tills in different moraine types and the deglaciation in a part of central Lappland. *Sveriges Geologiska Undersökning*, Serie C 754, Årsbok 72, No. 16, 82 pp.

Mitchell, C.W., 1973. *Terrain evaluation*. Longman, London.

Mohrig, D., Whipple, K.X., Hondzo, M., Ellis, C. and Parker, G., 1998. Hydroplaning of subaqueous debris flows. *Geological Society of America Bulletin*, 110, 387–394.

Mohrig, D., Elverhøi, A. and Parker, G., 1999. Experiments on the relative mobility of muddy subaqueous and subaerial debris flows, and their capacity to remobilize antecedent deposits. *Marine Geology*, 154, 117–129.

Mojski, J.E., 1995. Pleistocene glacial events in Poland. In J. Ehlers, S. Kozarski and P. Gibbard (eds), *Glacial deposits in north-east Europe*. Balkema, Rotterdam, pp. 287–292.

Mollard, J.D., 1983. The origin of reticulate and orbicular patterns on the floor of Lake Agassiz basin. In J.T. Teller and L. Clayton (eds), *Glacial Lake Agassiz*. Geological Association of Canada, Special Paper 26, 355–374.

Mollard, J.D., 2000. Ice-shaped ring forms in western Canada: their airphoto expression and manifold polygenetic origins. *Quaternary International*, 68, 187–198.

Möller, P., 1987. *Moraine morphology, till genesis, and deglaciation pattern in the Åsnen area, south-central Småland, Sweden*. Lundqua thesis, Lund University, 20, 146 pp.

Molnia, B.F., Carlson, P.R. and Levy, W.P., 1978. Holocene sediment volume and modern sediment yield, northeast Gulf of Alaska. *Bulletin of the American Association of Petroleum Geologists*, 62, 545.

Mooers, H.D., 1989a. On the formation of tunnel valleys of the Superior Lobe, central Minnesota. *Quaternary Research*, 32, 24–35.

Mooers, H.D., 1989b. Drumlin formation: a time transgressive model. *Boreas*, 18, 99–107.

Mooers, H.D., 1990. Ice-marginal thrusting of drift and bedrock: thermal regime, subglacial aquifers, and glacial surges. *Canadian Journal of Earth Sciences*, 27, 849–862.

Moran, K. and Fader, G.B.J., 1997. Glacial and glacimarine sedimentation: Halibut Channel, Grand Banks of Newfoundland. In T.A. Davies, T. Bell, A.K. Cooper *et al.* (eds), *Glaciated continental margins: an atlas of acoustic images*. Chapman and Hall, London, pp. 98–99

Moran, S.R., Clayton, L., Hooke, R.LeB., Fenton, M.M. and Andriashak, L.D., 1980. Glacier-bed landforms of the Prairie region of North America. *Journal of Glaciology*, 25, 457–476.

Morehead, M.D. and Syvitski, J.P.M., 1999. River-plume sedimentation modeling for sequence stratigraphy; application to the Eel margin, Northern California. *Marine Geology*, 154, 29–41.

Morgan, V.I., Jacka, T.H., Akerman, G.J. and Clarke, A.L., 1982. Outlet glacier and mass-budget studies in Enderby, Kemp and Mac Robertson lands, Antarctica. *Annals of Glaciology*, 3, 204–210.

Morlan, R.E., Nelson, D.E., Brown, T.A., Vogel, J.S. and Southon, J.R., 1990. Accelerator mass spectrometry dates on bones from Old Crow Basin, northern Yukon Territory. *Canadian Journal of Archaeology*, 14, 75–92.

Morris, S.E., 1981. Topoclimatic factors and the development of rock glacier facies, Sangre de Cristo Mountains, southern Colorado. *Arctic and Alpine Research*, 13, 329–338.

Mougeot, C.M., 1995. *The Quaternary Geology of the Vermilion – East Area, Alberta*. Unpublished MSc Thesis, University of Alberta.

Mulder, T. and Moran, K., 1995. Relationship among submarine instabilities, sea level variations, and the presence of an ice sheet on the continental shelf: An example from the Verrill Canyon Area, Scotian Shelf. *Paleoceanography*, 10, 137–154.

Mulder, T. and Syvitski, J.P.M., 1995. Turbidity currents generated at river mouths during exceptional discharges to the world oceans. *Journal of Geology*, 103, 285–299.

Mulder, T., Syvitski, J.P.M. and Skene, K.I., 1998. Modeling of erosion and deposition by turbidity currents generated at river mouths. *Journal of Sedimentary Research*, 68A, 124–137.

Mulholland, J.W., 1982. Glacial stagnation-zone retreat in New England: bedrock control. *Geology*, 10, 567–571.

Müller, B.U., 1999. Paraglacial sedimentation and denudation processes in an Alpine valley of Switzerland. An approach to the quantification of sediment budgets. *Geodinamica Acta*, 12(5), 291–301.

Müller, F., 1976. On the thermal regime of a high-Arctic valley glacier. *Journal of Glaciology*, 16, 119–133.

Müller, F., 1980. Present and late Pleistocene equilibrium line altitudes in the Mt Everest region – an application of the glacier inventory. *IASH Publication* 126 World Glacier Inventory, 75–94.

Müller, U., Rühberg, N. and Krienke, H.-D., 1995. The Pleistocene sequence in Mecklenburg-Vorpommern. In J. Ehlers, S. Kozarski and P. Gibbard (eds), *Glacial deposits in north-east Europe*. Balkema, Rotterdam, pp. 501–514.

Mullins, H.T., Eyles, N. and Hinchley, E.J., 1990. Seismic reflection investigation of Kalamalka Lake: a 'fiord lake' on the interior plateau of southern British Columbia. *Canadian Journal of Earth Sciences*, 27, 1225–1235.

Munro, M. and Shaw, J., 1997. Erosional origin of hummocky terrain in south-central Alberta, Canada. *Geology*, 25, 1027–1030.

Murray, T., Gooch, D.L. and Stuart, G.W., 1997. Structures within the surge front at Bakaninbreen, Svalbard, using ground-penetrating radar. *Annals of Glaciology*, 24, 122–129.

Naito, N., Nakawo, M., Kadota, T. and Raymond, C., 2000. Numerical simulation of recent shrinkage of Khumbu Glacier, Nepal Himalaya. In M. Nakawo, A. Fountain and C. Raymond (eds), *Debris-covered glaciers*. *IAHS Publication* 264, 245–254.

Nakawo, M. and Young, G.J., 1981. Field experiments to determine the effect of a debris layer on ablation of glacier ice. *Annals of Glaciology*, 2, 85–91.

Nakawo, M., Yabuki, H. and Sakai, A., 1999 Characteristics of Khumbu Glacier, Nepal Himalaya: recent changes in debris-covered area. *Annals of Glaciology*, 28, 118–122.

Nansen, F., 1904. The bathymetrical features of the north Polar Seas, with a discussion of the continental shelves and previous oscillations of the shore-line. In F. Nansen (ed.), *The Norwegian north polar Expedition 1893–1896 Scientific results*, vol. IV, 1-232.

Nemec, W., 1990. Aspects of sediment movement on steep delta slopes. In A. Colella and D.B. Prior (eds), *Coarse-grained deltas*. International Association of Sedimentologists, Special Publication, 10, 29–74.

Nesje, A. and Dahl, S.O., 2000. *Glaciers and Environmental Change*. Arnold, London, 203 pp.

Nesje, A. and Whillans, I.M., 1994. Erosion of Sognefjord, Norway. *Geomorphology*, 9, 33–45.

Newman, W.A. and Mickelson, D.M., 1994. Genesis of Boston Harbor Drumlins. *Sedimentary Geology*, 9, 333–343.

Nichols, R.J., Sparks, R.S.J. and Wilson, C.J.N., 1994. Experimental studies of the fluidization of layered sediments and the formation of fluid escape structures. *Sedimentology*, 41, 233–253.

Nicholson, L., 2000. *The geomorphological impact of climate change on glaciers in the Ngozumpa Valley, Khumbu Himal, Nepal*. Unpublished BSc Dissertation, University of Edinburgh.

Niewiarowski, W., 1963. Types of kames occurring within the area of the last glaciation in Poland as compared with kames from other regions. In J. Dylik (ed.), *INQUA, Report of the VIth International Congress on Quaternary*, 3, Warsaw, 1961, 475-485.

Nummedal, D., Hine, A.C. and Boothroyd, J.C., 1987. Holocene evolution of the south-central coast of Iceland. In D.M. FitzGerald and P.S. Rosen (eds), *Glaciated coasts*. Academic Press, San Diego, pp. 115–150.

Nürnberg, D., Wollenburg, I., Dethleff, D. *et al.*, 1994. Sediments in Arctic sea ice: Implications for entrainment, transport and release. *Marine Geology*, 119, 185–214.

Ó Cofaigh, C., 1998. Geomorphic and sedimentary signatures of early Holocene deglaciation in high arctic fiords, Ellesmere Island, Canada: implications for deglacial ice dynamics and thermal regime. *Canadian Journal of Earth Sciences*, 35, 437–452.

Ó Cofaigh, C. and Dowdeswell, J.A., 2001. Laminated sediments in glacimarine environments: diagnostic criteria for their interpretation. *Quaternary Science Reviews*, 20, 1411–1436.

Ó Cofaigh, C. and Evans, D.J.A., 2001a. Deforming bed conditions associated with a major ice stream of the last British ice sheet. *Geology*, 29, 795–798.

Ó Cofaigh, C. and Evans, D.J.A., 2001b. Sedimentary evidence for deforming bed conditions associated with a grounded Late Devensian Irish Sea glacier, southern Ireland. *Journal of Quaternary Science*, 16, 435–454.

Ó Cofaigh, C., Lemmen D.S., Evans, D.J.A. and Bednarski, J., 1999. Glacial landform/sediment assemblages in the Canadian High Arctic and their implications for late Quaternary glaciation. *Annals of Glaciology*, 28, 195–201.

Ó Cofaigh, C., England, J. and Zreda, M., 2000. Late Wisconsinan glaciation of southern Eureka Sound: evidence for extensive Innuitian ice in the Canadian high arctic during the last glacial maximum. *Quaternary Science Reviews*, 19, 1319–1341.

Ó Cofaigh, C., Dowdeswell, J.A. and Grobe, H., 2001. Holocene glacimarine sedimentation, inner Scoresby Sund, East Greenland: the influence of fast-flowing ice sheet outlet glaciers. *Marine Geology*, 175, 103–129.

Ó Cofaigh, C., Evans, D.J.A. and England, J., 2003. Ice-marginal terrestrial landsystems: sub-polar glacier margins of the Canadian and Greenland high arctic. In D.J.A. Evans (ed.), *Glacial landsystems*. Arnold, London.

O'Sullivan, P.E., 1983. Annually-laminated lake sediments and the study of Quaternary environmental changes – a review. *Quaternary Science Reviews*, 1, 243–313.

Ødegard, R.S., Hamran, S.E., Bø, P.H., Etzelmüller, B., Vatne, G. and Sollid, J.L., 1992. Thermal regime of a valley glacier, Erikbreen, northern Spitsbergen. *Polar Research*, 11, 69–79.

Oestreich, K., 1906. Die Täler des nordwestlichen Himalaya. *Petermans Mitteillungen*, 155(33).

Okko, M. and Perttunen, M., 1971. A moundfield in the second Salpausselkä ice-marginal belt at Kurhila, southern Finland. *Bulletin of the Geological Society of Finland*, 43, 47–54.

Oldale, R.N. and O'Hara, C.J., 1984. Glaciotectonic origin of the Massachusetts coastal end moraines and a fluctuating late Wisconsinan ice margin. *Geology*, 95, 61–74.

Ollier, C.D., 1977. Terrain classification, principles and applications. In J.R. Hails (ed.), *Applied geomorphology*. Elsevier, Amsterdam, 277–316.

Olszewski, A. and Weckwerth, P., 1997. The morphogenesis of kettles in the Höfðabrekkujökull forefield, Mýrdalssandur, Iceland. *Jökull*, 47, 71–88.

Orheim, O. and Elverhøi, A., 1981. Model for submarine glacial deposition. *Annals of Glaciology*, 2, 123–128.

Østrem, G., 1959. Ice melting under a thin layer of moraine and the existence of ice in moraine ridges. *Geografiska Annaler*, 41, 228–230.

Østrem, G., 1963. Comparative crystallographic studies on ice from ice-cored moraines, snow banks and glaciers. *Geografiska Annaler*, 45, 210–240.

Østrem, G., 1976. Sediment transport in glacial meltwater streams. In A.V. Jopling and B.C. McDonald (eds), *Glaciofluvial and glaciolacustrine sedimentation.* SEPM, Special Paper 23, 101–122.

Ottesen, D., Rise, L., Rokoengen, K. and Sættem, J., 2001. Glacial processes and large-scale morphology on the mid-Norwegian continental shelf. In O.J. Martinsen and T. Dreyer (eds), *Sedimentary environments offshore Norway – Paleozoic to Recent.* NPF Special Publication 10. Elsevier Science B.V., Amsterdam, pp. 441–449

Otvos, E.G., 2000. Beach ridges – definitions and significance. *Geomorphology*, 32, 83–108.

Ovenshine, A.T., 1970. Observations of iceberg rafting in Glacier Bay, Alaska, and the identification of ancient ice-rafted deposits. *Geological Society of America Bulletin*, 81, 891–894.

Owen, L.A., 1989. Terraces, uplift and climate in the Karakoram Mountains, northern Pakistan: Karakoram intermontane basin evolution. *Zeitschrift für Geomorphologie, Supplementband*, 76, 117–146.

Owen, L.A., 1991. Mass movement deposits in the Karakoram Mountains: their sedimentary characteristics, recognition and role in Karakoram landform evolution. *Zeitschrift für Geomorphologie*, 35, 401–424.

Owen, L.A., 1994. Glacial and non-glacial diamictons in the Karakoram Mountains and western Himalayas. In W.P. Warren and D.G. Croot (eds), *Formation and deformation of glacial deposits.* Balkema, Rotterdam, pp. 9–28.

Owen, L.A. and Derbyshire, E., 1989. The Karakoram glacial depositional system. *Zeitschrift für Geomorphologie Supplementband*, 76, 33–73.

Owen, L.A. and Derbyshire, E. 1993. Quaternary and Holocene intermontane basin sedimentation in the Karakoram Mountains. In J.F. Shroder (ed.), *Himalaya to the sea.* Routledge, London, pp. 108–131.

Owen, L.A., Derbyshire, E. and Scott, C.H., 2002. Contemporary sediment production and transfer in high-altitude glaciers. *Sedimentary Geology*, 155, 13–36.

Owen, L.A. and England, J., 1998. Observations on rock glaciers in the Himalayas and Karakoram Mountains of northern Pakistan and India. *Geomorphology*, 26, 199–214.

Owen, L.A. and Sharma, M.C., 1998. Rates and magnitudes of paraglacial fan formation in the Garwhal Himalaya: implications for landscape evolution. *Geomorphology*, 26, 171–184.

Owen, L.A., Benn, D.I., Derbyshire, E. *et al.*, 1995. The geomorphology and landscape evolution of the Lahul Himalaya, northern India. *Zeitschift für Geomorphologie*, 39, 145–174.

Palacios, D. and de Marcos, J., 1998. Glacial retreat and its geomorphological effects on Mexico's active volcanoes, 1994–95. *Journal of Glaciology*, 44, 63–67.

Palacios, D., Parilla, G. and Zamorano, J.J., 1999. Paraglacial and postglacial debris flows on a Little Ice Age terminal moraine: Jamapa Glacier, Pico de Orizabo (Mexico). *Geomorphology*, 28, 98–118.

Parizek, R.R., 1969. Glacial ice-contact rings and ridges. *Geological Society of America* Special Paper 123, 49–102.

Passchier, C.W. and Trouw, R.A.J., 1996. *Microtectonics.* Springer, Berlin, 289 pp.

Paterson, W.S.B., 1994. *The physics of glaciers.* Pergamon, Oxford.

Patterson, C.J., 1994. Tunnel-valley fans of the St. Croix moraine, east-central Minnesota, USA. In W.P. Warren and D.G. Croot (eds), *Formation and deformation of glacial deposits.* Balkema, Rotterdam, pp. 69–87.

Patterson, C.J., 1998. Laurentide glacial landscapes: the role of ice streams. *Geology*, 26, 643–646.

Patterson, C.J., 1997a. Surficial geology of southwestern Minnesota. In C.J. Patterson (ed.), *Contributions to Quaternary geology of southwestern Minnesota.* Minnesota Geological Survey Report of Investigation 47, 1–45.

Patterson, C.J., 1997b. Southern Laurentide ice lobes were created by ice streams: Des Moines lobe in Minnesota, USA. *Sedimentary Geology*, 111, 249–261.

Paul, M.A., 1983. The supraglacial landsystem. In N. Eyles (ed.), *Glacial geology.* Pergamon, Oxford, pp. 71–90.

Pelletier, B.R., 1966. Development of submarine physiography in the Canadian Arctic and its relation to crustal movement. *Royal Society of Canada, Special Publication*, 9, 77–101.

Pelto, M.S. and Warren, C.R., 1991. Relationship between tidewater calving velocity and water depths at the calving front. *Annals of Glaciology*, 15, 115–118.

Petrie, G. and Price, R.J., 1966. Photogrammetric measurements of the ice wastage and morphological changes near the Casement Glacier, Alaska. *Canadian Journal of Earth Sciences*, 3, 827–840.

Péwé, T.L., 1983. The periglacial environment in North America during Wisconsin time. In H.E. Wright and S.C. Porter (eds), *Late Quaternary environments of the United States*, University of Minnesota Press, Minneapolis, pp. 157-189.

Pfirman, S., Lange, M.A., Wollenburg, I. and Schlosser, P., 1989. Sea ice characteristics and the role of sediment inclusions in deep-sea deposition: Arctic and Antarctic comparison. In U. Bleil and J. Thiede (eds), *Geological history of the Polar Oceans: arctic versus antarctic*. NATO ASI ser. C308, 187–211.

Phillips, A.C., Smith, N.D. and Powell, R.D., 1991. Laminated sediments in prodeltaic deposits, Glacier Bay, Alaska. In J.B. Anderson and G.M. Ashley (eds), *Glacial marine sedimentation – paleoclimatic significance*. Geological Society America, Special Paper, 261, 51–60.

Picard, K., 1964. Der Einfluß der Tektonik auf das pleistozäne Geschehen in Schleswig-Holstein. *Schrifte des Naturwissenschaftlichen Vereins Schleswig-Holstein*, 35, 99–113.

Pickrill, R.A. and Irwin, J., 1983. Sedimentation in a deep glacier-fed lake – Lake Tekapo, New Zealand. *Sedimentology*, 30, 63–75.

Pickrill, R.A., 1993. Sediment yields in Fiordland. *Journal of Hydrology (New Zealand)*, 31, 39–55.

Piotrowski, J.A. and Kraus, A.M., 1997. Response of sediment to ice-sheet loading in northwestern Germany; effective stresses and glacier-bed stability. *Journal of Glaciology*, 43, 495–502.

Piotrowski, J.A. and Tulaczyk, S., 1999. Subglacial conditions under the last ice sheet in northwest Germany: ice-bed separation and enhanced basal sliding? *Quaternary Science Reviews*, 18, 737–751.

Piotrowski, J.A., Döring, U., Harder, A., Qadirie, R. and Wenghöfer, S., 1997. Deforming bed conditions on the Dänischer Wohld Peninsula, northern Germany: Comments. *Boreas*, 26, 73–77.

Piotrowski, J.A., Mickelson, D.M., Tulaczyk, S. and Krzyszkowski, D., 2001. Were deforming subglacial beds beneath past ice sheets really widespread? *Quaternary International*, 86, 139–150.

Piper, D.J.W., Letson, J.R.J., Delure, A.M. and Barrie, C.Q., 1983. Sediment accumulation in low-sedimentation, wave-dominated glacial inlets. *Sedimentary Geology*, 36, 195–215.

Piper, D.J.W., Mudie, P.J., Letson, J.R.J., Barnes, N.E. and Iulliucci, R.J., 1986. *The marine geology of the inner Scotian Shelf off the south shore, Nova Scotia*. Geological Survey of Canada, Paper, 85-19, 65 pp.

Plafker, G. and Addicott, W.O., 1976. Glaciomarine deposits of Miocene through Holocene age in the Yakataga Formation along the Gulf of Alaska margin, Alaska. In T.P. Miller (ed.), *Recent and ancient sedimentary environments in Alaska*. Proceedings of the Alaska Geological Symposium, pp. Q1–Q23.

Plassen, L., Vorren, T.O., 2002. Late Weichselian and Holocene sediment flux and sedimentation rates in Andfjord and Vågsfjord area, north Norway. *J Quaternary Science*, 17, 161–180.

Plink-Björklund, P. and Ronnert, L., 1999. Depositional processes and internal architecture of Late Weichselian ice-margin submarine fan and delta settings. *Sedimentology*, 46, 215–234.

Polyak, L., Edwards, M. H., Coakley, B.J. and Jakobsson, M., 2001. Ice shelves in the Pleistocene Arctic Ocean inferred from glaciogenic deep-sea bedforms. *Nature*, 410, 453–457.

Poole, D.A.R., Sættem, J. and Vorren, T.O., 1994. Foraminiferal stratigraphy, paleoenvironments and sedimentation of the glacigenic sequence south west of Bjørnøya. *Boreas*, 23, 122–138.

Porter, P.R., Murray, T. and Dowdeswell, J.A., 1997. Sediment deformation and basal dynamics beneath a glacier surge front: Bakaninbreen, Svalbard. *Annals of Glaciology*, 24, 21–26.

Porter, S.C. 1975 Equilibrium line altitudes of late Quaternary glaciers in the Sourthern Alps, New Zealand. *Quaternary Research* 5, 27–47.

Porter, S.C., 2001. Snowline depression in the tropics during the Last Glaciation. *Quaternary Science Reviews*, 20, 1067–1091.

Posamentier, H., Jervey, M. and Vail, P., 1988. Eustatic controls on clastic deposition; I – conceptual framework. In C.K. Wilgus, B. Hastings, C. Kendall, H. Posamentier, C. Ross and J. vanWagoner (eds), *Sea-level changes: an integrated approach*. SEPM, Special Publication 42, pp. 109–124.

Post, A.S., 1964. Effects of the March 1964 Alaska earthquake on glaciers. *United States Geological Survey Professional Paper* 544-D, 42 pp.

Post, A.S., 1969. Distribution of surging glaciers in western North America. *Journal of Glaciology*, 8, 229–240.

Postma, G., 1990 Depositional architecture and facies of river and fan deltas. In A. Colella and D.B. Prior (eds), *Coarse-grained deltas*. International Association of Sedimentologists, Special Publication 10, 13–27.

Powell, R.D., 1981a. A model for sedimentation by tidewater glaciers. *Annals of Glaciology*, 2, 129–134.

Powell, R.D., 1981b. Sedimentation conditions in Taylor Valley, Antarctica, inferred from textural analysis of DVDP cores. *AGU, Antarctic Research Series*, 33, 331–349.

Powell, R.D., 1983a. Discussion: Submarine flow tills at Victoria, British Columbia. *Canadian Journal of Earth Sciences*, 20, 509–510.

Powell, R.D., 1983b. Glacial-marine sedimentation processes and lithofacies of temperate tidewater glaciers, Glacier Bay, Alaska. In B.F. Molnia (ed.), *Glacial-marine sedimentation*. Plenum Publishing Co., N.Y., pp. 185–232.

Powell, R.D., 1984. Glacimarine processes and inductive lithofacies modelling of ice shelf and tidewater glacier sediments based on Quaternary examples. *Marine Geology*, 57, 1–52.

Powell, R.D., 1990. Glacimarine processes at grounding-line fans and their growth to ice-contact deltas. In J.A. Dowdeswell and J.D. Scourse (eds), *Glacimarine environments: processes and sediments*. Geological Society Special Publication 53, pp. 53–73.

Powell, R.D., 1991. Grounding-line systems as second order controls on fluctuations of temperate tidewater termini. In J.B. Anderson and G.M. Ashley (eds), *Glacial marine sedimentation – paleoclimatic significance*. Geological Society America, Special Paper, 261, pp. 75–94.

Powell, R.D. and Cooper, J.M., in press. A glacial sequence stratigraphic model for a temperate, glaciated continental shelf. In J.A. Dowdeswell and C. Ó Cofaigh (eds), *Glacier-influenced sedimentation on high-latitude continental margins*. Geological Society of London, Special Publication.

Powell, R.D. and Alley, R.B., 1997. Grounding-line systems: processes, glaciological inferences and the stratigraphic record. In P.F. Barker and A.C. Cooper (eds), *Geology and seismic stratigraphy of the Antarctic Margin, 2*. Antarctic Research Series, 71. AGU, Washington, DC, pp. 169–187.

Powell, R.D. and Domack, E.W., 1995. Glaciomarine environments, Chapter 13. In J. Menzies (ed.), *Glacial environments – processes, sediments and landforms*. Butterworth-Heinemann, Boston, pp. 445–486.

Powell, R.D. and Molnia, B.F., 1989. Glacimarine sedimentary processes, facies and morphology of the south-south-east Alaska Shelf and fjords. *Marine Geology*, 85, 359–390.

Powell, R.D., Dawber, M., McInnes, J.N. and Pyne, A.R., 1996. Observations of the grounding line area at a floating glacier terminus. *Annals of Glaciology*, 22, 217–223.

Powell, R.D., Hambrey, M.J. and Krissek, L.A., 1998. Quaternary and Miocene glacial and climatic history of the Cape Roberts drill site region, western Ross Sea, Antarctica. *Terra Antartica*, 5, 341–351.

Powell, R.D., Krissek, L.A. and van der Meer, J.J.M., 2000. Preliminary depositional environmental analysis of Cape Roberts 2/2A, Victoria Land Basin, Antarctica: palaeoglaciological and palaeoclimatic inferences. *Terra Antartica*, 7, 313–322.

Powell, R.D. and Cooper, J.M., 2003. A sequence stratigraphic model for temperate, glaciated continental shelves. In Dowdeswell, J.A. and Ó Cofaigh, C. (eds) Glacier-influenced sedimentation on high latitude continental margins: ancient and moderrn. Geological Society of London, Special Publication, 203: 215–244.

Praeg, D., 1997. Buried tunnel-valleys: 3D-seismic morphostratigraphy. In T.A. Davies, T. Bell, A.K. Cooper *et al.* (eds), *Glaciated continental margins: an atlas of acoustic images*. Chapman and Hall, London, pp. 68–69

Pratson, L.F., Imran, J., Hutton, E.W.H., Parker, G. and Syvitski, J.P.M., 2001. BANG1D, a one-dimensional, Lagrangian model of subaqueous turbid surges. *Computers and Geosciences*, 27, 701–716.

Prest, V.K., Grant, D.R. and Rampton, V.N., 1968. *Glacial map of Canada*. Geological Survey of Canada, Map 1253A, scale 1:5 000 000.

Price, R.J., 1964. *Landforms produced by the wastage of the Casement Glacier, southeast Alaska*. Institute of Polar Studies, Ohio State University, Report No. 9.

Price, R.J., 1965. The changing proglacial environment of the Casement Glacier, Glacier Bay, Alaska. *Transactions of the Institute of British Geographers*, 36, 107–116.

Price, R.J., 1966. Eskers near the Casement Glacier, Alaska. *Geografiska Annaler*, 48, 111–125.

Price, R.J., 1969. Moraines, sandar, kames and eskers near Breiðamerkurjökull, Iceland. *Transactions of the Institute of British Geographers*, 46, 17–43.

Price, R.J., 1970. Moraines at Fjallsjökull, Iceland. *Arctic and Alpine Research*, 2, 27–42.

Price, R.J., 1971. The development and destruction of a sandur, Breiðamerkurjökull, Iceland. *Arctic and Alpine Research*, 3, 225–237.

Price, R.J., 1973. *Glacial and fluvioglacial landforms*. Oliver and Boyd, Edinburgh.

Price, R.J., 1980. Rates of geomorphological changes in proglacial areas. In R.A. Cullingford, D.A. Davidson and J. Lewin (eds), *Timescales in geomorphology*. Wiley, Chichester, pp. 79–93.

Price, R.J., 1982. Changes in the proglacial area of Breiðamerkurjökull, southeastern Iceland: 1890–1980. *Jökull*, 32, 29–35.

Price, R.J. and Howarth, P.J., 1970. The evolution of the drainage system (1904–1965) in front of Breiðamerkurjökull, Iceland. *Jökull*, 20, 27–37.

Prior, D.B. and Bornhold, B.D., 1988. Submarine morphology and processes of fjord fan deltas and related high-gradient systems: modern examples from British Columbia. In W. Nemec and R.J. Steel (eds), *Fan deltas: sedimentology and tectonic settings*. Blackie, London, pp. 125–143.

Prior, D.B. and Bornhold, B.D., 1989. Submarine sedimentation on a developing Holocene fan delta. *Sedimentology*, 36, 1053–1076.

Prior, D.B. and Bornhold, B.D., 1990. The underwater development of Holocene fan deltas. In A. Colella and D.B. Prior (eds), *Coarse-grained deltas*. International Association of Sedimentologists, Special Publication, 10, 75–90.

Prior, D.B., Wiseman, W.J. and Bryant, W.R., 1981. Submarine chutes on the slopes of fjord deltas. *Nature*, 290, 326–328.

Prior, D.B., Bornhold, B.D. and Coleman, J.M., 1983. Geomorphology of a submarine landslide, Kitimat Arm, British Columbia. *Geological Survey Canada* Open File Report, 961.

Prior, D.B., Bornhold, B.D., Wiseman, W.J. and Lowe, D.R., 1987. Turbidity current activity in a British Columbia fjord. *Science*, 237, 1330–1333.

Punkari, M. 1993. Modelling of the dynamics of the Scandinavian Ice Sheet using remote sensing and GIS methods. In J. Aber (ed.), *Glaciotectonics and mapping glacial deposits*. Canadian Plains Research Center, University of Regina, 232–250.

Punkari, M., 1995. Function of the ice streams in the Scandinavian Ice Sheet: analyses of glacial geological data from southwestern Finland. Transactions of the Royal Society of Edinburgh; *Earth Sciences*, 85, 283–302.

Punkari, M., 1997. Glacial and glaciofluvial deposits in the interlobate areas of the Scandinavian Ice Sheet. *Quaternary Science Reviews*, 16, 741–753.

Quinterno, P., Carlson, P.R. and Molnia, B.F., 1980. Benthic foraminifers from the eastern Gulf of Alaska. In M.E. Field, A.H. Bouma, I.P. Colburn, R.G. Douglas and J.C. Ingle (eds), *Quaternary depositional environments of the Pacific Coast*. Proceedings of the Pacific Coast Paleogeography Symposium 4, 13–21.

Radbruch-Hall, D.H., 1978. Gravitational creep of rock masses on slopes. In B. Voight (ed.), *Rockslides and avalanches, Volume 1: Natural phenomena*. Elsevier, Amsterdam, pp. 607–657.

Radbruch-Hall, D.H., Varnes, D.J. and Savage, W.Z., 1976. Gravitational spreading of steep-sided ridges ('sackung') in western United States. *Bulletin of the International Association of Engineering Geology*, 14, 23–35.

Rafaelsen, B,. Andreassen, K., Kuilman, L.W., Lebesbye, E., Hogstad, K. and Midtbø, M., 2002. Geomorphology of buried glacigenic horizons in the Barents Sea from 3-dimensional seismic data. In J.A. Dowdeswell and C. Ó Cofaigh (eds), *Glacier-influenced sedimentation on high latitude continental margins*. Geological Society, London, Special Publication, 203, 259–276.

Rainio, H., Saarnisto, M. and Ekman, I., 1995. Younger Dryas end moraines in Finland and NW Russia. *Quaternary International*, 28, 179–192.

Rampton, V.N., 1974. The influence of ground ice and thermokarst upon the geomorphology of the Mackenzie-Beaufort region. In B.D. Fahey and R.D. Thompson (eds), *Research in polar and alpine geomorphology*. Proceedings of the 3rd Guelph Symposium on Geomorphology, pp. 43–59.

Rampton, V.N., 1988. *Quaternary Geology of the Tuktoyaktuk Coastlands, Northwest Territories*. Geological Survey of Canada, Memoir 423.

Rampton, V.N., 2001. Major end moraines of Younger Dryas age on Wollaston Peninsula, Victoria Island, Canadian Arctic: implications for paleoclimate and formation of hummocky moraine: discussion. *Canadian Journal of Earth Sciences*, 38, 1003–1006.

Rampton, V.N. and Walcott, R.I., 1974. Gravity profiles across ice-cored topography. *Canadian Journal of Earth Sciences*, 11, 110–122.

Rappol, M., 1984. Till in southeast Drente and origin of the Hondsrug complex, The Netherlands. *Eiszeitalter und Gegenwart*, 34, 7–27.

Rappol, M. and Stoltenberg, H.M.P., 1985. Compositional variability of Saalian till in The Netherlands and its origin. *Boreas*, 14, 33–50.

Rappol, M., Haldorsen, S., Jörgensen, P., Van der Meer, J.J.M. and Stoltenberg, H.M.P., 1989. Composition and origin of petrographically-stratified thick till in the northern Netherlands and a Saalian glaciation model for the North Sea basin. *Mededelingen van de Werkgroep voor Tertiaire en Kwartaire Geologie*, 26, 31–64.

Raukas, A., 1995. Glacial history of Estonia. In J. Ehlers, S. Kozarski and P. Gibbard (eds), *Glacial deposits in north-east Europe*. Balkema, Rotterdam, pp. 87–91.

Raymo, M.E. and Ruddiman, W.F., 1992. Tectonic forcing of late Cenozoic climate. *Nature*, 359, 117–122.

Raymond, C.F., 1987. How do glaciers surge? A review. *Journal of Geophysical Research*, 92, 9121–9134.

Raymond, C.F., Johannesson, T., Pfeffer, T. and Sharp, M., 1987. Propagation of a glacier surge into stagnant ice. *Journal of Geophysical Research*, 92, 9037–9049.

Rea, B.R. and Whalley, W.B., 1994. Subglacial observations from Øksfjordjøkelen, North Norway. *Earth Surface Processes and Landforms*, 19, 659–673.

Rea, B.R and Whalley, W.B., 1996. The role of bedrock topography, structure, ice dynamics and pre-glacial weathering in controlling subglacial erosion beneath a high latitude, maritime icefield. *Annals of Glaciology*, 22, 121–125.

Rea, B.R., Whalley, W.B. and Porter, E.M., 1996a. Rock weathering and the formation of summit blockfield slopes in Norway: examples and implications. In M.G. Anderson and S.M. Brooks (eds), *Advances in hillslope processes*, Wiley, Chichester, pp. 1257–1275

Rea, B.R., Whalley, W.B., Rainey, M. and Gordon, J.E., 1996b. Blockfields: old or new? Evidence and implications from some plateau blockfields in northern Norway. *Geomorphology*, 15, 109–121.

Rea, B.R., Whalley, W.B., Evans, D.J.A., Gordon, J.E. and McDougall, D.A., 1998. Plateau icefields: geomorphology and dynamics. *Quaternary Proceedings*, 6, 35–54.

Rea, B.R., Whalley, W.B., Dixon, T. and Gordon, J.E., 1999 Plateau icefields as contributing areas to valley glaciers and the potential impact on reconstructed ELAs: a case study from the Lyngen Alps, North Norway. *Annals of Glaciology*, 28, 97–102.

Rearic, D.M., Barnes, P.W. and Reimnitz, E., 1990. Bulldozing and resuspension of shallow-shelf sediment by ice keels: implication for Arctic sediment transport trajectories. *Marine Geology*, 91, 133–147.

Rebesco, M., Camerlenghi, A. and Zanolla, C., 1998. Bathymetri and morphogenesis of the Continental margin west of the Antarctic Peninsula. *Terra Antactica*, 5, 715–725.

Reeh, N., 1989. Dynamic and climatic history of the Greenland Ice Sheet. In R.J. Fulton (ed.), *Quaternary geology of Canada and Greenland*. Geological Survey of Canada, Geology of Canada no. 1, pp. 795–822.

Reimnitz, E. and Kempema, E.W., 1982. Dynamic ice-wallow relief on northern Alaska's nearshore. *Journal of Sedimentary Petrology*, 52, 451–461.

Reimnitz, E. and Kempema, E.W., 1987. Field observations of slush ice generated during freeze-up in arctic coastal waters. *Marine Geology*, 77, 219–231.

Reimnitz, E., Barnes, P., Forgatsech, T. and Rodeick, C., 1972. Influence of grounding ice on the Arctic shelf of Alaska. *Marine Geology*, 13, 323–334.

Reimnitz, E., Barnes, P.W. and Phillips, R.L., 1984. Geological evidence for 60 meter deep pressure-ridge keels in the Arctic Ocean. *Proceedings International Association Hydraulic Research, 7th International Symposium on Ice, Hamburg*. Volume 2, 189–206.

Reimnitz, E., Kempema, E.W. and Barnes, P.W., 1987. Anchor ice, sea bed freezing and sediment dynamics in shallow arctic seas. *Journal of Geophysical Research*, 92, 14671–14678.

Reimnitz, E., Toimil, L. and Barnes, P., 1978. Arctic continental shelf morphology related to sea-ice zonation, Beaufort Sea, Alaska. *Marine Geology*, 28, 179–210.

Retelle, M.J. and Bither, K.M., 1989. Late Wisconsinan glacial and glaciomarine sedimentary facies in the lower Androscoggin Valley, Topsham, Maine. In R.D. Tucker and R.G. Marvinney (eds), *Maine Geological Survey, Studies in Maine Geology*, 6, 33–51.

Reynolds, J.M., 2000. On the formation of supraglacial lakes on debris-covered glaciers. In M. Nakawo, A. Fountain and C. Raymond (eds), *Debris-covered glaciers*. IAHS Publication 264, 153–161.

Richards, B.W.M., Benn, D.I., Owen, L.A., Rhodes, E.J. and Spencer, J.Q., 2000. Timing of Late Quaternary glaciations south of Mount Everest in the Khumbu Himal, Nepal. *Geological Society of American Bulletin*, 112, 1621–1632.

Richardson, S.D. and Reynolds, J.M., 2000. An overview of glacial hazards in the Himalayas. *Quaternary International*, 65/66, 31–48.

Richter, W., Schneider, H. and Wager, R., 1951. Die Saaleeiszeitliche Stauchzone von Itterbeck-Uelsen (Grafschaft Bentheim). *Zeitschrift der Deutschen Geologischen Gesellschaft*, 76, 223–234.

Ridky, R.W. and Bindschadler, R.A., 1990. Reconstruction and dynamics of the Late Wisconsin 'Ontario' ice dome in the Finger Lakes region, New York. *Geological Society of America Bulletin*, 102, 1055–1064.

Riezebos, P.A., Boulton, G.S., van der Meer, J.J.M. *et al.*, 1986. Products and effects of modern eolian activity on a nineteenth century glacier-pushed ridge in West Spitsbergen, Svalbard. *Arctic and Alpine Research*, 18, 389–396.

Rijsdijk, K.F., Owen, G., Warren, W.P., McCarroll, D. and van der Meer, J.J.M., 1999. Clastic dykes in over-consolidated tills: evidence for subglacial hydrofracturing at Killiney Bay, eastern Ireland. *Sedimentary Geology*, 129, 111–126.

Ringberg, B., 1984. Cyclic lamination in proximal varves reflecting the length of summers during Late Weichsel in southernmost Sweden. In N. Mörner and W. Karlen (eds), *Climate changes on a yearly to millennial basis*. D. Reidel, Dordrecht, pp. 57–62.

Ringberg, B., 1991. Late Weichselian clay varve chronology and glaciolacustrine environment during deglaciation in southeast Sweden. *Sveriges Geologiska Undersökning* Ca 79, 42 pp.

Risberg, J., Sandgren, P., Teller, J., Last, W. 1999. Siliceous microfossils and mineral magnetic characteristics in a sediment core from Lake Manitoba, Canada. *Canadian Journal of Earth Sciences*, 36, 1299–1314.

Roberts, M.C. and Cunningham, F.F., 1992. Post-glacial loess deposition in a montane environment: South Thompson River Valley, British Columbia, Canada. *Journal of Quaternary Science*, 7, 291–301.

Roberts, M.J., Russell, A.J., Tweed, F.S. and Knudsen, O., 2000. Rapid sediment entrainment and englacial deposition during jokulhlaups. *Journal of Glaciology*, 46, 349–351.

Roberts, M.J., Russell, A.J., Tweed, F.S. and Knudsen, O., 2001. Controls on englacial sediment deposition during the November 1996 jokulhlaup, Skeiðarárjökull, Iceland. *Earth Surface Processes and Landforms*, 26, 935–952.

Robinson, P.H., 1984. Ice dynamics and thermal regime of Taylor Glacier, south Victoria Land, Antarctica. *Journal of Glaciology*, 30, 153–160.

Roed, M.A. and Waslyk, D.G., 1973. Age of inactive alluvial fans - Bow River Valley, Alberta. *Canadian Journal of Earth Sciences*, 10, 1834–1840.

Rosanova, C.E., Lucchitta, B.K. and Ferrigno, J.G., 1998. Velocities of Thwaites Glacier and smaller glaciers along Marie Byrd Land Coast, West Antarctica. *Annals of Glaciology*, 27, 47–53.

Rose, J., 1987. Drumlins as part of a glacier bedform continuum. In J. Menzies and J. Rose (eds), *Drumlin symposium*. Balkema, Rotterdam, 103–116.

Rose, J., 1989. Glacier stress patterns and sediment transfer associated with the formation of superimposed flutes. *Sedimentary Geology*, 62, 151–176.

Rose, J., 1991. Subaerial modification of glacial bedforms immediately following ice wastage. *Norsk Geografisk Tidsskrift*, 45, 143–153.

Rose, J., 1992. Boulder clusters in glacial flutes. *Geomorphology*, 6, 51–58.

Rose, J., Derbyshire, E., Hongwei, G. and Haizhou, M., 1998. Glaciation of Eastern Qilian Shan, Northwest China. *Quaternary Proceedings*, 6, 143–152.

Rose, K.E., 1979. Characteristics of ice flow in Marie Byrd Land, Antarctica. *Journal of Glaciology*, 24, 63–74.

Rosen, P.S. and Leach, K., 1987. Sediment accumulation forms, Thompson Island, Boston Harbour, Massachusetts. In D.M. FitzGerald and P.S. Rosen (eds), *Glaciated coasts*. Academic Press, San Diego, pp. 233–249.

Röthlisberger, F., 1986. *10,000 Jahre Gletschergeschichte der Erde*. Aarau, Verlag Sauerlander.

Röthlisberger, F., Haas, P., Holzhauser, H., Keller, W., Bircher, W. and Renner, F., 1980. Holocene climatic fluctuations – radiocarbon dating of fossil soils (fAh) and woods from moraines and glaciers in the Alps. *Geographica Helvetica*, 35, 21–52.

Röthlisberger, H., 1972. Water pressure in intra- and subglacial channels. *Journal of Glaciology*, 11, 177–203.

Rubulis, S., 1983. Deposition in a thermokarst sinkhole on a valley glacier, Mt. Tronodor, Argentina. In E.B. Evenson (ed.), *Tills and related deposits*. Balkema, Rotterdam, pp. 245–253.

Ruegg, G.H.J., 1981. Ice-pushed Lower and Middle Pleistocene deposits near Rhenen (Kwintelooijen): sedimentary/structural and lithological/granulometrical investigations. *Mededelingen Rijks Geologische Dienst*, 35, 165–177.

Rühberg, N., Schulz, W., Von Bülow, W. *et al.*, 1995. V. Mecklenburg-Vorpommern. In L. Benda (ed.), *Das Quartär Deutschlands*. Gebrüder Borntraeger, Berlin, pp. 95–115.

Russell, A.J. and Knudsen, O., 1999. Controls on the sedimentology of the November 1996 jokulhlaup deposits, Skeiðarársandur, Iceland. In N.D. Smith, J. Rogers and A.G. Plint (eds), *Advances in fluvial sedimentology*. International Association of Sedimentologists, Special Publication 28, 315–329.

Rust, B.R. and Romanelli, R., 1975. Late Quaternary subaqueous outwash deposits near Ottawa, Canada. In A.V. Jopling and B.C. McDonald (eds), *Glaciofluvial and glaciolacustrine sedimentation*. SEPM Special Publication 23, 177–192.

Ryder, J.M., 1971a. The stratigraphy and morphology of para-glacial alluvial fans in south-central British Columbia. *Canadian Journal of Earth Science*, 8, 279–298.

Ryder, J.M., 1971b. Some aspects of the morphometry of paraglacial alluvial fans in south-central British Columbia. *Canadian Journal of Earth Science*, 8, 1252–1264.

Ryder, J.M., Fulton, R.J. and Clague, J.J., 1991. The Cordilleran ice sheet and the glacial geomorphology of southern and central British Columbia. *Geographie Physique et Quaternaire*, 45, 365–377.

Sættem, J., 1990. Glaciotectonic forms and structures on the Norwegian continental shelf: observations, processes and implications. *Norsk Geologisk Tidsskrift*, 70, 81–94.

Sættem, J., 1991. *Glaciotectonics and glacial geology of the south western Barents Sea*. Universitet i Trondheim – Norges Tekniske Høgskole Doktoringeniøravhandling, 53.

Sættem, J., Poole, D.A.R., Ellingsen, L. and Sejrup, H.P., 1992. Glacial geology of outer Bjørnøyrenna, southwestern Barents Sea. *Marine Geology*, 103, 15–51.

Salaheldin, T.M., Imran, J., Chaudhry, M.H. and Reed, C., 2000. Role of fine-grained sediment in turbidity current flow dynamics and resulting deposits. *Marine Geology*, 171, 21–38.

Saunderson, H.C., 1975. Sedimentology of the Brampton esker and its associated deposits: an empirical test of theory. In A.V. Jopling and B.C. McDonald (eds), *Glaciofluvial and glaciolacustrine sedimentation*. SEPM Special Publication 23, 155–176.

Scambos, T.A. and Bindschadler, R., 1993. Complex ice stream flow revealed by sequential satellite imagery. *Annals of Glaciology*, 17, 177–182.

Scambos, T.A., Echelmeyer, K.A., Fahnestock, M.A. and Bindschadler, R.A., 1994. Development of enhanced ice flow at the southern margin of Ice Stream D, Antarctica. *Annals of Glaciology*, 20, 313–318.

Schafer, C.T. and Cole, F.E., 1986. Reconnaissance survey of benthic foraminifera from Baffin Island fiord environments. *Arctic*, 39, 232–239.
Schou, A., 1949. Landskabsformerne. In N. Nielsen (ed.), *Atlas over Danmark*. Hagerup, Copenhagen, 32 pp.
Schröder, E., 1978. Geomorphologische Untersuchungen im Hümmling. *Göttinger Geographische Abhandlungen*, 70, 1–113.
Schruben, P.G., Raymond, A.E. and Bawiec, W.J., 1999. Geology of the Conterminous United States at 1:2,500,000 Scale – a digital representation of the 1974 King and Beikman Map, U.S.G.S. Digital Data Series 11.
Schulz, W., 1963. Über einen Os bei Gellendin. *Geologie*, 12, 986–989.
Schulz, W., 1970. Über Oser und osähnliche Bildungen in der westlichen Prignitz. *Jahrbuch für Geologie*, 3, 411–420.
Schwerdtfeger, W., 1970. The climate of Antarctica. In S. Orvig (ed.), *Climates of polar regions*. Elsevier, Amsterdam, pp. 253–355.
Sejrup, H.P., King, E.L., Aarseth, I., Haflidason, H. and Elverhøi, A., 1996. Quaternary erosion and depositional processes: Western Norwegian fjords, Norwegian Channel and North Sea Fan. In M. DeBatist and P. Jacobs (eds), *Geology of siliciclastic shelf seas*. Geological Society Special Publication, 117, pp. 187–202.
Sejrup, H.P., Landvik, J.Y., Larsen, E., Janocko, J., Eiriksson, J. and King, E.L., 1998. The Jaeren area, a border zone of the Norwegian Channel ice stream. *Quaternary Science Reviews*, 17, 801–812.
Seramur, K.C., Powell, R.D. and Carlson, P.R., 1997. Evaluation of conditions along the grounding line of temperate marine glaciers: an example from Muir Inlet, Glacier Bay, Alaska. *Marine Geology*, 140, 307–327.
Sexton, D.J., Dowdeswell, J.A. Solheim, A. and Elverhøi, A., 1992. Seismic architecture and sedimentation in northwest Spitsbergen fjords. *Marine Geology*, 103, 53–68.
Shabtaie, S. and Bentley, C.R., 1987. West Antarctic ice streams draining into the Ross Ice Shelf: configuration and mass balance. *Journal of Geophysical Research*, 92(B2), 1311–1336.
Shabtaie, S., Whillans, I.M. and Bentley, C.R., 1987. The morphology of Ice Streams A, B, and C, West Antarctica, and their environs. *Journal of Geophysical Research*, 92, 8865–8883.
Shakesby, R.A. and Matthews, J.A., 1996. Glacial activity and paraglacial landsliding in the Devensian lateglacial: evidence from Craig Cerry-gleisiad and Fan Dringarth, Fforest Fawr (Brecon Beacons), South Wales. *Geological Journal*, 31, 143–157.
Sharp, M.J., 1982. Modification of clasts in lodgement tills by glacial erosion. *Journal of Glaciology*, 28, 475–481.
Sharp, M.J., 1984. Annual moraine ridges at Skalafellsjökull, south-east Iceland. *Journal of Glaciology*, 30, 82–93.
Sharp, M.J., 1985a. 'Crevasse-fill' ridges – a landform type characteristic of surging glaciers? *Geografiska Annaler*, 67A, 213–220.
Sharp, M.J., 1985b. Sedimentation and stratigraphy at Eyjabakkajökull – an Icelandic surging glacier. *Quaternary Research*, 24, 268–284.
Sharp, M.J., 1988. Surging glaciers: geomorphic effects. *Progress in Physical Geography*, 12, 533–559.
Sharp, M.J., Dowdeswell, J.A. and Gemmell, J.C., 1989. Reconstructing past glacier dynamics and erosion from glacial geomorphic evidence: Snowdon, North Wales. *Journal of Quaternary Science*, 4, 115–130.
Sharpe, D.R., 1988. Late Glacial landforms of Wollaston Peninsula, Victoria Island, Northwest Territories: product of ice-marginal retreat, surge, and mass stagnation. *Canadian Journal of Earth Sciences*, 25, 262–279.
Sharpe, D.R., 1992. *Quaternary Geology of Wollaston Peninsula, Victoria Island, Northwest Territories*. Geological Survey of Canada, Memoir 434.
Sharpe, D.R. and Cowan, W.R., 1990. Moraine formation in northwestern Ontario: product of subglacial fluvialand glaciolacustrine sedimentation. *Canadian Journal of Earth Sciences*, 27, 1478–1486.
Sharpe, D.R. and Nixon, F.M., 1989. *Surficial geology, Wollaston Peninsula, Victoria Island, District of Franklin, Northwest Territories*. Geological Survey of Canada, Map 1650A, scale 1:250 000.
Sharpe, D.R., Pullan, S.E. and Warman, T.A., 1992. A basin analysis of the Wabigoon area of Lake Agassiz, a Quaternary clay basin in northwestern Ontario. *Géographie physique et Quaternaire*, 46, 295–309.

Shaw, J., 1977a. Till body morphology and structure related to glacier flow. *Boreas*, 6, 189–201.
Shaw, J., 1977b. Tills deposited in arid polar environments. *Canadian Journal of Earth Science*, 14, 1239–1245.
Shaw, J., 1977c. Sedimentation in an alpine lake during deglaciation, Okanagan Valley, British Columbia, Canada. *Geografiska Annaler*, 59A, 221–240.
Shaw, J., 1982. Melt-out till in the Edmonton area, Alberta, Canada. *Canadian Journal of Earth Sciences*, 19, 1548–1569.
Shaw, J., 1996. A meltwater model for Laurentide subglacial landscapes. In S.B. McCann and D.C. Ford (eds), *Geomorphology Sans Frontières*. John Wiley and Sons, NY, pp. 181–236.
Shaw, J. and Forbes, D.L., 1992. Barriers, barrier platforms and spillover deposits in St George's Bay, Newfoundland: paraglacial sedimentation on the flanks of a deep coastal basin. *Marine Geology*, 105, 119–140.
Shaw, J. and Forbes, D.L., 1995. The postglacial relative sea-level lowstand in Newfoundland. *Canadian Journal of Earth Sciences*, 32, 1308–1330.
Shaw, J., Kvill, D. and Rains, R.B., 1989. Drumlins and catastrophic subglacial floods. *Sedimentary Geology*, 62, 177–202.
Shaw, J., Taylor, R.B. and Forbes, D.L., 1990. Coarse clastic barriers in eastern Canada: patterns of glaciogenic sediment dispersal with rising seas. *Journal of Coastal Research*, Special Issue, 9, 160–200.
Shaw, J., Taylor, R.B. and Forbes, D.L., 1993. Impact of the Holocene transgression on the coastline of Nova Scotia, Canada. *Géographie Physique et Quaternaire*, 47, 221–238.
Shevenelle, A..E., Domack, E.W. and Kernan, G.M., 1996. Record of Holocene palaeoclimate change along the Antarctic Peninsula: evidence from glacial marine sediments, Lallemand Fjord. Papers and Proceedings of the Royal Society of Tasmania, 130, pp. 55–64.
Shin, R.A. and Barron, E. 1989. Climate sensitivity to continental ice sheet size and configuration. *Journal of Climate*, 2, 1517-1537.
Shipp, S. and Anderson, J.B. 1997a Drumlin field on the Ross Sea continental shelf, Antarctica. In: Davies, T. A., Bell, T., Cooper, A. K., Josenhans, H., Polyak, L. Solheim, A., Stoker, M. S. and Stravers, J. A.(eds.): *Glaciated continental margins: an atlas of acoustic images*. pp. 52–53. Chapman & Hall.
Shipp, S. and Anderson, J.B., 1997b. Grounding zone wedges on the Antarctic continental shelf, Ross Sea. In T.A. Davies, T. Bell, A.K. Cooper *et al.* (eds), *Glaciated continental margins: an atlas of acoustic images*. Chapman and Hall, London, pp. 104–105.
Shipp, S., Anderson, J.B. and Domack, E.W., 1999. Late Pleistocene-Holocene retreat of the West Antarctic ice-sheet system in the Ross Sea; Part 1, Geophysical results. *Geological Society of America Bulletin*, 111, 1486–1516.
Shoemaker, E.M., 1992. Water sheet outburst floods from the Laurentide Ice Sheet. *Canadian Journal of Earth Sciences*, 29, 1250–1264.
Shroder, J.F., Bishop, M.P., Copland, L. and Sloan, V.F., 2000. Debris-covered glaciers and rock glaciers in the Nanga Parbat Himalaya, Pakistan. *Geografiska Annaler*, 82A, 17–31.
Sibrava, V., 1986. Correlation of European glaciations and their relation to the deep-sea record. *Quaternary Science Reviews*, 5, 433–441.
Sigurdsson, O. and Williams, R.S., 1991. Rockslides on the terminus of 'Jökulsárgilsjökull', southern Iceland. *Geografiska Annaler*, 73A, 129–140.
Sim, V.W., 1960. A preliminary account of late Wisconsin glaciation of Melville Peninsula. *Canadian Geographer*, 17, 21–34.
Sissons, J.B., 1974. A lateglacial ice cap in the central Grampians. *Transactions of the Institute of British Geographers*, 62, 95–114.
Sissons, J.B., 1977. The Loch Lomond Readvance in the northern mainland of Scotland. In J.M. Gray and J.J. Lowe (eds), *Studies in the Scottish Lateglacial environment*. Pergamon, Oxford, pp. 45–59.
Sissons, J.B., 1980. The Loch Lomond Advance in the Lake District, northern England. *Transactions of the Royal Society of Edinburgh: Earth Sciences*, 71, 13–27.

Sissons, J.B., 1981. Ice dammed lakes in Glen Roy and vicinity: a summary. In J. Neale and J. Flenley (eds), *The Quaternary in Britain*. Pergamon, Oxford, pp. 174–183.

Sjörberg, L.E., 1991. Fennoscandian uplift – an introduction. *Terra Nova*, 3, 356–357.

Sjørring, S., 1983. The glacial history of Denmark. In J. Ehlers (ed.), *Glacial deposits in north-west Europe*. Balkema, Rotterdam, pp. 163–179.

Skidmore, M.L. and Sharp, M.J., 1999. Drainage system behaviour of a high arctic polythermal glacier. *Annals of Glaciology*, 28, 209–215.

Skupin, K., Speetzen, E. and Zandstra, J.G., 1993. Die Eiszeit in Nordwestdeutschland – zur Vereisungsgeschichte der Westfälischen Bucht und angrenzender Gebiete. *Krefeld, Geologisches Landesamt Nordrhein-Westfalen*, 143 pp.

Slater, G., 1926. Glacial tectonics as reflected in disturbed drift deposits. *Proceedings of the Geologists Association*, 37, 392–400.

Slatt, R.M., 1971. Texture of ice-cored deposits from ten Alaskan valley glaciers. *Journal of Sedimentary Petrology*, 41, 828–834.

Sletten, K., Lyså, A. and Lønne, I., 2001. Formation and disintegration of a high-arctic ice-cored moraine complex, Scott Turnerbreen, Svalbard. *Boreas*, 30, 272–284.

Small, R.J., 1983. Lateral moraines of Glacier De Tsidjiore Nouve: form, development and implications. *Journal of Glaciology*, 29, 250–259.

Small, R.J., 1987a. Englacial and supraglacial sediment: transport and deposition. In A.M. Gurnell and M.J. Clark (eds), *Glacio-fluvial sediment transfer: an Alpine perspective*. Wiley, Chichester, pp. 111–145.

Small, R.J., 1987b. Moraine sediment budgets. In A.M. Gurnell and M.J. Clark (eds), *Glacio-fluvial sediment transfer: an Alpine perspective*. Wiley, Chichester, pp. 165–197.

Small, R.J., Clark, M.J. and Cawse, T.J.P., 1979. The formation of medial moraines on Alpine glaciers. *Journal of Glaciology*, 22, 43–52.

Smed, P. 1962. Studier over den fynske øgruppes glacial landskabformer. Meddelser Dansk Geologisk Forening 15, 1-74.

Smith, I.R., 1999. Late Quaternary glacial history of Lake Hazen Basin and eastern Hazen Plateau, northern Ellesmere Island, Nunavut, Canada. *Canadian Journal of Earth Sciences*, 36, 1547–1565.

Smith, I.R., 2000. Diamictic sediments within high arctic lake sediment cores: evidence for lake ice rafting along the lateral glacial margin. *Sedimentology*, 47, 1157–1179.

Smith, L.M. and Andrews, J.T., 2000. Sediment characteristics in iceberg dominated fjords, Kangerlussuaq region, East Greenland. *Sedimentary Geology*, 130, 11–25.

Smith, N.D., 1975. Sedimentary environments and Late Quaternary history of a 'low-energy' mountain delta. *Canadian Journal of Earth Sciences*, 12, 2004–2013.

Smith, N.D. and Ashley, G.M., 1985. Proglacial lacustrine environments. In G.M. Ashley, J. Shaw and N.D. Smith (eds), *Glacial sedimentary environments*. SEPM Short Course 16, 135–215.

Smith, N.D. and Ashley, G.M., 1996. A study of brash ice in the proximal marine zone of a sub-polar tidewater glacier. *Marine Geology*, 133, 75–87.

Smith, N D. and Syvitski, J.P.M., 1982. Sedimentation in a glacier-fed lake: the role of pelletisation on deposition of fine-grained suspensates. *Journal of Sedimentary Petrology*, 52, 503–513.

Smith, N.D., Vendl, M.A. and Kennedy, S.K., 1982. Comparison of sedimentation regimes in four glacier-fed lakes of western Alberta. In R. Davidson-Arnott, W. Nickling and B.D. Fahey (eds), *Research in glacial, glaciofluvial, and glaciolacustrine systems*. Geo Books, Norwich, pp. 203–238.

Smith, N.D., Phillips, A.C. and Powell, R.D., 1990. Tidal draw-down: A mechanism for producing cyclic sediment laminations in glacimarine deltas. *Geology*, 18, 10–13.

Socha, B.J., Colgan, P.M. and Mickelson, D.M., 1999. Ice-surface profiles and bed conditions of the Green Bay lobe from 13,000 to 11,000 14C-years BP. In D.M. Mickelson and J.W. Attig (eds), *Glacial processes past and present*. Geological Society of America, Special Paper 337, pp. 151–158.

Solheim, A., 1991. The depositional environment of surging sub-polar tidewater glaciers. *Norsk Polarinstitutt Skrifter*, 194, 97.

Solheim, A. and Kristoffersen, Y., 1984. Sediment distribution above the upper regional unconformity and the glacial history of western Barents Sea. *Norsk Polarinstitutt Skrifter*, 179(B), 26.

Solheim, A. and Pfirman, S.L., 1985. Sea-floor morphology outside a grounded, surging glacier: Bråsvellbreen, Svalbard. *Marine Geology*, 65, 127–143.

Solheim, A., Faleide, J.I., Andersen, E.S. *et al.*, 1998. Late Cenozoic seismic stratigraphy and glacial geological development of the East Greenland and Svalbard-Barents Sea continental margins. *Quaternary Science Reviews*, 17, 155–184.

Soller, D.R. and Packard, P.H., 1998. Digital representation of a map showing the thickness and character of Quaternary sediment in the glaciated United States east of the Rocky Mountains. US Geological Survey Digital Data Series DDS-38.

Sollid, J.L. and Sørbel, L., 1988. Influence of temperature conditions in formation of end moraines in Fennoscandia and Svalbard. *Boreas*, 17, 553–558.

Solomina, O.N., Savoskul, O.S. and Chirkinsky, A.E., 1994. Glacier variations, mudflow activity and landscape development in the Aksay Valley (Tien Shan) during the late Holocene. *The Holocene*, 4, 25–31.

Spedding, N.F., 2000. Hydrological controls on sediment transport pathways: implications for debris-covered glaciers. In M. Nakawo, A. Fountain and C. Raymond (eds), *Debris-covered glaciers*. IAHS Publication 264, 133–142.

Spedding, N. and Evans, D.J.A., 2002. Sediments and landforms at Kvíárjökull, southeast Iceland: a reappraisal of the glaciated valley landsystem. *Sedimentary Geology*, 149, 21–42.

Speight, J.G., 1963. Late Pleistocene historical geomorphology of the Lake Pukaki area, New Zealand. *New Zealand Journal of Geology and Geophysics*, 6, 160–188.

St Onge, D.A. and McMartin, I., 1995. *Quaternary Geology of the Inman River Area, Northwest Territories*. Geological Survey of Canada, Bulletin 446.

St Onge, D.A. and McMartin, I., 1999. La moraine du Lac Bluenose (Territoires du Nord-Ouest), une moraine a noyau de glace de glacier. *Geographie physique et Quaternaire*, 53, 287–295.

Stackebrandt, W., Ludwig, A.O. and Ostaficzuk, S., 2001. Base of Quaternary deposits of the Baltic Sea depression and adjacent areas (map 2). *Brandenburgische Geowissenschaftliche Beiträge*, 8, 13–19.

Stalker, A., 1960. *Ice-pressed drift forms and associated deposits in Alberta*. Geological Survey of Canada, Bulletin 57.

Stanford, S. and Mickelson, D.M., 1985. Till fabric and deformational structures in drumlins near Waukesha, Wisconsin, USA. *Journal of Glaciology*, 31, 220–228.

Stein R., Grobe, H., Hubberten, H., Marienfeld, P. and Nam, S., 1993. Latest Pleistocene to Holocene changes in glaciomarine sedimentation in Scoresby Sund and along the adjacent East Greenland continental margin: preliminary results. *Geo-Marine Letters*, 13, 9–16.

Stein, R. and Korolev, S., 1994. Shelf-to-basin sediment transport in the eastern Arctic Ocean. *Berichte zur Polarforschung*, 144, 87–100.

Stephan, H.-J., 1980. Some observations on flow tills in Schleswig-Holstein (North Germany). *Norsk Geografisk Tidsskrift*, 34, 99.

Stephan, H.-J., 1987. Moraine stratigraphy in Schleswig-Holstein and adjacent areas. In J.J.M. van der Meer (ed.), *Tills and glaciotectonics*. Balkema, Rotterdam, pp. 23–30.

Stephan, H.-J., 1995. I. Schleswig-Holstein. In L. Benda (ed.), *Das Quartär Deutschlands*. Gebrüder Borntraeger, Berlin, pp. 1–13.

Stephan, H.-J., 2002. Structural geology and sedimentology of the Heiligenhafen till section Northern Germany: comment. *Quaternary Science Reviews*, 21, 1112–1114.

Stephan, H.-J., Kabel, C. and Schlüter, G., 1983. Stratigraphic problems in the glacial deposits of Schleswig-Holstein. In J. Ehlers (ed.), *Glacial deposits in north-west Europe*. Balkema, Rotterdam, pp. 305–320.

Stewart, R.A., Bryant, D. and Sweat, M.J., 1988. Nature and origin of corrugated ground moraine of the Des Moines Lobe, Story County, Iowa. *Geomorphology*, 1, 111–130.

Stewart, T.G., 1991. Glacial marine sedimentation from tidewater glaciers in the Canadian high arctic. In J.B. Anderson and G.M. Ashley (eds), *Glacial marine sedimentation – paleoclimatic significance*. Geological Society of America, Special Paper, 261, pp. 95–105.

Stoker, M.S., 1995. The influence of glacigenic sedimentation on slope-apron development on the continental margin off Northwest Britain. In R.A. Scrutton, M.S. Stoker, G.B. Shimmield and A.W. Tudhope (eds), *The tectonics, sedimentation and Paleoceanography of the north Atlantic region*. Geological Society Special Publication 90. pp. 159–177.

Stoker, M.S., 1997. Submarine end-moraines on the West Shetland Shelf, north-west Britain. In T.A. Davies, T. Bell, A.K. Cooper *et al.* (eds), *Glaciated continental margins: an atlas of acoustic images*. Chapman and Hall, London, pp. 84–85.

Stoker, M. and Holmes, R., 1991. Submarine end-moraines as indicators of Pleistocene ice-limits off northwest Britain. *Journal of Geological Society London*, 148, 421–434.

Stokes, C.R. and Clark, C.D., 1999. Geomorphological criteria for identifying Pleistocene ice streams. *Annals of Glaciology*, 28, 67–74.

Stokes, C.R. and Clark, C.D., 2001. Palaeo-ice streams. *Quaternary Science Reviews*, 20(13), 1437–1457

Stokes, C.R. and Clark, C.D., 2002a. Ice stream shear margin moraines. *Earth Surface Processes and Landform*, 27, 547–558.

Stokes, C.R. and Clark, C.D., 2002b. Are long subglacial bedforms indicative of fast ice flow? *Boreas*, 31, 239–249.

Stokes, C.R. and Clark, C.D, 2003. The Dubawnt Lake Palaeo-ice Stream; evidence for dynamic ice sheet behaviour on the Canadian Shield and insights regarding controls on ice stream location and vigour. *Boreas*, 32, 263–279.

Stone, B.D. and Peper, J.D., 1982. Topographic control of the deglaciation of eastern Massachusetts: Ice lobation and the marine incursion. In G.J. Larsen and B.D. Stone (eds), *Late Wisconsinan glaciation of New England*. Kendall Hunt, Dubuque IA, pp. 145–166.

Stravers, J. A. and Powell, R. D., 1997. Glacial debris flow deposits on the Baffin Island shelf: seismic architecture of till-tongue-like deposits. *Marine Geology*, 143, 151–168.

Stravers, J.A. and Syvitski, J.M.P., 1991. Early Holocene land-sea correlations and deglacial evolution of the Cambridge Fjord basin. *Quaternary Research*, 35, 72–90.

Stravers, J.A., Syvitski, J.P.M. and Praeg, D.B., 1991. Application of size-sequence data to glacial-paraglacial sediment transport and sediment partitioning. In J.P.M. Syvitski (ed.), *Principles, methods and applications of particle-size analysis*. Cambridge University Press, Cambridge, pp. 293–310.

Strehl, E., 1998. Glazilimnische Kames in Schleswig-Holstein. *Eiszeitalter u. Gegenwart*, 48, 19–22.

Stuiver, M., Denton, G.H., Hughes, T.J. and Fastook, J.L., 1981. History of the marine ice sheet in West Antarctica during the last glaciation: a working hypothesis. In G.H. Denton and T.J.H. Hughes (eds), *The Last Great Ice Sheets*. Wiley, New York, pp. 319–436.

Sugden, D.E., 1968. The selectivity of glacial erosion in the Cairngorm Mountains, Scotland. *Transactions of the Institute of British Geographers*, 45, 79–92.

Sugden, D.E., 1974. Landscapes of glacial erosion in Greenland and their relationship to ice, topographic and bedrock conditions. *Institute of British Geographers*, Special Publication, 7, 177–95.

Sugden, D.E. and John, B.S., 1976. *Glaciers and landscape*. Edward Arnold, London, 376 pp.

Sugden, D.E. and Watts, S.H., 1977. Tors, felsenmeer, and glaciations in northern Cumberland Peninsula, Baffin Island. *Canadian Journal of Earth Science*, 14, 2817–2823

Svendson, H., Beszczynska-Møller, A., Ove Hagen, J., Lefauconnier, B., Tverberg, V., Gerland, S., Børre Ørbæk, J., *et al.*, 2002. The physical environment of Kongsfjorden-Krossfjorden, an Arctic fjord system in Svalbard. *Polar Research*, 21, 133–166.

Swinzow, G.K., 1962. Investigation of shear zones in the ice sheet margin, Thule area, Greenland. *Journal of Glaciology*, 4, 215–229.

Syverson, K.M., 2000. Morphology and sedimentology of ice-walled-lake plains in the Chippewa Moraine, western Wisconsin. *Geological Society of America Abstracts with Programs*, 32, A63.

Syvitski, J.P.M., 1980. Flocculation, agglomeration and zooplankton pelletization of suspended sediment in a fjord receiving glacial meltwater. In H.J. Freeland, D.H. Farmer and C.D. Levings (eds), *Fjord oceanography*. Plenum Press, New York, pp. 615–623.

Syvitski, J.P.M., 1989. On the deposition of sediments within glacier-influenced fjords: oceanographic controls. *Marine Geology*, 85, 301–330.

Syvitski, J.P.M., 1991. Towards an understanding of sediment deposition on glaciated continental shelves. *Continental Shelf Research*, 11, 897–937.

Syvitski, J.P.M., 1993. Glaciomarine environments in Canada: an overview. *Canadian Journal of Earth Sciences*, 30, 354–371.

Syvitski, J.P.M. and Hein F.J., 1991. Sedimentology of an Arctic basin: Itirbilung Fiord, Baffin Island, Northwest Territories. *Geological Survey Canada*, Paper 91-11, 66 pp.

Syvitski, J.P.M., Fader, G.B., Josenhans, H.W., Maclean, B. and Piper, D.J.W., 1983. Sea bed investigations of the Canadian east coast and arctic using Pisces TV. *Geoscience Canada*, 10, 59–68.

Syvitski, J.P.M., Burrell, D.C. and Skei, J.M., 1987. *Fjords: processes and products*. Springer-Verlag, New York, 379 pp.

Syvitski, J.P.M., Farrow, G.E., Atkinson, R.J.A., Moore, P.G. and Andrews, J.T., 1989. Baffin Island fjord macrobenthos: bottom communities and environmental significance. *Arctic*, 42, 232–247.

Syvitski, J.P.M. and Daughney, S., 1992. DELTA2; delta progradation and basin filling. *Computers and Geosciences*, 18, 839–897.

Syvitski, J.P.M. and Farrow, G.E., 1983. Structures and processes in bay head deltas: Knight and Butte Inlets, British Columbia. *Sedimentary Geology*, 30, 217–244.

Syvitski, J.P.M. and Lee, H.J., 1997. Postglacial sequence stratigraphy of Lake Melville, Labrador. *Marine Geology*, 143, 55–79.

Syvitski, J.P.M. and Praeg, D.B., 1989. Quaternary sedimentation in the St Lawrence Estuary and adjoining areas, eastern Canada: an overview based on high resolution seismo-stratigraphy. *Géographie Physique et Quaternaire*, 43, 291–310.

Syvitski, J.P.M. and Shaw, J., 1995. Sedimentology and geomorphology of fjords. In G.M.E. Perillo (ed.), *Geomorphology and sedimentology of estuaries. Developments in sedimentology 53*. Elsevier, Amsterdam, pp. 113–178.

Syvitski, J.P.M., Smith, J.N., Boudreau, B. and Calabrese, E.A., 1988. Basin sedimentation and the growth of prograding deltas. *Journal of Geophysical Research*, 93, 6895–6908.

Syvitski, J.P.M., William, K., LeBlanc, G. and Cranston, R.E., 1990. The flux and preservation of organic carbon in Baffin Island fjords. In J.A. Dowdeswell and J.D. Scourse (eds), *Glacimarine environments: processes and sediments*. Geological Society London, Special Publication, 53, pp. 177–199.

Syvitski, J.P.M., Andrews, J.T. and Dowdeswell, J.A., 1996. Sediment deposition in an iceberg-dominated glacimarine environment, East Greenland: basin fill implications. *Global and Planetary Change*, 12, 251–270.

Syvitski, J.P.M., Skene, K.I., Nicholson, M.K. and Morehead, M.D., 1998. PLUME1.1; deposition of sediment from a fluvial plume. *Computers and Geosciences*, 24, 159–171.

Syvitski, J.P.M., Stein, A.B., Andrews, J.T. and Milliman, J.D., 2001. Icebergs and the sea floor of the East Greenland (Kangerlussuaq) continental margin. *Arctic, Antarctic and Alpine Research*, 33, 52–61.

Tabor, R.W., 1971. Origin of ridge-top depressions by large-scale creep in the Olympic Mountains, Washington. *Geological Society of America Bulletin*, 82, 1811–1822.

Taylor, A., Judge, A. and Desrochers, D., 1983. Shoreline regression: its effect on permafrost and the geothermal regime, Canadian arctic archipelago. In *Permafrost: Fourth International Conference Proceedings*. National Academy Press, Washington DC, pp. 1239–1244.

Teller, J.T., 1976. Lake Agassiz deposits in the main offshore basin of southern Manitoba. *Canadian Journal of Earth Sciences*, 13, 27–43.

Teller, J.T., 1987. Proglacial lakes and the southern margin of the Laurentide Ice Sheet. In W.F. Ruddiman and H.E. Wright (eds), *North America and adjacent oceans during the last deglaciation*. Geological Society of America, Decade of North American Geology, K-3, pp. 39–69.

Teller, J.T., 2001. Formation of large beaches in an area of rapid differential isostatic rebound: a three-outlet model for Lake Agassiz. *Quaternary Science Reviews*, 20, 1649–1659.

Teller, J.T., 2003. Controls, history, outbursts, and impact of large late-Quarternary proglacial lakes in North America. Ch. 3 in A. Gillespie, S. Porter, and B. Atwater (eds), *The Quarternary Period in the United States*, (INQUA), Anniversary Volume, Elsevier.

Teller, J.T., and Clayton, L. (eds), 1983. *Glacial Lake Agassiz.* Geological Association of Canada, Special Paper 26, 451 pp.

Teller, J.T. and Kehew, A.E. (eds), 1994. Late glacial history of proglacial lakes and meltwater runoff along the Laurentide Ice Sheet. *Quaternary Science Reviews*, 13, 795–981.

Teller, J.T. and Mahnic, P., 1988. History of sedimentation in the northwestern Lake Superior basin and its relation to Lake Agassiz overflow. *Canadian Journal of Earth Sciences*, 25, 1660–1673.

Teller, J.T. and Thorleifson, L.H., 1983. The Lake Agassiz – Lake Superior connection. In J.T. Teller and L. Clayton (eds), *Glacial Lake Agassiz.* Geological Association of Canada, Special Paper 26, 261–290.

Teller, J.T., Risberg, J., Matile, G. and Zoltai, S., 2000. Postglacial history and paleoecology of Wampum, Manitoba, a former lagoon in the Lake Agassiz basin. *Geological Society of America Bulletin*, 112, 943–958.

Teller, J., Leverington, D. and Mann, J., 2002. Freshwater outbursts to the oceans from glacial Lake Agassiz and climate change during the last deglaciation. *Quaternary Science Reviews*, 21, 879–887.

ten Brink, U.S. and Schneider, C., 1995. Glacial morphology and depositional sequences of the Antarctic continental shelf. *Geology*, 23, 580–584.

ten Brink, U.S., Schneider, C. and Johnson, A.H., 1995. Morphology and stratal geometry of the Antarctic continental shelf; insights from models. *AGU, Antarctic Research Series*, 68, 1–24.

Ter Wee, M.W., 1983. The Elsterian Glaciation in the Netherlands. In J. Ehlers (ed.), *Glacial deposits in north-west Europe.* Balkema, Rotterdam, pp. 413–415.

Theakstone, W.H., 1982. Sediment fans and sediment flows generated by snowmelt: observations at Austerdalsisen, Norway. *Journal of Geology*, 90, 583–588.

Thomas, G.S.P., 1984a. A late Devensian glaciolacustrine fan-delta at Rhosesmor, Clwyd, North Wales. *Geological Journal*, 19, 125–141.

Thomas, G.S.P., 1984b. Sedimentation of a sub-aqueous esker-delta at Strabathie, Aberdeenshire. *Scottish Journal of Geology* 20, 9–20.

Thomas, G.S.P. and Connell, R.J., 1985. Iceberg drop, dump and grounding structures from Pleistocene glaciolacustrine sediments. *Journal of Sedimentary Petrology*, 55, 243–249.

Thompson, T.A. and Baedke, S.J., (1997) Strand-plain evidence for late Holocene lake-level variations in Lake Michigan. *Geological Society of America Bulletin*, 109, 666–682.

Thomson, M.H., Kirkbride, M.P. and Brock, B.W., 2000. Twentieth-century surface elevation change of the Miage Glacier, Italian Alps. In M. Nakawo, C. Raymond and A. Fountain (eds), *Debris-covered glaciers.* IAHS Publication 264, 219–225.

Thorarinsson, S., 1939. The ice-dammed lakes of Iceland, with particular reference to their values as indicators of glacier oscillations. *Geografiska Annaler*, 21, 216–242.

Thorarinsson, S., 1964. Sudden advance of Vatnajokull outlet glaciers 1930–1964. *Jökull*, 14, 76–89.

Thorarinsson, S., 1969. Glacier surges in Iceland, with special reference to the surges of Brúarjökull. *Canadian Journal of Earth Sciences*, 6, 875–882.

Thorp, P.W., 1991. Surface profiles and basal shear stresses of outlet glaciers from a Lateglacial mountain icefield in western Scotland. *Journal of Glaciology*, 37, 77–89.

Tipper, W.H., 1971. *The Glacial Geomorphology and Pleistocene History of Central British Columbia.* Geological Survey of Canada, Bulletin 196.

Tulaczyk, S.M., 1999. Ice sliding over weak, fine-grained tills: dependence of ice-till interactions on till granulometry. In D.M. Mickelson and J.V. Attig (eds), *Glacial processes: past and present.* Geological Society America, Special Paper 337, pp. 157–177.

Tulaczyk, S., Scherer, R.P. and Clark, C.D., 2001. A ploughing model for the origin of weak tills beneath ice streams: a qualitative treatment. *Quaternary International*, 86, 59–70.

Tulaczyk, S.M., Kamb, B. and Engelhardt, H.F., 2000. Basal mechanics of Ice Stream B, West Antarctica. I. Till mechanics. *Journal of Geophysical Research*, 105(B1), 463–481.

Tuthill, S.J., 1967. Late Pleistocene Mollusca of the Missouri Coteau district, North Dakota – a note and bibliography. *North Dakota Geological Survey Miscellaneous Series*, 30, 73–83.

Unstead, J.F., 1933. A system of regional geography. *Geography*, 18, 175–187.

Upham, W., 1894. The Madison type of drumlins. *American Geologist*, 14, 69–83.
van Bemmelen, R.W. and Rutten, M.G., 1955. *Tablemountains of Northern Iceland.* Leiden.
Van den Berg, M.W. and Beets, D.J., 1987. Saalian glacial deposits and morphology in The Netherlands. In J.J.M. Van der Meer (ed.), *Tills and glaciotectonics*. Balkema, Rotterdam, pp. 235–251.
Van der Meer J.J.M., Kjaer, K.H. and Krüger, J., 1999. Subglacial water-escape structures and till structures, Slettjökull, Iceland. *Journal of Quaternary Science*, 14, 191–205.
Van der Veen, C.J., 1996. Tidewater calving. *Journal of Glaciology*, 42, 375–385.
Van der Wateren, F.M., 1981. Glacial tectonics at the Kwintelooijen Sandpit, Rhenen, The Netherlands. *Mededelingen Rijks Geologische Dienst*, 35, 252–268.
Van der Wateren, F.M., 1985. A model of glacial tectonics, applied to the ice-pushed ridges in the central Netherlands. *Bulletin of the Geological Society of Denmark*, 34, 55–74.
Van der Wateren, F.M., 1987. Structural geology and sedimentation of the Dammer Berge push moraine, FRG. In J.J.M. Van der Meer (ed.), *Tills and glaciotectonics*. Balkema, Rotterdam, pp. 157–182.
Van der Wateren, F.M., 1994a. Processes of Glaciotectonism. In J. Menzies (ed.), *Glacial environments: processes, sediments and landforms*. Butterworth-Heinemann, Oxford, pp. 309–335.
Van der Wateren, F.M., 1994b. Proglacial subaquatic outwash fan and delta sediments in push moraines; indicators of subglacial meltwater activity. *Sedimentary Geology*, 91, 145–172.
Van der Wateren, F.M., 1995. Structural geology and sedimentology of push moraines. *Mededelingen Rijks Geologische Dienst* Nr. 54, 168.
Van der Wateren, F.M., 1999. Structural geology and sedimentology of Saalian tills near Heiligenhafen, Germany. *Quaternary Science Reviews*, 18, 1625–1639.
Van der Wateren, F.M., 2002a. Processes of Glaciotectonism. In J. Menzies (ed.), *Modern and past glacial environments – revised student edition*. Butterworth-Heinemann, Oxford, pp. 417–443.
Van der Wateren, F.M., 2002b. Structural geology and sedimentology of the Heiligenhafen till section Northern Germany: reply to comment by H.-J. Stephan. *Quaternary Science Reviews*, 21, 1114–1116.
Van der Wateren, F.M., Kluiving, S.J. and Bartek, L.R., 2000. Kinematic indicators of subglacial shearing. In A.J. Maltman, B.P. Hubbard and M.J. Hambrey (eds), *Deformation of glacial materials*. Geological Society Special Publication 176, pp. 259–278.
van Gijssel, K., 1987. A lithostratigraphic and glaciotectonic reconstruction of the Lamstedt Moraine, Lower Saxony (FRG). In J.J.M. Van der Meer (ed.), *Tills and glaciotectonics*. Balkema, Rotterdam, pp. 145–155.
van Heteren, S., FitzGerald, D.M., McKinlay, P.A. and Buynevich, I.V., 1998. Radar facies of paraglacial barrier systems: coastal New England, USA. *Sedimentology*, 45, 181–200.
van Husen, D., 1979. Verbreitung, Ursachen und Fülling glazial übertiefter Talabschnitte an Beispielen aus den Ostalpen. *Eiszeitalter und Gegenwart*, 29, 1–15.
van Tatenhove, F.G.M. and Huybrechts, P., 1996. Modelling of the thermal conditions at the Greenland ice sheet margin during Holocene deglaciation: boundary conditions for moraine formation. *Geografiska Annaler*, 78A, 83–99.
Vanneste, K., 1995. *A comparative seismostratigraphic study of large-scale Plio-Pleistocene glaciogenic depocenter along the polar North Atlantic margins*. University of Gent, 278 pp.
Vatne, G., Etzelmüller, B., Ødegard, R.S. and Sollid, J.L., 1996. Meltwater routing in a high arctic glacier, Hannabreen, northern Spitsbergen, *Norsk Geografisk Tidsskrift*, 50, 67–74.
Veatch, J.D., 1933. Agricultural land classification and land types of Michigan. *Michigan Agric. Exp. Stat.* Special Bulletin 231.
Veillette, J.J., 1994. Evolution and paleohydrology of glacial Lakes Barlow and Ojibway. *Quaternary Science Reviews*, 13, 945–971.
Veillette, J.J., Dyke, A.S. and Roy, M., 1999. Ice flow and ice divide history of the Labrador Sector of the Laurentide Ice Sheet from Wisconsinan inception to deglaciation and rare evidence of pre-Wisconsinan ice flow. *Quaternary Science Reviews*, 18, 993–1019.
Vere, D.M. and Benn, D.I., 1989. Structure and debris characteristics of medial moraines in Jotunheimen, Norway: implications for moraine classification. *Journal of Glaciology*, 35, 276–280.
Vere, D.M. and Matthews, J.A., 1985. Rock glacier formation from a lateral moraine at Bukkeholsbreen, Jotunheimen, Norway: a sedimentological approach. *Zeitschrift für Geomorphologie*, 29, 397–415.

Vincent, J.-S., 1982. The Quaternary history of Banks Island, NWT, Canada. *Géographie physique et Quaternaire*, 36, 209–232.

Vincent, J.-S., 1989. Quaternary geology of the northern Canadian interior plains. In R.J. Fulton (ed.), *Quaternary geology of Canada and Greenland*. Geological Survey of Canada, Geology of Canada, No. 1, pp. 100–137.

Vincent, J-S. and Hardy, L., 1979. *The evolution of glacial Lakes Barlow and Ojibway, Quebec and Ontario*. Geological Survey of Canada, Bulletin 316.

Vogt, P.R., 1986. Seafloor topography, sediments, and paleoenvironments. In B.G. Hurdle (ed.), *The Nordic Seas*. Springer-Verlag, New York, pp. 237–410.

Vogt, P.R. and Perry, R.K., 1978. Post-rifting accretion of continental margins in the Norwegian-Greenland and Labrador seas: morphological evidence. *EOS Trans. Am. Geophysical Union* 59, 1204.

Vogt, P.R., Crane, K. and Sundvor, E., 1993. Glacigenic mudflows on the Bear Island submarine fan. *EOS, Transaction of American Geophysical Union*, 74, 449, 452–453.

Vogt, P.R., Gardner, J. and Crane, K., 1999. The Norwegian-Barents-Svalbard (NBS) continental margin: introducing a natural laboratory of mass wasting, hydrates, and ascent of sediment, pore water, and methane. *Geo-Marine Letters*, 19, 2–21.

Vorren, T.O. and Kristoffersen, Y., 1986. Late Quaternary glaciation in the south-western Barents Sea. *Boreas*, 15, 51–59.

Vorren, T.O. and Laberg, J.S., 1996. Lateglacial air temperature, oceanographic and ice sheet interactions in the southern Barents Sea region. In J.T. Andrews, W.E.N. Austin, H. Bergsten and A.E. Jennings (eds), *Late Quaternary palaeoceanography of the North Atlantic Margins*. Geological Society of London Special Publication 111, pp. 303–321

Vorren, T.O. and Laberg, J.S., 1997. Trough mouth fans - palaeoclimate and ice-sheet monitors. *Quaternary Science Reviews*, 16, 865–881.

Vorren, T., Hald, M., Edvardsen, M. and Lind-Hansen, O.W., 1983. Glacigenic sediments and sedimentary environments on continental shelves: general principles with a case study from the Norwegian continental shelf. In J. Ehlers (ed.), *Glacial deposits in north-west Europe*. Balkema, Rotterdam, pp. 61–73.

Vorren, T.O., Hald, M. and Thomsen, E., 1984. Quaternary sediments and environments on the continental shelf off northern Norway. *Marine Geology*, 57, 229–257.

Vorren, T.O., Kristoffersen, Y. and Andreassen, K., 1986. Geology of the inner shelf west of North Cape, Norway. *Norsk Geologisk Tidsskrift*, 66, 99–105.

Vorren, T.O., Hald, M. and Lebesbye, E., 1988. Late Cenozoic environments in the Barents Sea. *Paleoceanography*, 3, 601–612.

Vorren, T.O., Lebesbye, E., Andreassen, K. and Larsen, K-B., 1989. Glacigenic sediments on a passive continental margin as exemplified by the Barents Sea. *Marine Geology*, 85, 251–272.

Vorren, T.O., Lebesbye, E. and Larsen, K-B., 1990. Geometry and genesis of the glacigenic sediments in the southern Barents Sea. In J.A. Dowdeswell and J.D. Scourse (eds), *Glacimarine environments: processes and sediments*. Geological Society of London Special Publication 53, pp. 309–328.

Vorren, T.O., Richardsen, G., Knutsen, S-M. and Henriksen, E., 1991. Cenozoic erosion and sedimentation in the western Barents Sea. *Marine and Petroleum Geology*, 8, 317–340.

Vorren, T.O., Rokoengen, K., Bugge, T. and Larsen, O.A., 1992. Tykkelse av kvartære sedimenter på Norsk kontinentalhylle [Thickness of Quaternary sediments on the Norwegian continental shelf]. Kart i 1:3,000,000 i Nasjonalatlas for Norge.

Vorren, T.O., Laberg, J.S., Blaume, F. *et al.*, 1998. The Norwegian-Greenland sea continental margins: morphology and late Quaternary sedimentary processes and environment. *Quaternary Science Reviews*, 17, 273–302.

Vorren, T.O., Plassen, L., 2002. Deglaciation and palaeoclimate of the Andfjorden-Vågsfjorden area, north Norway, *Boreas*, 31, 97–125.

Walker, R.G. and Plint, A.G., 1992. Wave- and storm-dominated shallow marine systems. In R.G. Walker and N.P. James (eds), *Facies models; response to sea level change*. Geological Association of Canada, 219–238.

Warburton, J., 1990. An alpine proglacial fluvial sediment budget. *Geografiska Annaler*, 72A, 261–272.
Warner, B.G. (ed), 1990. *Methods in Quaternary ecology*. Geoscience Canada Reprint Series 5, Geological Association of Canada, 170 pp.
Warren, C.R., Glasser, N.F., Harrison, S., Winchester, V., Kerr, A.R. and Rivera, A., 1995. Characteristics of tidewater calving at Glaciar San Rafael. *Journal of Glaciology*, 41, 273–289.
Washburn, A.L., 1979. *Geocryology*. Edward Arnold, London.
Watanabe, T., Dali, L. and Shiraiwa, T., 1998. Slope denudation and supply of debris to cones in Landtang Himal, Central Nepal Himalaya. *Geomorphology*, 26, 185–197.
Webb, P.-N., Harwood, D.M., Mabin, M.G.C. and McKelvey, B.C., 1996. A marine and terrestrial Sirius Group succession, middle Beardmore Glacier – Queen Alexandra Range, Transantarctic Mountains, Antarctica. *Marine Micropaleontology*, 27, 273–297.
Weertman, J., 1961. Mechanism for the formation of inner moraines found near the edge of cold ice caps and ice sheets. *Journal of Glaciology*, 3, 965–978.
Weidick, A., 1995. *Satellite image atlas of glaciers of the world: Greenland*. US Geological Survey Professional Paper 1386-C.
Welch, R., 1967. *The application of aerial photography to the study of a glacial area. Breiðamerkur, Iceland.* Unpublished PhD Thesis, University of Glasgow.
Welch, D.M., 1970. Substitution of space for time in a study of slope development. *Journal of Geology*, 78, 234–239.
Westergård, A.H., 1906. Platålera, en supramarin hvarfvig lera från Skåne. *Sveriges Geologiska Undersökning*, Series C, No. 201, 9 pp.
Whalley, W.B., Rea, B.R., Rainey, M.M., McAlister, J.J. 1997. Rock weathering in blockfields: some preliminary data from mountain plateaus in north Norway. In: Widdowson M (ed) Palaeosurfaces: recognition, reconstruction and interpretation. *Geological Society Special Publication* 129, 133–145.
Whalley, W.B., Gordon, J.E. Gellatly, A.F. and Hansom, J.D., 1995a. Plateau and valley glaciers in north Norway: responses to climate over the last 100 years. *Zeitschrift für Gletscherkunde und Glazialgeologie*, 31, 115–124.
Whalley, W.B., Gordon, J.E. and Thompson, D.L., 1981. Periglacial features on the margins of a receding plateau ice cap, Lyngen, North Norway. *Journal of Glaciology*, 27, 492–496.
Whalley, W.B., Hamilton, S.J., Palmer, C.F., Gordon, J.E. and Martin, H.E., 1995b. The dynamics of rock glaciers: data from Tröllaskagi, north Iceland. In O. Slaymaker (ed.), *Steepland Geomorphology*. John Wiley and Sons, Chichester, pp. 129–145.
Whillans, I.M. and Van der Veen, C.J., 1993. New and improved determinations of velocity of Ice Streams B and C, West Antarctica. *Journal of Glaciology*, 39, 483–490.
Whillans, I.M., Bolzan, J. and Shabtaie, S., 1987. Velocity of Ice Streams B and C, Antarctica. *Journal of Geophysical Research*, 92(B9), 8895–8902.
Whittecar, G.R. and Mickelson, D.M., 1977. Sequence of till deposition and erosion in drumlins. *Boreas*, 6, 213–217.
Whittecar, G.R. and Mickelson, D.M., 1979. Composition, internal structure, and a hypothesis of formation for drumlins, Waukesha County, Wisconsin, USA. *Journal of Glaciology*, 22, 357–371.
Wieczorek, G.F. and Jäger, S., 1996. Triggering mechanisms and depositional rates of postglacial slope movement processes in the Yosemite Valley, California. *Geomorphology*, 15, 17–31.
Wilke, H. and Ehlers, J., 1983. The thrust moraine of Hamburg-Blankenese. In J. Ehlers (ed.), *Glacial deposits in north-west Europe*. Balkema, Rotterdam, pp. 331–333.
Williams, R.S. and Ferrigno, J.G., 1988. *Satellite image atlas of glaciers of the world*. US Geological Survey Professional Paper 1386-A.
Wilson, G.S., Harwood, D.H., Askin, R.A. and Levy, R.H., 1998. Late Neogene Sirius Group strata in Reedy Valley, Antarctica: a multiple-resolution record of climate, ice-sheet and sea-level events. *Journal of Glaciology*, 44, 437–447.
Wilson, S.B. and Evans, D.J.A., 2000. Coire a' Cheud-chnoic, the 'hummocky moraine' of Glen Torridon. *Scottish Geographical Journal*, 116, 149–158.

Winters, G.V. and Syvitski, J.P.M., 1992. Suspended sediment character and distribution in McBeth Fiord, Baffin Island. *Arctic*, 45, 25–35.

Wohlfarth, B., Björck, S., Possnart, G. *et al.*, 1993. AMS dating Swedish varved clays of the last glacial-interglacial transition and the potential/difficulties of calibrating Late Weichselian 'absolute' chronologies. *Boreas*, 22, 113–128.

Woldstedt, P., 1925. Die großen Endmoränenzüge Norddeutschlands. *Zeitschrift der Deutsche Geologischen Gesellschaft*, 77, 172–184.

Woodworth-Lynas, C.M.T., 1996. Ice scour as an indicator of glaciolacustrine environments. In J. Menzies (ed.), *Post glacial environments, sediments, forms, and techniques*. Butterworth-Heinemann, Oxford, pp. 161–178.

Woodworth-Lynas, C.M.T. and Guigné, J.Y., 1990. Iceberg scours in the geological record: examples from glacial Lake Agassiz. In J.A. Dowdeswell and J.D. Scource (eds), *Glacimarine environments: processes and sediments*. Geological Society of London Special Publication 53, pp. 217–223.

Worsley, P., 1999. Context of relict Wisconsinan glacial ice at Angus Lake, SW Banks Island, western Canadian arctic and stratigraphic implications. *Boreas*, 28, 543–550.

Wright, H.E., Jr., 1973. Tunnel valleys, glacial surges, and subglacial hydrology of the Superior lobe, Minnesota. In R.F. Black, R.P. Goldthwait and H.B. Williams (eds), *The Wisconsinan Stage*. Geological Society of America Memoir 135, pp. 251–276.

Wright, H.E., Jr., 1980. Surge moraines of the Klutlan Glacier, Yukon Territory, Canada and application to the Late Glacial of Minnesota. *Quaternary Research*, 14, 2–17.

Wright, M.D., 1991. Pleistocene deposits of the South Wales coalfield and their engineering significance. In A. Forster, M.G. Culshaw, J.C. Cripps, J.A. Little and C.F. Moon (eds), *Quaternary engineering geology*. Geological Society Special Publication, 7, pp. 441–448.

Wyrwoll, K-H., 1977. Causes of rock-slope failure in a cold area: Labrador-Ungava. *Geological Society of America Reviews in Engineering Geology*, 3, 59–67.

Yamada, T., 1998. *Glacier lake and its outburst flood in the Nepal Himalaya*. Monograph No. 1, Data Center for Glacier Research, Japanese Society of Snow and Ice, 96 pp.

Zagwijn, W.H. and Doppert, J.W.C., 1978. Upper Cenozoic of the southern North Sea Basin: palaeoclimatic and palaeogeographic evolution. *Geologie en Mijnbouw*, 57, 577–588.

Zandstra, J.G., 1983. Fine gravel, heavy minerals and grain-size analyses of Pleistocene, mainly glacigenic deposits in the Netherlands. In J. Ehlers (ed.), *Glacial deposits in north-west Europe*. Balkema, Rotterdam, pp. 361–377.

Zeng, J. and Lowe, D.R., 1997a. Numerical simulation of turbidity current flow and sedimentation; I, Theory. *Sedimentology*, 44, 67–84.

Zeng, J. and Lowe, D.R., 1997b. Numerical simulation of turbidity current flow and sedimentation; II, Results and geological applications. *Sedimentology*, 44, 85–104.

Ziegler, P.A., 1990. *Geological atlas of western and central Europe*. Shell International Petroleum Maatscchappij, London, 239 pp.

Zielinski, T. and van Loon, A.J., 1996. Characteristics and genesis of moraine-derived flowtill varieties. *Sedimentary Geology*, 101, 119–143.

Zimmermann, M. and Haeberli, W., 1992. Climatic change and debris flow activity in high-mountain areas – a case study in the Swiss Alps. *Catena Supplement*, 22, 59–72.

INDEX

Note to index: page numbers in bold indicate tables and diagrams